Statistical Concepts

A Second Course

Statistical Concepts—A Second Course presents the last 10 chapters from *An Introduction to Statistical Concepts, Fourth Edition*. Designed for second and upper-level statistics courses, this book highlights how statistics work and how best to utilize them to aid students in the analysis of their own data and the interpretation of research results.

In this new edition, Hahs-Vaughn and Lomax discuss sensitivity, specificity, false positive and false negative errors. Coverage of effect sizes has been expanded upon and more organizational features (to summarize key concepts) have been included. A final chapter on mediation and moderation has been added for a more complete presentation of regression models.

This book acts as a clear and accessible instructional tool to help readers fully understand statistical concepts and how to apply them to data. It is an invaluable resource for students undertaking a course in statistics in any number of social science and behavioral science disciplines.

Debbie L. Hahs-Vaughn is Professor of Methodology, Measurement, and Analysis at the University of Central Florida, US. Her research primarily relates to methodological issues when analyzing, and applied research using, complex sample data.

Richard G. Lomax is Professor Emeritus of Educational and Human Ecology at The Ohio State University, US, and former Associate Dean for Research and Administration. His research primarily focuses on early literacy and statistics.

Statistical Concepts

A Second Course

Fifth Edition

Debbie L. Hahs-Vaughn
University of Central Florida

Richard G. Lomax
The Ohio State University

Routledge
Taylor & Francis Group
NEW YORK AND LONDON

Fifth edition published 2020
by Routledge
52 Vanderbilt Avenue, New York, NY 10017

and by Routledge
2 Park Square, Milton Park, Abingdon, Oxon, OX14 4RN

Routledge is an imprint of the Taylor & Francis Group, an informa business

First edition published by Longman Publishing Group 1992

Fourth edition published by Routledge 2012

Library of Congress Cataloging-in-Publication Data
A catalog record for this book has been requested

ISBN: 978-0-367-20404-4 (hbk)
ISBN: 978-0-367-20409-9 (pbk)
ISBN: 978-0-429-27782-5 (ebk)

Typeset in Palatino
by Apex CoVantage, LLC

Visit the companion website: www.routledge.com/cw/hahs-vaughn

This book is dedicated to our families and to all our former students.

You are statistically significant.

Contents

Preface

Approach

Many individuals have an aversion to statistics, which is quite unfortunate. Statistics is a tool that unleashes great power to the user—the potential to *really* make a difference. Being able to *understand* statistics means that you can critically evaluate empirical research conducted by others and thus better apply what others have found. Being able to *do* statistics means that you contribute to solving problems. We approach the writing of this text with the mindset that we want this text to be an instrument that contributes to your success as a researcher. With the help of this text, you will gain tools that can be used to make a positive contribution in your discipline. Perhaps this is the moment for which you have been created (Esther 4:14)! Consider the use of text as moving one step closer to making the world a better place.

This text is designed for a second course in statistics for students in any number of social science and behavioral disciplines—from education to business to communication to exercise science to psychology to sociology and more. The text assumes you have been introduced to descriptive statistics and basic inferential statistics (e.g., t tests and bivariate correlations). The text is designed for you to become a better prepared researcher and a more intelligent consumer *and* producer of research. We do not assume that you have extensive or recent training in mathematics. Perhaps you have only had algebra and one introductory statistics course, and perhaps that was some time ago. Rest assured; you will do fine.

We believe that a text should serve as an effective instructional tool. You should find this text to be more than a reference book. It is designed to help those who read it really understand statistical concepts, in what situations they can be applied, and how to apply them to data. With that said, there are several things that this text is *not*. This text is not a theoretical statistics book, nor is it a cookbook on computing statistics, nor a statistical software manual. Recipes suggest that there is one effective approach in all situations, and following that approach will produce the same results always. Additionally, recipes tend to be a crutch—followed without understanding how or why you obtain the desired product. As well, knowing how to run a statistics package without understanding the concepts or the output is not particularly useful. Thus, concepts drive the field of statistics, and that is the framework within which this text was approached.

Goals and Content Coverage

Our goals for this text are lofty, but the effort that you put forth in using it and its effects on your learning of statistics are more than worthwhile. First, the text provides comprehensive coverage of topics that could be included in an undergraduate or graduate second-course sequence in statistics. The text is flexible enough so that instructors can select those topics that they desire to cover as they deem relevant in their particular discipline. In other words, chapters and sections of chapters from this text can be included in a statistics course as the instructor sees fit. Most of the popular, as well as many of the lesser-known, procedures and models are described in the text. A particular feature is a thorough discussion of assumptions, the effects of their violation, and how to deal with their violation.

The text covers several inferential statistics, including all of the basic analysis of variance (ANOVA) models and various regression models.

Second, the text communicates a *conceptual, intuitive* understanding of statistics, which requires only a rudimentary knowledge of basic algebra, and emphasizes the important concepts in statistics. The most effective way to learn statistics is through the conceptual approach. Statistical concepts tend to be easy to learn because (a) concepts can be simply stated, (b) concepts can be made relevant thorough the use of real-life examples, (c) the same concepts are shared by many procedures, and (d) concepts can be related to one another. This is not to say that the text is void of mathematics, as understanding the math behind the technique blows apart the "black box" of statistics, particularly in a world where statistical software is so incredibly powerful. However, understanding the concepts is the first step in advancing toward a true understanding of statistics.

This text will help you to reach the goal of being a better consumer and producer of research. The following indicators may provide some feedback as to how you are doing. First, there will be a noticeable change in your attitude toward statistics. Thus, one outcome is for you to feel that "Statistics is not half bad," or "This stuff is OK." Second, you will feel comfortable using statistics in your own work. Finally, you will begin to "see the light." You will know when you have reached this highest stage of statistics development when suddenly in the middle of the night you wake up from a dream and say "Now I get it!" In other words, you will begin to *think* statistics rather than think of ways to get out of doing statistics.

Pedagogical Tools

The text contains several important pedagogical features to allow you to attain these goals. First, each chapter begins with a list of key concepts, which provide helpful landmarks within the chapter. Second, realistic examples from education and the behavioral sciences are used to illustrate the concepts and procedures covered in each chapter. Each example includes an initial vignette, an examination of the relevant procedures and necessary assumptions, how to run SPSS and **R** and develop an APA-style write-up, as well as tables, figures, and annotated SPSS and **R** output to assist you. Third, the text is based on the conceptual approach; that is, material is presented so that you obtain a good understanding of statistical concepts. *If you know the concepts, then you know statistics.* Finally, each chapter ends with three sets of problems, computational, conceptual, and interpretive. Pay particular attention to the conceptual problems as they provide the best assessment of your understanding of the concepts in the chapter. We strongly suggest using the example datasets and the computational and interpretive problems for additional practice through available statistical software. This will serve to reinforce the concepts covered. Answers to the odd-numbered problems are given at the end of each chapter.

New to This Edition

A number of changes have been made in this edition based on the suggestions of reviewers, instructors, teaching assistants, and students. These improvements have been made in order to better achieve the goals of the text. The changes include the following:

1. The content has been updated and numerous additional references have been provided.
2. The final chapter on mediation and moderation has been added for a more complete presentation of regression models.
3. To parallel generating statistics using SPSS (version 25), script and annotated output using **R** has been included to assist in the generation and interpretation of statistics.
4. Coverage of effect sizes has been expanded, including illustration of the use of online tools for computing effect sizes.
5. A discussion of sensitivity, specificity, false positive, and false negative has been included in the logistic regression chapter, along with using the receiver operator characteristic (ROC) curve to determine classification accuracy.
6. A discussion of the general linear model has been folded into the analysis of variance chapter to help readers understand how ANOVA and regression models are connected.
7. An expanded discussion of testing, and illustration of how to test for, interactions in factorial ANOVA.
8. More organizational features (e.g., boxes, tables, figures) have been included to summarize concepts and/or increase understanding of the material.
9. Additional end of chapter problems have been included.
10. The website for the text provides instructor-only access to a test bank (the website continues to offer the datasets, chapter outline, answers to the even-numbered problems, and PowerPoint slides for each chapter, with students granted access to the appropriate elements).

Acknowledgments

We have been blessed beyond measure, and are thankful for so many individuals who have played an important role in our personal and professional lives and, in some way, have shaped this text. Rather than include an admittedly incomplete listing, we just say "thank you" to all of you. A special thank you to all of the terrific students that we have had the pleasure of teaching at the University of Pittsburgh, the University of Illinois at Chicago, Louisiana State University, Boston College, Northern Illinois University, the University of Alabama, The Ohio State University, and the University of Central Florida. For all of your efforts, and the many lights that you have seen and shared with us, this book is for you.

Thanks also to so many wonderful publishing staff that we've had the pleasure of working along the way, first at Lawrence Erlbaum Associates and now at Routledge/Taylor & Francis. Additionally, we are most appreciative of the insightful suggestions provided by the reviewers of this text over the years.

For the users of this text, you are the reason we write. Thank you for bringing us along in your research and statistical journey. To those that have contacted us with questions, comments, and suggestions, we are very appreciative. We hope that you will continue to contact us to offer feedback (good and bad).

Last but not least, we extend gratitude to our families, in particular, to Lea and Kristen, and to Mark and Malani. Your unfailing love, understanding, and tolerance during the writing of this text allowed us to cope with such a major project. *You are statistically significant!* Thank you one and all.

DLHV & RGL

1

One-Factor Analysis of Variance—Fixed-Effects Model

Chapter Outline

Key Concepts

1. Between- and within-groups variability
2. Sources of variation
3. Partitioning the sums of squares
4. The ANOVA model
5. Expected mean squares

The first six chapters of this text are concerned with different analysis of variance (ANOVA) models. In this chapter, we consider the most basic ANOVA model, known as the *one-factor analysis of variance* model. Recall the independent t test where the means from two independent samples were compared. What if you wish to compare more than two means? The answer is to use the **analysis of variance**. At this point you may be wondering why the procedure is called the analysis of variance rather than the analysis of means, because the intent is to study possible mean differences. One way of comparing a set of means is to think in terms of the variability among those means. If the sample means are all the same, then the variability of those means would be zero. If the sample means are not all the same, then the variability of those means would be somewhat greater than zero. In general, the greater the mean differences are, the greater is the variability of the means. Thus, mean differences are studied by looking at the variability of the means; hence, the term analysis of variance is appropriate rather than analysis of means (further discussed in this chapter).

We use X to denote our single **independent variable**, which we typically refer to as a **factor**, and Y to denote our **dependent** (or **criterion**) **variable**. *Thus, the one-factor ANOVA is a bivariate, or two variable, procedure.* Our interest here is in determining whether mean differences exist on the dependent variable. Stated another way, the researcher is interested in the influence of the independent variable on the dependent variable (however, be cautious in inferring *causality* unless the design of your study allows that). For example, a researcher may want to determine the influence that method of instruction has on statistics achievement. The independent variable, or factor, would be method of instruction, and the dependent variable would be statistics achievement. Three different methods of instruction that might be compared are large lecture hall instruction, small-group instruction, and computer-assisted instruction. Students would be randomly assigned to one of the three methods of instruction and, at the end of the semester, evaluated as to their level of achievement in statistics. These results would be of interest to a statistics instructor in determining the most effective method of instruction (where "effective" is measured by student performance in statistics). Thus, the instructor may opt for the method of instruction that yields the highest mean achievement.

There are a number of new concepts introduced in this chapter as well as a refresher of concepts that you will likely remember from your previous statistics course. The concepts addressed in this chapter include the following: independent and dependent variables; between- and within-groups variability; fixed- and random-effects; the linear model; partitioning of the sums of squares; degrees of freedom, mean square terms, and F ratios; the ANOVA summary table; expected mean squares; balanced and unbalanced models; and alternative ANOVA procedures. Our objectives are that by the end of this chapter, you will be able to (a) understand the characteristics and concepts underlying a one-factor ANOVA, (b) generate and interpret the results of a one-factor ANOVA, and (c) understand and evaluate the assumptions of the one-factor ANOVA.

1.1 What One-Factor ANOVA Is and How It Works

Throughout the book, we will be following a group of superbly talented, creative, and energetic graduate research assistants (Challie Lenge, Ott Lier, Addie Venture, and Oso Wyse) working in their institution's statistics and research lab, fondly known as CASTLE

(Computing and Statistical Technology Laboratory). The group is supervised and mentored by a research methodology faculty member who empowers the group to lead their projects to infinity and beyond, so to speak. With each chapter, we will find the group, or a subset of members thereof, delving into a fantastical statistical journey. We now find Ott Lier assisting one of the region's leading sports psychologists in examining elite athletes and vulnerability to psychological distress based on the type of sport in which they participate.

The statistics and research lab at the university serve clients within the institution, such as faculty and staff, and outside the institution, including a multitude of diverse community partners. The lab is supervised by a research methodology faculty member and is staffed by the institution's best and brightest graduate students, Addie Venture, Oso Wyse, Challie Lenge, and Ott Lier. The research lab has been contracted to work with one of the leading sports psychologists in the region, Dr. Rhodes, and Ott Lier has the privilege of being assigned to the project. Dr. Rhodes is examining elite athletes and their vulnerability to psychological distress based on the sport in which they participate. Dr. Rhodes wants to determine if there is a difference in psychological stress based on type of sport (movement, target, fielding, or territory). Ott suggests the following research question is: *Is there a mean difference in psychological distress of elite athletes based on the type of sport in which they participate?* With one independent variable, Ott determines that a one-way ANOVA is the best statistical procedure to use to answer Dr. Rhodes's question. His next task is to collect and analyze the data to address this research question.

1.1.1 Characteristics

This section describes the distinguishing characteristics of the one-factor ANOVA model. Suppose you are interested in comparing the means of two independent samples. Here, the independent t test would be the method of choice (or perhaps Welch's t' test). What if your interest is in comparing the means of more than two independent samples? One possibility is to conduct multiple independent t tests on each pair of means. For example, if you wished to determine whether the means from five independent samples are the same, you could do all possible pairwise t tests. In this case, the following null hypotheses could be evaluated: $\mu_1 = \mu_2$, $\mu_1 = \mu_3$, $\mu_1 = \mu_4$, $\mu_1 = \mu_5$, $\mu_2 = \mu_3$, $\mu_2 = \mu_4$, $\mu_2 = \mu_5$, $\mu_3 = \mu_4$, $\mu_3 = \mu_5$, and $\mu_4 = \mu_5$. Thus, we would have to carry out 10 different independent t tests. The number of possible pairwise t tests that could be done for J means is equal to $\left[J(J-1)\right]/2$.

Is there a problem in conducting so many t tests? Yes; the problem has to do with the probability of making a Type I error (i.e., α), where the researcher incorrectly rejects a true null hypothesis. Although the α level for each t test can be controlled at a specified nominal α level that is set by the researcher, say .05, what happens to the overall α level for the entire set of tests? The overall α level for the entire set of tests (i.e., α_{total}), often called the **experiment-wise Type I error rate** (i.e., the Type I error across all experiments), is larger than the a level for each of the individual t tests.

In our example we are interested in comparing the means for 10 pairs of groups (again, these would be $\mu_1 = \mu_2$, $\mu_1 = \mu_3$, $\mu_1 = \mu_4$, $\mu_1 = \mu_5$, $\mu_2 = \mu_3$, $\mu_2 = \mu_4$, $\mu_2 = \mu_5$, $\mu_3 = \mu_4$, $\mu_3 = \mu_5$, and $\mu_4 = \mu_5$). A t test is conducted for each of the 10 pairs of groups at $\alpha = .05$. Although each test controls the a level at .05, the overall α level will be larger because the risk of a Type I error accumulates across the tests. For each test we are taking a risk; the more tests we do, the more risks we are taking. This can be explained by considering the risk you take each day you drive your car to school or work. The risk of an accident is small for any one day; however, over the period of a year, the risk of an accident is much larger.

For C independent (or **orthogonal**) tests, the experiment-wise error is as follows.

$$\alpha_{\text{total}} = 1 - (1 - \alpha)^C$$

Assume for the moment that our 10 tests are independent (although they are not, because within those 10 tests, each group is actually being compared to another group in four different instances). If we go ahead with our 10 t tests at $\alpha = .05$, then the experiment-wise error rate is

$$\alpha_{\text{total}} = 1 - (1 - .05)^{10} = 1 - .60 = .40$$

Although we are seemingly controlling our α level at the .05 level, the probability of making a Type I error across all 10 tests is .40. In other words, in the long run, if we conduct 10 independent t tests, 4 times out of 10 we will make a Type I error. For this reason, we do not want to do all possible t tests. Before we move on, the experiment-wise error rate for C dependent tests α_{total} (which would be the case when doing all possible pairwise t tests, as in our example) is more difficult to determine, so let us just say that

$$\alpha \le \alpha_{\text{total}} \le C\alpha$$

Are other options available to us where we can maintain better control over our experiment-wise error rate? The optimal solution, in terms of maintaining control over our overall α level as well as maximizing power, is to conduct *one overall test*, often called an **omnibus test**. Recall that power has to do with the probability of correctly rejecting a false null hypothesis. The omnibus test could assess the equality of all of the means simultaneously and is the one used in the analysis of variance. *The one-factor analysis of variance, then, represents an extension of the independent* t *test for two or more independent sample means, where the experiment-wise error rate is controlled.*

In addition, the one-factor ANOVA has *only one independent variable or factor* with two or more levels. The independent variable is a discrete or grouping variable, where each subject responds to only one level. The levels represent the different samples or groups or treatments whose means are to be compared. In an example, method of instruction is the independent variable with three levels: large lecture hall, small-group, and computer-assisted. There are two ways of conceptually thinking about the selection of levels. The **fixed-effects model** is one way to conceptualize the levels. As a result, generalizations can be made only about those particular levels of the independent variable that are actually selected. For instance, if a researcher is interested only in these three methods of instruction—large lecture hall, small-group, and computer-assisted—then only those levels are incorporated into the study. Generalizations about other methods of instruction cannot be made because no other methods were considered for selection. Other examples of fixed-effects independent variables might be socioeconomic status, sex, specific types of treatment, age group, weight, or marital status. Not all researchers agree on what constitutes fixed versus random effects, and some would argue that fixed effects are encountered only in experiments in which the researcher can manipulate the independent variable. We err on the side of defining fixed effects in this sense: *In the fixed-effects model, all levels that the researcher is interested in are included in the design and analysis for the study.* Should your independent variable meet this definition, then it is considered a fixed (not random) effect.

The **random-effects model** is the second way to conceptualize the levels. *In the random-effects model, the researcher randomly samples some levels of the independent variable from the population of levels.* As a result, generalizations can be made about *all* of the levels in the population, even those not actually sampled. For instance, a researcher interested in teacher effectiveness may have randomly sampled history teachers from the population of history teachers in a particular school district. Generalizations can then be made about other history teachers in that school district not actually sampled. The random selection of *levels* is much the same as the random selection of individuals or objects in the random sampling process. This is the nature of inferential statistics, where inferences are made about a population (of individuals, objects, or levels) from a sample. Other examples of random-effects independent variables might include randomly selected classrooms or time (e.g., hours, days), among others. The remainder of this chapter is concerned with the fixed-effects model. Chapter 5 discusses the random-effects model in more detail.

In the fixed-effects model, once the levels of the independent variable are selected, subjects (i.e., persons or objects) are randomly assigned to the levels of the independent variable. In certain situations, the researcher does not have control over which level a subject is assigned to. The groups may be preexisting—i.e., already in place when the research commences. For instance, students may be assigned to their classes at the beginning of the year by the school administration. Researchers typically have little input regarding class assignments. In another situation, it may be theoretically impossible to assign subjects to groups. For example, as much as we might like, researchers cannot randomly assign individuals to an age level. Thus, a distinction needs to be made about whether or not the researcher can control the assignment of subjects to groups. Although the analysis will not be altered, the interpretation of the results will differ depending on whether or not there is random assignment to groups. *When researchers have control over group assignments and exercise that control (e.g., in terms of random assignment to groups), the extent to which they can infer causality from their findings is greater than for those researchers who do not have such control.* For further information on the differences between **true experimental designs** (i.e., with random assignment) and **quasi-experimental designs** (i.e., without random assignment), take a look at Campbell and Stanley (1966), Cook and Campbell (1979), and Shadish, Cook, and Campbell (2002).

Moreover, in the model being considered here, each subject is exposed to only one level of the independent variable. Chapter 5 deals with models where a subject is exposed to multiple levels of an independent variable; these are known as **repeated-measures models**. For example, a researcher may be interested in observing a group of young children repeatedly over a period of several years. Thus, each child might be observed every 6 months from birth to 5 years of age. This would require a repeated-measures design because the observations of a particular child over time are obviously not independent observations.

One final characteristic is the **measurement scale** of the independent and dependent variables. In the analysis of variance, because this is a test of means, *a condition of the test is that the scale of measurement on the dependent variable is at the interval or ratio level.* If the dependent variable is measured at the ordinal level, then the nonparametric equivalent, the Kruskal-Wallis test, should be considered (discussed later in this chapter). If the dependent variable shares properties of both the ordinal and interval levels (e.g., grade point average), then both the ANOVA and Kruskal-Wallis procedures could be considered to cross-reference any potential effects of the measurement scale on the results.

As previously mentioned, *the independent variable is a grouping or discrete variable, so it can theoretically be measured on any scale*. However, there is one caveat to the measurement scale of the independent variable. Technically the condition is that the independent variable be a grouping or discrete variable. *Most often, ANOVAs are conducted with independent variables which are categorical—nominal or ordinal in scale*. ANOVAs can also be used in the case of interval or ratio values that are discrete. Recall that discrete variables are variables that can only take on certain values and that arise from the counting process. An example of a discrete variable that could be a good candidate for being an independent variable in an ANOVA model is number of children. What would make this a good candidate? The responses to this variable would likely be relatively limited (in the general population, it may be anticipated that the range would be from zero children to five or six, although outliers may be a possibility) and each discrete value would likely have multiple cases (with fewer cases having larger numbers of children). Applying this is obviously at the researcher's discretion; at some point the number of discrete values can become so numerous as to be unwieldy in an ANOVA model. Thus, while at first glance we may not consider it appropriate to use interval or ratio variables as independent variables in ANOVA models, there are situations where it is feasible and appropriate. While the minimum number of levels or categories of the independent variable is two, there is not a maximum number of categories. However, we have found that ANOVA models with independent variables that have more than five or six categories can become unwieldy—particularly in the case of factorial ANOVA, which we will study in a later chapter.

In summary, the characteristics of the one-factor analysis of variance fixed-effects model are as follows (also see Box 1.1):

(a) control of the experiment-wise error rate through an omnibus test;
(b) one independent variable with two or more levels;
(c) the levels of the independent variable are fixed by the researcher;
(d) subjects are randomly assigned to these levels;
(e) subjects are exposed to only one level of the independent variable; and
(f) the dependent variable is measured at least at the interval level, although the Kruskal-Wallis one-factor ANOVA can be considered for an ordinal-level dependent variable. In the context of experimental design, the one-factor analysis of variance is often referred to as the **completely randomized design**.

BOX 1.1 Characteristics of the One-Factor Analysis of Variance Fixed Effects

Model Feature	Characteristic
Variables	• One dependent variable • One independent variable
Measurement scale	• Dependent variable: At least interval in scale • Independent variable: Any scale of variable that is grouping (e.g., nominal or ordinal) or discrete (interval or ratio) but caution is given when considering independent variables with a large number of levels
Design features	• Subjects are randomly assigned to the levels of the independent variable • The levels of the independent variable are fixed by the researcher • Subjects are exposed to only one level of the independent variable
Statistical features	The experiment-wise error rate is controlled through an omnibus test

1.1.1.1 The Layout of the Data

Before we get into the theory and analysis of the data, let us examine one tabular form of the data, known as the layout of the data. We designate each observation as Y_{ij}, where the j subscript tells us what group or level the observation belongs to and the i subscript tells us the observation or identification number within that group. For instance, Y_{34} would mean this is the third observation in the fourth group, or level, of the independent variable. The first subscript ranges over $i = 1, \ldots, n$ and the second subscript ranges over $j = 1, \ldots, J$. Thus there are **J levels (or categories or groups) of the independent variable** and **n subjects in each group**, for a total of $Jn = N$ total observations. For now, presume there are n subjects (or cases or units) in each group in order to simplify matters; this is referred to as the **equal n's** or **balanced case**. Later on in this chapter, we consider the **unequal n's** or **unbalanced case**.

The layout of the data is shown in Table 1.1. Here we see that each column represents the observations for a particular group or level of the independent variable. At the bottom of each column are the sample group means ($\bar{Y}_{.j}$), with the overall sample mean ($\bar{Y}_{..}$) to the far right. In conclusion, the layout of the data is one form in which the researcher can think about the data.

1.1.1.2 ANOVA Theory

This section examines the underlying theory and logic of the analysis of variance, the sums of squares, and the ANOVA summary table. As noted previously, in the analysis of variance mean, differences are tested by looking at the variability of the means. Here we show precisely how this is done.

1.1.1.2.1 General Theory and Logic

We begin with the hypotheses to be tested in the analysis of variance. In the two-group situation of the independent t test, the null and alternative hypotheses for a two-tailed (i.e., nondirectional) test are as follows, where the null hypothesis is simply saying that the means of the two groups are the same.

$$H_0 : \mu_1 = \mu_2$$
$$H_1 : \mu_1 \neq \mu_2$$

TABLE 1.1

Layout for the One-Factor ANOVA Model

	Level of the Independent Variable					
	1	2	3	...	J	
	Y_{11}	Y_{12}	Y_{13}	...	Y_{1J}	
	Y_{21}	Y_{22}	Y_{23}	...	Y_{2J}	
	Y_{31}	Y_{32}	Y_{33}	...	Y_{3J}	
	.	.	.	.	.	
	.	.	.	.		
	Y_{n1}	Y_{n2}	Y_{n3}	...	Y_{nJ}	
Means	$\bar{Y}_{.1}$	$\bar{Y}_{.2}$	$\bar{Y}_{.3}$	...	$\bar{Y}_{.J}$	$\bar{Y}_{..}$

In the multiple-group situation (i.e., more than two groups), we have already seen the problem that occurs when multiple independent *t* tests are conducted for all pairs of population means (i.e., the problem is an increased likelihood of a Type I error). We concluded that the solution was to use an *omnibus test* where the equality of all of the means could be assessed simultaneously. The hypotheses for the omnibus analysis of variance test are as follows:

$$H_0: \mu_1 = \mu_2 = \mu_3 = \ldots = \mu_J$$
$$H_1: \textit{not all the } \mu_j \textit{ are equal}$$

Here H_1 is purposely written in a general form to cover the *multitude* of possible mean differences that could arise. These range from only two of the means being different to *all* of the means being different from one another. Thus, because of the way H_0 has been written, only a nondirectional alternative is appropriate. If H_0 were to be rejected, then the researcher might want to consider a multiple comparison procedure so as to determine which means or combination of means are significantly different (we cover this in greater detail in Chapter 2).

As was mentioned in the introduction to this chapter, the analysis of mean differences is actually carried out by looking at variability of the means. At first this seems strange. If one wants to test for mean differences, then do a test of means. If one wants to test for variance differences, then do a test of variances. These statements should make sense because logic pervades the field of statistics. And they do for the two-group situation. For the multiple-group situation, we already know things get a bit more complicated.

Say a researcher is interested in the influence of amount of daily study time on statistics achievement. Three groups were formed based on the amount of daily study time in statistics, 30 minutes, 1 hour, and 2 hours. Is there a differential influence of amount of time studied on subsequent mean statistics achievement (e.g., statistics final exam)? We would expect that the more one studied statistics, the higher the statistics mean achievement would be. *One possible situation in the population is where the amount of study time does not influence statistics achievement; here the population means will be equal.* That is, the null hypothesis of equal group means is actually *true*. Thus, the three groups are really three samples from the same population of students, with mean μ. The means are equal; thus there is no variability among the three group means. *A second possible situation in the population is where the amount of study time does influence statistics achievement; here the population means will not be equal.* That is, the null hypothesis is actually *false*. Thus, the three groups are not really three samples from the same population of students, but rather, each group represents a sample from a distinct population of students receiving that particular amount of study time, with mean μ_j. The means are not equal, so there is variability among the three group means. In summary, the statistical question becomes whether the difference between the sample means is due to the usual sampling variability expected from a single population, or the result of a true difference between the sample means from different populations.

We conceptually define **within-groups variability** *as the variability of the observations within a group combined across groups* (e.g., variability on test scores within children in the same proficiency level, such as low, moderate, and high, and then combined across all proficiency levels), and **between-groups variability** *as the variability between the groups* (e.g., variability among the test scores from one proficiency level to another proficiency level). In Figure 1.1, the columns represent low and high variability *within* the groups. The rows represent low and high variability *between* the groups.

In the upper left-hand plot of Figure 1.1, there is low variability both within and between the groups. That is, performance is very consistent, both within each group as well as across groups. We see that there is little variability *within* the groups, since the individual distributions are not very spread out, and little variability *between* the groups because the distributions are not very distinct, as they are nearly lying on top of one another. Here, within- and between-group variability are both low, and it is quite unlikely that one would reject H_0.

In the upper right-hand plot of Figure 1.1, there is high variability within the groups and low variability between the groups. That is, performance is very consistent across groups (i.e., the distributions largely overlap), but quite variable within each group. We see high variability *within* the groups because the spread of each individual distribution is quite large, and low variability *between* the groups because the distributions are lying so closely together. Here within-group variability exceeds between-group variability, and again it is quite unlikely that one would reject H_0.

In the lower left-hand plot of Figure 1.1, there is low variability within the groups and high variability between the groups. That is, performance is very consistent within each group, but quite variable across groups. We see low variability *within* the groups because each distribution is very compact with little spread to the data, and high variability *between* the groups because each distribution is nearly isolated from one another with very little overlap. Here between-group variability exceeds within-group variability, and it is quite likely that one would reject H_0.

In the lower right-hand plot of Figure 1.1, there is high variability both within and between the groups. That is, performance is quite variable within each group, as well as across the groups. We see high variability *within* groups because the spread of each individual distribution is quite large, and high variability *between* groups because of the minimal overlap

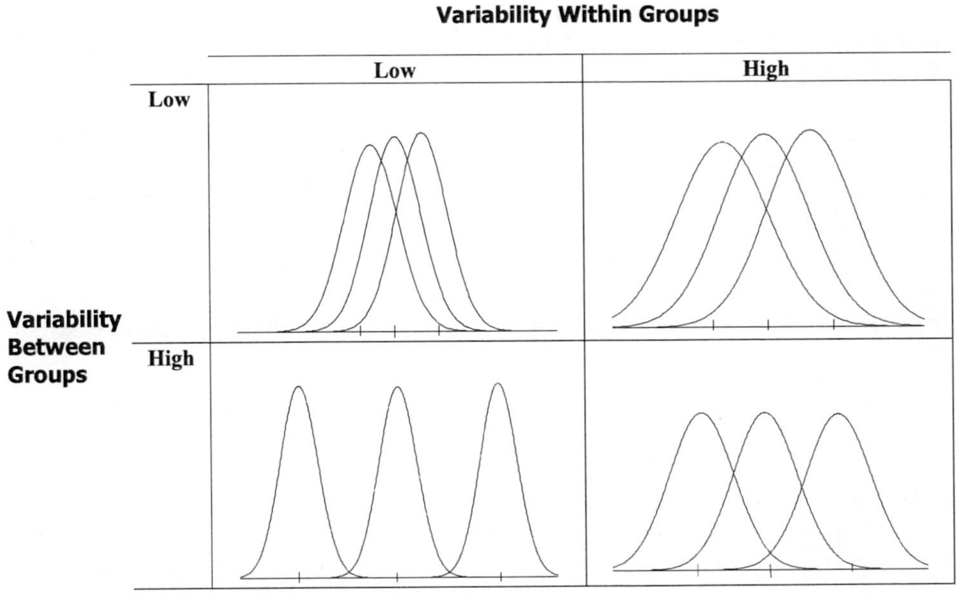

FIGURE 1.1
Conceptual look at between- and within-groups variability.

from one distribution to another. Here within- and between-group variability are both high, and depending on the relative amounts of between- and within-group variability, one may or may not reject H_0. *In summary, the optimal situation when seeking to reject H_0 is the one represented by high variability between the groups and low variability within the groups.*

1.1.1.2.2 General Linear Model

ANOVA and linear regression are both forms of the same **general linear model (GLM)**. In other words, ANOVA and linear regression are mathematically equal and thus correlational (as we will see later, we are provided ANOVA output when we generate regression). The differences between these methods is largely researcher-created. In practice, historically the methods were used in different disciplines. Multiple regression was used to examine natural variation in the biological and behavioral sciences, while ANOVA was used to study manipulated variation (i.e., experiments) in agricultural science (Cohen, 1968). Different algorithms, terminology, and uses have placed an artificial divide between ANOVA and regression; however, they are mathematically equivalent. Cohen (1968) illustrates the equivalence of the two systems. Thompson (2016) notes how other researchers have shown how other methods are subsumed in GLM. Indeed, most parametric procedures (e.g., *t* test, ANOVA, regression, multivariate ANOVA, structural equation modeling, and more) are part of the general linear model. Given their mathematical equivalence, whether you select ANOVA or regression is largely a consideration of convenience given your data. For example, should all or most of your predictors be continuous, single or multiple regression is more convenient. Should all or more of your predictors be categorical, one-way ANOVA or factorial ANOVA is more convenient.

Let's consider the basic case of the general linear model where one dependent variable is predicted from one or more independent variables. In fitting the linear model, weights or coefficients are computed for each independent variable. The dependent variable is the sum of three elements: the intercept (i.e., a mathematical constant), the sum of the weighted independent variables, and error. In regression, the intercept is the predicted value of the dependent variable when all independent variables have a value of zero. ANOVA in the GLM framework requires dummy coding of the categorical variables and inclusion of all but one of the dummy variables in the model. In ANOVA, therefore, the intercept becomes the mean of the category which was left out.

All statistical procedures that are subsumed under the GLM share several characteristics. These include partitioning variances and estimating ratio of those (e.g., eta squared, reliability coefficients) and yielding variance-accounted for effect size estimates (Thompson, 2016). One of the primary reasons this is important to understand is because it is the *design of your study* that allows claims to be made (e.g., causality) and not the statistical procedure that was used to analyze the data.

1.1.1.2.3 Partitioning the Sums of Squares

The **partitioning of the sums of squares** in ANOVA is a new concept in this chapter, which is also an important concept in regression analysis (see Chapters 7 and 8). In part this is because ANOVA and regression are both forms of the same GLM that we just learned about. Let us begin with the total **sum of squares in *Y***, denoted as SS_{total}. *The term SS_{total} represents the amount of total variation in Y*. The next step is to partition the total variation into **variation between the groups** (i.e., the categories or levels of the independent variable), denoted by SS_{betw}, and **variation within the groups** (i.e., units or cases within each category or level of the independent variable), denoted by SS_{with}. In the one-factor analysis of variance, we therefore partition the sum of square total, SS_{total}, into sum of squares between, SS_{betw}, and sum of squares within, SS_{with}, as follows:

$$SS_{\text{total}} = SS_{\text{betw}} + SS_{\text{with}}$$

Or

$$\sum_{i=1}^{n}\sum_{j=1}^{J}\left(Y_{ij} - \bar{Y}_{..}\right)^2 = \sum_{i=1}^{n}\sum_{j=1}^{J}\left(\bar{Y}_{.j} - \bar{Y}_{..}\right)^2 + \sum_{i=1}^{n}\sum_{j=1}^{J}\left(Y_{ij} - \bar{Y}_{.j}\right)^2$$

As a side note, algebraically, the sum of squares between can be calculated as follows, where the difference between the group mean and overall mean is weighted by the sample size of the group:

$$\sum_{i=1}^{n}\sum_{j=1}^{J}\left(Y_{ij} - \bar{Y}_{..}\right)^2 = \sum_{i=1}^{n} n_j\left(\bar{Y}_{.j} - \bar{Y}_{..}\right)^2 + \sum_{i=1}^{n}\sum_{j=1}^{J}\left(Y_{ij} - \bar{Y}_{.j}\right)^2$$

where

- SS_{total} is the total sum of squares due to variation among all of the observations without regard to group membership,
- SS_{betw} is the between-groups sum of squares due to the variation *between* the groups, and
- SS_{with} is the within-groups sum of squares due to the variation *within* the groups combined across groups.

We refer to this particular formulation of the partitioned sums of squares as the **definitional** (or **conceptual**) **formula**, because each term literally defines a form of variation.

Due to computational complexity and the likelihood of a computational error, the definitional formula is rarely used with real data. Instead, a **computational formula** for the partitioned sums of squares is used for hand computations. However, since nearly all data analysis at this level utilizes computer software, we defer to the software to actually perform an analysis of variance (SPSS and **R** examples are provided toward the end of this chapter). A complete example of the one-factor analysis of variance is also considered later in this chapter.

1.1.1.2.4 ANOVA Summary Table

An important result of the analysis is the **ANOVA summary table**. The purpose of the summary table is to literally summarize the analysis of variance. A general form of the summary table is shown in Table 1.2. *The first column lists the sources of variation in the model.* As we already know, in the one-factor model the total variation is partitioned into between-groups variation and within-groups variation. *The second column notes the sums of squares terms computed for each source (i.e.,* SS_{betw}, SS_{with}, *and* SS_{total}*).*

TABLE 1.2

Analysis of Variance Summary Table

Source	*SS*	*df*	*MS*	*F*
Between groups	SS_{betw}	$J - 1$	MS_{betw}	MS_{betw}/MS_{with}
Within groups	SS_{with}	$N - J$	MS_{with}	
Total	SS_{total}	$N - 1$		

The third column gives the degrees of freedom for each source. Recall that, in general, the degrees of freedom have to do with the number of observations that are free to vary. For example, if a sample mean and all of the sample observations except for one are known, then the final observation is not free to vary. That is, the final observation is predetermined to be a particular value. For instance, say the mean is 10 and there are three observations, 7, 11, and an unknown observation. Based on that information, first, the sum of the three observations must equal 30 for the mean to be 10. Second, the sum of the known observations is 18. Therefore, the unknown observation must be 12. Otherwise, the sample mean would not be exactly equal to 10.

For the **between-groups source**, the definitional formula is concerned with the deviation of each group mean from the overall mean. There are ***J* group means** (where J represents the number of groups or categories or levels of the independent variable), so the df_{betw} (also known as the *degrees of freedom numerator*) must be $J - 1$. Why? If we have J group means and we know the overall mean, then only $J - 1$ of the group means are free to vary. In other words, if we know the overall mean and all but one of the group means, then the final unknown group mean is predetermined.

For the **within-groups source**, the definitional formula is concerned with the deviation of each observation from its respective group mean. There are ***n* observations** (i.e., cases or units) in each group; consequently, there are $n - 1$ degrees of freedom in each group and J groups. Why are there $n - 1$ degrees of freedom in each group? If there are n observations in each group, then only $n - 1$ of the observations are free to vary. In other words, if we know one group mean and all but one of the observations for that group, then the final unknown observation for that group is predetermined. There are J groups, so the df_{with} (also known as the *degrees of freedom denominator*) is $J(n - 1)$, or more simply as $N - J$. Thus we lose one degree of freedom for each group.

For the **total source**, the definitional formula is concerned with the deviation of each observation from the overall mean. There are N total observations; therefore the df_{total} must be $N - 1$. Why? If there are N total observations and we know the overall mean, then only $N - 1$ of the observations are free to vary. In other words, if we know the overall mean, and all but one of the N observations, then the final unknown observation is predetermined.

Why is the number of degrees of freedom important in the analysis of variance? Suppose two researchers have conducted similar studies, except Researcher A uses 20 observations per group and Researcher B uses 10 observations per group. Each researcher obtains a SS_{with} value of 15. Would it be fair to say that this particular result for the two studies is the same? Such a comparison would be unfair because SS_{with} is influenced by the number of observations per group. A fair comparison would be to weight the SS_{with} terms by their respective number of degrees of freedom. Similarly, it would not be fair to compare the SS_{betw} terms from two similar studies based on different numbers of groups. A fair comparison would be to weight the SS_{betw} terms by their respective number of degrees of freedom. *The method of weighting a sum of squares term by the respective number of degrees of freedom on which it is based yields what is called a* **mean squares** *term*. Thus, $MS_{betw} = SS_{betw}/df_{betw}$ and $MS_{with} = SS_{with}/df_{with}$, as shown in the fourth column of Table 1.2. They are referred to as *mean squares* because they represent a summed quantity that is weighted by the number of observations used in the sum itself, like the mean. *The mean squares terms are also variance estimates* because they represent the sum of the squared deviations from a mean divided by their degrees of freedom, like the sample variance, s^2.

The last column in the ANOVA summary table, the F value (also known as the ***F ratio*** *), is the summary test statistic of the summary table.* The F value is computed by taking the ratio of the two mean squares or variance terms. Thus for the one-factor ANOVA fixed-effects model, the F value is computed as $F = MS_{betw}/MS_{with}$. When developed by Sir Ronald A. Fisher in the 1920s, this test statistic was originally known as the variance ratio because it represents the ratio of two variance estimates. Later, the variance ratio was renamed the F ratio by George W. Snedecor (who worked out the table of F values, discussed momentarily) in honor of Fisher (F for Fisher).

The F ratio tells us whether there is more variation *between* groups than there is *within* groups, which is required if we are to reject H_0. Thus, if there is more variation *between* groups than there is *within* groups, then MS_{betw} will be larger than MS_{with}. As a result of this, the F ratio of MS_{betw}/MS_{with} will be greater than 1. If, on the other hand, the amount of variation *between* groups is about the same as there is *within* groups, then MS_{betw} and MS_{with} will be about the same, and the F ratio will be approximately 1. Thus, we want to find large F values in order to reject the null hypothesis.

The F test statistic is then compared with the F critical value so as to make a decision about the null hypothesis. The critical value is found in the F table of Table A.4 as ${}_{\alpha}F_{(J-1,N-J)}$. Thus, the degrees of freedom are $df_{betw} = J - 1$ for the numerator of the F ratio and $df_{with} = N - J$ for the denominator of the F ratio. The significance test is a one-tailed test in order to be consistent with the alternative hypothesis. The null hypothesis is rejected if the F test statistic exceeds the F critical value. This is the **omnibus *F* test** which, again, simply provides evidence of the extent to which there is *at least one* statistically significant mean difference between the groups.

If the F test statistic exceeds the F critical value, and there are more than two groups, then it is not clear where the differences among the means lie. In this case, some **multiple comparison procedure** should be used to determine where the mean differences are in the groups; this is the topic of Chapter 2. When there are only two groups, it is obvious where the mean difference falls, that is, between groups 1 and 2. A researcher can simply look at the descriptive statistics to determine which group had the higher mean relative to the other group. *For the two-group situation*, it is also interesting to note that the F and t test statistics follow the rule of $F = t^2$, for a nondirectional alternative hypothesis in the independent t test. In other words, the one-way ANOVA with two groups and the independent t test will generate the same conclusion such that $F = t^2$. This result occurs when the numerator degrees of freedom for the F ratio is 1. In an actual ANOVA summary table (shown in the next section), except for the source of variation column, it is the values for each of the other entries generated from the data that are listed in the table. For example, instead of seeing SS_{betw}, we would see the computed value of SS_{betw}.

1.1.1.3 The ANOVA Model

In this section we introduce the analysis of variance linear model, cover the estimation of parameters of the model, effect size measures, confidence intervals, power, and an example, and finish up with expected mean squares.

1.1.1.3.1 The Model

The one-factor ANOVA fixed-effects model can be written in terms of population parameters as

$$Y_{ij} = \mu + \alpha_j + \varepsilon_{ij}$$

where Y is the observed score on the dependent (or criterion) variable for individual i in group j, μ is the overall or grand population mean (i.e., regardless of group designation), α_j is the group effect for group j, and ε_{ij} is the random residual error for individual i in group j. The residual error can be due to individual differences, measurement error, and/or other factors not under investigation (i.e., other than the independent variable X). The population group effect and residual error are computed as

$$\alpha_j = \mu_{.j} - \mu$$

and

$$\varepsilon_j = Y_{ij} - \mu_{.j}$$

respectively, and $\mu_{.j}$ is the population mean for group j, where the initial dot subscript indicates we have averaged across all i individuals in group j. That is, the group effect is equal to the difference between the population mean of group j and the overall population mean. The residual error is equal to the difference between an individual's observed score and the population mean of the group of which the individual is a member (i.e., group j). The group effect can also be thought of as the average effect of being a member of a particular group. A positive group effect implies a group mean greater than the overall mean, whereas a negative group effect implies a group mean less than the overall mean. Note that in a fixed-effects one-factor model, the population group effects sum to zero. The residual error in the analysis of variance represents that portion of Y not accounted for by X.

1.1.1.3.2 Estimation of the Parameters of the Model

Next, we need to estimate the parameters of the model μ, α_j, and ε_{ij}. The sample estimates are represented by $\bar{Y}_{..}$, a_j, and e_{ij}, respectively, where the latter two are computed as

$$a_j = \bar{Y}_{.j} - \bar{Y}_{..}$$

$$e_{ij} = Y_{ij} - \bar{Y}_{.j}$$

respectively. Note that $\bar{Y}_{..}$ represents the overall sample mean, where the double dot subscript indicates we have averaged across both the i and j subscripts, and $\bar{Y}_{.j}$ represents the sample mean for group j, where the initial dot subscript indicates we have averaged across all i individuals in group j.

1.1.1.3.3 Confidence Intervals

Confidence interval procedures are often useful in providing an interval estimate of a population parameter (i.e., mean or mean difference); these allow us to determine the accuracy of the sample estimate. One can form confidence intervals around any sample group mean from an ANOVA (provided in software such as SPSS), although confidence intervals for means have more utility for multiple comparison procedures, as discussed in Chapter 2. Confidence interval procedures have also been developed for several effect size measures (Fidler & Thompson, 2001; Smithson, 2001).

1.1.1.3.4 An Example

Consider now an example problem used throughout this chapter. Our dependent variable is psychological distress (a continuous score), whereas the independent variable is the type of sport in which an elite athlete competes. The researcher is interested in whether the type of sport in which an athlete competes influences their psychological distress. The types of sports are defined as follows:

- Sport 1, movement (e.g., gymnastics, dance);
- Sport 2, target (e.g., golf);
- Sport 3, fielding (e.g., baseball); and
- Sport 4, territory (e.g., football).

There were eight athletes in each sport, for a total of 32. In Table 1.3 we see the raw data and sample statistics (means and variances) for each sport and overall (far right).

The results are summarized in the ANOVA summary table as shown in Table 1.4. The test statistic, $F = 6.1877$, is compared to the critical value, ${}_{.05}F_{3,28} = 2.95$ obtained from Appendix Table A.4, using the .05 level of significance. To use the F table, find the numerator degrees of freedom, df_{betw}, which are represented by the columns, and then the denominator degrees of freedom, df_{with}, which are represented by the rows. The intersection of the two provides the F critical value. The test statistic exceeds the critical value, so we reject H_0 and conclude that type of sport is related to mean differences in psychological distress. The exact probability value (p value) given by SPSS is .001.

TABLE 1.3

Data and Summary Statistics for the Elite Athlete Example

	Psychological Distress by Type of Sport				
	Group 1: Movement (e.g., dance)	**Group 2: Target (e.g., golf)**	**Group 3: Fielding (e.g., baseball)**	**Group 4: Territory (e.g., football)**	**Overall**
	15	20	10	30	
	10	13	24	22	
	12	9	29	26	
	8	22	12	20	
	21	24	27	29	
	7	25	21	28	
	13	18	25	25	
	3	12	14	15	
Means	11.1250	17.8750	20.2500	24.3750	18.4063
Variances	30.1250	35.2679	53.0714	25.9821	56.4425

TABLE 1.4

Analysis of Variance Summary Table—Psychological Distress Example

Source	*SS*	*df*	*MS*	*F*
Between groups	738.5938	3	246.1979	6.8177*
Within groups	1,011.1250	28	36.1116	
Total	1,749.7188	31		

$^{*}\ {}_{.05}F_{3,28} = 2.95$

Next we examine the group effects and residual errors. The group effects are estimated as follows where the grand mean (irrespective of the group membership; here 18.4063) is subtracted from the group mean (e.g., 11.125 for group 1). The subscript of *a* indicates the level or group of the independent variable (e.g., 1 = movement; 2 = target; 3 = fielding; 4 = territory). A **negative group effect** indicates that group had a larger mean than the overall average and thus exerted a negative effect on the dependent variable (in our case, higher psychological distress). A **positive group effect** indicates that group had a smaller mean than the overall average and thus exerted a positive effect on the dependent variable (in our case, lower psychological distress).

$$a_1 = \bar{Y}_{.1} - \bar{Y}_{..} = 11.125 - 18.4063 = -7.2813$$
$$a_2 = \bar{Y}_{.2} - \bar{Y}_{..} = 17.875 - 18.4063 = -0.5313$$
$$a_3 = \bar{Y}_{.3} - \bar{Y}_{..} = 20.250 - 18.4063 = +1.8437$$
$$a_4 = \bar{Y}_{.4} - \bar{Y}_{..} = 24.375 - 18.4063 = +5.9687$$

Thus, group 4 (territory) has the largest *negative* group effect (i.e., highest psychological distress), while group 1 (movement) has the largest *positive* group effect (i.e., lowest psychological distress). In Chapter 2 we use the same data to determine which of these group means, or combination of group means, are statistically different. The residual errors (computed as the difference between the observed value and the group mean) for each individual by group are shown in Table 1.5 and discussed later in this chapter.

TABLE 1.5

Residuals for the Psychological Distress Example by Group

Group 1: Movement (e.g., dance)	Group 2: Target (e.g., golf)	Group 3: Fielding (e.g., baseball)	Group 4: Territory (e.g., football)
3.875	2.125	−10.250	5.625
−1.125	−4.875	3.750	−2.375
0.875	−8.875	8.750	1.625
−3.125	4.125	−8.250	−4.375
9.875	6.125	6.750	4.625
−4.125	7.125	0.750	3.625
1.875	0.125	4.750	0.625
−8.125	−5.875	−6.250	−9.375

1.1.1.3.5 Expected Mean Squares

There is one more theoretical concept called **expected mean squares** to introduce in this chapter. The notion of expected mean squares provides the basis for determining what the appropriate error term is when forming an F ratio (recall this ratio is $F = MS_{betw}/MS_{with}$). That is, when forming an F ratio to test a certain hypothesis, how do we know which source of variation to use as the error term in the denominator? For instance, in the one-factor fixed-effects ANOVA model, how did we know to use MS_{with} as the error term in testing for differences between the groups? There is a good rationale, as becomes evident.

Before we get into expected mean squares, consider the definition of an expected value. *An expected value is defined as the average value of a statistic that would be obtained with repeated sampling*. Using the sample mean as an example statistic, the expected value of the mean would be the average value of the sample means obtained from an infinite number of samples. *The expected value of a statistic is also known as the mean of the sampling distribution of that statistic*. In this case, the expected value of the mean is the mean of the sampling distribution of the mean.

An expected mean square for a particular source of variation represents the average mean square value for that source obtained if the same study were to be repeated an infinite number of times. For instance, the expected value of MS_{betw}, denoted by $E(MS_{betw})$, is the average value of MS_{betw} over repeated samplings. At this point you might be asking, "Why not only be concerned about the values of the mean square terms for my own little study?" Well, the mean square terms from your little study do represent a sample from a population of mean square terms. Thus, sampling distributions and sampling variability are as much a concern in the analysis of variance as they are in other situations previously described in this text.

Now we are ready to see what the expected mean square terms actually look like. Consider the two situations of H_0 actually being true and H_0 actually being false. If H_0 is actually *true*, such that there really are *no* differences between the population group means, then the *expected mean squares* [represented in statistical notation as either by $E(MS_{betw})$ or by $E(MS_{with})$] are as follows:

$$E(MS_{betw}) = \sigma_\varepsilon^2$$

$$E(MS_{with}) = \sigma_\varepsilon^2$$

and thus the ratio of expected mean squares is:

$$E(MS_{betw})/E(MS_{with}) = 1$$

where the expected value of F is then $E(F) = df_{with}/(df_{with} - 2)$, and σ_ε^2 is the population variance of the residual errors. This tells us the following: if H_0 is actually true, then each of the J samples really comes from the same population with mean μ.

If H_0 is actually *false*, such that there really *are* differences between the population group means, then the expected mean squares are as follows:

$$E(MS_{betw}) = \sigma_\varepsilon^2 + \frac{\left(n\sum_{j=1}^{J}\alpha_j^2\right)}{J-1}$$

$$E(MS_{with}) = \sigma_\varepsilon^2$$

and thus the ratio of the expected mean squares is as follows:

$$E(MS_{betw})/E(MS_{with}) > 1$$

where $E(F) > df_{with}/(df_{with} - 2)$. If H_0 is actually false, then the J samples do really come from different populations with different means μ_j.

There is a difference in the expected mean square between [i.e., E (MS_{betw})] when H_0 is actually true as compared to when H_0 is actually false, as in the latter situation there is a second term. The important part of this second term is $\sum_{j=1}^{J} \alpha_j^2$ which represents the **sum of the squared group effects**. The larger this part becomes, the larger MS_{betw} is, and thus the larger the F ratio becomes. In comparing the two situations, we also see that E (MS_{betw}) is the same whether H_0 is actually true or false, and thus represents a reliable estimate of σ_ε^2. This term is mean-free because it does not depend on group mean differences. Just to cover all of the possibilities, F could be less than one [or technically less than $df_{with}/(df_{with} - 2)$] due to sampling error, nonrandom samples, and/or assumption violations. For a mathematical proof of the E(MS) terms, see Kirk (2013).

Finally, let us try to put all of this information together. In general, the F ratio represents the following:

$$F = (systematic\ variability + error\ variability)/error\ variability$$

where, for the one-factor fixed-effects model, *systematic variability* is variability *between* the groups and *error variability* is variability *within* the groups. The F ratio is formed in a particular way because we want to isolate the systematic variability in the numerator. For this model, the only appropriate F ratio is MS_{betw}/MS_{with} because it does serve to isolate the systematic variability (i.e., the variability between the groups). That is, the appropriate error term for testing a particular effect (e.g., mean differences between groups) is the mean square that is identical to the mean square of that effect, except that it lacks a term due to the effect of interest. For this model, the appropriate error term to use for testing differences between groups is the mean square identical to the numerator MS_{betw}, except it lacks a term due to the between-groups etffect [i.e., $\left(n\sum_{j=1}^{J} \alpha_j^2\right)/(J-1)$]; this, of course, is MS_{with}. It should also be noted that the F ratio is a ratio of two independent variance estimates, here being MS_{betw} and MS_{with}.

1.1.1.4 The Unequal n*'s or Unbalanced Procedure*

Up to this point in the chapter, we have considered only the equal n's or balanced case where the number of observations is equal for each group. This was done to make things simple for presentation purposes. However, we do not need to assume that the n's must be equal (as some textbooks incorrectly do). This section briefly describes the **unequal *n*'s** or **unbalanced case**. For our purposes, the major statistical software can handle the analysis of this case for the one-factor ANOVA model without any special attention. Thus, interpretation of the analysis, the assumptions, and so forth are the same as with the equal n's case. However, once we get to factorial designs in Chapter 3, things become a bit more complicated for the unequal n's or unbalanced case.

1.1.1.5 Alternative ANOVA Procedures

There are several alternatives to the parametric one-factor fixed-effects ANOVA. These include the Kruskal-Wallis one-factor ANOVA (Kruskal & Wallis, 1952, 1953), the Welch test (Welch, 1951), the Brown-Forsythe procedure (Brown & Forsythe, 1974), and the James procedures (James, 1951). You may recognize the Welch and Brown-Forsythe procedures as similar alternatives to the independent t test.

1.1.1.5.1 Kruskal-Wallis Test

The Kruskal-Wallis test makes no normality assumption about the population distributions, although it assumes similar distributional shapes, but still assumes equal population variances across the groups (although heterogeneity does have some effect on this test, it is less than with the parametric ANOVA). When the normality assumption is met, or nearly so (i.e., with mild nonnormality), the parametric ANOVA is slightly more powerful than the Kruskal-Wallis test (i.e., less likelihood of a Type II error). Otherwise the Kruskal-Wallis test is more powerful.

The Kruskal-Wallis procedure works as follows. First, the observations on the dependent measure are rank ordered, regardless of group assignment (the ranking is done by the computer). That is, the observations are ranked from highest to lowest, disregarding group membership. *The procedure essentially tests whether the mean ranks are different across the groups such that they are unlikely to represent random samples from the same population.* Thus, according to the null hypothesis, the mean rank is the same for each group; whereas for the alternative hypothesis, the mean rank is not the same across groups. The test statistic is denoted by H and is compared to the critical value ${}_{\alpha}\chi^2_{J-1}$. The null hypothesis is rejected if the test statistic H exceeds the χ^2 critical value.

There are two situations to consider with this test. First, the χ^2 critical value is really only appropriate when there are at least three groups and at least five observations per group (i.e., the χ^2 is not an exact sampling distribution of H). The second situation is that when there are tied ranks, the sampling distribution of H can be affected. Typically a midranks procedure is used, which results in an overly conservative Kruskal-Wallis test. A correction for ties is commonly used. Unless the number of ties is relatively large, the effect of the correction is minimal.

Using the elite athlete data as an example, we perform the Kruskal-Wallis analysis of variance. The test statistic $H = 13.0610$ is compared with the critical value ${}_{.05}\chi^2_3 = 7.81$, from Appendix Table A.3, and the result is that H_0 is rejected ($p = .005$). Thus the Kruskal-Wallis result agrees with the result of the parametric analysis of variance. This should not be surprising because the normality assumption apparently was met. Thus, we would probably not have done the Kruskal-Wallis test for the example data. We merely provide it for purposes of explanation and comparison.

In summary, the Kruskal-Wallis test can be used as an alternative to the parametric one-factor analysis of variance under nonnormality and/or when data on the dependent variable are ordinal. Under normality and with interval/ratio dependent variable data, the parametric ANOVA is more powerful than the Kruskal-Wallis test, and thus is the preferred method.

1.1.1.5.2 Welch, Brown-Forsyth, and James Procedures

Next we briefly consider the following procedures for the heteroscedasticity condition: the Welch test (Welch, 1951); the Brown-Forsythe procedure (Brown & Forsythe, 1974); and the James first- and second-order procedures (James, 1951) (more fully desribed in sources such as Coombs, Algina, & Oltman, 1996; Myers, Lorch, & Well, 2010; Wilcox, 1996, 2003). These procedures do not require homogeneity. Research suggests that (a) under homogeneity the F test is slightly more powerful than any of these procedures, and (b) under heterogeneity each of these alternative procedures is more powerful than the F, although the choice among them depends on several conditions, making a recommendation amongst these alternative procedures somewhat complicated (e.g., Clinch & Keselman, 1982; Tomarken & Serlin, 1986; Coombs et al., 1996). The Kruskal-Wallis test is widely available in the major statistical software, and the Welch and Brown-Forsythe procedures are available in the SPSS one-way ANOVA module. Wilcox (1996) and Wilcox (2003) also provide assistance for these alternative procedures.

1.1.2 Power

As for power (the probability of correctly rejecting a false null hypothesis), one can consider either planned power (*a priori*) or observed power (post hoc). In the ANOVA context, we know that power is primarily a function of α, sample size, and effect size. For planned power, one inputs each of these components either into a statistical table or power chart (e.g., Cohen, 1988; Murphy, Myors, & Wolach, 2009), or into power software (such as G*Power). Planned power is most often used by researchers to determine adequate sample sizes in ANOVA models, which is highly recommended. Many disciplines recommend a minimum power value, such as .80. Thus, these methods are a useful way to determine the sample size that would generate a desired level of power. Observed power is determined by some statistics software, such as SPSS, and indicates the power that was actually observed in a completed study.

1.1.3 Effect Size

There are various effect size measures to indicate the strength of association between X and Y, that is, the relative strength of the group effect. Let us briefly examine η^2, ω^2, ε^2, and Cohen's (1988) f.

1.1.3.1 Eta Squared

First, η^2 (eta squared), ranging from zero to +1.00, is known as the correlation ratio (generalization of R^2) and represents the proportion of variation in Y explained by the group mean differences in X. An eta squared of zero suggests that *none* of the total variance in the dependent variable is due to differences between the groups. An eta squared of 1.00 indicates that *all* the variance in the dependent variable is due to the group mean differences. We find η^2 to be as follows (Olejnik & Algina, 2000):

$$\eta^2 = \frac{SS_{betw}}{SS_{total}}$$

It is well known that η^2 is a positively biased statistic (i.e., overestimates the association). The bias is most evident for n's (i.e., group sample sizes) less than 30. *In one-way ANOVA, eta squared and partial eta squared (which is reported in SPSS output) will be equal given there is just one independent variable.*

1.1.3.2 Omega Squared and Epsilon Squared

Other effect size measures are ω^2 (omega squared) and ε^2 (epsilon squared). Both are interpreted similarly to eta squared (specifically, the proportion of variation in *Y* explained by the group mean differences in *X*) but provide corrections that allow them to be less biased than η^2. Omega squared and epsilon squared will generally differ only slightly (Carroll & Nordholm, 1975). Both can provide negative estimates, and when that happens, the estimate is usually set to zero (Olejnik & Algina, 2000).

We determine **omega squared** through either of the following formulas (the first formula referenced in Olejnik & Algina, 2000):

$$\omega^2 = \frac{(df_{\text{betw}})(MS_{\text{betw}} - MS_{\text{with}})}{SS_{\text{total}} + MS_{\text{with}}}$$

$$\omega^2 = \frac{SS_{\text{betw}} - (J-1)MS_{\text{with}}}{SS_{\text{total}} + MS_{\text{with}}}$$

Epsilon squared is computed as (Olejnik & Algina, 2000):

$$\varepsilon^2 = \frac{(df_{\text{betw}})(MS_{\text{betw}} - MS_{\text{with}})}{SS_{\text{total}}}$$

1.1.3.3 Cohen's f

A final effect size measure is *f*, developed by Cohen (1988). The effect *f* can take on values from zero (when the means are equal) to an infinitely large positive value. This effect is interpreted as an approximate correlation index but can also be interpreted as the standard deviation of the standardized means (Cohen, 1988). We compute *f* through the following:

$$f = \sqrt{\frac{\eta^2}{1-\eta^2}}$$

We can also use *f* to compute the effect size *d*, which, you recall from the *t* test, is interpreted as the standardized mean difference. The formulas for translating *f* to *d* are dependent on whether there is minimum, moderate, or maximum variability between the means of the groups. Interested readers are referred to Cohen (1988).

1.1.3.4 Interpretation of Effect Size Values

Cohen's (1988) subjective standards can be used as follows to interpret these effect size values: small effect, $f = .1$, $\eta^2, \varepsilon^2, \omega^2 = .01$; medium effect, $f = .25$, $\eta^2, \varepsilon^2, \omega^2 = .06$; and large effect, $f = .40$, $\eta^2, \varepsilon^2, \omega^2 = .14$. Note that these are subjective standards developed for the behavioral sciences; your discipline may use other standards. For further discussion, see Keppel (1982), O'Grady (1982), Wilcox (1987), Cohen (1988), Keppel and Wickens (2004), and Murphy, Myors, and Wolach (2014, which includes software).

TABLE 1.6

Effect Sizes and Interpretations

Effect Size	Interpretation
Omega squared (ω^2), epsilon squared (ε^2), and eta squared (η^2)	Proportion of total variability in the dependent variable that is accounted for by the factor (i.e., independent variable) • Small effect = .01 • Medium effect = .06 • Large effect = .14
Cohen's *f*	Approximate correlation index but can also be interpreted as the standard deviation of the standardized means • Small effect = .1 • Medium effect = .25 • Large effect = .40

1.1.3.5 An Effect Size Example

Let's determine the effect size measures given the data on elite athletes. For illustrative purposes, all effect size measures that were previously discussed have been computed. In practice, only one effect size is usually computed and interpreted. First, eta squared, η^2, is computed as follows. Note that in the one-way ANOVA, eta squared will equal partial eta squared as output in SPSS.

$$\eta^2 = \frac{SS_{\text{betw}}}{SS_{\text{total}}} = \frac{738.5938}{1749.7188} = .4221$$

Next, omega squared, ω^2, is found to be the following (where either calculation will result in the same value). Note that in the one-way ANOVA, omega squared will equal partial omega squared as output in the online calculator that we will demonstrate shortly.

$$\omega^2 = \frac{(df_{\text{betw}})(MS_{\text{betw}} - MS_{\text{with}})}{SS_{\text{total}} + MS_{\text{with}}} = \frac{(3)(246.198 - 36.112)}{1749.7188 - 36.1116} = .3529$$

$$\omega^2 = \frac{SS_{\text{betw}} - (J-1)MS_{\text{with}}}{SS_{\text{betw}} + MS_{\text{with}}} = \frac{738.5938 - (4-1)36.1116}{1749.7188 + 36.1116} = .3529$$

Now we find epsilon squared, ε^2:

$$\varepsilon^2 = \frac{(df_{betw})(MS_{betw} - MS_{with})}{SS_{total}} = \frac{(3)(246.198 - 36.112)}{1749.7188} = .3602$$

Lastly, Cohen's *f* is computed as follows:

$$f = \sqrt{\frac{\eta^2}{1-\eta^2}} = \sqrt{\frac{.4221}{1-.4221}} = .8546$$

Recall Cohen's (1988) subjective standards that can be used to interpret these effect sizes. Based on these effect size measures, all measures lead to the same conclusion: *there is a large effect size for the influence of type of sport on psychological distress*. Examining ω^2, for example, we can also state that about 35% of the variation in *Y* (psychological distress) can be explained by X (type of sport in which the athlete competes). The other proportion of variance effect size indices provide similar interpretations. The effect *f* suggests a strong correlation.

In addition, if we rank the group means of the sport from movement (with the lowest mean) to territory (with the highest mean), we see that as physical contact of the sport increases (e.g., from movement to territory), the more psychological distress is reported by the athlete. While visual inspection of the means suggests descriptively there are differences in psychological distress by sport, we examine multiple comparison procedures with this same data in Chapter 2 to determine which groups are statistically significantly different from one another.

1.1.3.6 Confidence Intervals for Effect Size

As we know by this point, computing **confidence intervals** is valuable. The benefit in creating confidence intervals for effect size values is similar to that of creating confidence intervals for parameter estimates—*confidence intervals for the effect size provide an added measure of precision that is not obtained from knowledge of the effect size alone*. Computing confidence intervals for effect size indices, however, is not as straightforward as simply plugging in known values into a formula. Never fear; there are some nice online tools that can be used. One online calculator for computing many types of effect sizes and their confidence intervals is provided by Dr. David B. Wilson and is available through the Campbell Collaboration (see https://campbellcollaboration.org/research-resources/effect-size-calculator.htm). In the case of one-way ANOVA, this online calculator can be used with the *F* test when there are two groups with either a balanced or unbalanced design. Uanhoro's (2017) online calculator (available at https://effect-size-calculator.herokuapp.com/) uses the noncentral *F* method to compute confidence intervals for partial eta squared in fixed-effects ANOVA models that do not include covariates (i.e., ANCOVA, which we will study in a future chapter). As we see in Figure 1.2, only four inputs are required: *F*, numerator and denominator degrees of freedom, and confidence interval. Note that the default setting for the 90% confidence interval is equivalent to the 95% two-sided confidence interval since the *F* cannot be negative (Smithson, 2003)—thus, the recommendation on the site to "use the 90% CI if you have an alpha level of 5%." Partial eta squared is .422 with lower and upper confidence limits of .135 and .548, respectively. Putting this in context of our example, if multiple random samples were drawn from the population, 95% of the samples could expect about 14%, at minimum, and 55%, at maximum, of the proportion of the outcome to be explained by the independent variable.

1.1.3.7 Items to Consider

We will end our discussion on effect size with a few noteworthy items to consider as you compute and interpret effect sizes. Eta squared can be positively biased, overestimating the strength of the population relationship, and thus is best considered a descriptor of

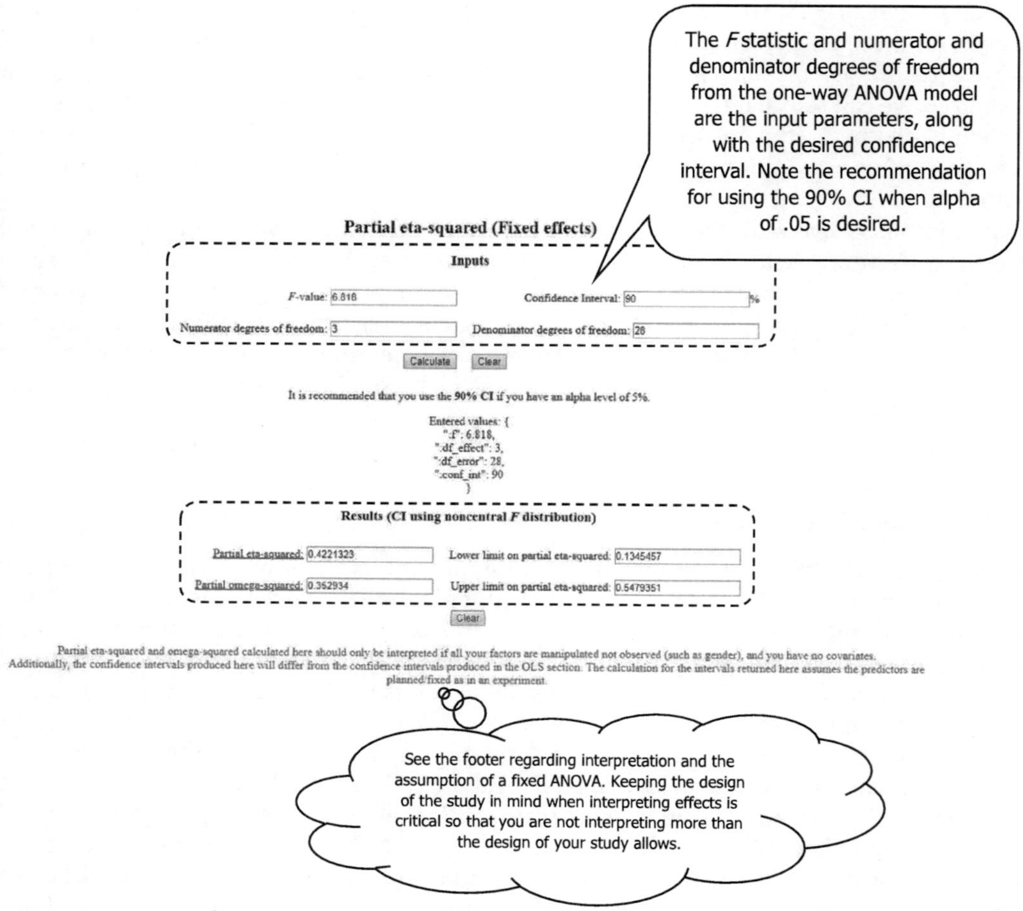

FIGURE 1.2
Confidence intervals for effect size.

proportion of variance in the dependent variable explained for a particular sample (Maxwell, Arvey, & Camp, 1981). Thus, many researchers discourage reporting eta squared or partial eta squared, although you will still see it widely reported given that it is the only effect size value that is output from SPSS. Both epsilon squared and omega squared introduce a correction to this problem and, generally, both will be quite similar in value (Carroll & Nordholm, 1975). If you find yourself in a situation where epsilon squared or omega squared are negative, the standard is simply to set the effect size to zero (Olejnik & Algina, 2000). Proportion of total variance effect size indices are not comparable across studies that incorporate different factors (Olejnik & Algina, 2000) or factors which are theoretically the same but measured differently. This is reasonable given that total variation is influenced by all the factors in the model.

Some researchers find interpreting proportion of variance effect sizes advantageous, as compared to standardized mean differences, given that the index range is from 0 to 1 (Rosenthal, 1994). However, even a large proportion of variance effect size values (e.g., .14+) suggest there is much variance that remains to be explained, and thus even large effects can be perceived as trivial (Rosenthal & Rubin, 1979).

Last but not least, we will touch on general reporting and interpretation considerations. Many researchers encourage interpreting effect size relative to other studies. However, several researchers (e.g., Fern & Monroe, 1996; Maxwell et al., 1981; O'Grady, 1982; Sechrest & Yeaton, 1982) have provided caution in doing this as effect size can be impacted by instrument reliability, heterogeneity of the populations that are compared, the levels or categories of the factors that are modeled, the strength of the treatments, and the range of treatments, all of which can lead to effect size comparisons that are misleading (Olejnik & Algina, 2000).

1.1.4 Assumptions

There are three standard assumptions made in analysis of variance models, which we are already familiar with from the independent *t* test. We see these assumptions often in the remainder of this text. The assumptions are concerned with **independence**, **homogeneity of variance**, and **normality**. We also mention some techniques appropriate to use in evaluating each assumption.

1.1.4.1 Independence

The first assumption is that observations are independent of one another (both within samples and across samples). In general, the assumption of independence for ANOVA designs can be met by (a) keeping the assignment of individuals to groups separate through the design of the experiment (specifically random assignment—not to be confused with random selection), and (b) keeping the individuals separate from one another through experimental control so that the scores on the dependent variable *Y* for group 1 do not influence the scores for group 2 and so forth for other groups of the independent variable. Zimmerman (1997) also stated that independence can be violated for supposedly independent samples due to some type of matching in the design of the experiment (e.g., matched pairs based on gender, age, and weight).

The use of independent random samples is crucial in the analysis of variance. The *F* ratio is very sensitive to violation of the independence assumption in terms of increased likelihood of a Type I and/or Type II error (e.g., Glass, Peckham, & Sanders, 1972). This effect can sometimes even be worse with larger samples (Keppel & Wickens, 2004). A violation of the independence assumption may affect the standard errors of the sample means and thus influence any inferences made about those means. One purpose of random assignment of individuals to groups is to achieve independence. If each individual is observed only once and individuals are randomly assigned to groups, then the independence assumption is usually met. If individuals work together during the study (e.g., through discussion groups or group work), then independence may be compromised. Thus, a carefully planned, controlled, and conducted research design is the key to satisfying this assumption.

What if your independent variable does not allow for random assignment, such as an observed characteristic or attribute (e.g., sex) or a preexisting group (e.g., self-selection into levels)? This does not prohibit the use of these variables as an independent variable in ANOVA. Indeed, for many disciplines, random assignment is rarely if ever possible, and it is a preexisting condition or already defined group that is of interest to examine as an independent variable. In these cases, it is particularly essential that evidence be examined to determine the extent to which the assumption of independence is met.

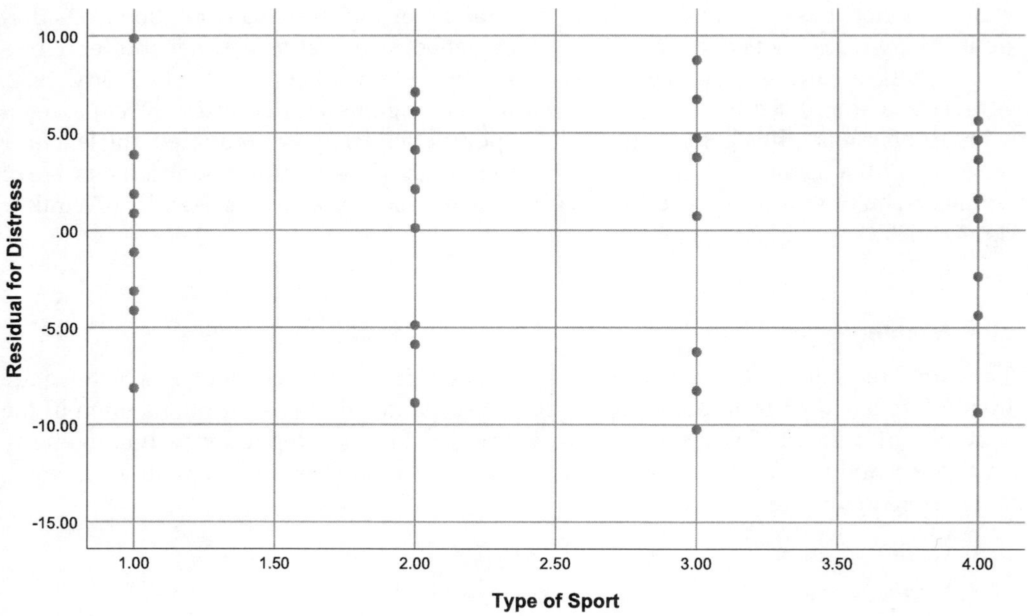

FIGURE 1.3
Residual plot by group for elite athlete example.

The simplest procedure for assessing independence is to examine residual plots by group. If the independence assumption is satisfied, then the residuals should fall into a random display of points for each group. If the assumption is violated, then the residuals will fall into some type of pattern. The Durbin-Watson statistic (Durbin & Watson, 1950, 1951, 1971) can be used to test for autocorrelation. Violations of the independence assumption generally occur in three situations: (1) when observations are collected over time; (2) when observations are made within blocks; or (3) when observation involves replication. For severe violations of the independence assumption, there is no simple "fix" (e.g., Scariano & Davenport, 1987). For the example data, a plot of the residuals by group is shown in Figure 1.3, and there does appear to be a random display of points for each group even though the units were not randomly assigned to group.

1.1.4.2 Homogeneity of Variance

The second assumption is that the variances of each population are equal. This is known as the assumption of **homogeneity of variance** or also sometimes referred to as **homoscedasticity**. In ANOVA models, the term *homogeneity of variance* is often used, while in regression, *homoscedasticity* is often used. Regardless, the terms *conceptually* relate to constant variance—i.e., the residuals are the same (i.e., *constant*) everywhere. A violation of the homogeneity of variance assumption can lead to bias in the SS_{with} term (recall that *within* measures variation of units within each group), as well as an increase in the Type I error rate (i.e., rejecting the null when it is really false) and possibly an increase in the Type II error rate.

Two sets of research studies have investigated violations of this assumption, classic work and more modern work. The classic work largely resulted from Box (1954) and

Glass et al. (1972). Their results indicated that the effect of the violation was small with equal or nearly equal n's across the groups. There is a more serious problem if the larger n's are associated with the smaller variances (actual observed α > nominal α, which is a liberal result; for example, if a researcher desires a nominal alpha of .05, the alpha actually observed will be greater than .05), or if the larger n's are associated with the larger variances (actual observed α < nominal α, which is a conservative result). [Note that Bradley's (1978) criterion is used in this text, where the actual α should not exceed 1.1 to 1.5 times the nominal alpha.] Thus, the suggestion from the classic work was that heterogeneity was a concern only when there were unequal n's. However, the classic work examined only minor violations of the assumption (the ratio of largest variance to smallest variance being relatively small), and unfortunately has been largely adapted in textbooks and by users.

There has been some research conducted since that time by researchers such as Brown and Forsythe (1974), and Wilcox (Wilcox, 1986, 1987, 1988, 1989), and nicely summarized by Coombs et al. (1996). In short, this work indicates that the effect of heterogeneity is more severe than previously thought (e.g., poor power; α can be greatly affected), even with equal n's (although having equal n's does reduce the magnitude of the problem). Thus F is not even robust to heterogeneity with equal n's (equal n's are sometimes referred to as a balanced design). However, heterogeneity is less problematic with a balanced design *and* when the assumption of normality holds (Wilcox, 2017).

Suggestions for dealing with such a violation include (a) using alternative procedures such as the Welch, Brown-Forsythe, and James procedures (Coombs et al., 1996; Glass & Hopkins, 1996; Keppel & Wickens, 2004; Myers et al., 2010; Wilcox, 1996, 2003), (b) reducing the alpha level and testing at a more stringent alpha level (e.g., alpha of .01 rather than the common .05) (e.g., Keppel & Wickens, 2004; Weinberg & Abramowitz, 2002), or (c) transforming Y (such as $\sqrt{Y}$, $1/Y$, or $\log Y$) (e.g., Keppel & Wickens, 2004; Weinberg & Abramowitz, 2002). The alternative procedures will be more fully described later in this chapter.

Examining the extent to which homogeneity has been met can be done visually. In a plot of residuals versus each value of X, the consistency of the variance of the conditional residual distributions may be examined simply by eyeballing the plot.

Another method for detecting violation of the homogeneity assumption is the use of formal statistical tests. The traditional homogeneity tests (e.g., Levene's test) are commonly available in statistical software but are not robust to nonnormality, and this is the only test for homogeneity currently available in SPSS. For the example data, the residual plot of Figure 1.2 shows similar variances across the groups, and Levene's test suggests the variances are not different [F (3, 28) = .905, p = .451]. A recent simulation study by Wang et al. (2017) studied the performance of 14 homogeneity tests on controlling Type I error and power in one-way ANOVA. They found that the Ramsey conditional, O'Brien, Brown-Forsythe, bootstrap Brown-Forsythe, and Levene with squared deviation tests maintained adequate control of Type I errors and performed better than others reviewed, including maintaining acceptable power, across the simulated conditions. Recommendations for selecting a test for homogeneity of variance based on average cell size include the following: (a) when cell size is less than 10, O'Brien is the recommended test for homogeneity of variance as it maintains adequate Type I error control; (b) when cell size is greater than 10 but less than 20, the Ramsey conditional test is recommended as it also maintains adequate Type I error control; and (c) when the cell size is more than 20, the Brown-Forsythe, bootstrap Brown-Forsythe, and Ramsey conditional test are recommended as these tests provide adequate Type I error control and greater power (around .80).

1.1.4.3 Normality

The third assumption is that each of the populations follows the normal distribution (i.e., there is normality of the dependent variable for each category or group or level of the independent variable). The *F* test is relatively robust to moderate violations of this assumption (i.e., in terms of Type I and Type II error rates). Specifically, effects of the violation will be minimal except for small *n*'s, for unequal *n*'s, and/or for extreme nonnormality. As noted in our earlier discussion of homogeneity of variance, when there are equal sample sizes and the assumption of normality is violated, the results from a *F* test will not be robust unless the distributions of the group are equal (e.g., each group has the same degree of skew) (Wilcox, 2017). Wilcox (2017) suggests that *F* is robust to Type I errors when the group distributions are equal (e.g., the same skew across all groups).

Violation of the normality assumption may be a result of outliers. The simplest outlier detection procedure is to look for observations that are more than two or three standard deviations from their respective group mean. We recommend (and will illustrate later) inspection of residuals for examination of evidence of normality. Formal procedures for the detection of outliers are now available in many statistical packages.

The following graphical techniques can be used to examine residuals and detect violations of the normality assumption: (a) the frequency distributions of the residuals for each group (through stem-and-leaf plots, boxplots, histograms, residual plots), (b) the normal probability or quantile (Q-Q) plot, or (c) a plot of group means versus group variances (which should be independent of one another). There are also several statistical procedures available for the detection of nonnormality including skewness and kurtosis as well as formal tests for normality (e.g., the Shapiro-Wilk test, Shapiro & Wilk, 1965).

As we've learned previously, sample statistics such as **skewness** and **kurtosis** of the residuals can be reviewed. Values within an absolute value of 2.0 suggest evidence of normality. We can also divide the skew and kurtosis values by their standard errors to get *standardized skew and kurtosis* values. We can review those values to a critical value (e.g., ±1.65 if alpha = .10; ±1.96 if alpha = .05; ±2.06 if alpha = .01) and determine if there is statistically significant skew and/or kurtosis. **D'Agostino's test** (D'Agostino, 1970) can be used to examine the null hypothesis that skewness equals zero, with a statistically significant D'Agostino's test indicating that there is statistically significant skewness. For kurtosis, we can use the **Bonett-Seier test for Geary's kurtosis** (Bonett & Seier, 2002). The null hypothesis states that data should have a Geary's kurtosis value equal to $\sqrt{2/\pi} = .7979$. Thus, a statistically significant Bonett-Seier test for Geary's kurtosis would indicate that there is statistically significant kurtosis. Thus, with these tests, as with Kolmogorov-Smirnov and Shapiro-Wilk, we do *not* want to find statistically significant results.

As is evident, many different tools can be used for testing the assumption of normality, and researchers should approach testing this assumption as collecting multiple forms of evidence to best understand the extent to which the assumption was met. A summary of several different types of evidence for examining normality is provided in Box 1.2.

Should you find yourself in a situation where there is a violation of normality, transformations can be used to normalize the data. For instance, a nonlinear relationship between *X* and *Y* may result in violations of the normality and/or homoscedasticity assumptions. Readers interested in learning more about potential data transformations are referred to sources such as Bradley (1982), Box and Cox (1964), or Mosteller and Tukey (1977).

In the example data, the residuals shown in Figure 1.3 appear to be somewhat normal in shape, especially considering the groups have fairly small *n*'s. This is suggested by the random display of points. In addition, as we will see later, for the residuals overall, skewness = –.2389 and kurtosis = –1.0191, indicating evidence of normality. Thus, it appears that all of our assumptions have been satisfied for the example data. We will delve

BOX 1.2 Evidence for Testing the Assumption of Normality

Evidence	Interpretation for Providing Evidence of Normality
Boxplot	Normality suggested when the quartiles are relatively evenly distributed with no outliers.
Histogram	Normality suggested with a relatively bell-shaped curve.
Skewness	Values within an absolute value of 2.0 suggest evidence of normality.
Kurtosis	Values within an absolute value of 2.0 suggest evidence of normality.
Standardized skew and standardized kurtosis	Divide the skew and kurtosis values by their standard errors to get *standardized skew and kurtosis* values. Review those values to a critical value (e.g., ±1.65 if alpha = .10; ±1.96 = if alpha = .05; ±2.06 if alpha = .01). Standardized skew and kurtosis that are less than the critical value suggest evidence of normality.
D'Agostino's test	Tests the null hypothesis that skewness equals zero, with a statistically significant D'Agostino's test indicating that there is statistically significant skewness.
Bonett-Seier test for Geary's kurtosis	Tests the null hypothesis that data should have a Geary's kurtosis value equal to $\sqrt{2/\pi} = .7979$. A statistically significant test indicates that there is statistically significant kurtosis.
Quantile-quantile (Q-Q) plots	Plots that depict quantiles of the sample distribution to quantiles of the theoretical normal distribution. Points that fall on or closely to the diagonal line of the Q-Q plot suggest evidence of normality.
Detrended quantile-quantile plot	Evidence of normality is provided when the points exhibit little or no pattern around zero (the horizontal line).

TABLE 1.7

Assumptions, Evidence to Examine, and Effects of Violations: One-Factor ANOVA Design

Assumption	Evidence to Examine	Effect of Assumption Violation
Independence	• Scatterplot of residuals by group	Increased likelihood of a Type I and/or Type II error in the *F* statistic; influences standard errors of means and thus inferences about those means.
Homogeneity of variance	• Scatterplot of residuals by X • Formal test of equal variances (e.g., Levene's test)	Bias in SS_{with}; increased likelihood of a Type I and/or Type II error; less effect with equal or nearly equal *n*'s when normality can be assumed; effect decreases as *n* increases.
Normality	• Graphs of residuals (or scores) by group (e.g., boxplots, histograms, stem-and-leaf plots) • Skewness and kurtosis of residuals • Q-Q plots of residuals • Formal tests of normality of residuals • Plot of group means by group variances	Minimal effect with moderate violation; effect less severe with large *n*'s, with equal or nearly equal *n*'s, and/or with homogeneously shaped distributions (e.g., all groups have the same degree of skew).

further into examination of assumptions later as we illustrate how to use SPSS to conduct a one-way ANOVA.

A summary of the assumptions and the effects of their violation for the one-factor analysis of variance design are presented in Table 1.7.

1.2 Computing Parametric and Nonparametric Models Using SPSS

Next we consider the use of SPSS for the elite athlete example. Instructions for determining the one-way ANOVA using SPSS are presented first, followed by additional steps for examining the assumptions for the one-way ANOVA. Next, instructions for computing the Kruskal-Wallis and Brown and Forsyth are presented.

1.2.1 One-Way Analysis of Variance

Note that SPSS needs the data to be in a specific form for any of the analyses below to proceed, which is different from the layout of the data in Table 1.1. For a one-factor ANOVA, the dataset must consist of at least two variables or columns (if there are more than two variables, only two of which will be used in the one-factor ANOVA) (see Figure 1.4). *One column or variable indicates the levels or categories of the independent variable, and the second is for the dependent variable.* Each *row* then represents one unit (e.g., individual), indicating the level or group within which that unit is a member of (1, 2, 3, or 4 in our example), and their score on the dependent variable. Thus we wind up with two long columns of group values and scores as shown in the screenshot (Figure 1.4).

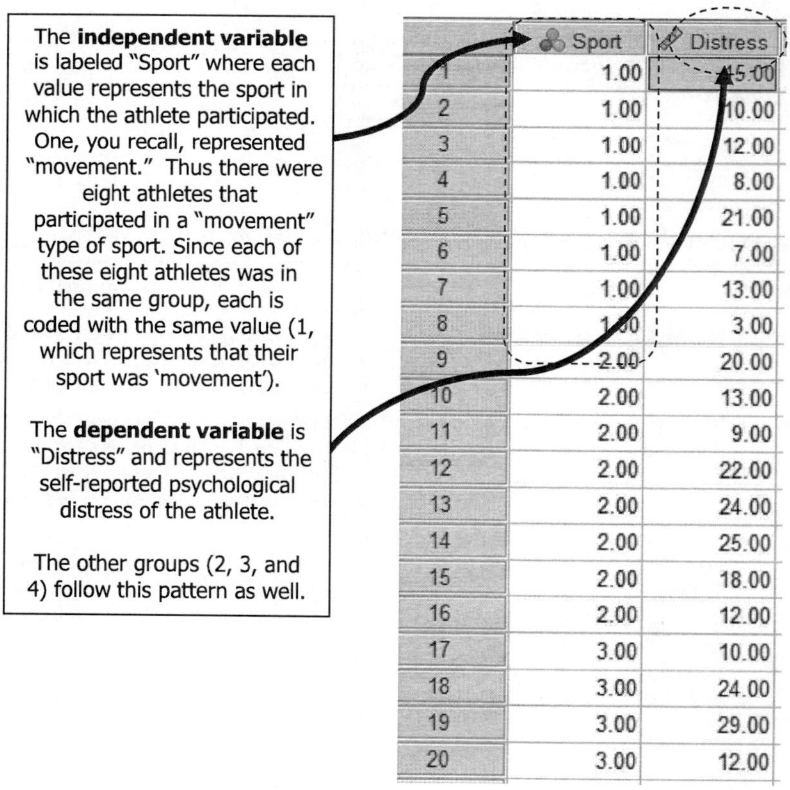

FIGURE 1.4
First 20 cases of ANOVA data.

Step 1. To conduct a one-way ANOVA, go to "Analyze" in the top pulldown menu, then select "General Linear Model," and then select "Univariate." Following the screenshot for Step 1 (shown in Figure 1.5) produces the Univariate dialog box.

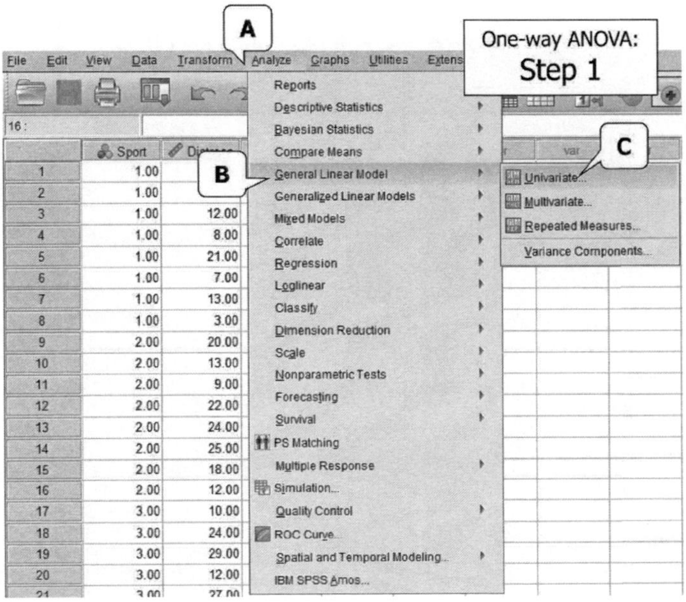

FIGURE 1.5
One-way ANOVA: Step 1.

Step 2. Click the dependent variable (e.g., psychological distress) and move it into the "Dependent Variable" box by clicking the arrow button. Click the independent variable (e.g., type of sport) and move it into the "Fixed Factors" box by clicking the arrow button. Next, click on "Options."

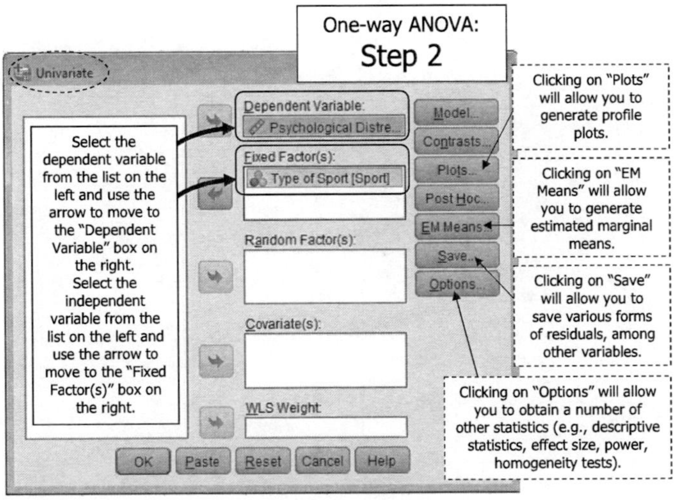

FIGURE 1.6
One-way ANOVA: Step 2.

Step 3. Clicking on "Options" will provide the option to select such information as "Descriptive statistics," "Estimates of effect size," "Observed power," and "Homogeneity tests." Click on "Continue" to return to the original dialog box.

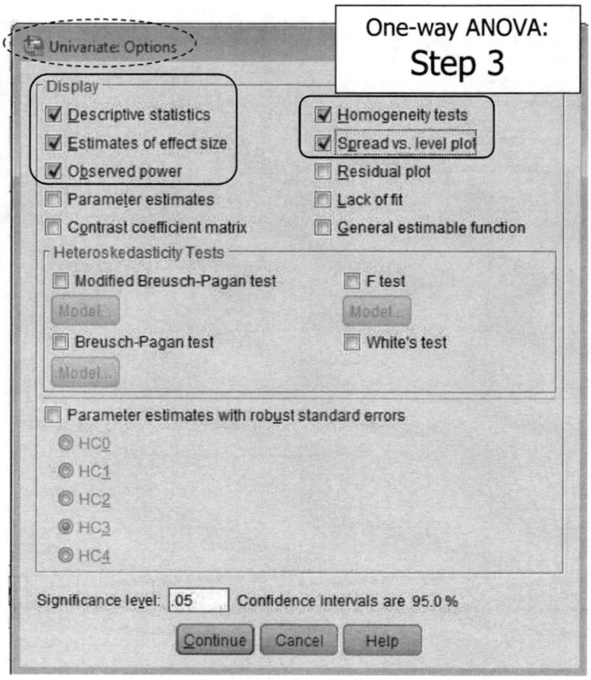

FIGURE 1.7
One-Way ANOVA: Step 3.

Step 4. Clicking on "EM Means" will provide the option to display the overall and factor means. Click on "Continue" to return to the original dialog box.

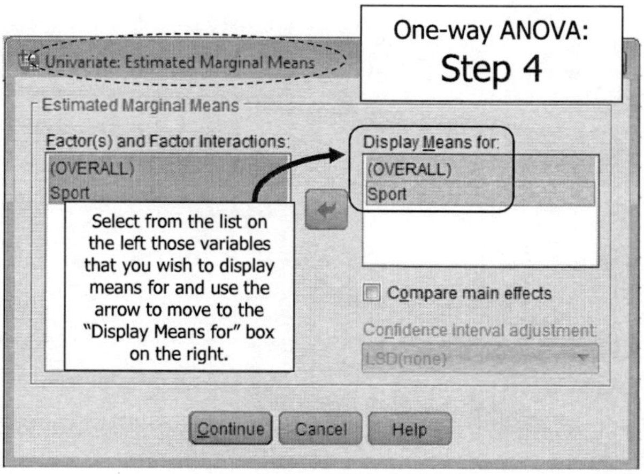

FIGURE 1.8
One-Way ANOVA: Step 4.

Step 5. From the Univariate dialog box, click on "Plots" to obtain a profile plot of means. Click the independent variable (e.g., type of sport, labeled as "Sport") and move it into the "Horizontal Axis" box by clicking the arrow button (see the screenshot for Step 5a in Figure 1.9). Then click on "Add" to move the variable into the "Plots" box at the bottom of the dialog box (see the screenshot for Step 5b in Figure 1.10). Click on "Continue" to return to the original dialog box.

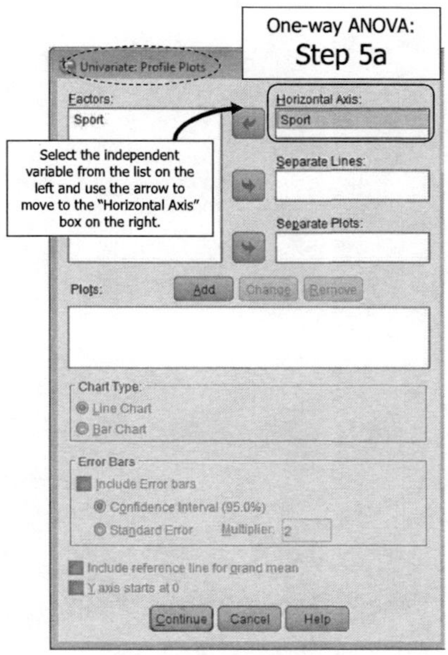

FIGURE 1.9
One-way ANOVA: Step 5a.

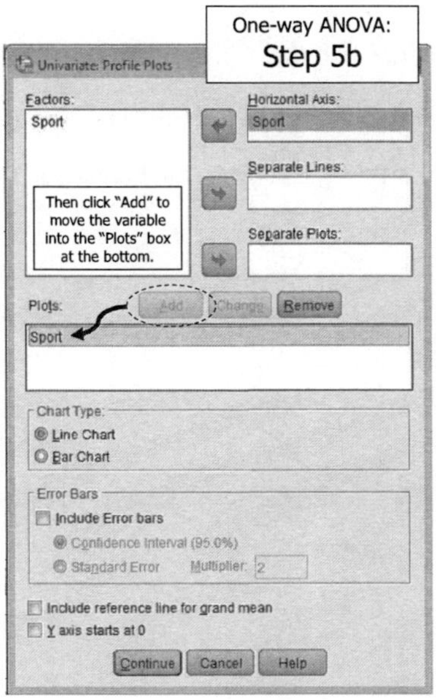

FIGURE 1.10
One-way ANOVA: Step 5b.

Step 6. From the Univariate dialog box, click on "Save" to select those elements that you want to save (in our case, we want to save the unstandardized residuals, which will be used later to examine the extent to which normality and independence are met). From the Univariate dialog box, click on "OK" to return to generate the output.

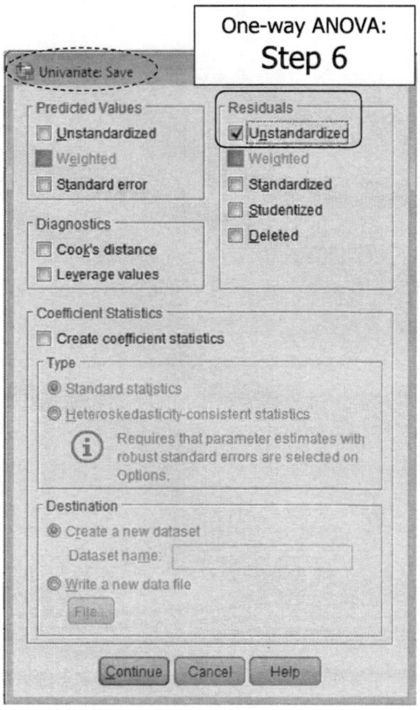

FIGURE 1.11
One-way ANOVA: Step 6.

1.2.1.1 Interpreting the Output for the One-Way Analysis of Variance

Annotated results are presented in Table 1.8 and the profile plot is shown in Figure 1.12.

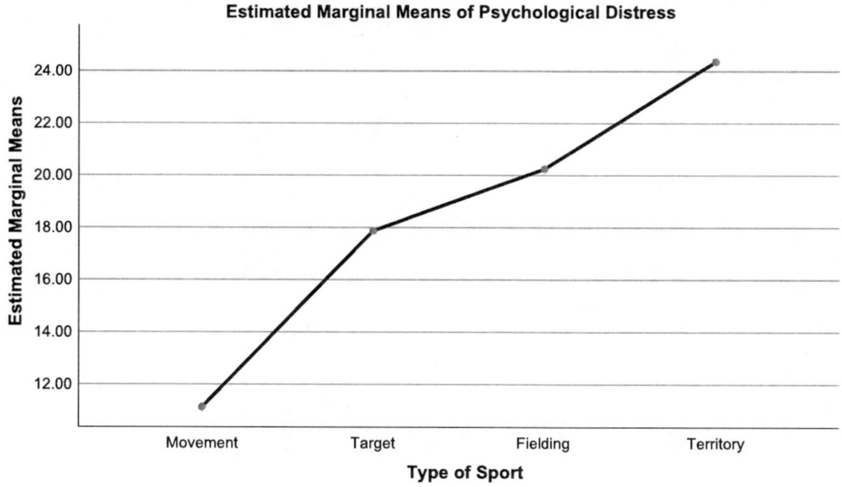

FIGURE 1.12
Profile plot for elite athlete example.

TABLE 1.8

Selected SPSS Results for the Psychological Distress Example

Between-Subjects Factors

		Value Label	N
Type of Sport	1.00	Movement	8
	2.00	Target	8
	3.00	Fielding	8
	4.00	Territory	8

The table labeled "Between-Subjects Factors" provides sample sizes for each of the categories of the independent variable (recall that the independent variable is the 'between subjects factor').

Descriptive Statistics

Dependent Variable: Psychological Distress

Type of Sport	Mean	Std. Deviation	N
Movement	11.1250	5.48862	8
Target	17.8750	5.93867	8
Fielding	20.2500	7.28501	8
Territory	24.3750	5.09727	8
Total	18.4062	7.51283	32

The table labeled "Descriptive Statistics" provides basic descriptive statistics (means, standard deviations, and sample sizes) for each group of the independent variable.

Levene's Test of Equality of Error Variances[a,b]

		Levene Statistic	df1	df2	Sig.
Psychological Distress	Based on Mean	.905	3	28	.451
	Based on Median	.604	3	28	.618
	Based on Median and with adjusted df	.604	3	26.059	.618
	Based on trimmed mean	.883	3	28	.462

Tests the null hypothesis that the error variance of the dependent variable is equal across groups.

a. Dependent variable: Psychological Distress

b. Design: Intercept + Sport

The F test (and associated p value) for Levene's Test for Equality of Error Variances is reviewed to determine if equal variances can be assumed. In this case, we meet the assumption (as p is greater than α). Note that *df1* is degrees of freedom for the numerator (calculated as $J - 1$) and *df2* are the degrees of freedom for the denominator (calculated as $N - J$).

(continued)

TABLE 1.8 (continued)

Selected SPSS Results for the Psychological Distress Example

The row labeled **"SPORT"** is the independent variable or between groups variable. The *between groups mean square* (246.198) tells how much the group means vary. The degrees of freedom for between groups is $J-1$ (3 in this example).

The omnibus *F* test is computed as:

$$F = \frac{MS_{betw}}{MS_{with}} = \frac{246.198}{36.112} = 6.818$$

The *p* value for the omnibus *F* test is .001. This indicates there is a statistically significant difference in the mean psychological distress based on the sport in which the athlete competes. The probability of observing these mean differences or more extreme mean differences by chance if the null hypothesis is really true (i.e., if the means really are equal) is substantially less than 1%. We reject the null hypothesis that all the population means are equal. For this example, this provides evidence to suggest that psychological distress differs based on the sport in which the athlete competes.

In one-way ANOVA, eta squared and partial eta squared are equal as there is just one independent variable. Thus, 'partial eta squared' on our one-way ANOVA output is really also eta squared, and is one measure of effect size computed as:

$$\eta_p^2 = \frac{SS_{betw}}{SS_{total}} = \frac{738.594}{1749.719} = .422$$

We can interpret this to mean that approximately 42% of the variation in the dependent variable (in this case, psychological distress) is accounted for by the sport in which the athlete competes.

Tests of Between-Subjects Effects

Dependent Variable: Psychological Distress

Source	Type III Sum of Squares	df	Mean Square	F	Sig.	Partial Eta Squared	Noncent. Parameter	Observed Power[b]
Corrected Model	738.594[a]	3	246.198	6.818	.001	.422	20.453	.956
Intercept	10841.281	1	10841.281	300.216	.000	.915	300.216	1.000
Sport	738.594	3	246.198	6.818	.001	.422	20.453	.956
Error	1011.125	28	36.112					
Total	12591.000	32						
Corrected Total	1749.719	31						

a. R Squared = .422 (Adjusted R Squared = .360)

b. Computed using alpha = .05

R squared is listed as a footnote underneath the table. *R* squared is the ratio of *sum of squares between* divided by *sum of squares total*:

$$R^2 = \frac{SS_{betw}}{SS_{total}} = \frac{738.594}{1749.719} = .422$$

and, in the case of one-way ANOVA, is also the simple bivariate Pearson correlation between the independent variable and dependent variable squared.

The row labeled **"Error"** is within groups. The *within groups mean square* tells us how much the observations within the groups vary (i.e., 36.112). The degrees of freedom for within groups is ($N-J$) or the total sample size minus the number of levels of the independent variable.

The row labeled "corrected total" is the *sum of squares total*. The degrees of freedom for the total is ($N-1$) or the total sample size minus 1.

Observed power tells whether our test is powerful enough to detect mean differences if they really exist. Power of .956 indicates that the probability of rejecting the null hypothesis if it is really false is about 96%; this represents strong power.

TABLE 1.8 (continued)

Selected SPSS Results for the Psychological Distress Example

Estimated Marginal Means

1. Grand Mean

Dependent Variable: Psychological Distress

Mean	Std. Error	95% Confidence Interval	
		Lower Bound	Upper Bound
18.406	1.062	16.230	20.582

The 'Grand Mean' (in this case, 18.406) represents the overall mean, regardless of group membership, of the dependent variable. The 95% CI represents the CI of the grand mean.

2. Type of Sport

Dependent Variable: Psychological Distress

Type of Sport	Mean	Std. Error	95% Confidence Interval	
			Lower Bound	Upper Bound
Movement	11.125	2.125	6.773	15.477
Target	17.875	2.125	13.523	22.227
Fielding	20.250	2.125	15.898	24.602
Territory	24.375	2.125	20.023	28.727

The table labeled **"Type of Sport"** provides descriptive statistics for each of the categories of the independent variable (notice that these are the same means reported previously). In addition to means, the *SE* and 95% CI of the means are reported.

Hold that thought! MCPs will be discussed in the next chapter.

Post Hoc Tests
Type of Sport

Multiple Comparisons

Dependent Variable: Psychological Distress

Tukey HSD

(I) Type of Sport	(J) Type of Sport	Mean Difference (I-J)	Std. Error	Sig.	95% Confidence Interval	
					Lower Bound	Upper Bound
Movement	Target	-6.7500	3.00465	.135	-14.9536	1.4536
	Fielding	-9.1250*	3.00465	.025	-17.3286	-.9214
	Territory	-13.2500*	3.00465	.001	-21.4536	-5.0464
Target	Movement	6.7500	3.00465	.135	-1.4536	14.9536
	Fielding	-2.3750	3.00465	.858	-10.5786	5.8286
	Territory	-6.5000	3.00465	.158	-14.7036	1.7036
Fielding	Movement	9.1250*	3.00465	.025	.9214	17.3286
	Target	2.3750	3.00465	.858	-5.8286	10.5786
	Territory	-4.1250	3.00465	.526	-12.3286	4.0786
Territory	Movement	13.2500*	3.00465	.001	5.0464	21.4536
	Target	6.5000	3.00465	.158	-1.7036	14.7036
	Fielding	4.1250	3.00465	.526	-4.0786	12.3286

Based on observed means.

The error term is Mean Square(Error) = 36.112.

*. The mean difference is significant at the 0.05 level.

(*continued*)

TABLE 1.8 (continued)

Selected SPSS Results for the Psychological Distress Example

Homogeneous Subsets

Psychological Distress

Tukey HSD[a,b]

		Subset	
Type of Sport	N	1	2
Movement	8	11.1250	
Target	8	17.8750	17.8750
Fielding	8		20.2500
Territory	8		24.3750
Sig.		.135	.158

Means for groups in homogeneous subsets are displayed.

Based on observed means.

The error term is Mean Square(Error) = 36.112.

a. Uses Harmonic Mean Sample Size = 8.000.

b. Alpha = 0.05.

Spread-versus-Level Plots

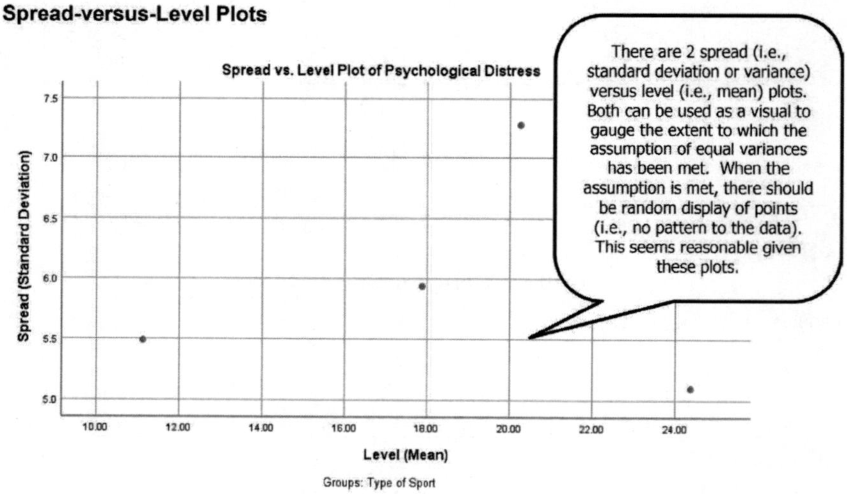

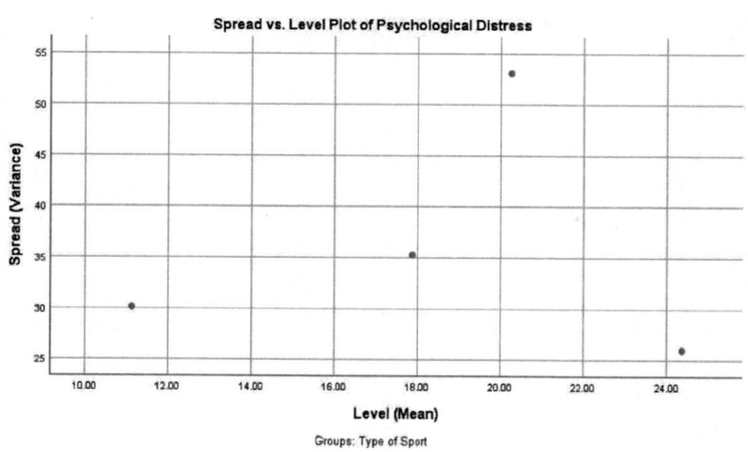

TABLE 1.8 (continued)

Selected SPSS Results for the Psychological Distress Example

Profile Plots

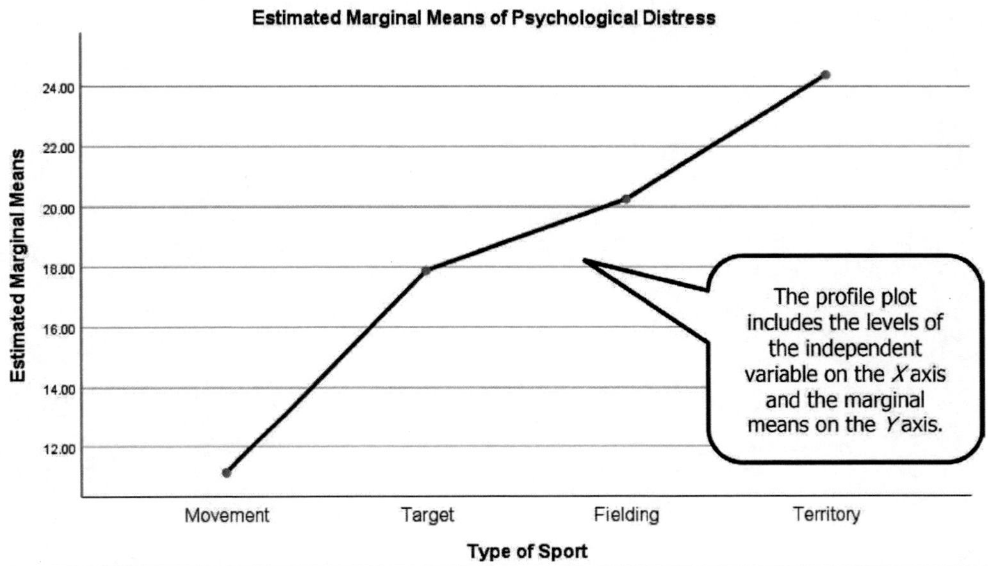

1.2.2 Nonparametric Procedures

Results from some of the recommended alternative procedures can be obtained from two other SPSS modules. Here we discuss the Kruskal-Wallis, Welch, and Brown-Forsythe procedures.

1.2.2.1 Kruskal-Wallis

Step 1. To conduct a Kruskal-Wallis test, go to "Analyze" in the top pulldown menu, then select "Nonparametric Tests," then select "Legacy Dialogs" and finally "K Independent Samples." Following the screenshot for Step 1 (Figure 1.13) produces the "Tests for Several Independent Samples" dialog box.

Step 2. Next, from the main "Tests for Several Independent Samples" dialog box, click the dependent variable (e.g., psychological distress) and move it into the "Test Variable List" box by clicking on the arrow button. Next, click the grouping variable (e.g., type of sport) and move it into the "Grouping Variable" box by clicking on the arrow button. You will notice that there are two question marks next to the name of your grouping variable. This is SPSS letting you know that you need to define (numerically) which categories of the grouping variable you want to include in the analysis (this must be done by identifying a range of values for all groups of interest). To do that, click on "Define Range." We have four groups or levels of our independent variable (labeled 1, 2, 3, and 4 in our raw data); thus enter 1 as the minimum and 4 as the maximum. In the lower left portion of the screen under "Test Type," check "Kruskal-Wallis H" to generate this nonparametric test. Then click on "OK" to generate the results, presented in Figure 1.14.

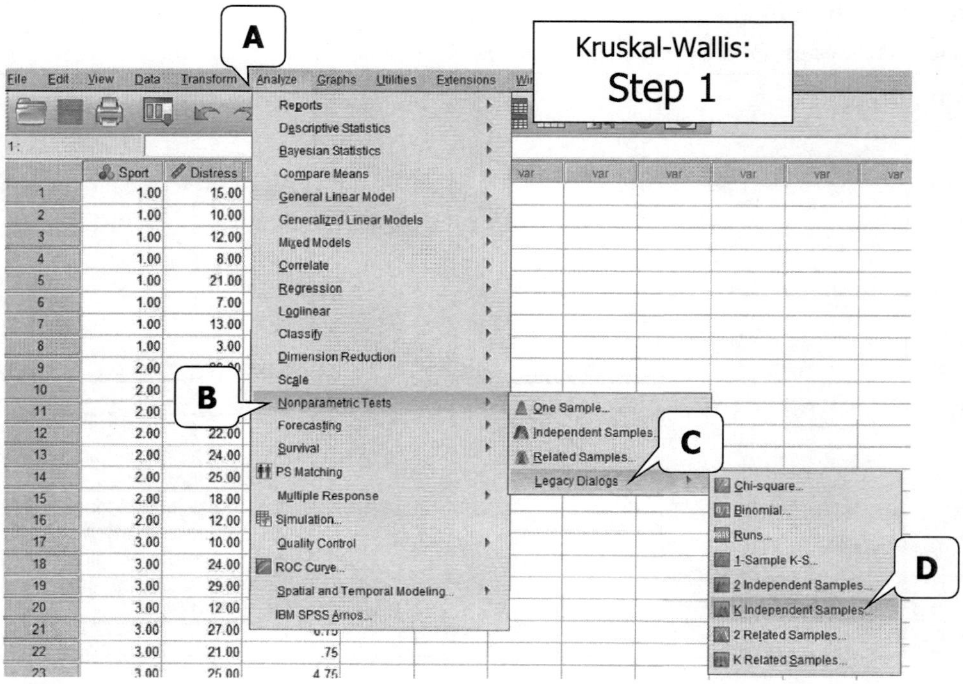

FIGURE 1.13
Kruskal-Wallis: Step 1.

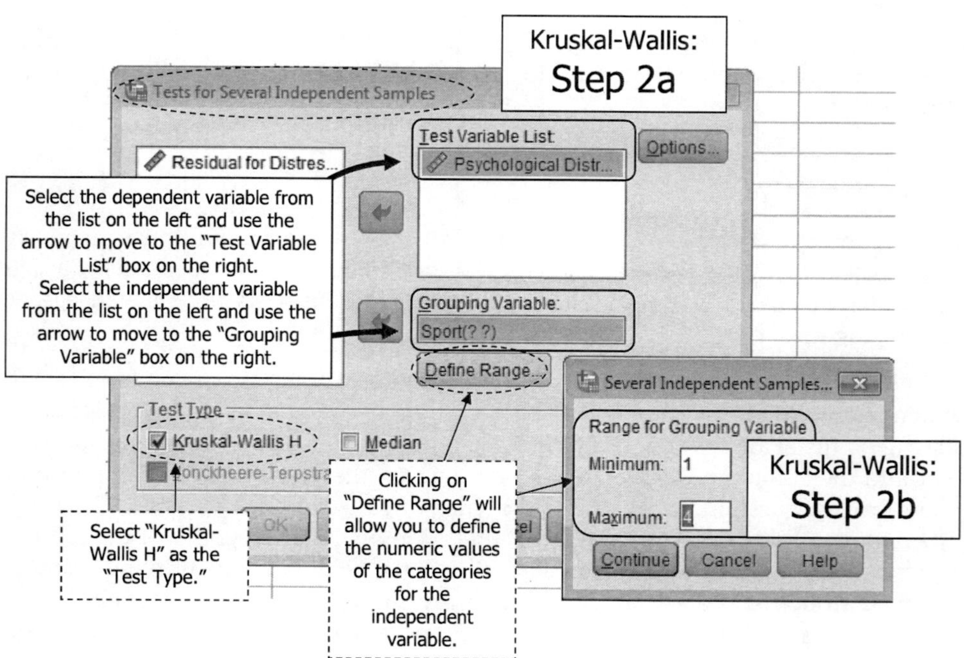

FIGURE 1.14
Kruskal-Wallis: Steps 2a and 2b.

1.2.2.1.1 Interpreting the Output for Kruskal-Wallis

The Kruskal-Wallis is literally an analysis of variance of ranks. Thus the null hypothesis is that the mean ranks of the groups of the independent variable will not be significantly different. In this example, the results (p = .005) suggest statistically significant differences in the mean ranks of the dependent variable by group of the independent variable.

Kruskal-Wallis Test

Ranks

	Type of Sport	N	Mean Rank
Psychological Distress	Movement	8	7.75
	Target	8	15.25
	Fielding	8	18.75
	Territory	8	24.25
	Total	32	

The mean rank is the rank order, from smallest to largest, of the means of the dependent variable (psychological distress) by sport (i.e., type of sport).

Test Statistics[a,b]

	Psychological Distress
Kruskal-Wallis H	13.061
df	3
Asymp. Sig.	.005

a. Kruskal-Wallis Test

b. Grouping Variable: Type of Sport

The p value (labeled "Asymp. Sig.") for the Kruskal-Wallis test is .005. This indicates there is a statistically significant difference in the mean ranks (i.e., rank order of the mean psychological distress by type of sport).

The probability of observing these mean ranks or more extreme mean ranks by chance if the null hypothesis is really true (i.e., if the mean ranks are really equal) is substantially less than 1%. We reject the null hypothesis that all the population mean ranks are equal. For the example, this provides evidence to suggest that psychological distress differs based on the type of sport in which the athlete participates.

FIGURE 1.15
Kruskal-Wallis results.

1.2.2.2 Welch and Brown-Forsythe

Step 1. To conduct the Welch and Brown-Forsythe procedures, go to the "Analyze" in the top pulldown menu, then select "Compare Means," and then select "One-way ANOVA." Following the screenshot for Step 1 (Figure 1.16) produces the One-way ANOVA dialog box.

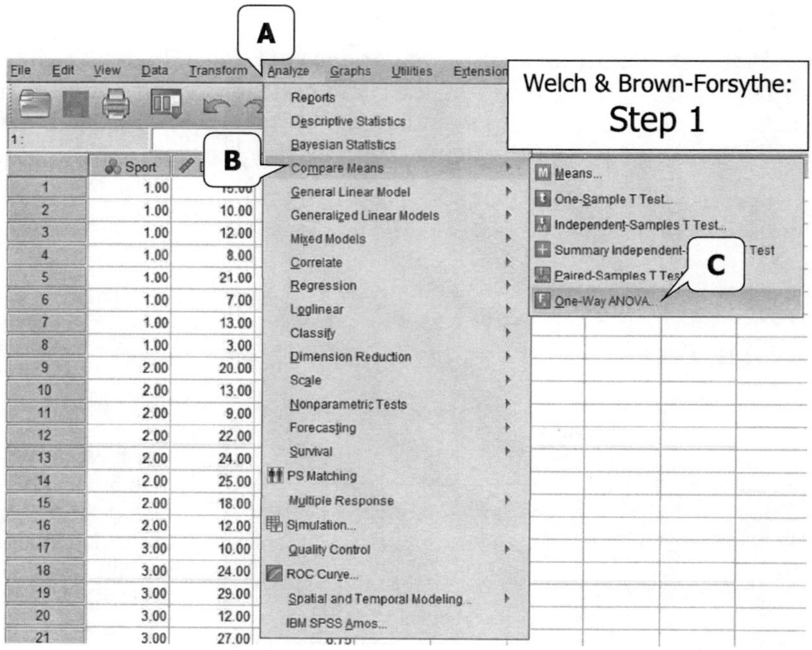

FIGURE 1.16
Welch and Brown-Forsythe: Step 1.

Step 2. Click the dependent variable (e.g., psychological distress) and move it into the "Dependent List" box by clicking the arrow button. Click the independent variable (e.g., type of sport) and move it into the "Factor" box by clicking the arrow button. Next, click on "Options."

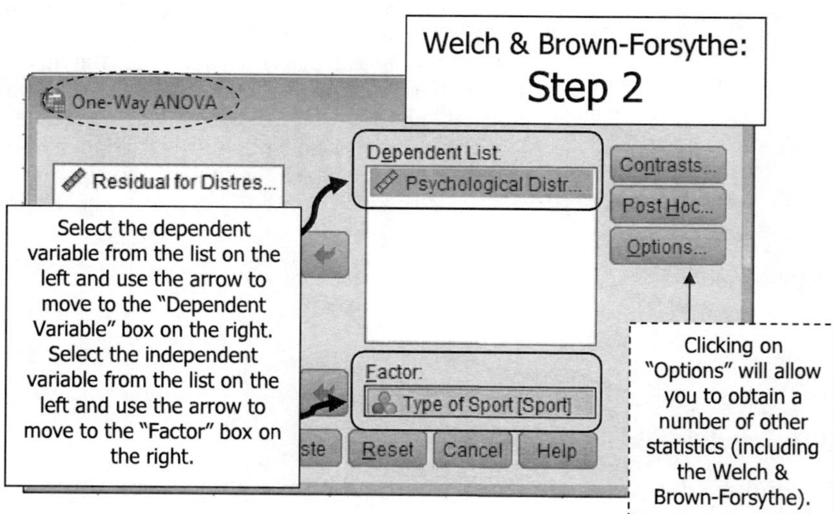

FIGURE 1.17
Welch and Brown-Forsythe: Step 2.

Step 3. Clicking on "Options" will provide the option to select such information as "Descriptive," "Homogeneity of variance test" (i.e., Levene's test for equal variances), "Brown-Forsythe," "Welch," and "Means plot." Click on "Continue" to return to the original dialog box. From the One-way ANOVA dialog box, click on "OK" to return and to generate the output.

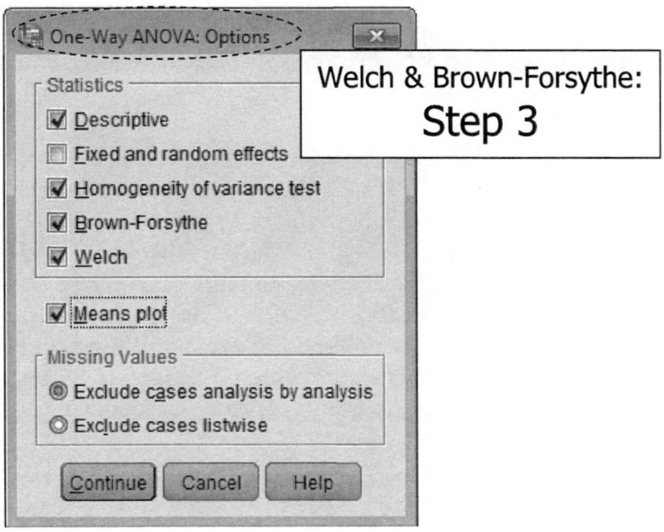

FIGURE 1.18
Welch and Brown-Forsythe: Step 3.

1.2.2.2.1 Interpreting the Output for the Welch and Brown-Forsythe

For illustrative purposes, and because the remainder of the one-way ANOVA results have been interpreted previously, only the results for the Welch and Brown-Forsythe procedures are displayed (Figure 1.19). Both tests suggest there are statistical differences between the groups in terms of the number of stats labs attended.

Robust Tests of Equality of Means

Psychological Distress

	Statistic[a]	df1	df2	Sig.
Welch	7.862	3	15.454	.002
Brown-Forsythe	6.818	3	25.882	.002

a. Asymptotically F distributed.

The *p* values for the Welch and Brown-Forsythe tests are .002. These indicate there is a statistically significant difference in mean psychological distress by type of sport in which the athlete participates.

The probability of observing the *F* statistics (7.862 and 6.818, respectively) or larger by chance if the means of the groups are really equal is substantially less than 1%. We reject the null hypothesis that all the population means are equal.

For this example, this provides evidence to suggest that psychological distress differs based on type of sport in which the athlete participates.

FIGURE 1.19
Welch and Brown-Forsythe results.

For further details on the use of SPSS for these procedures, be sure to examine books such as Page, Braver, and MacKinnon (2003), or Morgan, Leech, Gloeckner, and Barrett (2011).

1.3 Computing Parametric and Nonparametric Models Using R

1.3.1 Introduction to R

The purpose of this section is to briefly consider applications of **R** for the topics covered in this chapter. We will begin with a brief introduction to **R**, a freely accessible open source software environment that operates in both Windows and Mac platforms. *Open source* means that anyone can contribute to the environment, and thus there is a plethora of exciting tools that have been, and will continue to be, developed in **R**. We will interject here that we have a love-hate relationship with **R**. By being free and open source, **R** is broad and deep in terms of tools available, with more and more being added almost daily. Additionally, the **R** community is unparalleled for the help and support offered to users. On the other hand, **R** does not operate in a point-and-click environment. Rather, **R** operates using command language; users have to write "scripts" (aka code or syntax) to *tell* **R** what to do, and **R** is very finicky in its prose. At this point, you may be asking, "Why in the world would I want to subject myself to the torture of having to write commands to generate statistics when just learning statistics is hard enough?" Great question, and we've asked ourselves the same question! If you ask around to a few **R** users, what you'll often find is that many **R** users avoided **R** for as long as they could, but when they finally gave it a shot (or perhaps *had* to give it a shot, as the **R** environment was the only tool accessible for a particular statistic needed), were actually quite pleased—or at least were able to endure **R** sufficiently so that they saw the value in it and continued to use it. We've already mentioned a few benefits of **R**, and the fact that it's free, extremely powerful, open source, and overflowing with helpful users just waiting to lend support are really all that should be needed to convince you that **R** is a tool that you need in your toolkit. You may have heard the assertion, "Do something for 21 days and it becomes a habit."

(Pardon us for a momentary detour: This assertion came from a book published in 1960 by Dr. Maxwell Maltz, a plastic surgeon, who studied the number of days it took amputees to adjust to losing a limb. Dr. Maltz generalized these results to other major life events, and the assertion of 21 days for a habit was almost set in stone. More recently, there is research that actually says habit forming takes longer than 21 days and is quite varied depending on the task. For our purposes, we're going with Dr. Maltz! Now let's get back on track!)

We apply this principle to **R** and say "use **R** for 10 chapters and it becomes a habit" (okay, it's not 21, but all of the chapters in this text use **R**!) We encourage you to give **R** a shot throughout the entire book. By the time you've finished the book, or even sooner, we think you'll be an **R** convert. At the very least, you'll be able to say that you have used **R** and it is in your toolkit! That is no small feat!

All that being said, we're not necessarily saying that **R** will be easy to learn. Again, ask around to a few **R** users and you'll most likely quickly come to understand that there is a learning curve to **R**, one that is steeper for some than for others (we've ourselves experienced points at which it was nearly vertical). If you can get over the hump, so to speak, in using **R**, however, you're home free. (Remember our suggestion to try **R** for all chapters in which it's covered in this text?) Thus, just when you feel like throwing in the towel on **R**, don't do it. Stick with it. Take the hurdles in learning **R** as opportunities to connect to the **R** help community, and keep going. We have offered a number of excellent resources at the end of the chapter to help in learning more about **R**. We hope that the **R** sections in this textbook will provide a smooth transition into the **R** environment and will whet your appetite to learn more about **R**. We also want to remind you that what we have provided is an introduction to **R** for the various statistics generated in the text. Keep in mind that this text is *not* meant to serve as a comprehensive resource for all things **R**. There *are* resources that serve in that capacity, and we will offer a few of those at the conclusion of this chapter. However, this textbook is first and foremost a resource for learning about statistical concepts, which is supplemented with resources for computing statistics using both SPSS and **R**. With that introduction, let's get rolling in **R**!

1.3.1.1 R Basics

A few **R** basics need to be understood before we delve into writing commands. **R** is a base package. Similar to SPSS, Mplus, and many other software programs, there is a base package **R**, to which additional modules (called *packages* in the **R** environment) can be added. The packages written for **R** are stored in the Comprehensive **R** Archive Network (CRAN). There are identical versions of CRAN, called CRAN "mirrors," all around the world. Thus, when you first download **R**, you have to select from which mirror you want to download. There is no one correct location for you. Most **R** users select a CRAN location that is geographically close to them—or at least in their same time zone.

1.3.1.2 Downloading R and RStudio

R can be downloaded from www.r-project.org/. When you click on "download **R**," you will be asked from which CRAN mirror you prefer to download. Many **R** users work directly from the original **R** environment. We prefer using **R** from **RStudio**. The makers of RStudio claim that it "makes **R** easier to use" by providing a console within which to work, visualization space, debugging, and more—and we agree. We have used **R** (version 3.5.1) through RStudio (version 1.1.456) in a Windows platform throughout the text. To download RStudio, download **R** first and then visit www.rstudio.com/ and click on "download." The RStudio Desktop open source license can be downloaded for free. That is the version that has been used throughout the text.

When you use **R** through RStudio, you'll see that you will have access to four quadrants (see Figure 1.20). The top left quadrant is the **source editor window**. This is where you will write and execute scripts or commands (i.e., syntax or code). The bottom left quadrant is the **console**, and this is where the output will appear once you run a command (the exceptions are graphs and figures which display in the bottom right quadrant). When you open RStudio, the console will autopopulate with information—one very important piece of information is contained with the first line, and that is the version of **R** that is being used by RStudio. In Windows, the current version is the default. However, you may find yourself in a situation where you want to run an older version of **R**, as not all packages run in all versions of **R**. To override which version of **R** is being used, go to "Tools" in RStudio's top toolbar and select "Global options." If not already selected, click "general" in the left navigational menu. The version of **R** that is being used is displayed. Click on the button labeled "change"

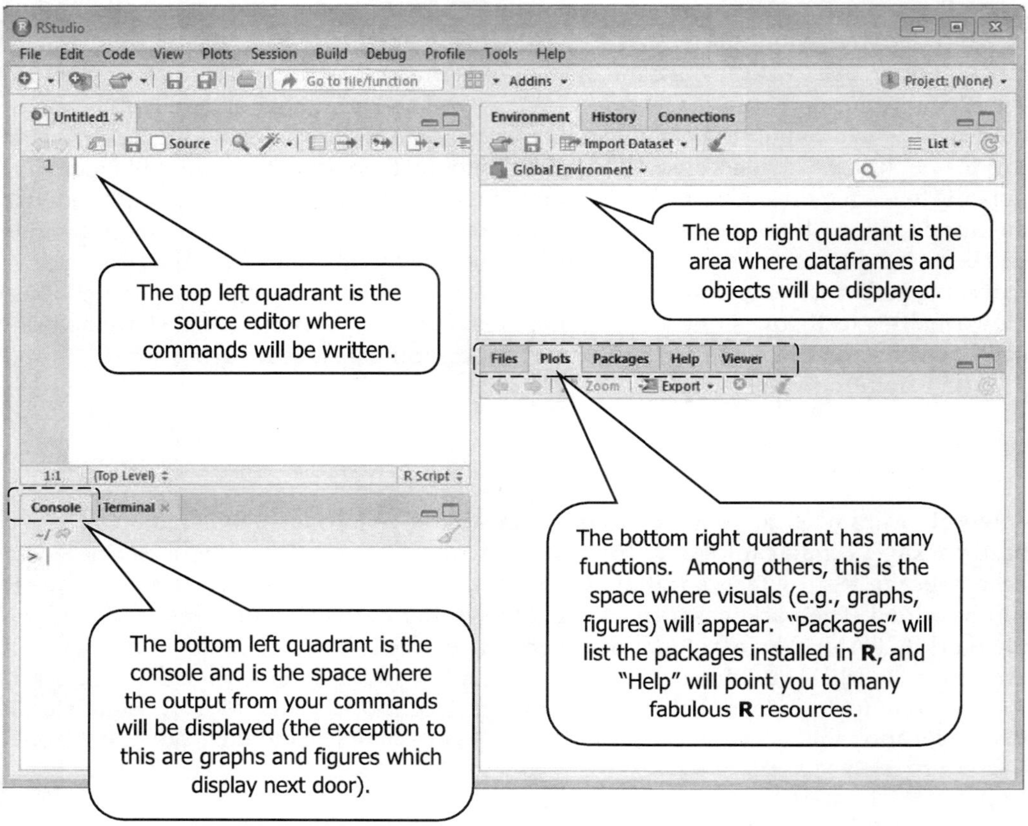

FIGURE 1.20
RStudio.

to display other options of **R** available on your computer and to select the specific version of **R** that you want to work. You can install previous releases of **R** by visiting the CRAN project website (https://cran.r-project.org/bin/windows/base/old/). The top right quadrant is where you will see the list of dataframes and objects (you'll learn about these soon) that you have called up to work with. The bottom right quadrant is the visualization space (i.e., graphs and plots that are generated will appear in this quadrant) and also where you can find the list of **R** packages installed, update packages, and links to various help tools.

1.3.1.3 Packages

Again, base **R** has a number of functionalities that are ready to use, but there are lots of great packages available as well that provide additional functionality. Accessing and using a package is a two-step process. The package first has to be installed (using the *install.packages("PackageNameHere")* command), and then the package has to be called into the library (using the *library(PackageNameHere)* command). Packages need be installed only once (i.e., once installed, always installed). However, they have to be called into the library each time they are used. Throughout the text, as a package is needed, we will provide the commands to both install it and call it into the library. It's not uncommon to get a warning when you install a package that it was created under an earlier version of **R**. Only in a rare instance will you encounter problems in continuing to use the package, so don't be too worried if you encounter this warning. Remember that packages, just like software, get updated from time to time. This is easy to do in RStudio using the *update* icon. **R** also gets updated from time to time. You can efficiently, and painlessly, update **R** by running the following script. You will want to copy all your packages to the new version of **R**.

```
install.packages("installr")
library(installr)
updateR()
```

Let us digress for a moment. Previously, we mentioned that **R** can be quite persnickety. Notice that quotation marks are used to enclose the name of the package in the command for installing the package but not in the command for calling it into the library. Yes, that's correct—you have to pay close attention to little details in **R** like quotation marks, commas, capitalization, etc.

1.3.1.4 Working in R

There are different ways to work in **R**, none of which are necessarily right or wrong and different users have different preferences. However, throughout the text we have guided your work in **R** so that you begin your work by establishing a **working directory**. Within that working directory, your files and scripts will be stored. We have found that to be the easiest way to keep track of your files and stay organized.

Throughout the text, we will refer to objects and functions. An **object** is something created from a **function**. Many times, a function will be running a statistical procedure (e.g., generating an ANOVA or generating a regression model), but a function could also be generating a table or creating a variable or more. An object is what results from that function (e.g., the results of the ANOVA model, the table, or the variable). Throughout the text, we will try to remind you what is the object versus the function, but generally this takes the form in **R** command language as the following: *object <- function*, where the object appears to the left of "<-" and the function appears to the right of "<-." Why do we need to know this? Creating objects from functions is not necessarily a requirement, but it can make life much easier when you want to extend the results from your function to something else. This is because rather than writing the entire function again (e.g., an entire ANOVA or regression model), you simply have to write the name of your object. As you will see, some functions are short and sweet, and some functions are long and tedious. Naming your function as an object is particularly helpful in the case of the latter!

Although data can be created in **R**, the data examples provided in the text use comma-separated (.csv) files. Command language is provided in the illustrations to bring the .csv file into the **R** environment. We have done this because it is usually the case that data that we work with already exist in spreadsheet form (e.g., Excel, SPSS, SAS). If the data do not already exist, we encourage you to use a spreadsheet tool to create the dataset and then bring it into the **R** environment. Once the data are brought into **R**, it is called a **dataframe.** There are lots of ways to work with data in dataframes (e.g., recoding variables, creating new variables), and you'll be introduced to quite a few of those in the examples throughout the text. If you manipulate your dataframe, you may want to save it and export out of the **R** environment. That's easy to do using the command *write.csv(DataframeName, "FileName.csv")*. This command, along with a few other "staples," are provided in Box 1.3. There are a number of time-saving shortcuts in RStudio. You can access these directly in RStudio by going to "Tools" in the top menu and then "keyboard shortcuts help." You'll find that some are the same as those to which you're accustomed (e.g., in Windows, Ctrl+O to open, Ctrl+S to save).

BOX 1.3 Need-to-Know Commands in R

Command	Functionality
install.packages("PackageNameHere")	Installs a package into **R**. Once a package is installed, it remains installed in **R**. However, each time it is used, the user needs to call it in using the library command. *Note: Quotation marks around the package name are required.*
library(PackageNameHere)	Calls a package into **R** so that it can be used. Each time a package is used, it must be called into the **R** environment using the library command.
getwd()	**R** is always directed to a directory on your computer. To find out which directory it's pointed to, run this *get working directory* command. We will assume that we need to change the working directory, and will use the next line of code to set the working directory to the desired path.
setwd("E:/Folder")	Establishes a working directory which points to a specific folder that is designated by the user. (Momentary detour: If you don't know where a file is located, right click on the file and go to "properties." The "location" in properties will provide the specific file location.) *Note: Quotation marks around the folder are required.*
DataframeName <- read.csv ("DatasetName.csv")	Renames the dataset to whatever is designated to the left of the "<-." *Note: Quotation marks around the file name are required.*
names(DataframeName)	Lists the names of the variables in the dataframe (this output is provided in the console).
View(DataframeName)	Calls the dataframe into RStudio (i.e., creates a tab in the source editor where the user can see the actual spreadsheet view of the data).
write.csv(DataframeName, "FileName.csv")	Exports the dataframe from **R** into a comma-separated file.

One additional tip as we're getting started: when you run script in **R**, do *not* highlight the command and then hit run. Rather, simply place your cursor anywhere in the command that you want to run and then hit the run icon (or Ctrl+Enter). This is especially helpful when you have very long lines of code, and it will prevent you from failing to highlight parts of it.

Now that you've been provided a segue into **R**, let's dive in! Next we consider **R** for the one-way ANOVA model. Note that the scripts are provided within the blocks with additional annotation to assist in understanding how the command works. Should you want to write reminder notes and annotation to yourself as you write the commands in **R** (and we highly encourage doing so), remember that any text that follows a hashtag (i.e., #) is annotation only and not part of the **R** script. Thus, you can write annotations directly into **R** with hashtags. We encourage this practice so that when you call up the commands in the future, you'll understand what the various lines of code are doing. You may think you'll remember what you did. However, trust us. There is a good chance that you won't. Thus, consider it best practice when using **R** to annotate heavily!

1.3.2 Reading Data Into R

```
getwd()
```

R is always pointed to a directory on your computer. The *get working directory* function can be used to determine to which directory **R** is pointed. We will assume that we need to change the working directory, and will use the next line of code to set the working directory to the desired path.

```
setwd("E:/FolderName")
```

We use the *setwd* function to establish the working directory. To set the working directory, change what is in quotation marks to your file location. Also, if you are copying the directory name from your properties, you will need to change the backslash (i.e., \) to a forward slash (i.e., /).

```
Ch1_distress <- read.csv("Ch1_distress.csv")
```

The *read.csv* function reads our data into **R**. What's to the left of the "<-" will be what the data will be called in **R**. In this example, we're calling the R dataframe "Ch1_distress." What's to the right of the "<-" tells **R** to find this particular csv file. In this example, our file is called "Ch1_distress.csv." Make sure the extension (i.e., .csv) is included in your script. Also note that the name of your file should be in quotation marks within the parentheses.

```
names(Ch1_distress)
```

The *names* function will produce a list of variable names for each dataframe as follows. This is a good check to make sure your data have been read in correctly.

```
[1] "Sport" "Distress"
```

```
View(Ch1_distress)
```

The *View* function will let you view the dataset in spreadsheet format in RStudio.

```
install.packages(car)
```

We will be using the *car* package for Levene's test. This function will install the package in **R**.

```
library(car)
```

The *library* function will load the *car* package in our library.

```
install.packages("compute.es")
```

We will use the *compute.es* package to compute effect sizes. The *install.packages* function will install the package in **R**.

```
library(compute.es)
```

The *library* function will load the *compute.es* package in our library.

```
Ch1_distress$SportF <- factor(Ch1_distress$Sport,
labels = c("movement", "target", "fielding", "territory"))
```

This command will create a new variable in our dataframe named "SportF." We use the *factor* function to define the variable Sport as nominal with the four groups defined here (i.e., *movement, target, fielding, territory*). What is to the left of "<-" in the script creates the new SportF variable in our dataframe.

```
summary(Ch1_distress)
```

The *summary* function will produce basic descriptive statistics on all the variables in your dataframe. This is a great way to quickly check to see if the data have been read in correctly and get a feel for your data, if you haven't already. The output from the summary statement for this dataframe looks like this. Because we defined *SportF* as a factor, we are provided only the frequencies for each category in that variable.

```
      Sport           Distress            SportF
Min.    :1.00    Min.    : 3.00    movement   :8
1st Qu. :1.75    1st Qu. :12.00    target     :8
Median  :2.50    Median  :20.00    fielding   :8
Mean    :2.50    Mean    :18.41    territory  :8
3rd Qu. :3.25    3rd Qu. :25.00
Max.    :4.00    Max.    :30.00
```

```
Levels(Ch1_distress$SportF)
```

The *levels* function will output the categories of our factor variable, a good way to double check your coding of the categories.

```
[1] "movement" "target" "fielding" "territory"
```

FIGURE 1.21
Reading data into **R**.

1.3.3 Generating the One-Way ANOVA Model

```
Ch1_ANOVA <- aov(Distress ~ SportF, data=Ch1_distress)
```

The *aov* function will generate the one-way ANOVA model with Distress as the dependent variable and *SportF* as the independent variable. The dataframe from which we are pulling the data is defined by the *data* function. We are calling this object 'Ch1_ANOVA.'

```
summary(Ch1_ANOVA)
```

The *summary* function will provide the output from our ANOVA model:

```
            Df  Sum Sq Mean Sq F value  Pr(>F)
SportF       3   738.6  246.20   6.818 0.00136 **
Residuals   28  1011.1   36.11
—
Signif. codes:  0 '***' 0.001 '**' 0.01 '*' 0.05 '.' 0.1 ' ' 1
```

FIGURE 1.22
Generating the one-way ANOVA.

```
summary.lm(Ch1_ANOVA)
```

The *summary.lm* function will produce additional output, including R^2 which, in one-way ANOVA, is also the same value as partial eta squared (recall that in one-way ANOVA, eta squared is equal to partial eta squared).

```
Call:
aov(formula = Distress ~ SportF, data = Ch1_distress)

Residuals:
     Min       1Q  Median      3Q     Max
-10.2500  -4.5000  0.8125  4.2500  9.8750

Coefficients:

                 Estimate Std. Error t value Pr(>|t|)
(Intercept)        11.125      2.125   5.236 1.45e-05 ***
SportFtarget        6.750      3.005   2.247 0.032741 *
SportFfielding      9.125      3.005   3.037 0.005125 **
SportFterritory    13.250      3.005   4.410 0.000139 ***
—
Signif. codes:  0 '***' 0.001 '**' 0.01 '*' 0.05 '.' 0.1 ' ' 1

Residual standard error: 6.009 on 28 degrees of freedom
Multiple R-squared: 0.4221, Adjusted R-squared: 0.3602
F-statistic: 6.818 on 3 and 28 DF, p-value: 0.001361
```

Homogeneity Tests

```
leveneTest(Ch1_distress$Distress, Ch1_distress$SportF, center=mean)
```

The *leveneTest* function can be used to generate Levene's test for homogeneity of variance. There are multiple ways to center Levene's. For this illustration, we centered on the mean (i.e., *center=mean*).

```
Levene's Test for Homogeneity of Variance (center = mean)
      Df F value Pr(>F)
group  3  0.9047 0.4513
      28
```

We read this output as $F(3,28) = .9047$, $p = .4513$, indicating we have met the assumption of equal variances.

```
leveneTest(Ch1_ANOVA)
```

We can also run the *leveneTest* function on the object *(Ch1_ANOVA)* of our one-way ANOVA model results to generate Levene's test with the default centering of the median, which may provide more robust results. These results still provide evidence of meeting the assumption of equal variances, with $p = .618$.

```
Levene's Test for Homogeneity of Variance (center = median)
      Df F value Pr(>F)
group  3  0.6039  0.618
      28
```

```
install.packages("onewaytests")
library(onewaytests)
```

Install the *onewaytests* package

FIGURE 1.22 (continued)
Generating the one-way ANOVA.

```
bf.test(Distress ~ SportF,
        Ch1_distress,
        alpha = .05,
        verbose = TRUE)
```

The *bf.test* function is used to generate the Brown-Forsythe test for equal variances. Within parentheses, we define the dependent variable, Distress, independent variable, SportF, and dataframe, *Ch1_distress*, along with alpha, and the final command of *verbose=TRUE* which tells **R** to print the output to the console.

```
Brown-Forsythe Test
-------------------------------------------------------------
  data : Distress and SportF

  statistic  : 6.817695
  num df     : 3
  denom df   : 25.88229
  p.value    : 0.001544356

  Result     : Difference is statistically significant.
-------------------------------------------------------------
```

```
install.packages("lawstat")
library(lawstat)
```

To install the *lawstat* package and load into the library.

```
levene.test(Ch1_distress$Distress,
            Ch1_distress$SportF,
            location = c("median"),
            bootstrap = TRUE,
            num.bootstrap = 1000,
            kruskal.test = FALSE,
            correction.method = c("zero.correction"))
```

```
      bootstrap modified robust Brown-Forsythe Levene-type
      test based on the absolute deviations from the median
      with modified structural zero removal method and
      correction factor

data: Ch1_distress$Distress
Test Statistic = 0.82624, p-value = 0.504
```

```
levene.test(Ch1_distress$Distress,
            Ch1_distress$SportF,
            location = c("median"),
            bootstrap = FALSE,
            kruskal.test = FALSE,
            correction.method = c("zero.correction"))
```

```
      modified robust Brown-Forsythe Levene-type test based
      on the absolute deviations from the median with
      modified structural zero removal method and correction
      factor

data: Ch1_distress$Distress
Test Statistic = 0.82624, p-value = 0.4924
```

FIGURE 1.22 (continued)
Generating the one-way ANOVA.

Effect Size

```
install.packages("sjstats")
library(sjstats)
```

The package *sjstats* can be used to generate multiple effect size indices in ANOVA. The *install.packages* and *library* functions will, respectively, install the package and then load it into our **R** library.

```
omega_sq(Ch1_ANOVA)
```

Using the object created from our ANOVA model, *Ch1_ANOVA*, we can generate omega squared with the *omega_sq* function.

```
    term omegasq
1 SportF   0.353
```

```
cohens_f(Ch1_ANOVA)
```

Using the object created from our ANOVA model, *Ch1_ANOVA*, we can generate Cohen's *f* with the *cohens_f* function.

```
    term  cohens.f
1 SportF 0.8546738
```

```
eta_sq(Ch1_ANOVA)
```

Using the object created from our ANOVA model, *Ch1_ANOVA*, we can generate eta squared with the *eta_sq* function.

```
    term etasq
1 SportF 0.422
```

```
Ch1_distress$unstandardizedResiduals <- residuals(Ch1_ANOVA)
```

We also want to save our unstandardized residuals to the dataframe. We use the *residuals* function to compute unstandardized residuals from our *Ch1_ANOVA* model. To the left of '<-' we will save the residuals as a variable named *unstandardizedResiduals* in our dataframe, *Ch1_distress*.

FIGURE 1.22 (continued)
Generating the one-way ANOVA.

1.3.4 Generating the Welch Test

```
oneway.test(Distress ~ Sport, data = Ch1_distress)
```

The *oneway.test* function produces Welch's test results, which are as follows:

```
        One-way analysis of means (not assuming equal variances)

data: Distress and Sport
F = 7.862, num df = 3.000, denom df = 15.454, p-value = 0.002055
```

FIGURE 1.23
Generating the Welch test in **R**.

1.3.5 Generating the Kruskal-Wallis Test

```
kruskal.test(Distress ~ Sport, data = Ch1_distress)
```

The *kruskal.test* function produces results for the Kruskal-Wallis test. We define our model with the dependent variable, "Distress," and independent variable, "Sport." The dataframe we use is *Ch1_distress*.

```
        Kruskal-Wallis rank sum test

data:  Distress by Sport
Kruskal-Wallis chi-squared = 13.061, df = 3, p-value =
0.004506
```

```
Ch1_distress$RankOrder <- rank(Ch1_distress$Distress)
```

This script produces a new variable in our dataframe called "RankOrder," which is the rank for each value of the variable *Distress* in our *Ch1_distress* dataframe.

```
by(Ch1_distress$RankOrder, Ch1_distress$Sport, mean)
```

The *by* function will produce the mean rank for each category of sport. The output looks like this:

```
Ch1_distress$Sport: movement
[1] 7.75
-------------------------------------
Ch1_distress$Sport: target
[1] 15.25
-------------------------------------
Ch1_distress$Sport: fielding
[1] 18.75
-------------------------------------
Ch1_distress$Sport: territory
[1] 24.25
```

FIGURE 1.24
Generating the Kruskal-Wallis test.

1.4 Data Screening

As noted earlier, there are three standard assumptions made in analysis of variance models and will see these assumptions often in the remainder of this text. The assumptions are concerned with *normality*, *independence*, and *homogeneity of variance* (also called *homoscedasticity*).

1.4.1 Normality

As alluded to earlier in the chapter, understanding the distributional shape, specifically the extent to which normality is a reasonable assumption, is important. For the one-way ANOVA, the distributional shape for the residuals should be a normal distribution. Recall that when we ran our ANOVA model, we saved the unstandardized residuals to our datafile (we could have also reviewed the standardized or studentized residuals—and will illustrate the use of those in later chapters). We can again use "Explore" to examine the extent to which the assumption of normality is met (see Figure 1.25). From the top tool bar in SPSS, to access Explore, go to 'Analyze' then 'Descriptive Statistics' and then 'Explore.' From the 'Explore' dialog box, click the residual and move it into the "Dependent List" box by clicking on the arrow button. The procedures for selecting normality statistics are as follows: click on "Plots" in the upper right corner. Place a checkmark in the boxes for "Normality plots with tests" and also for "Histogram." Then click "Continue" to return to the main Explore dialog box. Then click "OK" to generate the output. To identify normality by group, in the main dialog box, click the residual and move it into the Dependent List box, and click the independent variable and move to the "Factor List" box by clicking on the respective arrow buttons.

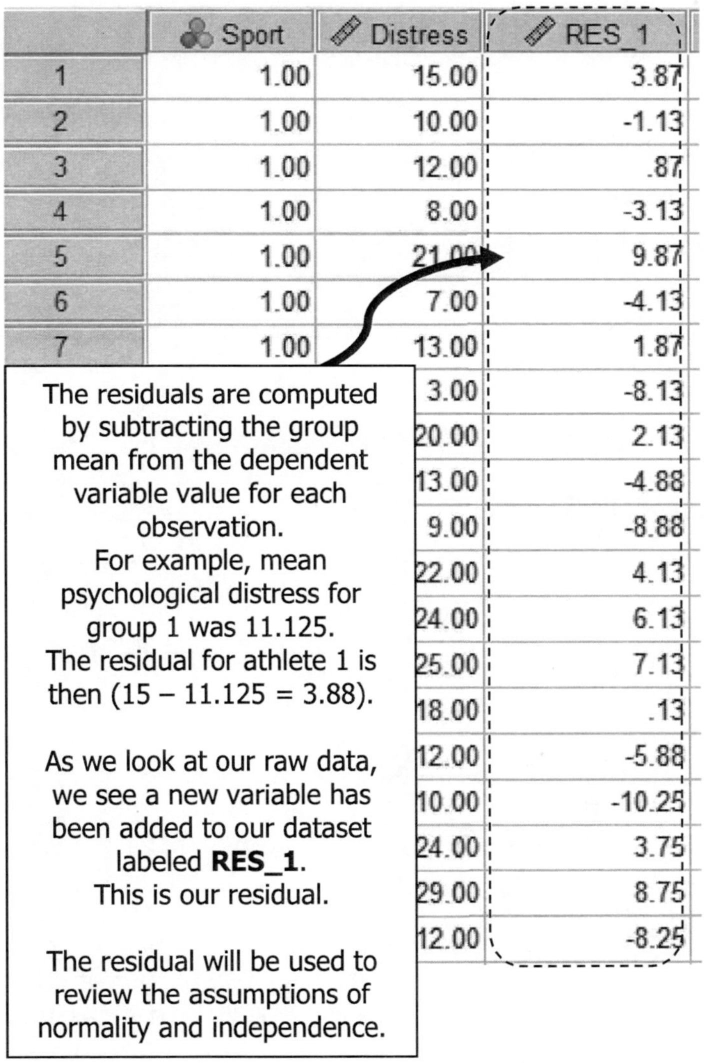

FIGURE 1.25
Generating normality evidence.

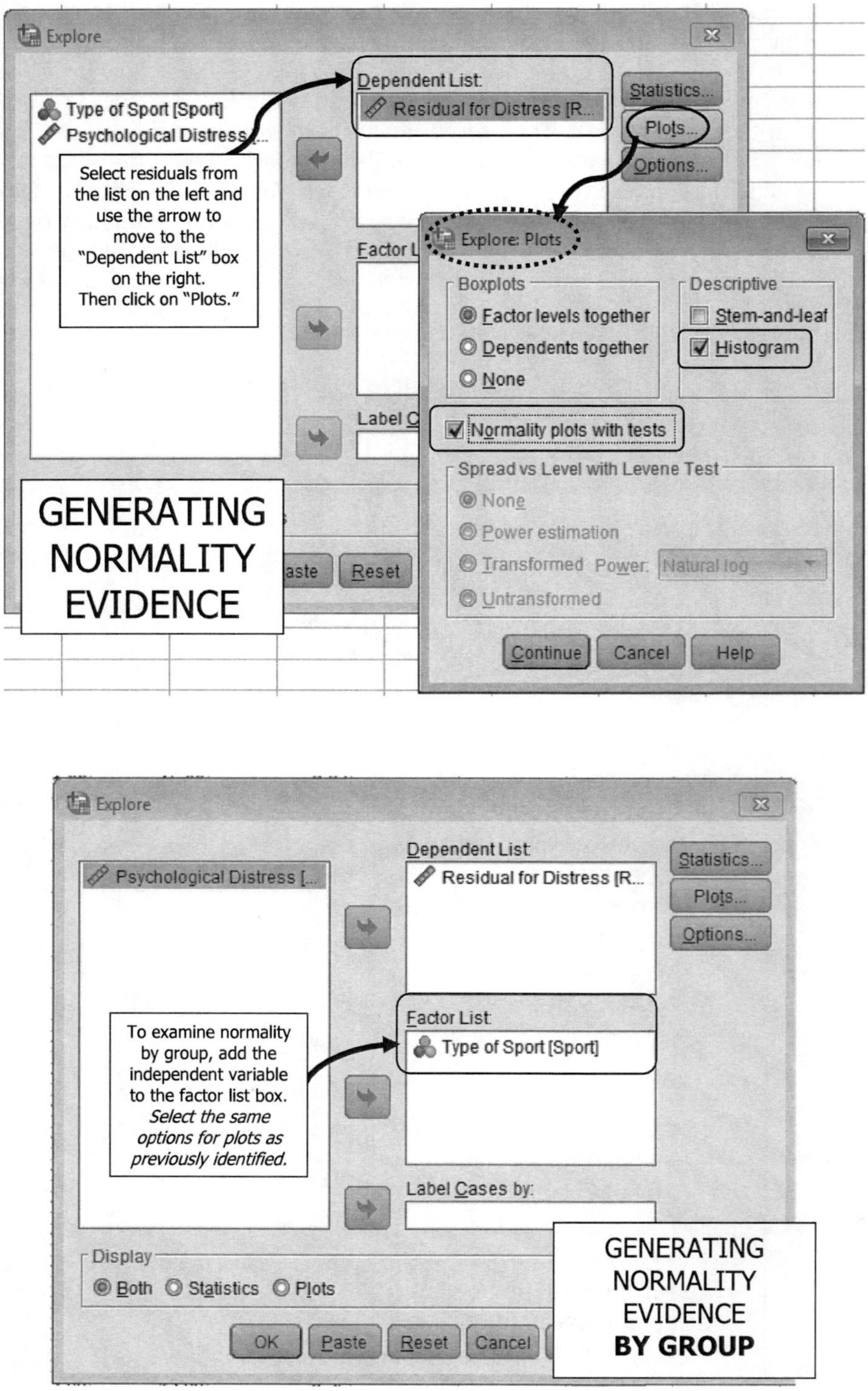

FIGURE 1.25 (continued)
Generating normality evidence.

1.4.1.1 Interpreting Normality Evidence

We have already developed a good understanding of how to interpret some forms of evidence of normality including skewness and kurtosis, histograms, and boxplots. The skewness and kurtosis statistics of the residuals, overall and by group, is within the range of an absolute value of 2.0, suggesting evidence of normality. Working in **R**, *D'Agostino's test* (D'Agostino, 1970) can be used to examine the null hypothesis that skewness equals zero. Thus, a statistically significant D'Agostino's test would indicate that there is statistically significant skewness. For kurtosis, we can use the *Bonett-Seier test for Geary's kurtosis* (Bonett & Seier, 2002) for data that are normally distributed. The null hypothesis states that data should have a Geary's kurtosis value equal to $\sqrt{2/\pi} = .7979$. Thus, a statistically significant Bonett-Seier test for Geary's kurtosis would indicate that there is statistically significant kurtosis. Thus, with these tests, as with Kolmogorov-Smirnov and Shapiro-Wilk, we do *not* want to find statistically significant results. Overall and by group, we find evidence of normality with p's > .05.

Descriptives

			Statistic	Std. Error
Residual for Distress	Mean		.0000	1.00959
	95% Confidence Interval for Mean	Lower Bound	−2.0591	
		Upper Bound	2.0591	
	5% Trimmed Mean		.0260	
	Median		.8125	
	Variance		32.617	
	Std. Deviation		5.71112	
	Minimum		−10.25	
	Maximum		9.87	
	Range		20.13	
	Interquartile Range		9.25	
	Skewness		−.239	.414
	Kurtosis		-1.019	.809

By group of the independent variable, we find the following, all indicating the normality assumption has been met.

	Type of Sport		Statistic	Std. Error
Residual for Distress	**Movement**	Skewness	.449	.752
		Kurtosis	.575	1.481
	Target	Skewness	-.315	.752
		Kurtosis	-1.522	1.481
	Fielding	Skewness	-.367	.752
		Kurtosis	-1.754	1.481
	Territory	Skewness	-.851	.752
		Kurtosis	.058	1.481

Working in R, we can generate various normality statistics as well.

FIGURE 1.26
Normality evidence.

```
install.packages("pastecs")
```

The *install.packages* function will install the *pastecs* package which we will use to generate various forms of normality evidence.

```
library(pastecs)
```

The *library* function will load the *pastecs* package.

```
stat.desc(Ch1_distress$unstandardizedResiduals,
          norm = TRUE)
```

The *stat.desc* function will generate normality indices on the variable *unstandardizedResiduals* in the dataframe *Ch1_distress* as follows. The *norm=TRUE* command will produce Shapiro- Wilk's results, which are displayed as *normtest.W* (which is the *S-W* statistic value) and *normtest.p* (which is the observed probability value).

Here, we see *S-W* = .958 and the related *p* = .240. We see skew (.249) and kurtosis (-.976) *(wait—these aren't the same values as what we found with SPSS; if that's what you're thinking, hold that thought!)* for the *unstandardizedResidual* variable.

Skew, kurtosis, and *S-W* all indicate the assumption of normality has been met. As we know, we can divide the skew and kurtosis values by their standard errors to get a standardized value that can be used to determine if the skew and/or kurtosis is statistically different from zero. Since this output provides '2SE', we would simply divide this value by 2 to arrive at the standard error.

Note: You may have noticed that the skewness and kurtosis value that we've just generated differs from what we found in SPSS, which was skew = -.239 and kurtosis = -1.019. *This is because there are different ways to calculate skewness and kurtosis.* Let's use another package in **R** to calculate these statistics with different algorithms.

```
      nbr.val      nbr.null        nbr.na           min           max
 3.200000e+01  0.000000e+00  0.000000e+00 -1.025000e+01  9.875000e+00
        range           sum        median          mean       SE.mean
 2.012500e+01 -4.329870e-15  8.125000e-01 -1.353084e-16  1.009594e+00
 CI.mean.0.95           var       std.dev      coef.var      skewness
 2.059080e+00  3.261694e+01  5.711124e+00 -4.220819e+16 -2.169593e-01
     skew.2SE      kurtosis      kurt.2SE    normtest.W    normtest.p
-2.617390e-01 -1.168535e+00 -7.218785e-01  9.578412e-01  2.395168e-01
```

```
install.packages("e1071")
```

The *install.packages* function will install the **e1071** package which we will use to generate skewness and kurtosis.

```
library(e1071)
```

The *library* function will load the **e1071** package.

```
skewness(Ch1_distress$unstandardizedResiduals, type=3)
skewness(Ch1_distress$unstandardizedResiduals, type=2)
skewness(Ch1_distress$unstandardizedResiduals, type=1)
```

The *skewness* function will generate skewness statistics on the variable(s) specified. The *type=* script defines how skewness is calculated. Specifying *type=2* will use the algorithm that is used by SPSS. Readers interested in

FIGURE 1.26 (continued)
Normality evidence.

learning more, including the algorithms for each of the three methods, are encouraged to review Joanes and Gill (1998). We see that using *type*=2, our skew is -.239, the same value as generated using SPSS.

```
# skewness(Ch1_distress$unstandardizedResiduals, type=3)
[1] -0.2169593

# skewness(Ch1_distress$unstandardizedResiduals, type=2)
[1] -0.2388885

# skewness(Ch1_distress$unstandardizedResiduals, type=1)
[1] -0.2275415
```

```
kurtosis(Ch1_distress$unstandardizedResiduals, type=3)
kurtosis(Ch1_distress$unstandardizedResiduals, type=2)
kurtosis(Ch1_distress$unstandardizedResiduals, type=1)
```

The *kurtosis* function will generate kurtosis statistics on the variable(s) we specify. The *type*= defines how kurtosis is calculated. Specifying *type*=2 will use the algorithm that is used by SPSS. Readers interested in learning more, including the algorithms for each of the three methods, are encouraged to review Joanes and Gill (1998). We see that using *type*=2, our kurtosis is -1.019, the same value as generated using SPSS.

```
# kurtosis(Ch1_distress$unstandardizedResiduals, type=3)
[1] -1.168535

# kurtosis(Ch1_distress$unstandardizedResiduals, type=2)
[1] -1.019064

# kurtosis(Ch1_distress$unstandardizedResiduals, type=1)
[1] -1.048471
```

Working in **R**, another way to test for normality is **Agostino's test** for skewness and the **Bonett-Seier test for Geary's kurtosis.**

```
install.packages("moments")
library(moments)
```

To conduct Agostino's test, we first have to install the *moments* package and then load it into our library. The null hypothesis for this test is that skewness equals zero. Thus, a statistically significant Agostino's test would indicate that there is statistically significant skewness.

```
agostino.test(Ch1_distress$unstandardizedResiduals)
```

The function *agostino.test* is generated using the variable *unstandardizedResiduals* from our *Ch11_distress* dataframe. The results suggest evidence of normality as p = .544, greater than alpha.

```
        D'Agostino skewness test

data:  Ch1_distress$unstandardizedResiduals
skew = -0.22754, z = -0.60681, p-value = 0.544
alternative hypothesis: data have a skewness
```

```
agostino.test(Ch1_distress$unstandardizedResiduals[Ch1_distress$Sport==1])
agostino.test(Ch1_distress$unstandardizedResiduals[Ch1_distress$Sport==2])
agostino.test(Ch1_distress$unstandardizedResiduals[Ch1_distress$Sport==3])
agostino.test(Ch1_distress$unstandardizedResiduals[Ch1_distress$Sport==4])
```

By group, the results for the D'Agostino test provide evidence of normality by group with all p's > .05.

FIGURE 1.26 (continued)
Normality evidence.

```
#agostino.test(Ch1_distress$unstandardizedResiduals[Ch1_distress$Sport==1])

        D'Agostino skewness test
data:   Ch1_distress$unstandardizedResiduals[Ch1_distress$Sport==1]
skew = 0.36014, z = 0.60580, p-value = 0.5446
alternative hypothesis: data have a skewness

#agostino.test(Ch1_distress$unstandardizedResiduals[Ch1_distress$Sport==2])

        D'Agostino skewness test
data:   Ch1_distress$unstandardizedResiduals[Ch1_distress$Sport==2]
skew = -0.25259, z = -0.42532, p-value = 0.6706
alternative hypothesis: data have a skewness

#agostino.test(Ch1_distress$unstandardizedResiduals[Ch1_distress$Sport==3])

        D'Agostino skewness test
data:   Ch1_distress$unstandardizedResiduals[Ch1_distress$Sport==3]
skew = -0.29418, z = -0.49518, p-value = 0.6205
alternative hypothesis: data have a skewness

#agostino.test(Ch1_distress$unstandardizedResiduals[Ch1_distress$Sport==4])

        D'Agostino skewness test
data:   Ch1_distress$unstandardizedResiduals[Ch1_distress$Sport==4]
skew = -0.6827, z = -1.1426, p-value = 0.2532
alternative hypothesis: data have a skewness
```

```
bonett.test((Ch1_distress$unstandardizedResiduals))
```

The *bonett.test* function, generated using the variable *unstandardizedResiduals* from our *Ch1_distress* dataframe, performs the Bonett-Seier test for Geary's kurtosis for data that is normally distributed. The null hypothesis states that data should have a Geary's kurtosis value equal to $\sqrt{2/\pi} = .7979$. The results suggest evidence of normality as $\pi = .1232$, greater than alpha.

```
        Bonett-Seier test for Geary kurtosis

data:   (Ch1_distress$unstandardizedResiduals)
tau = 4.8125, z = -1.5413, p-value = 0.1232
alternative hypothesis: kurtosis is not equal to sqrt(2/pi)
```

```
bonett.test((Ch1_distress$unstandardizedResiduals[Ch1_distress$Sport==1]))
bonett.test((Ch1_distress$unstandardizedResiduals[Ch1_distress$Sport==2]))
bonett.test((Ch1_distress$unstandardizedResiduals[Ch1_distress$Sport==3]))
bonett.test((Ch1_distress$unstandardizedResiduals[Ch1_distress$Sport==4]))
```

By group, the results for the Bonett-Seier test for Geary's kurtosis for data that is normally distributed provide evidence of normality by group with all p's > .05.

```
#bonett.test((Ch1_distress$unstandardizedResiduals[Ch1_distress$Sport==1]))

        Bonett-Seier test for Geary kurtosis
data:   (Ch1_distress$unstandardizedResiduals[Ch1_distress$Sport==1])
tau = 4.12500, z = -0.08177, p-value = 0.9348
alternative hypothesis: kurtosis is not equal to sqrt(2/pi)

#bonett.test((Ch1_distress$unstandardizedResiduals[Ch1_distress$Sport==2]))

        Bonett-Seier test for Geary kurtosis
data:   (Ch1_distress$unstandardizedResiduals[Ch1_distress$Sport==2])
```

FIGURE 1.26 (continued)
Normality evidence.

```
tau = 4.9062, z = -1.2053, p-value = 0.2281
alternative hypothesis: kurtosis is not equal to sqrt(2/pi)
#bonett.test((Ch1_distress$unstandardizedResiduals[Ch1_distress$Sport==3]))
        Bonett-Seier test for Geary kurtosis
data:   (Ch1_distress$unstandardizedResiduals[Ch1_distress$Sport==3])
tau = 6.1875, z = -1.5340, p-value = 0.125
alternative hypothesis: kurtosis is not equal to sqrt(2/pi)
#bonett.test((Ch1_distress$unstandardizedResiduals[Ch1_distress$Sport==4]))
        Bonett-Seier test for Geary kurtosis
data:   (Ch1_distress$unstandardizedResiduals[Ch1_distress$Sport==4])
tau = 4.03120, z = -0.68704, p-value = 0.4921
alternative hypothesis: kurtosis is not equal to sqrt(2/pi)
```

FIGURE 1.26 (continued)
Normality evidence.

The histogram of residuals, overall or by group, is not exactly what most researchers would consider a classic normally shaped distribution. Reviewing the residuals overall, it approaches a normal distribution and there is nothing to suggest normality may be an unreasonable assumption. By group, we will rely on other forms of normality evidence given the small group sizes make the histograms by group more difficult to visually evaluate.

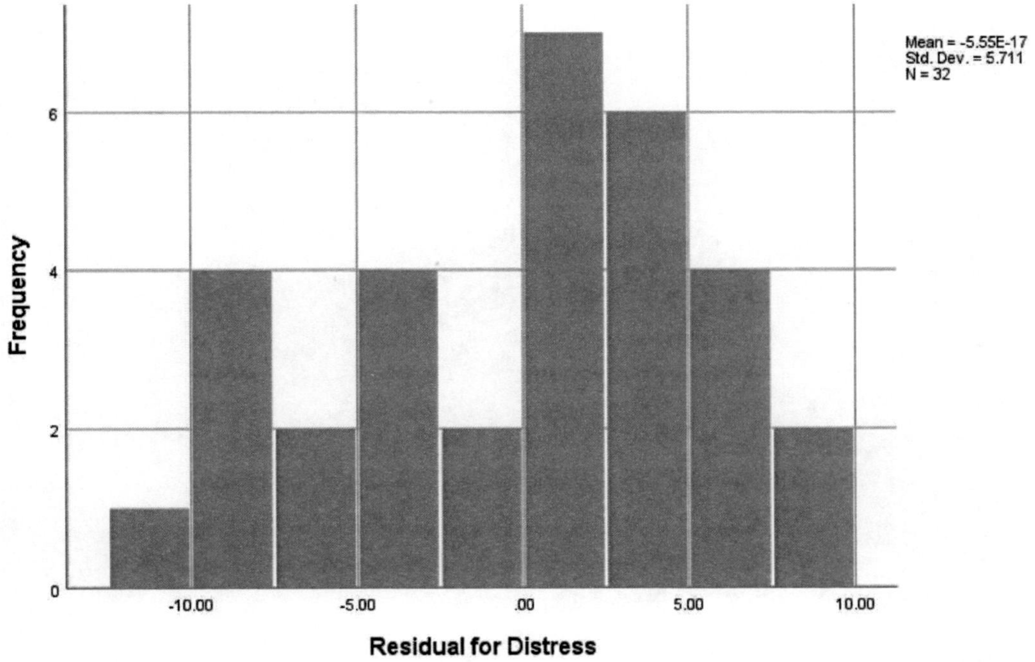

FIGURE 1.27
Histogram.

Two of the four histograms by group are presented. By group, the small group sizes are not conducive to suggesting normality. Thus, reviewing the normality evidence in aggregate will be helpful.

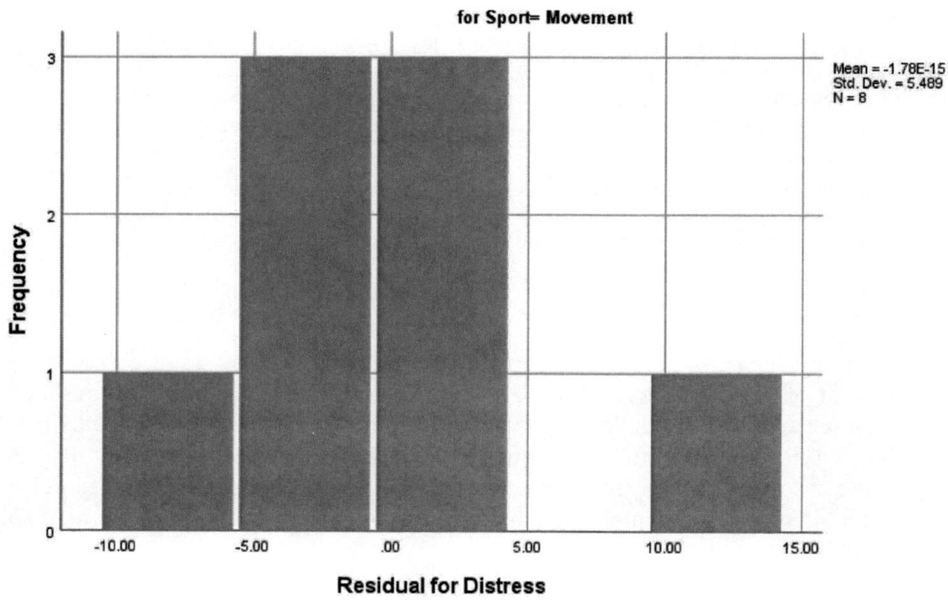

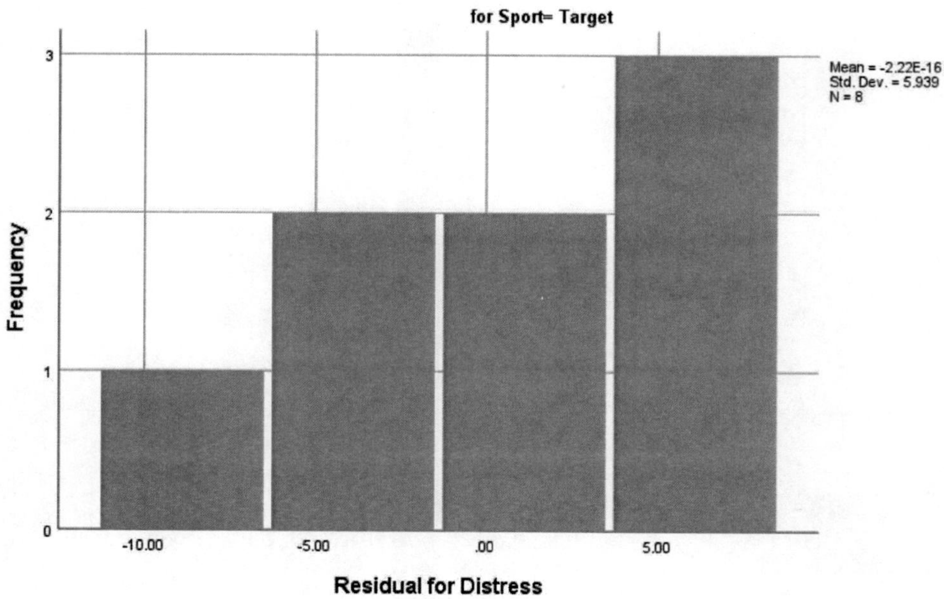

Working in **R**, we can generate a histogram using the *ggplot2* package.

FIGURE 1.27 (continued)
Histogram.

```
install.packages("ggplot2")
```

The *install.packages* function will install the *ggplot2* package which we can use to create various graphs and plots.

```
library(ggplot2)
```

The *library* function will load the *ggplot2* package.

```
qplot(Ch1_distress$unstandardizedResiduals,
geom="histogram",
binwidth=1,
main = "Histogram of Unstandardized Residuals",
xlab = "Unstandardized Residual", ylab = "Count",
fill=I("gray"),
col=I("white"))
```

Using the *gplot* function, we create a histogram (i.e., geom = "histogram") from our dataframe (i.e., *Ch1_distress*) using the variable "unstandardizedResiduals." We can add a few commands to change the width of the bars (i.e., *binwidth* = *1*), color of the bars (i.e., *fill=I("gray")*), and outline of the bars (i.e., *col=I("white")*). We can also add a title (i.e., *main* = *"Histogram of Unstandardized Residuals"*) and change the *X* and *Y* axes (*xlab* = *"Unstandardized Residual"*, *ylab* = *"Count"*).

```
hist(Ch1_distress$unstandardizedResiduals[Ch1_distress$Sport==1],
main="Histogram for Movement",
xlab="Unstandardized Residuals")
hist(Ch1_distress$unstandardizedResiduals[Ch1_distress$Sport==2],
main="Histogram for Target",
xlab="Unstandardized Residuals")
hist(Ch1_distress$unstandardizedResiduals[Ch1_distress$Sport==3],
main="Histogram for Fielding",
xlab="Unstandardized Residuals")
hist(Ch1_distress$unstandardizedResiduals[Ch1_distress$Sport==4],
main="Histogram for Territory",
xlab="Unstandardized Residuals")
```

Histograms by group can be created with these scripts, each one specifying one category of *Sport* as the variable with which to create the histogram of unstandardized residuals.

FIGURE 1.27 (continued)
Histogram.

There are a few other statistics that can be used to gauge normality. The formal test of normality, the Shapiro-Wilk test (*SW*) (Shapiro & Wilk, 1965), provides evidence of the extent to which our sample distribution is statistically different from a normal distribution. When testing the Kolmogorov-Smirnov (KS) and *SW* for normality, we do *not* want to find statistically significant results. Nonstatistically significant KS and *SW* results are interpreted to say that our distribution is *not* statistically significantly different from a normal distribution. The output for the Shapiro-Wilk test is presented in Figure 1.28

and suggests that our sample distribution for residuals overall is not statistically significantly different than what would be expected from a normal distribution ($SW = .958$, $df = 32$, $p = .240$), and the sample distribution for residuals by group is not statistically significantly different than what would be expected from a normal distribution (p's for all groups > .05).

Tests of Normality

	Kolmogorov-Smirnov[a]			Shapiro-Wilk		
	Statistic	df	Sig.	Statistic	df	Sig.
Residual for Distress	.112	32	.200*	.958	32	.240

* This is a lower bound of the true significance.

a Lilliefors Significance Correction

By group, we see evidence of normality as well.

Tests of Normality

	Type of Sport	Kolmogorov-Smirnova			Shapiro-Wilk		
		Statistic	df	Sig.	Statistic	df	Sig.
Residual for Distress	**Movement**	.116	8	.200*	.985	8	.982
	Target	.169	8	.200*	.932	8	.531
	Fielding	.197	8	.200*	.905	8	.320
	Territory	.174	8	.200*	.933	8	.548

* This is a lower bound of the true significance.

[a] Lilliefors Significance Correction

Working in **R**, we saw earlier how the *stat.desc* function from the *pastecs* package could be used to generate the Shapiro-Wilk test, along with many other statistics. Should we want to generate *just* the Shapiro-Wilk test, we can run the following script.

```
shapiro.test(Ch1_distress$unstandardizedResiduals)
```

```
        Shapiro-Wilk normality test

data:  Ch1_distress$unstandardizedResiduals
W = 0.95784, p-value = 0.2395
```

```
tapply(Ch1_distress$unstandardizedResiduals,
Ch1_distress$SportF, shapiro.test)
```

To generate the Shapiro-Wilk test by group, the *tapply* function can be used to apply the *shapiro.test* to the unstandardized residuals for all levels of the independent variable.

FIGURE 1.28
Shapiro-Wilk test.

Quantile-quantile (Q-Q) plots are also often examined to determine evidence of normality. Q-Q plots are graphs that plot quantiles of the theoretical normal distribution against quantiles of the sample distribution. Points that fall on or close to the diagonal line suggest evidence of normality. The Q-Q plot of residuals by group suggests relative normality.

Overall, the Q-Q plot suggests relative normality.

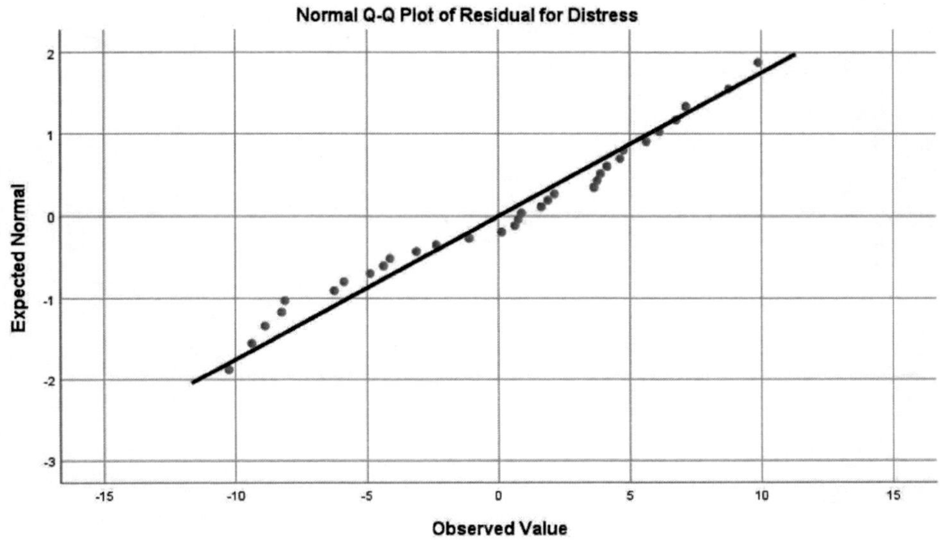

By group (for brevity, only two group graphs are presented), even with the small group sizes, there is general adherence to the diagonal line.

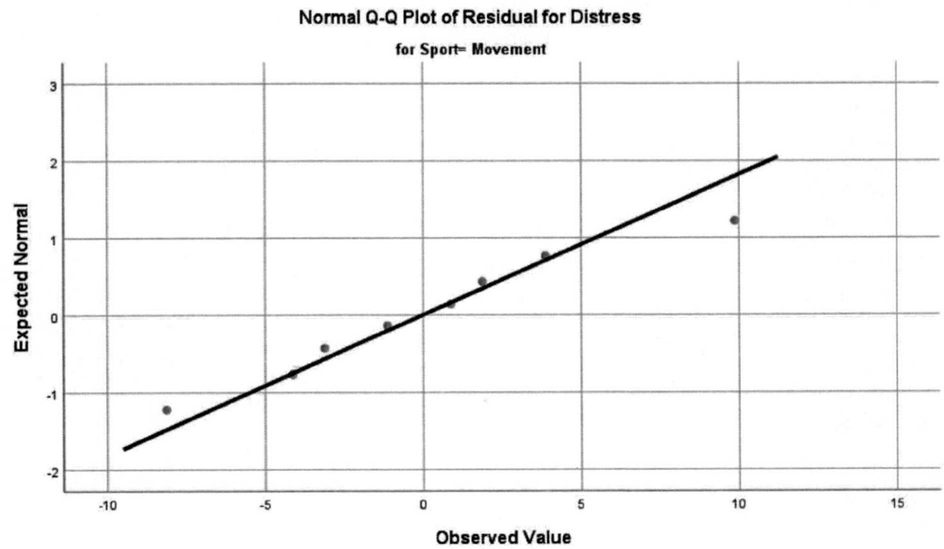

FIGURE 1.29
Q-Q plot.

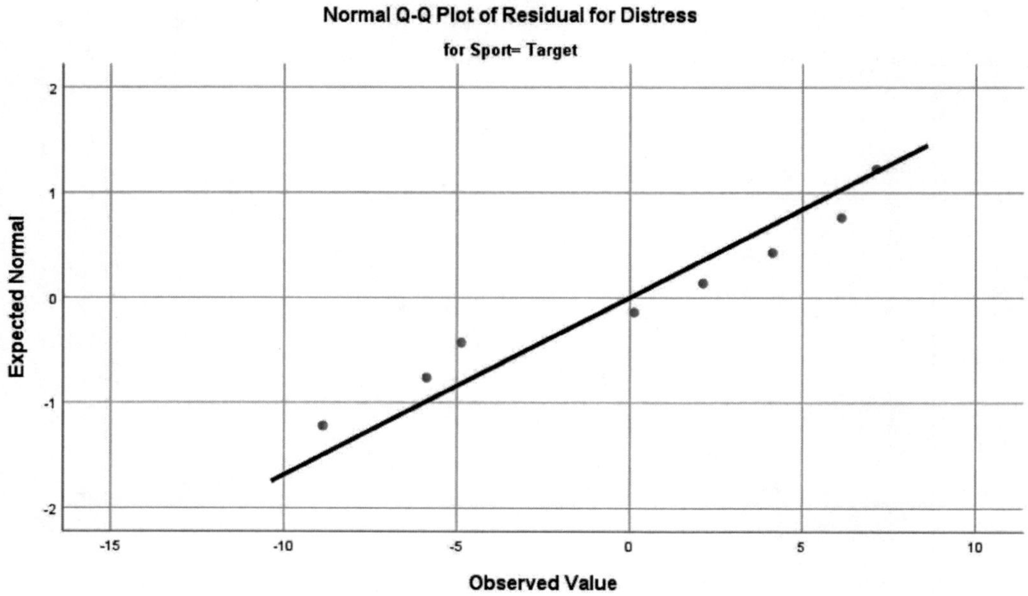

Working in **R**, we can use the *qplot* function to create a Q-Q plot of unstandardized residuals. The "data=" script defines the dataframe as "Ch1_distress."

```
qplot(sample=unstandardizedResiduals,
      data = Ch1_distress)
```

```
qqnorm(Ch1_distress$unstandardizedResiduals[Ch1_distress$Sport==1],
       main='movement')
qqnorm(Ch1_distress$unstandardizedResiduals[Ch1_distress$Sport==2],
       main='target')
qqnorm(Ch1_distress$unstandardizedResiduals[Ch1_distress$Sport==3],
       main='fielding')
qqnorm(Ch1_distress$unstandardizedResiduals[Ch1_distress$Sport==4],
       main='territory')
```

By group, Q-Q plots can be created with this script, with each command defining one category of the Sport variable.

FIGURE 1.29 (continued)
Q-Q plot.

Examination of the boxplot by group suggests a relatively normal distributional shape of residuals and no outliers.

Overall, the boxplot suggests normality.

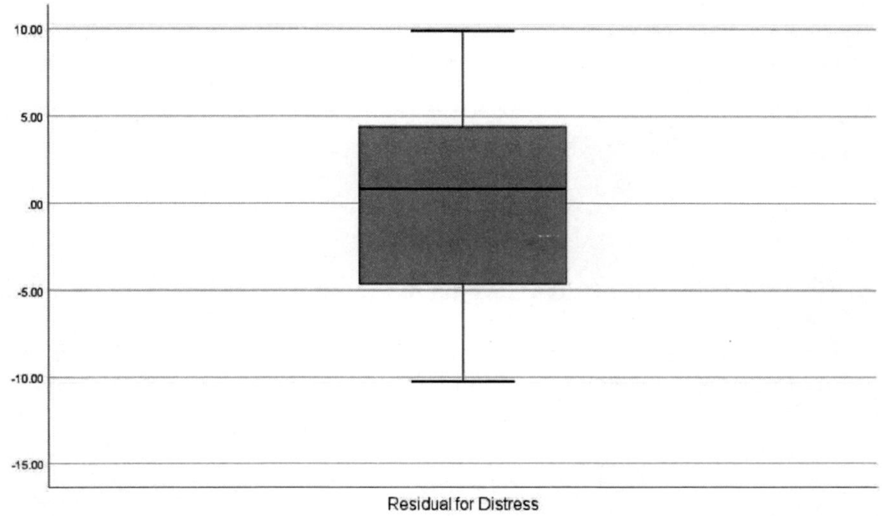

By group, even with the small group sizes, the distributions are generally acceptable in terms of normality (although fielding suggests more skew than the other groups) and do not suggest outliers.

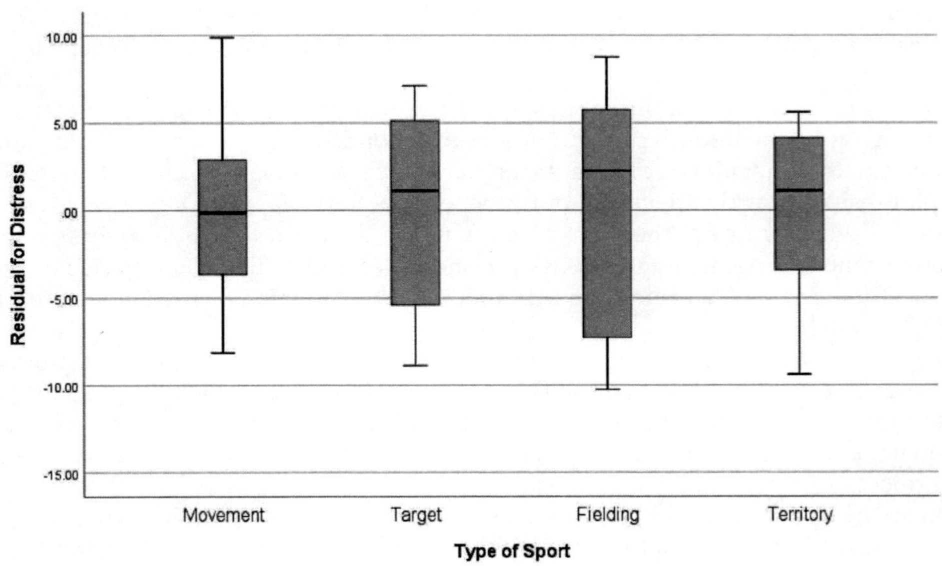

Working in **R**, we can generate a boxplot for unstandardized residuals using the *boxplot* function. To label the *Y* axis, we include the *ylab* command.

FIGURE 1.30
Boxplot.

```
boxplot(Ch1_distress$unstandardizedResiduals,
        ylab="Unstandardized Residuals")
```

Adding the independent variable to the script produces a boxplot by group. The command *xlab* will print "Sport" to identify the *X* axis.

```
boxplot(Ch1_distress$unstandardizedResiduals~Ch1_distress$SportF,
xlab="Sport", ylab="Unstandardized Residuals")
```

FIGURE 1.30 (continued)
Boxplot.

Considering the forms of evidence we have examined, skewness and kurtosis statistics, the Shapiro-Wilk test, the Q-Q plot, and the boxplot, all suggest normality by group is a reasonable assumption. We can be reasonably assured we have met the assumption of normality of the dependent variable for each group of the independent variable.

1.4.2 Independence

The only assumption we have not tested for yet is independence. If subjects have been randomly assigned to conditions (in other words, the different levels of the independent variable), the assumption of independence has been met. In this illustration, we have an observational study—athletes were not randomly assigned to the type of sport in which they participated, and thus we cannot assume that the assumption of independence was met. Had we randomly assigned units to the levels of the independent variable, we would have confidence in having met this assumption. The example we've been following, with athletes in types of sports, is common in that we often use independent variables that do not allow random assignment, such as preexisting characteristics. We can plot residuals against levels of our independent variable using a scatterplot to get an idea of whether or not there are patterns in the data and thereby provide an indication of whether we have met this assumption. Remember that these variables were added to the dataset by saving the unstandardized residuals when we generated the ANOVA model.

Please note that some researchers do not believe that the assumption of independence can be tested. If there is not random assignment to groups, then these researchers believe this assumption has been violated—period. The plot that we generate will give us a general idea of patterns, however, in situations where random assignment was not performed.

You are likely familiar with the general steps for generating a simple scatterplot (i.e., from 'Graphs' in the top tool bar of SPSS, go to 'Legacy Dialogs' then 'Scatter/Dot'). From the "Simple Scatterplot" dialog screen, click the residual variable and move it into the "Y Axis" box by clicking on the arrow. Click the independent variable (e.g., type of sport) and move it into the "X Axis" box by clicking on the arrow. Then click "OK."

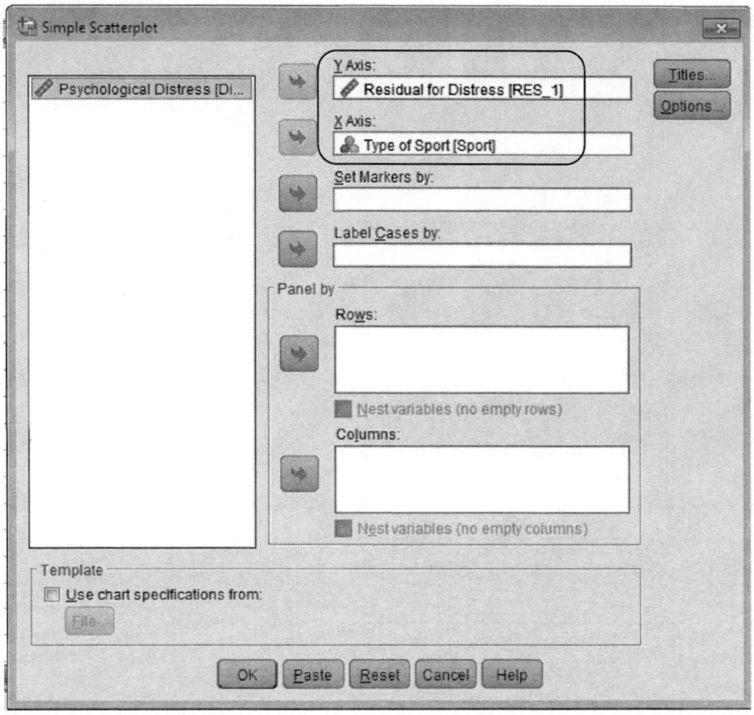

FIGURE 1.31
Generating a scatterplot.

Double click on the graph in the output to activate the chart editor. In the top toolbar within the chart editor, select "Options," then "Y Axis Reference Line."

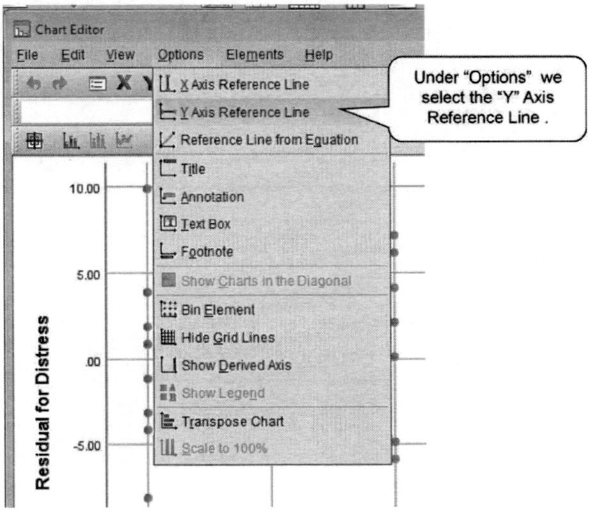

FIGURE 1.32
Chart editor.

Within the properties dialog box, we define the position for the *Y* axis reference line to be 0.

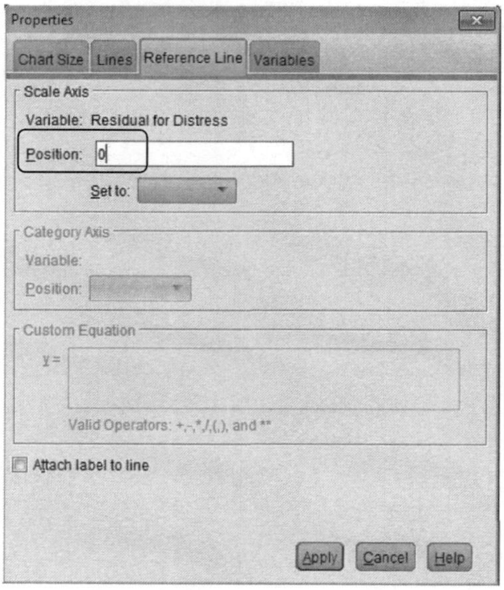

FIGURE 1.33
Adding a reference line.

1.4.2.1 Interpreting Independence Evidence

In examining the scatterplot for evidence of independence, the points should be falling relatively randomly above and below the reference line. In this example, our scatterplot suggests evidence of independence with a relatively random display of points above and below the horizontal line at zero. Thus, even though we had not met the assumption of independence through random assignment of cases to groups, this provides evidence that independence is a reasonable assumption.

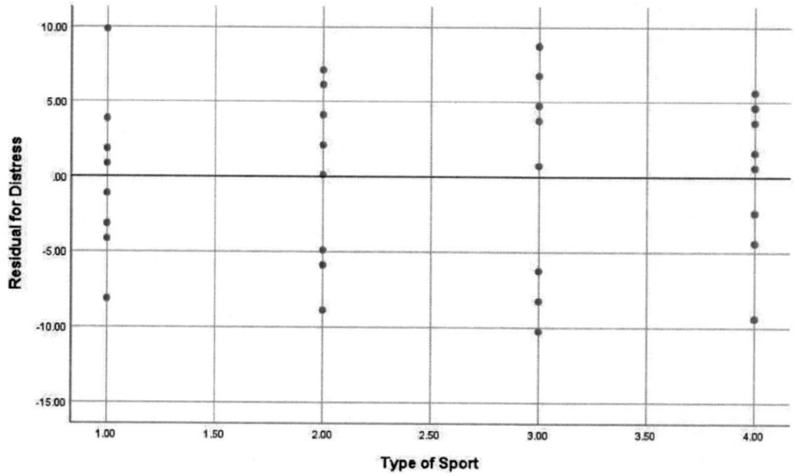

FIGURE 1.34
Scatterplot of residual by type of sport.

Working in **R**, we create a similar scatterplot.

```
plot(Ch1_distress$Sport,
     Ch1_distress$unstandardizedResiduals,
     xlab = "Sport",
     ylab = "Unstandardized Residual",
     main = "Scatterplot for independence")
```

Using the following *plot* function, with the first variable listed displaying on the X axis (e.g., "Ch1_distress$Sport"), and the second variable displaying on the Y axis (i.e., "Ch1_distress$unstandardizedResiduals"). Additional commands are provided to label the axes (*xlab* and *ylab*) and title the graph (*main*).

(Note that we are using our *Sport*, not *SportF*, variable in this script. Had we used *SportF*, the variable we defined as nominal, the plot generated would be a boxplot, not a scatterplot.)

```
plot(Ch1_ANOVA)
```

Using the *plot* function, additional plots (one of which is the Q-Q plot) that can be used for diagnostic purposes are created.

The residual versus fitted plot can be used to detect normality, unequal error variance and outliers. A random display of points, i.e., no patterns to the data, suggest assumptions of normality and equal variances have been met.

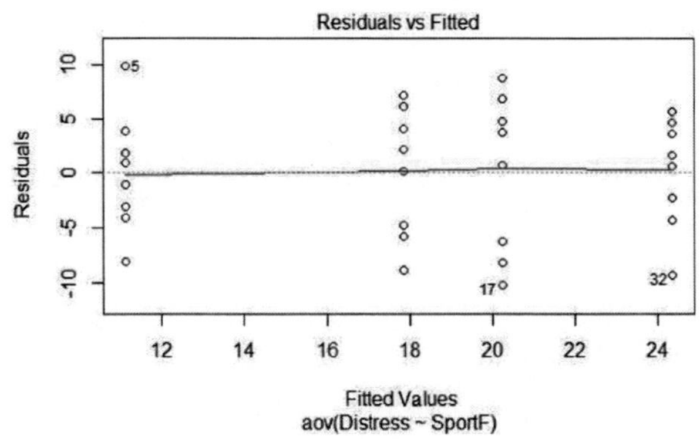

The normal Q-Q plot can be used to detect normality and outliers. Points that adhere closely to the diagonal line suggest the assumption of normality has been met.

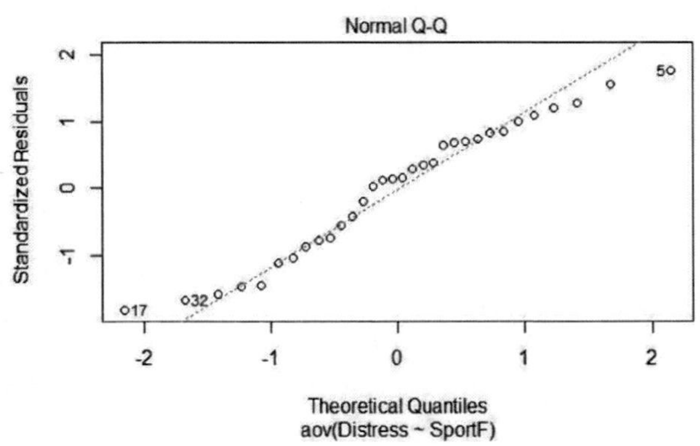

FIGURE 1.34 (continued)
Scatterplot of residual by type of sport.

The scale-location plot can be examined for evidence of equal variance. Relatively equally spaced points by group above and below a horizontal line (i.e., random and equal distribution of points and straight horizontal line) suggests evidence of meeting the assumption.

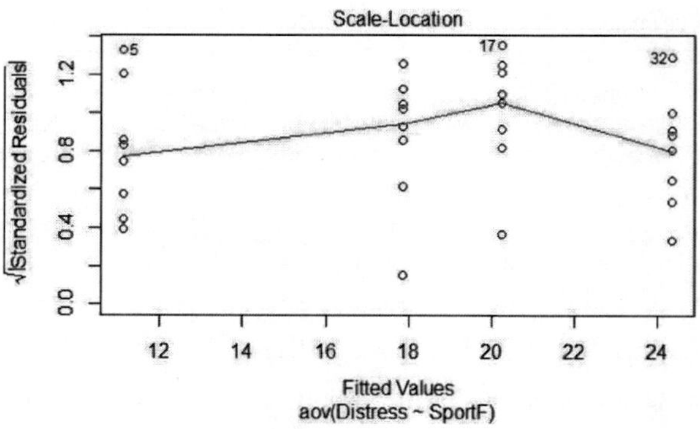

The constant leverage plot can be examined as evidence of normality as well to determine points that may exert influence.

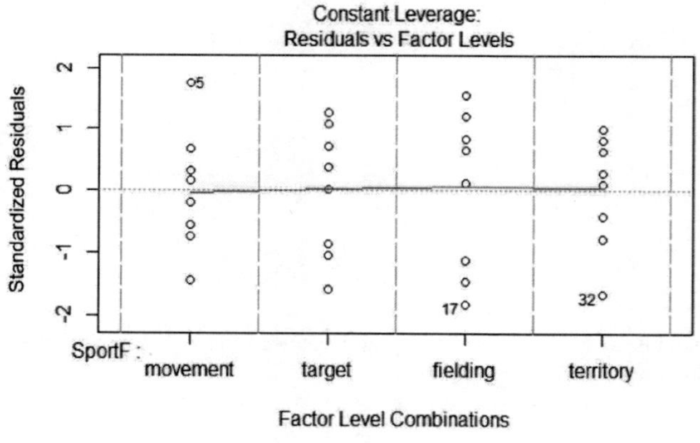

FIGURE 1.34 (continued)
Scatterplot of residual by type of sport.

1.4.3 Homogeneity of Variance

As we learned previously, another assumption to consider is that the variances of each population are equal. This is known as the assumption of *homogeneity of variance* or *homoscedasticity*. When generating ANOVA via SPSS, we requested Levene's test for examining homogeneity. Homogeneity tests using **R** were presented previously (see Figure 1.21).

1.5 Power Using G*Power

Using G*Power, post hoc power will be examined first. This will be following by an illustration of using G*Power to compute *a priori* power.

1.5.1 Post Hoc Power for the One-Way ANOVA Using G*Power

The first thing that must be done when using G*Power for computing post hoc power is to select the correct test family. In our case, we conducted a one-way ANOVA. To find the one-way ANOVA, we will select "Tests" in the top pulldown menu, then "Means," and then "Many groups: ANOVA: One-way (one independent variable)." Once that selection is made, the "Test family" automatically changes to "F tests."

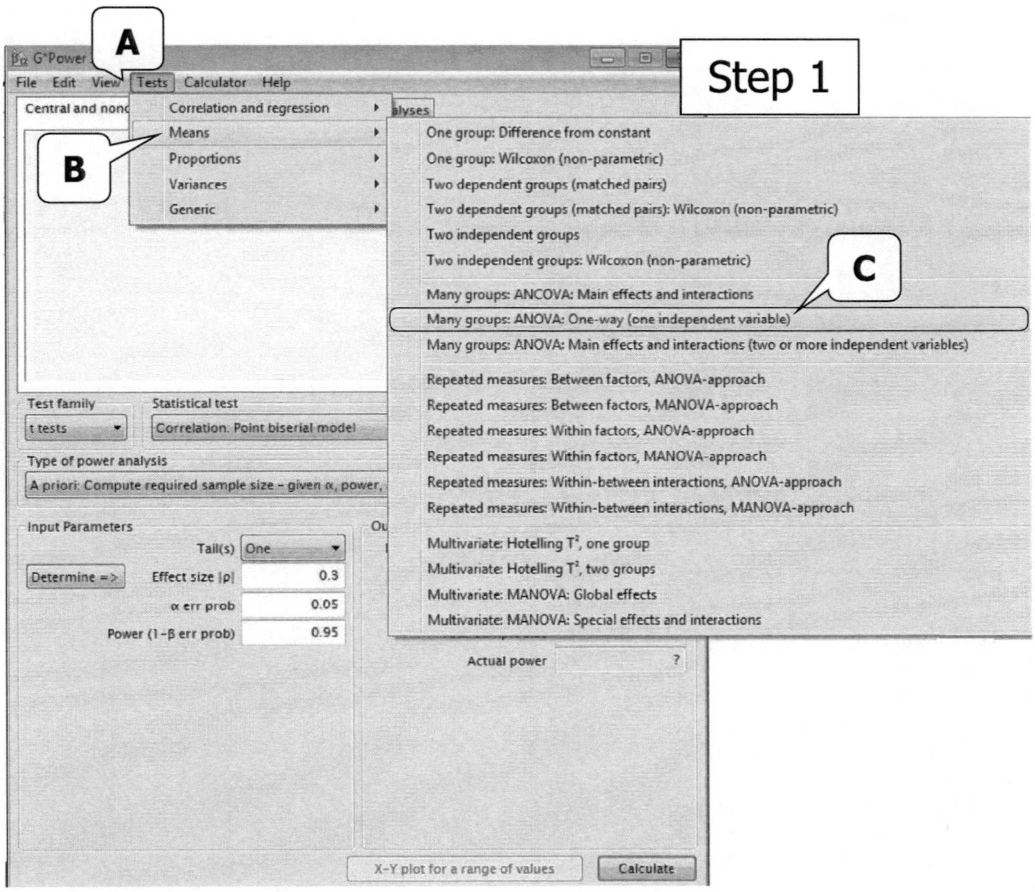

FIGURE 1.35
Power: Step 1.

The "Type of power analysis" desired then needs to be selected. To compute post hoc power, we need to select "Post hoc: Compute achieved power—given α, sample size, and effect size."

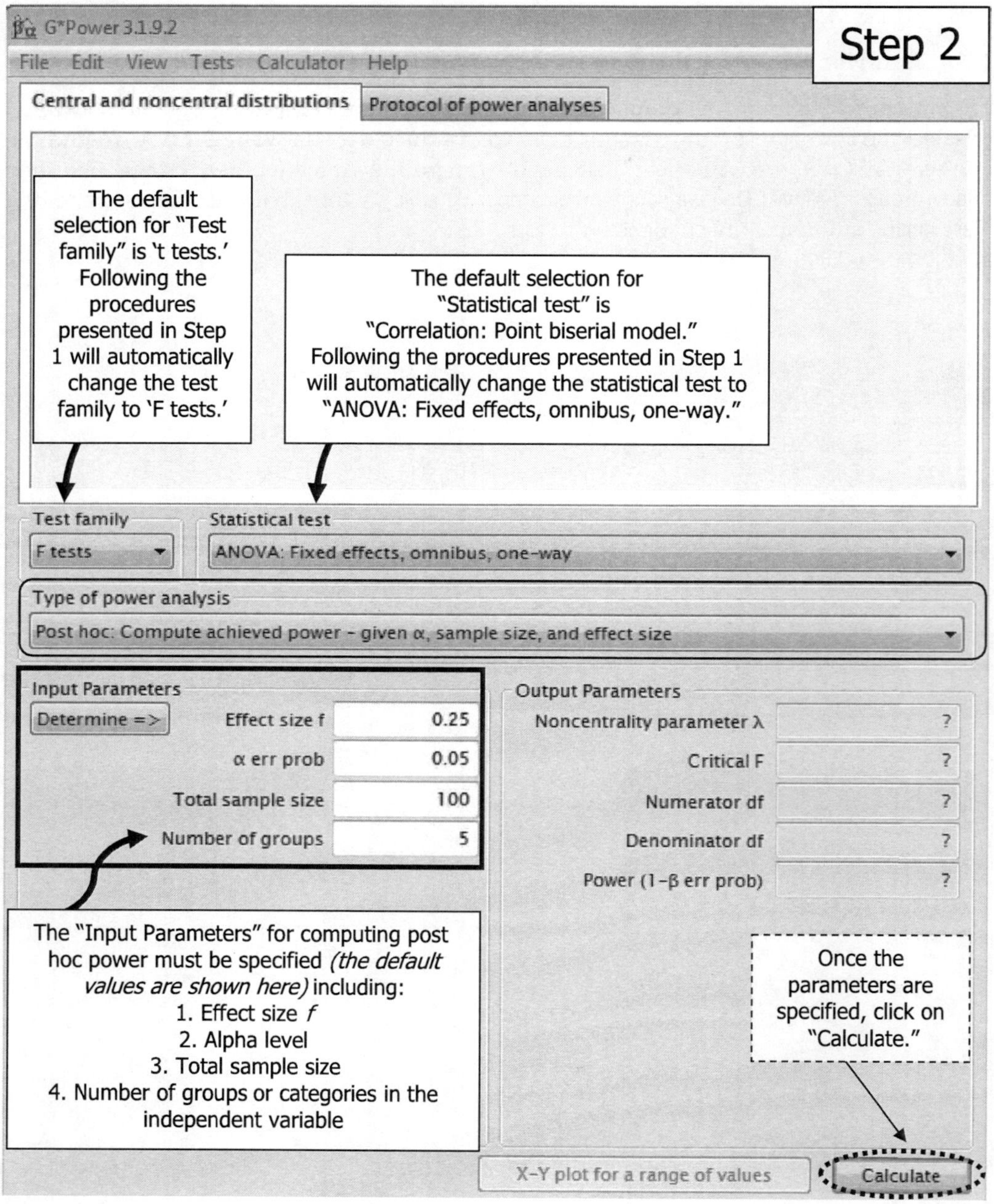

FIGURE 1.36
Power: Step 2.

The "Input Parameters" must then be specified. The first parameter is the effect size, *f*. In our example, the computed *f* effect size was .8546. The alpha level we used was .05, the total sample size was 32, and the number of groups (i.e., levels of the independent variable) was 4. Once the parameters are specified, click on "Calculate" to find the power statistics.

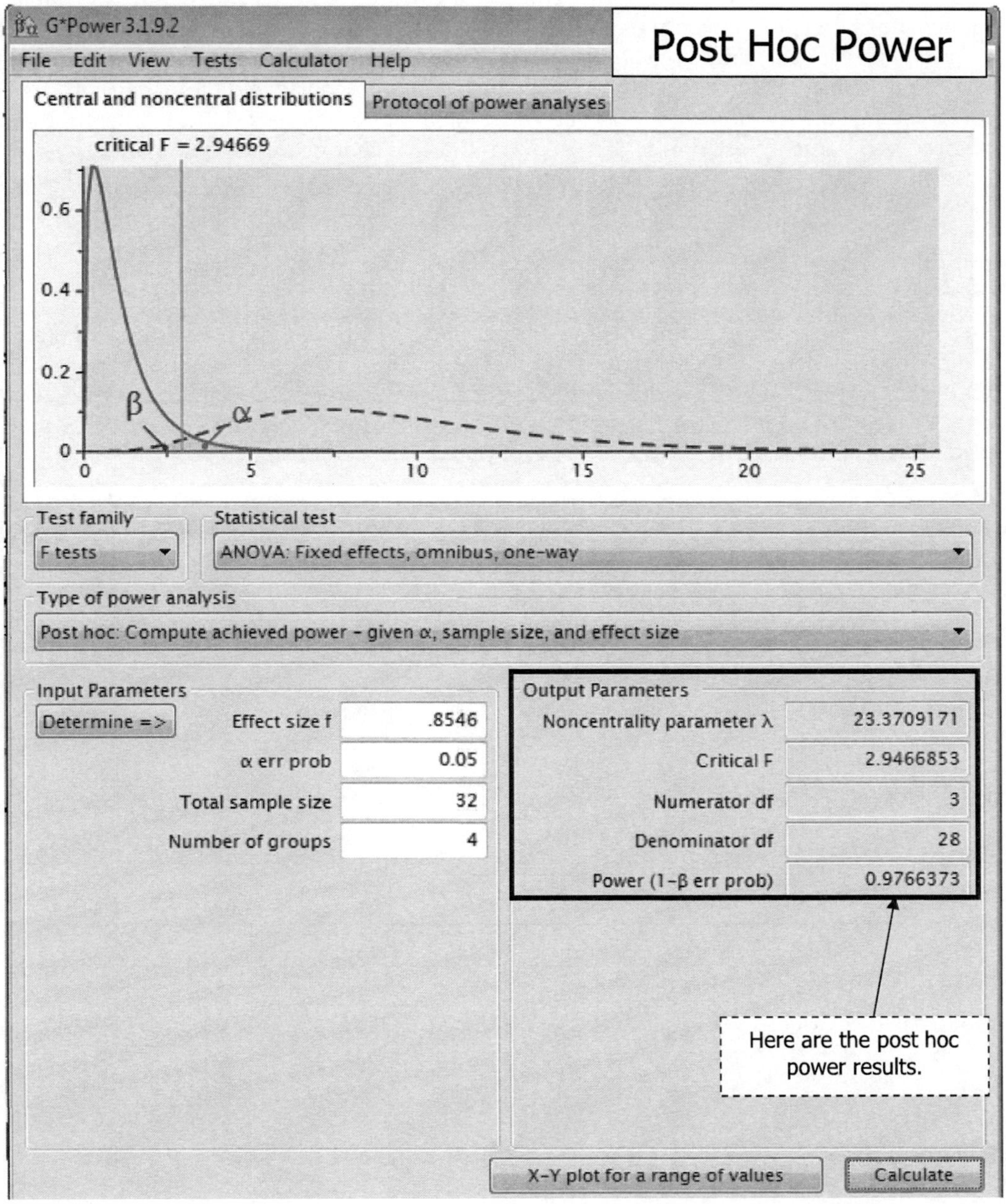

FIGURE 1.37
Post hoc power.

The "Output Parameters" provide the relevant statistics given the input just specified. In this example, we were interested in determining post hoc power for a one-way ANOVA with a computed effect size *f* of .8546, an alpha level of .05, total sample size of 32, and four groups (or categories) in our independent variable. Based on those criteria, the *post hoc power was .98*. In other words, with a one-way ANOVA, computed effect size *f* of .8546, alpha level of .05, total sample size of 32, and four groups (or categories) in our independent variable, the post hoc power of our test was .98—*the probability of rejecting the null hypothesis when it is really false (in this case, the probability that the means of the dependent variable would be equal for each level of the independent variable) was 98%*, which would be considered more than sufficient power (sufficient power is often .80 or above). Note that this value is slightly different than the observed value reported in SPSS. Keep in mind that conducting power analysis *a priori* is recommended so that you avoid a situation where, post hoc, you find that the sample size was not sufficient to reach the desired level of power (given the observed parameters).

1.5.2 *A Priori* Power for the One-Way ANOVA Using G*Power

For *a priori* power, we can determine the total sample size needed given an estimated effect size *f*, alpha level, desired power, and number of groups of our independent variable. In this example, had we estimated a moderate effect *f* of .25 (this is the default in G*Power), alpha of .05, desired power of .80, and four groups in the independent variable, we would need a total sample size of 180 (or 45 per group in a balanced design).

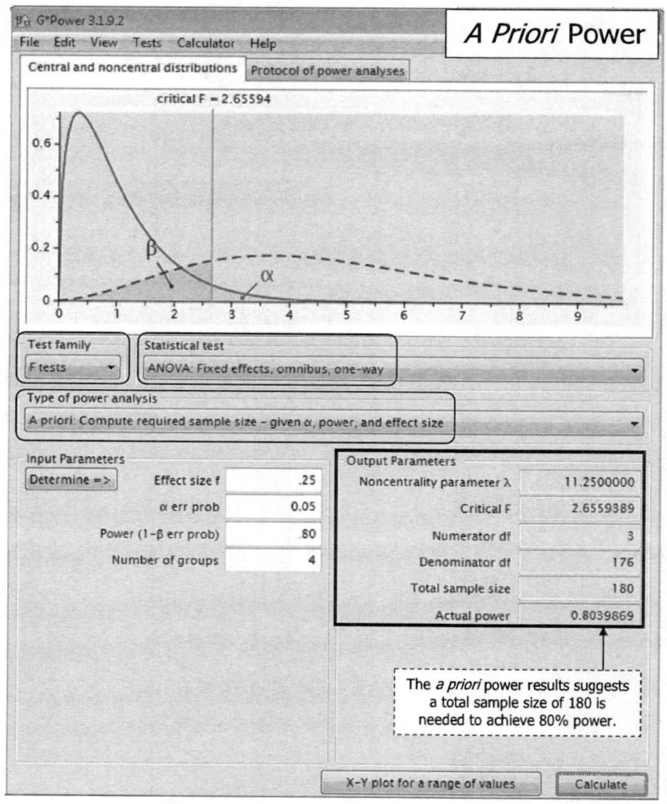

FIGURE 1.38
A priori power.

1.6 Research Question Template and Example Write-Up

Finally we come to an example paragraph of the results for the statistics lab example. Recall that Ott Lier was working with Dr. Rhodes, one of the leading sports psychologists in the region. Dr. Rhodes is examining elite athletes and their vulnerability to psychological distress based on the sport in which they participate. Ott suggested that the following research question is: *Is there a mean difference in psychological distress of elite athletes based on the type of sport in which they participate?* Ott then generated a one-way ANOVA to test the inference.

A template for writing a research question for a one-way ANOVA is presented below. Please note that it is important to ensure the reader understands the levels or groups of the independent variable. This may be done parenthetically in the actual research question, as an operational definition, or specified within the methods section. In this example, parenthetically we could have stated the following: Is there a mean difference in psychological distress of elite athletes based on the type of sport in which they compete (movement, target, fielding, territory)?

Is there a mean difference in [dependent variable] between [independent variable]?

It may be helpful to preface the results of the one-way ANOVA with information from an examination of the extent to which the assumptions were met (recall there are three assumptions: normality, homogeneity of variance, and independence). This assists the reader in understanding that you were thorough in data screening prior to conducting the test of inference.

A one-way analysis of variance (ANOVA) was conducted to determine if the mean psychological distress differed based on type of sport in which elite athletes compete. The assumptions of normality, homoscedasticity, and independence were reviewed.

The assumption of normality was tested and met via examination of the residuals. Review of the overall Shapiro-Wilk test for normality ($SW = .958$, $df = 32$, $p = .240$) and skewness (−.239) and kurtosis (−1.019) statistics suggested that normality was a reasonable assumption. Review of *SW*, skewness, and kurtosis by group also suggests normality [not presented for brevity]. Additional tests, including D'Agostino's test for skewness ($z = -.607$, $p = .544$) and the Bonett-Seier test for Geary's kurtosis ($z = -1.541$, $p = .123$) suggested evidence of normality overall as do the results for D'Agostino's and Bonnett-Seier by group [not presented for brevity]. The boxplots by group suggested a relatively normal distributional shape (with no outliers) of the residuals. The Q-Q plots and histograms by group suggested normality was reasonable.

According to Levene's test, the homogeneity of variance assumption was satisfied [F (3, 28) = .905, $p = .451$].

A scatterplot of residuals against the levels of the independent variable was reviewed. A random display of points around zero provided evidence that the assumption of independence was met. (*Note:* Had there been random assignment to groups, we could have added an additional statement such as: "Random assignment of individuals to groups helped ensure that the assumption of independence was met. Additionally, a random display of points around zero provided evidence that the assumption of independence was met.")

Here is an APA-style example paragraph of results for the one-way ANOVA (remember that this will be prefaced by the previous paragraph reporting the extent to which the assumptions of the test were met).

The one-way ANOVA is statistically significant ($F = 6.818$, $df = 3, 28$, $p = .001$). This suggests that the mean psychological distress differs by type of sport in which the athlete participates. Based on Tukey's HSD post hoc multiple comparison results, mean psychological distress for athletes in movement sports was statistically significantly lower than for athletes in fielding ($p = .025$) and territory sports ($p = .001$) [we've kept this here as a placeholder and will revisit multiple comparison procedures in the next chapter]. The means and standard deviations of psychological distress for each type of sport were as follows: 11.125 (*SD* = 5.489) for athletes competing in movement sports, 17.875 (*SD* = 5.939) for athletes competing in target sports, 20.250 (*SD* = 7.285) for athletes competing in fielding sports, and 24.375 (*SD* = 5.097) for athletes competing in territory sports.

The effect size is rather large ($\omega^2 = .35$; suggesting about 35% of the variance of psychological distress is due to differences in the type of sport in which an elite athlete competes), and observed power is quite strong (.956).

For completeness, we also conducted several alternative procedures. The Kruskal-Wallis test ($\chi^2 = 13.061$, $df = 3$, $p = .005$), the Welch procedure ($F_{Asymp} = 7.862$, $df1 = 3$, $df2 = 15.454$, $p = .002$), and the Brown-Forsythe procedure ($F_{Asymp} = 6.818$, $df1 = 3$, $df2 = 25.882$, $p = .002$) also indicated a statistically significant effect of type of sport on psychological distress.

1.7 Additional Resources

This chapter has provided a preview into conducting one-way ANOVA. However, there are a number of areas that space limitations prevent us from delving into. For more in-depth coverage of ANOVA models, see Maxwell et al. (2018). For readers interested in one-way ANOVA when there are censored data (i.e., when some data cannot be observed due to resource limitations, such as cost or time), see Celik and Senoglu (2018).

Problems

Conceptual Problems

1. Data for three independent random samples, each of size four, are analyzed by a one-factor analysis of variance fixed-effects model. If the values of the sample means are all equal, what is the value of MS_{betw}?
   a. 0
   b. 1
   c. 2
   d. 3

2. For a one-factor analysis of variance fixed-effects model, which of the following is always true?
   a. $df_{betw} + df_{with} = df_{total}$
   b. $SS_{betw} + SS_{with} = SS_{total}$
   c. $MS_{betw} + MS_{with} = MS_{total}$
   d. All of the above
   e. Both a and b
3. Suppose $n_1 = 19$, $n_2 = 21$, and $n_3 = 23$. For a one-factor ANOVA, the df_{with} would be
   a. 2
   b. 3
   c. 60
   d. 62
4. Suppose $n_1 = 19$, $n_2 = 21$, and $n_3 = 23$. For a one-factor ANOVA, the df_{betw} would be
   a. 2
   b. 3
   c. 60
   d. 62
5. Suppose $n_1 = 19$, $n_2 = 21$, and $n_3 = 23$. For a one-factor ANOVA, the df_{total} would be
   a. 2
   b. 3
   c. 60
   d. 62
6. Suppose $n_1 = 19$, $n_2 = 21$, and $n_3 = 23$. For a one-factor ANOVA, the *df* for the numerator of the *F* ratio would be which one of the following?
   a. 2
   b. 3
   c. 60
   d. 62
7. In a one-factor ANOVA, H_0 asserts that
   a. All of the population means are equal.
   b. The between-groups variance estimate and the within-groups variance estimate are both estimates of the same population residual variance.
   c. The within-groups sum of squares is equal to the between-groups sum of squares.
   d. Both a and b
8. For a one-factor ANOVA comparing three groups with $n = 10$ in each group, the *F* ratio has degrees of freedom equal to
   a. 2, 27
   b. 2, 29
   c. 3, 27
   d. 3, 29

9. For a one-factor ANOVA comparing five groups with $n = 50$ in each group, the F ratio has degrees of freedom equal to
   a. 4, 245
   b. 4, 249
   c. 5, 245
   d. 5, 249
10. Which of the following is not necessary in ANOVA?
   a. Observations are from random and independent samples.
   b. The dependent variable is measured on at least the interval scale.
   c. Populations have equal variances.
   d. Equal sample sizes are necessary.
11. If you find an F ratio of 1.0 in a one-factor ANOVA, it means that
   a. Between-groups variation exceeds within-groups variation.
   b. Within-groups variation exceeds between-groups variation.
   c. Between-groups variation is equal to within-groups variation.
   d. Between-groups variation exceeds total variation.
12. True or false? Suppose students in grades 7, 8, 9, 10, 11, and 12 were compared on absenteeism. If ANOVA were used rather than multiple t tests, then the probability of a Type I error will be less.
13. True or false? Mean square is another name for variance or variance estimate.
14. True or false? In ANOVA, each independent variable is known as a level.
15. True or false? A negative F ratio is impossible.
16. Suppose that for a one-factor ANOVA with $J = 4$ and $n = 10$, the four sample means are all equal to 15. I assert that the value of MS_{with} is necessarily equal to zero. Am I correct?
17. With $J = 3$ groups, I assert that if you reject H_0 in the one-factor ANOVA, you will necessarily conclude that all three group means are different. Am I correct?
18. True or false? The homoscedasticity assumption is that the populations from which each of the samples are drawn are normally distributed.
19. When analyzing mean differences among more than two samples, doing independent t tests on all possible pairs of means
   a. Decreases the probability of a Type I error
   b. Does not change the probability of a Type I error
   c. Increases the probability of a Type I error
   d. Cannot be determined from the information provided
20. Suppose for a one-factor fixed-effects ANOVA with $J = 5$ and $n = 15$, the five sample means are all equal to 50. I assert that the F test statistic cannot be significant. Am I correct?
21. True or false? The independence assumption in ANOVA is that the observations in the samples do not depend on one another.
22. True or false? For $J = 2$ and $\alpha = .05$, if the result of the independent t test is significant, then the result of the one-factor fixed-effects ANOVA is uncertain.

23. A statistician conducted a one-factor fixed-effects ANOVA and found the F ratio to be less than 0. I assert this means the between-groups variability is less than the within-groups variability. Am I correct?

24. Which of the following is *not* an alternative to the parametric one-factor fixed-effects ANOVA?
    a. Brown-Forsythe procedure
    b. Kruskal-Wallis test
    c. Levene's test
    d. Welch test

25. Which of the following is *not* a proportion of variance explained type of effect size measures that can be computed for one-way fixed-effects ANOVA?
    a. d
    b. e^2
    c. η^2
    d. ω^2

26. A researcher computes a one-way fixed-effects ANOVA and finds $\omega^2 = .07$. Using Cohen's subjective standards, how would this effect size be interpreted?
    a. Small
    b. Moderate
    c. Large
    d. Very large

27. Which of the following do *not* provide evidence of the assumption of normality?
    a. Levene's test
    b. Q-Q plot
    c. Shapiro-Wilk test
    d. Skewness and kurtosis

28. The assumption of homoscedasticity deals with which one of the following?
    a. Equal population variances
    b. Independence
    c. Linearity
    d. Normality

Answers to Conceptual Problems

1. **a** (if the sample means are all equal, then MS_{betw} is 0)
3. **c** (lose 1 df from each group; 63 – 3 = 60)
5. **d** (equals the $df_{betw} + df_{with} = df_{total}$; 60 + 2 = 62)
7. **d** (null hypothesis does not consider SS values)
9. **a** (for between source = 5 – 1 = 4 and for within source = 250 – 5 = 245)
11. **c** (an F ratio of 1.0 implies between- and within-groups variation are the same)
13. **True** (mean square is a variance estimate)

15. **True** (F ratio must be greater than or equal to 0)
17. **No** (rejecting the null hypothesis in ANOVA only indicates that there is some difference among the means, not that all of the means are different)
19. **c** (the more t tests conducted, the more likely a Type I error for the set of tests)
21. **True** (basically the definition of independence)
23. **No** (find a new statistician as a negative F value is not possible in this context)
25. **a** (effect size d is interpreted as a standardized mean difference, not proportion of variance)
27. **a** (Levene's test is used to examine the assumption of homogeneity of variances in ANOVA models, not normality)

Computational Problems

1. Complete the following ANOVA summary table for a one-factor analysis of variance, where there are four groups receiving different headache medications, each with 16 observations, and $\alpha = .05$.

Source	SS	df	MS	F	Critical Value and Decision
Between	9.75	–	–	–	
Within	–	–	–		
Total	18.75	–			

2. A social psychologist wants to determine if type of music has any effect on the number of beers consumed by people in a tavern. Four taverns are selected that have different musical formats. Five people are randomly sampled in each tavern and their beer consumption monitored for three hours. Complete the following one-factor ANOVA summary table using $\alpha = .05$.

Source	SS	df	MS	F	Critical Value and Decision
Between	–	–	7.52	5.01	
Within	–	–	–		
Total	–	–			

3. A psychologist would like to know whether the season (fall, winter, spring, and summer) has any consistent effect on people's overall mood. In the middle of each season, the psychologist selects a random sample of $n = 25$ students. Each individual is given a questionnaire that assesses their overall mood (and results in a continuous composite score). A one-factor ANOVA was used to analyze these data. Complete the following ANOVA summary table ($\alpha = .05$).

Source	SS	df	MS	F	Critical Value and Decision
Between	–	–	–	5.00	
Within	960	–	–		
Total	–	–			

4. The following five independent random samples are obtained from five normally distributed populations with equal variances. The dependent variable is the number of bank transactions in one month and the groups are five different banks.

Group 1	Group 2	Group 3	Group 4	Group 5
16	16	2	5	7
5	10	9	8	12
11	7	11	1	14
23	12	13	5	16
18	7	10	8	11
12	4	13	11	9
12	23	9	9	19
19	13	9	9	24

Use SPSS or **R** to conduct a one-factor analysis of variance to determine if the group means are equal using $\alpha = .05$.

5. The following three independent random samples are obtained from three normally distributed populations with equal variances. The dependent variable is starting hourly wage and the groups are the type of position (internship, co-op, work study).

Group 1: Internship	Group 2: Co-op	Group 3: Work Study
10	9	8
12	8	9
11	10	8
11	12	10
12	9	8
10	11	9
10	12	9
13	10	8

Conduct a one-factor analysis of variance to determine if the group means are equal using $\alpha = .05$. If needed, conduct Tukey's post hoc test. Report the extent to which the assumption of homogeneity of variances was met.

6. The following three independent random samples are obtained from three normally distributed populations with equal variances. The dependent variable is nurse's stress and the independent variable (denoted by group) is hospital size.

Group 1	Group 2	Group 3
4	5	6
2	4	8
3	5	6
3	5	7
4	4	6
3	5	7
3	5	6
4	5	6

Conduct a one-factor analysis of variance to determine if the group means are equal using $\alpha = .05$. If needed, conduct Tukey's post hoc test. Report the extent to which the assumption of homogeneity of variances was met.

Answers to Computational Problems

1. $df_{betw} = 3$, $df_{with} = 60$, $df_{total} = 63$, $SS_{with} = 9$, $MS_{with} = 3.25$, $MS_{with} = 0.15$, F = 21.6666, critical value = 2.76 (reject H_0).
3. $SS_{betw} = 150$, $SS_{total} = 1{,}110$, $df_{betw} = 3$, $df_{with} = 96$, $df_{total} = 99$, $MS_{betw} = 50$, $MS_{with} = 10$, critical value approximately 2.7 (reject H_0).
5. The one-way ANOVA was statistically significant, $F = 9.629$, $df = 2, 21$, $p < .001$. Based on Tukey's HSD, there were statistically significantly different wages for work study students relative to either interns or co-op students. The assumption of homogeneity of variances was met, based on Levene's test [F (2, 21) = 1.640, p = 2.18].

Interpretive Problems

1. Using the survey1 dataset, which is accessible from the website, use SPSS or **R** to conduct a one-factor fixed-effects ANOVA, where political view is the grouping variable (i.e., independent variable) ($J = 5$) and the dependent variable is an interval or ratio variable of your choice. Also compute effect size and test for assumptions. Then write an APA-style paragraph describing the results.
2. Using the survey1 dataset, which is accessible from the website, use SPSS or **R** to conduct a one-factor fixed-effects ANOVA, where hair color is the grouping variable (i.e., independent variable) ($J = 5$) and the dependent variable is an interval or ratio variable of your choice. Also compute effect size and test for assumptions. Then write an APA-style paragraph describing the results.
3. Use the IPEDS2017 dataset, which is accessible from the website, use SPSS or **R** to conduct a one-factor fixed-effects ANOVA. Select an appropriate independent variable (e.g., *land grant institution*, LANDGRNT) and appropriate dependent variable (e.g., *total dormitory capacity*, ROOMCAP). Also compute effect size and test for assumptions. Then write an APA-style paragraph describing the results.

2

Multiple Comparison Procedures

Chapter Outline

Key Concepts

1. Contrast
2. Simple and complex contrasts
3. Planned and post hoc comparisons
4. Contrast- and family-based Type I error rates
5. Orthogonal contrasts

In this chapter our concern is with multiple comparison procedures that involve comparisons among the group means. Recall from Chapter 1 the one-factor analysis of variance where the means from two or more samples were compared. What do we do if the omnibus F test leads us to reject H_0? First, consider the situation where there are only two samples (e.g., assessing the effectiveness of two types of medication), and H_0 has already been rejected in the omnibus test. Why was H_0 rejected? The answer should be obvious. Those two sample means must be significantly different, as there is no other way that the omnibus H_0 could have been rejected (e.g., one type of medication is significantly more effective than the other based on an inspection of the means).

Second, consider the situation where there are more than two samples (e.g., three types of medication), and H_0 has already been rejected in the omnibus test. Why was H_0 rejected? The answer is not so obvious. This situation is one where a **multiple comparison procedure (MCP)** would be quite informative. Thus, for situations where there are at least three groups and the analysis of variance (ANOVA) H_0 has been rejected, some sort of MCP is necessary to determine which means, or combination of means, are different. Third, consider the situation where the researcher is not even interested in the ANOVA omnibus test, but is only interested in comparisons involving particular means (e.g., certain medications are more effective than a placebo). This is a situation where a MCP is useful for evaluating those specific comparisons.

If the ANOVA omnibus H_0 has been rejected, why not do all possible independent t tests? First let us return to a similar question from Chapter 1. There we asked about doing all possible pairwise independent t tests rather than an ANOVA. The answer there was to do an omnibus F test. The reasoning was related to the probability of making a Type I error (i.e., α), where the researcher incorrectly rejects a true null hypothesis. Although the alpha level for each t test can be controlled at a specified nominal level, say .05, what would happen to the overall alpha level for the set of t tests? The overall α level for the set of tests, often called the **family-wise Type I error rate**, would be larger than the α level for each of the individual t tests. The optimal solution, in terms of maintaining control over our overall α level as well as maximizing power, is to conduct **one overall omnibus test** that assesses the equality of all of the means simultaneously.

Let us apply the same concept to the situation involving multiple comparisons. Rather than doing all possible pairwise independent t tests, where the family-wise error rate could be quite large, *one should use a procedure that controls the family-wise error rate in some way. This can be done with multiple comparison procedures.* As pointed out later in the chapter, there are two main methods for taking the Type I error rate into account.

This chapter is concerned with several important new concepts, such as a contrast, planned versus post hoc comparisons, the Type I error rate, and orthogonal contrasts. The remainder of the chapter consists of selected multiple comparison procedures, including when and how to apply them. The terms *comparison* and *contrast* are used here synonymously. *Also, MCPs are applicable only for comparing levels of an independent variable that are fixed, in other words, for fixed-effects independent variables, and not for random-effects independent variables.* Our objectives are that by the end of this chapter, you will be able to (a) understand the concepts underlying the MCPs, (b) select the appropriate MCP for a given research situation, and (c) determine and interpret the results of MCPs.

2.1 What Multiple Comparison Procedures Are and How They Work

In the previous chapter, Ott Lier, one of four graduate students who assist in the statistics lab, was embarking on a very exciting research adventure. He continues to work towards completion of this project. As you may recall, Ott was assisting Dr. Rhodes, one of the region's leading sports psychologists, in examining elite athletes and vulnerability to psychological distress based on type of sport in which they participate. Our graduate student team successfully analyzed the data and used (as we saw in a previous chapter) one-way ANOVA to answer a research question. As we will see in this chapter, Ott will be expanding on the analysis as it relates to examining the group means.

The research lab has been contracted to work with one of the leading sports psychologists in the region, Dr. Rhodes. Dr. Rhodes is examining elite athletes and their vulnerability to psychological distress based on the type of sport in which they participate. Dr. Rhodes wants to determine if there is a difference in psychological stress based on type of sport (movement, target, fielding, or territory). Ott suggests the following research question is: *Is there a mean difference in psychological distress of elite athletes based on the type of sport in which they participate?* With one independent variable, Ott conducted a one-way ANOVA to answer Dr. Rhodes's question, where he rejected the null hypothesis. Now his task is to determine which type of sport (recall there were four) were statistically different on the outcome (i.e., psychological distress).

2.1.1 Characteristics

This section describes the most important characteristics of the multiple comparison procedures. We begin by defining a contrast, and then move into planned versus post hoc contrasts, the Type I error rates, and orthogonal contrasts.

2.1.1.1 Contrasts

A **contrast** is a weighted combination of the means. For example, a researcher may want to form contrasts that examine the following combinations of means: (a) Group 1 with Group 2, and (b) the combination (or average) of Groups 1 and 2 with Group 3. Statistically, a contrast is defined as follows:

$$\psi_i = c_1\mu_{.1} + c_2\mu_{.2} + \ldots + c_J\mu_{.J}$$

Where psi, (ψ_i) is the particular contrast that is being investigated, the c_j are known as **contrast coefficients** (or **weights**), which are positive, zero, and negative values, and the μ_j are population group means. In other words, *a contrast is simply a particular combination of the group means, depending on which means the researcher is interested in comparing*. It should also be noted that to form a fair or legitimate contrast, $\sum c_j = 0$ for the equal n's or balanced case, and $\sum(n_j c_j) = 0$ for the unequal n's or unbalanced case.

For example, suppose we wish to compare the means of groups 1 and 3 for $J = 4$ groups or levels, and we call this contrast 1. The contrast would be written as follows, where the means of groups 2 and 4 are weighted as 0 since they are of no interest in this particular comparison:

$$\psi_1 = c_1\mu_{.1} + c_2\mu_{.2} + c_3\mu_{.3} + c_4\mu_{.4}$$
$$\psi_1 = (+1)\mu_{.1} + (0)\mu_{.2} + (-1)\mu_{.3} + (0)\mu_{.4}$$
$$\psi_1 = \mu_{.1} - \mu_{.3}$$

What hypotheses are we testing when we evaluate a contrast? The null and alternate hypotheses of any specific contrast can be written, respectively, simply as follows:

$$H_0 : \psi_i = 0$$

and

$$H_1 : \psi_i \neq 0$$

Thus we are testing whether a particular combination of means, as defined by the contrast coefficients, are different. How does this relate back to the omnibus F test? The null and alternate hypotheses for the omnibus F test can be written, respectively, in terms of contrasts as follows:

$$H_0 : all\ \psi_i = 0$$

and

$$H_1 : at\ least\ one\ \psi_i = 0$$

Here the omnibus test is used to determine whether any contrast that could be formulated for the set of J means is significant or not.

Contrasts can be divided into *simple or pairwise contrasts*, and *complex or nonpairwise contrasts*. A **simple or pairwise contrast** is a comparison involving only two means. Take as an example the situation where there are $J = 3$ groups. There are three possible distinct pairwise contrasts that could be formed: (a) $\mu_{.1} - \mu_{.2} = 0$ (comparing the mean of Group 1 to the mean of Group 2); (b) $\mu_{.1} - \mu_{.3} = 0$ (comparing the mean of Group 1 to the mean of Group 3); and (c) $\mu_{.2} - \mu_{.3} = 0$ (comparing the mean of Group 2 to the mean of Group 3). It should be obvious that a pairwise contrast involving Groups 1 and 2 is the same contrast whether it is written as $\mu_{.1} - \mu_{.2} = 0$, or as $\mu_{.2} - \mu_{.1} = 0$. In terms of *contrast coefficients*, these three contrasts for a simple or pairwise contrast could be written in the form of a table as in Table 2.1.

TABLE 2.1

Contract Coefficients for Simple or Pairwise Contrasts

	c_1	c_2	c_3
Ψ_1: $\mu_{.1} - \mu_{.2} = 0$	+1	−1	0
Ψ_2: $\mu_{.1} - \mu_{.3} = 0$	+1	0	−1
Ψ_3: $\mu_{.2} - \mu_{.3} = 0$	0	+1	−1

where each contrast (i.e., ψ_1, ψ_2, ψ_3) is read across the table (left to right) to determine its contrast coefficients (i.e., c_1, c_2, c_3). For example, the first contrast, ψ_1, does not involve Group 3 because that contrast coefficient is zero (see c_3 for ψ_1), but does involve Groups 1 and 2 because those contrast coefficients are not zero (see c_1 and c_2 for ψ_1). The contrast coefficients are +1 for Group 1 (see c_1) and −1 for Group 2 (see c_2); consequently we are interested in examining the difference between the means of Groups 1 and 2.

Written in long form so that we can see where the contrast coefficients come from, the three contrasts are as follows:

$$\psi_1 = (+1)\mu_{.1} + (-1)\mu_{.2} + (0)\mu_{.3} = \mu_{.1} - \mu_{.2}$$
$$\psi_2 = (+1)\mu_{.1} + (0)\mu_{.2} + (-1)\mu_{.3} = \mu_{.1} - \mu_{.3}$$
$$\psi_3 = (0)\mu_{.1} + (+1)\mu_{.2} + (-1)\mu_{.3} = \mu_{.2} - \mu_{.3}$$

An easy way to remember the number of possible unique pairwise contrasts that could be written is ½[(J)(J – 1)] or $\left[(J)(J-1)\right]\!\big/_{2}$. Thus for $J = 3$, the number of possible unique pairwise contrasts is 3, whereas for $J = 4$ the number of such contrasts is 6 (or ½[(4)(4 – 1)] = ½(4)(3) = ½(12) = 6).

A **complex contrast** is a comparison involving more than two means. Continuing with the example of $J = 3$ groups, we might be interested in testing the contrast of $\mu_{.1} - \left(\frac{1}{2}\right)(\mu_{.2} + \mu_{.3})$ [which could also be written as $\mu_{.1} - \left(\frac{\mu_{.2} + \mu_{.3}}{2}\right)$]. This contrast is a comparison of the mean for Group 1 (i.e., μ_1) with the average of the means for Groups 2 and 3 [i.e., $\left(\frac{\mu_{.2} + \mu_{.3}}{2}\right)$]. In terms of contrast coefficients, this contrast would be written as seen in Table 2.2.

TABLE 2.2

Complex Contract Coefficients

	c_1	c_2	c_3
$\psi_4 : \mu_{.1} - \frac{\mu_{.2}}{2} - \frac{\mu_{.3}}{2} = 0$	+1	−1/2	−1/2

Written in long form so that we can see where the contrast coefficients come from, this complex contrast is as follows:

$$\psi_4 = (+1)\mu_{.1} + \left(-\tfrac{1}{2}\right)\mu_{.2} + \left(-\tfrac{1}{2}\right)\mu_{.3}$$

$$\psi_4 = \mu_{.1} + \left(-\tfrac{1}{2}\right)\mu_{.2} + \left(-\tfrac{1}{2}\right)\mu_{.3}$$

$$\psi_4 = \mu_{.1} + \left(-\frac{\mu_{.2}}{2}\right) + \left(-\frac{\mu_{.3}}{2}\right) = 0$$

The number of unique complex contrasts is greater than ½[J (J – 1)], when J is at least 4. *In other words, the number of such contrasts that could be formed can be quite large when there are more than three groups.* It should be noted that the *total number of unique pairwise and complex contrasts is* $\left[1 + \left(\tfrac{1}{2}\right)(3^J - 1) - \mathbf{2}^J\right]$ (Keppel & Wickens, 2004. Thus for $J = 4$, one could form 25 total contrasts, $\left[1 + \left(\tfrac{1}{2}\right)(3^4 - 1) - 2^4\right] = 1 + {(81-1)}\big/_{2} - 16 = 1 + 40 - 16 = 25.$

Many of the multiple comparison procedures are based on the same test statistic, which we introduce here as the **standard *t***. The **standard *t* ratio for a contrast** is given as follows:

$$t = \frac{\psi'}{s_{\psi'}}$$

where $s_{\psi'}$ represents the standard error of the contrast as follows:

$$s_{\psi'} = \sqrt{MS_{error} \sum_{j=1}^{J} \left(\frac{c_j^2}{n_j}\right)}$$

where the prime (i.e.,′) indicates that this is a sample estimate of the population value of the contrast (i.e., based on sample data), and n_j refers to the number of observations in group j.

2.1.1.2 Planned Versus Post Hoc Comparisons

This section examines specific types of contrasts or comparisons. One way of classifying contrasts is whether the contrasts are formulated prior to the research or following a significant omnibus F test. **Planned contrasts** (also known as specific or *a priori* contrasts) involve particular comparisons that the researcher is interested in examining *prior* to data collection. These planned contrasts are generally based on theory, previous research, and/or specific hypotheses. Here the researcher is interested in certain specific contrasts *a priori*, where the number of such contrasts is usually small. *Planned contrasts are done without regard to the result of the omnibus F test (i.e., whether or not the overall F test is statistically significant).* In other words, the researcher is interested in certain specific contrasts, but not in the omnibus F test that examines all possible contrasts. In this situation the researcher could care less about the multitude of possible contrasts and need not even examine the overall F test; rather, the concern is only with a few contrasts of substantive interest. In addition, the researcher may not be as concerned with the family-wise error rate for planned comparisons because only a few of them will actually be carried out. Fewer planned comparisons are usually conducted (due to their specificity) than post hoc comparisons (due to their generality), so planned contrasts generally yield narrower confidence intervals, are more powerful, and have a higher likelihood of a Type I error than post hoc comparisons.

Post hoc contrasts are formulated such that the researcher provides no advance specification of the actual contrasts to be tested. This type of contrast is done *only* following a statistically significant omnibus F test. Post hoc is Latin for "after the fact," referring to contrasts tested after a statistically significant omnibus F in the ANOVA. Here the researcher may want to take the family-wise error rate into account somehow to achieve better overall Type I error protection. Post hoc contrasts are also known as *unplanned, a posteriori,* or *postmortem contrasts.* It should be noted that most MCPs (with the exception of post hoc contrasts) are not derived or based on finding a statistically significant F in the ANOVA.

2.1.1.3 The Type I Error Rate

The goal of multiple comparison procedures is to help ensure that some error rate is maintained and not exceeded so that we do not make a Type I error. **Type I error**, as you recall, is the probability of incorrectly rejecting a true null hypothesis. Type I error is sometimes referred to as **false positive**. Thus, when multiple comparisons are conducted, there is an increased chance of Type I error—or increased chance of false positives. The more multiple comparisons conducted, the higher the chance that there is a false positive—i.e., the higher the probability that a null comparison will be identified as statistically significant. The **false discovery rate** is the rate that comparisons identified as statistically significant are truly null (i.e., not statistically significant)—i.e., the ratio of the number of false positives to the number of total positives.

How does the researcher deal with the family-wise Type I error rate? Depending on the multiple comparison procedure selected, one may either *set alpha for each contrast* or *set alpha for a family of contrasts.* In the former category (i.e., alpha for each contrast), alpha is set for each individual contrast. The MCPs in this category are known as **contrast-based**. We designate the alpha level for contrast-based procedures as α_{pc}, as it represents the **per contrast** Type I error rate. Thus, alpha per contrast (i.e., α_{pc}) represents the *probability of making a Type I error (or false positive) for that particular contrast*.

In the latter category (i.e., alpha for a family of contrasts), alpha is set for a family or set of contrasts. The MCPs in this category are known as **family-wise**. Controlling for family-wise error controls for the probability of one or more false positives out of all comparison tests performed. We designate the alpha level for family-wise procedures as α_{fw}, as it represents the family-wise Type I error rate. Thus, α_{fw} represents the *probability of making at least one Type I error in the family or set of contrasts*. When *all* the null hypotheses are true, the false discovery rate equals the family-wise error rate. When not all the null hypotheses are true, controlling for the family-wise error rate also controls the false discovery rate.

For **orthogonal (or independent or unrelated) contrasts**, the following property holds:

$$\alpha_{fw} = 1 - \left(1 - \alpha_{pc}\right)^{c}$$

where $c = J - 1$ orthogonal contrasts (as defined in the next section). For **nonorthogonal (or related or oblique) contrasts**, this property is more complicated, so we simply say the following:

$$\alpha_{fw} \leq c\alpha_{pc}$$

These properties should be familiar from the discussion in Chapter 1, where we were looking at the probability of a Type I error in the use of multiple independent t tests.

2.1.1.4 Orthogonal Contrasts

Let us begin this section by defining orthogonal contrasts. A set of contrasts is **orthogonal** if they represent nonredundant and independent (if the usual ANOVA assumptions are met) sources of variation. For J groups, you will only be able to construct $J - 1$ orthogonal contrasts in a set. However, more than one set of orthogonal contrasts may exist. Note that although the contrasts *within* each set are orthogonal, contrasts *across* such sets may not be orthogonal.

For purposes of simplicity, we first consider the **equal *n*'s or balanced case** (in other words, the sample sizes are the same for each group). *With equal observations per group, two contrasts are defined to be orthogonal if the products of their contrast coefficients sum to zero*. That is, two contrasts are orthogonal if the following holds:

$$\sum_{j=1}^{J}\left(c_j c_{j'}\right) = c_1 c_{1'} + c_2 c_{2'} + \ldots + c_J c_{J'} = 0$$

where j and j' represent two distinct contrasts. Thus we see that orthogonality depends on the contrast coefficients, the c_j, and *not* the group means, the $\mu_{.j}$.

For example, if $J = 3$, then we can form a set of two orthogonal contrasts. One such set is in Table 2.3. In this set of contrasts, the first contrast (ψ_1) compares the mean of Group 1 ($c_1 = +1$) to the mean of Group 2 $\left(c_2 = -1\right)$. The second contrast (ψ_2) compares the average of the means of Group 1 $\left(c_1 = +\frac{1}{2}\right)$ and Group 2 $\left(c_2 = +\frac{1}{2}\right)$ to the mean of Group 3 $\left(c_3 = -1\right)$.

TABLE 2.3
Orthogonal Contrast

	c_1	c_2	c_3	
ψ_1: $\mu_{.1} - \mu_{.2} = 0$	+1	−1	0	
ψ_2: $\frac{1}{2}\mu_{.1} + \frac{1}{2}\mu_{.2} - \mu_{.3} = 0$	+½	+½	−1	
$\sum_{j=1}^{J}(c_j c_{j'}) =$	+½	−½	0	= 0

Thus, plugging these values into our equation produces the following:

$$\sum_{j=1}^{J}(c_j c_{j'}) = c_1 c_{1'} + c_2 c_{2'} + c_3 c_{3'}$$

$$\sum_{j=1}^{J}(c_j c_{j'}) = (+1)(+\tfrac{1}{2}) + (-1)(+\tfrac{1}{2}) + (0)(-1) = (+\tfrac{1}{2}) + (-\tfrac{1}{2}) + 0 = 0$$

If the sum of the contrast coefficient products for a set of contrasts is equal to zero, then we define this as an orthogonal set of contrasts.

A set of two contrasts that are *not* orthogonal is in Table 2.4, where we see that the set of contrasts does *not* sum to zero.

TABLE 2.4
Nonorthogonal Contrasts

	c_1	c_2	c_3	
ψ_3: $\mu_{.1} - \mu_{.2} = 0$	+1	−1	0	
ψ_4: $\mu_{.1} - \mu_{.3} = 0$	+1	0	−1	
$\sum_{j=1}^{J}(c_j c_{j'}) =$	+1	0	0	= +1

Thus, plugging these values into our equation produces the following, where we see that the product of the contrasts also does not sum to zero.

$$\sum_{j=1}^{J}(c_j c_{j'}) = c_1 c_{1'} + c_2 c_{2'} + c_3 c_{3'}$$

$$\sum_{j=1}^{J}(c_j c_{j'}) = (+1)(+1) + (-1)(0) + (0)(-1) = (+1) + 0 + 0 = +1$$

Consider a situation (Table 2.5) where there are three groups and we decide to form three pairwise contrasts, knowing full well that they cannot all be orthogonal to one another.

For this set of contrasts, the first contrast (ψ_1) compares the mean of Group 1 ($c_1 = +1$) to the mean of Group 2 ($c_2 = -1$). The second contrast (ψ_2) compares the mean of Group 2 ($c_2 = +1$) to the mean of Group 3 ($c_3 = -1$), and the third contrast compares the mean of Group 1 ($c_1 = +1$) to the mean of Group 3 ($c_3 = -1$).

TABLE 2.5

Nonorthogonal Contrasts

	c_1	c_2	c_3
$\psi_1: \mu_{.1} - \mu_{.2} = 0$	+1	−1	0
$\psi_2: \mu_{.2} - \mu_{.3} = 0$	0	+1	−1
$\psi_3: \mu_{.1} - \mu_{.3} = 0$	+1	0	−1

Say that the group population means are $\mu_{.1} = 30$, $\mu_{.2} = 24$, and $\mu_{.3} = 20$. We find $\psi_1 = 6$ for the first contrast (i.e., $\psi_1: \mu_{.1} - \mu_{.2} = 30 - 24 = 6$), and $\psi_2 = 4$ for the second contrast (i.e., $\psi_2: \mu_{.2} - \mu_{.3} = 24 - 20 = 4$). Because these three contrasts are not orthogonal and contain totally redundant information about these means, $\psi_3 = 10$ for the third contrast by definition (i.e., $\psi_3: \mu_{.1} - \mu_{.3} = 30 - 20 = 10$). Thus, the third contrast contains no additional information beyond that contained in the first two contrasts.

Finally, for the unequal n's or unbalanced case, two contrasts are orthogonal if the following holds:

$$\sum_{j=1}^{J}\left(\frac{c_j c_{j'}}{n_j}\right) = 0$$

The denominator n_j makes it more difficult to find an orthogonal set of contrasts that is of any interest to the applied researcher (see Pedhazur, 1997, for an example).

2.1.2 Selected Multiple Comparison Procedures

This section considers a selection of multiple comparison procedures (MCP). These represent the "best" procedures in some sense, in terms of ease of utility, popularity, and control of Type I and Type II error rates. Other procedures are briefly mentioned. In the interest of consistency, each procedure is discussed in the hypothesis testing situation based on a test statistic. Most, but not all, of these procedures can also be formulated as confidence intervals (sometimes called a **critical difference**), although these will not be discussed here. The first few procedures discussed are for planned comparisons, whereas the remainder of the section is devoted to post hoc comparisons. For each MCP, we describe its major characteristics, and then present the test statistic with an example using the data from Chapter 1.

Unless otherwise specified, each MCP makes the standard assumptions of normality, homogeneity of variance, and independence of observations. Some of the procedures do have additional restrictions, such as equal n's per group. Throughout this section we also presume that a two-tailed alternative hypothesis is of interest, although some of the MCPs can also be used with a one-tailed alternative hypothesis. In general, the MCPs are fairly robust to nonnormality (but not for extreme cases), but are not as robust to departures from homogeneity of variance or from independence (e.g., Pavur, 1988).

2.1.2.1 Planned Analysis of Trend

Trend analysis is a *planned MCP* useful when the groups represent different quantitative levels of a factor (i.e., an interval or ratio level independent variable). Examples of such a factor might be age, drug dosage, and different amounts of instruction, practice, or trials. Here, the researcher is interested in whether the sample means vary with a change in the amount of the independent variable. *We define trend analysis in the form of orthogonal polynomials, and assume that the levels of the independent variable are equally spaced (i.e., same distances between the levels of the independent variable, such as 100, 200, 300, and 400cc), and that the number of observations per group is the same.* This is the standard case; other cases are briefly discussed at the end of this section.

Orthogonal polynomial contrasts use the standard t test statistic, which is compared to the critical values of $\pm_{\alpha/2} t_{df(error)}$ obtained from the t table in Appendix Table A.2. The form of the contrasts is a bit different and requires a bit of discussion. *Orthogonal polynomial contrasts incorporate two concepts, orthogonal contrasts (recall these are unrelated or independent contrasts) and polynomial regression.* For J groups, there can be only $J-1$ orthogonal contrasts in a set. In *polynomial regression*, we have terms in the model for a linear trend, a quadratic trend, a cubic trend, and so on. For example, linear trend is represented by a straight line (no bends), quadratic trend by a curve with one bend (e.g., U or upside-down U shapes), and cubic trend by a curve with two bends (e.g., S shape).

Now put those two ideas together. *A set of orthogonal contrasts can be formed where the first contrast evaluates a linear trend, the second a quadratic trend, the third a cubic trend, and so forth.* Thus for J groups, the highest order polynomial that can be formed is $J-1$. With four groups, for example, one could form a set of three orthogonal contrasts to assess linear, quadratic, and cubic trend.

You may be wondering just how these contrasts are formed. For $J=4$ groups, the contrast coefficients for the linear, quadratic, and cubic trends are found in Table 2.6.

TABLE 2.6

Orthogonal Polynomial Contrasts

	c_1	c_2	c_3	c_4
ψ_{linear}	−3	−1	+1	+3
$\psi_{quadratic}$	+1	−1	−1	+1
ψ_{cubic}	−1	+3	−3	+1

Where the contrasts can be written out as follows:

$$\psi_{linear} = (-3)\mu_{.1} + (-1)\mu_{.2} + (+1)\mu_{.3} + (+3)\mu_{.4}$$

$$\psi_{quadratic} = (+1)\mu_{.1} + (-1)\mu_{.2} + (-1)\mu_{.3} + (+1)\mu_{.4}$$

$$\psi_{cubic} = (-1)\mu_{.1} + (+3)\mu_{.2} + (-3)\mu_{.3} + (+1)\mu_{.4}$$

These contrast coefficients, for a number of different values of J, can be found in Appendix Table A.6. If you look in the table of contrast coefficients for values of J greater than 6, you see that the coefficients for the higher-order polynomials are not included. As an example, for J = 7, coefficients only up through a quintic trend are included. Although they could easily be derived and tested, these higher-order polynomials are usually not of interest to the researcher. In fact, it is rare to find anyone interested in polynomials beyond the cubic because they are difficult to understand and interpret (although statistically sophisticated, they say little to the applied researcher as the results must be interpreted in values that are highly complex). The contrasts are typically tested sequentially beginning with the linear trend and proceeding to higher-order trends (cubic then quadratic).

Using the example data on the elite athletes from Chapter 1, let us test for linear, quadratic, and cubic trends. While trend analysis may not be relevant for this data because the groups do not represent different quantitative levels of type of sport, we'll walk through the example for illustrative purposes and assume the levels are appropriate for trend analysis. Because J = 4, we can use the contrast coefficients given previously. The following are the computations, based on these mean values, to test the trend analysis. The critical values (where df_{error} is calculated as $N - J$ or 32 − 4 = 28) are determined to be as follows: $\pm_{\alpha/2} t_{df(error)} = \pm_{.025} t_{28} = \pm 2.048$. The standard error for *linear trend* is computed as follows (where n_j = 8 for each of the J = 4 groups; MS_{error} was computed in the previous chapter and found to be 36.1116). Recall that the contrast equation for the linear trend is

$$\psi_{linear} = (-3)\mu_{.1} + (-1)\mu_{.2} + (+1)\mu_{.3} + (+3)\mu_{.4}$$

and thus these are the c_j values in the equation below (−3, −1, +1, and +3, respectively).

$$s_{\psi'} = \sqrt{MS_{error} \sum_{j=1}^{J} \left(\frac{c_j^2}{n_j} \right)}$$

$$s_{\psi'} = \sqrt{36.1116 \left(\frac{(-3)^2}{8} + \frac{(-1)^2}{8} + \frac{(+1)^2}{8} + \frac{(+3)^2}{8} \right)} = \sqrt{36.1116 \left(\frac{9}{8} + \frac{1}{8} + \frac{1}{8} + \frac{9}{8} \right)} = 9.5015$$

The standard error for *quadratic trend* is determined similarly. Recall that the contrast equation for the quadratic trend is

$$\psi_{quadratic} = (+1)\mu_{.1} + (-1)\mu_{.2} + (-1)\mu_{.3} + (+1)\mu_{.4}$$

and thus these are the c_j values in the equation below (+1, −1, −1, and +1, respectively).

$$s_{\psi'} = \sqrt{MS_{error} \sum_{j=1}^{J} \left(\frac{c_j^2}{n_j} \right)}$$

$$s_{\psi'} = \sqrt{36.1116 \left(\frac{(+1)^2}{8} + \frac{(-1)^2}{8} + \frac{(-1)^2}{8} + \frac{(+1)^2}{8} \right)} = \sqrt{36.1116 \left(\frac{1}{8} + \frac{1}{8} + \frac{1}{8} + \frac{1}{8} \right)} = 4.2492$$

The standard error for *cubic trend* is computed similarly. Recall that the contrast equation for the cubic trend is

$$\psi_{cubic} = (-1)\mu_{.1} + (+3)\mu_{.2} + (-3)\mu_{.3} + (+1)\mu_{.4}$$

and thus these are the c_j values in the equation below (–1, +3, –3, and +1, respectively).

$$s_{\psi'} = \sqrt{MS_{error} \sum_{j=1}^{J} \left(\frac{c_j^2}{n_j} \right)}$$

$$s_{\psi'} = \sqrt{36.1116 \left(\frac{(-1)^2}{8} + \frac{(+3)^2}{8} + \frac{(-3)^2}{8} + \frac{(+1)^2}{8} \right)} = \sqrt{36.1116 \left(\frac{1}{8} + \frac{9}{8} + \frac{9}{8} + \frac{1}{8} \right)} = 9.5015$$

Recall the following means for each group (as presented in the previous chapter; Table 2.7).

TABLE 2.7

Data and Summary Statistics for the Elite Athlete Example

	Psychological Distress by Type of Sport				
	Group 1: Movement (e.g., dance)	**Group 2: Target (e.g., golf)**	**Group 3: Fielding (e.g., baseball)**	**Group 4: Territory (e.g., football)**	**Overall**
	15	20	10	30	
	10	13	24	22	
	12	9	29	26	
	8	22	12	20	
	21	24	27	29	
	7	25	21	28	
	13	18	25	25	
	3	12	14	15	
Means	11.1250	17.8750	20.2500	24.3750	18.4063
Variances	30.1250	35.2679	53.0714	25.9821	56.4425

Thus, using the *contrast coefficients* (represented by the constant *c* values in the numerator of each term) and the values of the means for each of the four groups (represented by $\bar{Y}_{.1}$, $\bar{Y}_{.2}$, $\bar{Y}_{.3}$, $\bar{Y}_{.4}$), the test statistics are computed as follows:

$$t_{linear} = \frac{\psi_{linear}}{s_{\psi'}} = \frac{-3\bar{Y}_{.1} - 1\bar{Y}_{.2} + 1\bar{Y}_{.3} + 3\bar{Y}_{.4}}{s_{\psi'}}$$

$$t_{linear} = \frac{-3(11.1250) - 1(17.8750) + 1(20.2500) + 3(24.3750)}{9.5015} = 4.4335$$

$$t_{cubic} = \frac{\psi_{cubic}}{s_{\psi'}} = \frac{-1\bar{Y}_{.1} + 3\bar{Y}_{.2} - 3\bar{Y}_{.3} + 1\bar{Y}_{.4}}{s_{\psi'}}$$

$$t_{cubic} = \frac{-1(11.1250) + 3(17.8750) - 3(20.2500) + 1(24.3750)}{9.5015} = 0.6446$$

$$t_{quadratic} = \frac{\psi_{quadratic}}{s_{\psi'}} = \frac{1\bar{Y}_{.1} - 1\bar{Y}_{.2} - 1\bar{Y}_{.3} + 1\bar{Y}_{.4}}{s_{\psi'}}$$

$$t_{quadratic} = \frac{1(11.1250) - 1(17.8750) - 1(20.2500) + 1(24.3750)}{4.2492} = -0.6178$$

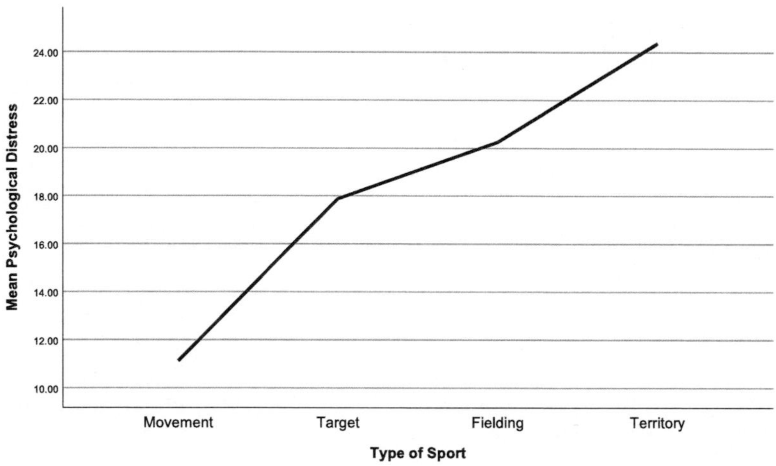

FIGURE 2.1
Profile plot for psychological distress example.

The t test statistic for the linear trend exceeds the t critical value. Thus, we see that there is a statistically significant *linear trend* in the means, but no significant *higher-order* trend (in other words, no significant quadratic or cubic trend). This should not be surprising as shown in the profile plot of the means of Figure 2.1, where there is a very strong linear trend, and that is about it. In other words, there is a steady increase in mean psychological distress as the type of sport increases from movement to target, fielding, and territory. Always plot the means so that you can interpret the results of the contrasts.

Let us make some final points about orthogonal polynomial contrasts. First, be particularly careful about extrapolating beyond the range of the levels investigated. The trend may or may not be the same outside of this range; that is, given only those sample means, we have no way of knowing what the trend is outside of the range of levels investigated. Second, in the unequal n's or unbalanced case, it becomes difficult to formulate a set of orthogonal contrasts that make any sense to the researcher. See the discussion in the next section on planned orthogonal contrasts, as well as Kirk (2013). Third, when the levels are not equally spaced, this needs to be taken into account in the contrast coefficients (Kirk, 2013).

2.1.2.2 Planned Orthogonal Contrasts

Planned orthogonal contrasts (POC) are MCPs where the contrasts are defined ahead of time by the researcher (i.e., planned) and the set of contrasts are orthogonal (or unrelated). The POC method is a **contrast-based procedure** where the researcher is not concerned with control of the family-wise Type I error rate across the set of contrasts. The set of contrasts are *orthogonal,* so the number of contrasts should be small, and concern with the family-wise error rate is lessened.

Computationally, planned orthogonal contrasts use the standard t test statistic that is compared to the critical values of $\pm_{\alpha/2} t_{df(error)}$ obtained from the t table in Appendix Table A.2. Using the example dataset from Chapter 1, let us find a set of orthogonal contrasts and complete the computations. Since $J = 4$, we can find at most a set of three (or $J - 1$) orthogonal contrasts. One orthogonal set that seems reasonable for these data is in Table 2.8.

TABLE 2.8

Planned Orthogonal Contrast

	c_1	c_2	c_3	c_4
$\psi_1: \left(\frac{\mu_{.1}+\mu_{.2}}{2}\right)-\left(\frac{\mu_{.3}+\mu_{.4}}{2}\right)=0$	+ ½	+ ½	−½	−½
$\psi_2: \mu_{.1}-\mu_{.2}=0$	+1	−1	0	0
$\psi_3: \mu_{.3}-\mu_{.4}=0$	0	0	+1	−1

Here we see that the first contrast compares the average of the first two groups (i.e., Movement and Target) with the average of the last two groups (i.e., Fielding and Territory), the second contrast compares the means of the first two groups (i.e., Movement and Target), and the third contrast compares the means of the last two groups (Fielding and Territory). Note that the design is balanced (i.e., the equal *n*'s case as all groups had a sample size of 8). What follows are the computations. The critical values are: $\pm_{\alpha/2}t_{df(error)} = \pm_{.025}t_{28} = \pm 2.048$.

The standard error for contrast 1 is computed as follows (where n_j = 8 for each of the J = 4 groups; MS_{error} was computed in the previous chapter and found to be 36.1116). The equation for contrast one is $\psi_1: \left(\frac{\mu_{.1}+\mu_{.2}}{2}\right)-\left(\frac{\mu_{.3}+\mu_{.4}}{2}\right)=0$ and thus the c_j values in the equation below (+1/2, +1/2, −1/2, −1/2, respectively, and these values are then squared which results in the value of .25).

$$s_{\psi'} = \sqrt{MS_{error}\sum_{j=1}^{J}\left(\frac{c_j^2}{n_j}\right)}$$

$$s_{\psi'} = \sqrt{36.1116\left(\frac{.25}{8}+\frac{.25}{8}+\frac{.25}{8}+\frac{.25}{8}\right)} = 2.1246$$

Similarly, the standard errors for contrasts 2 and 3 are computed below:

$$s_{\psi'} = \sqrt{MS_{error}\sum_{j=1}^{J}\left(\frac{c_j^2}{n_j}\right)} = \sqrt{36.1116\left(\frac{1}{8}+\frac{1}{8}\right)} = 3.0046$$

The **test statistics** are computed as follows:

$$t_1 = \frac{(1/2)\bar{Y}_{.1}+(1/2)\bar{Y}_{.2}-(1/2)\bar{Y}_{.3}-(1/2)\bar{Y}_{.4}}{s_{\psi'}}$$

$$t_1 = \frac{(1/2)(11.1250)+(1/2)(17.8750)-(1/2)(20.2500)-(1/2)(24.3750)}{2.1246} = -3.6772$$

$$t_2 = \frac{\bar{Y}_{.1}-\bar{Y}_{.2}}{s_{\psi'}} = \frac{11.1250-17.8750}{3.0046} = -2.2466$$

$$t_3 = \frac{\bar{Y}_{.3} - \bar{Y}_{.4}}{s_{\psi'}} = \frac{20.2500 - 24.3750}{3.0046} = -1.3729$$

The result for contrast 1 is that the combined first two groups (Movement and Target) have statistically significantly lower psychological distress, on average, than the combined last two groups (Fielding and Territory). The result for contrast 2 is that Movement and Target groups are statistically significantly different from one another, on average. The result for contrast 3 is that the means of Fielding and Territory are not statistically significantly different from one another.

There is a practical problem with this procedure because (a) the contrasts that are of interest to the researcher may not necessarily be orthogonal, or (b) the researcher may not be interested in all of the contrasts of a particular orthogonal set. Another problem already mentioned occurs when the design is unbalanced, where an orthogonal set of contrasts may be constructed at the expense of meaningful contrasts. Our advice is simple.

1. If the contrasts you are interested in are not orthogonal, then use another MCP.
2. If you are not interested in all of the contrasts of an orthogonal set, then use another MCP.
3. If your design is not balanced and the orthogonal contrasts formed are not meaningful, then use another MCP.

In each case you need a different *planned* MCP. We recommend using one of the following procedures discussed later in this chapter: the Dunnett, Dunn (Bonferroni), or Dunn-Sidak procedure.

We defined the POC as a *contrast-based procedure*. One could also consider an alternative family-wise method where the α_{pc} level is divided among the contrasts in the set. This procedure is defined by $\alpha_{pc} = {}^{\alpha_{fw}}/_{c}$, where c is the number of orthogonal contrasts in the set (i.e., $c = J - 1$). As we show later, this borrows a concept from the Dunn (Bonferroni) procedure. If the variances are not equal across the groups, several approximate solutions have been proposed that take the individual group variances into account (Kirk, 2013).

2.1.2.3 Planned Contrasts With Reference Group: Dunnett Method

A third method of planned comparisons is attributed to Dunnett (1955) and thus referred to as the Dunnett method. It is designed to test pairwise contrasts where a reference group (e.g., a control or baseline group) is compared to each of the other $J - 1$ groups. Thus, a family of *prespecified* pairwise contrasts is to be evaluated. The Dunnett method is a *family-wise MCP* and is slightly more powerful than the Dunn procedure (another planned family-wise MCP). The test statistic is the standard t except that the standard error is simplified as follows:

$$s_{\psi'} = \sqrt{MS_{error}\left(\frac{1}{n_c} + \frac{1}{n_j}\right)}$$

where c is the reference group and j is the group to which it is being compared. The test statistic is compared to the critical values $\pm_{\alpha/2} t_{df(error), J-1}$, obtained from the Dunnett table located in Appendix Table A.7.

Using the example dataset, compare Group 1, the movement sport (used as a reference or baseline group), to each of the other three types of sports. The contrasts are found in Table 2.9.

TABLE 2.9

Planned Contrasts With Reference Group: Dunnett Method

	c_1	c_2	c_3	c_4
$\psi_1: \mu_{.1} - \mu_{.2} = 0$	+1	−1	0	0
$\psi_2: \mu_{.1} - \mu_{.3} = 0$	+1	0	−1	0
$\psi_4: \mu_{.1} - \mu_{.4} = 0$	+1	0	0	−1

The critical values are as follows: $\pm_{\alpha/2} t_{df(error),J-1} = \pm_{.025} t_{28,3} \approx \pm 2.48$

The standard error is computed as follows (where $n_c = 8$ for the reference group; $n_j = 8$ for each of the other groups; MS_{error} was computed in the previous chapter and found to be 36.1116).

$$s_{\psi'} = \sqrt{MS_{error}\left(\frac{1}{n_c} + \frac{1}{n_j}\right)} = \sqrt{36.1116\left(\frac{1}{8} + \frac{1}{8}\right)} = 3.00$$

The test statistics for the three contrasts (i.e., Group 1 to Group 2; Group 1 to Group 3; and Group 1 to Group 4) are computed as follows:

$$Movement\ to\ Target: t_1 = \frac{\bar{Y}_{.1} - \bar{Y}_{.2}}{s_{\psi'}} = \frac{11.1250 - 17.8750}{3.0046} = -2.2466$$

$$Movement\ to\ Target: t_2 = \frac{\bar{Y}_{.1} - \bar{Y}_{.3}}{s_{\psi'}} = \frac{11.1250 - 20.2500}{3.0046} = -3.0370$$

$$Movement\ to\ Fielding: t_3 = \frac{\bar{Y}_{.1} - \bar{Y}_{.4}}{s_{\psi'}} = \frac{11.1250 - 24.3750}{3.0046} = -4.4099$$

Comparing the test statistics to the critical values, we see that the second group (i.e., Target) is not statistically significantly different from group one (i.e., Movement), but the third (Fielding) and fourth (Territory) groups are significantly different from group one (i.e., Movement).

If the variance of the reference group is different from the variances of the other $J - 1$ groups, then a modification of this method is described in Dunnett (1964). For related procedures that are less sensitive to unequal group variances, see Wilcox (1987) or Wilcox (1996) (e.g., variation of the Dunnett T3 procedure).

2.1.2.4 Other Planned Contrasts: Dunn (or Bonferroni) and Dunn-Sidak Methods

The Dunn (1961) procedure (commonly attributed to Dunn as the developer is unknown), also often called the **Bonferroni procedure** (because it is based on the Bonferroni inequality), is a planned family-wise MCP. *It is designed to test either pairwise or complex contrasts for balanced or unbalanced designs.* Thus this MCP is very flexible and may be used to test any planned contrast of interest. Dunn's method uses the standard t test statistic with one important exception. The alpha level is split up among the set of planned contrasts. Typically the per contrast alpha level (denoted as α_{pc}) is set at α/c, where c is the number of contrasts. That is, $\alpha_{pc} = \alpha_{fw}/c$. According to this rationale, the family-wise Type I error rate (denoted as α_{fw}) will be maintained at alpha. For example, if $\alpha_{fw} = .05$ is desired and there are five contrasts to be tested, then each contrast would be tested at the .01 level of significance (.05/5 = .01). We are reminded that alpha need not be distributed equally among the set of contrasts, as long as the sum of the individual α_{pc} terms is equal to α_{fw} (Keppel & Wickens, 2004; Rosenthal & Rosnow, 1985).

Computationally, the Dunn method uses the standard t test statistic, which is compared to the critical values of $\pm_{\alpha/c} t_{df(error)}$ for a two-tailed test obtained from the table in Appendix Table A.8. The table takes the number of contrasts into account without requiring you to physically split up the α. Using the example dataset from Chapter 1, for comparison purposes, let us test the same set of three orthogonal contrasts we evaluated with the POC method. These contrasts are in Table 2.10.

TABLE 2.10

Dunn Method

	c_1	c_2	c_3	c_4
$\psi_1: \left(\frac{\mu_{.1}+\mu_{.2}}{2}\right)-\left(\frac{\mu_{.3}+\mu_{.4}}{2}\right)=0$	+ ½	+ ½	− ½	− ½
$\psi_2: \mu_{.1}-\mu_{.2}=0$	+1	−1	0	0
$\psi_3: \mu_{.3}-\mu_{.4}=0$	0	0	+1	−1

Below are the computations; the critical values include:

$$\pm_{\alpha/c} t_{df(error)} = \pm_{.05/3} t_{28} = \pm 2.539$$

The standard error for contrast 1 is computed as follows:

$$s_{\psi'} = \sqrt{MS_{error} \sum_{j=1}^{J} \left(\frac{c_j^2}{n_j}\right)} = \sqrt{36.1116\left(\frac{.25}{8}+\frac{.25}{8}+\frac{.25}{8}+\frac{.25}{8}\right)} = 2.1246$$

Similarly, the standard error for contrasts 2 and 3 is computed below:

$$s_{\psi'} = \sqrt{MS_{error} \sum_{j=1}^{J} \left(\frac{c_j^2}{n_j}\right)} = \sqrt{36.1116\left(\frac{1}{8}+\frac{1}{8}\right)} = 3.0046$$

The test statistics are computed as follows:

$$t_1 = \frac{(1/2)\bar{Y}_{.1} + (1/2)\bar{Y}_{.2} - (1/2)\bar{Y}_{.3} - (1/2)\bar{Y}_{.4}}{s_{\psi'}}$$

$$t_1 = \frac{(1/2)(11.1250) + (1/2)(17.8750) - (1/2)(20.2500) - (1/2)(24.3750)}{2.1246} = -3.6772$$

$$t_2 = \frac{\bar{Y}_{.1} + \bar{Y}_{.2}}{s_{\psi'}} = \frac{11.1250 - 17.8750}{3.0046} = -2.2466$$

$$t_3 = \frac{\bar{Y}_{.3} + \bar{Y}_{.4}}{s_{\psi'}} = \frac{20.2500 - 24.3750}{3.0046} = -1.3729$$

Notice that the test statistic values have not changed from the POC, but the critical value *has* changed. For this set of contrasts, then, we see the same results as were obtained via the POC procedure with the exception of contrast 2, which is now nonsignificant (i.e., only contrast 1 is significant). The reason for this difference lies in the critical values used, which were ±2.048 for the POC method and ±2.539 for the Dunn method. Here we see the conservative nature of the Dunn procedure because the critical value is larger than with the POC method, thus making it a bit more difficult to reject H_0.

The Dunn procedure is slightly conservative (i.e., not as powerful) in that the true α_{fw} may be less than the specified nominal α level. For example, if the nominal alpha (specified by the researcher) is .05, then the true alpha may be less than .05. *Thus when using the Dunn, you may be less likely to reject the null hypothesis (i.e., less likely to find a statistically significant contrast).* A less conservative (i.e., more powerful) modification is known as the **Dunn-Sidak procedure** (Dunn, 1974; Sidak, 1967), and uses slightly different critical values. For more information see Kirk (2013), Wilcox (1987), and Keppel and Wickens (2004). The Bonferroni modification can also be applied to other MCPs.

2.1.2.5 Complex Post Hoc Contrasts: Scheffé and Kaiser-Bowden Methods

Another early MCP due to Scheffé (1953) is quite versatile. The Scheffé procedure can be used for any possible type of comparison, orthogonal or nonorthogonal, pairwise or complex, planned or post hoc, where the family-wise error rate is controlled. The Scheffé method is so general that the tests are quite conservative (i.e., less powerful), particularly for the pairwise contrasts. This is so because the family of contrasts for the Scheffé method consists of all possible linear comparisons. To control the Type I error rate for such a large family, the procedure has to be conservative (i.e., making it less likely to reject the null hypothesis if it is really true). *Thus we recommend the Scheffé method only for complex post hoc comparisons.*

The Scheffé procedure is the only MCP that is necessarily consistent with the results of the F ratio in the analysis of variance. If the F ratio is statistically significant, then this means that at least one contrast in the entire family of contrasts will be significant with the Scheffé method. Do not forget, however, that this family can be quite large and you may not even be interested in the contrast(s) that wind up being significant. If the F ratio is not statistically significant, then none of the contrasts in the family will be significant with the Scheffé method.

The test statistic for the Scheffé method is the standard t again. This is compared to the critical value $\sqrt{(J-1)\left({}_{\alpha}F_{J-1,df(error)}\right)}$ taken from the F table in Appendix Table A.4. In other words, the square root of the F critical value is adjusted by $J - 1$, which serves to increase the Scheffé critical value and make the procedure a more conservative one.

Consider a few example contrasts with the Scheffé method. Using the example dataset from Chapter 1, for comparison purposes we test the same set of three orthogonal contrasts that were evaluated with the POC method. These contrasts are again as follows (Table 2.11).

TABLE 2.11

Scheffé Method

	c_1	c_2	c_3	c_4
$\psi_1 : \left(\frac{\mu_{.1}+\mu_{.2}}{2}\right)-\left(\frac{\mu_{.3}+\mu_{.4}}{2}\right)=0$	+ ½	+ ½	− ½	− ½
$\psi_2 : \mu_{.1}-\mu_{.2}=0$	+1	−1	0	0
$\psi_3 : \mu_{.3}-\mu_{.4}=0$	0	0	+1	−1

Below are the computations with the following critical value:

$$\sqrt{(J-1)\left({}_{\alpha}F_{J-1,df(error)}\right)}=\sqrt{(4-1)\left({}_{.05}F_{3,28}\right)}=\sqrt{(3)(2.95)}=2.97$$

Standard error for contrast 1:

$$s_{\psi'}=\sqrt{MS_{error}\sum_{j=1}^{J}\left(\frac{c_j^2}{n_j}\right)}=\sqrt{36.1116\left(\frac{.25}{8}+\frac{.25}{8}+\frac{.25}{8}+\frac{.25}{8}\right)}=2.1246$$

Standard error for contrasts 2 and 3:

$$s_{\psi'}=\sqrt{MS_{error}\left(\frac{1}{n_j}+\frac{1}{n_j}\right)}=\sqrt{36.1116\left(\frac{1}{8}+\frac{1}{8}\right)}=3.0046$$

The test statistics are computed as follows:

$$t_1=\frac{(1/2)\bar{Y}_{.1}+(1/2)\bar{Y}_{.2}-(1/2)\bar{Y}_{.3}-(1/2)\bar{Y}_{.4}}{s_{\psi'}}$$

$$t_1=\frac{(1/2)(11.1250)+(1/2)(17.8750)-(1/2)(20.2500)-(1/2)(24.3750)}{2.1246}=-3.6772$$

$$t_2=\frac{\bar{Y}_{.1}+\bar{Y}_{.2}}{s_{\psi'}}=\frac{11.1250-17.8750}{3.0046}=-2.2466$$

$$t_3=\frac{\bar{Y}_{.3}+\bar{Y}_{.4}}{s_{\psi'}}=\frac{20.2500-24.3750}{3.0046}=-1.3729$$

Using the Scheffé method, these results are precisely the same as those obtained via the Dunn procedure. There is somewhat of a difference in the critical values, which were 2.97 for the Scheffé method, 2.539 for the Dunn method, and 2.048 for the POC method. Here we see that the Scheffé procedure is even more conservative than the Dunn procedure, thus making it a bit more difficult to reject H_0.

For situations where the group variances are unequal, a modification of the Scheffé method less sensitive to unequal variances has been proposed by Brown and Forsythe (1974). Kaiser and Bowden (1983) found that the Brown-Forsythe procedure may cause the actual α level to exceed the nominal α level, and thus we recommend the Kaiser-Bowden modification. For more information see Kirk (2013), Wilcox (1987), and Wilcox (1996).

2.1.2.6 Simple Post Hoc Contrasts: Tukey HSD, Tukey-Kramer, Fisher LSD and Fisher-Hayter Tests

Tukey's (1953) honestly significant difference (HSD) test is one of the most popular post hoc MCPs. The HSD test is a family-wise procedure and is most appropriate for considering all pairwise contrasts with equal n's per group (i.e., a balanced design). The HSD test is sometimes referred to as the **studentized range test** because it is based on the sampling distribution of the studentized range statistic developed by William Sealy Gossett (forced to use the pseudonym "Student" by his employer, the Guinness brewery). For the traditional approach, the first step in the analysis is to rank order the means from largest ($\bar{Y}_{.1}$) to smallest ($\bar{Y}_{.J}$). The test statistic, or **studentized range statistic**, is computed as follows:

$$q_i = \frac{\bar{Y}_{.j} - \bar{Y}_{.j'}}{s_{\psi'}}$$

where

$$s_{\psi'} = \sqrt{\frac{MS_{error}}{n}}$$

and where i identifies the specific contrast, j and j' designate the two group means to be compared, and n represents the number of observations per group (equal n's per group is required). The test statistic is compared to the critical value $\pm\,{}_{\alpha}q_{df(error),\,J'}$ where df_{error} is equal to $J\,(n-1)$. The table for these critical values is given in Appendix Table A.9.

The first contrast involves a test of the largest pairwise difference in the set of J means (q_1) *(i.e., largest vs. smallest means).* If these means are *not* statistically significantly different, then the analysis stops because no other pairwise difference could be significant. If these means *are* statistically significantly different, then we proceed to test the second pairwise difference involving the largest mean (i.e., q_2). Contrasts involving the largest mean are continued until a nonsignificant difference is found. Then the analysis picks up with the *second largest mean* and compares it with the smallest mean. Contrasts involving the second largest mean are continued until a nonsignificant difference is detected. The analysis continues with the *third largest mean* and the smallest mean, and so on, until it is obvious that no other pairwise contrast could be significant.

Finally, consider an example using the HSD procedure with the elite athlete data. Below are the computations, with the following critical value: $\pm_{\alpha} q_{df(error),J} = \pm_{.05} q_{28,4} \approx \pm 3.87$. The standard error is computed as follows where n represents the sample size per group:

$$s_{\psi'} = \sqrt{\frac{MS_{error}}{n}} = \sqrt{\frac{36.1116}{8}} = 2.1246$$

The test statistics are computed as follows:

Territory to Movement: $$q_1 = \frac{\bar{Y}_{.4} - \bar{Y}_{.1}}{s_{\psi'}} = \frac{24.3750 - 11.1250}{2.1246} = 6.2365$$

Territory to Target: $$q_2 = \frac{\bar{Y}_{.4} - \bar{Y}_{.2}}{s_{\psi'}} = \frac{24.3750 - 17.8750}{2.1246} = 3.0594$$

Fielding to Movement: $$q_3 = \frac{\bar{Y}_{.3} - \bar{Y}_{.1}}{s_{\psi'}} = \frac{20.2500 - 11.1250}{2.1246} = 4.2949$$

Fielding to Target: $$q_4 = \frac{\bar{Y}_{.3} - \bar{Y}_{.2}}{s_{\psi'}} = \frac{20.2500 - 17.8750}{2.1246} = 1.1179$$

Target to Movement: $$q_5 = \frac{\bar{Y}_{.2} - \bar{Y}_{.1}}{s_{\psi'}} = \frac{17.8750 - 11.1250}{2.1246} = 3.1771$$

Comparing the test statistic values to the critical value, these results indicate that the group means are significantly different for Groups 1 (Movement) and 4 (Territory) and for Groups 1 (Movement) and 3 (Fielding). Just for completeness, we examine the final possible pairwise contrast involving Groups 3 and 4. However, we already know from the results of previous contrasts that these means cannot possibly be significantly different. The test statistic result for this contrast is as follows:

Territory to Fielding: $$q_6 = \frac{\bar{Y}_{.4} - \bar{Y}_{.3}}{s_{\psi'}} = \frac{24.3750 - 20.2500}{2.1246} = 1.9415$$

Occasionally researchers need to summarize the results of their pairwise comparisons. Table 2.12 shows the results of Tukey's HSD contrasts for the example data. For ease of interpretation, the means are ordered from lowest to highest. The first row consists of the results for those contrasts that involve Group 1. Thus the mean for Group 1 (Movement) is statistically different from those of Groups 3 (Fielding) and 4 (Territory) only. None of the other pairwise contrasts were shown to be significant. Such a table could also be developed for other pairwise MCPs.

TABLE 2.12

Tukey HSD Contrast Test Statistics and Results

	Group 1: Movement	Group 2: Target	Group 3: Fielding	Group 4: Territory
Group 1 (mean = 11.1250)	–	3.1771	4.2949*	6.2365*
Group 2 (mean = 17.8750)		–	1.1179	3.0594
Group 3 (mean = 20.2500)			–	1.9415
Group 4 (mean = 24.3750)				–

* $p < .05$; ${}_{.05}q_{28,4} = 3.87$

The HSD test has exact control of the family-wise error rate assuming normality, homogeneity, and equal n's (better than Dunn or Dunn-Sidak). The HSD procedure is more powerful than the Dunn (aka Bonferroni) or Scheffé procedures for testing all possible pairwise contrasts, although Dunn is more powerful for less than all possible pairwise contrasts. The HSD technique is the recommended MCP as a pairwise method in the equal n's situation and when there is homoscedasticity. The HSD test is reasonably robust to non-normality, but not in extreme cases, and is not as robust as the Scheffé MCP.

There are several alternatives to the HSD for the unequal n's case. These include the Tukey-Kramer modification (Kramer, 1957; Tukey, 1953), which assumes normality and homogeneity. The Tukey-Kramer test statistic is the same as the Tukey HSD except that the standard error is computed as follows.

$$s_{\psi'} = \sqrt{MS_{error}\left[\frac{1}{2}\left(\frac{1}{n_1}+\frac{1}{n_2}\right)\right]}$$

The critical value is determined in the same way as with the Tukey HSD procedure.

If you are using SPSS to compute Tukey's MCP, you will find there are two options: Tukey and Tukey's b. Tukey's HSD is operationalized within SPSS as "Tukey," while "Tukey b" is a variation developed by Tukey that does not control the experiment-wise error rate (and thus is not recommended).

Fisher's (1942) **least significant difference (LSD) test**, also known as the protected t test, was the first MCP developed and is a pairwise post hoc procedure. It is a sequential procedure where a significant ANOVA F is followed by the LSD test in which all (or perhaps some) pairwise t tests are examined. The standard t test statistic is compared with the critical values of $\pm_{\alpha/2}t_{df(error)}$. The LSD test has precise control of the family-wise error rate for the three-group situation, assuming normality and homogeneity; but, as noted by Levin, Serlin, and Seaman (1994), for more than three groups, the protection deteriorates rather rapidly. In that case, a modification due to Hayter (1986) is suggested for more adequate protection.

The **Fisher-Hayter test** (also referred to as the Hayter-Fisher method) is a two-step procedure that was originally devised for equal sample sizes but also shown to work well with unbalanced designs when a modification is applied (Hayter, 1986). The Fisher-Hayter commences with a significant ANOVA F and then is followed by all pairwise comparisons using the studentized range distribution, with the comparisons treated as if there is one group less in the comparison, thereby being more powerful than the Tukey HSD.

In the case of unequal sample sizes, the Tukey-Kramer (relative to Fisher-Hayter) may have more power to detect the largest pairwise difference but is less powerful than the Fisher-Hayter in detecting all pairwise differences. The Fisher-Hayter can be applied in balanced and unbalanced designs and has excellent control of family-wise error (Keppel & Wickens, 2004).

2.1.2.7 Simple Post Hoc Contrasts for Unequal Variances: Games-Howell, Dunnett T3, and C Tests

When the group variances are unequal, several alternative procedures are available. These alternatives include the Games-Howell (Games & Howell, 1976), Dunnett T3, and Dunnett C (Dunnett, 1980) procedures. **Dunnett T3** is recommended for small sample sizes, $n < 50$ and **Games-Howell** for larger sample sizes, $n > 50$ (Maxwell, Delaney, & Kelley, 2018; Wilcox, 1995, 2003b). Games-Howell has been found to be slightly liberal, with an experiment-wise error rate above the nominal alpha, with smaller samples (Dunnett, 1980). **Dunnett C** performs about the same as Games-Howell (Wilcox, 1995, 2003b). For further details on these methods, please consult additional references (e.g., Benjamini & Hochberg, 1995; Hochberg, 1988; Kirk, 2013; Maxwell et al., 2018; Wilcox, 1987, 1995, 2003).

2.1.2.8 Follow-Up Tests to Kruskal-Wallis

Recall from Chapter 1 the nonparametric equivalent to the analysis of variance, the Kruskal-Wallis test. Several post hoc procedures are available to follow up a statistically significant overall Kruskal-Wallis test. The procedures discussed here are the nonparametric equivalents to the Scheffé and Tukey HSD methods. One may form pairwise or complex contrasts as in the parametric case. The test statistic is Z and computed as follows:

$$Z = \frac{\psi_i'}{s_{\psi'}}$$

where the standard error in the denominator is computed as:

$$s_{\psi'} = \sqrt{\left(\frac{N(N+1)}{12}\right)\sum_{j=1}^{J}\left(\frac{c_j^2}{n_j}\right)}$$

and where N is the total number of observations. For the Scheffé method, the test statistic Z is compared to the critical value $\sqrt{{}_{\alpha}\chi_{J-1}}$ obtained from the χ^2 table in Appendix Table A.3. For the Tukey HSD procedure, the test statistic Z is compared to the critical value $\left({}_{\alpha}q_{df(error),J}\right)/\sqrt{2}$ obtained from the table of critical values for the studentized range statistic in Appendix Table A.9.

Let us use the psychological distress data to illustrate. Do not forget that we use the ranked data as described in Chapter 1. The rank means for the groups are as follows: Group 1 (Movement) = 7.7500; Group 2 (Target) = 15.2500; Group 3 (Fielding) = 18.7500; and Group 4 (Territory) = 24.2500. Here we examine only two contrasts and then compare the results for both the Scheffé and Tukey HSD methods. The first contrast compares the first two types of sports to each other (i.e., Groups 1 and 2, Movement and Target), whereas the second contrast compares the first two types of sports (i.e., Movement and Target in aggregate) with the last two types of sports (i.e., Groups 3 and 4, Fielding and Territory in aggregate). In other words, we examine a pairwise contrast and a complex contrast, respectively. The results are given here. The critical values are as follows:

$$\textit{Scheffé}\ \sqrt{{}_{\alpha}\chi_{J-1}} = \sqrt{{}_{.05}\chi_{3}} = \sqrt{7.8147} = 2.7955$$

$$\textit{Tukey}\ \frac{{}_{\alpha}q_{df(error),J}}{\sqrt{2}} = \frac{{}_{.05}q_{28,4}}{\sqrt{2}} = \frac{3.87}{\sqrt{2}} = 2.7365$$

The standard error for contrast 1 is computed as:

$$s_{\psi'} = \sqrt{\left(\frac{N(N+1)}{12}\right)\sum_{j=1}^{J}\left(\frac{c_j^2}{n_j}\right)} = \sqrt{\left(\frac{32(32+1)}{12}\right)\left(\frac{1}{8}+\frac{1}{8}\right)} = 4.6904$$

The standard error for contrast 2 is calculated as follows:

$$s_{\psi'} = \sqrt{\left(\frac{N(N+1)}{12}\right)\sum_{j=1}^{J}\left(\frac{c_j^2}{n_j}\right)}$$

$$= \sqrt{\left(\frac{32(32+1)}{12}\right)\left(\frac{.25}{8}+\frac{.25}{8}+\frac{.25}{8}+\frac{.25}{8}\right)} = 3.3166$$

The test statistics are computed as follows:

$$Z_1 = \frac{\bar{Y}_{.1} - \bar{Y}_{.2}}{s_{\psi'}} = \frac{7.75 - 15.25}{4.6904} = -1.5990$$

$$Z_2 = \frac{(1/2)\bar{Y}_{.1} + (1/2)\bar{Y}_{.2} - (1/2)\bar{Y}_{.3} - (1/2)\bar{Y}_{.4}}{s_{\psi'}}$$

$$Z_2 = \frac{\left(\frac{1}{2}\right)(7.75) + \left(\frac{1}{2}\right)(15.25) - \left(\frac{1}{2}\right)(18.75) - \left(\frac{1}{2}\right)(24.25)}{3.3166} = -3.0151$$

For both procedures we find a statistically significant difference with the second contrast, but not with the first. These results agree with most of the other parametric procedures for these particular contrasts. That is, the first two groups are not statistically significantly different (only statistically significant with POC), whereas the first two groups in aggregate (Movement and Target) are statistically significantly different from the last two groups in aggregate (Fielding and Territory) (significant with all procedures). One could also devise nonparametric equivalent MCPs for methods other than the Scheffé and Tukey procedures.

2.1.3 Selecting the Proper Multiple Comparison Procedure

This chapter has attempted to summarize some of the most common MCPs. Box 2.1 is provided to assist in understanding typologies of MCPs, including simple versus complex contrasts and planned versus post hoc contrasts. Box 2.2 is provided to assist in summarizing some of the primary advantages and disadvantages of various multiple comparison procedures.

BOX 2.1 MCP Typologies

Typology	Definition
Simple versus complex comparison	
Simple or pairwise contrast	Comparison involving only two means
Complex or non-pairwise contrast	Comparison involving more than two means
Planned versus post hoc comparison	
Planned contrast (also known as specific or *a priori* contrast)	Comparisons that are determined without regard to the outcome of the omnibus F test
Post hoc comparison	Comparisons that are conducted only when the outcome of the omnibus F test is statistically significant

BOX 2.2 Advantages and Disadvantages of MCPs

	Advantage	Disadvantage
Planned Contrasts		
Trend analysis (planned polynomial MCP)	Can be used when groups represent different quantitative levels of a factor, allowing the examination of whether the sample means vary with a change in the amount of the independent variable	• Assumes equidistance between levels of the independent variable • Assumes equal n's per group
Planned orthogonal contrast	• Contrasts are determined *a priori* • Unconcerned with control of family-wise Type I error rate across the set of contrasts	• The number of contrasts should be small • The contrasts that are of interest may not be orthogonal • Not all the contrasts of a particular orthogonal set may be of interest • Assumes equal n's per group
Dunnett method (planned orthogonal contrast with reference group)	• Allows examination of pairwise contrasts where a reference group is compared to each of the other J-1 groups • More powerful than the Dunn MCP	• A modification of the procedure is needed in the presence of heteroscedasticity

(*continued*)

(continued)

Dunn method (also known as the Bonferonni method; planned orthogonal contrast)	• Can test pairwise or complex contrasts • Can deal with unbalanced groups	• Conservative (more difficult to reject the null hypothesis) relative to other MCPs • A modification (the Dunn-Sidak) is more powerful than the Dunn
Post hoc contrasts		
Scheffé (complex post hoc contrast)	• Can test orthogonal or nonorthogonal • Can test pairwise or complex contrasts • Can test planned or post hoc comparisions • Controls the family-wise error rate • The only MCP that is consistent with the results of the omnibus *F* test in ANOVA	• Even more conservative (more difficult to reject the null hypothesis) than the Dunn • The family of contrasts is potentially large and the contrast(s) that are statistically significant may not be of interest • Recommended only for complex post hoc comparisons • A modification of the procedure is needed in the presence of heteroscedasticity
Tukey's honestly significant difference (HSD) (simple post hoc comparison)	• Family-wise procedure with exact control of the family-wise error rate in the following conditions: normality (and is relatively robust to non-normality) and homogeneity are assumed and a balanced design • Most appropriate when considering all possible pairwise contrasts with equal *n*'s per group • More powerful than Dunn or Scheffé for testing all possible pairwise contrasts	• Not robust to extreme non-normality • Assumes a balanced design (Tukey-Kramer can be used for unbalanced designs) • Assumes homogeneity of variances
Fisher's least significance difference (LSD) test (pairwise post hoc comparison)	Precise control of the family-wise error rate for the three group situation	• Assumes normality and homogeneity of variances • Family-wise error rate deteriorates quickly with more than three groups
Fisher-Hayter test (pairwise post hoc comparison)	• More powerful than Tukey's HSD • Excellent control of family-wise error rate • Can be used with balanced or unbalanced designs	With unequal sample sizes, the Tukey-Kramer is more powerful in detecting the *largest* pairwise difference but is less powerful than the Fisher-Hayter in detecting *all* pairwise differences
Dunnett T3 (simple post hoc contrasts for unequal variances)	• Appropriate in the presence of heteroscedasticity • Recommended when $n < 50$	Not recommended for larger samples
Dunnett C tests (simple post hoc contrasts for unequal variances)	• Appropriate in the presence of heteroscedasticity • Recommended when $n > 50$	Not recommended for smaller samples
Games-Howell (simple post hoc contrasts for unequal variances)	• Appropriate in the presence of heteroscedasticity • Recommended when $n > 50$	Not recommended for smaller samples

Figure 2.2 is a flowchart to assist you in making decisions about which MCP to use. Not every statistician will agree with every decision on the flowchart as there is not total consensus about which MCP is appropriate in every single situation. Nonetheless, this is simply a guide. Whether you use it in its present form, or adapt it for your own needs, we hope you find the figure to be useful in your own research.

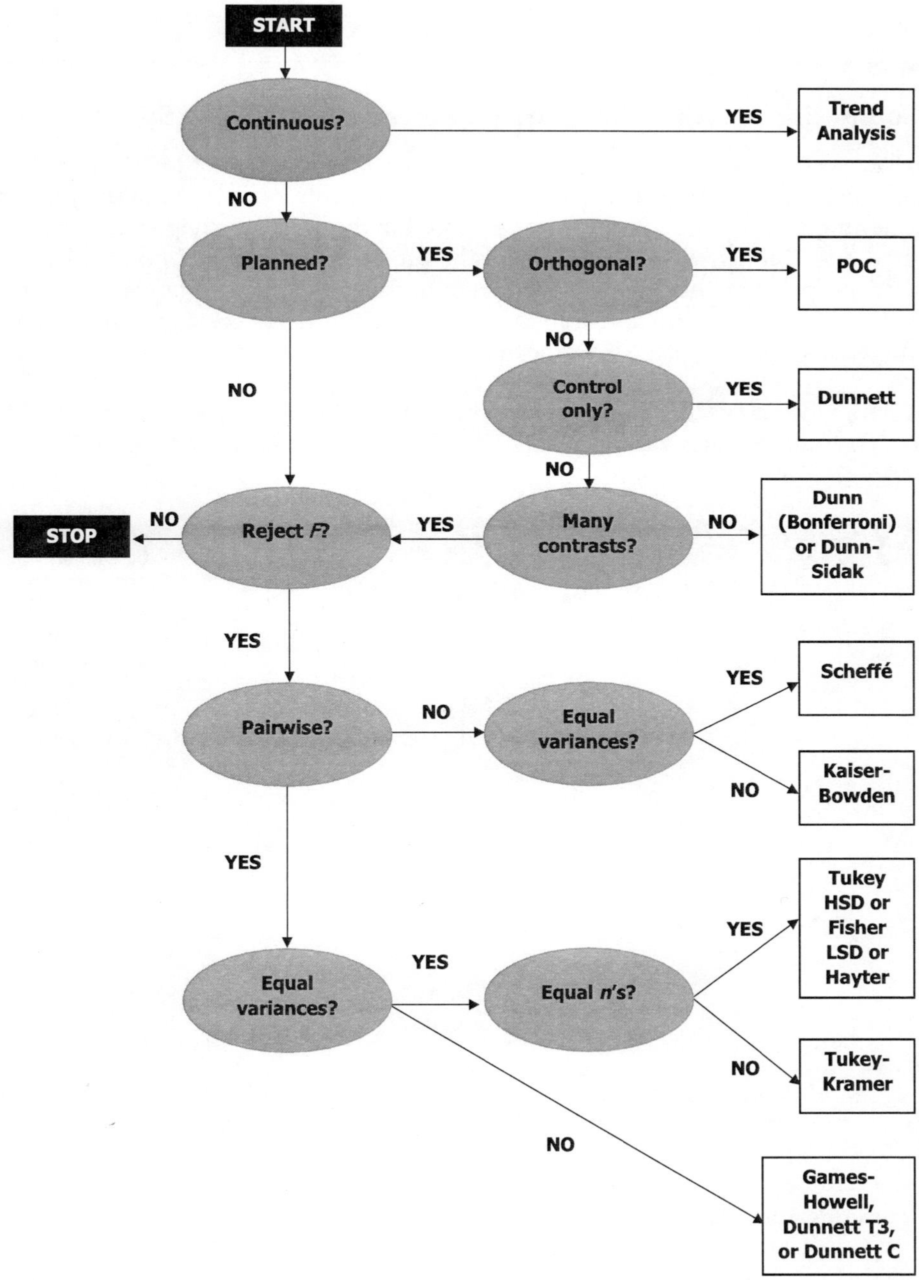

FIGURE 2.2
Flowchart of recommended MCPs.

At this point you should have met the following objectives: (a) be able to understand the concepts underlying the MCPs, (b) be able to select the appropriate MCP for a given research situation, and (c) be able to determine and interpret the results of MCPs. Chapter 3 returns to the analysis of variance again and discusses models for which there is more than one independent variable.

2.2 Computing Multiple Comparison Procedures Using SPSS

In our last section, we examined what SPSS has to offer in terms of MCPs. Here we use the GLM module (although the one-way ANOVA module can also be used). The steps for requesting a one-way ANOVA were presented in the previous chapter and will not be reiterated here. Rather, we will assume all the previously mentioned options have been selected. The last step, therefore, is selection of one or more planned (*a priori*) or post hoc MCPs. For purposes of this illustration, the Tukey will be selected. However, you are encouraged to examine other MCPs for this dataset.

Step 1. From the Univariate dialog box, click on "Post Hoc" to select various post hoc MCPs or click on "Contrasts" to select various planned MCPs (see the screenshot for Step 1, Figure 2.3).

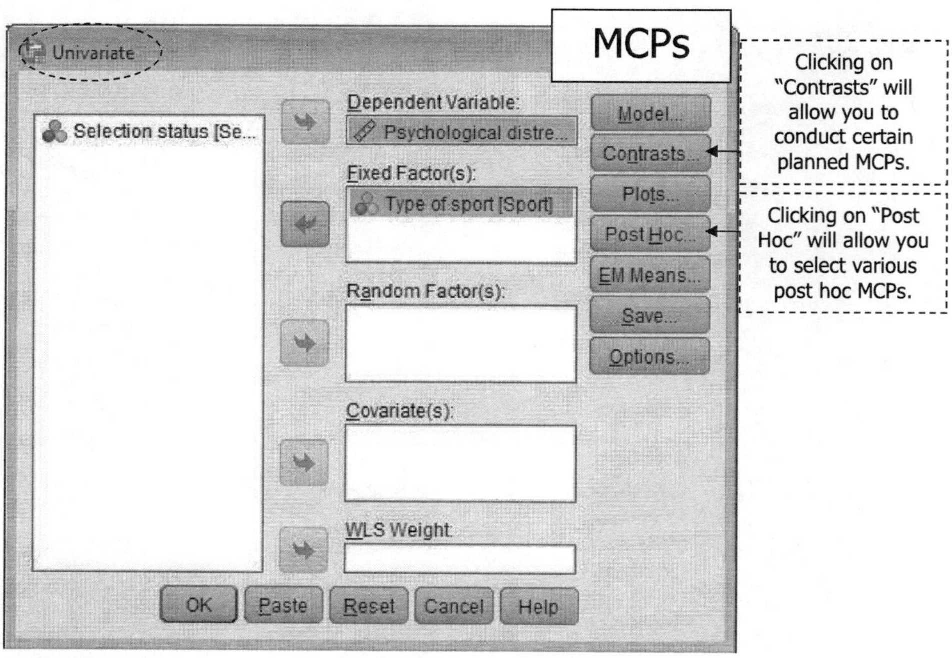

FIGURE 2.3
Multiple comparison procedure in SPSS.

Step 2 (Post hoc MCP). Click on the name of independent variable in the "Factor(s)" list box in the top left and move to the "Post Hoc Tests for" box in the top right by clicking on the arrow key. Check an appropriate MCP for your situation by placing a checkmark in the box next to the desired MCP. In this example, we will select "Tukey." Recall that SPSS operationalizes Tukey's HSD as *Tukey* within the ANOVA procedure. Click on "Continue" to return to the original dialog box. Click on "OK" to return to generate the output.

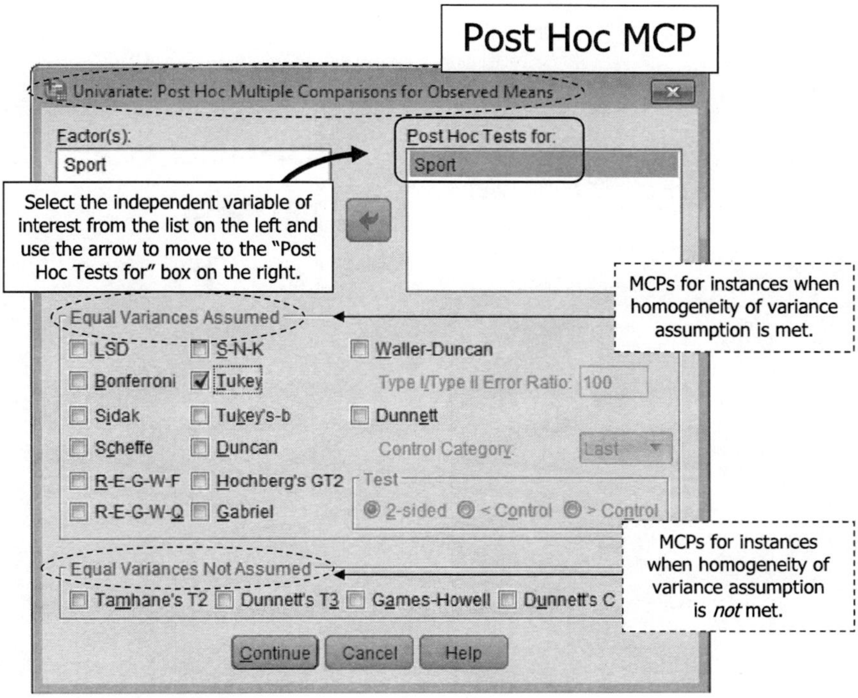

FIGURE 2.4
Post hoc MCP.

Step 3a (Planned MCP). To obtain trend analysis contrasts, click the "Contrasts" button from the Univariate dialog box (see the screenshot for Step 1, Figure 2.3). From the Contrasts dialog box, click the "Contrasts" pulldown and scroll down to "Polynomial."

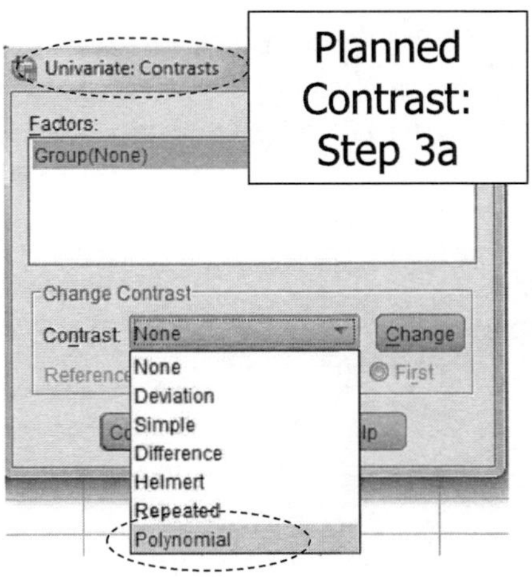

FIGURE 2.5
Planned contrast: Step 3a.

Step 3b. Click "Change" to select Polynomial and move it to be displayed in parentheses next to the independent variable. Recall that this type of contrast will allow testing of linear, quadratic, and cubic contrasts. Other specific planned contrasts are also available. Then click "Continue" to return to the Univariate dialog box.

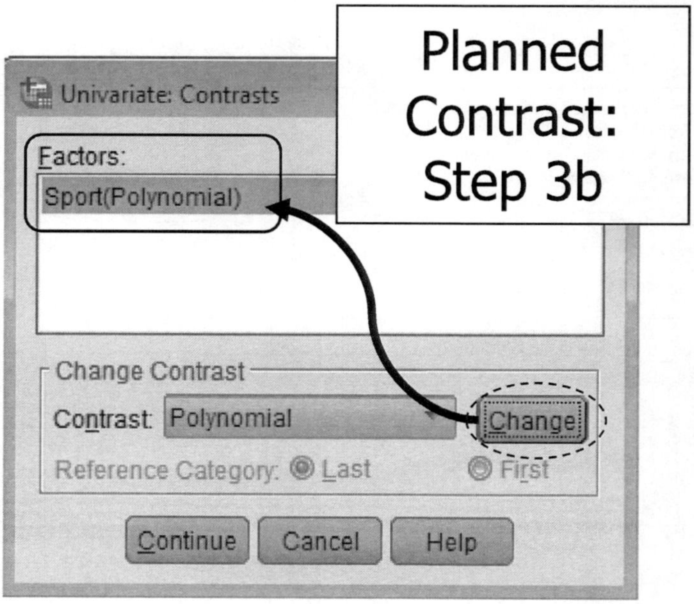

FIGURE 2.6
Planned contrast: Step 3b.

Interpreting the output. Annotated results from the Tukey HSD procedure, as one example of MCP, are shown in Table 2.13. Note that confidence intervals around a mean difference of zero are given to the right for each contrast.

TABLE 2.13
Tukey HSD SPSS Results for Psychological Distress Example

Descriptive Statistics

Dependent Variable: Psychological Distress

Type of Sport	Mean	Std. Deviation	N
Movement	11.1250	5.48862	8
Target	17.8750	5.93867	8
Fielding	20.2500	7.28501	8
Territory	24.3750	5.09727	8
Total	18.4062	7.51283	32

Recall the means of the groups as presented in the previous chapter.

TABLE 2.13 (continued)

Tukey HSD SPSS Results for Psychological Distress Example

Note that our selection of *Tukey* as the post hoc operationalizes to **Tukey's HSD**

"Mean difference" is simply the difference between the means of the two groups compared. For example, the mean difference of group 1 and group 2 is calculated as 11.1250 – 17.8750 = -6.7500

Multiple Comparisons

Dependent Variable: Psychological Distress

Tukey HSD

(I) Type of Sport	(J) Type of Sport	Mean Difference (I-J)	Std. Error	Sig.	95% Confidence Interval Lower Bound	Upper Bound
Movement	Target	–6.7500	3.00465	.135	–14.9536	1.4536
	Fielding	–9.1250*	3.00465	.025	–17.3286	–.9214
	Territory	–13.2500*	3.00465	.001	–21.4536	–5.0464
Target	Movement	6.7500	3.00465	.135	–1.4536	14.9536
	Fielding	–2.3750	3.00465	.858	–10.5786	5.8286
	Territory	–6.5000	3.00465	.158	–14.7036	1.7036
Fielding	Movement	9.1250*	3.00465	.025	.9214	17.3286
	Target	2.3750	3.00465	.858	–5.8286	10.5786
	Territory	–4.1250	3.00465	.526	–12.3286	4.0786
Territory	Movement	13.2500*	3.00465	.001	5.0464	21.4536
	Target	6.5000	3.00465	.158	–1.7036	14.7036
	Fielding	4.1250	3.00465	.526	–4.0786	12.3286

Based on observed means.

The error term is Mean Square(Error) = 36.112.

* The mean difference is significant at the .05 level.

The standard error calculated in SPSS uses the harmonic mean (Tukey-Kramer modification) where n_j and n_k are the sample sizes for the two groups whose means are being compared (Toothaker, 1993):

$$s_{\Psi'} = \sqrt{MS_{error}\left(\frac{1}{n_j}+\frac{1}{n_k}\right)}$$

$$s_{\Psi'} = \sqrt{36.112\left(\frac{1}{8}+\frac{1}{8}\right)} = \sqrt{6.04275} = 3.00466$$

"Sig." denotes the observed *p* value and provides the results of the contrasts. There are only two statistically significant contrasts. There is a statistically significant mean difference between: (1) group 1 (Movement) and group 3 (Fielding); and (2) between group 1 (Movement) and group 4 (Territory). Note that there are only 6 unique contrast results:

$$\tfrac{1}{2}[J(J-1)] = \tfrac{1}{2}[4\,(4-1)] = \tfrac{1}{2}(12) = 6.$$

However there are redundant results presented in the table. For example, the comparison of groups 1 and 2 (presented in results row 1) is the same as the comparison of groups 2 and 1 (presented in results row 2).

Toothaker, L. E. (1993). *Multiple comparison procedures.* Newbury Park, CA: Sage.

Homogeneous Subsets

Psychological Distress

Tukey HSD[a,b]

Type of Sport	N	Subset 1	Subset 2
Movement	8	11.1250	
Target	8	17.8750	17.8750
Fielding	8		20.2500
Territory	8		24.3750
Sig.		.135	.158

Means for groups in homogeneous subsets are displayed.

Based on observed means.

The error term is Mean Square(Error) = 36.112.

a. Uses Harmonic Mean Sample Size = 8.000.

b. Alpha = 0.05.

For each requested post hoc test that provides homogenous subset results, the groups are listed in order of *ascending* means. The means that are listed under each subset comprise a set of means that are ***not*** significantly different from each other. Movement is statistically different from fielding and territory as they do not appear in the same subset together.

NOTE: Tests available in the MCP table generally has better properties than the homogenous subset tests and are the preferred focus for post hoc analysis.

2.3 Computing Multiple Comparison Procedures Using R

Next we consider **R** for multiple comparison procedures. Note that the scripts are provided within the blocks with additional annotation to assist in understanding how the command works. Should you want to write reminder notes and annotation to yourself as you write the commands in **R** (and we highly encourage doing so), remember that any text that follows a hashtag (i.e., #) is annotation only and not part of the **R** script. Thus, you can write annotations directly into **R** with hashtags. We encourage this practice so that when you call up the commands in the future, you'll understand what the various lines of code are doing. You may think you'll remember what you did. However, trust us. There is a good chance that you won't. Thus, consider it best practice when using **R** to annotate heavily!

2.3.1 Reading Data Into R

```
getwd()
```

R is always pointed to a directory on your computer. The *get working directory* function can be used to determine which directory **R** is pointed. We will assume that we need to change the working directory, and will use the next line of code to set the working directory to the desired path.

```
setwd("E:/FolderName")
```

We use the *setwd* function to establish the working directory. To set the working directory, change what is in quotation marks to your file location. Also, if you are copying the directory name from your properties, you will need to change the backslash (i.e., \) to a forward slash (i.e., /).

```
Ch1_distress <- read.csv("Ch1_distress.csv")
```

The *read.csv* function reads our data into **R**. What's to the left of the "<-" will be what the data will be called in **R**. In this example, we're calling the **R** dataframe "Ch1_distress." What's to the right of the "<-" tells **R** to find this particular csv file. In this example, our file is called "Ch1_distress.csv." Make sure the extension (i.e., .csv) is included in your script. Also note that the name of your file should be in quotation marks within the parentheses.

```
names(Ch1_distress)
```

The *names* function will produce a list of variable names for each dataframe as follows. This is a good check to make sure your data have been read in correctly.

```
[1] "Sport" "Distress"
```

```
View(Ch1_distress)
```

The *View* function will let you view the dataset in spreadsheet format in RStudio.

FIGURE 2.7
Reading data into **R**.

```
Ch1_distress$SportF <- factor(Ch1_distress$Sport,
labels = c("movement", "target", "fielding", "territory"))
```

This script will create a new variable in our dataframe named "SportF." We use the *factor* function to define the variable *Sport* as nominal with the four groups defined here (i.e., movement, target, fielding, territory). What is to the left of "<-" in the script creates the new *SportF* variable in our dataframe.

```
summary(Ch1_distress)
```

The *summary* function will produce basic descriptive statistics on all the variables in your dataframe. This is a great way to quickly check to see if the data have been read in correctly and to get a feel for your data, if you haven't already. The output from the summary statement for this dataframe looks like this. Because we defined SportF as a factor, we are provided only the frequencies for each category in that variable.

```
    Sport            Distress            SportF
Min.   :1.00    Min.   : 3.00     movement :8
1st Qu.:1.75    1st Qu.:12.00     target   :8
Median :2.50    Median :20.00     fielding :8
Mean   :2.50    Mean   :18.41     territory:8
3rd Qu.:3.25    3rd Qu.:25.00
Max.   :4.00    Max.   :30.00
```

FIGURE 2.7 (continued)
Reading data into **R**.

2.3.2 Generating the One-Way ANOVA

```
Ch1_ANOVA <- aov(Distress ~ SportF, data=Ch1_distress)
```

The *aov* function will generate the one-way ANOVA model with "Distress" as the dependent variable and "SportF" as the independent variable. The dataframe from which we are pulling the data is defined by the *data* function. We are calling this object "Ch1_ANOVA."

FIGURE 2.8
Generating the one-way ANOVA.

2.3.3 Generating Tukey's Multiple Comparison Procedure

```
install.packages("multcomp")
```

The *install.packages* function will be used to install the *multcomp* package that is needed for the post hoc tests.

```
library(multcomp)
```

The *library* function will call up the package into our library.

```
PostHoc1<-glht(Ch1_ANOVA, linfct=mcp(SportF="Tukey"))
```

The *glht* function will generate Tukey's HSD post hoc analysis and name the object "PostHoc1." We could replace "Tukey" with "Dunnett" if we had wanted to run the Dunnett MCP rather than Tukey's.

```
summary(PostHoc1)
```

The *summary* function will output the results of the Tukey post hoc analysis from the previous command.

FIGURE 2.9
Generating Tukey's multiple comparison procedure.

```
         Simultaneous Tests for General Linear Hypotheses
Multiple Comparisons of Means: Tukey Contrasts

Fit: aov(formula = Distress ~ SportF, data = Ch1_distress)

Linear Hypotheses:
                         Estimate Std. Error t value  Pr(>|t|)
target-movement == 0       6.750     3.005   2.247     0.1356
fielding-movement == 0     9.125     3.005   3.037     0.0246
territory-movement == 0   13.250     3.005   4.410    <0.001
fielding-target == 0       2.375     3.005   0.790     0.8581
territory-target == 0      6.500     3.005   2.163     0.1585
territory-fielding == 0    4.125     3.005   1.373     0.5261
target-movement == 0

fielding-movement == 0 *
territory-movement == 0 ***
fielding-target == 0
territory-target == 0
territory-fielding == 0
---
Signif. codes: 0 '***' 0.001 '**' 0.01 '*' 0.05 '.' 0.1 ' ' 1
(Adjusted p values reported-single-step method)
```

As we already know from the *F* test, there were differences between at least some of the types of sport. Tukey's post hoc tells us there are statistically significant differences between movement and fielding and between movement and territory.

```
confint(PostHoc1)
```

The *confint* function will output confidence intervals of the post hoc results. The lower confidence limit is labeled "lwr" and the upper confidence interval is "upr." The confidence intervals that do not contain zero suggest statistically significant differences in the outcome between those groups.

```
Simultaneous Confidence Intervals
Multiple Comparisons of Means: Tukey Contrasts

Fit: aov(formula = Distress ~ SportF, data = Ch1_distress)

Quantile = 2.7325
95% family-wise confidence level

Linear Hypotheses:
                         Estimate lwr      upr
target-movement == 0       6.7500 -1.4601 14.9601
fielding-movement == 0     9.1250  0.9149 17.3351
territory-movement == 0   13.2500  5.0399 21.4601
fielding-target == 0       2.3750 -5.8351 10.5851
territory-target == 0      6.5000 -1.7101 14.7101
territory-fielding == 0    4.1250 -4.0851 12.3351
```

FIGURE 2.9 (continued)
Generating Tukey's multiple comparison procedure.

2.3.4 Generating Trend Analysis

```
contrasts(Ch1_distress$SportF)<-contr.poly(4)
```

To conduct a trend analysis, we use the *contr.poly* function. This defines "SportF" as the variable to conduct the contrast, using the Ch1_distress dataframe. This also defines four categories in that variable. Note that the categories need to be in ascending order in order to detect a meaningful trend. Our categories range from movement (1) to territory (4). Our categories are arguably not the best for trend analysis as they are not explicitly ordinal; however, for the sake of illustration, we will go with it!

```
Ch2trend <- aov(Distress ~ SportF, data=Ch1_distress)
```

We run our ANOVA model again.

```
summary.lm(Ch2trend)
```

Then we generate the output using the *summary.lm* function on the "Ch2trend" object. In the coefficient table, we see "Sport.L." The "L" refers to the linear trend. "Q" refers to the quadratic trend. "C" refers to the cubic trend. We see the only statistically significant trend is for the linear model.

```
Call:
aov(formula = Distress ~ Sport, data = Ch1_distress)

Residuals:
     Min       1Q    Median       3Q      Max
-10.2500  -4.5000   0.8125   4.2500   9.8750

Coefficients:
            Estimate Std. Error t value Pr(>|t|)
(Intercept)   18.406      1.062  17.327   < 2e-16 ***
Sport.L        9.419      2.125   4.433   0.00013 ***
Sport.Q       -1.313      2.125  -0.618   0.54172
Sport.C 1.370 2.125 0.645 0.52441
—
Signif. codes:
0 '***' 0.001 '**' 0.01 '*' 0.05 '.' 0.1 ' ' 1

Residual standard error: 6.009 on 28 degrees of freedom
Multiple R-squared: 0.4221, Adjusted R-squared: 0.3602
F-statistic: 6.818 on 3 and 28 DF, p-value: 0.001361
```

FIGURE 2.10
Generating trend analysis.

2.3.5 Generating Other MCPs

```
pairwise.t.test(Ch1_distress$Distress,
                Ch1_distress$Sport,
                paired = FALSE,
                p.adjust.method = "bonferroni")
```

Bonferonni and other methods can be computed using the pairwise *t* test function, *pairwise.t.test*, available in **R**. We first define the outcome (*Ch1_distress$Distress*) and independent variable (*Ch1_distress$Sport*). We indicate *paired* = *FALSE* as we do not actually have a matched sample—we want to conduct an independent *t* test. For the method, we will illustrate the Bonferroni; however, we could have used "BH" (Benjamin-Hochberg), "hochberg," "holm," "hommel," or "none" (the last of which is not recommended as it does not adjust the *p* value for multiple tests).

```
Pairwise comparisons using t tests with pooled SD

data: Ch1_distress$Distress and Ch1_distress$Sport

          movement target fielding
target    0.19645- -
fielding  0.03075  1.00000 -
territory 0.00083  0.23522 1.00000

P value adjustment method: bonferroni
```

The results provide all possible pairwise comparisons. The values represent *p* values for the cells where the variables intersect. We see the same results with Bonferonni as we did with Tukey's post hoc results. More specifically, *movement* is statistically different from *fielding* (p = .03075) and *territory* (p = .00083).

FIGURE 2.11
Generating other MCPs.

2.4 Research Question Template and Example Write-Up

In terms of an APA-style write-up, the MCP results for Tukey's HSD test for the statistics lab example are as follows.

Recall that our graduate research assistant, Ott, was working with one of the leading sports psychologists in the region, Dr. Rhodes. Dr. Rhodes is examining elite athletes and their vulnerability to psychological distress based on type of sport in which they participate. His research question was: *Is there a mean difference in psychological distress of elite athletes based on the type of sport in which they participate?* Ott then generated a one-way ANOVA as the test of inference. The APA-style example paragraph of results for the one-way ANOVA, prefaced by the extent to which the assumptions of the test were met, was presented in the previous chapter. Thus, only the results of the MCP (specifically the Tukey HSD) are presented here.

Post hoc analyses were conducted given the statistically significant omnibus ANOVA *F* test. Specifically, given the balanced design and equal variances, Tukey HSD multiple comparison tests were conducted on all possible pairwise contrasts. The following pairs of groups were found to be significantly different: Movement (M = 11.125, SD = 5.4886) and Fielding (M = 20.2500, SD = 7.2850) (p = .025); and Movement and Territory (M = 24.3750, SD = 5.0973) (p = .001). In other words, athletes who participate in sports related to Movement have statistically significantly lower psychological distress than athletes who participate in Fielding or Territory types of sports.

Problems

Conceptual Problems

1. True or false? The Tukey HSD procedure requires equal *n*'s and equal means.
2. Applying the Dunn procedure, given a nominal family-wise error rate of .10 and two contrasts, what is the per contrast alpha?
   a. .01
   b. .05
   c. .10
   d. .20
3. Which of the following linear combinations of population means is not a legitimate contrast?
   a. $\frac{(\mu_{.1}+\mu_{.2}+\mu_{.3})}{3}-\mu_{.4}$
   b. $\mu_{.1}-\mu_{.4}$
   c. $\frac{(\mu_{.1}+\mu_{.2})}{2}-(\mu_{.3}+\mu_{.4})$
   d. $\mu_{.1}-\mu_{.2}+\mu_{.3}-\mu_{.4}$
4. When a one-factor fixed-effects ANOVA results in a significant *F* ratio for $J = 2$, one should follow the ANOVA with which one of the following procedures?
   a. Tukey HSD method
   b. Scheffé method
   c. Fisher-Hayter method
   d. None of the above
5. If a family-based error rate for alpha is desired, and hypotheses involving all pairs of means are to be tested, which method of multiple comparisons should be selected?
   a. Tukey HSD
   b. Scheffé
   c. Planned Orthogonal Contrasts
   d. Trend analysis
   e. None of the above
6. *A priori* comparisons are which one of the following?
   a. Planned in advance of the research
   b. Often arise out of theory and prior research
   c. May be done without examining the *F* ratio
   d. All of the above
7. True or false? For planned contrasts involving the control group, the Dunn procedure is most appropriate.
8. Which is not a property of planned orthogonal contrasts?
   a. The contrasts are independent.
   b. The contrasts are post hoc.

c. The sum of the cross-products of the contrast coefficients = 0.

d. If there are J groups, there are $J - 1$ orthogonal contrasts.

9. Which multiple comparison procedure is most flexible in the contrasts that can be tested?

a. Planned orthogonal contrasts

b. Newman-Keuls

c. Dunnett

d. Tukey HSD

e. Scheffé

10. Post hoc tests are necessary after an ANOVA given which one of the following?

a. H_0 is rejected with two groups.

b. Fail to reject the null hypothesis and there are more than two groups.

c. H_0 is rejected and there are more than two groups.

d. You should always do post hoc tests after an ANOVA.

11. True or false? Post hoc tests are done after ANOVA to determine why H_0 was *not* rejected.

12. True or false? Holding the α level and the number of groups constant, as the df_{error} increases, the critical value of the q decreases.

13. True or false? The Tukey HSD procedure maintains the family-wise Type I error rate at α.

14. True or false? The Dunnett procedure assumes equal numbers of observations per group.

15. For complex post hoc contrasts with unequal group variances, which of the following MCPs is most appropriate?

a. Kaiser-Bowden

b. Dunnett

c. Tukey HSD

d. Scheffé

16. The number of levels of the independent variable is six. How many orthogonal contrasts can be tested?

a. 1

b. 3

c. 5

d. 6

17. A researcher is interested in testing the following contrasts in a $J = 6$ study: Group 1 vs. 2; Group 3 vs. 4; and Group 5 vs. 6. I assert that these contrasts are orthogonal. Am I correct?

18. I assert that rejecting H_0 in a one-factor fixed-effects ANOVA with $J = 3$ indicates that all three pairs of group means are necessarily statistically significantly different using the Scheffé procedure. Am I correct?

19. For complex post hoc contrasts with equal group variances, which of the following MCPs is most appropriate?

a. Planned orthogonal contrasts

b. Dunnett

c. Tukey HSD

d. Scheffé

20. A researcher finds a statistically significant omnibus F test. For which one of the following will there be at least one statistically significant MCP?

a. Kaiser-Bowden

b. Dunnett

c. Tukey HSD

d. Scheffé

21. Suppose all $J = 4$ of the sample means are equal to 100. I assert that it is possible to find a significant contrast with some MCP. Am I correct?

22. True or false? In contrast-based multiple comparison procedures, alpha is set for each individual contrast.

23. When alpha is established for a family of contrasts, which of the following does it represent?

a. Contrast based alpha

b. Per contrast alpha

c. Probability of making a Type I error for a particular contrast

d. Probability of making at least one Type I error in a set of contrasts

24. Contrasts can be divided into which two of the following types?

a. Contrast and complex

b. Nonpairwise and complex

c. Pairwise and nonpairwise

d. Simple and pairwise

25. A trend analysis is evaluated in terms of which one of the following?

a. Oblique higher-order terms

b. Orthogonal polynomials

c. Pairwise contrast

d. Simple contrast

26. MCPs that apply trend analysis are usually conducted sequentially in which order?

a. Cubic, linear, quadratic

b. Linear, quadratic, cubic

c. Linear, cubic, quadratic

d. Quadratic, cubic, linear

27. A researcher has conducted a one-way ANOVA with an independent variable with 5 categories. How many orthogonal contrasts can be tested?

a. 2

b. 3

c. 4

d. 5

Answers to Conceptual Problems

1. **False** (requires equal n = s and equal variances; we hope the means are different)
3. **c** (c is not legitimate as the contrast coefficients do not sum to 0)
5. **a** (see flowchart of MCPs in Figure 2.2)
7. **False** (use Dunnett procedure)
9. **e** (Scheffé is most flexible of all MCPs; can test simple and complex contrasts)
11. **False** (post hoc tests are conducted after ANOVA to determine why null *has* been rejected; post hoc tests are not needed when the null is not rejected)
13. **True** (see characteristics of Tukey HSD)
15. **a** (see Figure 2.2)
17. **Yes** (each contrast is orthogonal to the others as they rely on independent information)
19. **d** (see Figure 12.2)
21. **No** (with equal sample means, the numerator of any t will be zero; thus nothing can possibly be significant).
23. **d** (family-wise alpha represents the probability of making at least one Type I error in a set, or family, of contrasts)
25. **b** (trend analysis is defined in the form of orthogonal polynomials)
27. **c** (c (the number of orthogonal contrasts is one less than the number of levels or groups of the independent variable; in this case $J - 1 = 5 - 1 = 4$)

Computational Problems

1. A one-factor fixed-effects analysis of variance is performed on data for 10 groups of unequal sizes and H_0 is rejected at the .01 level of significance. Using the Scheffé procedure, test the following contrast:

$$\bar{Y}_{.2} - \bar{Y}_{.5} = 0$$

at the .01 level of significance given the following information: $df_{with} = 40$, $\bar{Y}_{.2} = 10.8$, $n_2 = 8$, $\bar{Y}_{.5} = 15.8$, $n_2 = 8$, and $MS_{with} = 4$.

2. A one-factor fixed-effects ANOVA is performed on data from three groups of equal size ($n = 10$) and H_0 is rejected at the .01 level. The following values were computed: $MS_{with} = 40$ and the sample means are $\bar{Y}_{.1} = 4.5$, $\bar{Y}_{.2} = 12.5$, and $\bar{Y}_{.3} = 13.0$. Use the Tukey HSD method to test all possible pairwise contrasts.
3. A one-factor fixed-effects ANOVA is performed on data from three groups of equal size ($n = 20$) and H_0 is rejected at the .05 level. The following values were computed: $MS_{with} = 60$ and the sample means are $\bar{Y}_{.1} = 50$, $\bar{Y}_{.2} = 70$, and $\bar{Y}_{.3} = 85$. Use the Tukey HSD method to test all possible pairwise contrasts.
4. Consider the situation where there are $J = 4$ groups of subjects. Answer the following questions:
   a. Construct a set of orthogonal contrasts and show that they are orthogonal.
   b. Is the following contrast legitimate? Why or why not?

   $$H_0: \mu_{.1} - (\mu_{.2} + \mu_{.3} + \mu_{.4})$$

   c. Using the same means, how might the contrast in part (b) be altered to yield a legitimate contrast?

5. Using the following data, conduct a one-factor fixed-effects ANOVA and perform Tukey's HSD using SPSS or **R**. Indicate which means are statistically significantly different based on Tukey's HSD.

Group	Outcome
1	10
1	13
1	12
1	11
1	10
2	15
2	16
2	14
2	17
2	16
3	17
3	18
3	16
3	17
3	16
4	21
4	22
4	20
4	21
4	22

6. Using the following data, conduct a one-factor fixed-effects ANOVA and perform Tukey's HSD using SPSS or **R**. Indicate which means are statistically significantly different based on Tukey's HSD.

Group	Outcome
1	36
1	45
1	32
1	57
1	46
1	60
1	23
1	32
1	60
1	45
2	57

2	47
2	32
2	42
2	42
2	53
2	60
2	33
2	64
2	37
3	23
3	61
3	58
3	52
3	28
3	52
3	43
3	64
3	47
3	62

Selected Answers to Computational Problems

1. Contrast = –5; standard error = 1; t = –5; critical values are 5.10 and –5.10; fail to reject.
3. Standard error = $\sqrt{60/20} = \sqrt{3} = 1.7321$
   - $q_1 = (85-50)/1.7321 = 20.2073$
   - $q_2 = (85-70)/1.7321 = 8.6603$
   - $q_3 = (70-50)/1.7321 = 11.5470$
   - Critical values approximately 3.39 and –3.39; all contrasts are statistically significant.
5. Based on the one-factor fixed-effects ANOVA and Tukey's HSD (see the following table), there are statistically significant mean differences between the following groups:
   a. Group 1 and Group 2
   b. Group 1 and Group 3
   c. Group 1 and Group 4
   d. Group 2 and Group 4
   e. Group 3 and Group 4

Multiple Comparisons

Dependent Variable: outcome
Tukey HSD

(I) group	(J) group	Mean Difference (I-J)	Std. Error	Sig.	95% Confidence Interval Lower Bound	95% Confidence Interval Upper Bound
1.00	2.00	**−4.4000***	**.66332**	**.000**	−6.2978	−2.5022
	3.00	**−5.6000***	**.66332**	**.000**	−7.4978	−3.7022
	4.00	**−10.0000***	**.66332**	**.000**	−11.8978	−8.1022
2.00	1.00	4.4000*	.66332	.000	2.5022	6.2978
	3.00	−1.2000	.66332	.305	−3.0978	.6978
	4.00	**−5.6000***	**.66332**	**.000**	−7.4978	−3.7022
3.00	1.00	5.6000*	.66332	.000	3.7022	7.4978
	2.00	1.2000	.66332	.305	−.6978	3.0978
	4.00	**−4.4000***	**.66332**	**.000**	−6.2978	−2.5022
4.00	1.00	10.0000*	.66332	.000	8.1022	11.8978
	2.00	5.6000*	.66332	.000	3.7022	7.4978
	3.00	4.4000*	.66332	.000	2.5022	6.2978

Based on observed means.
The error term is Mean Square(Error) = 1.100.
* The mean difference is significant at the 0.05 level.

6. Based on the one-factor fixed-effects ANOVA and Tukey's HSD (see the following table), there are statistically significant mean differences between *none* of the groups.

Multiple Comparisons

Dependent Variable: Outcome
Tukey HSD

(I) Group	(J) Group	Mean Difference (I-J)	Std. Error	Sig.	95% Confidence Interval Lower Bound	95% Confidence Interval Upper Bound
1.00	2.00	−3.1000	5.73262	.852	−17.3136	11.1136
	3.00	−5.4000	5.73262	.619	−19.6136	8.8136
2.00	1.00	3.1000	5.73262	.852	−11.1136	17.3136
	3.00	−2.3000	5.73262	.915	−16.5136	11.9136
3.00	1.00	5.4000	5.73262	.619	−8.8136	19.6136
	2.00	2.3000	5.73262	.915	−11.9136	16.5136

Based on observed means.
The error term is Mean Square(Error) = 164.315.

Interpretive Problems

1. For the interpretive problem you selected in Chapter 1 (using the survey1 dataset accessible from the website), select an *a priori* MCP, apply it using SPSS, and write an APA-style paragraph describing the results.
2. For the interpretive problem you selected in Chapter 1 (using the survey1 dataset accessible from the website), select a *post hoc* MCP, apply it using SPSS, and write an APA-style paragraph describing the results.
3. For the interpretive problem you selected in Chapter 1 (using the IPEDS2017 dataset accessible from the website), select a *post hoc* MCP, apply it using SPSS, and write an APA-style paragraph describing the results.

3

Factorial Analysis of Variance—Fixed-Effects Model

Chapter Outline

Key Concepts

1. Main effects
2. Interaction effects
3. Partitioning the sums of squares
4. The ANOVA model
5. Main effects contrasts, simple and complex interaction contrasts
6. Nonorthogonal designs

The last two chapters have dealt with the one-factor analysis of variance (ANOVA) model and various multiple comparison procedures (MCPs) for that model. In this chapter, we continue our discussion of analysis of variance models by extending the one-factor case to the two- and three-factor models. This chapter seeks an answer to the question: What should we do if there are multiple factors for which we want to make comparisons of the means? In other words, the researcher is interested in the effect of two or more independent variables or factors on the dependent (or criterion) variable. This chapter is most concerned with two- and three-factor models, but the extension to more than three factors, when warranted, is fairly simple.

For example, suppose that a researcher is interested in the effects of textbook choice and time of day on statistics achievement. Thus, one independent variable would be the textbook selected for the course, and the second independent variable would be the time of day the course was offered. The researcher hypothesizes that certain texts may be more effective in terms of achievement than others, and that student learning may be greater at certain times of the day. For the time-of-day variable, one might expect that students would not do as well in an early morning section or a late evening section than at other times of the day. In the example study, say that the researcher is interested in comparing three textbooks (A, B, and C) and three times of the day (early morning, mid-afternoon, and evening sections). Students would be randomly assigned to sections of statistics based on a combination of textbook and time of day. One group of students might be assigned to the section offered in the evening using textbook A. These results would be of interest to statistics instructors for selecting a textbook and optimal time of the day. This is just one example, but it should allow you to see how multiple independent variables can be applied within one model.

Most of the concepts used in this chapter are the same as those covered in Chapters 1 and 2. In addition, new concepts include main effects, interaction effects, multiple comparison procedures for main and interaction effects, and nonorthogonal designs. Our objectives are that by the end of this chapter, you will be able to (a) understand the characteristics and concepts underlying factorial ANOVA, (b) determine and interpret the results of factorial ANOVA, and (c) understand and evaluate the assumptions of factorial ANOVA.

3.1 What Two-Factor ANOVA Is and How It Works

Our very talented group of graduate students has been performing amazing statistical feats that have garnered rave reviews from those with which they have worked. We now find Ott Lier assisting one of the region's leading sports psychologists in examining elite athletes and vulnerability to psychological distress following selection procedures and player status (selection or deselection to remain on their team). Our graduate student team successfully analyzed the data (as we saw in a previous chapter) using one-way ANOVA to answer one research question using this data. As we will see in this chapter, Ott will be extending analysis to include an additional independent variable.

The research lab has been contracted to work with one of the leading sports psychologists in the region, Dr. Rhodes. Ott Lier, one of our very capable graduate students, has the pleasure of being selected to work with Dr. Rhodes. Dr. Rhodes is examining elite athletes and their vulnerability to psychological distress after selection procedures in which athletes are either selected or deselected for their team. Dr. Rhodes wants to determine if there is a difference in psychological stress based on type of sport (movement, target, fielding, or territory) and selection status (selected or deselected). Ott suggests the following research question is: *Is there a mean difference in psychological distress of elite athletes based on the type of sport and selection status?* With two independent variables, Ott determines that a factorial ANOVA is the best statistical procedure to use to answer Dr. Rhodes's question. His next task is to collect and analyze the data to address this research question.

This section describes the distinguishing characteristics of the two-factor ANOVA model, the layout of the data, the linear model, main effects and interactions, assumptions of the model and their violation, partitioning the sums of squares, the ANOVA summary table, multiple comparison procedures, effect size measures, confidence intervals, power, an example, and expected mean squares.

3.1.1 Characteristics

The first characteristic of the two-factor ANOVA model should be obvious by now; this model considers the effect of *two factors or independent variables* on one dependent variable. Each factor consists of two or more levels (or categories). This yields what we call a **factorial design** because more than a single factor is included. We see then that the two-factor ANOVA is an extension of the one-factor ANOVA. Why would a researcher want to complicate things by considering a second factor? Three reasons come to mind. First, the researcher may have a genuine interest in studying the second factor and, more specifically, how the second factor operates on the outcome in the presence of another factor. Rather than studying each factor separately in two analyses, the researcher includes both factors in the same analysis. This allows a test not only of the effect of each individual factor, known as **main effects**, but of the effect of both factors *collectively*. This latter effect is known as an **interaction effect** and provides information about whether the two factors are operating independent of one another (i.e., no interaction exists) or whether the two factors are operating *together* to produce some additional impact (i.e., an interaction exists). If two separate analyses were conducted, one for each independent variable, no information would be obtained about the interaction effect. As becomes evident, assuming a factorial ANOVA with two independent variables, the researcher will test three hypotheses: one for each factor or main effect individually and a third for the interaction between the factors. *Factorial ANOVA models with more than two independent variables will, accordingly, test for additional main effects and interactions*. This chapter spends considerable time discussing interactions.

A second reason for including an additional factor is an attempt to **reduce the error (or within-groups) variation**, which is variation that is unexplained by the first factor. The use of a second factor provides a more precise estimate of error variance. *For this reason, a two-factor design is generally more powerful than two one-factor designs, as the second factor and the interaction serve to control for additional extraneous variability*.

A third reason for considering two factors simultaneously is to provide *greater generalizability* of the results and to provide a more efficient and economical use of observations and resources. Thus the results can be generalized to more situations, and the study will be more cost efficient in terms of time and money.

For the two-factor ANOVA, every level of the first factor (hereafter known as factor A) is paired with every level of the second factor (hereafter known as factor B). In other words, *every combination of factors A and B is included in the design of the study*, yielding what is referred to as a **fully crossed design**. If some combinations are not included, then the design is not fully crossed and may form some sort of a nested design (see Chapter 6). Units (e.g., individuals or objects) are randomly assigned to one combination of the two factors. In other words, each individual responds to only one combination of the factors. If individuals respond to more than one combination of the factors, this would be some sort of repeated measures design, which we examine in Chapter 5. In this chapter we consider only models where all factors are fixed. Thus, the overall design is known as a **fixed-effects model**. If one or both factors are random, then the design is not a fixed-effects model, which we discuss in Chapter 5. It is also a condition for factorial ANOVA that the dependent variable is measured at least at the interval level and the independent variables are categorical (either nominal or ordinal).

In this section of the chapter, for simplicity's sake, we impose the restriction that the number of observations is the same for each factor combination (i.e., equal or balanced n's). This yields what is known as an **orthogonal design**, where the effects due to the factors (separately and collectively) are independent or unrelated. We leave the discussion of nonorthogonal (i.e., unequal n's or unbalanced) factorial ANOVA until later in this chapter. In addition, there must be at least two observations per factor combination so as to have within-groups variation.

In summary, the characteristics of the two-factor analysis of variance fixed-effects model are as follows: (a) two independent variables (both of which are categorical) each with two or more levels, (b) the levels of both independent variables are fixed by the researcher, (c) subjects are randomly assigned to only one combination of these levels, (d) the two factors are fully crossed, and (e) the dependent variable is measured at least at the interval level. In the context of experimental design, the two-factor analysis of variance is often referred to as the **completely randomized factorial design**.

3.1.1.1 The Layout of the Data

Before we get into the theory and analysis of the data, let us examine one form in which the data can be placed, known as the layout of the data. We designate each observation as Y_{ijk}, where the j subscript tells us what level (or category) of factor A (i.e., independent variable 1) the observation belongs to, the k subscript tells us what level of factor B (i.e., independent variable 2) the observation belongs to, and the i subscript tells us the observation or identification number within that combination of factor A and factor B. For instance, Y_{321}

would mean that this is the third observation in the second level of factor A and the first level of factor B. The first subscript ranges over $i = 1, \ldots, n$, the second subscript ranges over $j = 1, \ldots, J$, and the third subscript ranges over $k = 1, \ldots, K$. Note also that the latter two subscripts denote the cell of an observation. Using the same example, we are referring to the third observation in the 21 cell. Thus, there are J levels of factor A, K levels of factor B, and n subjects in each cell, for a total of $JKn = N$ observations. For now, we consider the case where there are n subjects in each cell in order to simplify matters; this is referred to as the equal n's case. Later in this chapter we consider the unequal n's case.

The layout of the sample data is shown in Table 3.1. Here we see that each row represents the observations for a particular level of factor A (independent variable 1), and that each

TABLE 3.1

Layout for the Two-Factor ANOVA

	Level of Factor B				
Level of Factor A	**1**	**2**	**...**	***K***	**Row Mean**
1	Y_{111}	Y_{112}	...	Y_{11K}	$\bar{Y}_{.1.}$
	.	.	...		
	.	.	...		
	.	.	...		
	Y_{n11}	Y_{n12}	...	Y_{n1K}	
	—	—		—	
	$\bar{Y}_{.11}$	$\bar{Y}_{.12}$	...	$\bar{Y}_{.1K}$	
2	Y_{121}	Y_{122}	...	Y_{12K}	$\bar{Y}_{.2.}$
	.	.	...	.	
	.	.	...	.	
	.	.	...	.	
	Y_{n21}	Y_{n22}	...	Y_{n2K}	
	—	—		—	
	$\bar{Y}_{.21}$	$\bar{Y}_{.22}$	...	$\bar{Y}_{.2K}$	
.	.	.	...	.	.
	Y_{1J1}	Y_{1J2}	...	Y_{1JK}	
J	.	.	...	.	$\bar{Y}_{.J.}$
	.	.	...	.	
	Y_{nJ1}	Y_{nJ2}	...	Y_{nJK}	
	—	—		—	
	$\bar{Y}_{.J1}$	$\bar{Y}_{.J2}$	...	$\bar{Y}_{.JK}$	
Column Mean	$\bar{Y}_{..1}$	$\bar{Y}_{..2}$		$\bar{Y}_{..K}$	$\bar{Y}_{...}$

column represents the observations for a particular level of factor B (independent variable 2). At the bottom of each column are the column means $\left(\bar{Y}_{..k}\right)$, to the right of each row are the row means $\left(\bar{Y}_{.j.}\right)$, and in the lower right-hand corner is the overall mean $\left(\bar{Y}_{...}\right)$. We also need the cell means $\left(\bar{Y}_{.jk}\right)$, which are shown at the bottom of each cell. Thus, the layout is one form in which to think about the data.

3.1.1.2 The ANOVA Model

This section introduces the analysis of variance linear model, as well as estimation of the parameters of the model. The two-factor analysis of variance model is a form of the **general linear model,** like the one-factor ANOVA model of Chapter 1. The two-factor ANOVA fixed-effects model can be written in terms of **population parameters** as follows:

$$Y_{ijk} = \mu + \alpha_j + \beta_k + (\alpha\beta)_{jk} + \varepsilon_{ijk}$$

where Y_{ijk} is the observed score on the criterion (i.e., dependent variable) variable for individual i in level j of factor A (i.e., independent variable 1) and level k of factor B (i.e., independent variable 2) (or in the jk cell), μ is the overall or grand population mean (i.e., regardless of cell designation), α_j is the main effect for level j of factor A (row or effect of independent variable 1), β_k is the main effect for level k of factor B (column or effect of independent variable 2), $(\alpha\beta)_{jk}$ is the interaction effect for the combination of level j of factor A and level k of factor B, and ε_{ijk} is the random residual error for individual i in cell jk. The residual error can be due to individual differences, measurement error, and/or other factors not under investigation.

The population effects and residual error can be computed as follows:

$$\alpha_j = \mu_{.j.} - \mu$$

$$\beta_k = \mu_{..k} - \mu$$

$$(\alpha\beta)_{jk} = \mu_{.jk} - \left(\mu_{.j.} + \mu_{..k} - \mu\right)$$

$$\varepsilon_{ijk} = Y_{ijk} - \mu_{.jk}$$

That is, the **row effect,** α_j, is equal to the difference between the population mean of level j of factor A (i.e., one particular group or category of independent variable 1, $\mu_{.j.}$) and the overall population mean, μ. The **column effect,** β_k, is equal to the difference between the population mean of level k of factor B (i.e., one particular group or category of independent variable 2, $\mu_{..k}$) and the overall population mean, μ. The **interaction effect,** $(\alpha\beta)_{jk}$, is the difference between the population cell mean $\left(\mu_{.jk}\right)$ and the sum of the population mean of level j of factor A (i.e., one particular group or category of independent variable 1, $\mu_{.j.}$) and the population mean of level k of factor B (i.e., one particular group or category of independent variable 2, $\mu_{..k}$) subtracted from the overall population mean, μ. The **residual error,** ε_{ijk}, is equal to the difference between an individual's observed score, Y_{ijk}, and the population mean of cell jk, $\mu_{.jk}$.

The row, column, and interaction effects can also be thought of as the average effect of being a member of a particular row (i.e., a unit assigned to group or category A, B, or C of independent variable 1), *column* (i.e., a unit assigned to group or category X, Y, or Z of independent variable 2), *or cell* (e.g., a unit assigned to group A of independent variable 1 and group Y of independent variable 2), respectively. It should also be noted that the sum of the row effects is equal to zero, the sum of the column effects is equal to zero, and the sum of the interaction effects is equal to zero (both across rows and across columns). This implies, for example, that if there are any *nonzero* row effects, then the row effects will balance out around zero with some positive and some negative effects. Likewise for column and interaction effects.

You may be wondering why the interaction effect looks a little different from the main effects. We have given you the version that is solely a function of population means. A more intuitively convincing **conceptual version of the interaction effect** is as follows:

$$(\alpha\beta)_{jk} = \mu_{.jk} - \alpha_j - \beta_k - \mu$$

which is written in similar fashion to the row and column effects. Here we see that the interaction effect $\left[(\alpha\beta)_{jk}\right]$ is equal to the population cell mean $(\mu_{.jk})$ minus the following: (a) the row effect, (α_j); (b) the column effect, (β_k); and (c) the overall population mean, (μ). *In other words, the interaction is solely a function of cell means without regard to, or controlling for, its row effect, column effect, or the overall mean.*

To estimate the parameters of the model [μ, α_j, β_k, $(\alpha\beta)_{jk}$, and ε_{ijk}], the least squares method of estimation is used as the most appropriate for general linear models (e.g., regression, ANOVA). These sample estimates are represented by $\bar{Y}_{...}$, a_j, b_j, $(ab)_{jk}$, and e_{ijk}, respectively, where the latter four are computed as follows, respectively:

$$a_j = \bar{Y}_{.j.} - \bar{Y}_{...}$$
$$b_k = \bar{Y}_{..k} - \bar{Y}_{...}$$
$$(ab)_{jk} = \bar{Y}_{.jk} - \left(\bar{Y}_{.j.} + \bar{Y}_{..k} - \bar{Y}_{...}\right)$$
$$e_{ijk} = Y_{ijk} - \bar{Y}_{.jk}$$

Note that $\bar{Y}_{...}$ represents the overall sample mean, $\bar{Y}_{.j.}$ represents the sample mean for level *j* of factor A (independent variable 1), $\bar{Y}_{..k}$ represents the sample mean for level *k* of factor B (independent variable 2), and $\bar{Y}_{.jk}$ represents the sample mean for cell *jk* (the interaction of factor A and factor B).

For the two-factor ANOVA model, there are three sets of hypotheses, one for each of the main effects, and one for the interaction effect. The null and alternative hypotheses, respectively, for testing the main effect of factor A (independent variable 1) are as follows:

$$H_{01}\text{: } \mu_{.1.} = \mu_{.2.} = \ldots \mu_{.J.}$$
$$H_{11}\text{: not all the } \mu_{.j.} \text{ are equal}$$

The hypotheses for testing the main effect of factor B (independent variable 2) are noted as:

$$H_{02}\text{: } \mu_{..1} = \mu_{..2} = \ldots = \mu_{..K}$$
$$H_{12}\text{: not all the } \mu_{..k} \text{ are equal}$$

Finally, the hypotheses for testing the interaction effect (independent variable 1 with independent variable 2) are as follows:

$$H_{03}\text{: } \left(\mu_{.jk} - \mu_{.j.} - \mu_{..k} + \mu\right) = 0 \text{ for all } j \text{ and } k$$

$$H_{13}\text{: not all the} \left(\mu_{.jk} - \mu_{.j.} - \mu_{..k} + \mu\right) \text{ are equal}$$

The null hypotheses can also be written in terms of row, column and interaction effects (which may make more intuitive sense to you) as follows:

$$H_{01}\text{: } \alpha_1 = \alpha_2 = \ldots = \alpha_J = 0$$
$$H_{02}\text{: } \beta_1 = \beta_2 = \ldots = \beta_K = 0$$
$$H_{03}\text{: } (\alpha\beta)_{jk} = 0 \text{ for all } j \text{ and } k$$

As in the one-factor model, all of the alternative hypotheses are written in a general form to cover the multitude of possible mean differences that could arise. These range from only two of the means being different to all of the means being different from one another. Also, because of the way the alternative hypotheses have been written, only a nondirectional alternative is appropriate. If one of the null hypotheses is rejected, then consider a multiple comparison procedure so as to determine which means, or combination of means, are significantly different (this is discussed later).

3.1.1.3 Main Effects and Interaction Effects

Finally, we come to a formal discussion of main effects and interaction effects. A **main effect** of factor A (independent variable 1) is defined as the effect of factor A, averaged across the levels of factor B (independent variable 2), on the dependent variable Y. More precisely, it represents the unique effect of factor A on the outcome Y, controlling statistically for factor B. A similar statement may be made for the main effect of factor B.

As far as the concept of interaction is concerned, things are a bit more complex. An **interaction** can be defined in any of the following ways: An interaction is said to exist if (a) certain combinations of the two factors produce effects *beyond* the effects of the two factors when those two factors are considered separately; (b) the mean differences among the levels of factor A are not constant across, and thus depend on, the levels of factor B; (c) there is *a joint effect* of factors A and B on Y; or (d) there is a *unique effect* that could not be predicted from knowledge of only the main effects.

Let us mention two fairly common examples of interaction effects. The first is known as an aptitude-treatment interaction (ATI). This means that the effectiveness of a particular treatment depends on the aptitude of the individual. In other words, some treatments are more effective for individuals with a high aptitude, and other treatments are more effective for those with a low aptitude. A second example is an interaction between treatment and sex. Here some treatments may be more effective for males and others may be more effective for females. This is often considered in gender studies research.

For some graphical examples of main and interaction effects, take a look at the various plots in Figure 3.1. Each plot represents the graph of a particular set of *cell means* (the mean of the dependent variable for a cell—the combination of a particular category of factor A and a particular category of factor B), sometimes referred to as a **profile plot**. On the **X axis** are the levels of factor A, the **Y axis** provides the cell means on the dependent variable Y, and the **separate lines** in the body of the plot represent the levels of factor B (although the specific placement of the two factors here is arbitrary; alternatively factor B could be plotted on the X axis and factor A as the separate lines). *Profile plots provide information about the possible existence of a main effect for A, a main effect for B, and/or an interaction effect.* A *main effect for factor A* can be examined by taking the means for each level of A and averaging them across the levels of B. If these marginal means for the levels of A are the same or nearly so, this would indicate no main effect for factor A. A *main effect for factor B* can be assessed by taking the means for each level of B and averaging them across the levels of A. If these marginal means for the levels of B are the same or nearly so, this would imply no main effect for factor B. An *interaction effect* is determined by whether the cell means for the levels of A are constant across the levels of B (or vice versa). This is easily viewed in a profile plot by checking to see whether or not the lines are parallel. *Parallel lines indicate no interaction, whereas nonparallel lines suggest that an interaction may exist.* Of course, the statistical significance of the main and interaction effects is a matter to be determined by the F test statistics (which we will soon learn). The profile plots give you only a rough idea as to the possible existence of the effects. For instance, lines that are nearly parallel will probably not show up as a significant interaction. It is suggested that the plot can be simplified if the factor with the most levels is shown on the X axis. This cuts down on the number of lines drawn.

The plots shown in Figure 3.1 represent the eight different sets of results possible for a two-factor design, that is, from no effects to all three effects being evident. To simplify matters, only two levels of each factor are used. Figure 3.1a indicates that there is no main effect for either factor A or B, and there is no interaction effect. The lines are horizontal (no A effect), lie nearly on top of one another (no B effect), and are parallel (no interaction effect). Figure 3.1b suggests the presence of an effect due to factor A only (the lines are not horizontal because the mean for A_1 is greater than the mean for A_2), but the lines are nearly on top of one another (no B effect) and are parallel (no interaction). In Figure 3.1c we see a separation between the lines for the levels of B (B_1 being greater than B_2); thus a main effect for B is likely, but the lines are horizontal (no A effect), and are parallel (no interaction).

For Figure 3.1d there are no main effects (the means for the levels of A are the same, and the means for the levels of B are the same), but an interaction is indicated by the lack of parallel lines. Figure 3.1e suggests a main effect for both factors, as shown by mean differences (A_1 less than A_2, and B_1 greater than B_2), but no interaction (the lines are parallel). In Figure 3.1f we see a main effect for A (A_1 less than A_2) and an interaction effect, but no main effect for B (little separation between the lines for factor B). For Figure 3.1g there appears to be a main effect for B (B_1 greater than B_2) and an interaction, but no main effect for A. Finally, in Figure 3.1h we see the likelihood of two main effects (A_1 less than A_2, and B_1 greater than B_2), and an interaction. Although these are clearly the only possible outcomes from a two-factor design, the precise pattern will differ depending on the obtained cell means. In other words, if your study yields a significant effect only for factor A, your profile plot need not look exactly like Figure 3.1b, but it will retain the same general pattern and interpretation.

In many statistics texts, a big deal is made about the type of interaction shown in the profile plot. They make a distinction between an ordinal interaction and a disordinal interaction. An **ordinal interaction** is said to exist when the lines are not parallel and they do not cross; ordinal here means the same relative order of the cell means is maintained across the levels of one of the factors. For example, the means for level 1 of factor B are always greater than the means for level 2 of B, regardless of the level of factor A. A **disordinal interaction** is said to exist when the lines are not parallel and they do cross. For example, the mean for B_1 is greater than the mean for B_2 at A_1, but the opposite is true at A_2. Dwelling on the distinction between the two types of interaction is not recommended as it can depend on how the plot is drawn (i.e., which factor is plotted on the X axis). That is, when factor A is plotted on the X axis, a disordinal interaction may be shown, and when factor B is plotted on the X axis, an ordinal interaction may be shown. The purpose of the profile plot is to simplify interpretation of the results; worrying about the type of interaction may merely serve to confuse that interpretation.

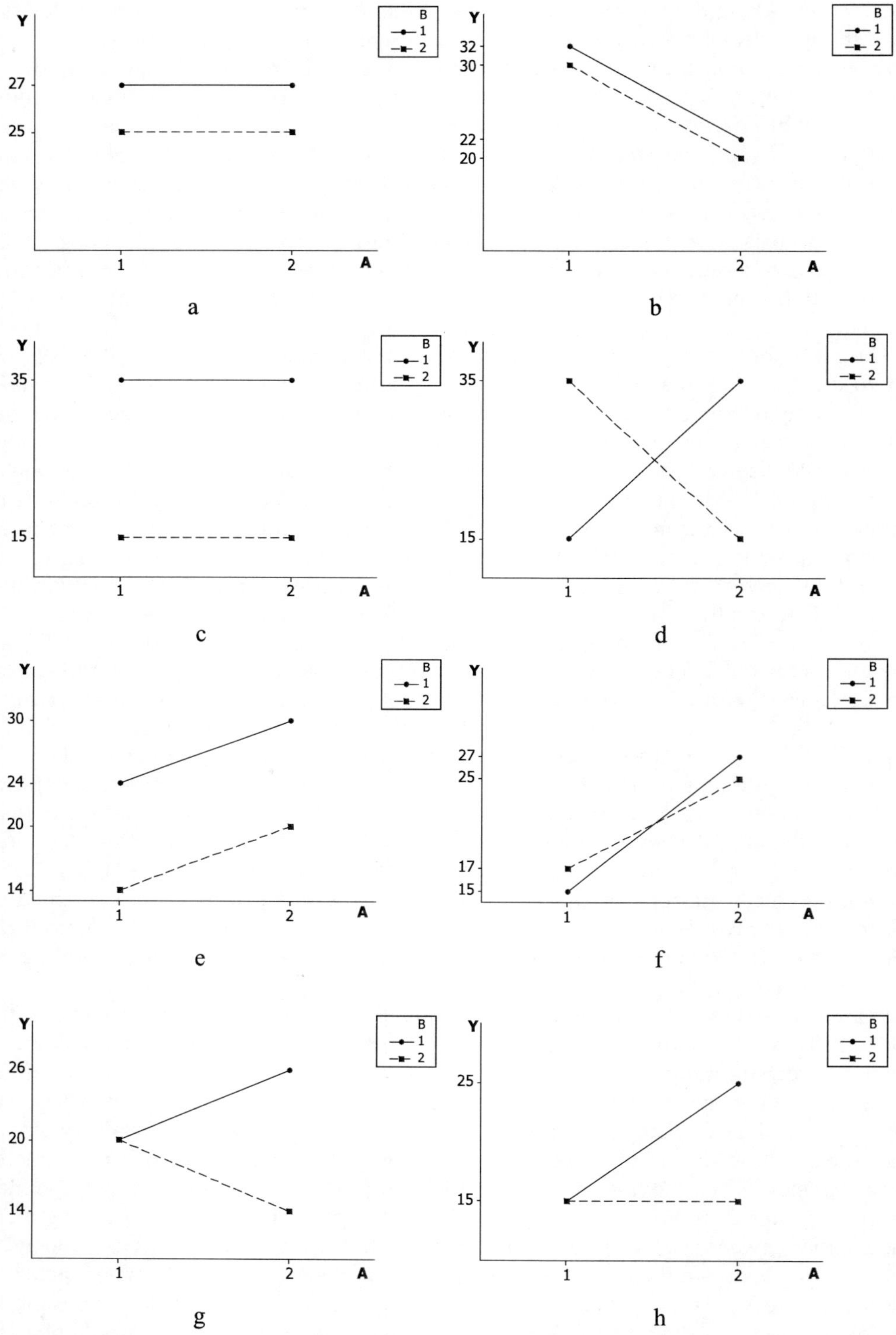

FIGURE 3.1
Display of possible two-factor ANOVA effects.

Let us take a moment to discuss how to deal with an interaction effect. Consider two possible situations, one where there is a significant interaction effect and one where there is no such effect. If there is *no* significant interaction effect, then the findings regarding the main effects can be generalized with greater confidence. In this situation, the main effects are known as **additive effects**, and an additive linear model with no interaction term could actually be used to describe the data. For example, the results might be that for factor A, the level 1 means always exceed those of level 2 by 10 points, across all levels of factor B. *Thus, we can make a blanket statement about the constant added benefits of* A_1 *over* A_2*, regardless of the level of factor B*. In addition, for the no-interaction situation, the main effects are statistically independent of one another; that is, *each of the main effects serves as an independent predictor* of Y.

If there *is* a significant interaction effect, then the findings regarding the main effects cannot be generalized with such confidence. In this situation, the main effects are not additive and the interaction term must be included in the linear model. For example, the results might be that (a) the mean for A_1 is greater than A_2 when considering B_1, but (b) the mean for A_1 is less than A_2 when considering B_2. *Thus, we cannot make a blanket statement about the constant added benefits of* A_1 *over* A_2*, because it depends on the level of factor B*. In addition, for the interaction situation, the main effects are not statistically independent of one another; that is, *each of the main effects does not serve as an independent predictor of* Y. In order to predict Y well, information is necessary about the levels of factors A and B. *Thus, in the presence of a significant interaction, generalizations about the main effects must be qualified.* A profile plot should be examined so that a proper graphical interpretation of the interaction and main effects can be made. A significant interaction serves as a warning that one cannot generalize statements about a main effect for A over all levels of B. If you obtain a significant interaction, this is an important result. Do not ignore it and go ahead to interpret the main effects.

3.1.1.4 Partitioning the Sums of Squares

As pointed out in Chapter 1, partitioning the sums of squares is an important concept in the analysis of variance. We will illustrate with a two-factor model, but this can be extended to more than two factors. Let us begin with the **total sum of squares in** Y, denoted here as SS_{total}. The term SS_{total} represents the amount of total variation among all of the observations without regard to row, column, or cell membership. The next step is to partition the total variation into variation between the levels of factor A (denoted by SS_A), variation between the levels of factor B (denoted by SS_B), variation due to the interaction of the levels of factors A and B (denoted by SS_{AB}), and variation within the cells combined across cells (denoted by SS_{with}). In the two-factor analysis of variance, then, we can partition SS_{total} into the following:

$$SS_{total} = SS_A + SS_B + SS_{AB} + SS_{with}$$

Then computational formulas are used by statistical software to actually compute these sums of squares.

3.1.1.5 The ANOVA Summary Table

The next step is to assemble the ANOVA summary table. The purpose of the summary table is to simply summarize the analysis of variance. A general form of the summary table for the two-factor model is shown in Table 3.2. The first column lists the sources of variation

TABLE 3.2

Two-Factor Analysis of Variance Summary Table

Source	SS	df	MS	F
A	SS_A	$J-1$	MS_A	MS_A/MS_{with}
B	SS_B	$K-1$	MS_B	MS_B/MS_{with}
AB	SS_{AB}	$(J-1)(K-1)$	MS_{AB}	MS_{AB}/MS_{with}
Within	SS_{with}	$N-JK$	MS_{with}	
Total	SS_{total}	$N-1$		

in the model. We note that the total variation is divided into a within-groups source, and a general between-groups source, which is then subdivided into sources due to A, B, and the AB interaction. This is in keeping with the spirit of the one-factor model, where total variation was divided into a between-groups source (just one effect because there is only one factor and no interaction term) and a within-groups source. The second column provides the computed sums of squares.

The third column gives the degrees of freedom for each source. As always, degrees of freedom have to do with the number of observations that are free to vary in a particular context. Because there are J levels of factor A, the number of degrees of freedom for the A source is equal to $J-1$. As there are J means and we know the overall mean, then only $J-1$ of the means are free to vary. This is the same rationale we have been using throughout this text. As there are K levels of factor B, there are $K-1$ degrees of freedom for the B source. For the AB interaction source, we take the product of the degrees of freedom for the main effects. Thus we have as degrees of freedom for AB the product $(J-1)(K-1)$. The degrees of freedom within groups is equal to the total number of observations minus the number of cells, $N-JK$. Finally, the degrees of freedom total can be written simply as $N-1$.

The fourth column provides the mean squares terms. In this column, the sum of squares terms are weighted by the appropriate degrees of freedom to generate the mean squares terms. Thus, for instance, $MS_S = SS_A/df_A$.

Finally, in the last column of the ANOVA summary table, we have the F values, which represent the summary statistics for the analysis of variance. There are three hypotheses that we are interested in testing, one for each of the two main effects and one for the interaction effect, so there will be three F test statistics. For the factorial fixed-effects model, each F value is computed by taking the MS for the source that you are interested in testing and dividing it by MS_{with}. Thus for each hypothesis, the same error term is used in forming the F ratio (i.e., MS_{with}). We return to the two-factor model for cases where the effects are not fixed in Chapter 5.

Each of the F test statistics is then compared with the appropriate F critical value so as to make a decision about the relevant null hypothesis. These critical values are found in the F table of Appendix Table A.4 as follows: for the test of factor A as ${}_{\alpha}F_{J-1,N-JK}$; for the test of factor B as ${}_{\alpha}F_{K-1,N-JK}$; and for the test of the interaction as ${}_{\alpha}F_{(J-1)(K-1),N-JK}$. Thus, with a two-factor model, testing two main effects and one interaction, there are three F tests and three decisions that must be made. Each significance test is one-tailed so as to be consistent with the alternative hypothesis. The null hypothesis is rejected if the F test statistic exceeds the F critical value

Recall that these F tests are **omnibus tests** that tell only if there is an *overall* main effect or interaction effect. If the F test statistic does exceed the F critical value, and there is more than one degree of freedom for the source being tested, then it is not clear precisely why the null hypothesis was rejected. For example, if there are three levels of factor A and the null hypothesis for A is rejected, then we are not sure where the mean differences lie among the levels of A. In this case, some multiple comparison procedure should be used to determine where the mean differences are; this is the topic of the next section.

3.1.1.6 Multiple Comparison Procedures

In this section, we extend the concepts related to multiple comparison procedures (MCPs) covered in Chapter 2 to the two-factor ANOVA model. This model includes main and interaction effects; consequently you can examine contrasts of both main and interaction effects. In general, the procedures described in Chapter 2 can be applied to the two-factor situation. Things become more complicated as we have row and column means (i.e., marginal means), and cell means. Thus we have to be careful about which means are being considered.

Let us begin with contrasts of the *main effects*. If the effect for factor A is significant, and there are more than two levels of factor A, then we can form contrasts that compare the levels of factor A ignoring factor B. Here we would be comparing the means for the levels of factor A, which are marginal means as opposed to cell means. Considering each factor separately is strongly advised; considering the factors simultaneously is to be avoided. Some statistics texts suggest that you consider the design as a one-factor model with JK levels when using MCPs to examine main effects. This is inconsistent with the design and the intent of separating effects, and is not recommended.

For contrasts involving the *interaction*, our recommendation is to begin with a complex interaction contrast if there are more than four cells in the model. Thus, for example, in a 4 × 4 design that consists of four levels of factor A and four levels of factor B, one possibility is to test both 4 x 2 complex interaction contrasts. An example of one such contrast is as follows [where $(\bar{Y}_{.11} + \bar{Y}_{.21} + \bar{Y}_{.31} + \bar{Y}_{.41})$, for example, is the sum of the cell means of each level of factor A for level 1 of factor B and $(\bar{Y}_{.12} + \bar{Y}_{.22} + \bar{Y}_{.32} + \bar{Y}_{.42})$, is the sum of the cell means of each level of factor A for level 2 of factor B]:

$$\psi' = \frac{(\bar{Y}_{.11} + \bar{Y}_{.21} + \bar{Y}_{.31} + \bar{Y}_{.41})}{4} - \frac{(\bar{Y}_{.12} + \bar{Y}_{.22} + \bar{Y}_{.32} + \bar{Y}_{.42})}{4}$$

with a standard error of the following:

$$s_{\psi'} = \sqrt{MS_{\text{with}}\left(\sum_{j=1}^{J}\sum_{k=1}^{K}\frac{c_{jk}^2}{n_{jk}}\right)}$$

where n_{jk} is the number of observations in cell jk. This contrast would examine the interaction between the four groups in factor A and the first two groups in factor B. A second complex interaction contrast could consider the interaction between the four groups in factor A and the other two groups in factor B.

If the complex interaction contrast is significant, then follow this up with a simple interaction contrast that involves only four cell means. This is a single degree of freedom contrast because it involves only two levels of each factor (known as a **tetrad difference**). An example of such a contrast is the following:

$$\psi' = \left(\bar{Y}_{.11} - \bar{Y}_{.21}\right) - \left(\bar{Y}_{.12} - \bar{Y}_{.22}\right)$$

with a similar standard error term. Using the same example, this contrast would examine the interaction between the first two groups in factor A and the first two groups in factor B.

Most of the MCPs described in Chapter 2 can be used for testing main effects and interaction effects (although there is some debate about the appropriate use of interaction contrasts; see Boik, 1979; Marascuilo and Levin, 1970, 1976). Keppel and Wickens (2004) consider interaction contrasts in much detail. Finally, some statistics texts suggest the use of simple main effects in testing a significant interaction. These involve comparing, for example, the levels of factor A at a particular level of factor B, and are generally conducted by further partitioning the sums of squares. However, the simple main effects sums of squares represent a portion of a main effect plus the interaction effect. Thus, the simple main effect does not really help us to understand the interaction, and it is not recommended here.

3.1.1.7 Expected Mean Squares

As we asked in Chapter 1 for the one-factor fixed-effects model, for the two-factor fixed-effects model being considered here, we again ask the question, "How do we know which source of variation to use as the error term in the denominator?" That is, for the two-factor fixed-effects ANOVA model, how did we know to use MS_{with} as the error term in testing for the main effects and the interaction effect? As we learned in Chapter 1, an expected mean square for a particular source of variation represents the average mean square value for that source obtained if the same study were to be replicated an infinite number of times. For instance, the expected value of MS_A, denoted by $E(MS_A)$, is the average value of MS_A over repeated samplings.

Let us examine what the expected mean square terms actually look like for our two-factor fixed-effects model. Consider the two situations of (a) all of the H_0 actually being true and (b) all of the H_0 actually being false. If all of the H_0 are actually *true*, such that there really are no main effects or an interaction effect, then the expected mean squares are:

$$E(MS_A) = \sigma_\varepsilon^2$$
$$E(MS_B) = \sigma_\varepsilon^2$$
$$E(MS_{AB}) = \sigma_\varepsilon^2$$
$$E(MS_{with}) = \sigma_\varepsilon^2$$

and thus using MS_{with} as the error term will produce F values around 1.

If all of the H_0 are actually *false*, such that there really are main effects and an interaction effect, then the expected mean squares are as follows:

$$E(MS_A) = \sigma_\varepsilon^2 + \left(nK\sum_{j=1}^{J}\alpha_j^2\right)\Big/(J-1)$$

$$E(MS_B) = \sigma_\varepsilon^2 + \left(nJ\sum_{k=1}^{K}\beta_k^2\right)\Big/(K-1)$$

$$E(MS_{AB}) = \sigma_\varepsilon^2 + \left(n\sum_{j=1}^{J}\sum_{k=1}^{K}(\alpha\beta)_{jk}^2\right)\Big/(J-1)(K-1)$$

$$E(MS_{\text{with}}) = \sigma_\varepsilon^2$$

and thus using MS_{with} as the error term will produce F values greater than 1.

There is a difference in the main and interaction effects between when H_0 is actually true as compared to when H_0 is actually false because in the latter situation there is a second term. The important parts of this second term are, α, β, and $\alpha\beta$, which represent the effects for A, B and AB, respectively. The larger this part becomes, the larger the F ratio becomes. In comparing the two situations, we also see that $E(MS_{with})$ is the same whether H_0 is actually true or false, and thus it represents a reliable estimate of σ_ε^2. This term is *mean-free* because it does not depend on any mean differences.

Finally let us put all of this information together. In general, the ***F* ratio** represents

$$F = \frac{(systematic\ variability + error\ variability)}{error\ variability}$$

where, for the two-factor fixed-effects model, systematic variability is variability due to the main or interaction effects (i.e., between sources) and error variability is variability within. The F ratio is formed in a particular way because we want to isolate the systematic variability in the numerator. For this model, the only appropriate error term to use for each F ratio is MS_{with} because it does serve to isolate the systematic variability.

3.1.1.8 An Example

Consider the following illustration of the two-factor design. Here we expand on the example presented in Chapter 1 by adding a second factor to the model. Our dependent variable will again be psychological distress, factor A is type of sport in which the sampled athlete participates, and factor B is selection status (i.e., whether the athlete is deselected or selected to continue to compete on their team). Thus, the researcher is interested in whether the type of sport in which the athlete participates, the selection status (i.e., whether they are deselected or selected to continue to participate on their team), or the interaction of type of sport and selection status influences psychological distress. The categories of type of sport are defined again as (a) movement, (b) target, (c) fielding, and (d) territory. Selection status is defined as (a) deselected and (b) selected. This is not a manipulated design; i.e., this is truly an observational study where athletes were *not* randomly assigned to either type of sport or selection status. There were four athletes in each cell and eight cells (four levels of type of sport and two categories of selection status, thus 4 × 2 or 8 combinations of type of sport and selection status) for a total of 32 observations. Table 3.3 depicts the raw data and sample means for each cell (given beneath each cell), column, row, and overall.

The results are summarized in the ANOVA summary table as shown in Table 3.4. The F test statistics are compared to the following critical values obtained from Appendix Table A.4 ($\alpha = .05$): ${}_{.05}F_{3,24} = 3.01$ for the A (i.e., type of sport) and AB (i.e., type of sport by selection status) effects; and ${}_{.05}F_{1,24} = 4.26$ for the B (i.e., selection status) effect. The test statistics exceed the critical values for the A and B effects only, so we can reject these H_0 and conclude that both the type of sport and selection status are related to mean differences in psychological distress. The interaction was shown not to be a significant effect. If you would like to see an example of a two-factor design where the interaction is significant, take a look at the end of chapter problems, computational problem 6.

TABLE 3.3

Data for the Elite Athlete Example: Psychological Distress by Type of Sport and Selection

	Selection (B)		
Sport (A)	**Deselected**	**Selected**	**Row Mean**
Movement (e.g., gymnastics, dance)	15	10	11.1250
	12	8	
	21	7	
	13	3	
	15.2500	7.0000	
Target (e.g., golf)	20	13	17.8750
	22	9	
	24	18	
	25	12	
	22.7500	13.0000	
Fielding (e.g., baseball)	24	10	20.2500
	29	12	
	27	21	
	25	14	
	26.2500	14.2500	
Territory (e.g., football)	30	22	24.3750
	26	20	
	29	25	
	28	15	
	28.2500	20.5000	
Column mean	23.1250	13.6875	18.4063 *(Overall mean)*

TABLE 3.4

Two-Factor Analysis of Variance Summary Table—Elite Athlete Example

Source	*SS*	*df*	*MS*	*F*
A	738.5938	3	246.1979	21.3504*
B	712.5313	1	712.5313	61.7911**
AB	21.8438	3	7.2813	0.6314*
Within	276.7500	24	11.5313	
Total	1749.7188	31		

$^{*}{}_{.05}F_{3,24} = 3.01$

$^{**}{}_{.05}F_{1,24} = 4.26$

Next we estimate the main and interaction effects. The *main effects for the levels of A* (i.e., type of sport) are estimated to be:

$Movement: a_1 = \bar{Y}_{.1.} - \bar{Y}_{...} = 11.1250 - 18.4063 = -7.2813$

$Target: a_2 = \bar{Y}_{.2.} - \bar{Y}_{...} = 17.8750 - 18.4063 = -0.5313$

$Fielding: a_3 = \bar{Y}_{.3.} - \bar{Y}_{...} = 20.2500 - 18.4063 = 1.8437$

$Territory: a_4 = \bar{Y}_{.4.} - \bar{Y}_{...} = 24.3750 - 18.4063 = 5.9687$

The *main effects for the levels of B* (selection status) are estimated to be:

$Deselected: b_1 = \bar{Y}_{..1} - \bar{Y}_{...} = 23.1250 - 18.4063 = 4.7187$

$Selected: b_2 = \bar{Y}_{..2} - \bar{Y}_{...} = 13.6875 - 18.4063 = -4.7187$

Finally, the *interaction effects for the combinations of the levels of factors A (type of sport) and B (selection status)* are:

$$(ab)_{11} = \bar{Y}_{.11} - (\bar{Y}_{.1.} + \bar{Y}_{..1} - \bar{Y}_{...}) = 15.2500 - (11.1250 + 23.1250 - 18.4063) = -0.5937$$

$$(ab)_{12} = \bar{Y}_{.12} - (\bar{Y}_{.1.} + \bar{Y}_{..2} - \bar{Y}_{...}) = 7.0000 - (11.1250 + 13.6875 - 18.4063) = 0.5938$$

$$(ab)_{21} = \bar{Y}_{.21} - (\bar{Y}_{.2.} + \bar{Y}_{..1} - \bar{Y}_{...}) = 22.7500 - (17.8750 + 23.1250 - 18.4063) = 0.1563$$

$$(ab)_{22} = \bar{Y}_{.22} - (\bar{Y}_{.2.} + \bar{Y}_{..2} - \bar{Y}_{...}) = 13.0000 - (17.8750 + 13.6875 - 18.4063) = -0.1562$$

$$(ab)_{31} = \bar{Y}_{.31} - (\bar{Y}_{.3.} + \bar{Y}_{..1} - \bar{Y}_{...}) = 26.2500 - (20.2500 + 23.1250 - 18.4063) = 1.2813$$

$$(ab)_{41} = \bar{Y}_{.41} - (\bar{Y}_{.4.} + \bar{Y}_{..1} - \bar{Y}_{...}) = 28.2500 - (24.3750 + 23.1250 - 18.4063) = -0.8437$$

$$(ab)_{42} = \bar{Y}_{.42} - (\bar{Y}_{.4.} + \bar{Y}_{..2} - \bar{Y}_{...}) = 20.5000 - (24.3750 + 13.6875 - 18.4063) = 0.8438$$

The profile plot shown in Figure 3.2 graphically depicts these effects. The main effect for type of sport (factor A) was statistically significant and has more than two levels, so let us consider one example of a multiple comparison procedure, Tukey's HSD test. Recall from Chapter 2 that the HSD test is a family-wise procedure most appropriate for considering all pairwise contrasts with a balanced design (which is the case for these data). The following are the computations:

The critical value (obtained from Appendix Table A.9):

$${}_{\alpha}q_{df(with),\,J} = {}_{.05}q_{24,4} = 3.901$$

The standard error:

$$s_{\psi'} = \sqrt{\frac{MS_{with}}{n}} = \sqrt{\frac{11.5313}{8}} = 1.2006$$

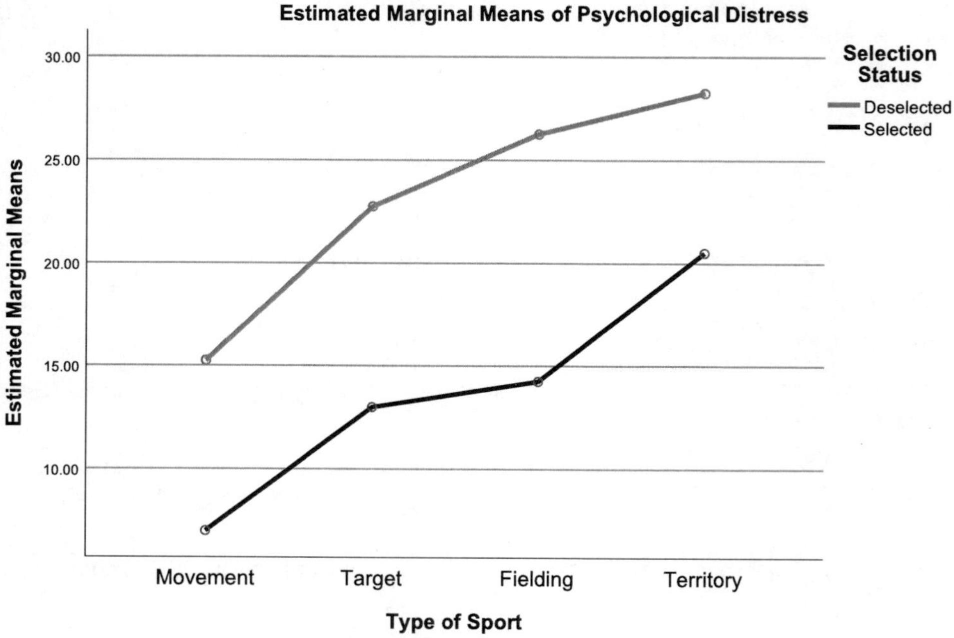

FIGURE 3.2
Profile plot for elite athlete example data.

The test statistics:

$$q_1 = \frac{\bar{Y}_{.4.} - \bar{Y}_{.1.}}{s_{\psi'}} = \frac{24.3750 - 11.1250}{1.2006} = \mathbf{11.0361*}$$

$$q_2 = \frac{\bar{Y}_{.4.} - \bar{Y}_{.2.}}{s_{\psi'}} = \frac{24.3750 - 17.8750}{1.2006} = \mathbf{5.4140*}$$

$$q_3 = \frac{\bar{Y}_{.4.} - \bar{Y}_{.3.}}{s_{\psi'}} = \frac{24.3750 - 20.2500}{1.2006} = 3.4358$$

$$q_4 = \frac{\bar{Y}_{.3.} - \bar{Y}_{.1.}}{s_{\psi'}} = \frac{20.2500 - 11.1250}{1.2006} = \mathbf{7.6004*}$$

$$q_5 = \frac{\bar{Y}_{.3.} - \bar{Y}_{.2.}}{s_{\psi'}} = \frac{20.2500 - 17.8750}{1.2006} = 1.9782$$

$$q_6 = \frac{\bar{Y}_{.2.} - \bar{Y}_{.1.}}{s_{\psi'}} = \frac{17.8750 - 11.1250}{1.2006} = \mathbf{5.6222*}$$

Recall that we compare the test statistic value to the critical value to make our hypothesis testing decision. If the test statistic value exceeds the critical value, we reject the null hypothesis and conclude that those means differ. For these tests, the results indicate that the means for the levels of factor A (type of sport) are statistically significantly different for levels 1 and 4 (i.e., the test statistic value is 11.0361 and the critical value is 3.901), 2 and 4, 1 and 3, and 1 and 2 (see equation results in bold). Thus, level 1 (movement) is significantly different from the other three types of sports, and levels 2 and 4 (target and fielding) are also significantly different. The only levels that are not statistically different are levels 2 and 3 $(q_5 = 1.9782)$ and levels 3 and 4 $(q_3 = 3.4358)$.

These results are somewhat different from those found with the one-factor model in Chapters 1 and 2 (where the significantly different levels were only 1 vs. 4 and 1 vs. 3). The MS_{with} has been reduced with the introduction of the second factor from 36.1116 to 11.5313 because SS_{with} has been reduced from 1,011.1250 to 276.7500. Although the *SS* and *MS* for the type of sport factor remain unchanged, this resulted in the *F* test statistic being considerably larger (increased from 6.8177 to 21.3504), although observed power was quite high in both models. Recall that this is one of the benefits we mentioned earlier about the use of additional factors in the model. Also, although the effect of factor B (selection status) was significant, there are only two levels, and thus we need not carry out any multiple comparisons (psychological distress is higher for deselected athletes). Finally, since the interaction was not significant, it is not necessary to consider any related contrasts.

3.1.2 Power

As mentioned in Chapter 1, power can be determined either in the planned (*a priori*) or observed (post hoc) power context. For planned power, we typically use tables or power charts (e.g., Cohen, 1988; Murphy, Myors, & Wolach, 2009) or software (e.g., G*Power). These are particularly useful in terms of determining adequate sample sizes when designing a study. Observed power is reported by statistics software, such as SPSS, to indicate the actual power in a given study.

3.1.3 Effect Size

Various measures of effect size have been proposed. Let us examine some commonly used measures, which assume equal variances across the cells, and that are presented in Olejnik and Algina (2000). The formulas presented assume a two-factor design (factors A and B and the interaction of AB); however, these can easily be extended to additional factorial designs.

3.1.3.1 Proportion of Total Variance Effect Size

We begin with *proportion of total variance effect size* indices. *These effect size indices are interpreted as the proportion of total variability in the dependent variable that is accounted for by a factor (e.g., A, B, or the interaction AB).* One such effect size measure is the **omega squared statistic**, ω^2. We can determine ω^2 as follows and will offer two different ways to calculate the index, both of which should yield the same value:

$$\omega_A^2 = \frac{SS_A - (J-1)(MS_{with})}{SS_{total} + MS_{with}} = \frac{df_A(MS_A - MS_{with})}{SS_{total} + MS_{with}}$$

$$\omega_B^2 = \frac{SS_B - (K-1)(MS_{with})}{SS_{total} + MS_{with}} = \frac{df_B(MS_B - MS_{with})}{SS_{total} + MS_{with}}$$

$$\omega_{AB}^2 = \frac{SS_{AB} - (J-1)(K-1)(MS_{with})}{SS_{total} + MS_{with}} = \frac{df_{AB}(MS_{AB} - MS_{with})}{SS_{total} + MS_{with}}$$

Epsilon squared, ε^2, is another proportion of total variance effect size and can be computed as follows:

$$\varepsilon_A^2 = \frac{df_A(MS_A - MS_{with})}{SS_{total}}$$

$$\varepsilon_B^2 = \frac{df_B(MS_B - MS_{with})}{SS_{total}}$$

$$\varepsilon_{AB}^2 = \frac{df_{AB}(MS_{AB} - MS_{with})}{SS_{total}}$$

Eta squared, η^2, is the last proportion of total variance effect size that we'll discuss, and it can be computed as follows:

$$\eta_A^2 = \frac{SS_A}{SS_{total}}$$

$$\eta_B^2 = \frac{SS_B}{SS_{total}}$$

$$\eta_{AB}^2 = \frac{SS_{AB}}{SS_{total}}$$

3.1.3.2 Proportion of Partial Variance Effect Size

There are also **proportion of partial variance effect size** indices. These are referred to as *partial* because the effect size is computed by excluding other factors in the model when computing the effect size, and in this way controls other factors in the model. They are generally interpreted as the proportion of variation in the dependent variable, Y, explained by the effect of interest (i.e., by factor A, or factor B, or the AB interaction) that is *not* explained by other variables in the model. Generally, a proportion of partial variance effect size for a factor will be larger than a proportion of total variance for that same factor (Olejnik & Algina, 2000).

We can determine **partial omega squared** as follows:

$$\omega_{A,partial}^2 = \frac{df_A(MS_A - MS_{with})}{(df_A)(MS_A) + (N - df_A)(MS_{with})}$$

$$\omega_{B,partial}^2 = \frac{df_B(MS_B - MS_{with})}{(df_B)(MS_B) + (N - df_B)(MS_{with})}$$

$$\omega_{AB,partial}^2 = \frac{df_{AB}(MS_{AB} - MS_{with})}{(df_{AB})(MS_{AB}) + (N - df_{AB})(MS_{with})}$$

Partial epsilon squared is another proportion of partial variance effect size and can be computed as follows:

$$\varepsilon^2_{A,partial} = \frac{df_A(MS_A - MS_{with})}{SS_A + SS_{with}}$$

$$\varepsilon^2_{B,partial} = \frac{df_B(MS_B - MS_{with})}{SS_B + SS_{with}}$$

$$\varepsilon^2_{AB,partial} = \frac{df_{AB}(MS_{AB} - MS_{with})}{SS_{AB} + SS_{with}}$$

Partial eta squared is the estimate of effect size that can be requested when using SPSS for computing factorial ANOVA. We determine $\eta^2_{partial}$ as follows:

$$\eta^2_{A,partial} = \frac{SS_A}{SS_A + SS_{with}}$$

$$\eta^2_{B,partial} = \frac{SS_B}{SS_B + SS_{with}}$$

$$\eta^2_{AB,partial} = \frac{SS_{AB}}{SS_{AB} + SS_{with}}$$

3.1.3.3 Interpreting Effect Size

Using Cohen's (1988) subjective standards, these effect sizes, whether they are the proportion of total variance or proportion of partial variance, can be interpreted as follows: small effect, ω^2, ε^2, or η^2 = .01; medium effect, ω^2, ε^2, or η^2 = .06; large effect, ω^2, e^2, or η^2 = .14. See Table 3.5. Researchers interested in further discussion on effect size in

TABLE 3.5

Effect Sizes and Interpretations

Effect Size	Interpretation
Proportion of Total Variability Accounted For	
Omega squared (ω^2), epsilon squared (ε^2), and eta squared (η^2)	Proportion of total variability in the dependent variable that is accounted for by a factor (e.g., A, B, or AB) • Small effect = .01 • Medium effect = .06 • Large effect = .14
Proportion of Partial Variability Accounted For	
Partial omega squared ($\omega^2_{A,partial}$), partial epsilon squared ($\varepsilon^2_{A,partial}$), and partial eta squared ($\eta^2_{A,partial}$)	Proportion of total variability in the dependent variable that is accounted for by a factor (e.g., A, B, or AB) that is *not* explained by other variables in the model • Small effect = .01 • Medium effect = .06 • Large effect = .14

factorial designs are encouraged to review any number of resources (e.g., Cohen, 1988; Fidler & Thompson, 2001; Keppel & Wickens, 2004; Murphy et al., 2009; O'Grady, 1982; Wilcox, 1987).

3.1.3.4 Additional Effect Size Considerations

We will end our discussion on effect size with a few noteworthy items to consider as you compute and interpret effect sizes. Eta squared can be positively biased, overestimating the strength of the population relationship, and thus is best considered a descriptor of proportion of variance in the dependent variable explained for a particular sample (Maxwell, Arvey, & Camp, 1981). Thus, many researchers discourage reporting eta squared or partial eta squared, although you will still see it widely reported given that it is the only effect size value that is output from SPSS. Both epsilon squared and omega squared introduce a correction to this problem, and generally, both will be quite similar in value (Carroll & Nordholm, 1975). If you find yourself in a situation where epsilon squared or omega squared are negative, the standard is simply to set the effect size to zero (Olejnik & Algina, 2000).

One cautionary note in using omega squared, both the total and partial, is that the computation uses variance components from the expected mean squares for the source of variation, and the expected mean square assumes a balanced design (Olejnik & Algina, 2000). When sample sizes are not equal, researchers may wish to report a different measure of effect.

Also consider that proportion of total variance effect size indices are not comparable across studies that incorporate different factors (Olejnik & Algina, 2000). This is reasonable given that total variation is influenced by all the factors in the model. An even more stringent stumbling block pertains to proportion of partial variance effect size measures. Because the denominator for each factor and/or interaction differs when computing the proportion of partial variance effect size, these effects cannot be compared within the same study (Olejnik & Algina, 2000). Additionally, because the denominators differ in proportion of partial variance effect size, the sum of the partial measures of effect may total more than one, and this may occur even if the factors are orthogonal (i.e., balanced) (Olejnik & Algina, 2000).

Some researchers find interpreting proportion of variance effect sizes advantageous, as compared to standardized mean differences, given that the index ranges from 0 to 1 (Rosenthal, 1994). However, even large proportion of variance effect size values (e.g., .14+) suggest there is much variance that remains to be explained, and thus even large effects can be perceived as trivial (Rosenthal & Rubin, 1979).

Last but not least, we will touch on *general reporting and interpretation recommendations for effect size*. First, reporting effect size values for omnibus tests are rarely meaningful as the omnibus test is usually not the hypothesis test of interest (Rosnow & Rosenthal, 1988). Reporting effect size measures for factors and contrasts (e.g., A, B, and AB, as the computations provided here allow) are encouraged as those are likely where the real interest (and hypotheses of interest) lie (Olejnik & Algina, 2000). Second, many researchers encourage interpreting effect size relative to other studies. However, several researchers (e.g., Fern & Monroe, 1996; Maxwell et al., 1981; O'Grady, 1982; Sechrest & Yeaton, 1982) have provided caution in doing this as effect size can be impacted by instrument reliability, heterogeneity of the populations that are compared, the levels or categories of the factors that are modeled, the strength of the treatments, and the range of treatments, all of which can lead to effect size comparisons that are misleading (Olejnik & Algina, 2000). In our perspective, this doesn't mean that you should avoid interpreting effect size relative to other studies. Rather, recognizing these limitations and pointing out differences that are known when making those interpretations are important.

3.1.3.5 Effect Size Example

Let us estimate effect size given the elite athlete example and results that are presented in Table 3.9. The partial η^2 are determined to be the following:

$$\eta_A^2 = \frac{SS_A}{SS_A + SS_{with}} = \frac{738.5938}{738.5938 + 276.7500} = 0.7274$$

$$\eta_B^2 = \frac{SS_B}{SS_B + SS_{with}} = \frac{712.5313}{712.5313 + 276.7500} = 0.7203$$

$$\eta_{AB}^2 = \frac{SS_{AB}}{SS_{AB} + SS_{with}} = \frac{21.8438}{21.8438 + 276.7500} = 0.0732$$

We calculate ω^2 to be the following:

$$\omega_A^2 = \frac{SS_A - (J-1)(MS_{with})}{SS_{total} + MS_{with}} = \frac{738.5938 - (4-1)(11.5313)}{1749.7188 + 11.5313} = 0.3997$$

$$\omega_B^2 = \frac{SS_B - (K-1)(MS_{with})}{SS_{total} + MS_{with}} = \frac{712.5313 - (2-1)(11.5313)}{1749.7188 + 11.5313} = 0.3980$$

$$\omega_{AB}^2 = \frac{SS_{AB} - (J-1)(K-1)(MS_{with})}{SS_{total} + MS_{with}} = \frac{21.8438 - (4-1)(2-1)(11.5313)}{1749.7188 + 11.5313} = -0.007$$

Based on these effect size measures, using Cohen's subjective standards, one would conclude that there is a large effect for type of sport and for selection status, but very little effect for the type of interaction of sport and selection status. An example of interpretation is the following: Partial eta squared for the main effect for type of sport tells us that the proportion of variation in psychological distress explained by the type of sport in which the athlete participates that is *not* explained by selection status is about 73%. Omega squared for the main effect for type of sport tells us that proportion of total variability in the dependent variable that is accounted for by type of sport is about 40%. Interpretations for selection status and the interaction of sport by selection status can be made similarly.

3.1.3.6 Confidence Intervals for Effect Size

To refresh our memory, computing **confidence intervals** is valuable. As mentioned in Chapter 1, confidence intervals can be used for providing interval estimates of a population mean or mean difference; this gives us information about the accuracy of a sample estimate. In the case of the two-factor model, we can form confidence intervals for row means, column means, cell means, the overall mean, as well as any possible contrast formed through a multiple comparison procedure. Note also that confidence intervals have been developed for effect sizes. The benefit in creating confidence intervals for effect size values is similar to that of creating confidence intervals for parameter estimates—*confidence intervals for the effect size provide an added measure of precision that is not obtained from knowledge of the effect size alone.* Computing confidence intervals for effect size indices, however, is not as straightforward as simply plugging in known values into a formula. Never fear; there are some nice online tools that can be used. For factorial ANOVA, Uanhoro's (2017) online calculator, available at https://effect-size-calculator.herokuapp.com/, uses the noncentral F method to compute confidence intervals for partial eta squared in fixed-effects ANOVA models that do not include covariates (i.e., ANCOVA, which we will study in a future chapter). As we see in Figure 3.3, and as we saw with one-way ANOVA, only four inputs are required: F, numerator and denominator degrees of freedom, and confidence interval. Note that because the F cannot be negative, the default setting for the 90% confidence interval is equivalent

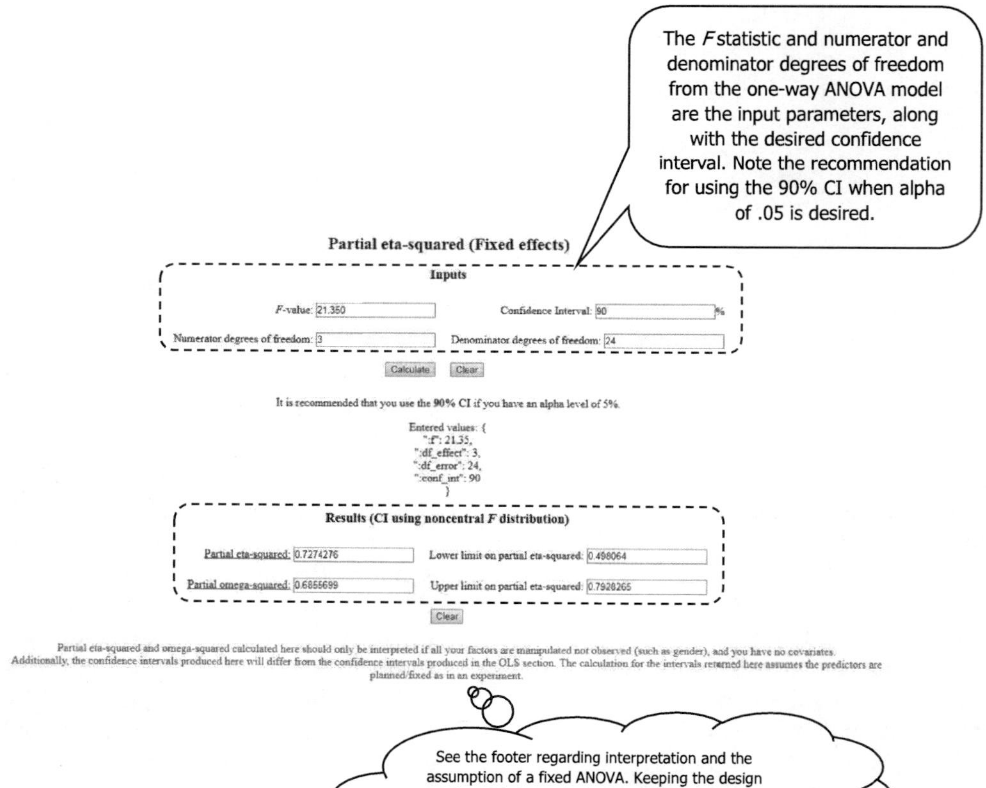

FIGURE 3.3
Confidence intervals for effect size.

to the 95% two-sided confidence interval (Smithson, 2003). Therefore, the site for the online calculator recommends that you "use the 90% CI if you have an alpha level of 5%."

We will use the results from our elite athlete data (see Table 3.9) and will illustrate with the main effect for type of sport; however, confidence intervals for results for all main effects and interactions can be computed similarly (see Figure 3.3). With an F of 21.350 and numerator and denominator degrees of freedom of 3 and 24, respectively, partial eta squared is .727 with lower and upper confidence limits of .498 and .793, respectively. Putting this in context of our example, if multiple random samples were drawn from the population, 95% of the samples could expect about 50%, at minimum, and 79%, at maximum, of the proportion of the outcome to be explained by the independent variable type of sport that is not explained by other variables in the model (specifically selection status).

3.1.4 Assumptions

In Chapter 1 we described in detail the assumptions for the one-factor analysis of variance. In the two-factor model, the assumptions are again concerned with **independence, homogeneity of variance**, and **normality**. A summary of the effects of their violation is provided in Table 3.6. The same methods for detecting violations described in Chapter 1 can be used for this model.

There are only two different wrinkles for the two-factor model as compared to the one-factor model. First, as the effect of heterogeneity is small with balanced designs (equal n's per cell) or nearly balanced designs, and/or with larger n's, this is a reason to strive for such a design. Unfortunately, there is very little research on this problem, except the classic (Box, 1954) article for a no-interaction model with one observation per cell. There are limited solutions for dealing with a violation of the homogeneity assumption, such as the Welch (1951) test, the Johansen (1980) procedure, and variations described by Wilcox (1996 or 2003). Transformations are not usually used, as they may destroy an additive linear model and create interactions that did not previously exist. Nonparametric techniques are not commonly used with the two-factor model, although see the description of the Brunner, Dette, and Munk (1997) procedure in Wilcox (2003). Second, the effect of nonnormality seems to be the same as heterogeneity (Miller, 1997).

TABLE 3.6

Assumptions and Effects of Violations for the Two-Factor ANOVA Design

Assumption	Effect of Assumption Violation
Independence	• Increased likelihood of a Type I and/or Type II error in the F statistic • Influences standard errors of means and thus inferences about those means
Homogeneity of variance	• Bias in SS_{with} • Increased likelihood of a Type I and/or Type II error • Less effect with balanced or nearly balanced design • Effect decreases as n increases
Normality	• Minimal effect with moderate violation • Minimal effect with balanced or nearly balanced design • Effect decreases as n increases

3.2 What Three-Factor and Higher-Order ANOVA Models Are and How They Work

3.2.1 Characteristics

All of the characteristics we discussed for the two-factor model apply to the three-factor model, with one obvious exception. There are three factors rather than two. This will result in three main effects (one for each factor, known as A, B, and C), three two-way interactions (known as AB, AC, and BC), and one three-way interaction (known as ABC). The only new concept is the three-way interaction, which may be stated as follows: "Is the AB interaction constant across all levels of factor C?" This may also be stated as "AC across the levels of B" or as "BC across the levels of A." These each have the same interpretation as there is only one way of testing the three-way interaction. In short, the three-way interaction can be thought of as the two-way interaction behaving differently across the levels of the third factor.

We do not explicitly consider models with more than three factors (compare Keppel & Wickens, 2004; Marascuilo & Serlin, 1988; Myers & Well, 1995). However, be warned that such models do exist, and that they will necessitate more main effects, more two-way interactions, more three-way interactions, as well as higher-order interactions—and thus more complex interpretations. Conceptually, the only change is to add these additional effects to the model.

3.2.2 The ANOVA Model

The model for the three-factor design is

$$Y_{ijkl} = \mu + \alpha_j + \beta_k + \gamma_l + (\alpha\beta)_{jk} + (\alpha\gamma)_{jl} + (\beta\gamma)_{kl} + (\alpha\beta\gamma)_{jkl} + \varepsilon_{ijkl}$$

where Y_{ijkl} is the observed score on the criterion (i.e., dependent) variable for individual i in level j of factor A, level k of factor B, and level l of factor C (or in the jkl cell), μ is the overall or grand population mean (i.e., regardless of cell designation), α_j is the effect for level j of factor A, β_k is the effect for level k of factor B, γ_l is the effect for level l of factor C, $\alpha\beta_{jk}$ is the interaction effect for the combination of level j of factor A and level k of factor B, $(\alpha\gamma)_{jl}$ is the interaction effect for the combination of level j of factor A and level l of factor C, $(\beta\gamma)_{kl}$ is the interaction effect for the combination of level k of factor B and level l of factor C, $(\alpha\beta\gamma)_{jkl}$ is the interaction effect for the combination of level j of factor A, level k of factor B, and level l of factor C, and ε_{ijkl} is the random residual error for individual i in cell jkl. Given that there are three main effects, three two-way interactions, and one three-way interaction, there will be an accompanying null and alternative hypothesis for each of these effects. At this point in your statistics career, the hypotheses should be obvious (simply expand on the hypotheses at the beginning of this chapter).

3.2.3 The ANOVA Summary Table

The ANOVA summary table for the three-factor model is shown in Table 3.7, with the usual columns for sources of variation, sums of squares, degrees of freedom, mean squares, and

TABLE 3.7

Three-Factor Analysis of Variance Summary Table

Source	*SS*	*df*	*MS*	*F*
Between				
A	SS_A	$J-1$	MS_A	MS_A/MS_{with}
B	SS_B	$K-1$	MS_B	MS_B/MS_{with}
C	SS_C	$L-1$	MS_C	MS_C/MS_{with}
AB	SS_{AB}	$(J-1)(K-1)$	MS_{AB}	MS_{AB}/MS_{with}
AC	SS_{AC}	$(J-1)(L-1)$	MS_{AC}	MS_{AC}/MS_{with}
BC	SS_{BC}	$(K-1)(L-1)$	MS_{BC}	MS_{BC}/MS_{with}
ABC	SS_{ABC}	$(J-1)(K-1)(L-1)$	MS_{ABC}	MS_{ABC}/MS_{with}
Within	SS_{with}	$N-JKL$	MS_{with}	
Total	SS_{total}	$N-1$		

F. A quick three-factor example dataset and the resulting ANOVA summary table from SPSS are shown in Table 3.8. Note that the only statistically significant effects are the main effect for B and the AC interaction ($p < .01$).

3.2.4 The Triple Interaction

Everything else about the three-factor design follows from the two-factor model. The assumptions are the same, MS_{with} is the error term used for testing each of the hypotheses in the fixed-effects model, and the multiple comparison procedures are easily utilized. The main new feature is the three-way interaction. If this interaction is significant, then this means that the two-way interaction is different across the levels of the third factor. This result will need to be taken into account prior to interpreting the two-way interactions and the main effects.

Although the inclusion of additional factors in the design should result in a reduction in MS_{with}, there is a price to pay for the study of additional factors. Although the analysis is simple for the computer, you must consider the possibility of significant higher-order interactions. If you find, for example, that the four-way interaction is significant, how do you deal with it? First you have to interpret this interaction, which could be difficult if it is unexpected. Then you may have difficulty in dealing with the interpretation of your other effects. Our advice is simple. *Do not include additional factors just because they sound interesting. Include only those factors that are theoretically or empirically important*. Then, if a significant higher-order interaction occurs, you will be in a better position to understand it because you will have already thought about its consequences. Reporting that an interaction is significant, but not interpretable, is not sound research. For additional discussion on this topic, see Keppel and Wickens (2004).

TABLE 3.8

Three-Factor Analysis of Variance Example—Raw Data and SPSS ANOVA Summary Table

Raw Data:

$A_1B_1C_1$: 8, 10, 12, 9

$A_1B_1C_1$: 23, 17, 21, 19

$A_1B_1C_1$: 22, 19, 16, 24

$A_1B_2C_2$: 33, 31, 27, 30

$A_2B_1C_1$: 16, 19, 21, 24

$A_2B_1C_2$: 6, 8, 11, 13

$A_2B_2C_1$: 27, 30, 31, 33

$A_2B_2C_2$: 16, 19, 21, 25

SPSS ANOVA Summary Table:

The row labeled "A" is the first independent variable or factor or between-groups variable. The *between-groups mean square* for factor A (.031) provides an indication of the variation in the dependent variable attributable to factor A.

The degrees of freedom for the sum of squares between groups for factor A is $J-1$ ($df = 1$ in this example indicating 2 levels for factor A).

Similar interpretations are made for the other main effects and interactions.

The omnibus *F* test for the main effect for factor A (and computed similarly for the other main effects and interactions) is computed as follows, where MS_{with} is the mean square of the error term.

$$F = \frac{MS_A}{MS_{with}} = \frac{0.31}{8.698} = .004$$

The *p* value for the omnibus *F* test of the main effect for factor A is .953. This indicates there is not a statistically significant difference in the dependent variable based on factor A, averaged across the levels of factors B and C. In other words, there is not a unique effect of factors A on the dependent variable, controlling for factors B and C. The probability of observing these mean differences or more extreme mean differences by chance if the null hypothesis is really true (i.e., if the population means really are equal) is about 95%.

We fail to reject the null hypothesis that the population means of factor A are equal. For this example, this provides evidence to suggest that the dependent variable does not differ, on average, across the levels of factor A, when controlling for factors B and C.

Tests of Between-Subjects Effects

Dependent Variable: OUTCOME

Source	Type III Sum of Squares	df	Mean Square	F	Sig.
Corrected Model	1702.219[a]	7	243.174	27.958	.000
Intercept	12840.031	1	12840.031	1476.219	.000
A	.031	1	.031	.004	.953
B	871.531	1	871.531	100.200	.000
C	.031	1	.031	.004	.953
A * B	.031	1	.031	.004	.953
A * C	830.281	1	830.281	95.457	.000
B * C	.031	1	.031	.004	.953
A * B * C	.281	1	.281	.032	.859
Error	208.750	24	8.698		
Total	14751.000	32			
Corrected Total	1910.969	31			

a. R Squared = .891 (Adjusted R Squared = .859)

The row labeled **"Error"** is within groups. The within groups sum of squares tells us how much variation there is within the cells combined across the cells (i.e., 208.750). The degrees of freedom for the sum of squares within groups is ($N - JKL$) or the sample size minus the number of levels of the independent variables [i.e., 32 – (2)(2)(2) = 24].

The row labeled **"corrected total"** is the sum of squares total. The degrees of freedom for the total is ($N-1$) or the sample size minus one.

3.3 What the Factorial ANOVA With Unequal *n*'s Is and How It Works

Up until this point in the chapter, we have considered only the equal *n*'s or **balanced design**. *That is, the model used was where the number of observations in each cell was equal.* This served to make the formulas and equations easier to deal with. However, we do not need to assume that the *n*'s are equal. In this section we discuss ways to deal with the unequal *n*'s (or unbalanced) case for the two-factor model, although these notions can be transferred to higher-order models as well.

When *n*'s are unequal, things become a bit trickier as the main effects and the interaction effect are not orthogonal. In other words, the sums of squares cannot be partitioned into independent effects and thus the individual *SS* do not necessarily add up to the SS_{total}. As a result, several computational approaches have been developed. In the old days, prior to the availability of high-speed computers, the standard approach was to use unweighted means analysis. This is essentially an analysis of means, rather than raw scores, which are unweighted by cell size. This approach is only an approximate procedure. Due to the availability of quality statistical software, the unweighted means approach is no longer necessary. A rather silly approach, and one that we do not condone, is to delete enough data until you have an equal *n*'s model.

There are three more modern approaches to this case. Each of these approaches really test different hypotheses and thus may result in different results and conclusions: (a) the **sequential approach** (also known as the hierarchical sums of squares approach), (b) the **partially sequential approach** (also known as the partially hierarchical, or experimental design, or method of fitting constants approach), and (c) the **regression approach** (also known as the marginal means or unique approach). There has been considerable debate over the years about the relative merits of each approach (e.g., Applebaum & Cramer, 1974; Carlson & Timm, 1974; Cramer & Applebaum, 1980; Overall, Lee, & Hornick, 1981; Overall & Spiegel, 1969; Timm & Carlson, 1975). Below we describe what each approach is actually testing.

In the **sequential approach,** the effects being tested are:

$$\alpha \mid \mu$$
$$\beta \mid \mu, \alpha$$
$$\alpha\beta \mid \mu, \alpha, \beta$$

This indicates, for example, that the effect for factor B (β) is adjusted or controls for (as denoted by the vertical line) the overall mean (m) and the main effect due to factor A (a). Thus, each effect is adjusted for prior effects in the sequential order given (i.e., α, β $\alpha\beta$). Here the α effect is given theoretical or practical priority over the β effect. In SAS and SPSS, this is the **Type I sum of squares** method.

In the **partially sequential approach,** the effects being tested are:

$$\alpha \mid \mu, \beta$$
$$\beta \mid \mu, \alpha$$
$$\alpha\beta \mid \mu, \alpha, \beta$$

There is difference here because each main effect controls for the other main effect, but not for the interaction effect. In SAS and SPSS, this is the **Type II sum of squares** method. This is the only one of the three methods where the sums of squares will add up to the total sum of squares. Notice in the sequential and partially sequential approaches that the interaction is not taken into account in estimating the main effects, which is fine only if there is no interaction effect.

In the **regression approach,** the effects being tested are:

$$\alpha \mid \mu, \beta, \alpha\beta$$
$$\beta \mid \mu, \alpha, \alpha\beta$$
$$\alpha\beta \mid \mu, \alpha, \beta$$

In this approach, each effect controls for each of the other effects. In SAS and SPSS this is the **Type III sum of squares** method (and is the default selection in SPSS). Many statisticians (Glass & Hopkins, 1996; Keppel & Wickens, 2004; Mickey, Dunn, & Clark, 2004), including the authors of this text, recommend exclusive use of the regression approach because each effect is estimated taking the other effects into account. The hypotheses tested in the sequential and partially sequential approaches are seldom of interest and are difficult to interpret (Carlson & Timm, 1974; Kirk, 2013; Overall, 1981; Timm & Carlson, 1975). The regression approach seems to be conceptually closest to the traditional analysis of variance in that each effect is estimated controlling for all other effects. When the n's are equal, each of these three approaches tests the same hypotheses and yields the same results.

3.4 Computing Factorial ANOVA Using SPSS

In this section we take a look at SPSS for the elite athlete example. As already noted in Chapter 1, SPSS needs the data to be in a specific form for the analysis to proceed, which is different from the layout of the data in Table 3.1. For a two-factor ANOVA, the dataset must consist of three variables or columns, one for the level of factor A, one for the level of factor B, and the third for the dependent variable. Each row still represents one individual, indicating the levels of factors A and B within which the individual is a member, and their score on the dependent variable. As seen in the screenshot (Figure 3.4), for a two-factor ANOVA, the SPSS data are in the form of two columns that represent the group values (i.e., the two independent variables) and one column that represents the scores or values of the dependent variable.

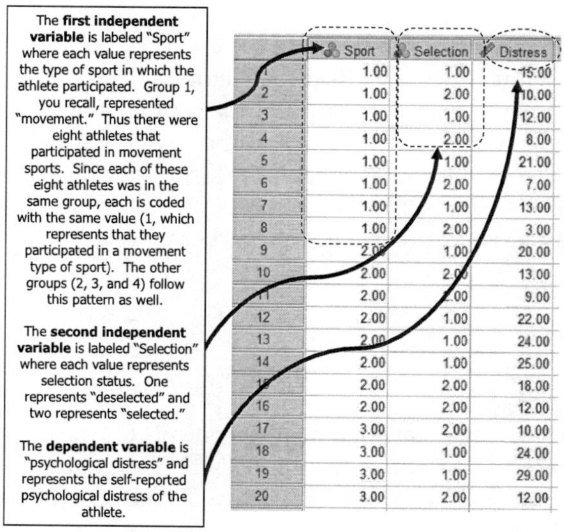

FIGURE 3.4
First 20 cases of the factorial ANOVA data.

Step 1. To conduct a factorial ANOVA, go to "Analyze" in the top pulldown menu, then select "General Linear Model," and then select "Univariate." Following the screenshot for Step 1 (Figure 3.5) produces the Univariate dialog box.

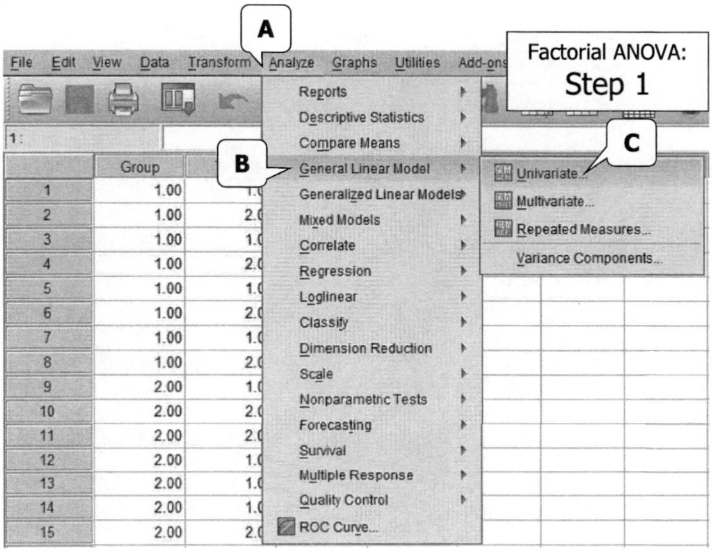

FIGURE 3.5
Factorial ANOVA: Step 1.

Step 2. Click the dependent variable (e.g., psychological distress) and move it into the "Dependent Variable" box by clicking the arrow button. Click the first independent variable (e.g., type of sport) and move it into the "Fixed Factors" box by clicking the arrow button. Follow this same step to move the second independent variable into the Fixed Factors box. Next, click on "Options."

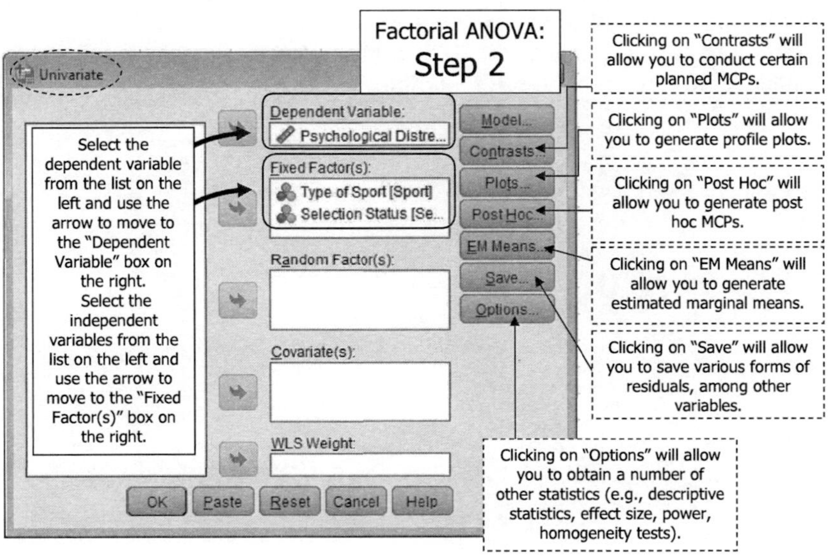

FIGURE 3.6
Factorial ANOVA: Step 2.

Step 3. Clicking on "Options" will provide the option to select such information as "Descriptive statistics," "Estimates of effect size," "Observed power," "Homogeneity tests" (i.e., Levene's test for equal variances), and "Spread vs. level plot." Click on "Continue" to return to the original dialog box.

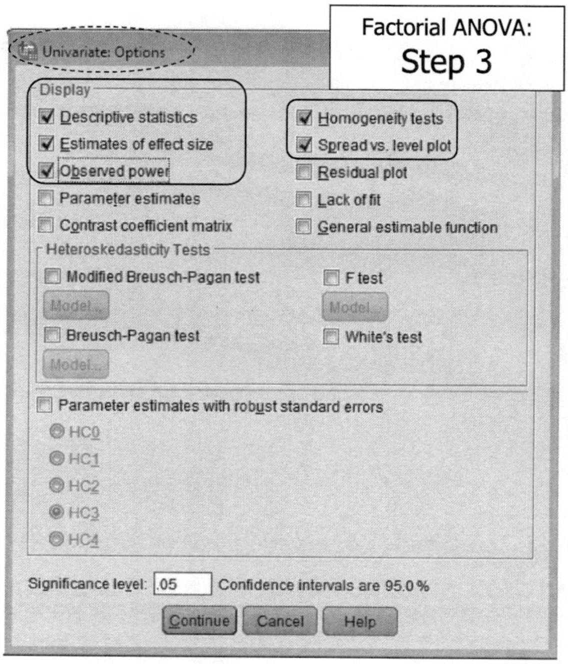

FIGURE 3.7
Factorial ANOVA: Step 3.

Step 4. Clicking on "EM Means" (see the main dialog box in Step 2, Figure 3.6) will provide the option to display overall and marginal means. Move the items that are listed in the "Factor(s) and Factor Interactions" box into the "Display Means for" box to generate adjusted means. Click on "Continue" to return to the original dialog box.

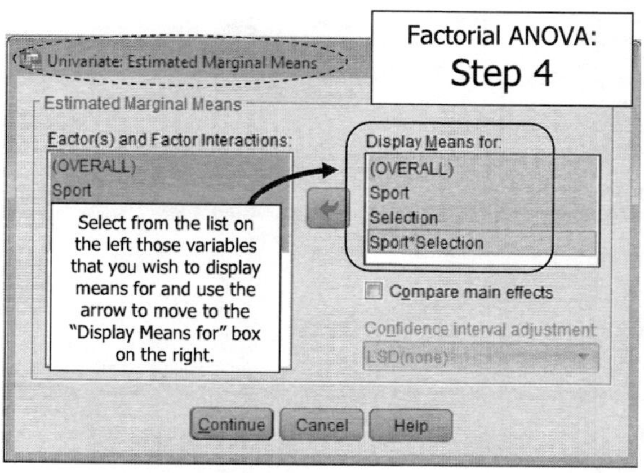

FIGURE 3.8
Factorial ANOVA: Step 4.

Step 5. From the Univariate dialog box, click on "Plots" to obtain a profile plot of means. Click the independent variable (e.g., type of sport labeled as "Sport") and move it into the "Horizontal Axis" box by clicking the arrow button (see screenshot for Step 5a, Figure 3.9). *(Tip: Placing the independent variable that has the most categories or levels on the horizontal axis of the profile plots will make for easier interpretation of the graph; however, this is really personal preference.)* Then click the second independent variable (e.g., "Selection") and move it into the "Separate Lines" box by clicking the arrow button (see Figure 3.9). Then click on "Add" to move the variable into the "Plots" box at the bottom of the dialog box (see the screenshot for Step 5b, Figure 3.10). Click on "Continue" to return to the original dialog box.

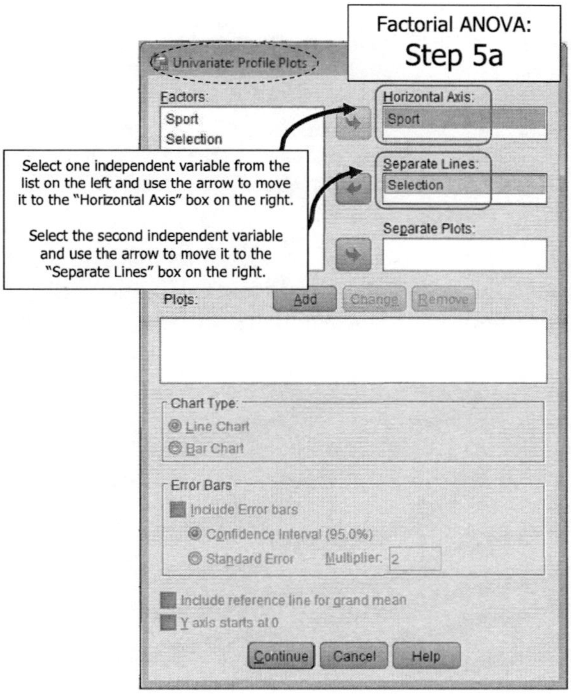

FIGURE 3.9
Factorial ANOVA: Step 5a.

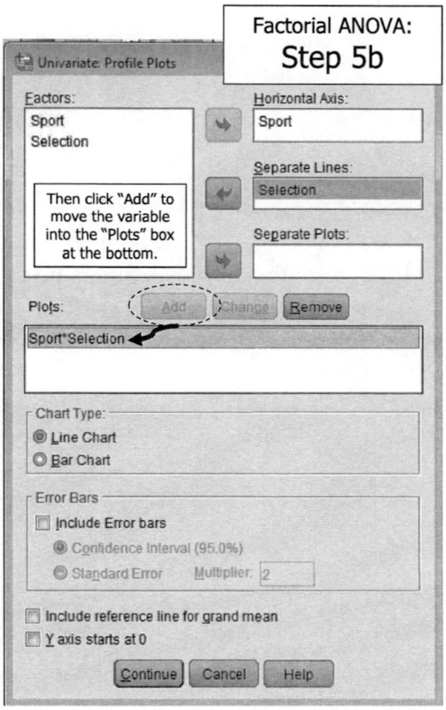

FIGURE 3.10
Factorial ANOVA: Step 5b.

Step 6. From the Univariate dialog box, click on "Post Hoc" to select various post hoc MCPs, or click on "Contrasts" to select various planned multiple comparison procedures (MCPs) (see main dialog box in screenshot for Step 2, Figure 3.6). From the "Post Hoc Multiple Comparisons for Observed Means" dialog box, click on the names of the independent variables in the "Factor(s)" list box in the top left (e.g., "Sport" and "Selection") and move them to the "Post Hoc Tests for" box in the top right by clicking on the arrow key. Check an appropriate MCP for your situation by placing a checkmark in the box next to the desired MCP. In this example, we will select Tukey, which is operationalized within SPSS as Tukey's HSD. Click on "Continue" to return to the original dialog box.

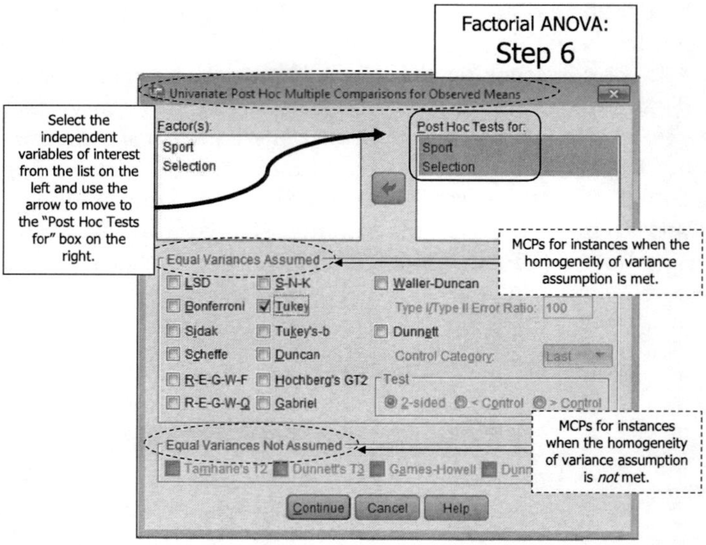

FIGURE 3.11
Factorial ANOVA: Step 6.

Step 7. From the Univariate dialog box, click on "Save" to select those elements that you want to save. For this illustration, we want to save the unstandardized residuals which will be used later to examine the extent to which normality and independence are met. From the Univariate dialog box, click on "OK" to return to generate the output.

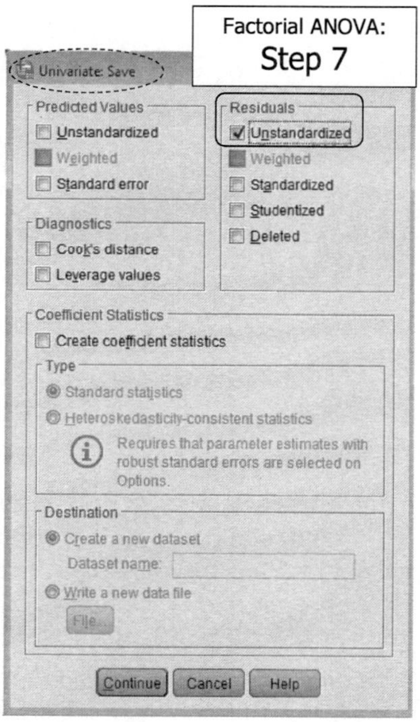

FIGURE 3.12
Factorial ANOVA: Step 7.

Interpreting the output. Annotated results are presented in Table 3.9 and the profile plot is shown in Figure 3.2. Note also that the SPSS ANOVA summary table will include additional sources of variation that we find not to be useful (i.e., corrected model, intercept, total); thus they are not annotated in Table 3.9.

TABLE 3.9

Selected SPSS Results for the Elite Athlete Psychological Distress Example

Between-Subjects Factors

		Value Label	N
Type of Sport	1.00	Movement	8
	2.00	Target	8
	3.00	Fielding	8
	4.00	Territory	8
Selection Status	1.00	Deselected	16
	2.00	Selected	16

The table labeled "Between-Subjects Factors" provides sample sizes for each of the categories of the independent variables (recall that the independent variables are the "between-subjects factors").

Descriptive Statistics

Dependent Variable: Psychological Distress

Type of Sport	Selection Status	Mean	Std. Deviation	N
Movement	Deselected	15.2500	4.03113	4
	Selected	7.0000	2.94392	4
	Total	11.1250	5.48862	8
Target	Deselected	22.7500	2.21736	4
	Selected	13.0000	3.74166	4
	Total	17.8750	5.93867	8
Fielding	Deselected	26.2500	2.21736	4
	Selected	14.2500	4.78714	4
	Total	20.2500	7.28501	8
Territory	Deselected	28.2500	1.70783	4
	Selected	20.5000	4.20317	4
	Total	24.3750	5.09727	8
Total	Deselected	23.1250	5.65538	16
	Selected	13.6875	6.09611	16
	Total	18.4062	7.51283	32

The table labeled "Descriptive Statistics" provides basic descriptive statistics (means, standard deviations, and sample sizes) for each cell of the design.

Levene's Test of Equality of Error Variances[a,b]

		Levene Statistic	df1	df2	Sig.
Psychological Distress	Based on Mean	.579	7	24	.766
	Based on Median	.417	7	24	.882
	Based on Median and with adjusted df	.417	7	15.398	.877
	Based on trimmed mean	.524	7	24	.807

Tests the null hypothesis that the error variance of the dependent variable is equal across groups.

a. Dependent variable: Psychological Distress

b. Design: Intercept + Sport + Selection + Sport * Selection

The *F* test (and associated *p* value) for Levene's Test for Equality of Error Variances is reviewed to determine if equal variances can be assumed. In this case, we meet the assumption (as *p* is greater than α). Note that *df1* is calculated as $(JK - 1)$ and *df2* is calculated as $(N - JK)$.

(continued)

TABLE 3.9 (continued)

Selected SPSS Results for the Elite Athlete Psychological Distress Example

The omnibus *F* test for the main effect for "Sport" (i.e., type of sport) (and computed similarly for the other main effects and interactions) is computed as $F = \frac{MS_A}{MS_{with}} = \frac{246.198}{11.531} = 21.350$

The *p* value for the omnibus *F* test for the main effect for "sport" is .000. This indicates there is astatistically significant difference in the dependent variable based on type of sport, averaged across selection status (i.e., deselected or selected).

In other words, there is a unique effect of type of sport on psychological distress, controlling for selection status. The probability of observing these mean differences or more extreme mean differences by chance if the null hypothesis is really true (i.e., if the population means are really equal) is less than 1%. We reject the null hypothesis that the population means of type of sport are equal. For our example, this provides evidence to suggest that psychological distress differs, on average, across type of sport in which an athlete

Tests of Between-Subjects Effects

Dependent Variable: Psychological Distress

Source	Type III Sum of Squares	df	Mean Square	F	Sig.	Partial Eta Squared	Noncent. Parameter	Observed Power[b]
Corrected Model	1472.969[a]	7	210.424	18.248	.000	.842	127.737	1.000
Intercept	10841.281	1	10841.281	940.165	.000	.975	940.165	1.000
Sport	738.594	3	246.198	21.350	.000	.727	64.051	1.000
Selection	712.531	1	712.531	61.791	.000	.720	61.791	1.000
Sport * Selection	21.844	3	7.281	.631	.602	.073	1.894	.162
Error	276.750	24	11.531					
Total	12591.000	32						
Corrected Total	1749.719	31						

a. R Squared = .842 (Adjusted R Squared = .796)

b. Computed using alpha = .05

R^2 is listed as a footnote underneath the table. R^2 is the ratio of sum of squares between (i.e., combined *SS* for main effects and for the interaction) divided by sum of squares total:

$$R^2 = \frac{SS_{betw}}{SS_{total}}$$

$$R^2 = \frac{738.594 + 712.531 + 21.844}{1749.719} = .842$$

The row labeled "Error" is for within groups. The within groups sum of squares tells us how much variation there is within the cells combined across the cells (i.e., 276.750). The degrees of freedom for within groups is ($N - JK$) or the sample size minus the number of levels of the independent variables [i.e., 32 – (4)(2) = 24].

The row labeled **"Corrected Total"** is the sum of squares total. The degrees of freedom for the total is ($N-1$) or the total sample size minus 1.

Observed power tells whether our test is powerful enough to detect mean differences if they really exist. Power of 1.000 indicates the maximum probability of rejecting the null hypothesis if it is really false (i.e., very strong power).

TABLE 3.9 (continued)

Selected SPSS Results for the Elite Athlete Psychological Distress Example

Estimated Marginal Means

1. Type of Sport

Dependent Variable: Psychological Distress

Type of Sport	Mean	Std. Error	95% Confidence Interval Lower Bound	Upper Bound
Movement	11.125	1.201	8.647	13.603
Target	17.875	1.201	15.397	20.353
Fielding	20.250	1.201	17.772	22.728
Territory	24.375	1.201	21.897	26.853

The table labeled **"Type of Sport"** provides descriptive statistics for each of the categories of the first independent variable. In addition to means, the *SE* and 95% CI of the means are reported.

2. Selection Status

Dependent Variable: Psychological Distress

Selection Status	Mean	Std. Error	95% Confidence Interval Lower Bound	Upper Bound
Deselected	23.125	.849	21.373	24.877
Selected	13.688	.849	11.935	15.440

The table labeled **"Selection Status"** provides descriptive statistics for each of the categories of the second independent variable. In addition to means, the *SE* and 95% CI of the means are reported.

3. Type of Sport * Selection Status

Dependent Variable: Psychological Distress

Type of Sport	Selection Status	Mean	Std. Error	95% Confidence Interval Lower Bound	Upper Bound
Movement	Deselected	15.250	1.698	11.746	18.754
	Selected	7.000	1.698	3.496	10.504
Target	Deselected	22.750	1.698	19.246	26.254
	Selected	13.000	1.698	9.496	16.504
Fielding	Deselected	26.250	1.698	22.746	29.754
	Selected	14.250	1.698	10.746	17.754
Territory	Deselected	28.250	1.698	24.746	31.754
	Selected	20.500	1.698	16.996	24.004

The table labeled **"Type of Sport * Selection Status"** provides descriptive statistics for each of the categories of the first independent variable by the second independent variable (i.e., cell means) (notice that these are the same means reported previously). In addition to means, the *SE* and 95% CI of the means are reported.

(continued)

TABLE 3.9 (continued)

Selected SPSS Results for the Elite Athlete Psychological Distress Example

Post Hoc Tests

Type of Sport

When requested, post hoc tests are conducted for main effects that have more than two levels

"Mean Difference" is simply the difference between the means of the two levels of type of sport being compared. For example, the mean difference of Movement and Target is calculated as 11.1250 – 17.8750 = -6.7500.

Multiple Comparisons

Dependent Variable: Psychological Distress

Tukey HSD

(I) Type of Sport	(J) Type of Sport	Mean Difference (I-J)	Std. Error	Sig.	95% Confidence Interval Lower Bound	95% Confidence Interval Upper Bound
Movement	Target	–6.7500*	1.69788	.003	–11.4338	–2.0662
	Fielding	–9.1250*	1.69788	.000	–13.8088	–4.4412
	Territory	–13.2500*	1.69788	.000	–17.9338	–8.5662
Target	Movement	6.7500*	1.69788	.003	2.0662	11.4338
	Fielding	–2.3750	1.69788	.512	–7.0588	2.3088
	Territory	–6.5000*	1.69788	.004	–11.1838	–1.8162
Fielding	Movement	9.1250*	1.69788	.000	4.4412	13.8088
	Target	2.3750	1.69788	.512	–2.3088	7.0588
	Territory	–4.1250	1.69788	.098	–8.8088	.5588
Territory	Movement	13.2500*	1.69788	.000	8.5662	17.9338
	Target	6.5000*	1.69788	.004	1.8162	11.1838
	Fielding	4.1250	1.69788	.098	–.5588	8.8088

Based on observed means.

The error term is Mean Square(Error) = 11.531.

*. The mean difference is significant at the .05 level.

The standard error calculated in SPSS uses the harmonic mean (Tukey-Kramer modification) where n_j and n_k are the sample sizes for the two groups whose means are being compared (Toothaker, 1993):

$$s_{\psi'} = \sqrt{MS_{error}\left(\frac{1}{n_j}+\frac{1}{n_k}\right)}$$

$$s_{\psi'} = \sqrt{11.531\left(\frac{1}{8}+\frac{1}{8}\right)}$$

$$s_{\psi'} = \sqrt{2.88275} = 1.69766$$

"Sig." denotes the observed p values and provides the results of the contrasts. There are four statistically significant mean differences between: 1) Movement and Target; 2) Movement and Fielding; 3) Movement and Territory; and 4) Target and Territory

Note that there are only ***6 unique contrast*** results:

$$\frac{1}{2}\left[J(J-1)\right] = \frac{1}{2}\left[4(4-1)\right] = \frac{1}{2}(12) = 6$$

Thus there are redundant results presented in the table. For example, the comparison of Movement and Target (presented in results row 1) is the same as the comparison of Target and Movement (presented in results row 2).

TABLE 3.9 (continued)

Selected SPSS Results for the Elite Athlete Psychological Distress Example

Homogeneous Subsets

Psychological Distress

Tukey HSD[a,b]

Type of Sport	N	Subset 1	Subset 2	Subset 3
Movement	8	11.1250		
Target	8		17.8750	
Fielding	8		20.2500	20.2500
Territory	8			24.3750
Sig.		1.000	.512	.098

Means for groups in homogeneous subsets are displayed.

Based on observed means.

The error term is Mean Square(Error) = 11.531.

a. Uses Harmonic Mean Sample Size = 8.000.

b. Alpha = .05.

This table displays the means for the types of sports that are *not* statistically significantly different. *We read the table by columns.* For example, in **subset 1,** the only mean displayed is Movement. *This indicates that Movement is statistically significantly different than all other sports.*

In **subset 2,** the means for Target and Fielding are displayed, indicating that those group means are "homogeneous" or *not* significantly different. The means for Movement and Territory are not displayed in subset 2, which indicates those means *are* statistically different from each other.

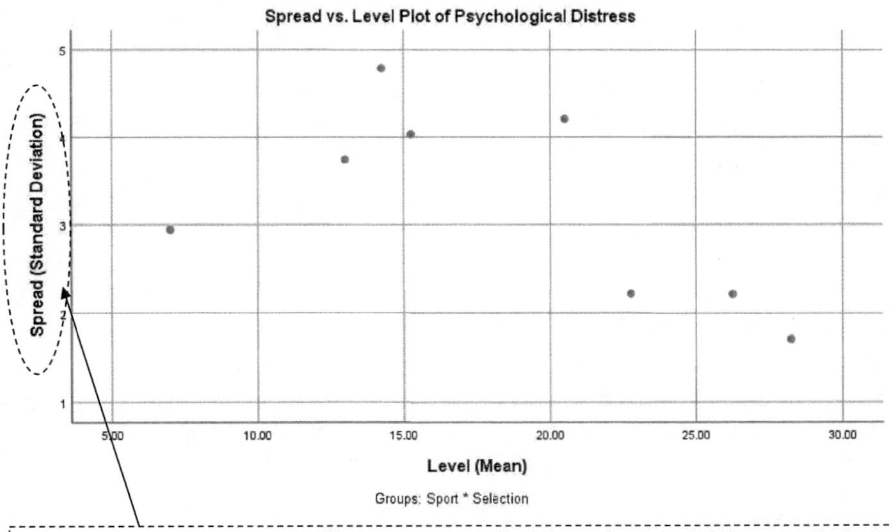

Spread vs. level plots are plots of the *dependent variable standard deviations (or variances)* against the *cell means.* These plots can be used to determine what to do when the homogeneity of variance assumption has been violated (remember, we already have evidence of meeting the homogeneity of variance assumption). *In addition to Levene's test, homogeneity is suggested when the spread vs. level plots provide a random display of points (i.e., no systematic pattern).*

If the plot suggests a linear relationship between the standard deviation and mean, transforming the data by taking the log of the dependent variable values may be a solution to the heterogeneity (since the calculation of logarithms requires positive values, this assumes all the data values are positive). If there is a linear relationship between the variance and mean, transforming the data by taking the square root of the dependent variable values may be a solution to the heterogeneity (since the calculation of square roots requires positive values, this assumes all the data values are positive).

Note: This plot displays the standard deviations. The plot for variances is not displayed for brevity; however, you will find the variance plot looks nearly identical with the exception of the scale of the *Y* axis.

(continued)

TABLE 3.9 (continued)

Selected SPSS Results for the Elite Athlete Psychological Distress Example

Profile Plots

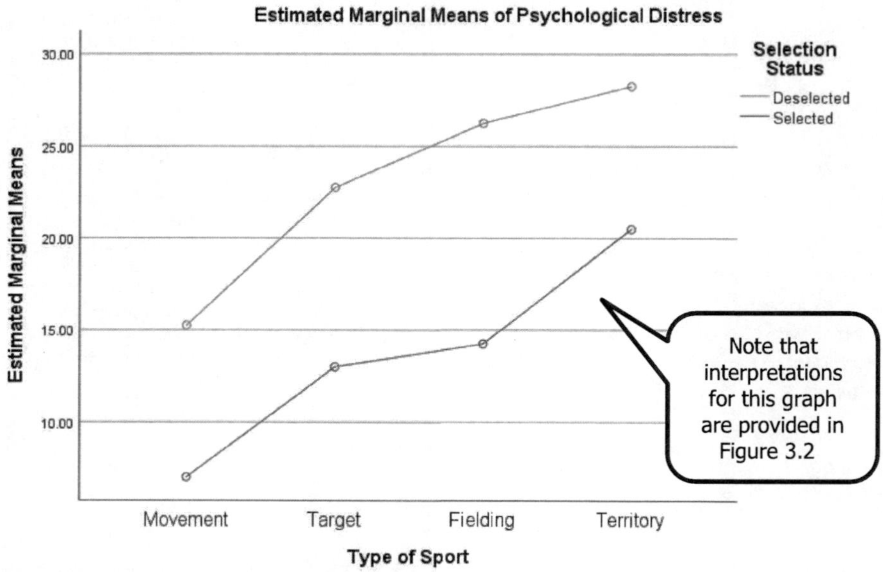

3.4.1 Testing a Statistically Significant Interaction

In this illustration, the interaction was not statistically significant. However, if you find yourself in a situation with a statistically significant interaction, there are some additional steps that you need to take to examine the **simple effects**. We will go back to our model, which should still be defined within SPSS, and click on "EM Means" (see the main dialog box in Step 2, Figure 3.6).

From the EM Means dialog box, we can define the factors and factor interactions for which we want to compute simple effects. We are interested in the interaction of type of sport and selection status (Sport*Selection) but to induce the option to "Compare main effects," we must include at least one main effect (i.e., one independent variable) in the "Display Means for" box. We will check the "Compare main effects" box. We will leave the default option of LSD, which is Fisher's Least Significant Difference (i.e., unadjusted probabilities). We could have selected Bonferroni; just remember that Bonferroni will be conservative if there are many comparisons that are made. Other options to which we are accustomed (e.g., Tukey's) are not available through the EM Means tool. Then click "Continue" to return to the main Univariate dialog box.

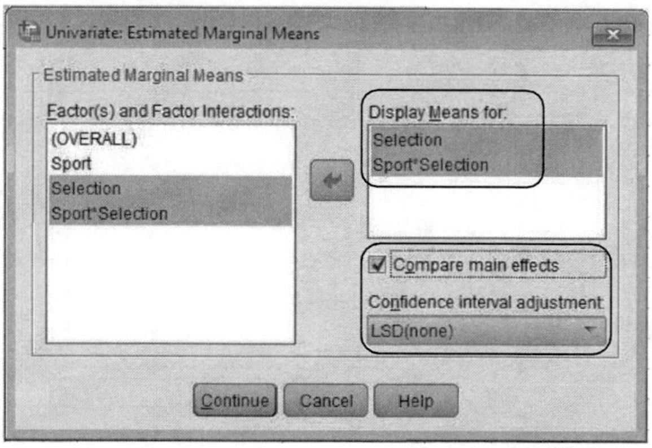

FIGURE 3.13
EM Means dialog box.

Interaction Step 2. Rather than clicking on "OK," we will click on "Paste" (see Figure 3.6). This will open a syntax box with the following script (for illustrative purposes, we have removed all other commands that are not necessary at this point):

```
UNIANOVA Distress BY Sport Selection
/METHOD=SSTYPE(3)
/INTERCEPT=INCLUDE
/EMMEANS=TABLES(Selection) COMPARE ADJ(LSD)
/EMMEANS=TABLES(Sport*Selection)
/CRITERIA=ALPHA(0.05)
/DESIGN=Sport Selection Sport*Selection.
```

The line of code that reads /EMMEANS=TABLES(Sport*Selection) is the syntax that we will adjust by adding the syntax, COMPARE ADJ(LSD) and including the name of the factor of interest in parentheses in this syntax (i.e., **COMPARE (Sport) ADJ (LSD)** to our syntax). This line of syntax now reads:

```
/EMMEANS=TABLES(Sport*Selection) COMPARE (Sport) ADJ (LSD)
```

We will simply copy and paste this line of code back into the previous syntax and remove the unnecessary line of code for comparing the main effect for selection (if we choose). This will result in the following syntax (where ADJ specifies the adjustment to the p value for each pairwise comparison):

```
UNIANOVA Distress BY Sport Selection
/METHOD=SSTYPE(3)
/INTERCEPT=INCLUDE
/EMMEANS=TABLES(Sport*Selection) COMPARE (Sport) ADJ(LSD)
/CRITERIA=ALPHA(0.05)
/DESIGN=Sport Selection Sport*Selection.
```

TABLE 3.10

Results of Simple Effects Tests for Statistically Significant Interactions

Pairwise Comparisons

Dependent Variable: Psychological Distress

Selection Status	(I) Type of Sport	(J) Type of Sport	Mean Difference (I-J)	Std. Error	Sig.[b]	95% Confidence Interval for Difference[b] Lower Bound	Upper Bound
Deselected	Movement	Target	−7.500*	2.401	.005	−12.456	−2.544
		Fielding	−11.000*	2.401	.000	−15.956	−6.044
		Territory	−13.000*	2.401	.000	−17.956	−8.044
	Target	Movement	7.500*	2.401	.005	2.544	12.456
		Fielding	−3.500	2.401	.158	−8.456	1.456
		Territory	−5.500*	2.401	.031	−10.456	−.544
	Fielding	Movement	11.000*	2.401	.000	6.044	15.956
		Target	3.500	2.401	.158	−1.456	8.456
		Territory	−2.000	2.401	.413	−6.956	2.956
	Territory	Movement	13.000*	2.401	.000	8.044	17.956
		Target	5.500*	2.401	.031	.544	10.456
		Fielding	2.000	2.401	.413	−2.956	6.956
Selected	Movement	Target	−6.000*	2.401	.020	−10.956	−1.044
		Fielding	−7.250*	2.401	.006	−12.206	−2.294
		Territory	−13.500*	2.401	.000	−18.456	−8.544
	Target	Movement	6.000*	2.401	.020	1.044	10.956
		Fielding	−1.250	2.401	.607	−6.206	3.706
		Territory	−7.500*	2.401	.005	−12.456	−2.544
	Fielding	Movement	7.250*	2.401	.006	2.294	12.206
		Target	1.250	2.401	.607	−3.706	6.206
		Territory	−6.250*	2.401	.016	−11.206	−1.294
	Territory	Movement	13.500*	2.401	[illegible]00	8.544	18.456
		Target	7.500*	2.401	[illegible]05	2.544	12.456
		Fielding	6.250*	2.[illegible]	[illegible]16	1.294	11.206

Based on estimated marginal means

*Mean difference is significant at the 0.05 level.

b. Adjustment for multiple comparisons: **Least Significant Difference** (equivalent to no adjustments).

When players are *deselected,* there is no statistically significant difference in psychological distress between athletes in fielding as compared to territory sports. However, when players are *selected,* there is a statistically significant difference in psychological distress. More specifically, athletes in territory sports have statistically significantly more psychological distress when *selected* as compared to athletes in fielding sports (mean difference = -6.250, p = .016) but athletes in these sports are similar in psychological distress when *deselected.*

We paste this into our syntax window, highlight these six lines of code, and then click the green triangle to run it (or click "Run," then "Selection" in the top toolbar of the syntax window) and generate our output, which now includes a test of simple effects.

Table 3.10 includes the new table of simple effects that is generated to test the interactions. Recall that an interaction means that the effect of one factor depends on the level of another factor (and vice versa). In the elite athlete example, that would mean that the effect of selection status depends on the type of sport (and vice versa). In other words, the interaction means that psychological distress depends on the type of sport in which the athlete participates and selection status. Conversely, the effect of selection status on psychological distress depends on the type of sport in which the athlete participates. The simple effects, for which we just generated output, reveal the degree to which one independent variable is differentially effective at every level of the second independent variable.

Although we see lots of statistically significant results, we are specifically looking for comparisons where the results for a factor level are statistically significant in one level of a second factor but *not* statistically significant for the other level of that second factor. We see, for example, that *movement* is statistically significantly different than all other types of sports, regardless of whether the play is *deselected* or *selected*. In other words, we are not seeing an interaction between selection status and type of sport when considering the comparisons of *movement* to the other types of sport.

Our results for the simple effects of the interaction generate *one statistically significant simple effect*. Let's interpret the interaction of selection status (selected, deselected) with type of sport when the sports of *fielding* and *territory* are considered. When players are *deselected*, there is no statistically significant difference in psychological distress between athletes in fielding as compared to territory sports. However, when players are *selected*, there is a statistically significant difference in psychological distress. More specifically, athletes in territory sports have statistically significantly more psychological distress when *selected* as compared to athletes in fielding sports (mean difference = −6.250, p = .016), but athletes in these sports are similar in psychological distress when *deselected*. Because our omnibus F test for the interaction was not statistically significant, we will not interpret this comparison later in our write-up. However, it should provide an illustration that will assist you should you find a statistically significant interaction in your own research.

3.5 Computing Factorial ANOVA Using R

Next we consider **R** for factorial ANOVA. Note that the scripts are provided within the blocks with additional annotation to assist in understanding how the command works. Should you want to write reminder notes and annotation to yourself as you write the commands in **R** (and we highly encourage doing so), remember that any text that follows a hashtag (i.e., #) is annotation only and not part of the **R** script. Thus, you can write annotations directly into **R** with hashtags. We encourage this practice so that when you call up the commands in the future, you'll understand what the various lines of code are doing. You may think you'll remember what you did. However, trust us. There is a good chance that you won't. Thus, consider it best practice when using **R** to annotate heavily!

3.5.1 Reading Data Into R

```
getwd()
```

R is always pointed to a directory on your computer. The *get working directory* function can be used to determine to which directory **R** is pointed. We will assume that we need to change the working directory, and will use the next line of code to set the working directory to the desired path.

```
setwd("E:/FolderName")
```

We use the *setwd* function to establish the working directory. To set the working directory, change what is in quotation marks to your file location. Also, if you are copying the directory name from your properties, you will need to change the backslash (i.e., \) to a forward slash (i.e., /).

```
Ch3_distress <- read.csv("Ch3_psychdistress.csv")
```

The *read.csv* function reads our data into **R**. What's to the left of the "<-" will be what the data will be called in **R**. In this example, we're calling the R dataframe "Ch3_distress." What's to the right of the "<-" tells **R** to find this particular csv file. In this example, our file is called "Ch3_psychdistress.csv." Make sure the extension (i.e., .csv) is included in your script. Also note that the name of your file should be in quotation marks within the parentheses.

```
names(Ch3_distress)
```

The *names* function will produce a list of variable names for each dataframe as follows. This is a good check to make sure your data have been read in correctly.

```
[1] "Sport" "Selection" "Distress"
```

```
View(Ch3_distress)
```

The *View* function will let you view the dataset in spreadsheet format in RStudio.

```
install.packages(car)
```

We will be using the *car* package for Levene's test. This function will install the package in **R**.

```
library(car)
```

The *library* function will load the *car* package in our library.

```
install.packages("compute.es")
```

We will use the *compute.es* package to compute effect sizes. The *install.packages* function will install the package in **R**.

```
library(compute.es)
```

The *library* function will load the *compute.es* package in our library.

```
Ch3_distress$SportF <- factor(Ch3_distress$Sport,
labels = c("movement", "target", "fielding", "territory"))
```

The *factor* function will create a new variable in our dataframe named "SportF." We use the *factor* function to define the variable *Sport* as nominal with the four groups defined here (i.e., movement, target, fielding, territory). What is to the left of "<-" in the script creates the new *SportF* variable in our dataframe.

FIGURE 3.14
Reading data into **R**.

```
Ch3_distress$SelectionF <- factor(Ch3_distress$Selection,
labels = c("deselected","selected"))
```

The *factor* function will create a new variable in our dataframe named "SelectionF." We use the *factor* function to define the variable "Selection" as nominal with the two groups defined here (i.e., deselected, selected). What is to the left of "<-" in the script creates the new *SelectionF* variable in our dataframe.

```
summary(Ch3_distress)
```

The *summary* function will produce basic descriptive statistics on all the variables in your dataframe. This is a great way to quickly check to see if the data have been read in correctly and to get a feel for your data, if you haven't already. The output from the summary statement for this dataframe looks like this. Because we defined *SportF* and *SelectionF* as factors, we are provided only the frequencies for each category in those variables.

```
    Sport         Selection       Distress          SportF        SelectionF
Min.   :1.00   Min.   :1.0   Min.   : 3.00   movement :8   deselected:16
1st Qu.:1.75   1st Qu.:1.0   1st Qu.:12.00   target   :8   selected  :16
Median :2.50   Median :1.5   Median :20.00   fielding :8
Mean   :2.50   Mean   :1.5   Mean   :18.41   territory:8
3rd Qu.:3.25   3rd Qu.:2.0   3rd Qu.:25.00
Max.   :4.00   Max.   :2.0   Max.   :30.00
```

FIGURE 3.14 (continued)
Reading data into **R**.

3.5.2 Generating the Factorial ANOVA

```
Ch3_2way <- aov(Distress ~ SportF*SelectionF, data=Ch3_distress)
```

The *aov* function will generate the factorial ANOVA model with "Distress" as the dependent variable and "SportF" and "SelectionF" as the independent variables. The main effects and the interaction of these variables will be generated with the command *Sport*Selection*. Had we included more independent variables, we would have simply continued adding them to this command line with asterisks such as A*B*C. We are using data from the Ch3_distress dataframe, and we are calling this object "Ch3_2way."

```
                  Df  Sum Sq  Mean Sq  F value    Pr(>F)
SportF             3   738.6    246.2   21.350  5.86e-07 ***
SelectionF         1   712.5    712.5   61.791  4.30e-08 ***
SportF:SelectionF  3    21.8      7.3    0.631     0.602
Residuals         24   276.7     11.5
---
Signif. codes: 0 '***' 0.001 '**' 0.01 '*' 0.05 '.' 0.1 ' ' 1
```

```
summary.lm(Ch3_2way)
```

The *summary.lm* function will produce additional output, including R^2.

```
Call:
aov(formula = Distress ~ SportF * SelectionF, data = Ch3_distress)

Residuals:
   Min      1Q  Median     3Q    Max
-5.500  -2.250  -0.250  1.562  6.750
```

FIGURE 3.15
Generating factorial ANOVA.

```
Coefficients:
                                    Estimate  Std. Error  t value    Pr(>|t|)
(Intercept)                           15.250       1.698    8.982    3.83e-09 ***
SportFtarget                           7.500       2.401    3.123     0.00462 **
SportFfielding                        11.000       2.401    4.581     0.00012 ***
SportFterritory                       13.000       2.401    5.414    1.46e-05 ***
SelectionFselected                    -8.250       2.401   -3.436     0.00216 **
SportFtarget:SelectionFselected       -1.500       3.396   -0.442     0.66264
SportFfielding:SelectionFselected     -3.750       3.396   -1.104     0.28041
SportFterritory:SelectionFselected     0.500       3.396    0.147     0.88417
---
Signif. codes: 0 '***' 0.001 '**' 0.01 '*' 0.05 '.' 0.1 ' ' 1

Residual standard error: 3.396 on 24 degrees of freedom
Multiple R-squared: 0.8418, Adjusted R-squared: 0.7957
F-statistic: 18.25 on 7 and 24 DF, p-value: 3.476e-08
```

```
Ch3_distress$unstandardizedResiduals <- residuals(Ch3_2way)
```

We also want to save our unstandardized residuals to the dataframe. We use the *residuals* function to compute unstandardized residuals from our Ch3_2way model. To the left of "<-" we will save the residuals as a variable named "unstandardizedResiduals" in our *Ch3_distress* dataframe.

```
install.packages("sjstats")
library(sjstats)
```

Installing and loading into the library the *sjstats* package will provide another great function for generating the ANOVA summary table, along with multiple effect size indices.

```
anova_stats(Ch3_2way)
```

```
                 term  df    sumsq   meansq statistic p. value etasq partial.etasq
1              SportF   3  738.594  246.198    21.350    0.000 0.422         0.727
2          SelectionF   1  712.531  712.531    61.791    0.000 0.407         0.720
3   SportF:SelectionF   3   21.844    7.281     0.631    0.602 0.012         0.073
4           Residuals  24  276.750   11.531        NA       NA    NA            NA

  omegasq partial.omegasq cohens.f  power
1   0.400           0.656    1.634  1.000
2   0.398           0.655    1.605  1.000
3  -0.007          -0.036    0.281  0.183
4      NA              NA       NA     NA
```

FIGURE 3.15 (continued)
Generating factorial ANOVA.

3.5.3 Generating Tests for Homogeneity of Variance

```
install.packages("car")
library(car)
```

We use the *car* package to run Levene's Test, so we will install using the *install.packages* function (if not already installed; if you have already installed this package, you can skip the install step) and then load into our library using the *car* package.

```
LeveneTest(Ch3_distress$Distress,
interaction(Ch3_distress$SportF,
Ch3_distress$SelectionF),
           center=mean)
```

```
Levene's Test for Homogeneity of Variance (center = mean)
      Df F value Pr(>F)
group  7 0.5785 0.7663
      24
```

We read this output as F(7,24) = .5785, p = .7663, indicating we have met the assumption of equal variances.

```
leveneTest(Ch3_2way)
```

We can also run the *leveneTest* function on the object (Ch3_2way) of our factorial ANOVA model results to generate Levene's test with the default centering of the median, which may provide more robust results. These results still provide evidence of meeting the assumption of equal variances, with $p = .882$.

```
Levene's Test for Homogeneity of Variance (center = median)
      Df F value Pr(>F)
group  7 0.4172 0.882
      24
```

FIGURE 3.16
Generating homoscedasticity tests.

3.5.4 Generating Post Hoc Tests

```
Ch3_tukey <- TukeyHSD(Ch3_2way)
```

The *TukeyHSD* function will generate Tukey's HSD post hoc analysis on our factorial ANOVA model, Ch3_2way, and will name the object "Ch3_tukey."

```
Ch3_tukey
```

This will output the results of the Tukey's HSD post hoc analysis from the previous command.

```
   Tukey multiple comparisons of means
     95% family-wise confidence level

Fit: aov(formula = Distress ~ SportF * SelectionF, data = Ch3_distress)

$SportF
                       diff        lwr       upr     p adj
target-movement       6.750  2.0662003  11.4338  0.0029427
fielding-movement     9.125  4.4412003  13.8088  0.0000901
territory-movement   13.250  8.5662003  17.9338  0.0000003
fielding-target       2.375 -2.3087997   7.0588  0.5122072
territory-target      6.500  1.8162003  11.1838  0.0042196
territory-fielding    4.125 -0.5587997   8.8088  0.0982713
```

FIGURE 3.17
Generating post hoc comparisons.

As we already know from the *F* test, there were differences between at least some of the types of sport. Tukey's post hoc tells us there are statistically significant differences between *movement* and all other sports as well as differences between *territory* and *target*.

```
$SelectionF
                       diff        lwr       upr  p adj
selected-deselected -9.4375  -11.91539 -6.959613      0
```

As we already know from the *F* test, there is a statistically significant difference between the selection status groups. This confirms that finding again.

```
$'SportF:SelectionF'
                                        diff         lwr         upr
target:deselected-movement:deselected   7.50  -0.4524709  15.4524709
fielding:deselected-movement:deselected 11.00  3.0475291  18.9524709
territory:deselected-movement:deselected 13.00 5.0475291  20.9524709
```

For brevity, and because there was not a statistically significant omnibus interaction, the complete results for the post hoc comparisons of the "sport by selection" interaction are not presented here.

FIGURE 3.17 (continued)
Generating post hoc comparisons.

3.5.5 Computing Effect Size

```
install.packages("sjstats")
library(sjstats)
```

The package *sjstats* can be used to generate multiple effect size indices in ANOVA. The *install.packages* and *library* functions will, respectively, install the package and then load it into our **R** library.

```
omega_sq(Ch3_2way)
```

Using the object created from our ANOVA model, Ch3_2way, we can generate omega squared with the *omega_sq* function. We see that ω^2 for type of sport (i.e., A) is approximately .40, for selection status (i.e., B) is also about .40, and for the interaction of type of sport and selection status is about −.007.

```
               term  omegasq
1            SportF    0.400
2        SelectionF    0.398
3 SportF:SelectionF   -0.007
```

```
omega_sq(Ch3_2way, partial = TRUE)
```

Using the object created from our ANOVA model, Ch3_2way, we can generate partial omega squared with the *omega_sq* function, defining *partial = TRUE*. The partial variance effect size is interpreted as the proportion of variation in the dependent variable, *Y*, explained by the effect of interest (i.e., by factor A, or factor B, or the AB interaction) that is *not* explained by other variables in the model.

```
               term  partial.omegasq
1            SportF            0.656
2        SelectionF            0.655
3 SportF:SelectionF           -0.036
```

FIGURE 3.18
Generating effect size.

```
cohens_f(Ch3_2way)
```

Using the object created from our ANOVA model, Ch3_2way, we can generate Cohen's f with the *cohens_f* function. The effect f can take on values from zero (when the means are equal) to an infinitely large positive value. This effect is interpreted as an approximate correlation index but can also be interpreted as the standard deviation of the standardized means (Cohen, 1988). Small effects for $f = .1$, moderate $f = .25$, and large effect $f = .40$.

```
               term  cohens.f
1            SportF  1.633650
2        SelectionF  1.604568
3 SportF:SelectionF  0.280944
```

```
eta_sq(Ch3_2way)
```

Using the object created from our ANOVA model, Ch3_2way, we can generate eta squared with the *eta_sq* function.

```
               term  etasq
1            SportF  0.422
2        SelectionF  0.407
3 SportF:SelectionF  0.012
```

```
eta_sq(Ch3_2way, partial = TRUE)
```

Using the object created from our ANOVA model, Ch3_2way, we can generate partial eta squared with the *eta_sq* function, defining *partial = TRUE*.

```
               term  partial.etasq
1            SportF          0.727
2        SelectionF          0.720
3 SportF:SelectionF          0.073
```

```
anova_stats(Ch3_2way)
```

The *anova_stats* function can be used with our ANOVA model, Ch3_2way, to present a comprehensive summary, including effect size measures and power.

```
               term df   sumsq  meansq statistic p. value
1            SportF  3 738.594 246.198    21.350    0.000
2        SelectionF  1 712.531 712.531    61.791    0.000
3 SportF:SelectionF  3  21.844   7.281     0.631    0.602
4         Residuals 24 276.750  11.531        NA       NA

  etasq partial.etasq omegasq partial.omegasq cohens.f
1 0.422         0.727   0.400           0.656    1.634
2 0.407         0.720   0.398           0.655    1.605
3 0.012         0.073  -0.007          -0.036    0.281
4    NA            NA      NA              NA       NA

  power
1 1.000
2 1.000
3 0.183
4    NA
```

FIGURE 3.18 (continued)
Generating effect size.

3.6 Data Screening

3.6.1 Normality

We will use the residuals (which were requested and created through the "Save" option when generating our factorial ANOVA) to examine the extent to which normality was met.

As alluded to earlier in the chapter, understanding the distributional shape, specifically the extent to which normality is a reasonable assumption, is important. For factorial ANOVA, the distributional shape for the residuals by group should be a normal distribution.

The general steps for accessing "Explore" have been presented in previous chapters and will not be repeated here. Click the residual and move it into the "Dependent List" box by clicking on the arrow button. For dependent variable by group, click the factor variables (in this case, "Sport" and "Selection") and move to the "Factor List" box. The procedures for selecting normality statistics include: click on "Plots" in the upper right corner. Place a checkmark in the boxes for "Normality plots with tests" and also for "Histogram." Then click "Continue" to return to the main Explore dialog box. Then click "OK" to generate the output.

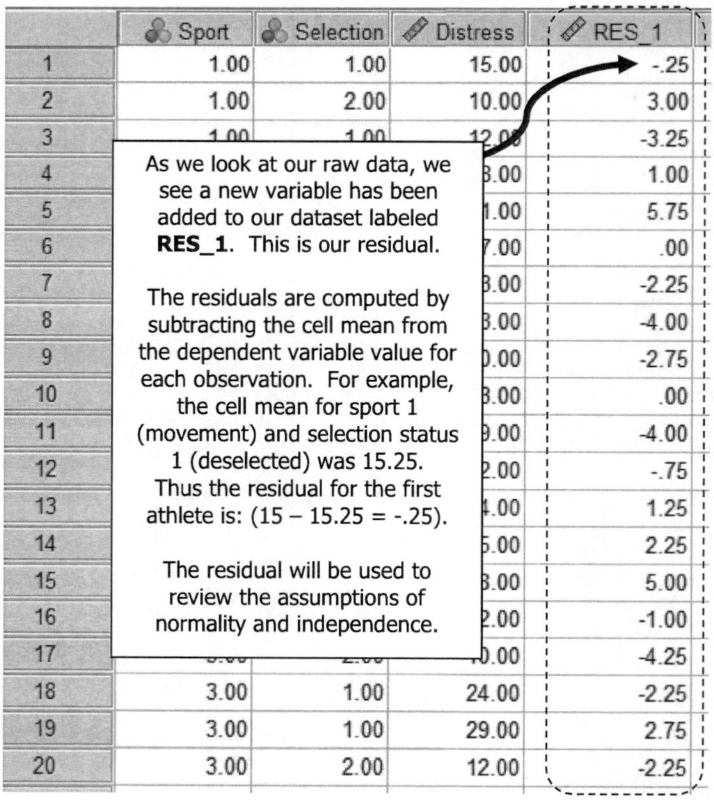

FIGURE 3.19
First 20 cases of residual data.

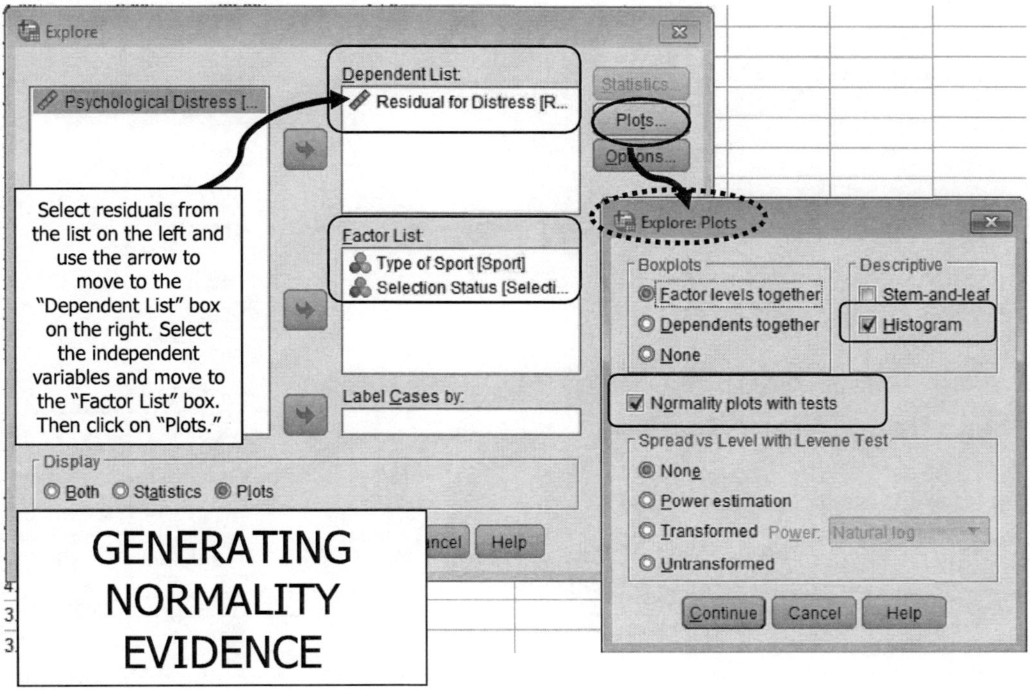

FIGURE 3.20
Generating normality evidence.

3.6.1.1 Interpreting Normality Evidence

We have already developed a good understanding of how to interpret some forms of evidence of normality including skewness and kurtosis, histograms, and boxplots. The skewness statistic of the residuals, overall, is .400 and kurtosis is –.162—both within the range of an absolute value of 2.0, suggesting some evidence of normality. By group, skewness and kurtosis are all within this range as well. Working in **R**, **D'Agostino's test** (D'Agostino, 1970) can be used to examine the null hypothesis that skewness equals zero. Thus, a statistically significant D'Agostino's test would indicate that there is statistically significant skewness. For kurtosis, we can use the **Bonett-Seier test for Geary's kurtosis** (Bonett & Seier, 2002) for data that are normally distributed. The null hypothesis states that data should have a Geary's kurtosis value equal to $\sqrt{2/\pi} = .7979$. Thus, a statistically significant Bonett-Seier test for Geary's kurtosis would indicate that there is statistically significant kurtosis. Thus, with these tests, as with Kolmogorov-Smirnov and Shapiro-Wilk, we do *not* want to find statistically significant results—and that's exactly what we found (i.e., $p >$ alpha).

Descriptives

			Statistic	Std. Error
Residual for Distress	Mean		.0000	.52819
	95% Confidence Interval for Mean	Lower Bound	-1.0772	
		Upper Bound	1.0772	
	5% Trimmed Mean		-.0747	
	Median		-.2500	
	Variance		8.927	
	Std. Deviation		2.98788	
	Minimum		-5.50	
	Maximum		6.75	
	Range		12.25	
	Interquartile Range		3.94	
	Skewness		.400	.414
	Kurtosis		-.162	.809

By group of the independent variable, we find the following, all indicating the normality assumption has been met.

Statistics

Residual for Distress

Movement	**Deselected**	N	Valid	4
			Missing	0
		Skewness		**1.469**
		Std. Error of Skewness		1.014
		Kurtosis		**2.031**
		Std. Error of Kurtosis		2.619
	Selected	N	Valid	4
			Missing	0
		Skewness		**-.941**
		Std. Error of Skewness		1.014
		Kurtosis		**1.500**
		Std. Error of Kurtosis		2.619
Target	**Deselected**	N	Valid	4
			Missing	0
		Skewness		**-.482**
		Std. Error of Skewness		1.014

FIGURE 3.21
Normality evidence.

		Kurtosis		**-1.700**
		Std. Error of Kurtosis		2.619
	Selected	N	Valid	4
			Missing	0
		Skewness		**.764**
		Std. Error of Skewness		1.014
		Kurtosis		**1.500**
		Std. Error of Kurtosis		2.619
Fielding	**Deselected**	N	Valid	4
			Missing	0
		Skewness		**.482**
		Std. Error of Skewness		1.014
		Kurtosis		**-1.700**
		Std. Error of Kurtosis		2.619
	Selected	N	Valid	4
			Missing	0
		Skewness		**1.333**
		Std. Error of Skewness		1.014
		Kurtosis		**1.910**
		Std. Error of Kurtosis		2.619
Territory	**Deselected**	N	Valid	4
			Missing	0
		Skewness		**-.753**
		Std. Error of Skewness		1.014
		Kurtosis		**.343**
		Std. Error of Kurtosis		2.619
	Selected	N	Valid	4
			Missing	0
		Skewness		**-.646**
		Std. Error of Skewness		1.014
		Kurtosis		**.707**
		Std. Error of Kurtosis		2.619

Working in **R**, we can generate various normality statistics as well.

```
install.packages("pastecs")
```

The *install.packages* function will install the *pastecs* package which we will use to generate various forms of normality evidence.

FIGURE 3.21 (continued)
Normality evidence.

```
library(pastecs)
```

The *library* function will load the *pastecs* package.

```
stat.desc(Ch3_distress$unstandardizedResiduals,
          norm = TRUE)
```

The *stat.desc* function will generate normality indices on the variable "unstandardizedResiduals" in the dataframe *Ch3_distress* as follows. The *norm=TRUE* command will produce Shapiro-Wilk results (*SW*), which are displayed as *normtest.W* (which is the *SW* statistic value) and *normtest.p* (which is the observed probability value).

Here, we see $SW = .977$ and the related $p = .701$. We see skew (.363) and kurtosis (−.485) for the *unstandardizedResidual* variable.

Skew, kurtosis, and *SW* all indicate the assumption of normality has been met. As we know, we can divide the skew and kurtosis values by their standard errors to get a standardized value that can be used to determine if the skew and/or kurtosis is statistically different from zero. Since this output provides "2SE," we would simply divide this value by 2 to arrive at the standard error.

Note: You may have noticed that the skewness and kurtosis value that we've just generated differs from what we found in SPSS, which was skew = .400 and kurtosis = −.162. This is because there are different ways to calculate skewness and kurtosis. Let's use another package in **R** to calculate these statistics with different algorithms.

```
      nbr.val      nbr.null        nbr.na           min
 3.200000e+01  0.000000e+00  0.000000e+00 -5.500000e+00

          max         range           sum        median
 6.750000e+00  1.225000e+01 -1.318390e-15 -2.500000e-01

         mean       SE.mean  CI.mean.0.95           var
-4.119968e-17  5.281873e-01  1.077245e+00  8.927419e+00

      std.dev      coef.var      skewness     skew.2SE
 2.987879e+00 -7.252189e+16  3.633272e-01  4.383167e-01

     kurtosis     kurt.2SE     normtest.W    normtest.p
-4.847138e-01 -2.994385e-01  9.767450e-01  7.009705e-01
```

```
install.packages("e1071")
```

The *install.packages* function will install the e1071 package which we will use to generate skewness and kurtosis.

```
library(e1071)
```

The *library* function will load the e1071 package.

```
skewness(Ch3_distress$unstandardizedResiduals, type=3)
skewness(Ch3_distress$unstandardizedResiduals, type=2)
skewness(Ch3_distress$unstandardizedResiduals, type=1)
```

The *skewness* function will generate skewness statistics on the variable(s) specified. The "type=" script defines how skewness is calculated. Specifying "type=2" will use the algorithm that is used by SPSS. Readers interested in learning more, including the algorithms for each of the three methods, are encouraged to review Joanes and Gill (1998). We see that using type=2, our skew is .400, the same value as generated using SPSS.

```
# skewness(Ch3_distress$unstandardizedResiduals, type=3)
[1] 0.3633272
```

FIGURE 3.21 (continued)
Normality evidence.

```
# skewness(Ch3_distress$unstandardizedResiduals, type=2)
[1] 0.4000506

# skewness(Ch3_distress$unstandardizedResiduals, type=1)
[1] 0.3810486
```

```
kurtosis(Ch3_distress$unstandardizedResiduals, type=3)
kurtosis(Ch3_distress$unstandardizedResiduals, type=2)
kurtosis(Ch3_distress$unstandardizedResiduals, type=1)
```

The *kurtosis* function will generate kurtosis statistics on the variable(s) we specify. The "type=" script defines how kurtosis is calculated. Specifying "type=2" will use the algorithm that is used by SPSS. Readers interested in learning more, including the algorithms for each of the three methods, are encouraged to review Joanes and Gill (1998). We see that using "type=2," our kurtosis is −.162, the same value as generated using SPSS.

```
# kurtosis(Ch3_distress$unstandardizedResiduals, type=3)
[1] -0.4847138

# kurtosis(Ch3_distress$unstandardizedResiduals, type=2)
[1] -0.162271

# kurtosis(Ch3_distress$unstandardizedResiduals, type=1)
[1] -0.3198199
```

Working in **R**, another way to test for normality is D'Agostino's test for skewness and the Bonett-Seier test for Geary's kurtosis.

```
install.packages("moments")
library(moments)
```

To conduct D'Agostino's test, we first have to install the *moments* package and then load it into our library. The null hypothesis for this test is that skewness equals zero. Thus, a statistically significant D'Agostino's test would indicate that there is statistically significant skewness.

```
agostino.test(Ch3_distress$unstandardizedResiduals)
```

The function *agostino.test* is generated using the variable "unstandardizedResiduals" from our Ch3_distress dataframe. The results suggest evidence of normality for the residuals overall as p = .3154, greater than alpha.

```
        D'Agostino skewness test

data: Ch3_distress$unstandardizedResiduals
skew = 0.38105, z = 1.00390, p-value = 0.3154
alternative hypothesis: data have a skewness
```

```
agostino.test(Ch3_distress$unstandardizedResiduals[Ch3_distress$Sport==1])
agostino.test(Ch3_distress$unstandardizedResiduals[Ch3_distress$Sport==2])
agostino.test(Ch3_distress$unstandardizedResiduals[Ch3_distress$Sport==3])
agostino.test(Ch3_distress$unstandardizedResiduals[Ch3_distress$Sport==4])
agostino.test(Ch3_distress$unstandardizedResiduals[Ch3_distress$Selection==1])
agostino.test(Ch3_distress$unstandardizedResiduals[Ch3_distress$Selection==2])
```

By group, the results for the D'Agostino test provide evidence of normality by group with all p's > .05. For brevity, only the results from "Selection" are presented.

```
# agostino.test(Ch3_distress$unstandardizedResiduals[Ch3_distress$Selection==1])
```

FIGURE 3.21 (continued)
Normality evidence.

```
        D'Agostino skewness test
data: Ch3_distress$unstandardizedResiduals[Ch3_distress$Selection == 1]
skew = 0.68393, z = 1.37170, p-value = 0.1702
alternative hypothesis: data have a skewness

# agostino.test(Ch3_distress$unstandardizedResiduals[Ch3_distress$Selection==2])

        D'Agostino skewness test
data: Ch3_distress$unstandardizedResiduals[Ch3_distress$Selection == 2]
skew = 0.26226, z = 0.54189, p-value = 0.5879
alternative hypothesis: data have a skewness
```

```
bonett.test((Ch3_distress$unstandardizedResiduals))
```

The *bonett.test* function, generated using the variable "unstandardizedResiduals" from our Ch3_distress dataframe, performs the Bonett-Seier test for Geary's kurtosis for data that are normally distributed. The null hypothesis states that data should have a Geary's kurtosis value equal to $\sqrt{2/\pi} = .7979$. The results suggest evidence of normality as $p = .7488$, greater than alpha.

```
        Bonett-Seier test for Geary kurtosis

data: (Ch3_distress$unstandardizedResiduals)
tau = 2.31250, z = 0.32019, p-value = 0.7488
alternative hypothesis: kurtosis is not equal to sqrt(2/pi)
```

```
bonett.test((Ch3_distress$unstandardizedResiduals[Ch3_distress$Sport==1]))
bonett.test((Ch3_distress$unstandardizedResiduals[Ch3_distress$Sport==2]))
bonett.test((Ch3_distress$unstandardizedResiduals[Ch3_distress$Sport==3]))
bonett.test((Ch3_distress$unstandardizedResiduals[Ch3_distress$Sport==4]))
bonett.test((Ch3_distress$unstandardizedResiduals[Ch3_distress$Selection==1]))
bonett.test((Ch3_distress$unstandardizedResiduals[Ch3_distress$Selection==2]))
```

By group, the results for the Bonett-Seier test for Geary's kurtosis for data that are normally distributed provide evidence of normality by group with all p's > .05. For brevity, only the results from "Selection" are presented.

```
# bonett.test((Ch3_distress$unstandardizedResiduals[Ch3_distress$Selection==1]))

        Bonett-Seier test for Geary kurtosis

data: (Ch3_distress$unstandardizedResiduals[Ch3_distress$Selection == 1])
tau = 1.90630, z = -0.38564, p-value = 0.6998
alternative hypothesis: kurtosis is not equal to sqrt(2/pi)

# bonett.test((Ch3_distress$unstandardizedResiduals[Ch3_distress$Selection==2]))

        Bonett-Seier test for Geary kurtosis

data: (Ch3_distress$unstandardizedResiduals[Ch3_distress$Selection == 2])
tau = 2.71880, z = 0.16969, p-value = 0.8653
alternative hypothesis: kurtosis is not equal to sqrt(2/pi)
```

FIGURE 3.21 (continued)
Normality evidence.

As suggested by the skewness statistic, the histograms of residuals, overall, is slightly positively skewed, but it approaches a normal distribution and there is nothing to suggest normality may be an unreasonable assumption. Similarly, the histograms by group suggest some skew (for brevity, only residuals by selection status are presented). Additional normality indices will be reviewed to better understand the extent that normality may be reasonable.

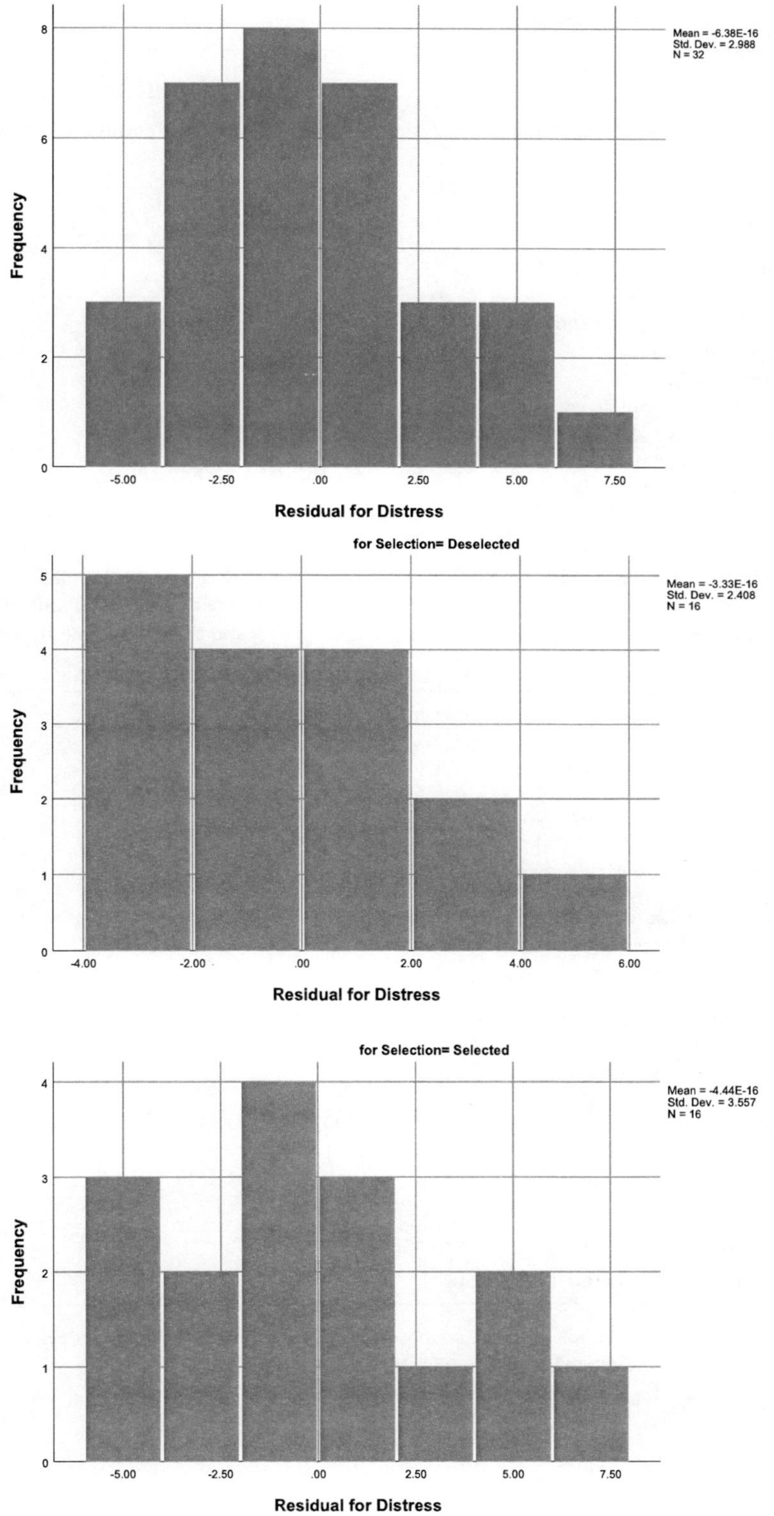

FIGURE 3.22
Histogram.

Working in **R**, we can generate a histogram using the *ggplot2* package.

```
install.packages("ggplot2")
```

The *install.packages* function will install the *ggplot2* package which we can use to create various graphs and plots.

```
library(ggplot2)
```

The *library* function will load the *ggplot2* package.

```
qplot(Ch3_distress$unstandardizedResiduals,
      geom="histogram",
      binwidth=1,
      main = "Histogram of Unstandardized Residuals",
      xlab = "Unstandardized Residual", ylab = "Count",
      fill=I("gray"),
      col=I("white"))
```

Using the *gplot* function, we create a histogram (i.e., geom = "histogram") from our dataframe (i.e., Ch3_distress) using the variable "unstandardizedResiduals." We can add a few commands to change the width of the bars (i.e., *binwidth=1*), color of the bars (i.e., *fill=I("gray")*), and outline of the bars (i.e., *col=I("white")*). We can also add a title (i.e., *main = "Histogram of Unstandardized Residuals"*) and change the *X* and *Y* axes (*xlab = "Unstandardized Residual", ylab = "Count"*).

```
hist(Ch3_distress$unstandardizedResiduals[Ch3_distress$Sport==1],
     main="Histogram for Movement",
     xlab="Unstandardized Residuals")

hist(Ch3_distress$unstandardizedResiduals[Ch3_distress$Sport==2],
     main="Histogram for Target",
     xlab="Unstandardized Residuals")

hist(Ch3_distress$unstandardizedResiduals[Ch3_distress$Sport==3],
     main="Histogram for Fielding",
     xlab="Unstandardized Residuals")

hist(Ch3_distress$unstandardizedResiduals[Ch3_distress$Sport==4],
     main="Histogram for Territory",
     xlab="Unstandardized Residuals")

hist(Ch3_distress$unstandardizedResiduals[Ch3_distress$Selection==1],
     main="Histogram for Deselected",
     xlab="Unstandardized Residuals")

hist(Ch3_distress$unstandardizedResiduals[Ch3_distress$Selection==2],
     main="Histogram for Selected",
     xlab="Unstandardized Residuals")
```

Histograms by group can be created with these scripts, each one specifying one category of Sport or Selection as the variable with which to create the histogram of unstandardized residuals.

FIGURE 3.22 (continued)
Histogram.

There are a few other statistics that can be used to gauge normality. The formal test of normality, the Shapiro-Wilk test (*SW*) (Shapiro & Wilk, 1965), provides evidence of the extent to which our sample distribution is statistically different from a normal distribution. The output for the Shapiro-Wilk test is presented in Figure 3.23 and suggests that our sample distribution for overall residuals is not statistically significantly different than what would be expected from a normal distribution ($SW = .977$, $df = 32$, $p = .701$), nor are the residuals by group statistically significantly different than what would be expected from a normal distribution.

Tests of Normality

	Kolmogorov-Smirnov[a]			Shapiro-Wilk		
	Statistic	df	Sig.	Statistic	df	Sig.
Residual for Distress	.094	32	.200*	.977	32	.701

*. This is a lower bound of the true significance.

a. Lilliefors Significance Correction

Tests of Normality

	Type of Sport	Kolmogorov-Smirnov[a]			Shapiro-Wilk		
		Statistic	df	Sig.	Statistic	df	Sig.
Residual for Distress	**Movement**	.130	8	.200*	.957	8	.785
	Target	.125	8	.200*	.982	8	.970
	Fielding	.164	8	.200*	.931	8	.529
	Territory	.183	8	.200*	.966	8	.863

*. This is a lower bound of the true significance.

a. Lilliefors Significance Correction

Tests of Normality

	Selection Status	Kolmogorov-Smirnov[a]			Shapiro-Wilk		
		Statistic	df	Sig.	Statistic	df	Sig.
Residual for Distress	**Deselected**	.137	16	.200*	.945	16	.413
	Selected	.125	16	.200*	.963	16	.708

*. This is a lower bound of the true significance.

a. Lilliefors Significance Correction

Working in **R**, we saw earlier how the *stat.desc* function from the *pastecs* package could be used to generate the Shapiro-Wilk test, along with many other statistics. Should we want to generate *just* the Shapiro-Wilk test, we can run the following script.

```
shapiro.test(Ch1_distress$unstandardizedResiduals)
```

```
        Shapiro-Wilk normality test
data:  Ch3_distress$unstandardizedResiduals
W = 0.97674, p-value = 0.701
```

FIGURE 3.23
Shapiro-Wilk test of normality.

```
tapply(Ch3_distress$unstandardizedResiduals,
Ch3_distress$SportF, shapiro.test)
tapply(Ch3_distress$unstandardizedResiduals,
Ch3_distress$SelectionF, shapiro.test)
```

To generate the Shapiro-Wilk test by group, the *tapply* function can be used to apply the *shapiro.test* to the unstandardized residuals for all levels of the independent variable.

FIGURE 3.23 (continued)
Shapiro-Wilk test of normality.

Quantile-quantile (Q-Q) plots are also often examined to determine evidence of normality. Q-Q plots are graphs that plot quantiles of the theoretical normal distribution against quantiles of the sample distribution. Points that fall on or close to the diagonal line suggest evidence of normality. The Q-Q plot of residuals, overall and by group (for brevity, only one group is presented) suggests relative normality.

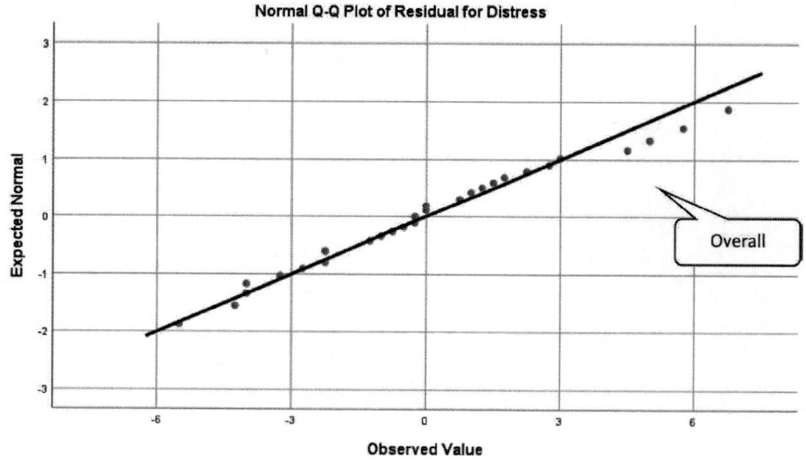

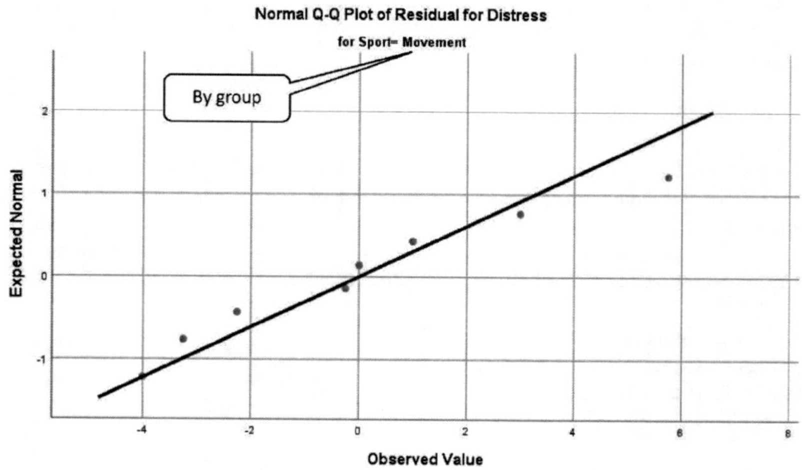

FIGURE 3.24
Normal Q-Q plot.

Working in **R**, we can use the *gplot* function to create a Q-Q plot of unstandardized residuals. The "data=" script defines the dataframe as "Ch3_distress."

```
qplot(sample=unstandardizedResiduals,
data = Ch3_distress)
```

```
qqnorm(Ch3_distress$unstandardizedResiduals[Ch3_distress$Sport==1],
        main='movement')
qqnorm(Ch3_distress$unstandardizedResiduals[Ch3_distress$Sport==2],
        main='target')
qqnorm(Ch3_distress$unstandardizedResiduals[Ch3_distress$Sport==3],
        main='fielding')
qqnorm(Ch3_distress$unstandardizedResiduals[Ch3_distress$Sport==4],
        main='territory')
qqnorm(Ch3_distress$unstandardizedResiduals[Ch3_distress$Selection==1],
        main='deselected')
qqnorm(Ch3_distress$unstandardizedResiduals[Ch3_distress$Selection==2],
        main='selected')
```

By group, Q-Q plots can be created with this script, with each command defining one category of the *Sport* and *Selection* variables.

FIGURE 3.24 (continued)
Normal Q-Q plot.

Examination of the boxplot suggests a relatively normal distributional shape of residuals, overall and by group, and no outliers.

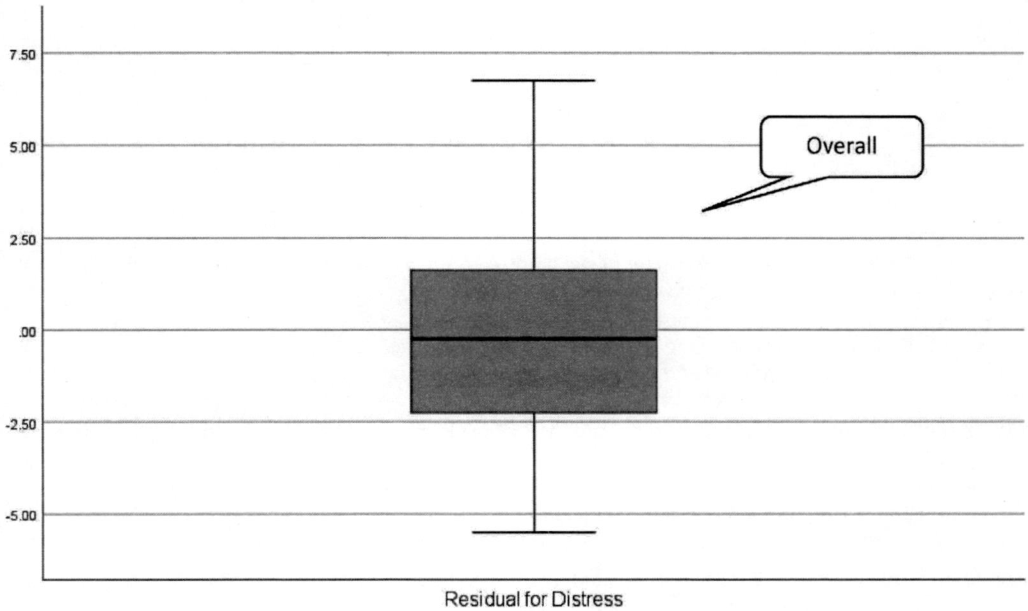

FIGURE 3.25
Boxplot.

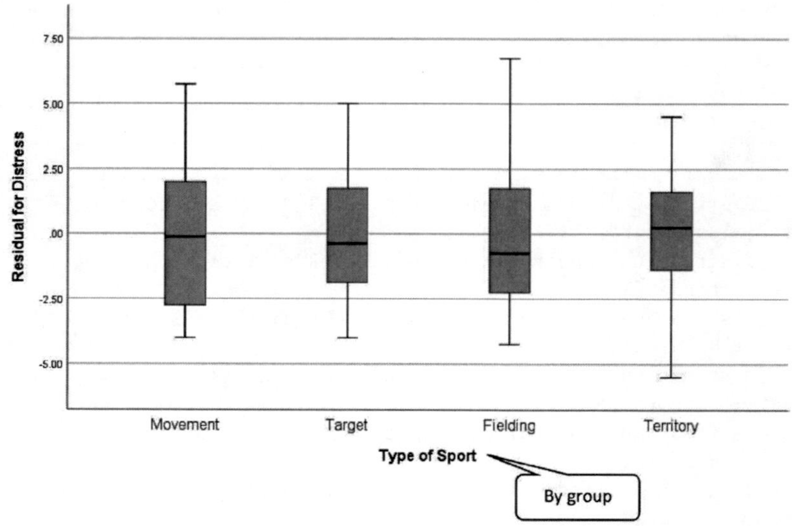

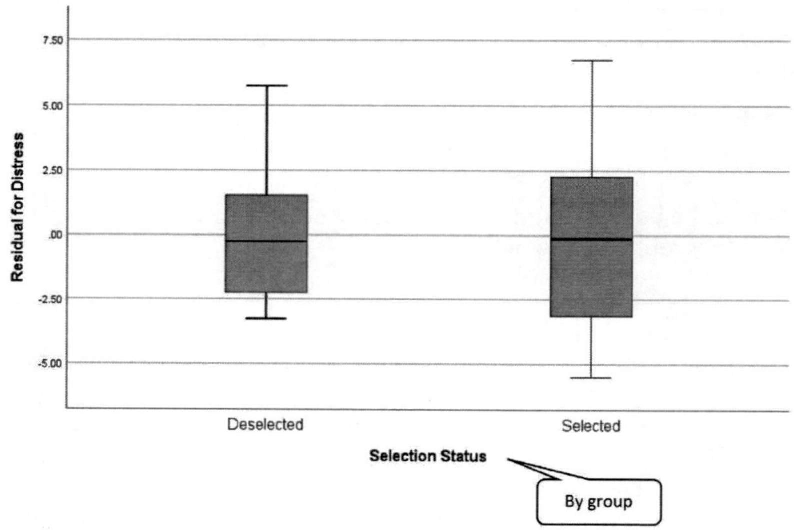

Working in **R**, we can generate a boxplot for unstandardized residuals using the *boxplot* function. To label the *Y* axis, we include the *ylab* command.

```
boxplot(Ch3_distress$unstandardizedResiduals,
        ylab="Unstandardized Residuals")
```

Adding the independent variable to the script produces a boxplot by group. The command *xlab* will print "Sport" to identify the *X* axis.

```
boxplot(Ch3_distress$unstandardizedResiduals~Ch3_distress$SportF,
        xlab="Sport", ylab="Unstandardized Residuals")
boxplot(Ch3_distress$unstandardizedResiduals~Ch3_distress$SelectionF,
        xlab="Selection", ylab="Unstandardized Residuals")
```

FIGURE 3.25 (continued)
Boxplot.

Considering the forms of evidence we have examined, skewness and kurtosis statistics, the Shapiro-Wilk test, the Q-Q plot, and the boxplot, all suggest normality by group is a reasonable assumption. We can be reasonably assured that we have met the assumption of normality of the dependent variable for each group of the independent variable.

3.6.2 Independence

The only assumption we have not tested for yet is independence. As we discussed in reference to the one-way ANOVA, if subjects have been randomly assigned to conditions (or to the different combinations of the levels of the independent variables in a factorial ANOVA), the assumption of independence has been met. In this illustration, athletes were randomly assigned to neither type of sport nor selection status; thus we cannot assume that the assumption of independence was met. We often use independent variables such as this that do not allow random assignment, such as preexisting characteristics. We can plot residuals against levels of our independent variables in a scatterplot to get an idea of whether or not there are patterns in the data and thereby provide an indication of whether we have met this assumption. Given we have multiple independent variables in the factorial ANOVA, we will split the scatterplot by levels of one independent variable ("Sport") and then generate a bivariate scatterplot for "Selection" by residual. Remember that the residual was added to the dataset by saving it when we generated the factorial ANOVA model.

Please note that some researchers do not believe that the assumption of independence can be tested. If there is not random assignment to groups, then these researchers believe this assumption has been violated—period. The plot that we generate will give us a general idea of patterns, however, in situations where random assignment was not performed or not possible.

Splitting the file. The first step is to split our file by the levels of one of our independent variables (e.g., "Sport"). To do that, go to "Data" in the top pulldown menu and then select "Split File."

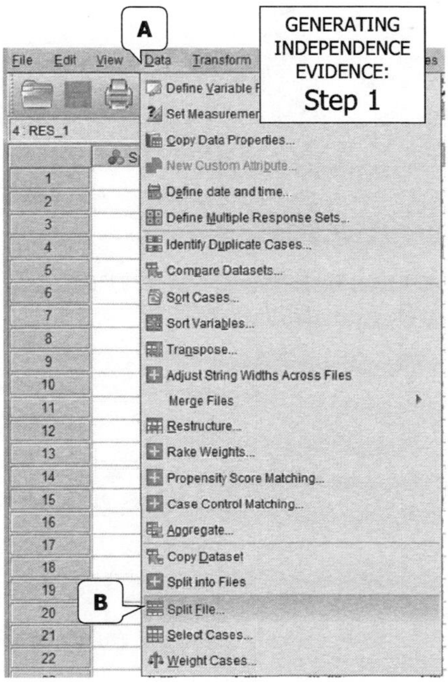

FIGURE 3.26
Generating independence evidence: Step 1.

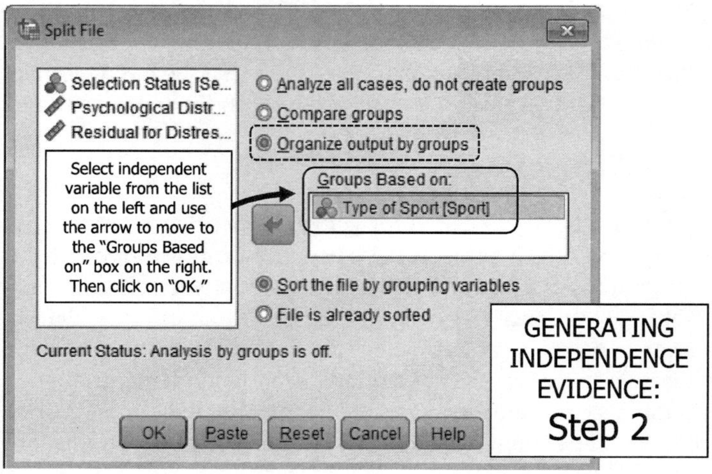

FIGURE 3.27
Generating independence evidence: Step 2.

Generating the scatterplot. The general steps for generating a simple scatterplot through "Scatter/dot" are likely not new to you (from the top toolbar in SPSS, go to Graphs -> Legacy Dialogs -> Scatter/Dot). From the "Simple Scatterplot" dialog screen, click the residual variable and move it into the "Y Axis" box by clicking on the arrow. Click the independent variable that was not used to split the file (e.g., "Selection Status") and move it into the "X Axis" box by clicking on the arrow. Then click "OK."

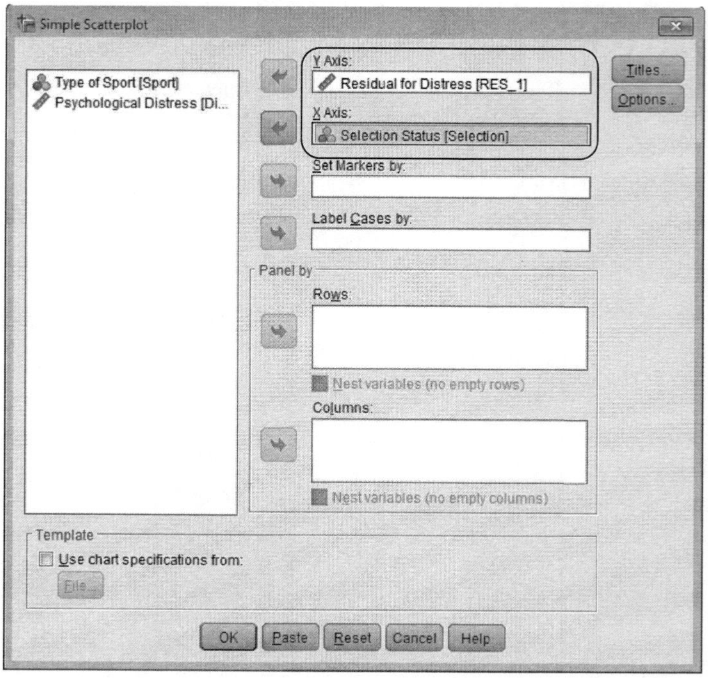

FIGURE 3.28
Generating a scatterplot.

3.6.2.1 Interpreting Independence Evidence

In examining the scatterplots for evidence of independence, the points should fall relatively randomly above and below a horizontal line at zero. (You may recall in Chapter 1 that we added a reference line to the graph using Chart Editor. To add a reference line, double click on the graph in the output to activate the chart editor. Select "Options" in the top pulldown menu, then "Y axis reference line." This will bring up the "Properties" dialog box. Change the value of the position to be "0." Then click on "Apply" and "Close" to generate the graph with a horizontal line at zero.)

In this example, our scatterplot for each type of sport generally suggests evidence of independence with a relatively random display of residuals above and below the horizontal line at zero for each category of time. Thus, even though we have not met the assumption of independence through random assignment of cases to groups, this provides evidence that independence is a reasonable assumption.

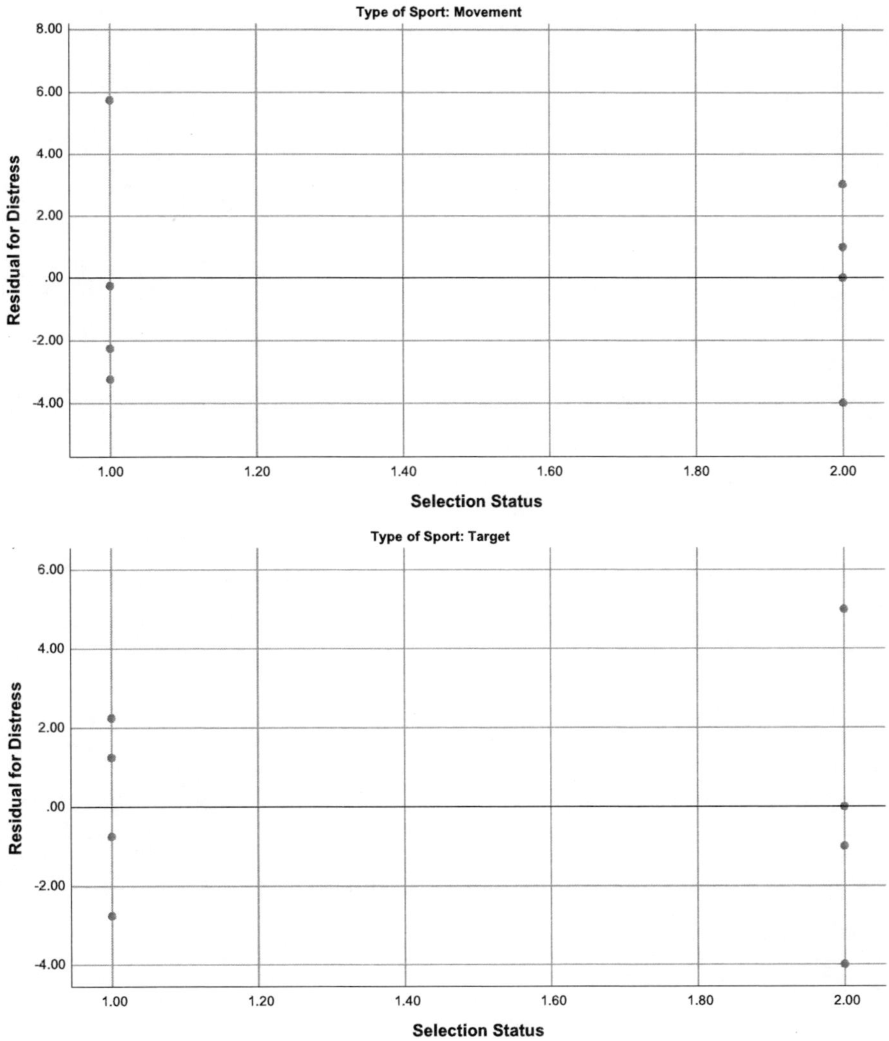

FIGURE 3.29
Residual plots.

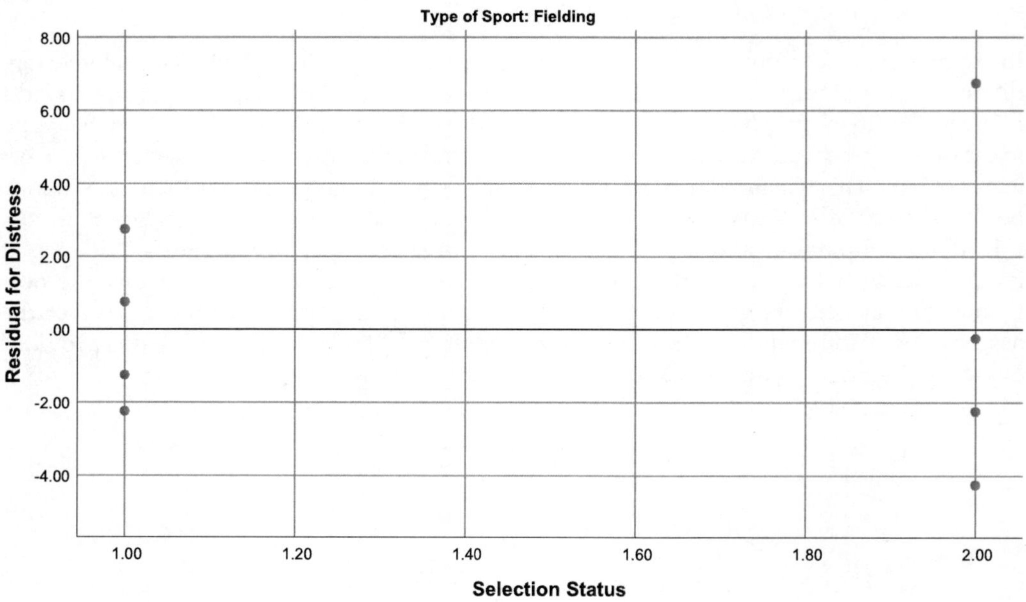

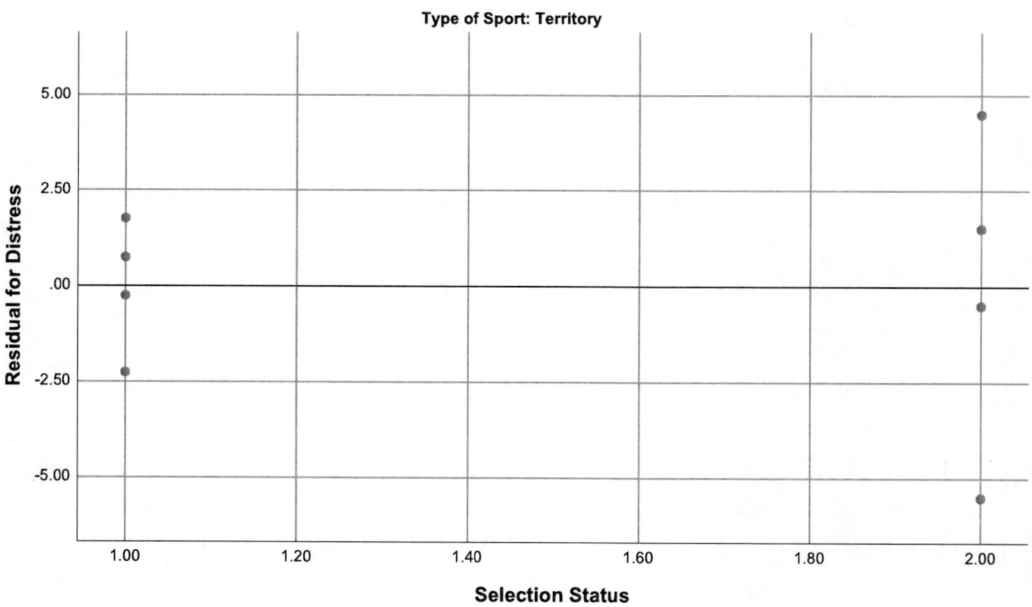

FIGURE 3.29 (continued)
Residual plots.

Working in **R**, we create similar scatterplots.

```
plot(Ch3_distress$unstandardizedResiduals ~ Ch3_distress$Selection,
      xlab="Selection",
      ylab="Unstandardized Residual",
      Ch3_distress[Ch3_distress$Sport == 1, ])
plot(Ch3_distress$unstandardizedResiduals ~ Ch3_distress$Selection,
      xlab="Selection",
      ylab="Unstandardized Residual",
      Ch3_distress[Ch3_distress$Sport == 2, ])
plot(Ch3_distress$unstandardizedResiduals ~ Ch3_distress$Selection,
      xlab="Selection",
      ylab="Unstandardized Residual",
      Ch3_distress[Ch3_distress$Sport == 3, ])
plot(Ch3_distress$unstandardizedResiduals ~ Ch3_distress$Selection,
      xlab="Selection",
      ylab="Unstandardized Residual",
      Ch3_distress[Ch3_distress$Sport == 4, ])
```

Using the *plot* function, we can create a scatterplot of unstandardized residuals, "Ch3_distress$unstandardized Residuals," by selection status, "Ch3_distress$Selection," for each sport, "Ch3_distress[Ch3_distress$Sport == 1,]," where there are four plots, one for each category of sport. Note that we are using "Sport" and "Selection," the original numeric variables in our dataframe (i.e., *not* the new factor variables created from these variables). Doing so provides us the scatterplot. Had we used the factor variables in our script, we would have generated boxplots.

```
plot(Ch3_2way)
```

Using the *plot* function, additional plots that can be used for diagnostic purposes are created.

The residual versus fitted plot can be used to detect normality, unequal error variance, and outliers. A random display of points, i.e., no patterns to the data, suggest assumptions of normality and equal variances have been met.

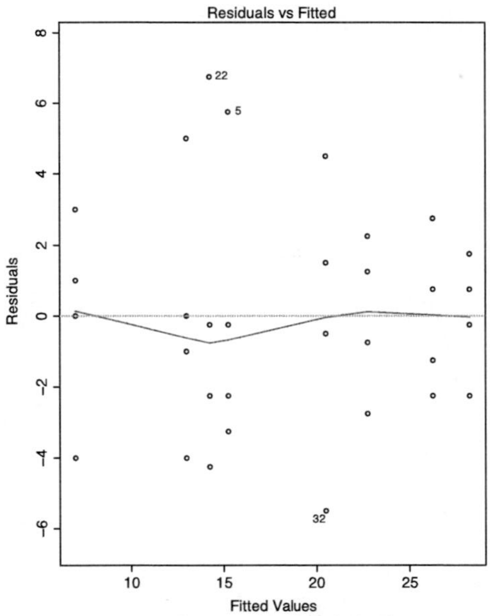

The normal Q-Q plot can be used to detect normality and outliers. Points that adhere closely to the diagonal line suggest the assumption of normality has been met.

FIGURE 3.29 (continued)
Residual plots.

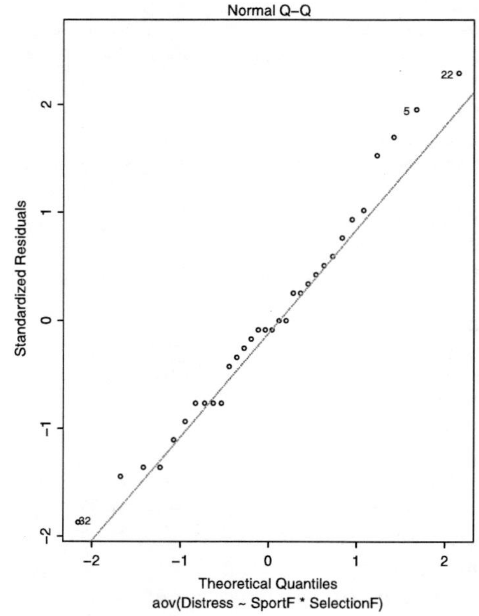

The scale-location plot can be examined for evidence of equal variance. Relatively equally spaced points by group above and below a horizontal line (i.e., random and equal distribution of points and straight horizontal line) suggests evidence of meeting the assumption.

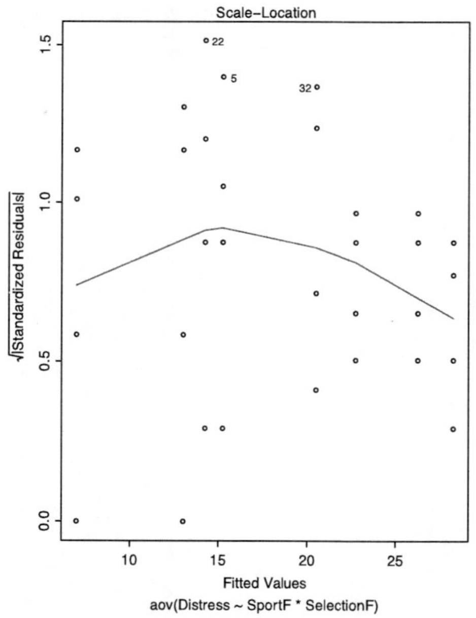

The constant leverage plot can be examined as evidence of normality as well to determine points that may exert influence.

FIGURE 3.29 (continued)
Residual plots.

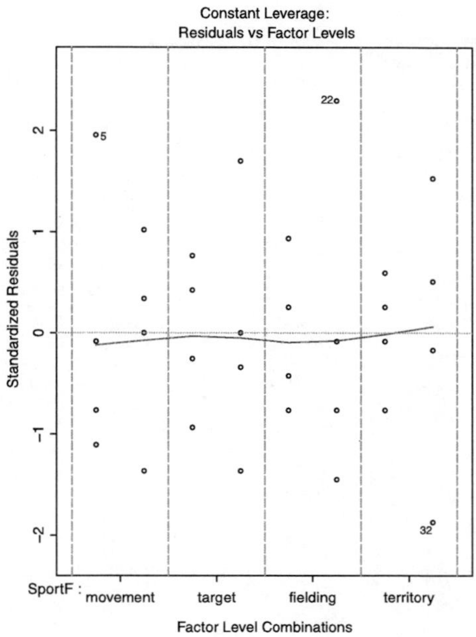

FIGURE 3.29 (continued)
Residual plots.

3.6.3 Homogeneity of Variance

As we learned previously, another assumption to consider is that the variances of each population are equal. This is known as the assumption of **homogeneity of variance** or **homoscedasticity**. When generating factorial ANOVA via SPSS, we requested a number of tests for examining homogeneity. Homogeneity tests using **R** were presented previously (see Figure 3.16).

3.7 Power Using G*Power

3.7.1 Post Hoc Power for Factorial ANOVA Using G*Power

*When there are multiple independent variables, G*Power must be calculated for each main effect and for each interaction.* We will illustrate computing post hoc power for the **main effect** for type of sport, but note that computing power for the other main effect(s) and interaction(s) are similarly obtained.

The first thing that must be done when using G*Power for computing post hoc power is to select the correct test family. In our case, we conducted a factorial ANOVA. To find the factorial ANOVA, we select "Tests" in the top pulldown menu, then "Means," and then "Many groups: ANOVA: Main effects and interactions (two or more independent variables)." Once that selection is made, the "Test family" automatically changes to "F tests."

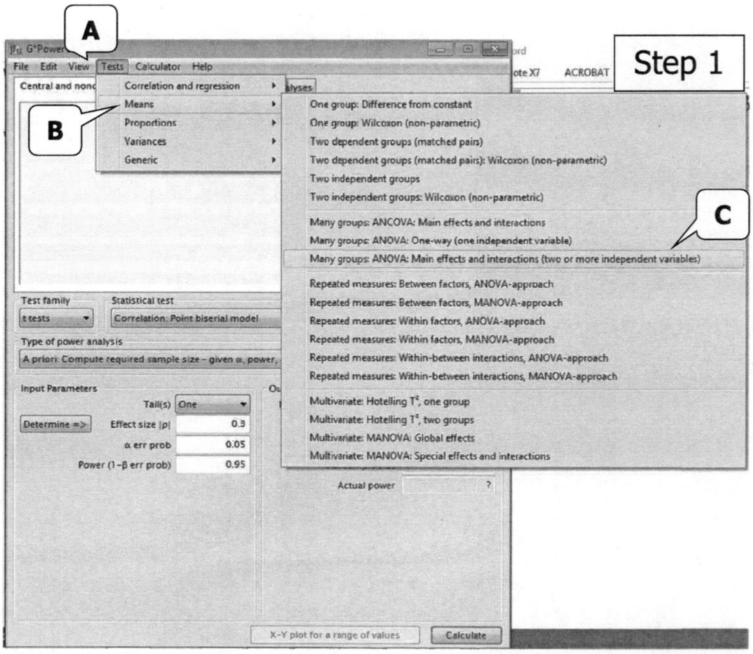

FIGURE 3.30
Power: Step 1.

The "Type of power analysis" desired then needs to be selected. To compute post hoc power, we need to select "Post hoc: Compute achieved power—given α, sample size, and effect size."

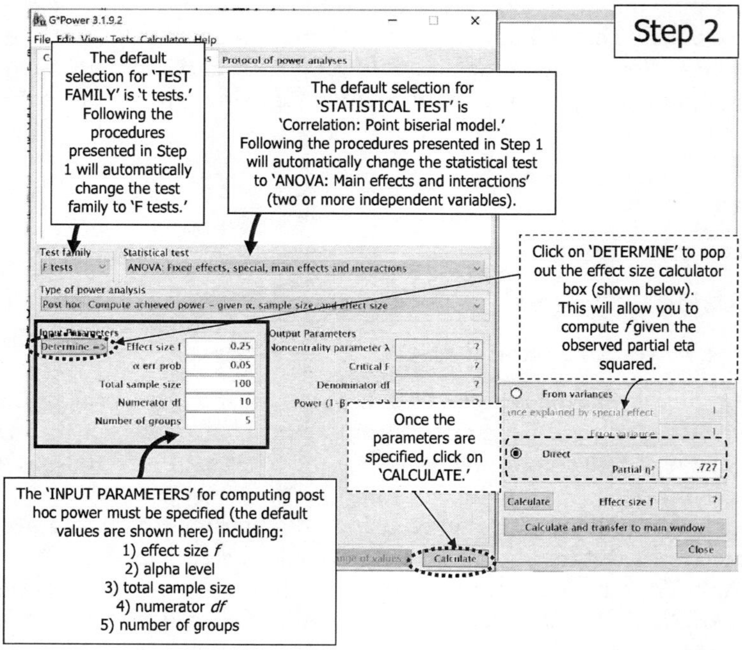

FIGURE 3.31
Power: Step 2.

The "Input Parameters" must then be specified. We compute the effect size *f* last, so skip that for the moment. In our example, the alpha level we used was .05 and the total sample size was 32. The numerator *df* for type of sport (recall that we are computing post hoc power for the main effect for type of sport here) is equal to the number of categories of this variable (i.e., 4) minus 1; thus there are three degrees of freedom for type of sport. The *number of groups* is equal to the product of the number of levels or categories of the independent variables, or (*J*)(*K*). In this example, the number of groups or cells then equals (*J*)(*K*) = (4)(2) = 8.

We skipped filling in the first parameter, the effect size *f*, for a reason. SPSS provided only a partial eta squared effect size. Thus, we will use the pop out effect size calculator in G*Power to compute the effect size *f* (we saved this parameter for last as the calculation is based on the previous values just entered). To pop out the effect size calculator, click on "Determine" which is displayed under "Input Parameters." In the pop out effect size calculator, click on the radio button for "Direct" and then enter the partial eta squared value for type of sport that was calculated in SPSS (i.e., .727). Clicking on "Calculate" in the pop out effect size calculator will calculate the effect size *f*. Then click on "Calculate and transfer to main window" to transfer the calculated effect size (i.e., 1.6318712) to the "Input Parameters." Once the parameters are specified, click on "Calculate" to find the power statistics.

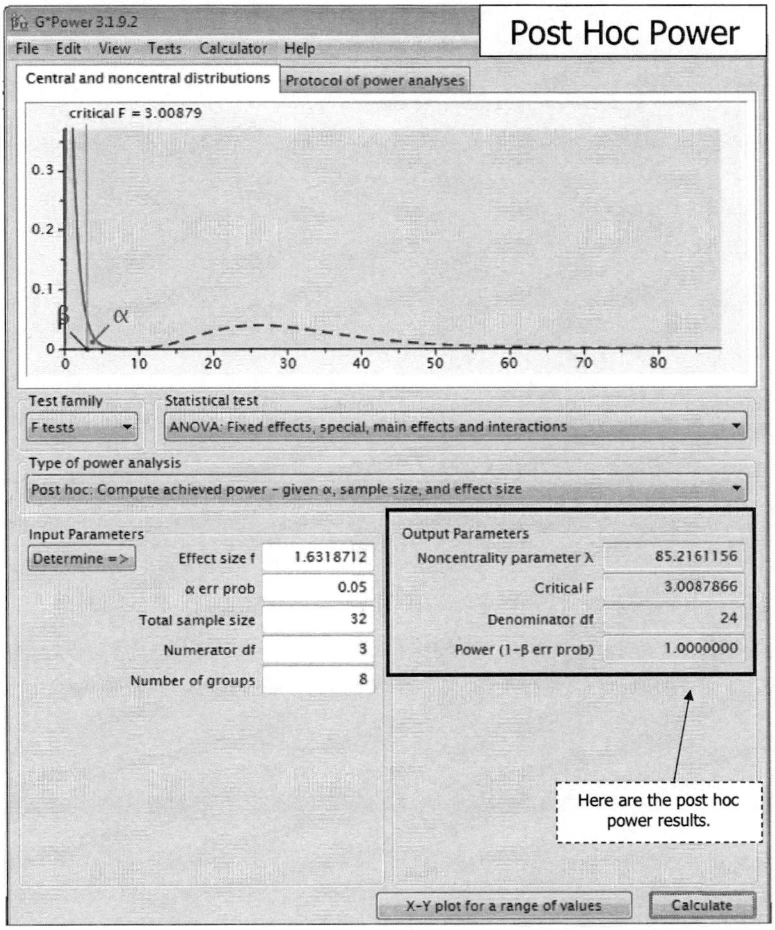

FIGURE 3.32
Post hoc power results.

The "Output Parameters" provide the relevant statistics given the input just specified. In this example, we were interested in determining post hoc power for a two-factor ANOVA with a computed effect size *f* of 1.632, an alpha level of .05, total sample size of 32, numerator degrees of freedom of three and eight groups or cells. Based on those criteria, the post hoc power for the main effect of type of sport was 1.00. In other words, with the input parameters we defined, the post hoc power of our main effect was 1.00—the probability of rejecting the null hypothesis when it is really false (in this case, the probability that the means of the dependent variable would be equal for each level of the independent variable) was 1.00, which would be considered maximum power (sufficient power is often .80 or above). Note that this value is the same as that reported in SPSS. Keep in mind that conducting power analysis *a priori* is recommended so that you avoid a situation where, post hoc, you find that the sample size was not sufficient to reach the desired level of power (given the observed parameters).

3.7.1.1 Power for Interactions

Calculation of power for interactions is conducted similarly. Calculating *f* from the input of .727 for partial eta squared results in the following output for interaction power. The post hoc power of the interaction effect for this test was .204—the probability of rejecting the null hypothesis when it is really false (in this case, the probability that the means of the dependent variable would be equal for each cell) was about 20%, which would be considered very low power (sufficient power is often .80 or above). Note that this value is not the same as that reported in SPSS.

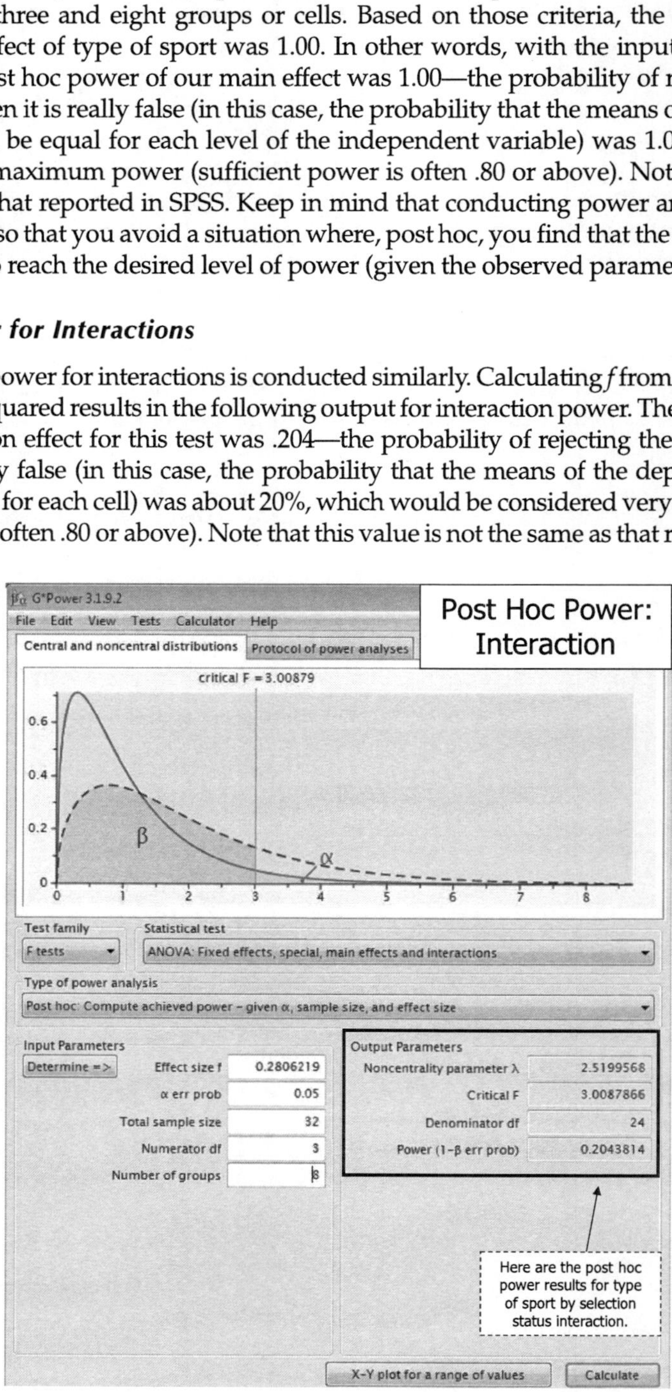

FIGURE 3.33
Post hoc power interaction results.

3.7.2 *A Priori* Power for Factorial ANOVA Using G*Power

For *a priori* power, we can determine the total sample size needed for the main effects and/or interactions given an estimated effect size *f*, alpha level, desired power, numerator degrees of freedom (i.e., number of categories of our independent variable or interaction, depending on which *a priori* power is of interest), and number of groups or cells (i.e., the product of the number of levels of the independent variables). We follow Cohen's (1988) conventions for effect size (i.e., small $f = .10$; moderate $f = .25$; large $f = .40$). In this example, had we estimated a moderate effect *f* of .25, alpha of .05, desired power of .80, numerator degrees of freedom of three [four types of sports, two categories in selection status, thus $(4 - 1)(2 - 1) = 3$], and number of groups of eight (i.e., four types of sports, two categories in selection status, thus $4 \times 2 = 8$), we would need a total sample size of 179 (or about 22 or 23 individuals per cell).

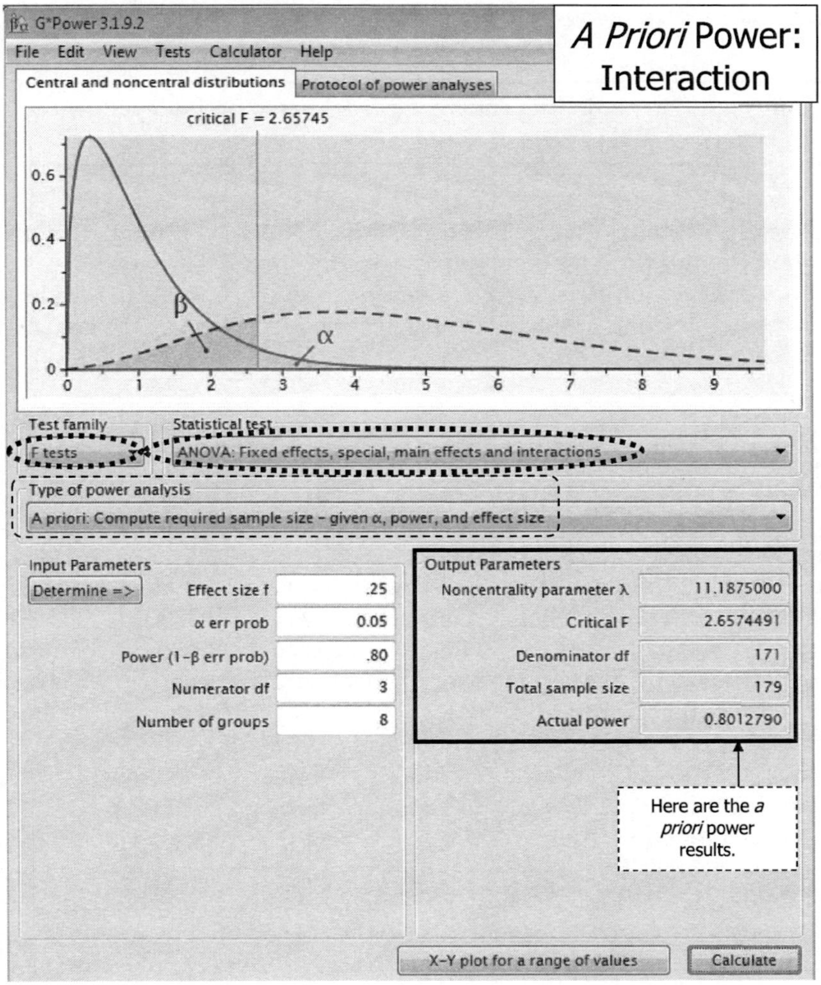

FIGURE 3.34
A priori power interaction results.

3.8 Research Question Template and Example Write-Up

Finally we come to an example paragraph of the results for the two-factor elite athlete example. Recall that our graduate research assistant, Ott, was working with Dr. Rhodes, one of the leading sports psychologists in the region. Dr. Rhodes was interested in examining elite athletes and their vulnerability to psychological distress after selection procedures in which athletes are either selected or deselected for their team and/or to continue in their athletic field. Ott then generated a factorial ANOVA as the test of inference. A template for writing a research question for a factorial ANOVA is presented in this section. This is illustrated assuming a two-factor model, but it can easily be extended to more than two factors. As we noted in Chapter 1, it is important to ensure the reader understands the levels or groups of the independent variables. This may be done parenthetically in the actual research question, as an operational definition, or specified within the methods section. In this example, parenthetically we could have stated the following: *Is there a mean difference in psychological distress of elite athletes based on the type of sport in which they participate (movement, target, fielding, or territory) and selection status (deselected or selected)?*

Is there a mean difference in [dependent variable] based on [independent variable 1] and [independent variable 2]?

It may be helpful to preface the results of the factorial ANOVA with information on an examination of the extent to which the assumptions were met (recall there are three assumptions: normality, homogeneity of variance, and independence). This assists the reader in understanding that you were thorough in data screening prior to conducting the test of inference.

A factorial analysis of variance (ANOVA) was conducted to determine if the mean psychological distress of elite athletes differed based on type of sport in which they participated (i.e., movement, target, fielding, or territory) and selection status (deselected or selected). The assumptions of normality, homoscedasticity, and independence were reviewed.

The assumption of normality was tested and met via examination of the residuals. Review of the overall Shapiro-Wilk test for normality ($SW = .977$, $df = 32$, $p = .701$), and skewness (.400) and kurtosis (−.162) statistics, suggested that normality of the residuals overall was a reasonable assumption. Normality of residuals by group was also reasonable, as results for the Shapiro-Wilk test for normality by group were all nonstatistically significant. Skewness and kurtosis of residuals by group were all within the range of an absolute value of 2.0, suggesting normality by group was reasonable. Additional tests, including the overall D'Agostino's test for skewness ($z = 1.00$, $p = .315$) and the Bonett-Seier test for Geary's kurtosis ($z = .320$, $p = .749$) suggested evidence of normality of residuals overall. The results for D'Agostino's test for skewness and the Bonett-Seier test for Geary's kurtosis by group were all nonstatistically significant, suggesting normality is reasonable. The boxplots of residuals, overall and by group, suggested a relatively normal distributional shape (with no outliers). The Q-Q plots and histograms generally suggested normality was reasonable, although the plots by group reflected some nonnormality. In aggregate, the results suggest normality by group is reasonable.

According to Levene's test, the homogeneity of variance assumption was satisfied [F (7, 24) = .579, p = .766].

Scatterplots of residuals against the levels of the independent variables were reviewed. A random display of points around zero provided evidence that the assumption of independence was met in the absence of random assignment to groups. [*Note:* Had there been random assignment to groups, we could have also stated, "Random assignment of individuals to groups helped ensure that the assumption of independence was met."]

Here is an APA-style example paragraph of results for the factorial ANOVA (remember that this will be prefaced by the previous paragraph reporting the extent to which the assumptions of the test were met).

The interaction of type of sport by selection status is not statistically significant ($F_{sport*selection}$ = 21.844, df = 3,24, p = .602), but there are statistically significant main effects for both type of sport (F_{sport} = 21.350, df = 3,24, p = .001) and selection status ($F_{selection}$ = 61.791, df = 1,24, p = .001).

Effect sizes are large for both type of sport status (partial $\eta^2_{sport} = .727$, $\omega^2_{sport} = .400$) and selection status (partial $\eta^2_{selection} = .720$; $\omega^2_{\text{selection}} = .398$) but very small for the interaction of sport by selection status (partial $\eta^2_{sport*selection} = -.007$; $\omega^2_{sport*selection} = .007$). Partial eta squared for the main effect for type of sport tells us that the proportion of variation in psychological distress explained by the type of sport in which the athlete participates that is *not* explained by selection status is about 73%. Partial eta squared for the main effect for selection status tells us that the proportion of variation in psychological distress explained by whether the athlete is selected or deselected that is *not* explained by the type of sport in which they participate is about 72%. Omega squared for the main effect for type of sport tells us that proportion of total variability in the dependent variable that is accounted for by type of sport is about 40%. Omega squared for the main effect for selection status tells us that proportion of total variability in the dependent variable that is accounted for by selection status is about 40%.

Observed power for type of sport and selection status is maximal (i.e., 1.000). However, the test of the interaction of sport by selection status was underpowered, with observed power of only .162.

Post hoc analyses were conducted given the statistically significant omnibus ANOVA F tests for the main effects. The profile plot (Figure 3.2) summarizes these differences. Tukey's HSD tests were conducted on all possible pairwise contrasts. For the main effect of type of sport, Tukey's HSD post hoc comparisons revealed that athletes in "movement" sports had statistically significantly lower psychological distress than all the other types of sports, and that athletes in "target" sports had statistically significantly lower psychological distress than the athletes in "territory" types of sports. More specifically, the following pairs of types of sports were found to be significantly different ($p < .05$):

- Movement (M = 11.125, SD = 5.4886) and Target (M = 17.875, SD = 1.201);
- Movement and Fielding (M = 20.2500, SD = 7.2850);
- Movement and Territory (M = 24.3750, SD = 5.0973); and
- Target and Territory.

In other words, athletes enrolled in "movement" types of sports had statistically significantly less psychological distress than athletes who participated in any of the three other types of sports (i.e., target, fielding, and territory).

For the main effect of selection status, a comparison of means revealed that athletes who were deselected for their sport ($M = 23.125$, $SD = .849$) had statistically significantly higher psychological distress than athletes who were selected ($M = 13.688$, $SD = .849$).

3.9 Additional Resources

This chapter has provided a preview into conducting factorial ANOVA. However, there are a number of areas that space limitations prevent us from delving into. For more in-depth coverage of ANOVA models, see Maxwell, Delaney, and Kelley (2018) and Keppel and Wickens (2004), among others.

Problems

Conceptual Problems

1. You are given a two-factor design with the following cell means (cell 11 = 25; cell 12 = 75; cell 21 = 50; cell 22 = 50; cell 31 = 75; cell 32 = 25). Assume that the within-cell variation is small. Which one of the following conclusions seems most probable?
   a. The row means are significantly different.
   b. The column means are significantly different.
   c. The interaction is significant.
   d. All of the above.
2. In a two-factor ANOVA, one independent variable has five levels and the second has four levels. If each cell has seven observations, what is df_{with}?
   a. 20
   b. 120
   c. 139
   d. 140
3. In a two-factor ANOVA, one independent variable has three levels or categories and the second has three levels or categories. What is df_{AB}, the interaction degrees of freedom?
   a. 3
   b. 4
   c. 6
   d. 9

4. Which of the following conclusions would result in the greatest generalizability of the main effect for factor A across the levels of factor B? The interaction between the independent variables A and B was
   a. Not significant at the .25 level.
   b. Significant at the .05 level.
   c. Significant at the .01 level.
   d. Significant at the .001 level.
5. In a two-factor fixed-effects ANOVA tested at an alpha of .05, the following p values were found: main effect for factor A, $p = .06$; main effect for factor B, $p = .09$; interaction AB, $p = .02$. What can be interpreted from these results?
   a. There is a statistically significant main effect for factor A.
   b. There is a statistically significant main effect for factor B.
   c. There is a statistically significant main effect for factors A and B.
   d. There is a statistically significant interaction effect.
6. In a two-factor fixed-effects ANOVA, $F_A = 2$, $df_A = 3$, $df_B = 6$, $df_{AB} = 18$, and $df_{with} = 56$. The null hypothesis for factor A can be rejected
   a. At the .01 level
   b. At the .05 level, but not at the .01 level
   c. At the .10 level, but not at the .05 level
   d. None of the above
7. In ANOVA, the interaction of two factors is certainly present when
   a. The two factors are positively correlated.
   b. The two factors are negatively correlated.
   c. Row effects are not consistent across columns.
   d. Main effects do not account for all of the variation in Y.
   e. Main effects do account for all of the variation in Y.
8. For a design with four factors, how many *total* interactions will there be?
   a. 4
   b. 8
   c. 11
   d. 12
   e. 16
9. Degrees of freedom for the AB interaction are equal to which one of the following?
   a. $df_A - df_B$
   b. $(df_A)(df_B)$
   c. $df_{with} - (df_A + df_B)$
   d. $df_{total} - df_{with}$
10. A two-factor experiment means that the design necessarily includes which one of the following?
   a. Two independent variables
   b. Two dependent variables

   c. An interaction between the independent and dependent variables
   d. Exactly two separate groups of subjects

11. Two independent variables are said to interact when which one of the following occurs?
   a. Both variables are equally influenced by a third variable.
   b. These variables are differentially affected by a third variable.
   c. Each factor produces a change in the subjects' scores.
   d. The effect of one variable depends on the second variable.
12. True or false? If there is an interaction between the independent variables textbook and time of day, this means that the textbook used has the same effect at different times of the day.
13. True or false? If the AB interaction is significant, then at least one of the two main effects must be significant.
14. For a two-factor fixed-effects model, if the degrees of freedom for testing factor A = 2,24, then I assert that the degrees of freedom for testing factor B will necessarily be = 2,24. Am I correct?

Questions 15 through 17 are based on the following ANOVA summary table (fixed-effects):

Source	*df*	*MS*	*F*
A	2	45	4.5
B	1	70	7.0
AB	2	170	17.0
Within	60	10	

15. For which source of variation is the null hypothesis rejected at the .01 level of significance?
   a. A
   b. B
   c. AB
   d. All of the above
16. How many cells are there in the design?
   a. 1
   b. 2
   c. 3
   d. 5
   e. None of the above
17. The total sample size for the design is which one of the following?
   a. 66
   b. 68
   c. 70
   d. None of the above

Questions 18 through 20 are based on the following ANOVA summary table (fixed-effects):

Source	*df*	*MS*	*F*
A	2	164	5.8
B	1	80	2.8
AB	2	68	2.4
Within	9	28	

18. For which source of variation is the null hypothesis rejected at the .01 level of significance?

a. A

b. B

c. AB

d. All of the above

19. How many cells are there in the design?

a. 1

b. 2

c. 3

d. 6

e. None of the above

20. The total sample size for the design is which one of the following?

a. 10

b. 15

c. 20

d. 25

21. Which of the following assumptions is applicable in ANOVA but not in factorial ANOVA? Select all that apply?

a. Equal variances

b. Independent

c. Normality

d. All of the above

e. None of the above

22. In the absence of random assignment to groups, which one of the following can be used to examine the extent to which the assumption of independence has been met?

a. Boxplot of residuals by levels of factor A

b. Scatterplot of residuals to categories of the independent variables

c. Shapiro-Wilk test

d. Spread versus level plots

23. The following table is provided in your output. Which groups have statistically significantly different means on the outcome based on Tukey's MCP?

DEPENDENT VARIABLE

Tukey HSD[a,b,c]

Group	N	Subset 1	Subset 2
A	1000	1.6524	
B	1000		1.8304
C	1000		1.8490
D	1000		1.9691

Means for groups in homogeneous subsets are displayed.

Based on observed means.

a. Group A is statistically different from Groups B, C, and D.
b. Groups B, C, and D are statistically different from each other.
c. Groups A and B differ, but Groups C and D are not statistically different.
d. Groups C and D differ, but Groups A and B are not statistically different.

24. A researcher finds a p value of .04 for Levene's test. Which of the following can be concluded if the alpha level is .05?
a. The assumption of equal variances has been violated.
b. The assumption of normality has been met.
c. There is a statistically significant main effect for this factor.
d. There is a statistically significant omnibus test.

25. An inappropriate way to deal with factorial ANOVA with unequal n's includes which one of the following?
a. Deleting data until n's are equal
b. Partially sequential approach
c. Regression approach
d. Sequential approach

Answers to Conceptual Problems

1. c (a plot of the cell means reveals an interaction)
3. **b** (product of the number of degrees of freedom for each main effect; $(J-1)(K-1) = (2)(2) = 4$)
5. **d** (p less than alpha only for the interaction term)
7. c (c is one definition of an interaction)
9. **b** (interaction df = product of main effects df)

11. **d** (the effect of one factor depends on the second factor; see definition of interaction)
13. **False** (when the interaction is significant, this implies nothing about the main effects)
15. **c** (check *F* table for critical values; only reject for interaction)
17. **a** (as $df_{total} = 65$, then total sample size = 66)
19. **d** (3 levels of A, 2 levels of B, thus 6 cells)
21. **e** (all assumptions of ANOVA are also applicable to factorial ANOVA; these include independence, homogeneity of variance, and normality)
23. **a** (interpreting the subset columns, Group A is statistically different from Groups B, C, and D; however, Groups B, C, and D are not statistically different from each other)
25. **a** (with factorial ANOVA with unequal *n*'s, the inappropriate way to deal with the unbalance is to delete data until the sample sizes of the groups are the same)

Computational Problems

1. Complete the following ANOVA summary table for a two-factor fixed-effects analysis of variance, where there are two levels of factor A (drug) and three levels of factor B (dosage). Each cell includes 26 patients and $\alpha = .05$.

Source	*SS*	*df*	*MS*	*F*	Critical Value	Decision
A	6.15	–	–	–	–	–
B	10.60	–	–	–	–	–
AB	9.10	–	–	–	–	–
Within	–	–	–			
Total	250.85	–				

2. Complete the following ANOVA summary table for a two-factor fixed-effects analysis of variance, where there are three levels of factor A (program) and two levels of factor B (gender). Each cell includes four individuals and $\alpha = .01$.

Source	*SS*	*df*	*MS*	*F*	Critical Value	Decision
A	3.64	–	–	–	–	–
B	.57	–	–	–	–	–
AB	2.07	–	–	–	–	–
Within	–	–	–			
Total	8.18	–				

3. Complete the following ANOVA summary table for a two-factor fixed-effects analysis of variance, where there are two levels of factor A (undergraduate versus graduate) and two levels of factor B (treatment). Each cell includes four students and $\alpha = .05$.

Source	*SS*	*df*	*MS*	*F*	Critical Value	Decision
A	14.06	–	–	–	–	–
B	39.06	–	–	–	–	–
AB	1.56	–	–	–	–	–
Within	–	–	–			
Total	723.43	–				

4. Conduct a two-factor fixed-effects ANOVA to determine if there are any effects due to A (task type), B (task difficulty), or the AB interaction ($\alpha = .01$). Conduct Tukey's HSD post hoc comparisons, if necessary. The following are the scores from the individual cells of the model:

   A_1B_1: 41, 39, 25, 25, 37, 51, 39, 101

   A_1B_2: 46, 54, 97, 93, 51, 36, 29, 69

   A_1B_3: 113, 135, 109, 96, 47, 49, 68, 38

   A_2B_1: 86, 38, 45, 45, 60, 106, 106, 31

   A_2B_2: 74, 96, 101, 124, 48, 113, 139, 131

   A_2B_3: 152, 79, 135, 144, 52, 102, 166, 155

5. An experimenter is interested in the effects of strength of reinforcement (factor A), type of reinforcement (factor B), and sex of the adult administering the reinforcement (factor C) on children's behavior. Each factor consists of two levels. Thirty-two children are randomly assigned to 8 cells (i.e., 4 per cell), one for each of the factor combinations. Using the scores from the individual cells of the model that follow, conduct a three-factor fixed-effects analysis of variance ($\alpha = .05$). If there are any significant interactions, graph and interpret the interactions.

   $A_1B_1C_1$: 3, 6, 3, 3

   $A_1B_1C_2$: 4, 5, 4, 3

   $A_1B_2C_1$: 7, 8, 7, 6

   $A_1B_2C_2$: 7, 8, 9, 8

   $A_2B_1C_1$: 1, 2, 2, 2

   $A_2B_1C_2$: 2, 3, 4, 3

   $A_2B_2C_1$: 5, 6, 5, 6

   $A_2B_2C_2$: 10, 10, 9, 11

6. A replication study dataset of the example from this chapter is given below (A = type of sport, B = selection status; same levels). Using the scores from the individual cells of the model that follow, conduct a two-factor fixed-effects analysis of variance ($\alpha = .05$). Are the results different as compared to the original dataset?

   A_1B_1: 10, 8, 7, 3

   A_1B_2: 15, 12, 21, 13

   A_2B_1: 13, 9, 18, 12

   A_2B_2: 20, 22, 24, 25

   A_3B_1: 24, 29, 27, 25

   A_3B_2: 10, 12, 21, 14

A_4B_1: 30, 26, 29, 28

A_4B_2: 22, 20, 25, 15

7. Using the following data, conduct a two-factor fixed-effects ANOVA to determine if there are any effects due to A (intervention), B (group), or the AB interaction (a = .05). Conduct Tukey HSD post hoc comparisons, if necessary.

Intervention	Group	Outcome
1	1	69
1	1	67
1	1	55
1	1	72
1	1	70
1	1	59
1	1	63
1	1	64
1	2	69
1	2	74
1	2	71
1	2	67
2	2	69
2	2	64
2	2	54
2	2	58
2	3	58
2	3	62
2	3	56
2	3	60
1	3	64
1	3	56
1	3	55
1	3	64

8. Using the following data, conduct a two-factor fixed-effects ANOVA to determine if there are any effects due to A (intervention), B (group), or the AB interaction (a = .05). Conduct Tukey HSD post hoc comparisons, if necessary.

Intervention	Group	Outcome
1	1	69
1	1	72
1	1	63
1	1	74
1	2	67

(continued)

(continued)

Intervention	Group	Outcome
1	2	70
1	2	64
1	2	71
1	3	55
1	3	59
1	3	69
1	3	67
2	1	69
2	1	58
2	1	56
2	1	56
2	2	64
2	2	58
2	2	60
2	2	55
2	3	54
2	3	62
2	3	64
2	3	64

Answers to Computational Problems

1. $SS_{with} = 225$; $df_A = 1$; $df_B = 2$; $df_{AB} = 2$; $df_{with} = 150$; $df_{total} = 155$; $MS_A = 6.15$; $MS_B = 5.30$; $MS_{AB} = 4.55$; $MS_{with} = 1.50$; $F_A = 4.10$; $F_B = 3.5333$; $F_{AB} = 3.0333$; critical value for A is approximately 3.91, thus reject H_0 for A; critical value for B and AB approximately 3.06, thus reject H_0 for B and fail to reject H_0 for AB.
3. See completed table below:

Source	*SS*	*df*	*MS*	*F*	Critical Value	Decision
A	14.06	1	14.06	.25	4.75	Fail to reject H_0
B	39.06	1	39.06	.70	4.75	Fail to reject H_0
AB	1.56	1	1.56	.03	4.75	Fail to reject H_0
Within	668.75	12	55.73			
Total	723.43	15				

5. $F_A = 4.0541$; $F_B = 210.1622$; $F_C = 31.7838$; $F_{AB} = 7.9459$; $F_{AC} = 13.1351$; $F_{BC} = 10.3784$; $F_{ABC} = 4.0541$; all but ABC and A are significant.
7. $F_{intervention} = 3.723$, $p = .069$; $F_{group} = 3.185$, $p = .064$; $F_{intervention*group} = 2.666$, $p = .119$; fail to reject H_0 for intervention, group, and the intervention by group interaction. No post hoc comparisons are needed given there were no hypotheses rejected.

Interpretive Problems

1. Building on the interpretive problem from Chapter 1, utilize the survey1 dataset, which is accessible from the website. Use SPSS or **R** to conduct a two-factor fixed-effects ANOVA, including effect size, where political view is factor A ($J = 5$), gender is factor B ($K = 2$), and the dependent variable is the same one that you used for interpretative problem #1 in Chapter 1. Then write an APA-style paragraph summarizing the results.
2. Building on the interpretive problem from Chapter 1, use the survey1 dataset, which is accessible from the website. Use SPSS or **R** to conduct a two-factor fixed-effects ANOVA, including effect size, where hair color is factor A (i.e., one independent variable) ($J = 5$), gender is factor B (a new factor, $K = 2$), and the dependent variable is an interval or ratio variable of your choice. Then write an APA-style paragraph describing the results.
3. Building on the interpretive problem from Chapter 1, use the IPEDS2017 dataset, which is accessible from the website. Use SPSS or **R** to conduct a factorial ANOVA. To the model that you created in Chapter 1, add a second independent variable. Compute the results of the factorial ANOVA. Also compute effect size and test for assumptions. Then write an APA-style paragraph describing the results.

4

Introduction to Analysis of Covariance: The One-Factor Fixed-Effects Model With a Single Covariate

Chapter Outline

Key Concepts

1. Statistical adjustment
2. Covariate
3. Adjusted means

4. Homogeneity of regression slopes
5. Independence of the covariate and the independent variable

We have now considered several different analysis of variance (ANOVA) models. As we moved through Chapter 3, we saw that the inclusion of additional factors helped to reduce the residual or uncontrolled variation. These additional factors served as experimental design controls, in that their inclusion in the design helped to reduce the uncontrolled variation. In fact, this could be the reason an additional factor is included in a factorial design.

In this chapter a new type of variable, known as a **covariate**, is incorporated into the analysis. Rather than serving as an "experimental design control," the covariate serves as a "statistical control" where uncontrolled variation is reduced statistically in the analysis. Thus a model where a covariate is used is known as **analysis of covariance (ANCOVA)**. We are most concerned with the one-factor fixed-effects model here, although this model can be generalized to any of the other ANOVA designs considered in this text. That is, any of the ANOVA models discussed in the text can also include a covariate, and thus become an ANCOVA model. Additionally, multiple covariates can be included in a model, although our discussion will focus on the inclusion of just one covariate.

Most of the concepts used in this chapter have already been covered in the text. In addition, new concepts include statistical adjustment, covariate, adjusted means, and two important assumptions, homogeneity of regression slopes and independence of the covariate and the independent variable. Our objectives are that by the end of this chapter, you will be able to (a) understand the characteristics and concepts underlying ANCOVA; (b) determine and interpret the results of ANCOVA, including adjusted means and multiple comparison procedures; and (c) understand and evaluate the assumptions of ANCOVA.

4.1 What ANCOVA Is and How It Works

We have been following a superbly talented group of graduate students. We now find Addie Venture assisting with experimental data from her institution's Exercise Physiology and Wellness Institute. As we will see in this chapter, Addie will be examining data generated from an experiment of athletes.

Addie Venture and her group of graduate researchers have been extremely successful in providing support to researchers in a number of areas. We now find Addie assisting Dr. Waung, the university's director of the Exercise Physiology and Wellness Institute, with an experimental study to determine if there was a mean difference in self-rated physical performance based on the use of caffeine in an attempt to facilitate improved athletic performance. Twelve athletes were randomly assigned to ingest either a caffeinated (treatment) or decaffeinated (control) beverage prior to physical activity. Prior to random assignment to sections, participants were also measured on mental fatigue. After random assignment, participants completed a 2000-meter self-paced jog and were then asked to self-rate their physical performance. Addie is now ready to examine this data. Addie's research question that she recommends to Dr. Waung is: *Is there a mean difference in self-rated physical performance based on caffeine ingestion, controlling for mental fatigue?* With one independent variable and one covariate for which to control, Addie determines that an analysis of covariance (ANCOVA) is the best statistical procedure to use to answer the question. Her next task is to analyze the data to address the research question.

In this section, we describe the distinguishing characteristics of the one-factor fixed-effects ANCOVA model. However, before we begin an extended discussion of these characteristics, consider the following example (a situation similar to which we find Addie Venture with her project with Dr. Waung).

4.1.1 Characteristics

Imagine a situation where a statistics professor is scheduled to teach two sections of introductory statistics. The professor, being a cunning researcher, decides to perform a little experiment where Section 1 is taught using the traditional lecture method and Section 2 is taught with more innovative methods using extensive graphics, computer simulations, and computer-assisted and calculator-based instruction, as well as using mostly small-group and self-directed instruction. The professor is interested in which section performs better in the course.

Before the study/course begins, the professor thinks about whether there are other variables related to statistics performance that should somehow be taken into account in the design. An obvious one is ability in quantitative methods. From previous research and experience, the professor knows that ability in quantitative methods is highly correlated with performance in statistics and decides to give a measure of quantitative ability in the first class and use that as a covariate in the analysis. A *covariate* (e.g., quantitative ability) is defined as a source of variation not controlled for in the design of the experiment, but that the researcher believes to affect the dependent variable (e.g., course performance). The covariate is used to *statistically adjust* the dependent variable. For instance, if Section 1 has higher quantitative ability than Section 2 going into the study, then it would be wise to take this into account in the analysis. Otherwise, Section 1 might outperform Section 2 due to their higher quantitative ability rather than due to the method of instruction. This is precisely the point of the analysis of covariance. Some of the more typical examples of covariates in education and the behavioral sciences, depending on the study of course, are pretest (where the dependent variable is the posttest), prior achievement, weight, IQ, aptitude, age, experience, previous training, motivation, and grade point average.

Let us now begin with the characteristics of the ANCOVA model. The first set of characteristics is obvious because they carry over from the one-factor fixed-effects ANOVA model. *There is a single independent variable or factor with two or more levels or categories (thus the independent variable continues to be either nominal or ordinal in measurement scale)*. The *levels of the independent variable are fixed* by the researcher rather than randomly sampled from a population of levels. Once the levels of the independent variable are selected, *subjects or individuals are somehow assigned to these levels or groups*. Each subject is then exposed to only one level of the independent variable (although ANCOVA with repeated measures is also possible, but is not discussed here). In our example, method of statistics instruction is the independent variable with two levels or groups, the traditional lecture method and the cutting-edge method.

Situations where the researcher is able to randomly assign subjects to groups are known as **true experimental designs**. Situations where the researcher does not have control over which level a subject is assigned to are known as **quasi-experimental designs**. This lack of control may occur for one of two reasons. First, the groups may be already in place when the researcher arrives on the scene; these groups are referred to as **intact groups** (e.g., based on class assignments made by students at the time of registration). Second, it may be theoretically impossible for the researcher to assign subjects to groups (e.g., income level). Thus a distinction is typically made about whether or not the researcher can control the assignment of subjects to groups. The distinction between the use of ANCOVA in true and quasi-experimental situations has been quite controversial over the past few decades; we look at it in more detail later in this chapter. For further information on true experimental designs and quasi-experimental designs, we suggest you consider Campbell and Stanley (1966), Cook and Campbell (1979), and Shadish, Cook, and Campbell (2002). In our example again, if assignment to groups is random, then we have a true experimental design. If assignment to groups is not random, perhaps already assigned at registration, then we have a quasi-experimental design.

One final item in the first set of characteristics has to do with the measurement scales of the variables. In the analysis of covariance, it is assumed the *dependent variable is measured at the interval level or better*. If the dependent variable is measured at the ordinal level, then nonparametric procedures described toward the end of this chapter should be considered. It is also assumed that the *covariate is measured at the interval level or better*. Lastly, as indicated previously, the *independent variable must be a grouping or categorical variable.*

The remaining characteristics have to do with the uniqueness of the analysis of covariance. As already mentioned, the analysis of covariance is a form of statistical control developed specifically to reduce unexplained error variation. The covariate (sometimes known as a *concomitant variable*, as it accompanies or is associated with the dependent variable), is a source of variation not controlled for in the design of the experiment, but believed to affect the dependent variable. In a factorial design, for example, a factor could be included to reduce error variation. However, this represents an experimental design form of control as it is included as a factor in the model.

In ANCOVA, the dependent variable is adjusted statistically to remove the effects of the portion of uncontrolled variation represented by the covariate. The group means on the dependent variable are adjusted so that they now represent groups with the same means on the covariate. The analysis of covariance is essentially an analysis of variance on these "adjusted means." This needs further explanation. Consider first the situation of the randomized true experiment where there are two groups. Here it is unlikely that the two groups will be statistically different on any variable related to the dependent measure. The two groups should have roughly equivalent means on the covariate, although 5% of the time we would expect a significant difference due to chance at $\alpha = .05$. Thus, we typically do not see preexisting differences between the two groups on the covariate in a true experiment—that is the value and beauty of random assignment, especially as it relates to ANCOVA. The advantage of ANCOVA in randomized studies, as compared to other types of statistical designs, is *increased precision* and *unbiased estimates* of treatment effects. However, the relationship between the covariate and the dependent variable is important. If these variables are linearly related (discussed later), then the use of the covariate in the analysis will serve to reduce the unexplained variation in the model. The greater the magnitude of the correlation, the more uncontrolled variation can be removed, as shown by a reduction in mean square error.

Let us divert for a moment to ensure we understand statistical precision and bias. **Precision** refers to the size of deviations from an estimate (e.g., mean) that occurs when the same sampling procedures using the same sampling frame and sample size are repeated. Estimates that are more precise, for example, have more narrow confidence intervals. **Bias** refers to the difference between an estimate's expected value and the true value of the parameter. Thus, bias basically means that an estimate is systematically "off," either overestimating or underestimating the true population parameter.

BOX 4.1 Precision and Bias and the Relationship to ANCOVA

Term	Definition
Precision	The size of deviations from an estimate (e.g., mean) that occurs when the same sampling procedures using the same sampling frame and sample size are repeated.
Bias	The difference between an estimate's expected value and the true value of the parameter and basically means that an estimate is systematically "off," either overestimating or underestimating the true population parameter.
How these concepts relate to ANCOVA	The advantage of ANCOVA in randomized studies, as compared to other types of statistical designs, is *increased precision* and *unbiased estimates* of treatment effects through the inclusion of the covariate.

Consider next the situation of the **quasi-experiment**, that is, *without randomization to groups or levels of the independent variable*. Here it is more likely that the two groups will be statistically different on the covariate as well as other variables related to the dependent variable. Thus, there may indeed be a preexisting difference between the two groups on the covariate. If the groups do differ on the covariate and we ignore it by conducting an ANOVA, our ability to get a precise estimate of the group effects will be reduced as the group effect will be confounded with the effect of the covariate. For instance, if a significant group difference is revealed by the ANOVA, we would not be certain if there was truly a group effect or whether the effect was due to preexisting group differences on the covariate, or some combination of group and covariate effects. The analysis of covariance takes the covariate mean difference into account as well as the linear relationship between the covariate and the dependent variable.

Thus, the covariate is used to (a) reduce error variation, (b) take any preexisting group mean difference on the covariate into account, (c) take into account the relationship between the covariate and the dependent variable, and (d) yield a more precise and less biased estimate of the group effects. If error variation is reduced, the analysis of covariance will be more powerful and require smaller sample sizes than the analysis of variance (Keppel & Wickens, 2004; Mickey, Dunn, & Clark, 2004; Myers, Lorch, & Well, 2010). If error variation is not reduced, the analysis of variance is more powerful. A more extensive comparison of ANOVA versus ANCOVA is given in Chapter 6. In addition, as shown later, one degree of freedom is lost from the error term for each covariate used. This results in a larger critical value for the F test and makes it a bit more difficult to find a statistically significant F test statistic. This is the major cost of using a covariate. If the covariate is not effective in reducing error variance, then we are worse off than if we had ignored the covariate. Importance references on ANCOVA include Elashoff (1969) and Huitema (2011).

4.1.1.1 The Layout of the Data

Before we get into the theory and subsequent analysis of the data, let us examine the layout of the data. We designate each observation on the dependent or criterion variable as Y_{ij}, where the j subscript tells us what group or level the observation belongs to and the i subscript tells us the observation or identification number within that group. The first subscript ranges over $i = 1, \ldots, n_j$ and the second subscript ranges over $j = 1, \ldots, J$. Thus,

TABLE 4.1

Layout for the One-Factor ANCOVA

Level of the Independent Variable						
1		**2**		**...**	**J**	
Y_{11}	X_{11}	Y_{12}	X_{12}	...	Y_{1j}	X_{1j}
Y_{21}	X_{21}	Y_{22}	X_{22}	...	Y_{2j}	X_{1j}
...	...	...	...	...	...	...
Y_{n1}	X_{n1}	Y_{n2}	X_{n2}	...	Y_{nJ}	Y_{nJ}
$\bar{Y}_{.1}$	$\bar{X}_{.1}$	$\bar{Y}_{.2}$	$\bar{X}_{.2}$	...	$\bar{Y}_{.J}$	$\bar{X}_{.J}$

there are J levels of the independent variable and n_j subjects in group j. We designate each observation on the covariate as X_{ij}, where the subscripts have the same meaning.

The layout of the data is shown in Table 4.1. Here we see that each pair of columns represents the observations for a particular group or level of the independent variable on the dependent variable (i.e., Y) and the covariate (i.e., X). At the bottom of the pair of columns for each group j are group means $(\bar{Y}_{.j}, \bar{X}_{.j})$. Although the table shows there are n observations for each group, we need not make such a restriction, as this was done only for purposes of simplifying the table.

4.1.1.2 The ANCOVA Model

The analysis of covariance model is a form of the general linear model much like the models shown in the last few chapters of this text. The **one-factor ANCOVA fixed-effects model** can be written in terms of population parameters as follows:

$$Y_{ij} = \mu_Y + \alpha_j + \beta_w \left(X_{ij} - \mu_X\right) + \varepsilon_{ij}$$

where Y_{ij} is the observed score on the dependent variable for individual i in group j, μ_Y is the overall or grand population mean (i.e., regardless of group designation) for the dependent variable Y, α_j is the group effect for group j, β_w is the within-groups regression slope from the regression of Y on X (i.e., the covariate), X_{ij} is the observed score on the covariate for individual i in group j, μ_X is the overall or grand population mean (i.e., regardless of group designation) for the covariate X, and ε_{ij} is the random residual error for individual i in group j. The residual error can be due to individual differences, measurement error, and/or other factors not under investigation. As you would expect, the least squares sample estimators for each of these parameters are as follows: $\bar{Y}$ for μ_Y, $\bar{X}$ for μ_X, a_j for α_j, b_w for β_w, and e_{ij} for ε_{ij}. Just like in the analysis of variance, the sum of the group effects is equal to zero. This implies that if there are any nonzero group effects, then the group effects will balance out around zero with some positive and some negative effects.

The hypotheses consist of testing the equality of the adjusted means (defined by $\mu'_{.j}$ and discussed later) as follows:

$$H_0 : \mu'_{.1} = \mu'_{.2} = \ldots = \mu'_{.J}$$
$$H_1 : not\ all\ the\ \mu'_{.j}\ are\ equal$$

4.1.1.3 The ANCOVA Summary Table

We turn our attention to the familiar summary table, this time for the one-factor ANCOVA model. A general form of the summary table is shown in Table 4.2. Under the first column you see the following sources: adjusted between-groups variation, adjusted within-groups variation, variation due to the covariate, and total variation. The second column notes the sums of squares terms for each source (i.e., $SS_{betw(adj)}$, $SS_{with(adj)}$, SS_{cov}, SS_{total}). Recall that the *between* source represents the independent variable being systematically studied and the *within* source represents the error or residual.

The third column gives the degrees of freedom for each source. For the adjusted between-groups source (i.e., the independent variable controlling for the covariate), because there are J group means, the $df_{betw(adj)}$ is $J - 1$, the same as in the one-factor ANOVA model. For the adjusted within-groups source, because there are N total observations and J groups, we would expect the degrees of freedom within to be $N - J$, because that was the case in the one-factor ANOVA model. However, as we pointed out earlier in the characteristics of the ANCOVA model, a price is paid for the use of a covariate. The price here is that we lose one degree of freedom from the within term for single covariate, so that $df_{with(adj)}$ is $N - J - 1$. For multiple covariates, we lose one degree of freedom for each covariate used (see later discussion). This degree of freedom has gone to the covariate source such that df_{cov} is equal to 1. Finally, for the total source, as there are N total observations, the df_{total} is the usual $N - 1$.

The fourth column gives the mean squares for each source of variation. As always, the mean squares represent the sum of squares weighted by their respective degrees of freedom. Thus $\left[MS_{betw(adj)} = SS_{betw(adj)} / (J-1)\right]$, $\left[MS_{with(adj)} = SS_{with(adj)} / (N-J-1)\right]$, and $\left[MS_{cov} = SS_{cov} / 1\right]$. The last column in the ANCOVA summary table is for the F values. Thus for the one-factor fixed-effects ANCOVA model, the F value tests for differences between the adjusted means (i.e., to test for differences in the mean of the dependent variable based on the levels of the independent variable when controlling for the covariate) and is computed as $F = MS_{betw(adj)} / MS_{with(adj)}$. A second F value, which is obviously not included in the ANOVA model, is the test of the covariate. To be specific, this F statistic is actually testing the hypothesis of H_0: $\beta_w = 0$. If the slope is equal to zero, then the covariate and the dependent variable are unrelated. This F value is equal to $F = MS_{cov} / MS_{with(adj)}$. If the F test for the covariate is *not* statistically significant (and has a negligible effect size), the researcher may want to consider removing that covariate from the model.

The critical value for the test of difference between the adjusted means is ${}_{\alpha}F_{J-1,N-J-1}$. The critical value for the test of the covariate is ${}_{\alpha}F_{1,N-J-1}$. The null hypotheses in each case are

TABLE 4.2

One-Factor Analysis of Covariance Summary Table

Source	SS	*df*	*MS*	F
Covariate	SS_{cov}	1	MS_{cov}	$MS_{cov}/MS_{with(adj)}$
Between adjusted (i.e., independent variable)	$SS_{betw(adj)}$	$J - 1$	$MS_{betw(adj)}$	$MS_{betw(adj)}/MS_{with(adj)}$
Within adjusted (i.e., Error)	$SS_{with(adj)}$	$N - J - 1$	$MS_{with(adj)}$	
Total	SS_{total}	$N - 1$		

rejected if the F test statistic exceeds the F critical value. The critical values are found in the F table of Appendix Table A.4.

If the F test statistic for the adjusted means exceeds the F critical value, and there are more than two groups, then it is not clear exactly how the means are different. In this case, some multiple comparison procedure may be used to determine which means are different (see later discussion). For the test of the covariate (i.e., the within-groups regression slope), we hope that the F test statistic *does* exceed the F critical value. Otherwise, the power and precision of the test of the adjusted means in ANCOVA will be lower than the test of the unadjusted means in ANOVA because the covariate is not significantly related to the dependent variable. (As stated previously, if the F test for the covariate is *not* statistically significant and has a negligible effect size, the researcher may want to consider removing that covariate from the model).

4.1.1.4 Partitioning the Sum of Squares

As seen already, the partitioning of the sums of squares is the backbone of all general linear models, whether we are dealing with an ANOVA model, an ANCOVA model, or a linear regression model. As always, the first step is to partition the total variation into its relevant parts or sources of variation. As we have learned from the previous section, the sources of variation for the one-factor ANCOVA model are adjusted between groups (i.e., the independent variable), adjusted within groups (i.e., error), and the covariate. This is written as follows:

$$SS_{total} = SS_{betw(adj)} + SS_{with(adj)} + SS_{cov}$$

From this point the statistical software is used to handle the remaining computations.

4.1.1.5 Adjusted Means and Related Procedures

In this section we formally define the adjusted mean and briefly examine several multiple comparison procedures. We have spent considerable time already discussing the analysis of the adjusted means. Now it is time to define them. The **adjusted mean** is denoted by $\bar{Y}'_{.j}$ and estimated by

$$\bar{Y}'_{.j} = \bar{Y}_{.j} - b_w\left(\bar{X}_{.j} - \bar{X}_{..}\right)$$

Here it should be noted that the adjusted mean is simply equal to the unadjusted mean (i.e., $\bar{Y}_{.j}$) minus the adjustment [i.e., $b_w\left(\bar{X}_{.j} - \bar{X}_{..}\right)$]. The adjustment is a function of the within-groups regression slope (i.e., b_w) and the difference between the group mean and the overall mean for the covariate (i.e., the difference being the group effect, $\bar{X}_{.j} - \bar{X}_{..}$). No adjustment will be made if (a) $b_w = 0$ (i.e., X and Y are unrelated), or (b) the group means on the covariate are all the same. Thus, in both cases $\bar{Y}_{.j} = \bar{Y}'_{.j}$. In all other cases, at least some adjustment will be made for some of the group means (although not necessarily for all of the group means).

You may be wondering how this adjustment actually works. Let us assume the covariate and the dependent variable are positively correlated such that b_w is also positive, and there are two treatment groups with equal n's that differ on the covariate. If Group 1 has a higher mean on *both* the covariate and the dependent variable than Group 2, then the adjusted means will be closer together than the unadjusted means. For our first example, we have the following conditions:

$$b_w = 1, \bar{Y}_{.1} = 50, \bar{Y}_{.2} = 30, \bar{X}_{.1} = 20, \bar{X}_{.2} = 10, \bar{X}_{..} = 15$$

The adjusted means are determined as follows:

$$\bar{Y}'_{.1} = \bar{Y}_{.1} - b_w(\bar{X}_{.1} - \bar{X}_{..}) = 50 - 1(20 - 15) = 45$$

$$\bar{Y}'_{.2} = \bar{Y}_{.2} - b_w(\bar{X}_{.2} - \bar{X}_{..}) = 30 - 1(10 - 15) = 35$$

This is shown graphically in Figure 4.1a. In looking at the covariate X, we see that Group 1 has a higher mean $(\bar{X}_{.1} = 20)$ than Group 2 $(\bar{X}_{.2} = 10)$ by 10 points. The vertical line represents the overall mean on the covariate $(\bar{X}_{..} = 15)$. In looking at the dependent variable Y, we see that Group 1 has a higher mean $(\bar{Y}_{.1} = 50)$ than Group 2 $(\bar{Y}_{.2} = 30)$ by 20 points. The diagonal lines represent the regression lines for each group, with b_w = 1.0. The points at which the regression lines intersect (or cross) the vertical line $(\bar{X}_{..} = 15)$ represent on the Y scale the values of the adjusted means. Here we see that the adjusted mean for Group 1 $\bar{Y}'_{.1} = 45$ is larger than the adjusted mean for Group 2 $\bar{Y}'_{.2} = 35$ by 10 points. Thus, because of the preexisting difference on the covariate, the adjusted means here are somewhat closer together than the unadjusted means (10 points vs. 20 points, respectively).

If Group 1 has a higher mean on the covariate and a lower mean on the dependent variable than Group 2, then the adjusted means will be further apart than the unadjusted means. As a second example, we have the following slightly different conditions:

$$b_w = 1, \bar{Y}_{.1} = 30, \bar{Y}_{.2} = 50, \bar{X}_{.1} = 20, \bar{X}_{.2} = 10, \bar{X}_{..} = 15$$

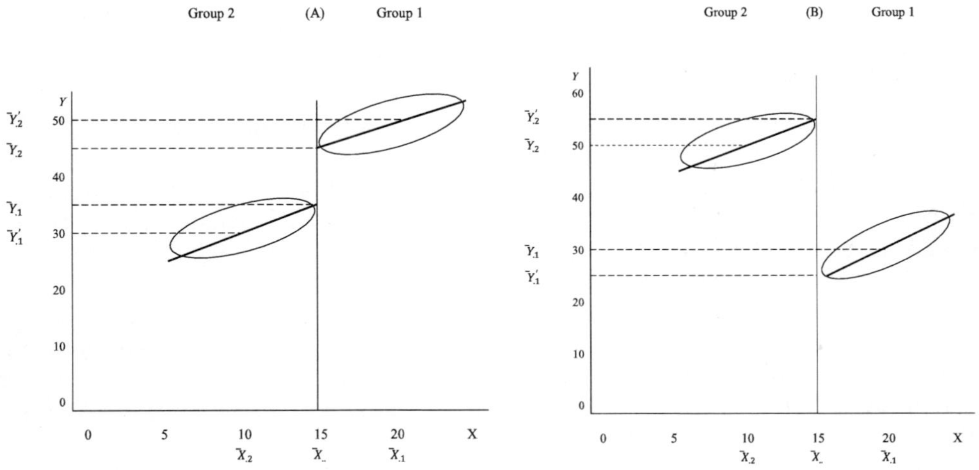

FIGURE 4.1
Graphs of ANCOVA adjustments.

Then the adjusted means become as follows:

$$\bar{Y}'_{.1} = \bar{Y}_{.1} - b_w\left(\bar{X}_{.1} - \bar{X}_{..}\right) = 30 - 1(20 - 15) = 25$$

$$\bar{Y}'_{.2} = \bar{Y}_{.2} - b_w\left(\bar{X}_{.2} - \bar{X}_{..}\right) = 50 - 1(10 - 15) = 55$$

This is shown graphically in Figure 4.1b where the unadjusted means differ by 20 points and the adjusted means differ by 30 points. There are obviously other possible situations.

Let us briefly examine multiple comparison procedures (MCPs) for use in the analysis of covariance situation. Most of the procedures described in Chapter 2 can be adapted for use with a covariate, although a few procedures are not mentioned here as critical values do not currently exist. The adapted procedures involve a different form of the standard error of a contrast. The contrasts are formed based on adjusted means, of course. Let us briefly outline just a few procedures. Each of the test statistics has as its numerator the contrast ψ', such as $\psi' = \bar{Y}'_{.1} - \bar{Y}'_{.2}$. The standard errors do differ somewhat depending on the specific MCP, just as they do in ANOVA.

The example procedures briefly described here are easily translated from the ANOVA context into the ANCOVA context. Dunn's (or the Bonferroni) method is appropriate to use for a small number of planned contrasts (still utilizing the critical values from Appendix Table A.8). Scheffé's procedure can be used for unplanned complex contrasts with equal group variances (again based on the *F* table in Appendix Table A.4). Tukey's HSD test is most desirous for unplanned pairwise contrasts with equal *n*'s per group. There has been some discussion in the literature about the appropriateness of this test in ANCOVA. Most statisticians currently argue that the procedure is only appropriate when the covariate is fixed, when in fact it is almost always random. As a result the Bryant-Paulson (Bryant & Paulson, 1976) generalization of the Tukey procedure has been developed for the random covariate case. The test statistic is compared to the critical value ${}_{\alpha}q_{X,df(error),J}$ taken from Appendix Table A.10, where *X* is the number of covariates. If the group sizes are unequal, the harmonic mean can be used in ANCOVA (Huitema, 2011). A generalization of the Tukey-Bryant procedure for unequal *n*'s ANCOVA was developed by Hochberg and Varon-Salomon (1984) (see also Hochberg & Tamhane, 1987; Miller, 1997).

4.1.1.6 An Example

Consider the following illustration of what we have covered in this chapter. Our dependent variable is self-rated physical performance (with a maximum possible score of 6), the covariate is self-rated mental fatigue assessed prior to random assignment (with a maximum possible score of 10), and the independent variable is the assigned group (where Group 1 ingests a decaffeinated beverage and Group 2 ingests a caffeinated beverage prior to a jog). Thus, the researcher is interested in whether caffeine influences athletes' physical performance, controlling for mental fatigue (assume we have developed a measure that is relatively error-free). Athletes are randomly assigned to one of the two groups prior to random assignment when the measure of mental fatigue is administered. There are 6 athletes in each group for a total of 12. The layout of the data is shown in Table 4.3, where we see the data and sample statistics (means, variances, slopes, and correlations).

The results are summarized in the ANCOVA summary table as shown in the top panel of Table 4.5. The ANCOVA test statistics are compared to the critical value ${}_{.05}F_{1,9} = 5.12$ obtained from Appendix Table A.4, using the .05 level of significance. Both test statistics exceed the critical value, so we reject H_0 in each case. We conclude that (a) physical performance means do differ for the two groups when adjusted (or controlling) for mental fatigue (i.e., the between adjusted test of the independent variable controlling for the covariate), and (b) the slope of the regression of Y (i.e., dependent variable) on X (i.e., covariate) is statistically significantly different from zero (i.e., the test of the covariate). Just to be complete, the results for the analysis of variance (ANOVA) on Y are shown in the bottom panel of Table 4.4. We see that in the analysis of the unadjusted means (i.e., the ANOVA), there is

TABLE 4.3

Data and Summary Statistics for the Physical Performance Example

	Group 1 (Decaffeinated)		Group 2 (Caffeinated)		Overall	
Statistic	Physical Performance (Y)	Mental Fatigue (X)	Physical Performance (Y)	Mental Fatigue (X)	Physical Performance (Y)	Mental Fatigue (X)
	1	4	1	1		
	2	3	2	3		
	3	5	4	2		
	4	6	5	4		
	5	7	6	5		
	6	9	6	7		
Means	3.5000	5.6667	4.0000	3.6667	3.7500	4.6667
Variances	3.5000	4.6667	4.4000	4.6667	3.6591	5.3333
b_{YX}	0.8143		0.8143		0.5966	
r_{YX}	0.9403		0.8386		0.7203	
Adjusted means	2.6857		4.8143			

TABLE 4.4

One-Factor ANCOVA and ANOVA Summary Tables—Statistics Instruction Example

Source	*SS*	*df*	*MS*	*F*
ANCOVA				
Covariate	20.8813	1	20.8813	21.9641*
Adjusted between (i.e., independent variable)	10.8127	1	10.8127	11.3734*
Adjusted within (i.e., error)	8.5560	9	0.9507	
Total	40.2500	11		
ANOVA				
Between	0.7500	1	0.7500	0.1899**
Within	39.5000	10	3.9500	
Total	40.2500	11		

* $_{.05}F_{1,9} = 5.12$
** $_{.05}F_{1,10} = 4.96$

no significant group difference. Thus the adjustment (i.e., ANCOVA which controlled for the covariate, mental fatigue) yielded a different statistical result. The covariate also "did its thing" in that a reduction in MS_{with} resulted due to the strong relationship between the covariate and the dependent variable (i.e., $r_{XY} = 0.7203$ overall).

Let us next examine the group physical performance means, as shown previously in Table 4.3. Here we see that with the *unadjusted* physical performance means (i.e., prior to controlling for the covariate), there is a 0.5000-point difference in favor of Group 2 (the group that ingested caffeine prior to a self-paced jog), whereas for the *adjusted* physical performance means (i.e., the ANCOVA results which controlled for mental fatigue), there is a 2.1286-point difference in favor of Group 2. In other words, the adjustment (i.e., controlling for mental fatigue) in this case resulted in a greater difference between the adjusted physical performance means than between the unadjusted physical performance means. Since there are only two groups, a multiple comparison procedure is unnecessary (although we illustrate this in the SPSS section).

4.1.1.7 ANCOVA Without Randomization

As referenced previously in the discussion of assumptions, there has been a great deal of discussion and controversy over the years, particularly in education and the behavioral sciences, about the use of the analysis of covariance in situations where randomization is not conducted. **Randomization** is defined as an experiment where individuals are randomly assigned to groups (or cells in a factorial design). In the Campbell and Stanley (1966) system of experimental design, these designs are known as **true experiments**. (Do not confuse random assignment with random selection, the latter of which deals with how the cases are sampled from the population.)

In certain situations, randomization either has not occurred or is not possible due to circumstances in the study. The best example is the situation where there are **intact groups**, which are groups that have been formed prior to the researcher arriving on the scene. Either the researcher chooses not to randomly assign these individuals to groups through a reassignment (e.g., it is just easier to keep the groups in their current form), or the researcher cannot randomly assign them (legally, ethically, or otherwise). When randomization does not occur, the resulting designs are known as **quasi-experimental**. For instance, in classroom research, the researcher is almost never able to come into a school and randomly assign students to classrooms. Once students are given their class assignments at the beginning of the year, this cannot be altered. On occasion, the researcher might be able to pull a few students out of several classrooms, randomly assign them to small groups, and conduct a true experiment. In general, this is possible only on a very small scale and for short periods of time.

Let us briefly consider the issues as it relates to ANCOVA, as not all statisticians agree. In *true experiments* (i.e., with randomization), there is no cause for concern (except for dealing with the statistical assumptions). The analysis of covariance is more powerful and has greater precision for true experiments than for quasi-experiments. So if you have a choice, go with a true experimental situation (which is a big *if*). In a true experiment, the probability that the groups differ on the covariate or any other concomitant variable is equal to α. That is, the likelihood that the group means will be different on the covariate is small, and thus the adjustment in the group means may be small. The payoff is in the possibility that the error term will be greatly reduced.

In *quasi-experiments*, as it relates to ANCOVA, there are several possible causes for concern. Although this is the situation where the researcher needs the most help, this is also the situation where less help is available. Here it is more likely that there will be statistically significant differences among the group means on the covariate. Thus the adjustment in the group means can be substantial (assuming that b_w is different from zero). Because there are significant mean differences on the covariate, any of the following may occur: (a) it is likely that the groups may be different on other important characteristics as well, which have not been controlled for either statistically or experimentally; (b) the homogeneity of regression slopes assumption is less likely to be met; (c) adjusting for the covariate may remove part of the treatment effect; (d) equating groups on the covariate may be an extrapolation beyond the range of possible values that occur for a particular group (e.g., the examples on trying to equate men and women (Lord, 1960, 1967) or trying to equate mice and elephants (Ferguson & Takane, 1989); these groups should not be equated on the covariate because their distributions on the covariate do not overlap); (e) although the slopes may be equal for the range of X's obtained, when extrapolating beyond the range of scores, the slopes may not be equal; (f) the standard errors of the adjusted means may increase, making tests of the adjusted means not significant; and (g) there may be differential growth in the groups confounding the results (e.g., adult vs. child groups).

Although one should be cautious about the use of ANCOVA in quasi-experiments, this is not to suggest that ANCOVA should never be used in such situations. Schneider, Avivi-Reich, and Mozuraitis (2015) provide recommendations for conducting ANCOVA in various situations, including experimental designs where there is both random selection and random assignments, as well as quasi-experimental designs where it cannot be assumed that the covariate is equal across the population. Just be extra careful and do not go too far in terms of interpreting your results. If at all possible, replicate your study. For further discussion, see Huitema (2011), Porter and Raudenbush (1987), or Schneider et al. (2015).

4.1.1.8 More Complex ANCOVA Models

The one-factor ANCOVA model can be extended to more complex models in the same way as we expanded the one-factor ANOVA model. Thus, we can consider ANCOVA designs that involve any of the following characteristics: (a) factorial designs (i.e., having more than one factor or independent variable); (b) fixed-, random-, and mixed-effects designs; (c) repeated measures and split-plot (mixed) designs; (d) hierarchical designs; and (e) randomized block designs. Conceptually, there is nothing new for these types of ANCOVA designs, and you should have no trouble getting a statistical package to do such analyses. For further information on these designs, or for information on how one can also utilize multiple covariates in an analysis of covariance design, see any number of excellent references (Huitema, 2011; Keppel & Wickens, 2004; Kirk, 2014; Myers et al., 2010; Page, Braver, & MacKinnon, 2003).

4.1.1.9 Nonparametric ANCOVA Procedures

In situations where the assumptions of normality, homogeneity of variance, and/or linearity have been seriously violated, one alternative is to consider nonparametric ANCOVA procedures. Some rank ANCOVA procedures have been proposed by Quade (1967), Puri and Sen (1969), Conover and Iman (1982), Rutherford (1992), Mansouri and Zhang (2018). For a description of such procedures, see these references as well as Huitema (2011), Harwell (1992), Harwell (2003), or Wilcox (2003).

4.1.2 Sample Size

As you have likely gauged in the discussion of sample size in other chapters, there are not suggested sample size guidelines that we will offer in consideration of computing ANCOVA. We know there are many elements that work together to impact sample size. These include the alpha level (where smaller alphas require larger sample size), power (where increased power requires larger sample size), effect size (where larger effect size estimates decrease sample size), and variation in the data (where increased variance increases required sample size). Rather than attempt to suggest criteria for the number of cases required in ANCOVA, we encourage researchers to compute required sample size based on power analysis, as we will illustrate later. We also encourage researchers to consider work that has been done in this area. For example, a two-step method for computing sample size with ANCOVA using one covariate was proposed by Borm, Fransen, and Lemmens (2007). Shan and Ma (2014) proposed an exact approach that produces power closer to the pre-specified power when the correlation between the dependent variable and covariate is large. Researchers interested in sample size determination with more complex ANCOVA models that include multiple covariates are encouraged to review Shieh (2017).

4.1.3 Power

Given a fixed sample size, ANCOVA is more powerful than ANOVA, and this has been demonstrated (e.g., Egbewale, Lewis, & Sim, 2014; Van Breukelen, 2006). Approaching power from a slightly different angle, a smaller sample size is needed in ANCOVA to obtain the same power in ANOVA (Maxwell, Delaney, & Kelley, 2018). How large the sample size needs to be in ANCOVA to achieve a desired power is best gauged by conducting a power analysis using power tables (e.g., Cohen, 1988) or appropriate software (e.g., G*Power).

4.1.4 Effect Size

For the one-factor ANCOVA model, effect size works exactly the same as in the factor-ANOVA model, except that they are based on adjusted means (Cohen, 1988), and as we will see in SPSS, partial eta squared is still the effect size reported in SPSS. The effect size representing the *standardized difference between adjusted means for a two-group design* (i.e., two groups or categories in the independent variable) can be computed as follows (Maxwell et al., 2018):

$$d = \frac{\bar{Y}_1' - \bar{Y}_2'}{\sqrt{MS_{with(adj)}}}$$

Where $\bar{Y}_1'$ and $\bar{Y}_2$ represent the adjusted means of the two groups in the study.

Effect size values for the proportion of total variance (specifically for eta squared, epsilon squared, and omega squared) for the omnibus test can be computed as follows. In effect size indices that represent the proportion of total variance, the variance due to the independent variable (i.e., the "effect of interest") is expressed as a proportion of the sum of the error variance and the total variance (i.e., the variance due to all factors) (Olejnik & Algina, 2000, p. 268). Because effect size values that represent the proportion of total variance are influenced by all the factors in the model, these effect size indices cannot be compared across models that incorporate different factors or different research designs (Olejnik & Algina, 2000).

Eta squared for group effects in ANCOVA, representing the proportion of total variance, can be computed as follows (Olejnik & Algina, 2000), where SS_{effect} corresponds to our notation of $SS_{betw(adj)}$:

$$\eta^2 = \frac{SS_{effect}}{SS_{total}} = \frac{SS_{betw(adj)}}{SS_{total}}$$

Epsilon squared for group effects in ANCOVA, representing the proportion of total variance in the dependent variable that is explained by the independent variable after the effects of the covariate have been removed, can be computed as follows (Olejnik & Algina, 2000):

$$\varepsilon^2 = \frac{df_{effect}\left(MS_{effect} - MS_{error}\right)}{SS_{total}} = \frac{df_{betw(adj)}\left(MS_{betw(adj)} - MS_{with(adj)}\right)}{SS_{total}}$$

Omega squared for group effects in ANCOVA, representing the proportion of total variance in the dependent variable that is explained by the independent variable after the effects of the covariate have been removed, can be computed as follows (Olejnik & Algina, 2000):

$$\omega^2 = \frac{df_{effect}\left(MS_{effect} - MS_{error}\right)}{SS_{total} + MS_{error}} = \frac{df_{betw(adj)}\left(MS_{betw(adj)} - MS_{with(adj)}\right)}{SS_{total} + MS_{with(adj)}}$$

Where df_{effect} corresponds to our notation of $df_{betw(adj)}$ and MS_{effect} corresponds to our notation of $MS_{betw(adj)}$. MS_{error} corresponds to our notation of $MS_{with(adj)}$. Using the example data and results (presented in Table 4.7), we find omega squared to be as follows:

$$\omega^2 = \frac{df_{betw(adj)}\left(MS_{betw(adj)} - MS_{with(adj)}\right)}{SS_{total} + MS_{with(adj)}} = \frac{1(10.812 - .951)}{40.250 + .951} = \frac{9.861}{21.201} = .465$$

Epsilon squared and omega squared will yield similar values (Carroll & Nordholm, 1975). Both epsilon squared and omega squared can result in negative values when F is less than one. In the event this occurs, setting the effect size value to zero is typical practice (Olejnik & Algina, 2000).

We will leave our discussion of effect size measures with a few cautionary notes discussed in Olejnik and Algina (2000). Omega squared effect size indices are derived from expected mean squares variance components. Expected means squares assume a balanced design, and in the absence of balance, omega squared is not recommended (Vaughan & Corballis, 1969). Small sample sizes (e.g., N = 15 and N = 30) can detrimentally impact epsilon squared and omega squared by producing large standard errors for these effects (Carroll & Nordholm, 1975).

TABLE 4.5
Effect Sizes and Interpretations

Effect Size	Interpretation
Standardized difference between adjusted means for a two-group design (d)	Standardized mean difference controlling for the covariate • Small effect, $d = .20$ • Medium effect, $d = .50$ • Large effect, $d = .80$
Eta squared (η^2)	Proportion of total variability in the dependent variable that is accounted for by the independent variable after controlling for the covariate • Small effect, $\eta^2 = .01$ • Medium effect, $\eta^2 = .06$ • Large effect, $\eta^2 = .14$
Epsilon squared (ε^2)	Proportion of total variability in the dependent variable that is accounted for by the independent variable after controlling for the covariate • Small effect, $\varepsilon^2 = .01$ • Medium effect, $\varepsilon^2 = .06$ • Large effect, $\varepsilon^2 = .14$
Omega squared (ω^2)	Proportion of total variability in the dependent variable that is accounted for by the independent variable after controlling for the covariate • Small effect, $\omega^2 = .01$ • Medium effect, $\omega^2 = .06$ • Large effect, $\omega^2 = .14$

4.1.5 Assumptions

The introduction of a covariate requires several assumptions beyond the traditional ANOVA assumptions. For the familiar assumptions (e.g., independence of observations, homogeneity, and normality), the discussion is kept to a minimum as these have already been described in Chapters 1 and 3. The new assumptions are as follows: (a) linearity, (b) independence of the covariate and the independent variable, (c) the covariate is measured without error, and (d) homogeneity of the regression slopes. In this section, we describe each assumption, how each assumption can be evaluated, the effects that a violation of the assumption might have, and how one might deal with a serious violation. Later in the chapter, when we illustrate how to use SPSS and **R** to generate ANCOVA, we will specifically test for the assumptions of independence of observations, homogeneity of variance, normality, linearity, independence of the covariate and the independent variable, and homogeneity of regression slopes.

4.1.5.1 Independence

As we learned previously, the assumption of independence of observations can be met by (a) keeping the assignment of individuals to groups (i.e., to the levels or categories of the independent variable) separate through the design of the experiment (specifically random assignment—not to be confused with random selection), and (b) keeping the individuals separate from one another through experimental control so that the scores on the dependent variable Y are independent across subjects (both within and across groups).

As in previous ANOVA models, the use of independent random samples is also crucial in the analysis of covariance. The F ratio is very sensitive to violation of the independence assumption in terms of increased likelihood of a Type I and/or Type II error. A violation of the independence assumption may affect the standard errors of the sample adjusted means and thus influence any inferences made about those means. One purpose of random assignment of individuals to groups is to achieve independence. If each individual is observed only once and individuals are randomly assigned to groups, then the independence assumption is usually met. Random assignment is important for valid interpretation of both the F test and multiple comparison procedures. Otherwise, the F test and adjusted means may be biased.

The simplest procedure for assessing independence is to examine residual plots by group. If the independence assumption is satisfied, then the residuals should fall into a random display of points. If the assumption is violated, then the residuals will fall into some type of cyclical pattern. As discussed in Chapter 1, the Durbin-Watson statistic (Durbin & Watson, 1950, 1951, 1971) can be used to test for autocorrelation. Violations of the independence assumption generally occur in the three situations we mentioned in Chapter 1: time series data, observations within blocks, or replication. For severe violations of the independence assumption, there is no simple "fix," such as the use of transformations or nonparametric tests (Scariano & Davenport, 1987).

4.1.5.2 Homogeneity of Variance

The second assumption is that the variances of each population are the same, known as the homogeneity of variance or homoscedasticity assumption. A violation of this assumption may lead to bias in the SS_{with} term, as well as an increase in the Type I error rate, and possibly an increase in the Type II error rate. A summary of Monte Carlo research on ANCOVA assumption violations by Harwell (2003) indicates that the effect of the violation is negligible with equal or nearly equal n's across the groups. There is a more serious problem if the larger n's are associated with the smaller variances (actual or observed $\alpha >$ nominal or stated α selected by the researcher, which is a liberal result), or if the larger n's are associated with the larger variances (actual $\alpha <$ nominal α, which is a conservative result).

In a plot of Y versus the covariate X for each group, the variability of the distributions may be examined for evidence of the extent to which this assumption is met. Another method for detecting violation of the homogeneity assumption is the use of formal statistical tests for homoscedasticity (e.g., Levene's), as discussed in Chapter 1 and as we illustrate using SPSS and **R** later in this chapter. Several solutions are available for dealing with a violation of the homogeneity assumption. These include the use of variance stabilizing transformations or other ANCOVA models that are less sensitive to unequal variances, such as nonparametric ANCOVA procedures (described at the end of this chapter).

4.1.5.3 Normality

The third assumption is that each of the populations follows the normal distribution. Based on the classic work by Box and Anderson (1962) and Atiqullah (1964), as well as the summarization of modern Monte Carlo work by Harwell (2003), the F test is relatively robust to nonnormal Y distributions, "minimizing the role of a normally distributed X" (Harwell, 2003, p. 62). Thus we need only really be concerned with serious nonnormality (although "serious nonnormality" is a subjective call made by the researcher).

We will examine residuals for this assumption, and the following graphical techniques can be used to detect violation of the normality assumption: (a) frequency distributions (such as stem-and-leaf plots, boxplots, or histograms), or (b) normal probability plots. There are also several statistical procedures available for the detection of nonnormality [e.g., the Shapiro-Wilk test (Shapiro & Wilk, 1965)]. If the assumption of normality is violated, transformations can also be to normalize the data, as previously discussed in Chapter 1. In addition, nonparametric ANCOVA has been shown to be robust to nonnormality, have reasonable power, and preserve the nominal alpha level (Wu & Lai, 2015), and one can use one of the rank ANCOVA procedures previously mentioned.

4.1.5.4 Linearity

The next assumption is that the regression of Y (i.e., the dependent variable) on X (i.e., the covariate) is linear. If the relationship between Y and X is not linear, then use of the usual ANCOVA procedure is not appropriate, just as linear regression (see Chapter 7) would not be appropriate in cases of nonlinearity. In ANCOVA (as well as in correlation and linear regression), we fit a straight line to the data points in a scatterplot. When the relationship is nonlinear, a straight line will not fit the data particularly well. In addition, the magnitude of the linear correlation will be smaller. If the relationship is not linear, the estimate of the group effects will be biased, and the adjustments made in SS_{with} and SS_{betw} will be smaller.

Violations of the linearity assumption can generally be detected by looking at scatterplots of Y versus X, overall and for each group or category of the independent variable. Once a serious violation of the linearity assumption has been detected, two alternatives can be used: transformations and nonlinear ANCOVA. Transformations on one or both variables can be used to achieve linearity (Keppel & Wickens, 2004). The second option is to use nonlinear ANCOVA methods as described by Huitema (2011) and Keppel and Wickens (2004).

4.1.5.5 Fixed Independent Variable

The fifth assumption states that the levels of the independent variable are fixed by the researcher. This results in a fixed-effects model rather than a random-effects model. As in the one-factor ANOVA model, the one-factor ANCOVA model is the same computationally in the fixed- and random-effects cases. The summary of Monte Carlo research by Harwell (2003) indicates that the impact of a random effect on the F test is minimal.

4.1.5.6 Independence of the Covariate and the Independent Variable

A condition of the ANCOVA model (although not an assumption) requires that the covariate and the independent variable be independent. That is, the covariate is not influenced by the independent or treatment variable. If the covariate is affected by the treatment itself, then the use of the covariate in the analysis either (a) may remove part of the treatment effect or produce a spurious (inflated) treatment effect, or (b) may alter the covariate scores as a result of the treatment being administered prior to obtaining the covariate data. The obvious solution to this potential problem is to obtain the covariate scores prior to the administration of the treatment. In other words, be alert prior to the study for possible covariate candidates. Thus, in a true experiment, the treatment (i.e., independent variable) and covariate are not related by default of random assignment and thereby the assumption of independence of the covariate and independent variable are met. If randomization is not possible, closely matching participants on the covariate may also help to ensure the assumption is not violated.

Let us consider an example where this condition is obviously violated. A psychologist is interested in which of several hypnosis treatments is most successful in reducing or eliminating cigarette smoking. A group of heavy smokers is randomly assigned to the hypnosis treatments. After the treatments have been completed, the researcher suspects that some patients are more susceptible to hypnosis (i.e., are more suggestible) than others. By using suggestibility as a covariate after the study is completed, the researcher would not be able to determine whether group differences were a result of hypnosis treatment, suggestibility, or some combination. Thus, the measurement of suggestibility after the hypnosis treatments have been administered would be ill-advised. An extended discussion of this condition is given in Maxwell et al. (2018).

Evidence of the extent to which this assumption is met can be done by examining mean differences on the covariate across the levels of the independent variable. If the independent variable has only two levels, an independent t test would be appropriate. If the independent variable has more than two categories, a one-way ANOVA would suffice. If the groups are not statistically different on the covariate, then that lends evidence that the assumption of independence of the covariate and the independent variable has been met. If the groups are statistically different on the covariate, then the groups are not likely to be equivalent.

4.1.5.7 Covariate Measured Without Error

An assumption that we have not yet discussed in this text is that the covariate is measured without error. This is of special concern in education and the behavioral sciences, where variables are often measured with considerable measurement error. In the presence of measurement error, in randomized experiments, b_w (i.e., the within-groups regression slope from the regression of the dependent variable, Y, on the covariate, X) will be underestimated so that less of the covariate effect is removed from the dependent variable (i.e., the adjustments will be smaller). In addition, the reduction in the unexplained variation will not be as great and the F test will not be as powerful when there is measurement error. The F test is generally conservative in terms of Type I error (the actual observed alpha will be less than the nominal alpha which was selected by the researcher—the nominal alpha is often .05). However, the treatment effects will not be biased. In quasi-experimental designs, b_w will also be underestimated with similar effects. However, the treatment effects may be seriously biased. A method by Porter (1967) is suggested for this situation.

There is considerable discussion about the effects of measurement error (Cohen, Cohen, West, & Aiken, 2003; Keppel & Wickens, 2004; Lord, 1960, 1967, 1969; Mickey et al., 2004; Pedhazur, 1997; Porter, 1967; Reichardt, 1979; Weisberg, 1985). Obvious violations of this assumption can be detected by computing the reliability of the covariate prior to the study or from previous research. This is the minimum that should be done. One may also want to consider the *validity* of the covariate as well, where validity may be defined as the extent to which an instrument measures what it was intended to measure. While this is the first mention in the text of measurement error, it is certainly important that all measures included in a model—regardless of which statistical procedure is being conducted—are measured such that the scores provide high reliability and validity.

4.1.5.8 Homogeneity of Regression Slopes

The final assumption puts forth that the slope of the regression line between the dependent variable and covariate is the same for each category of the independent variable. Here we assume that $\beta_1 = \beta_2 = \ldots = \beta_J$. This is an important assumption because it allows us to use b_w, the sample estimator of β_w, as the within-groups regression slope, and some researchers have noted that this is *the most important assumption* (Shieh, 2017). Assuming that the group slopes are parallel allows us to test for group intercept differences, *which is all we are really doing when we test for differences among the adjusted means*. Without this assumption of homogeneity of regression slopes, groups can differ on *both* the regression slope and intercept, and β_w cannot legitimately be used. If the slopes differ, then the regression lines interact in some way. As a result, the size of the group differences in Y (i.e., the dependent variable) will depend on the value of X (i.e., the covariate). For example, Treatment 1 may be most effective on the dependent variable for low values of the covariate, Treatment 2 may be most effective on the dependent variable for middle values of the covariate, and Treatment 3 may be most effective on the dependent variable for high values of the covariate. Thus, we do not have constant differences on the dependent variable between the groups of the independent variable across the values of the covariate. A straightforward interpretation is not possible, which is the same situation in factorial ANOVA when the interaction between factor A and factor B is found to be significant. *Thus, unequal slopes in ANCOVA represent a type of interaction.*

There are other potential outcomes if this assumption is violated. Without homogeneous regression slopes, the use of β_w can yield biased adjusted means and can affect the F test. Earlier simulation studies by Peckham (1968) and Glass, Peckham, and Sanders (1972) suggest that for the one-factor fixed-effects model, the effects will be minimal. Later analytical research by Rogosa (1980) suggests that there is little effect on the F test for balanced designs with equal variances, but the F is less robust for mild heterogeneity. However, a summary of modern Monte Carlo work by Harwell (2003) indicates that the effect of slope heterogeneity on the F test is (a) negligible with equal n's and equal covariate means (randomized studies), (b) modest with equal n's and unequal covariate means (nonrandomized studies), and (c) modest with unequal n's.

A formal statistical procedure is often conducted to test for homogeneity of slopes using statistical software (we will illustrate this later in this chapter), although the eyeball method (i.e., see if the slopes look about the same by reviewing scatterplots of the dependent variable and covariate for each category of the independent variable) can be a good starting point. Some alternative tests for equality of slopes when the variances are unequal are provided by Tabatabai and Tan (1985).

Several alternatives are available if the homogeneity of slopes assumption is violated. The first is to use the concomitant variable not as a covariate but as a blocking variable. This will work because this assumption is not made for the randomized block design (see Chapter 6). A second option, and not a very desirable one, is to analyze each group separately with its own slope or subsets of the groups having equal slopes. A third possibility is to utilize interaction terms between the covariate and the independent variable and conduct a regression analysis (see Agresti, 2018). A fourth option is to use the Johnson-Neyman (Johnson & Neyman, 1936) technique, whose purpose is to determine the values of X (i.e., the covariate) that are related to significant group differences on Y (i.e., the dependent variable). Interested readers are referred to Huitema (2011) or Wilcox (1987). A fifth option is use more modern robust methods (e.g., Maxwell et al., 2018; Wilcox, 2003).

A summary of the ANCOVA assumptions is presented in Table 4.6.

TABLE 4.6

Assumptions and Effects of Violations—One-Factor ANCOVA

Assumption	Effect of Assumption Violation
Independence	• Increased likelihood of a Type I and/or Type II error in *F* • Affects standard errors of means and inferences about those means
Homogeneity of variance	• Bias in SS_{with}; increased likelihood of a Type I and/or Type II error • Negligible effect with equal or nearly equal *n*'s • Otherwise more serious problem if the larger *n*'s are associated with the smaller variances (increased α) or larger variances (decreased α)
Normality	• *F* test relatively robust to nonnormal *Y*, minimizing the role of nonnormal *X*
Linearity	• Reduced magnitude of r_{XY} • Straight line will not fit data well • Estimate of group effects biased • Adjustments made in *SS* smaller
Fixed-effect	• Minimal impact
Covariate and factor are independent	• May reduce/increase group effects; may alter covariate scores

Covariate measured without error	• True experiment:
	  • b_W underestimated
	  • Adjustments smaller
	  • Reduction in unexplained variation smaller
	  • *F* less powerful
	  • Reduced likelihood of Type I error
	• Quasi-experiment:
	  • b_W underestimated
	  • Adjustments smaller
	  • Group effects seriously biased

Homogeneity of slopes	• Negligible effect with equal *n*'s in true experiment • Modest effect with equal *n*'s in quasi-experiment • Modest effect with unequal *n*'s

4.2 Computing ANCOVA Using SPSS

Next we consider SPSS for the *physical performance* (i.e., dependent variable) example that includes *treatment group* as the independent variable and *mental fatigue* as the covariate (Ch4_fatigue.sav). As noted in previous chapters, SPSS needs the data to be in a specific form for the analysis to proceed, which is different from the layout of the data in Table 4.1. For a one-factor ANCOVA with a single covariate, the dataset must contain three variables or columns: one for the level of the factor or independent variable, one for the covariate, and a third for the dependent variable. The screenshot in Figure 4.2 presents an example of the dataset for the physical performance example. Each row still represents one individual, displaying the level of the factor (or independent variable) for which they are a member, as well as their scores on the covariate and the scores for the dependent variable.

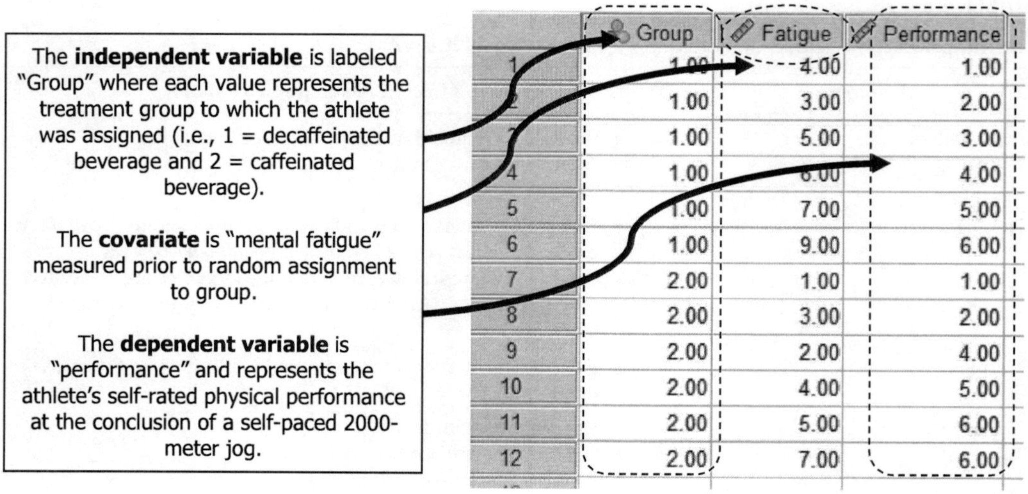

FIGURE 4.2
Data.

Step 1. To conduct an ANCOVA, go to "Analyze" in the top pulldown menu, then select "General Linear Model," and then select "Univariate." Following the screenshot for Step 1 (Figure 4.3) produces the Univariate dialog box.

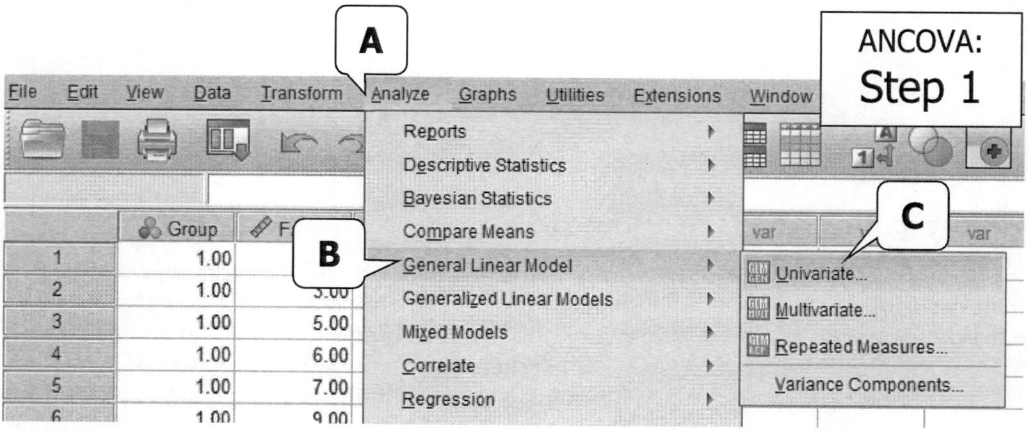

FIGURE 4.3
ANCOVA: Step 1.

Step 2. From the Univariate dialog box (see Step 2, shown in Figure 4.4), click the dependent variable (e.g., self-rated physical performance) and move it into the "Dependent Variable" box by clicking the arrow button. Click the independent variable (e.g., group) and move it into the "Fixed Factor(s)" box by clicking the arrow button. Click the covariate (e.g., fatigue) and move it into the "Covariate(s)" box by clicking the arrow button. Next, click on "Options."

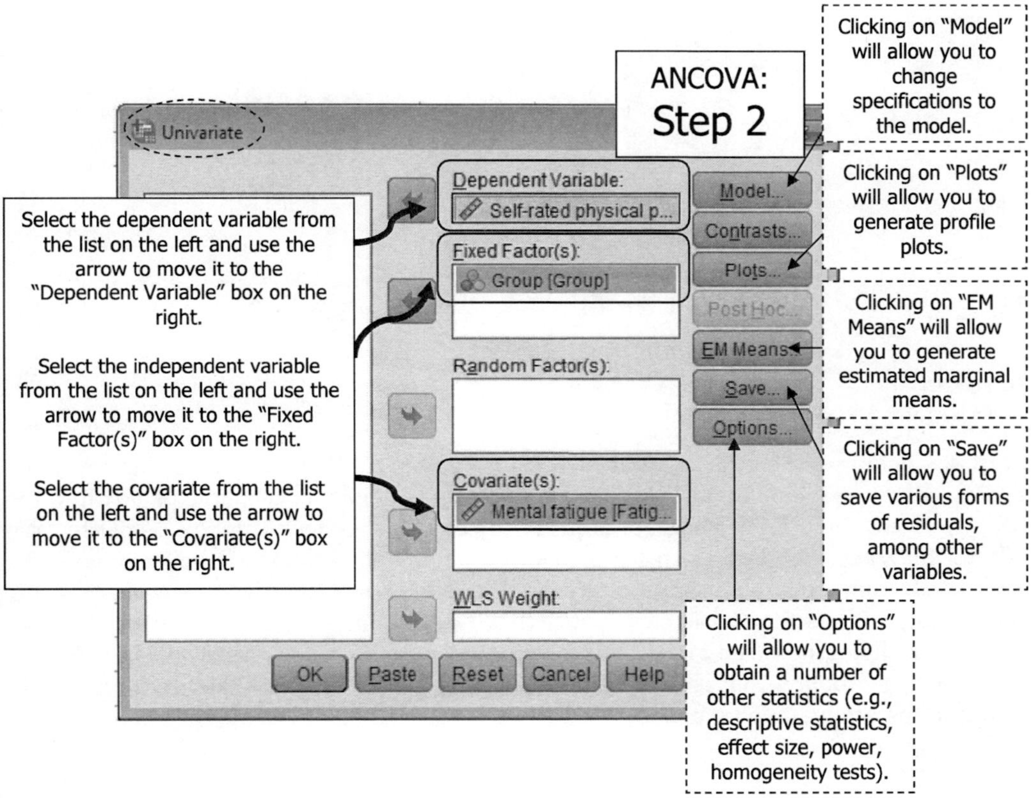

FIGURE 4.4
ANCOVA: Step 2.

Step 3. Clicking on "Options" will provide the option to select such information as "Descriptive statistics," "Estimates of effect size," "Observed power," and "Homogeneity tests." Click on "Continue" to return to the original dialog box.

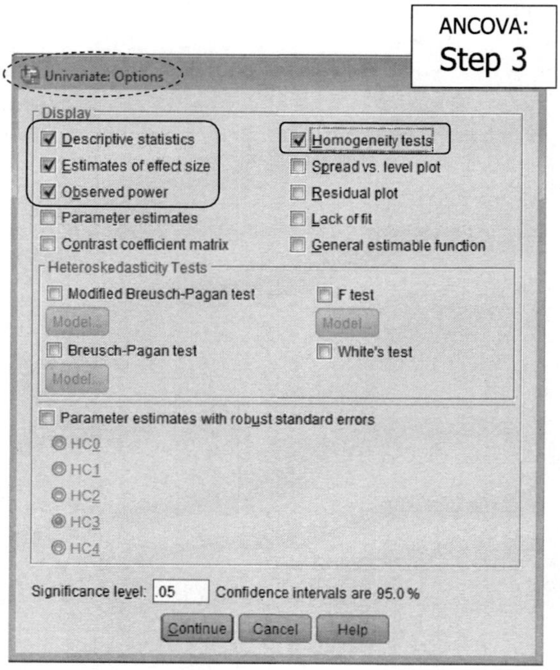

FIGURE 4.5
ANCOVA: Step 3.

Step 4. Clicking on "EM Means" will provide the option to display overall and marginal means. Move the items that are listed in the "Factor(s) and Factor Interactions" box into the "Display Means for" box to generate adjusted means. Also, check the box "Compare main effects," then click the pulldown for "Confidence interval adjustment" to choose among the LSD, Bonferroni, or Sidak multiple comparison procedures of the adjusted means. For this illustration, we select the Bonferonni. Notice that the "Post Hoc" option button from the main Univariate dialog box (see Figure 4.4) is not active; thus you are restricted to the three MCPs just mentioned that are accessible from this Options screen. Click on "Continue" to return to the original dialog box.

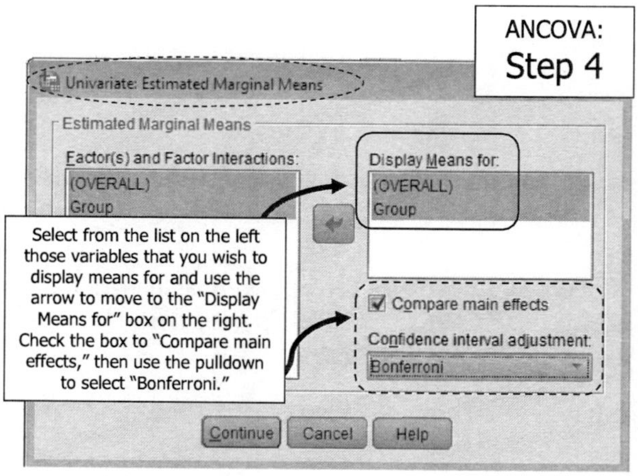

FIGURE 4.6
ANCOVA: Step 4.

Step 5. From the Univariate dialog box (see Figure 4.4), click on "Plots" to obtain a profile plot of means. Click the independent variable (e.g., statistics course section, "Group") and move it into the "Horizontal Axis" box by clicking the arrow button (see screenshot for Step 5a, Figure 4.7). Then click on "Add" to move the variable into the Plots box at the bottom of the dialog box (see screenshot for Step 5b, shown in Figure 4.8). Click on "Continue" to return to the original dialog box.

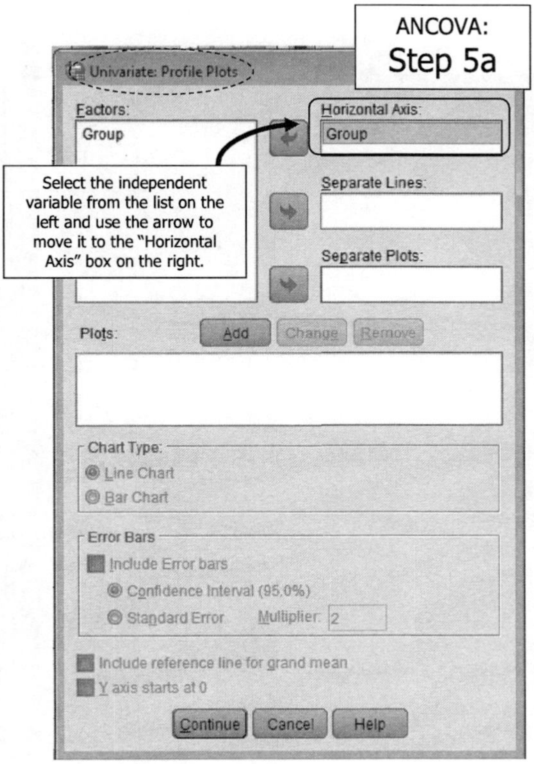

FIGURE 4.7
ANCOVA: Step 5a.

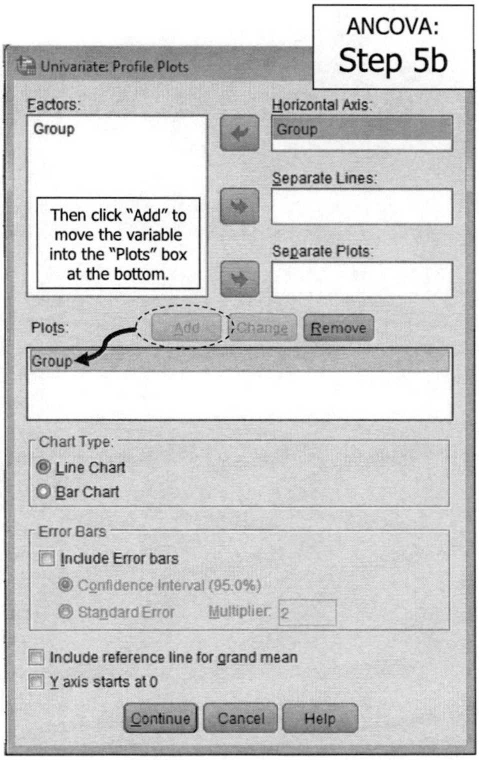

FIGURE 4.8
ANCOVA: Step 5b.

Step 6. Finally, in order to generate the appropriate sources of variation and results as recommended in this chapter, from the main Univariate dialog box (see Step 2, Figure 4.4), you need to click on the "Model" button. Then select "Type I" from the "Sum of squares" pull-down menu. Click on "Continue" to return to the original dialog box.

You may be asking yourself why we need to utilize the Type I sum of squares, as up until this point in the text we have always recommended the Type III (which is the default in SPSS). In a study conducted by Li and Lomax (2011), the following were confirmed with SPSS (as well as with SAS). First, when generating the Type I sum of squares, the covariate is extracted first, then the treatment is estimated controlling for the covariate. The Type I sum of squares will also correctly add up to the total sum of squares. Second, when generating the Type III sum of squares, each effect is estimated controlling for each of the other effects. In other words, the covariate is computed controlling for the treatment, and the treatment is determined controlling for the covariate. The former is not of interest as the treatment is administered after the covariate has been measured; thus no such control is necessary. Also, the Type III sum of squares will not add up to the total sum of squares, as the covariate sum of squares will be different than when using Type I. Thus, you do not want to estimate the covariate controlling for the treatment, and thus you want to use the Type I, not Type III, in the ANCOVA context. In other words, Type I sum of squares is sequential, with each term adjusted for the term that precedes it in the model.

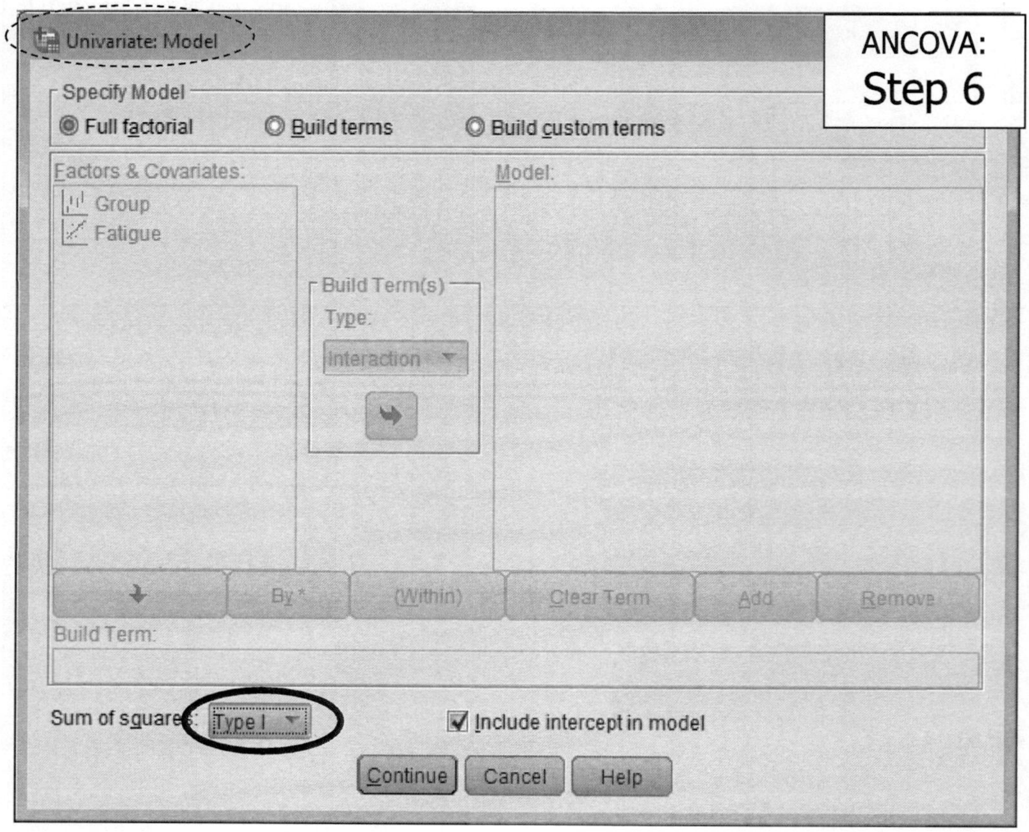

FIGURE 4.9
ANCOVA: Step 6.

Step 7. From the Univariate dialog box (see Step 2, Figure 4.4), click on "Save" to select those elements that you want to save (here we want to save the unstandardized residuals for later use in order to examine the extent to which normality and independence are met). Click on "Continue" to return to the original dialog box. From the Univariate dialog box, click on "OK" to return to generate the output.

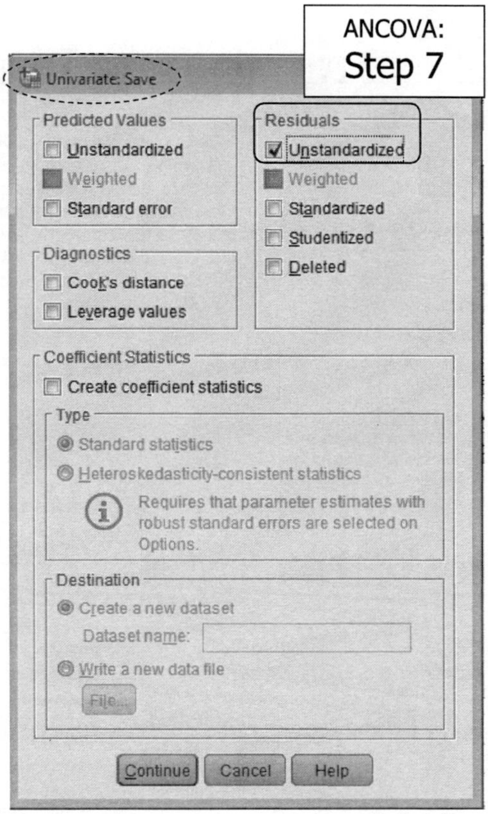

FIGURE 4.10
ANCOVA: Step 7.

Interpreting the output. Annotated results are presented in Table 4.7.

TABLE 4.7
SPSS Results for the Physical Performance Example

Between-Subjects Factors

		Value Label	N
Group	1.00	Control (decaffeinated beverage)	6
	2.00	Treatment (caffeinated beverage)	6

The table labeled "Between-Subjects Factors" provides sample sizes for each of the categories of the independent variable (recall that the independent variable is the 'between subjects factor').

Descriptive Statistics

Dependent Variable: Self-rated physical performance

Group	Mean	Std. Deviation	N
Control (decaffeinated beverage)	3.5000	1.87083	6
Treatment (caffeinated beverage)	4.0000	2.09762	6
Total	3.7500	1.91288	12

The table labeled "Descriptive Statistics" provides basic descriptive statistics (means, standard deviations, and sample sizes) for each group of the independent variable.

(continued)

TABLE 4.7 (continued)

SPSS Results for the Physical Performance Example

Levene's Test of Equality of Error Variances[a]

Dependent Variable: Self-rated physical performance

F	df1	df2	Sig.
6.768	1	10	.026

Tests the null hypothesis that the error variance of the dependent variable is equal across groups.

a. Design: Intercept + Fatigue + Group

The *F* test (and associated *p* value) for Levene's Test for Equality of Error Variances is reviewed to determine if equal variances are assumed. In this case, we do *not* meet the assumption (as *p* is less than alpha). Note that *df1* is degrees of freedom for the numerator (calculated as $J-1$) and *df2* are the degrees of freedom for the denominator (calculated as $N-J$).

The row labeled **"GROUP"** is the independent variable or between groups variable. The *between groups mean square* (10.812) tells how much individual observations should vary if the null hypothesis is true. The degrees of freedom for the sum of squares between groups is $J-1$ (or 2-1 = 1 in this example).

The omnibus *F* test is computed as:

$$F = \frac{MS_{betw(adj)}}{MS_{with(adj)}} = \frac{10.812}{.951} = 11.37$$

The *p* value for the independent variable's *F* test is .008. This indicates there is a statistically significant difference in physical performance based on treatment group, controlling for mental fatigue. The probability of observing these mean differences or more extreme mean differences by chance if the null hypothesis is really true (i.e., if the means really are equal) is substantially less than 1%. We reject the null hypothesis that all the population means are equal. The *p* value for the covariate's *F* test is .001. This indicates there is a statistically significant relationship between the covariate (mental fatigue) and physical performance.

Partial eta squared is one measure of effect size:

$$\eta_p^2 = \frac{SS_{betw(adj)}}{SS_{betw(adj)} + SS_{error}}$$

$$\eta_p^2 = \frac{10.812}{10.812 + 8.557} = .558$$

We can interpret this to say that approximately 56% of the variation in the dependent variable (in this case, statistics quiz score) is accounted for by the instructional method when controlling for aptitude.

Tests of Between-Subjects Effects

Dependent Variable: Self-rated physical performance

Source	Type I Sum of Squares	df	Mean Square	F	Sig.	Partial Eta Squared	Noncent. Parameter	Observed Power[b]
Corrected Model	31.693[a]	2	15.846	16.667	.001	.787	33.333	.993
Intercept	168.750	1	168.750	177.483	.000	.952	177.483	1.000
Fatigue *(covariate)*	20.881	1	20.881	21.961	.001	.709	21.961	.986
Group *(ind. variable)*	10.812	1	10.812	11.372	.008	.558	11.372	.850
Error	8.557	9	.951					
Total	209.000	12						
Corrected Total	40.250	11						

a. R Squared = .787 (Adjusted R Squared = .740)

b. Computed using alpha = .05

R^2 is listed as a footnote underneath the table. R^2 is the ratio of $SS_{betw(adj)}$ and SS_{cov} divided by SS_{total}:

$$R^2 = \frac{SS_{betw(adj)} + SS_{cov}}{SS_{total}}$$

$$R^2 = \frac{10.812 + 20.881}{40.250} = .787$$

The row labeled **"Error"** is within groups. The within groups mean square tells us how much the observations within the groups really vary (i.e., .951). The degrees of freedom for the sum of squares within groups is (*N*-*J*-1) or the sample size minus the number of levels of the independent variable minus one.

The row labeled "corrected total" is the sum of squares total. The degrees of freedom for the total is (*N*-1) or the sample size minus one.

Observed power tells whether our test is powerful enough to detect mean differences if they really exist. Power of .850 indicates that the probability of rejecting the null hypothesis if it is really false is about 85%, strong power.

TABLE 4.7 (continued)

SPSS Results for the Physical Performance Example

Estimated Marginal Means

1. Grand Mean

Dependent Variable: Self-rated physical performance

Mean	Std. Error	95% Confidence Interval Lower Bound	Upper Bound
3.750[a]	.281	3.113	4.387

a. Covariates appearing in the model are evaluated at the following values: Mental fatigue = 4.6667.

The 'Grand Mean' (in this case, 3.750) represents the overall mean, regardless of group membership in the independent variable. The 95% CI represents the CI of the grand mean.
Here, 95% of the time, the true grand mean will be between 3.113 and 4.387.

2. Group

Estimates

Dependent Variable: Self-rated physical performance

Group	Mean	Std. Error	95% Confidence Interval Lower Bound	Upper Bound
Control (decaffeinated beverage)	2.686[a]	.423	1.729	3.642
Treatment (caffeinated beverage)	4.814[a]	.423	3.858	5.771

a. Covariates appearing in the model are evaluated at the following values: Mental fatigue = 4.6667.

The table labeled **"Group"** provides descriptive statistics for each of the categories of the independent variable, controlling for the covariate (notice that these are NOT the same means reported previously; also note the table footnote). In addition to means, the *SE* and 95% CI of the means are reported.

'Mean difference' is simply the difference between the adjusted group means of the two groups compared. For example, the mean difference of group 1 and group 2, controlling for the covariate, is calculated as 2.686 – 4.814 = -2.12.
Because there are only two groups to our independent variable, the values in the table are the same (in absolute value) for row 1 as compared to row 2 (the exception is that the CI for the difference is switched).

Pairwise Comparisons

Dependent Variable: Self-rated physical performance

(I) Group	(J) Group	Mean Difference (I-J)	Std. Error	Sig.[b]	95% Confidence Interval for Difference[b] Lower Bound	Upper Bound
Control (decaffeinated beverage)	Treatment (caffeinated beverage)	-2.129*	.631	.008	-3.556	-.701
Treatment (caffeinated beverage)	Control (decaffeinated beverage)	2.129*	.631	.008	.701	3.556

Based on estimated marginal means

*. The mean difference is significant at the .05 level.

b. Adjustment for multiple comparisons: Bonferroni.

'Sig.' denotes the observed *p* value and provides the results of the Bonferroni post hoc procedure. There is a statistically significant adjusted mean difference in the outcome between groups of the independent variable (controlling for the covariate).

Because we had only two groups, requesting post hoc results really was not necessarily. We could have reviewed the *F* test and then the adjusted means to determine which group had the higher adjusted mean. The pairwise comparison results will become more valuable when the ANCOVA includes independent variables with more than two categories.

Note there are redundant results presented in the table.
The comparison of group 1 and 2 (presented in results row 1) is the same as the comparison of group 2 and 1 (presented in results row 2).

(*continued*)

TABLE 4.7 (continued)

SPSS Results for the Physical Performance Example

The table labeled **"Univariate Tests"** is simply an omnibus *F* test. In the case of one independent variable, the row labeled "Contrast" provides the same results for the independent variable as that presented in the summary table previously. The results from this table suggest there is a statistically significant difference in adjusted mean physical performance based on treatment group when controlling for mental fatigue.

Univariate Tests

Dependent Variable: Self-rated physical performance

	Sum of Squares	df	Mean Square	F	Sig.	Partial Eta Squared	Noncent. Parameter	Observed Power[a]
Contrast	10.812	1	10.812	11.372	.008	.558	11.372	.850
Error	8.557	9	.951					

The F tests the effect of Group. This test is based on the linearly independent pairwise comparisons among the estimated marginal means.

a. Computed using alpha = .05

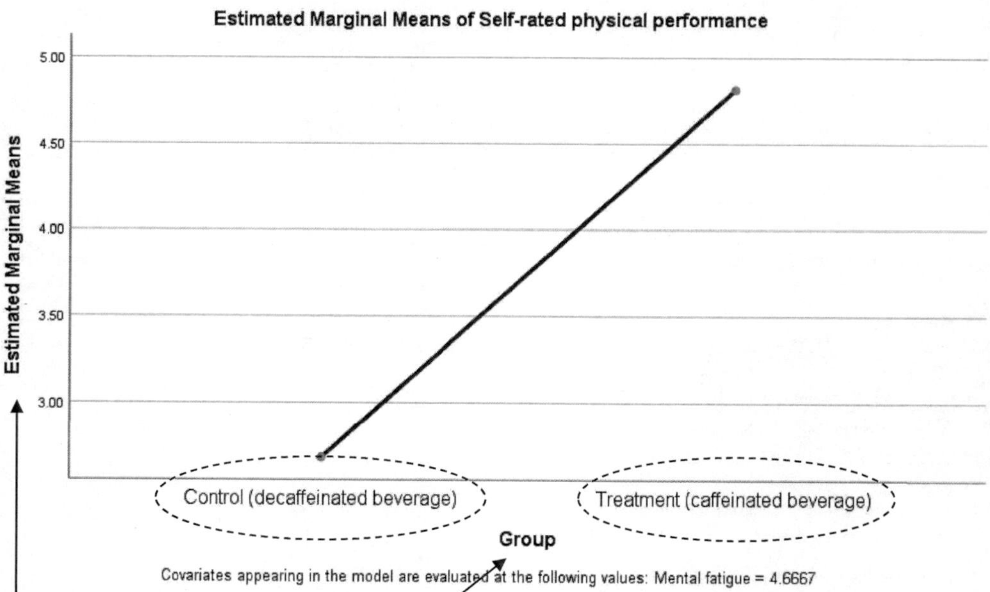

The graph is a plot of the adjusted means (i.e., controlling for the covariate) against the categories of the independent variable. This provides visual representation of the extent to which the physical performance means differ by treatment group when controlling for mental fatigue.

4.3 Computing ANCOVA Using R

Next we consider **R** for the ANCOVA model. Note that the scripts are provided within the blocks with additional annotation to assist in understanding how the command works. Should you want to write reminder notes and annotation to yourself as you write the commands in **R** (and we highly encourage doing so), remember that any text that follows a hashtag (i.e., #) is annotation only and not part of the **R** script. Thus, you can write annotations directly into **R** with hashtags. We encourage this practice so that when you call up the commands in the future, you'll understand what the various lines of code are doing. You may think you'll remember what you did. However, trust us. There is a good chance that you won't. Thus, consider it best practice when using **R** to annotate heavily!

4.3.1 Reading Data Into R

```
getwd()
```

R is always pointed to a directory on your computer. The *get working directory* function can be used to determine to which directory **R** is pointed. We will assume that we need to change the working directory, and will use the next line of code to set the working directory to the desired path.

```
setwd("E:/FolderName")
```

We use the *setwd* function to establish the working directory. To set the working directory, change what is in quotation marks to your file location. Also, if you are copying the directory name from your properties, you will need to change the backslash (i.e., \) to a forward slash (i.e., /).

```
Ch4_fatigue <- read.csv("Ch4_fatigue.csv")
```

The *read.csv* function reads our data into **R**. What's to the left of the "<-" will be what the data will be called in **R**. In this example, we're calling the **R** dataframe "Ch4_fatigue." What's to the right of the "<-" tells **R** to find this particular csv file. In this example, our file is called "Ch4_fatigue.csv." Make sure the extension (i.e., .csv) is included in your script. Also note that the name of your file should be in quotation marks within the parentheses.

```
names(Ch4_fatigue)
```

The *names* function will produce a list of variable names for each dataframe as follows. This is a good check to make sure your data have been read in correctly.

```
[1] "Group"    "Fatigue"    "Performance"
```

```
View(Ch4_fatigue)
```

The *View* function will let you view the dataset in spreadsheet format in RStudio.

```
Ch4_fatigue$GroupF <- factor(Ch4_fatigue$Group,
labels = c("control",
"treatment"))
```

FIGURE 4.11
Reading data into **R**.

This command will create a new variable in our dataframe named "GroupF." We use the *factor* function to define the variable "Group" as nominal with the two groups defined here (i.e., control, treatment). What is to the left of "<-" in the script creates the new GroupF variable in our dataframe.

```
summary(Ch4_fatigue)
```

The *summary* function will produce basic descriptive statistics on all the variables in your dataframe. This is a great way to quickly check to see if the data have been read in correctly and to get a feel for your data, if you haven't already. The output from the summary statement for this dataframe looks like this. Because we defined GroupF as a factor, we are provided only the frequencies for each category in that variable.

```
    Group         Fatigue          Performance           GroupF
Min.   :1.0   Min.   :1.000   Min.   :1.00   control  :6
1st Qu.:1.0   1st Qu.:3.000   1st Qu.:2.00   treatment:6
Median :1.5   Median :4.500   Median :4.00
Mean   :1.5   Mean   :4.667   Mean   :3.75
3rd Qu.:2.0   3rd Qu.:6.250   3rd Qu.:5.25
Max.   :2.0   Max.   :9.000   Max.   :6.00
```

FIGURE 4.11 (continued)
Reading data into **R**.

4.3.2 Generating the ANCOVA Model

```
ANCOVA_fatigue <- lm(Performance ~ Fatigue+GroupF, data=Ch4_fatigue)
```

The *lm* function will generate the ANCOVA model with "Performance" as the dependent variable and "Group" as the independent variable, with "Fatigue" as the covariate. The dataframe from which we are pulling the data is defined by the *data* function. We are calling this object "ANCOVA_fatigue."

```
anova(ANCOVA_fatigue)
```

This command will output the results, which we see here:

```
Analysis of Variance Table

Response: Performance
          Df  Sum Sq  Mean Sq F value  Pr(>F)
Fatigue    1  20.8807 20.8807 21.961 0.001142 **
GroupF     1  10.8122 10.8122 11.372 0.008228 **
Residuals  9  8.5571  0.9508
—
Signif. codes:
0 '***' 0.001 '**' 0.01 '*' 0.05 '.' 0.1 ' ' 1
```

```
summary(ANCOVA_fatigue)
```

The *summary* function will provide additional output from our ANCOVA model:

```
Call:
lm(formula = Performance ~ Fatigue + GroupF, data = Ch4_fatigue)

Residuals:
    Min       1Q  Median      3Q     Max
-1.4571 -0.7429  0.1357 0.6857 1.3571
```

FIGURE 4.12
Generating ANCOVA in **R**.

```
Coefficients:
                 Estimate  Std. Error  t value  Pr(>|t|)
(Intercept)       -1.1143      0.9015   -1.236  0.247732
Fatigue            0.8143      0.1427    5.705  0.000293 ***
GroupFtreatment    2.1286      0.6312    3.372  0.008228 **
---
Signif. codes: 0 '***' 0.001 '**' 0.01 '*' 0.05 '.' 0.1 ' ' 1

Residual standard error: 0.9751 on 9 degrees of freedom
Multiple R-squared: 0.7874, Adjusted R-squared: 0.7402
F-statistic: 16.67 on 2 and 9 DF, p-value: 0.000942
```

```
install.packages("sjstats")
```

The *install.packages* function will install *sjstats*.

```
library(sjstats)
```

The *library* function will call the *sjstats* package into the library.

```
anova_stats(ANCOVA_fatigue)
```

The *anova_stats* function, applied to our ANCOVA model (i.e., ANCOVA_fatigue) will produce the ANOVA summary table, along with multiple effect size indices.

```
term          df  sumsq  meansq  statistic  p. value  etasq
1 Fatigue      1 20.881  20.881     21.961     0.001  0.519
2 GroupF       1 10.812  10.812     11.372     0.008  0.269
3 Residuals 9     8.557   0.951         NA        NA     NA

 partial.etasq  omegasq  partial.omegasq  cohens.f  power
1        0.709    0.484            0.636     1.562  0.996
2        0.558    0.239            0.464     1.124  0.911
3           NA       NA               NA        NA     NA
```

```
omega_sq(ANCOVA_fatigue)
omega_sq(ANCOVA_fatigue, partial = TRUE)
cohens_f(ANCOVA_fatigue)
eta_sq(ANCOVA_fatigue)
eta_sq(ANCOVA_fatigue, partial = TRUE)
```

Had we wanted a specific effect size value, we could have generated the above code.

```
# omega_sq(ANCOVA_fatigue)
     term   omegasq
1 Fatigue     0.484
2  GroupF     0.239

# omega_sq(ANCOVA_fatigue, partial = TRUE)
     term partial.omegasq
1 Fatigue           0.636
2  GroupF           0.464
```

FIGURE 4.12 (continued)
Generating ANCOVA in **R**.

```
# cohens_f(ANCOVA_fatigue)
      term cohens.f
1 Fatigue  1.562097
2  GroupF  1.124067

# eta_sq(ANCOVA_fatigue)
      term etasq
1 Fatigue  0.519
2  GroupF  0.269

# eta_sq(ANCOVA_fatigue, partial = TRUE)
     term partial.etasq
1 Fatigue         0.709
2  GroupF         0.558
```

```
Ch4_fatigue$unstandardizedResiduals <- residuals(ANCOVA_fatigue)
```

We also want to save our unstandardized residuals to the dataframe. We use the *residuals* function to compute unstandardized residuals from our *ANCOVA_fatigue* model. To the left of "<-" we will save the residuals as a variable named "unstandardizedResiduals" in our dataframe, *Ch4_fatigue*.

FIGURE 4.12 (continued)
Generating ANCOVA in **R**.

4.4 Data Screening

The assumptions that we will test for in our ANCOVA model include: (a) independence of observations; (b) homogeneity of variance (this was previously generated; thus you can examine Table 4.6 for this assumption as it will not be reiterated here); (c) normality; (d) linearity; (e) independence of the covariate and the independent variable; and (f) homogeneity of regression slopes. We will examine the assumptions after generating the ANCOVA results. This is because many of the tests for assumptions are based on examination of the residuals, which were requested when generating the ANCOVA.

4.4.1 Independence

If subjects have been randomly assigned to conditions (in other words, the different levels of the independent variable), the assumption of independence has been met. In this illustration, students were randomly assigned to group (i.e., ingest caffeinated versus decaffeinated beverage), and thus the assumption of independence was met. As we have learned in previous chapters, however, we often use independent variables that do not allow random assignment (e.g., intact groups). We can plot residuals against levels of the independent variable in a scatterplot to get an idea of whether or not there are patterns in the data and thereby provide an indication of the extent to which we have met this assumption. Remember that these variables were added to the dataset by saving the unstandardized residuals when we generated the ANCOVA model.

Note that some researchers do not believe that the assumption of independence can be tested. If there is not random assignment to groups, then these researchers believe this assumption has been violated—period. The plot that we generate will give us a general idea of patterns, however, in situations where random assignment was not performed.

The general steps for generating a simple scatterplot through "Scatter/dot" have been presented in Chapter 10 of the previous volume, and they will not be reiterated here. From the "Simple Scatterplot" dialog screen, click the residual variable and move it into the "Y Axis" box by clicking on the arrow. Click the independent variable (e.g., group) and move it into the "X Axis" box by clicking on the arrow. Then click "OK."

4.4.1.1 Interpreting Independence Evidence

In examining the scatterplot for evidence of independence, the points should fall relatively randomly above and below the horizontal reference line at zero. In this example, the scatterplot does suggest evidence of independence with relative randomness of points above and below the horizontal line at zero.

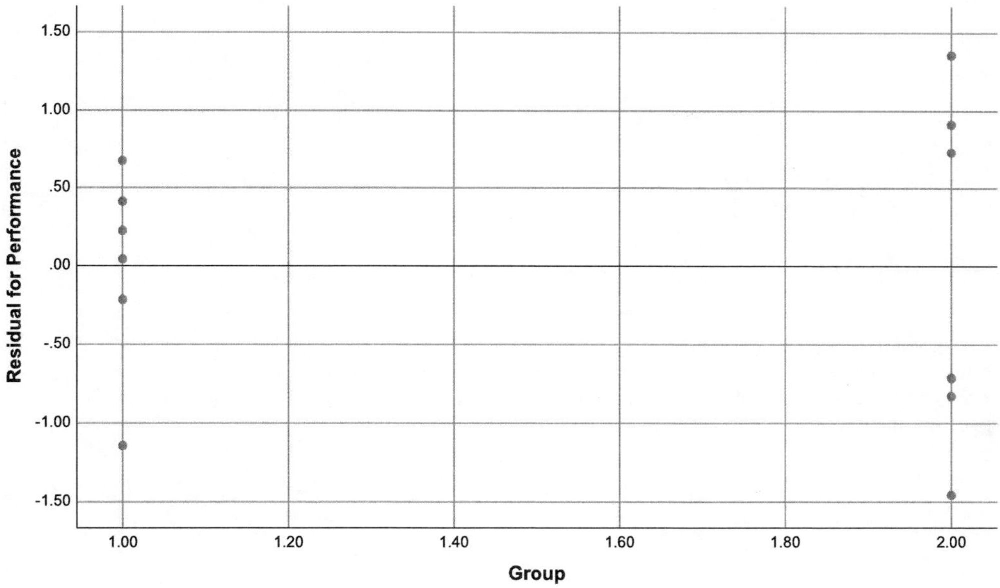

Working in **R**, we create a similar scatterplot.

```
plot(Ch4_fatigue$Group,
     Ch4_fatigue$unstandardizedResiduals,
     xlab = "Group",
     ylab = "Unstandardized Residual",
     main = "Scatterplot for Independence")
```

Using the following *plot* function, with the first variable listed displaying on the *X* axis (e.g., "Ch4_fatigue$Group"), and the second variable displaying on the *Y* axis (i.e., "Ch4_fatigue$unstandardized Residuals"). Additional commands are provided to label the axes (*xlab* and *ylab*) and title the graph (*main*).

(Note that we are using our "Group" variable, not GroupF, in this script. Had we used GroupF, the variable we defined as nominal, the plot generated would be a boxplot, not a scatterplot.)

FIGURE 4.13
Independence evidence.

```
plot(ANCOVA_fatigue)
```

Using the *plot* function, additional plots that can be used for diagnostic purposes are created.

The **residual versus fitted plot** can be used to detect normality, unequal error variance, and outliers. A random display of points, i.e., no patterns to the data, suggest assumptions of normality and equal variances have been met.

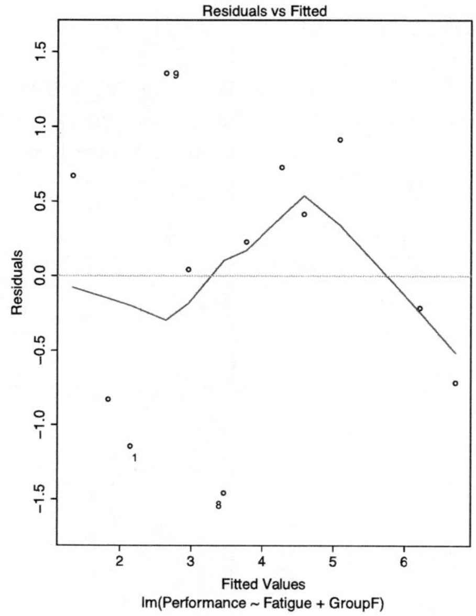

The **normal Q-Q plot** can be used to detect normality and outliers. Points that adhere closely to the diagonal line suggest the assumption of normality has been met. In this case, there are a few points that may be suggestive of outliers.

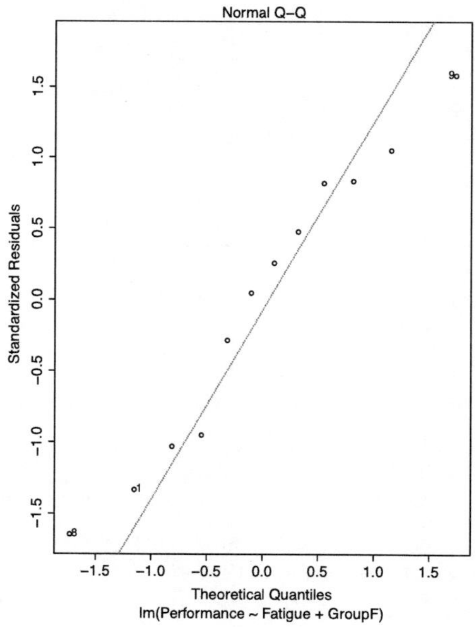

FIGURE 4.13 (continued)
Independence evidence.

The **scale-location plot** can be examined for evidence of equal variance. Relatively equally spaced points by group above and below a horizontal line (i.e., random and equal distribution of points and straight horizontal line) suggests evidence of meeting the assumption. There is some evidence in this graph to suggest heteroscedasticity.

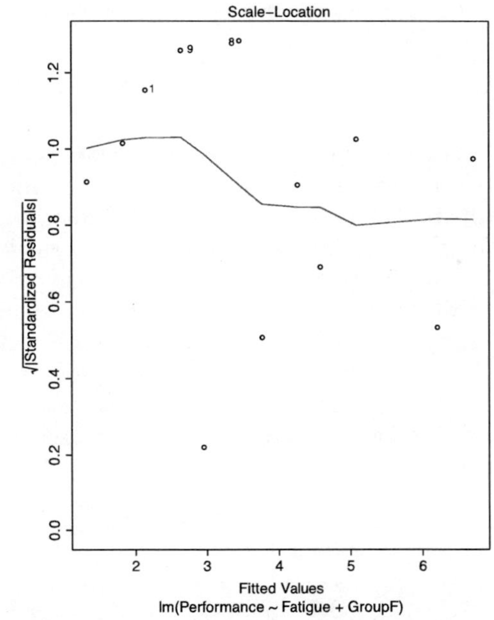

The **constant leverage plot** can be examined as evidence of normality as well to determine points that may exert influence.

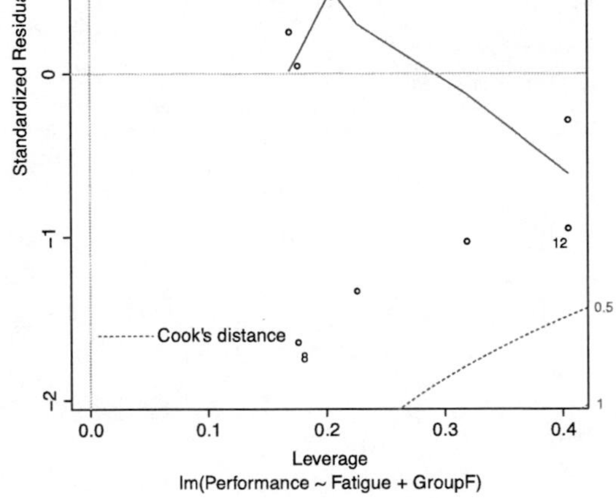

FIGURE 4.13 (continued)
Independence evidence.

4.4.2 Homogeneity of Variance

As we learned previously, another assumption to consider is that the variances of each population are equal. This is known as the assumption of **homogeneity of variance**. When generating factorial ANOVA via SPSS, we requested Levene's test (see Figure 4.5).

4.4.3 Normality

As alluded to earlier in the chapter, understanding the distributional shape, specifically the extent to which normality is a reasonable assumption, is important. For the ANCOVA, the distributional shape for the residuals should be a normal distribution. We can again use "Explore" to examine the extent to which the assumption of normality is met.

The general steps for accessing Explore have been presented in previous chapters and will not be repeated here. From the Explore dialog menu (see the screenshot in Figure 4.14), click the residual and move it into the "Dependent List" box by clicking on the arrow button. The procedures for selecting normality statistics include the following: click on "Plots" in the upper right corner. Place a checkmark in the boxes for "Normality plots with tests" and also for "Histogram." Then click "Continue" to return to the main Explore dialog box. Then click "OK" to generate the output.

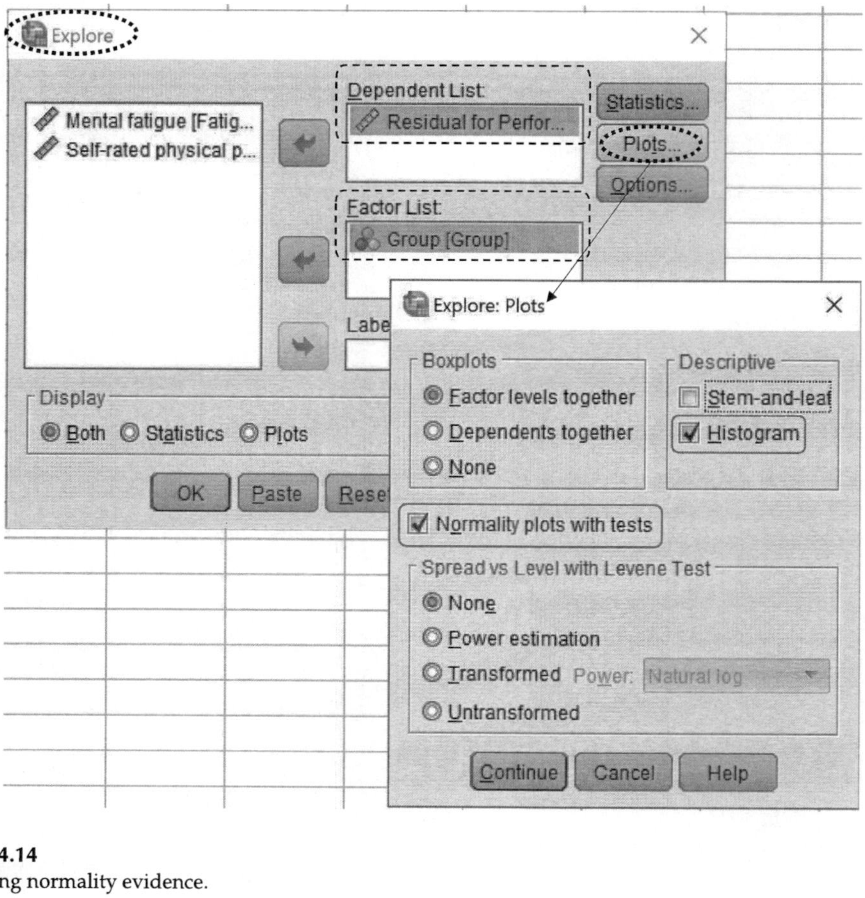

FIGURE 4.14
Generating normality evidence.

4.4.3.1 Interpreting Normality Evidence

We have already developed a good understanding of how to interpret some forms of evidence of normality including skewness and kurtosis, histograms, and boxplots. Here we examine the output for these statistics again. The skewness statistic of the residuals, overall and by group, is within the range of an absolute value of 2.0 suggesting some evidence of normality. There is slight non-normality based on kurtosis (treatment skew = −.076, kurtosis = −2.303; control skew = −1.296, kurtosis = 2.015) (see the "Descriptives" output in Figure 4.15).

Descriptives

			Statistic	Std. Error
Residual for Performance	Mean		.0000	.25461
	95% Confidence Interval for Mean	Lower Bound	-.5604	
		Upper Bound	.5604	
	5% Trimmed Mean		.0056	
	Median		.1357	
	Variance		.778	
	Std. Deviation		.88200	
	Minimum		-1.46	
	Maximum		1.36	
	Range		2.81	
	Interquartile Range		1.51	
	Skewness		-.237	.637
	Kurtosis		-1.024	1.232

Statistics

Residual for Performance

Control (decaffeinated beverage)	N	Valid	6
		Missing	0
	Skewness		**-1.296**
	Std. Error of Skewness		.845
	Kurtosis		**2.015**
	Std. Error of Kurtosis		1.741
Treatment (caffeinated beverage)	N	Valid	6
		Missing	0
	Skewness		**-.076**
	Std. Error of Skewness		.845
	Kurtosis		**-2.303**
	Std. Error of Kurtosis		1.741

FIGURE 4.15
Normality evidence.

Working in **R**, we can generate various normality statistics as well.

```
install.packages("pastecs")
```

The *install.packages* function will install the *pastecs* package which we will use to generate various forms of normality evidence.

```
library(pastecs)
```

The *library* function will load the *pastecs* package.

```
stat.desc(Ch4_fatigue$unstandardizedResiduals,
          norm = TRUE)
```

The *stat.desc* function will generate normality indices on the variable "unstandardizedResiduals" in the dataframe Ch4_fatigue as follows. The *norm=TRUE* command will produce Shapiro-Wilk results (*SW*), which are displayed as *normtest.W* (which is the *SW* statistic value) and *normtest.p* (which is the observed probability value).

Here, we see *SW* = .965 and the related *p* = .854. We see skew (–.181) and kurtosis (–.141) for the "unstandardizedResidual" variable.

Skew, kurtosis, and *SW* all indicate the assumption of normality has been met. As we know, we can divide the skew and kurtosis values by their standard errors to get a standardized value that can be used to determine if the skew and/or kurtosis is statistically different from zero. Since this output provides "2SE," we would simply divide this value by 2 to arrive at the standard error.

Note: You may have noticed that the overall skewness and kurtosis value that we've just generated differs from what we found in SPSS, which was skew = –.237 and kurtosis = –1.024. This is because there are different ways to calculate skewness and kurtosis. Let's use another package in **R** to calculate these statistics with different algorithms.

```
     nbr.val       nbr.null          nbr.na           min           max
1.200000e+01   0.000000e+00    0.000000e+00 -1.457143e+00  1.357143e+00
       range            sum          median          mean       SE.mean
2.814286e+00  -1.276756e-15    1.357143e-01 -1.064009e-16  2.546112e-01
CI.mean.0.95            var         std.dev      coef.var      skewness
5.603954e-01   7.779221e-01    8.819989e-01 -8.289394e+15 -1.813642e-01
    skew.2SE       kurtosis        kurt.2SE    normtest.W    normtest.p
-1.422906e-01 -1.408656e+00   -5.715804e-01  9.651737e-01  8.543019e-01
```

```
install.packages("e1071")
```

The *install.packages* function will install the **e1071** package which we will use to generate skewness and kurtosis.

```
library(e1071)
```

The *library* function will load the **e1071** package.

```
skewness(Ch4_fatigue$unstandardizedResiduals, type=3)
skewness(Ch4_fatigue$unstandardizedResiduals, type=2)
skewness(Ch4_fatigue$unstandardizedResiduals, type=1)
```

FIGURE 4.15 (continued)
Normality evidence.

The *skewness* function will generate skewness statistics on the variable(s) specified. The "type=" script defines how skewness is calculated. Specifying "type=2" will use the algorithm that is used by SPSS. Readers interested in learning more, including the algorithms for each of the three methods, are encouraged to review Joanes and Gill (1998). We see that using type=2, our skew is −.237, the same value as generated using SPSS.

```
# skewness(Ch4_fatigue$unstandardizedResiduals, type=3)
[1] -0.1813642

#skewness(Ch4_fatigue$unstandardizedResiduals, type=2)
[1]   -0.2374222

# skewness(Ch4_fatigue$unstandardizedResiduals, type=1)
[1] -0.2066495
```

```
kurtosis(Ch4_fatigue$unstandardizedResiduals, type=3)
kurtosis(Ch4_fatigue$unstandardizedResiduals, type=2)
kurtosis(Ch4_fatigue$unstandardizedResiduals, type=1)
```

The *kurtosis* function will generate kurtosis statistics on the variable(s) we specify. The "type=" script defines how kurtosis is calculated. Specifying "type=2" will use the algorithm that is used by SPSS. Readers interested in learning more, including the algorithms for each of the three methods, are encouraged to review Joanes and Gill (1998). We see that using "type=2," our kurtosis is −1.024, the same value as generated using SPSS.

```
# kurtosis(Ch4_fatigue$unstandardizedResiduals, type=3)
[1] -1.408656

# kurtosis(Ch4_fatigue$unstandardizedResiduals, type=2)
[1]   -1.024246

# kurtosis(Ch4_fatigue$unstandardizedResiduals, type=1)
[1] -1.106169
```

Working in **R**, another way to test for normality is D'Agostino's test for skewness and the Bonett-Seier test for Geary's kurtosis.

```
install.packages("moments")
library(moments)
```

To conduct D'Agostino's test, we first have to install the *moments* package and then load it into our library. The null hypothesis for this test is that skewness equals zero. Thus, a statistically significant D'Agostino's test would indicate that there is statistically significant skewness.

```
agostino.test(Ch4_fatigue$unstandardizedResiduals)
```

The function *agostino.test* is generated using the variable "unstandardizedResiduals" from our Ch4_fatigue dataframe. The results suggest evidence of normality as $p = .6967$, greater than alpha.

```
        D'Agostino skewness test

data: Ch4_fatigue$unstandardizedResiduals
skew = -0.20665, z = -0.38974, p-value = 0.6967
alternative hypothesis: data have a skewness
```

FIGURE 4.15 (continued)
Normality evidence.

```
agostino.test(Ch4_fatigue$unstandardizedResiduals[Ch4_fatigue$Group==1])
agostino.test(Ch4_fatigue$unstandardizedResiduals[Ch4_fatigue$Group==2])
```

This test can also be generated by group of the independent variable. Given the small sample size by group in this illustration, however, results are not available.

```
bonett.test((Ch4_fatigue$unstandardizedResiduals))
```

The *bonett.test* function, generated using the variable "unstandardizedResiduals" from our Ch4_fatigue dataframe, performs the Bonett-Seier test for Geary's kurtosis for data that are normally distributed. The null hypothesis states that data should have a Geary's kurtosis value equal to $\sqrt{2/\pi} = .7979$. The results suggest evidence of normality as $p = .293$, greater than alpha.

```
        Bonett-Seier test for Geary kurtosis

data:  (Ch4_fatigue$unstandardizedResiduals)
tau = 0.72619, z = -1.05160, p-value = 0.293
alternative hypothesis: kurtosis is not equal to sqrt(2/pi)
```

```
bonett.test((Ch4_fatigue$unstandardizedResiduals[Ch4_fatigue$Group==1]))
bonett.test((Ch4_fatigue$unstandardizedResiduals[Ch4_fatigue$Group==2]))
```

This test can also be generated by group of the independent variable. By group, the results for the Bonett-Seier test for Geary's kurtosis for data that are normally distributed provide evidence of normality by group with both p's > .05.

```
# bonett.test((Ch4_fatigue$unstandardizedResiduals[Ch4_fatigue$Group==1]))

        Bonett-Seier test for Geary kurtosis
data:  (Ch4_fatigue$unstandardizedResiduals[Ch4_fatigue$Group == 1])
tau = 0.45238, z = 0.26847, p-value = 0.7883
alternative hypothesis: kurtosis is not equal to sqrt(2/pi)

# bonett.test((Ch4_fatigue$unstandardizedResiduals[Ch4_fatigue$Group==2]))

        Bonett-Seier test for Geary kurtosis

data:  (Ch4_fatigue$unstandardizedResiduals[Ch4_fatigue$Group == 2])
tau = 1.0000, z = -1.9487, p-value = 0.05133
alternative hypothesis: kurtosis is not equal to sqrt(2/pi)
```

FIGURE 4.15 (continued)
Normality evidence.

The histogram of residuals, overall and by group, is not what most would consider normal in shape, and this is largely an artifact of the small sample size. Because of this, we will rely more heavily on the other forms of normality evidence.

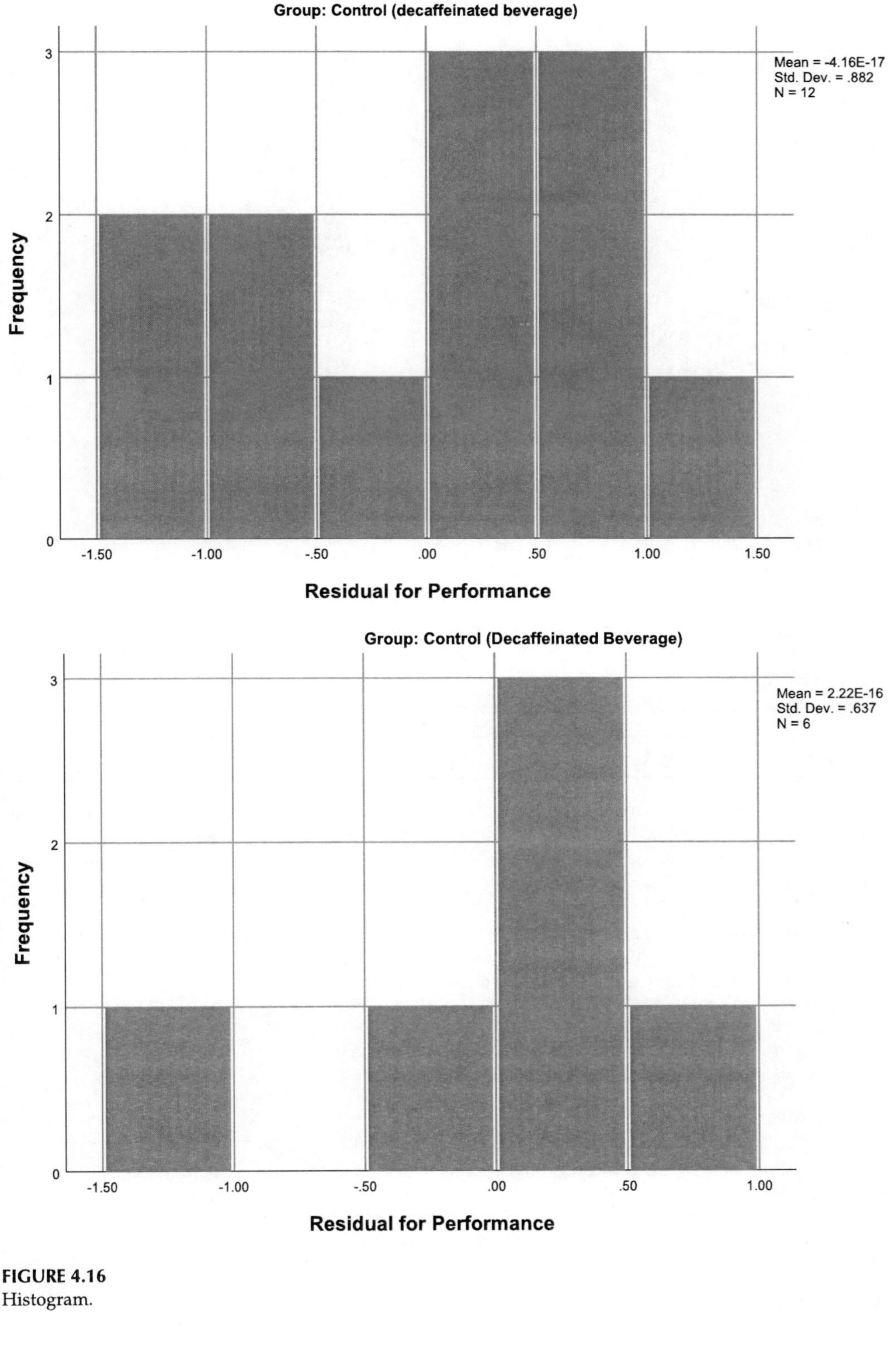

FIGURE 4.16
Histogram.

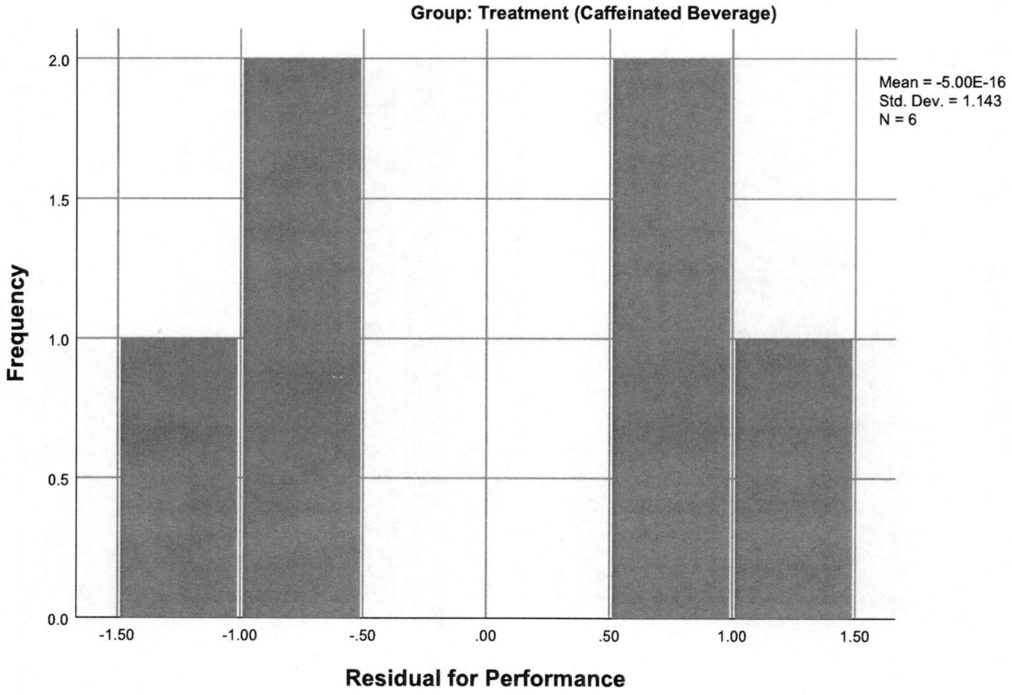

Working in **R**, we can generate a histogram using the *ggplot2* package.

```
install.packages("ggplot2")
```

The *install.packages* function will install the *ggplot2* package which we can use to create various graphs and plots. If you have already installed *ggplot2* previously, there is no need to run this script again.

```
library(ggplot2)
```

The library function will load the ggplot2 package.

```
qplot(Ch4_fatigue$unstandardizedResiduals,
      geom="histogram",
      binwidth=.5,
      main = "Histogram of Unstandardized Residuals",
      xlab = "Unstandardized Residual", ylab = "Count",
      fill=I("gray"),
      col=I("white"))
```

Using the *qplot* function, we create a histogram (i.e., geom = "histogram") from our dataframe (i.e., Ch4_fatigue) using the variable "unstandardizedResiduals." We can add a few commands to change the width of the bars (i.e., *binwidth=.5*), color of the bars (i.e., *fill=I("gray")*), and outline of the bars (i.e., *col=I("white")*). We can also add a title (i.e., *main = "Histogram of Unstandardized Residuals"*) and change the X and Y axes (*xlab = "Unstandardized Residual", ylab = "Count"*).

FIGURE 4.16 (continued)
Histogram.

```
hist(Ch4_fatigue$unstandardizedResiduals[Ch4_fatigue$Group==1],
     main="Histogram for Control",
     xlab="Unstandardized Residuals")
```

```
hist(Ch4_fatigue$unstandardizedResiduals[Ch4_fatigue$Group==2],
     main="Histogram for Treatment",
     xlab="Unstandardized Residuals")
```

Histograms by group can be created with these scripts, each one specifying one category of "Group" as the variable with which to create the histogram of unstandardized residuals.

FIGURE 4.16 (continued)
Histogram.

There are a few other statistics that can be used to gauge normality. The formal test of normality, the Shapiro-Wilk test (*SW*) (Shapiro & Wilk, 1965), provides evidence of the extent to which our sample distribution is statistically different from a normal distribution. The output for the Shapiro-Wilk test is presented in Figure 4.17 and suggests that our sample distribution for residuals is not statistically significantly different than what would be expected from a normal distribution (overall $SW = .965$, $df = 12$, $p = .854$; control $SW = .911$, $p = .4426$; treatment $SW = .902$, $p = .3832$).

Tests of Normality

	Kolmogorov-Smirnov[a]			Shapiro-Wilk		
	Statistic	df	Sig.	Statistic	df	Sig.
Residual for Performance	.124	12	.200*	.965	12	.854

*. This is a lower bound of the true significance.

a. Lilliefors Significance Correction

Tests of Normality

		Kolmogorov-Smirnov[a]			Shapiro-Wilk		
	Group	Statistic	df	Sig.	Statistic	df	Sig.
Residual for Performance	Control (decaffeinated beverage)	.202	6	.200*	.911	6	.443
	Treatment (caffeinated beverage)	.238	6	.200*	.902	6	.383

*. This is a lower bound of the true significance.

a. Lilliefors Significance Correction

FIGURE 4.17
Shapiro-Wilk test of normality.

```
shapiro.test(Ch4_fatigue$unstandardizedResiduals)
```

Working in **R**, had we wanted to generate only the Shapiro-Wilk test, the *shapiro.test* function could be used.

```
        Shapiro-Wilk normality test

data:  Ch4_fatigue$unstandardizedResiduals
W = 0.96517, p-value = 0.8543
```

```
tapply(Ch4_fatigue$unstandardizedResiduals,
       Ch4_fatigue$GroupF, shapiro.test)
```

To generate the Shapiro-Wilk test by group, the *tapply* function can be used to apply the *shapiro.test* to the unstandardized residuals for all levels of the independent variable.

```
$control

      Shapiro-Wilk normality test
data:  X[[i]]
W = 0.91093, p-value = 0.4426

$treatment

      Shapiro-Wilk normality test
data:  X[[i]]
W = 0.90156, p-value = 0.3832
```

FIGURE 4.17 (continued)
Shapiro-Wilk test of normality.

Quantile-quantile (Q-Q) plots are also often examined to determine evidence of normality. Q-Q plots are graphs that plot quantiles of the theoretical normal distribution against quantiles of the sample distribution. Points that fall on or close to the diagonal line suggest evidence of normality. The Q-Q plots of residuals by group shown below suggests relative normality.

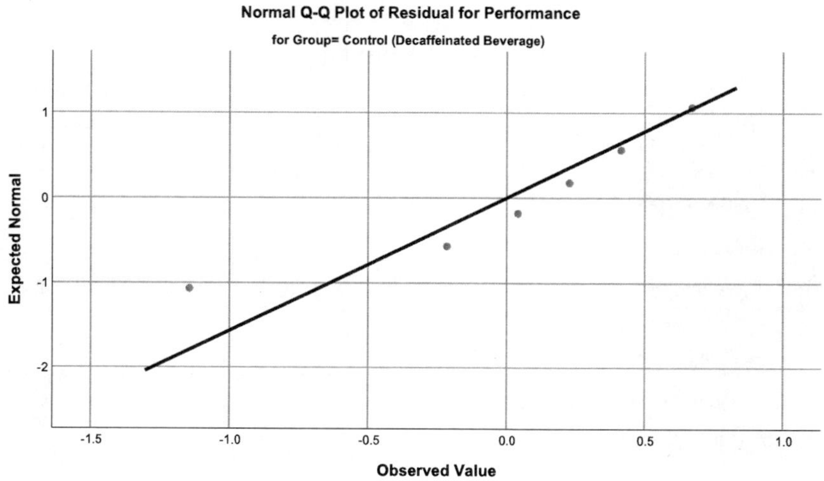

FIGURE 4.18
Normal Q-Q plot

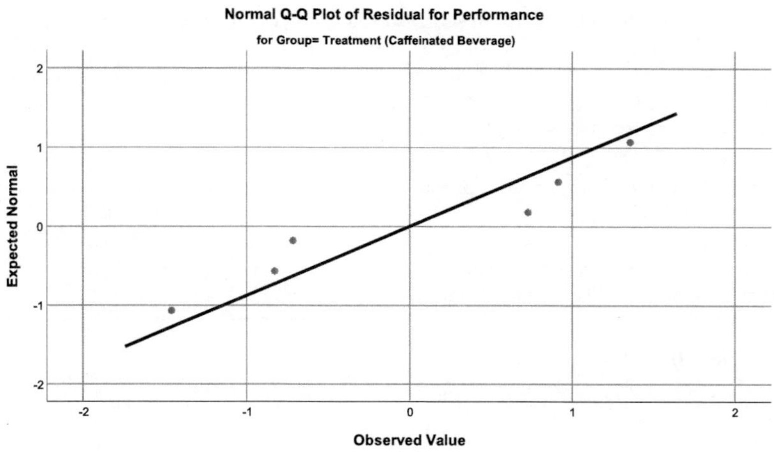

Working in **R**, we can use the *qplot* function to create a Q-Q plot of unstandardized residuals. The "data=" script defines the dataframe as "Ch4_fatigue."

```
qplot(sample=unstandardizedResiduals,
      data = Ch4_fatigue)
```

```
qqnorm(Ch4_fatigue$unstandardizedResiduals[Ch4_fatigue$Group==1],
       main='control')

qqnorm(Ch4_fatigue$unstandardizedResiduals[Ch4_fatigue$Group==2],
       main='treatment')
```

By group, Q-Q plots can be created with this script, with each command defining one category of the *Group* variable.

FIGURE 4.18 (continued)
Normal Q-Q plot

Examination of the boxplot by group in Figure 4.19 suggests a relatively normal distributional shape of residuals and no outliers for both groups.

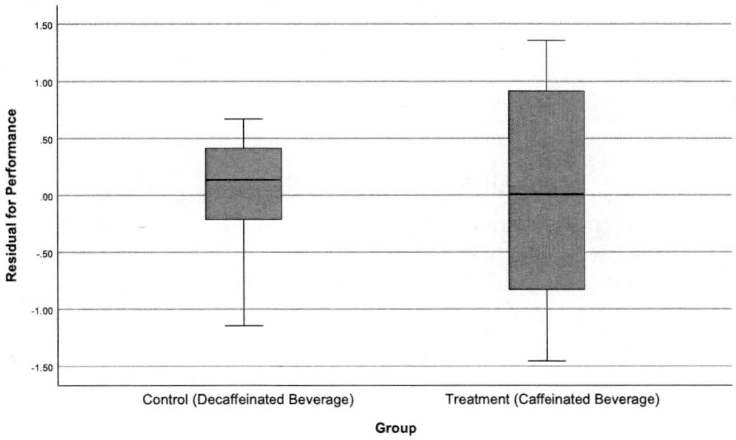

FIGURE 4.19
Boxplot.

Working in **R**, we can generate a boxplot for unstandardized residuals using the *boxplot* function. To label the *Y* axis, we include the *ylab* command.

```
boxplot(Ch4_fatigue$unstandardizedResiduals,
        ylab="Unstandardized Residuals")
```

Adding the independent variable to the script produces a boxplot by group. The command *xlab* will print "Group" to identify the *X* axis.

```
boxplot(Ch4_fatigue$unstandardizedResiduals~Ch4_fatigue$GroupF,
        xlab="Group", ylab="Unstandardized Residuals")
```

FIGURE 4.19 (continued)
Boxplot.

Considering the forms of evidence we have examined, skewness and kurtosis statistics, histogram, the Shapiro-Wilk test, the Q-Q plot, and the boxplot, all suggest normality is a reasonable assumption. We can be reasonably assured we have met the assumption of normality of the dependent variable for each group of the independent variable.

4.4.4 Linearity

Recall that the assumption of linearity means that the regression of the dependent variable (i.e., "physical performance" in this illustration) on the covariate (i.e., "mental fatigue") is linear. Evidence of the extent to which this assumption is met can be done by examining scatterplots of the dependent variable versus the covariate—both overall and also for each category or group of the independent variable.

4.4.4.1 Overall Linearity Evidence

The general steps for generating a simple scatterplot through "Scatter/dot" have been presented in Chapter 10 of the previous volume, and they will not be reiterated here. To generate the overall scatterplot, from the "Simple Scatterplot" dialog screen, click the dependent variable (i.e., performance) and move it into the "Y Axis" box by clicking on the arrow. Click the covariate (i.e., fatigue) and move it into the "X Axis" box by clicking on the arrow. Then click "OK."

4.4.4.1.1 Interpreting Overall Linearity Evidence

In examining the scatterplot for overall evidence of linearity, the points should fall relatively linearly (in other words, we should not be seeing a curvilinear or some other nonlinear relationship). In this example, our scatterplot suggests we have evidence of overall linearity as there is a relatively clear pattern of points which suggest a positive and linear relationship between the dependent variable and covariate.

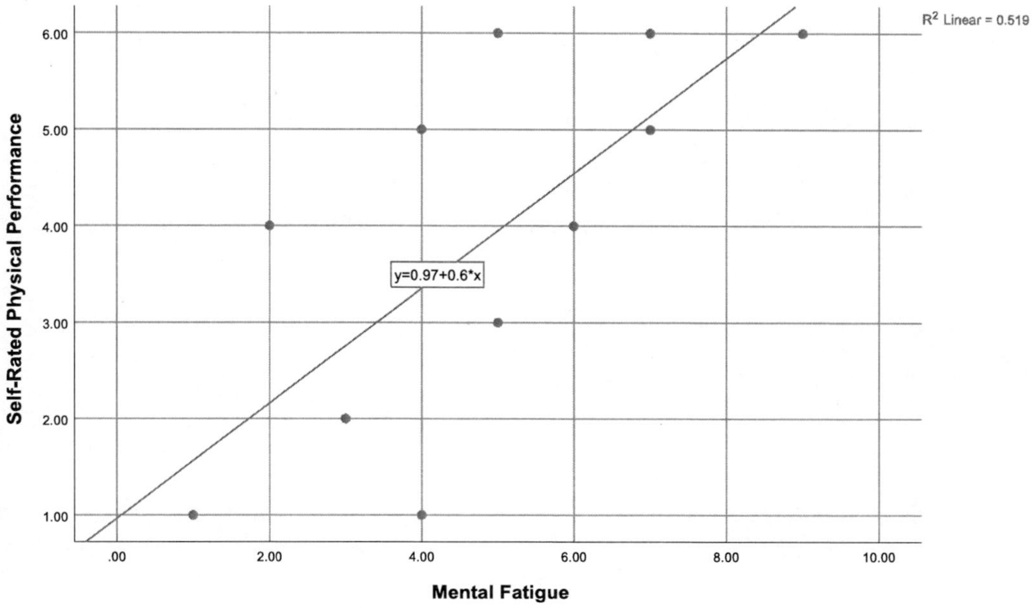

Working in **R**, we can generate a similar scatterplot.

```
plot(Ch4_fatigue$Fatigue,
     Ch4_fatigue$Performance,
     xlab = "Fatigue",
     ylab = "Performance",
     main = "Scatterplot for Independence")
```

Using the *plot* function, with the first variable listed displaying on the *X* axis (e.g., "Ch4_fatigue$Fatigue"), and the second variable displaying on the *Y* axis (i.e., "Ch4_fatigue$Performance"). Additional commands are provided to label the axes (*xlab* and *ylab*) and title the graph (*main*).

FIGURE 4.20
Scatterplot.

4.4.4.2 Linearity Evidence by Group

To generate the scatterplot of the dependent variable and covariate for each group of the independent variable, we must first split the data file. To do this, go to "Data" in the top pulldown menu. Then select "Split File."

From the Split File dialog screen, select the radio button for "Organize output by groups," then click the independent variable and move it into the "Groups Based on" box by clicking on the arrow. Then click "OK."

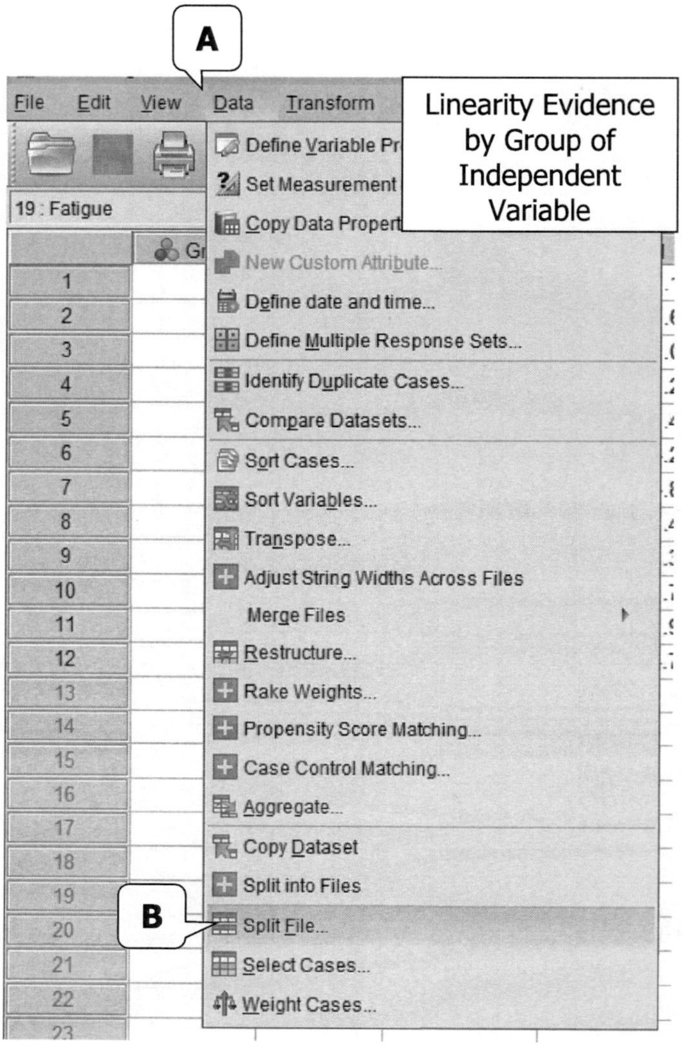

FIGURE 4.21
Linearity by group of independent variable.

After splitting the file, the next step is to generate the scatterplot of the dependent variable by covariate. Because we have split the file, there will be two scatterplots generated: one for the treatment (i.e., ingestion of caffeinated beverage) and one for the control (i.e., ingestion of decaffeinated beverage). Because we have just generated the overall scatterplot, the selections made previously will remain, and thus from the "Simple Scatterplot" dialog screen, simply click "OK" to generate the output.

4.4.4.2.1 Interpreting Evidence of Linearity Evidence by Group

In examining the scatterplot for evidence of linearity by group of the independent variable, our interpretation should remain the same: the points should fall relatively linearly (in

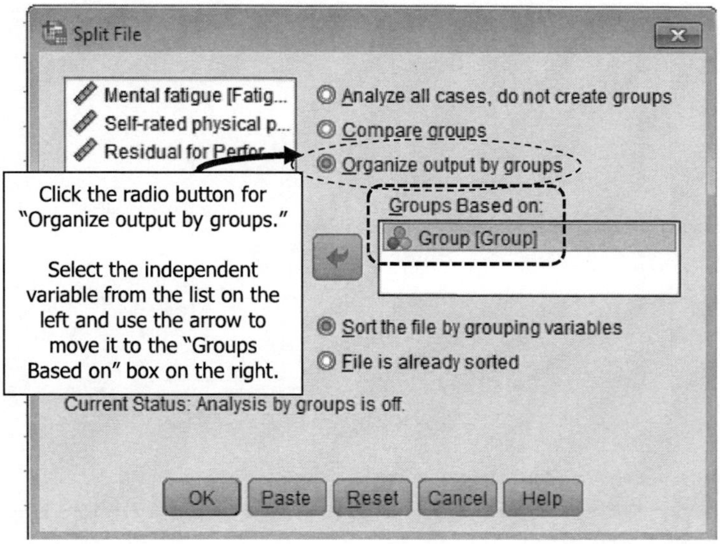

FIGURE 4.22
Split file.

other words, we should not see a curvilinear or some other nonlinear relationship). In this example, our scatterplots suggest we have evidence of linearity by group of the independent variable as there is a relatively clear pattern of points which suggest a positive and linear relationship between the dependent variable and covariate for each group of the independent variable.

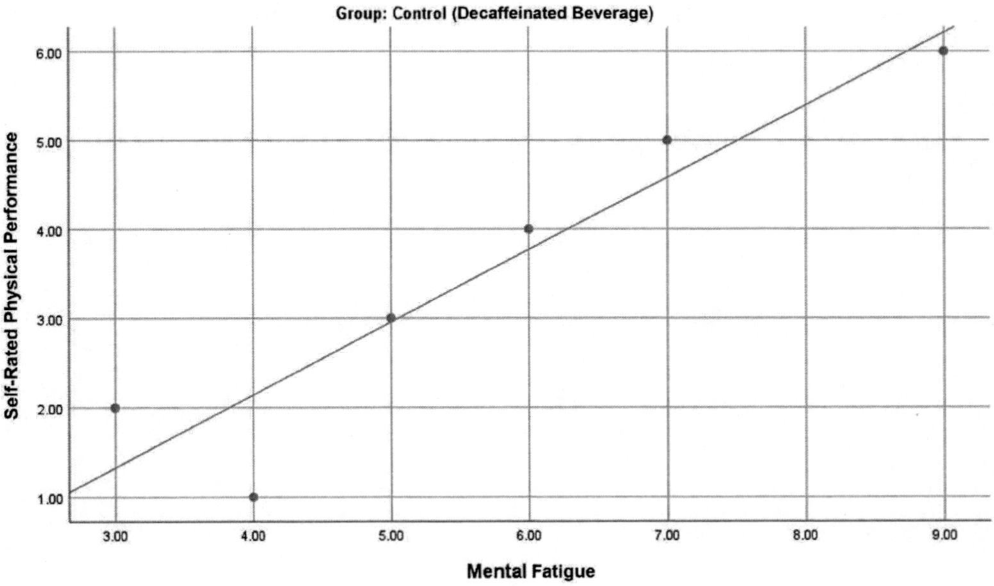

FIGURE 4.23
Scatterplot by group of independent variable.

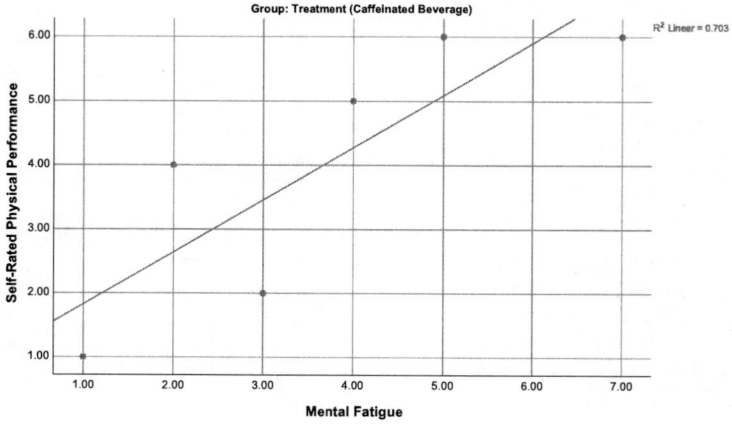

Working in **R**, we create a similar plot.

```
install.packages("ggplot2")
library(ggplot2)
```

For this plot, we use the *ggplot2* package.

```
ggplot(Ch4_fatigue,
       aes(x=Fatigue, y=Performance, shape=factor(GroupF))) +
   geom_point(size=4) +
   geom_smooth(method=lm, se=FALSE, fullrange=TRUE)
```

Using the *ggplot* function, we can create a scatterplot of the outcome, *Performance,* by covariate, *Fatigue,* for each group, *GroupF*. This will create one plot with different shapes for each group of the independent variable. The regression line, by group, is added with *geom.smooth*. The standard error shading is removed with *se=FALSE,* and the regression line is extended to both sides of the graph with *fullrange=TRUE.*

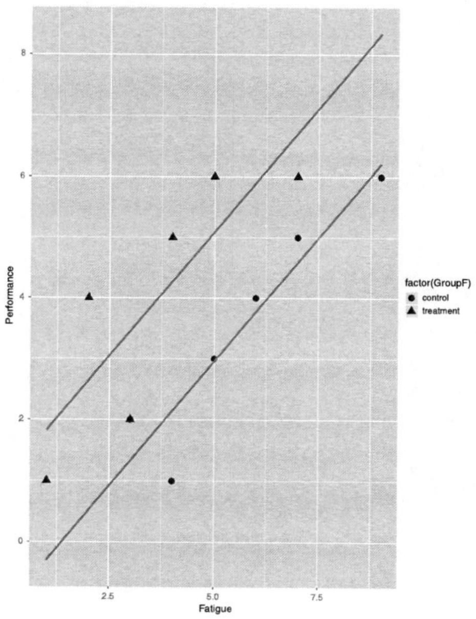

FIGURE 4.24
Scatterplot by group of independent variable.

4.4.5 Independence of Covariate and Independent Variable

Recall the assumption of independence of the covariate and independent variable. In other words, the levels of the independent variable should not differ on the covariate. If subjects have been randomly assigned to conditions (in other words, the different levels of the independent variable), the assumption of independence of the covariate and independent variable has likely been met. In this illustration, athletes were randomly assigned to treatment group (i.e., ingestion of caffeinated or decaffeinated beverage), and thus the assumption of independence of the covariate and independent variable was likely met. As we have learned in previous chapters, however, we often use independent variables that do not allow random assignment. Evidence of the extent to which this assumption is met can be done by examining mean differences on the covariate based on the independent variable. If the independent variable has only two levels, an independent *t* test would be appropriate. If the independent variable has more than two categories, a one-way ANOVA would suffice. If the groups are not statistically different on the covariate, then that lends evidence that the assumption of independence of the covariate and the independent variable has been met.

We have two levels of our independent variable, thus we will generate an independent *t* test. The general steps for generating an independent *t* test have been presented in Chapter 8 of the previous volume, and they will not be reiterated here. From the "Independent Samples *T* Test" dialog screen, click the covariate (e.g., mental fatigue) and move it into the "Test Variable(s)" box by clicking on the arrow. Click the independent variable (e.g., treatment group) and move it into the "Grouping Variable" box by clicking on the arrow. Click the "Define Groups" box and enter "1" for "Group 1" and "2" for "Group 2." Then click "Continue" to return to the main the "Independent Samples *T* Test" dialog screen and click on "OK" to generate the output.

4.4.5.1 Interpreting Evidence of Independence of Covariate and Independent Variable

In examining the independent *t* test results, evidence of independence of the covariate and independent variable is provided when the test results are *not* statistically significant. In this example, our results suggest we have evidence of independence of the covariate and independent variable as the results are *not* statistically significant, $t(10)$, = 1.604, p = .140. Thus, we have likely met this assumption through random assignment of cases to groups, and this provides further confirmation that we have not violated the assumption of independence of the covariate and independent variable.

Independent Samples Test

		Levene's Test for Equality of Variances		t-test for Equality of Means					95% Confidence Interval of the Difference	
		F	Sig.	t	df	Sig. (2-tailed)	Mean Difference	Std. Error Difference	Lower	Upper
Mental fatigue	Equal variances assumed	.000	1.000	1.604	10	.140	2.00000	1.24722	-.77898	4.77898
	Equal variances not assumed			1.604	10.000	.140	2.00000	1.24722	-.77898	4.77898

FIGURE 4.25
Independent *t* test results.

Working in **R**, we can compute a *t* test when we have two groups or an ANOVA if there are more than two groups. For illustrative purposes, we'll run an ANOVA.

```
Ch4_independence <- aov(Ch4_fatigue$Fatigue ~ Ch4_fatigue$GroupF)
```

The *aov* function will generate the ANOVA model with *Fatigue* as the dependent variable and *GroupF* as the independent variable. We are using data from the *Ch4_fatigue* dataframe, and we are calling this object "Ch4_independence." Recall that in a two-group situation, $F = t^2$ or $\sqrt{F} = t$. Thus, $\sqrt{F} = \sqrt{2.571} = t$, we find $t = 1.603$, which is roughly equivalent (likely due to rounding) that we found using SPSS.

```
                   Df Sum Sq Mean Sq F value Pr(>F)
Ch4_fatigue$GroupF  1  12.00  12.000   2.571   0.14
Residuals          10  46.67   4.667
```

FIGURE 4.25 (continued)
Independent *t* test results.

4.4.6 Homogeneity of Regression Slopes

Step 1. In order the test the homogeneity of slopes assumption, you will need to rerun the ANCOVA analysis. Keep every screen the same as before, *with one exception*. Return to the main Univariate dialog box (see Step 2, Figure 4.4) and click on "Model." From the Model dialog box, click on the "Build terms" radio button to build a custom model to include the interaction between the independent and covariate variables. To do this, under the "Build Term(s)" pulldown in the middle of the dialog box, select "Main effects."

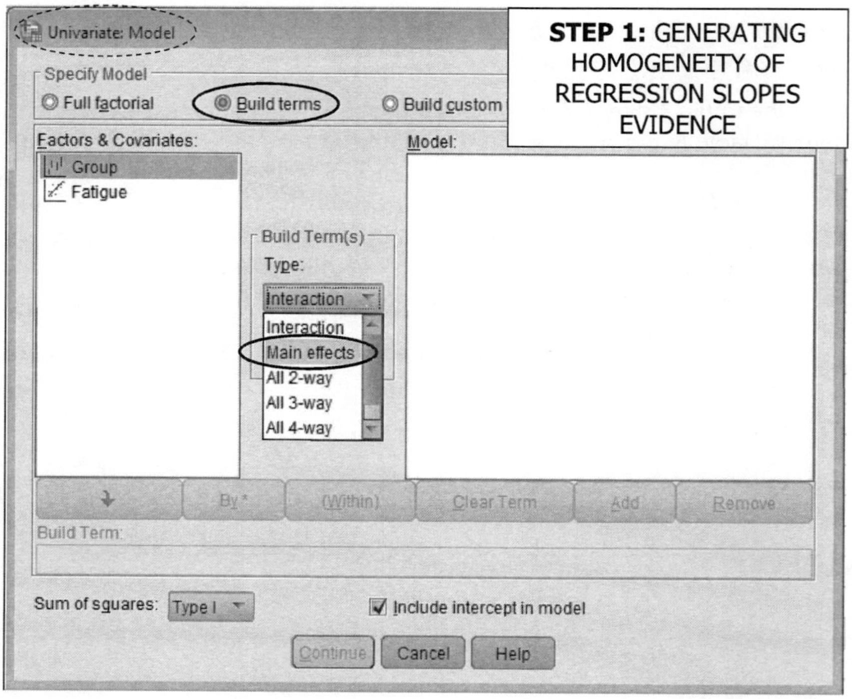

FIGURE 4.26
Homogeneity of regression slopes: Step 1.

Step 2. Click the independent variable and move it into the Model box by clicking on the arrow button. Next, click the covariate and move it into the Model box by clicking on the arrow button. This will place "Group" and "Fatigue" in the Model box on the right of the screen.

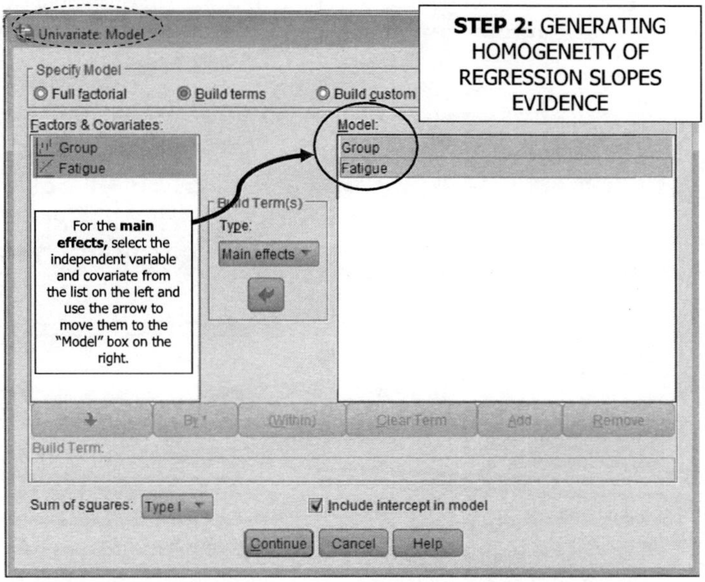

FIGURE 4.27
Homogeneity of regression slopes: Step 2.

Step 3. Then, from the Build Term(s) pulldown menu, select "Interaction."

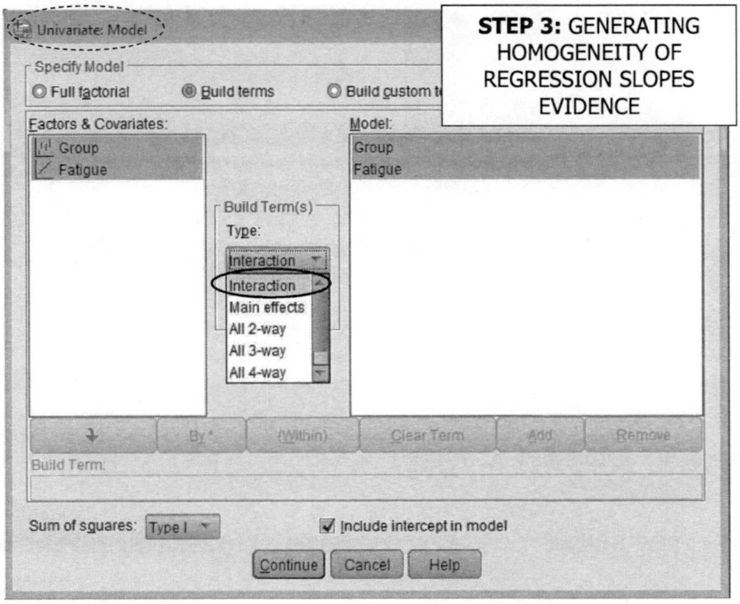

FIGURE 4.28
Homogeneity of regression slopes: Step 3.

Step 4. From the left "Factors & Covariates" box, click both variables at the same time (e.g., using the shift key) and use the arrow key to move the interaction of Fatigue*Group into the Model box on the right. There should now be *three terms* in the Model box: the interaction and two main effects. Then click "Continue" to return to the main Univariate dialog box. Then click "OK" to generate the output.

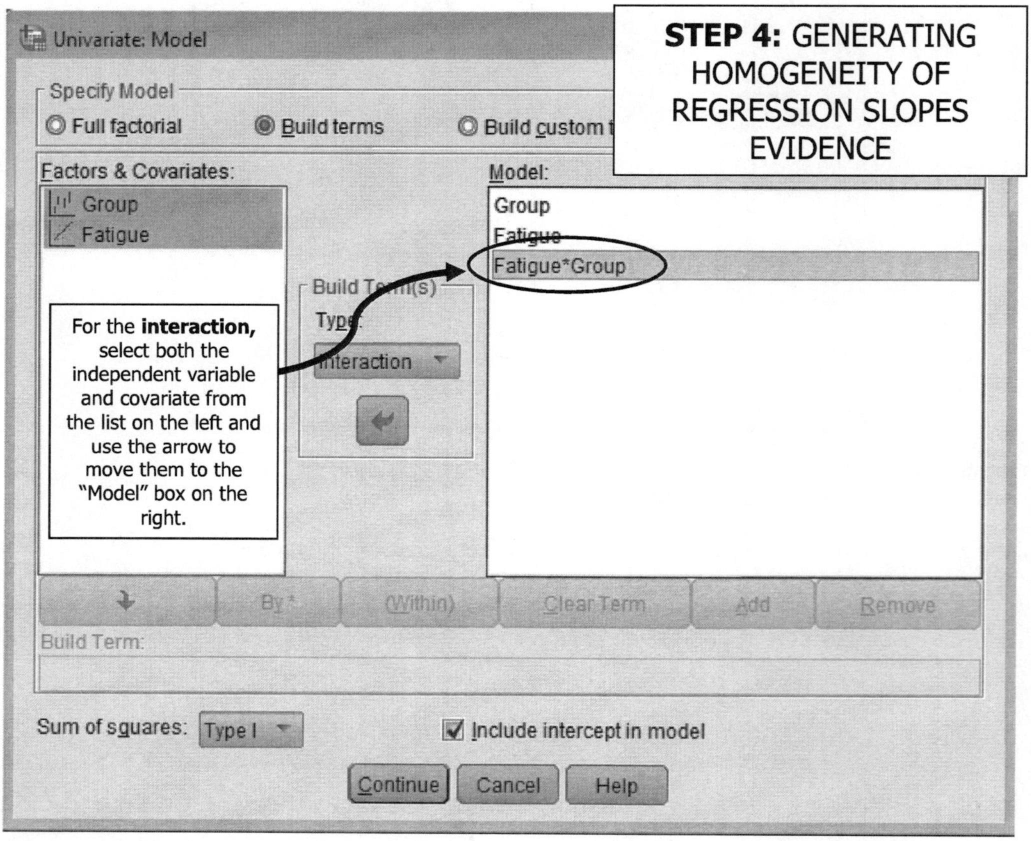

FIGURE 4.29
Homogeneity of regression slopes: Step 4.

4.4.6.1 Interpreting Evidence of Homogeneity of Regression Slopes

Selected results, specifically the ANCOVA summary table which presents the results for the homogeneity of slopes test, are presented as follows. Here the only thing that we care about is the test of the interaction, which we want to be nonsignificant, and we find this to be the case: $F\ (1, 8) = .000$, $p = 1.000$. This indicates that we have met the homogeneity of regression slopes assumption.

Tests of Between-Subjects Effects

Dependent Variable: Self-rated physical performance

Source	Type I Sum of Squares	df	Mean Square	F	Sig.	Partial Eta Squared	Noncent. Parameter	Observed Power[b]
Corrected Model	31.693[a]	3	10.564	9.876	.005	.787	29.629	.955
Intercept	168.750	1	168.750	157.763	.000	.952	157.763	1.000
Group	.750	1	.750	.701	.427	.081	.701	.115
Fatigue	30.943	1	30.943	28.928	.001	.783	28.928	.997
Group * Fatigue	.000	1	.000	.000	1.000	.000	.000	.050
Error	8.557	8	1.070					
Total	209.000	12						
Corrected Total	40.250	11						

a. R Squared = .787 (Adjusted R Squared = .708)

b. Computed using alpha = .05

Working in **R**, we can examine homogeneity of regression slopes by building in an interaction term in the model.

```
HRS <- aov(Performance ~ Fatigue + GroupF + Fatigue:GroupF,
         data=Ch4_fatigue)
```

We use the *aov* function to generate our model, which takes the form of *dependent variable ~ covariate + independent variable + covariate:independent variable interaction*. We name this function *HRS* (i.e., homogeneity of regression slopes).

```
Anova(HRS, type="II")
```

We use the *Anova* function on our object, defining Type II sum of squares as *type = "II"*.

```
Anova Table (Type II tests)
```

```
Response: Performance
                Sum Sq Df F value    Pr(>F)
Fatigue         30.9429  1  28.928 0.0006628 ***
GroupF          10.8122  1  10.108 0.0130104 *
Fatigue:GroupF   0.0000  1   0.000 1.0000000
Residuals        8.5571  8
—
Signif. codes:  0 '***' 0.001 '**' 0.01 '*' 0.05 '.' 0.1 ' ' 1
```

FIGURE 4.30
Homogeneity of regression slopes evidence.

4.5 Power Using G*Power

Generating power analysis for ANCOVA models follows similarly to that for ANOVA and factorial ANOVA. In particular, if there is more than one independent variable, we must test for main effects and interactions separately. Because we have only one independent variable for our ANCOVA model, our illustration assumes only one main effect. If there were additional independent variables and/or interactions, we would have followed these steps for those as well.

4.5.1 Post Hoc Power for ANCOVA Using G*Power

The first thing that must be done when using G*Power for computing post hoc power is to select the correct test family. In our case, we conducted an ANCOVA. To find ANCOVA, we will select "Tests" in the top pulldown menu, then "Means," and then "Many groups: ANCOVA: Main effects and interactions." Once that selection is made, the "Test family" automatically changes to "F tests."

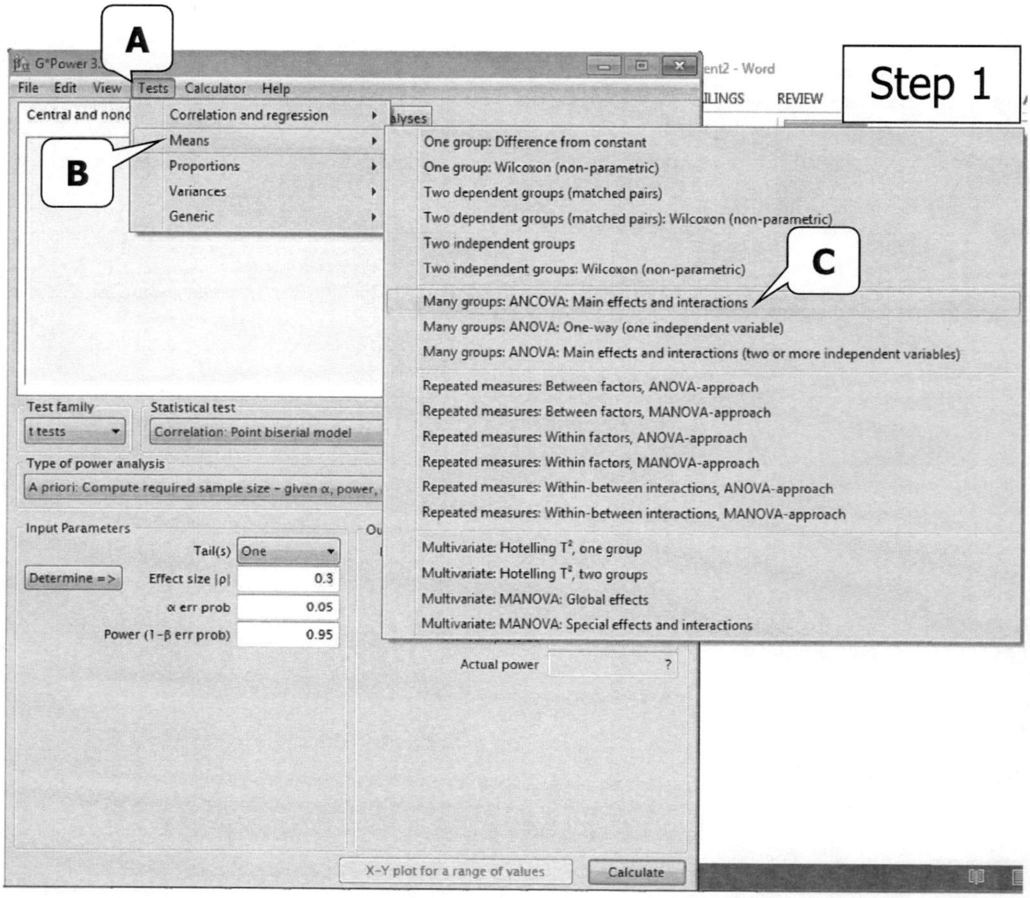

FIGURE 4.31
Power: Step 1.

The "Type of power analysis" desired then needs to be selected. To compute post hoc power, we need to select "Post hoc: Compute achieved power—given α, sample size, and effect size."

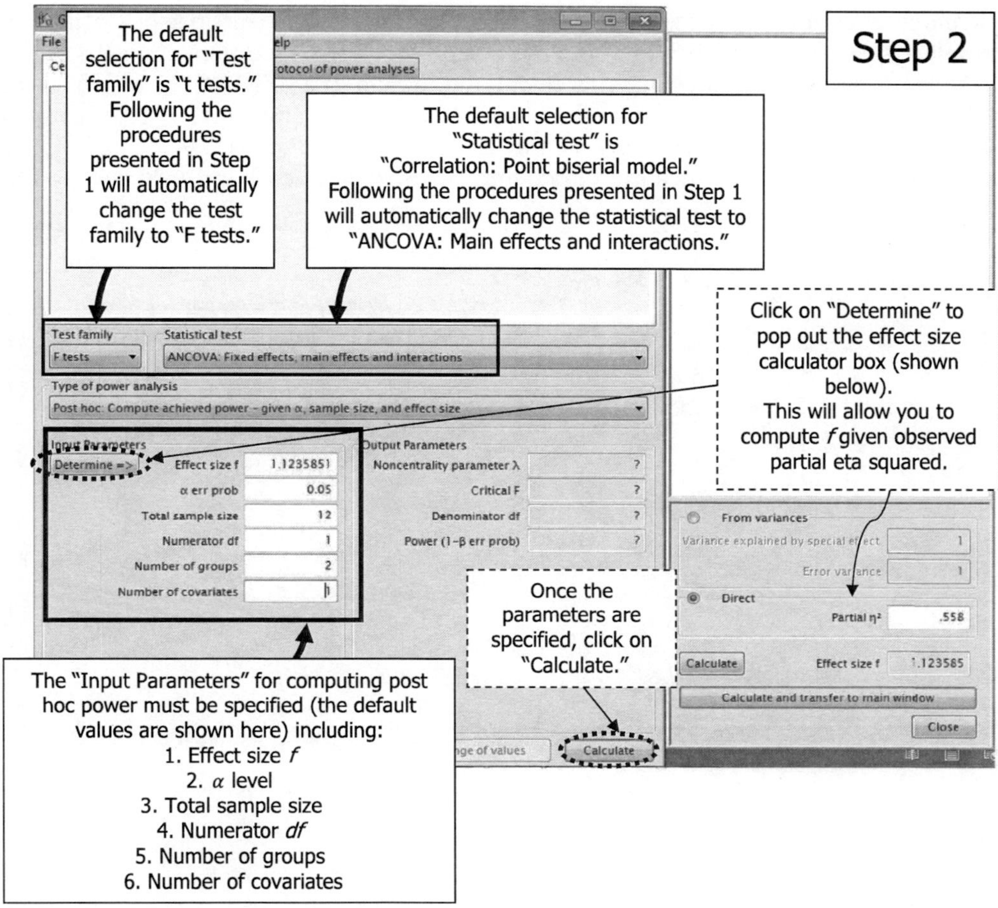

FIGURE 4.32
Power: Step 2.

The "Input Parameters" must then be specified. We will compute the effect size *f* last so we skip that for the moment. In our example, the alpha level we used was .05 and the total sample size was 12. The *numerator degrees of freedom* for group (our independent variable) is equal to the number of categories of this variable (i.e., 2) minus 1; thus there is one degree of freedom for the numerator. The *number of groups* equals, in the case of an ANCOVA with multiple independent variables, the product of the number of levels or categories of the independent variables or (*J*)(*K*). In this example, we have only one independent variable. Thus the number of groups when there is only one independent variable is equal to the number of categories of this independent variable (i.e., 2). The last parameter that must be inputted is the number of covariates. In this example, we have only one covariate; thus we enter 1 in this box.

We skipped filling in the first parameter, the effect size *f*, for a reason. SPSS provides only a partial eta squared measure of effect size. Thus we will use the pop out effect size calculator in G*Power to compute the effect size *f* (we saved this parameter for last as the calculation is based on the previous values just entered). To pop out the effect size calculator, click on "Determine" which is displayed under "Input Parameters." In the pop out effect size calculator, click on the radio button for "Direct" and then enter the partial eta squared value for group that was calculated in SPSS (i.e., .558). Clicking on "Calculate" in the pop out effect size calculator will calculate the effect size *f*. Then click on "Calculate and transfer to main window" to transfer the calculated effect size (i.e., 1.1235851) to the "Input Parameters." Once the parameters are specified, click on "Calculate" to find the power statistics.

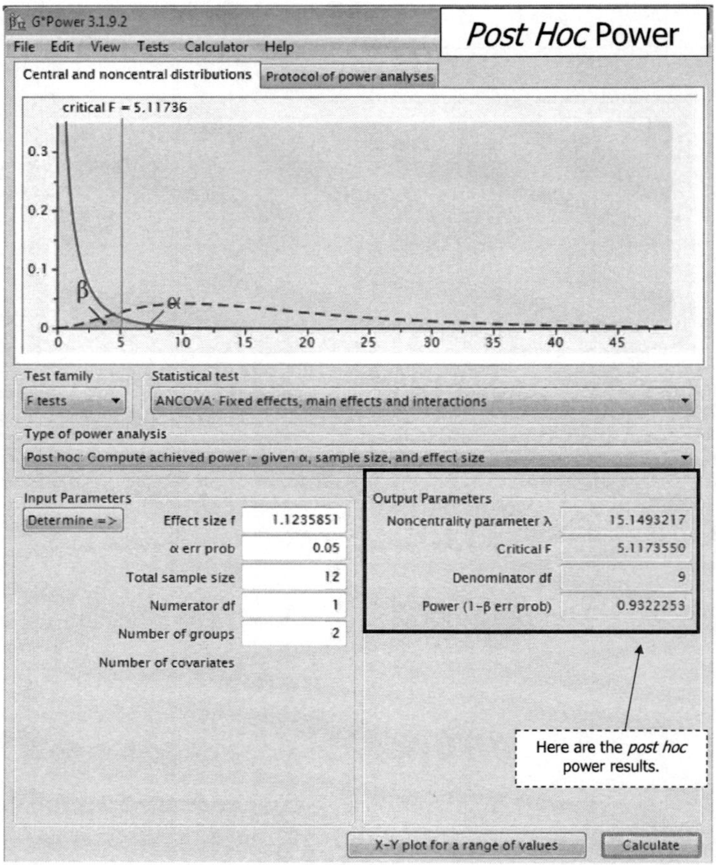

FIGURE 4.33
Post hoc power.

The "Output Parameters" provide the relevant statistics given the input just specified. In this example, we were interested in determining post hoc power for an ANCOVA with a computed effect size *f* of 1.1235851, an alpha level of .05, total sample size of 12, numerator degrees of freedom of one, two groups, and one covariate.

Based on those criteria, the post hoc power for the main effect of treatment group (i.e., our only independent variable) was .93. In other words, with an ANCOVA, computed effect size *f* of 1.124, alpha level of .05, total sample size of 12, numerator degrees of freedom of one, two groups, and one covariate, the post hoc power of our main effect for this test was .93—the probability of rejecting the null hypothesis when it is really false (in this case, the probability that the adjusted means of the dependent variable would be equal for each level of the independent variable, controlling for the covariate) was about 93%, which would be considered more than sufficient power (sufficient power is often .80 or above). Note that this value differs slightly than that reported in SPSS. Keep in mind that conducting power analysis *a priori* is recommended so that you avoid a situation where, post hoc, you find that the sample size was not sufficient to reach the desired level of power (given the observed parameters).

4.5.2 *A Priori* Power for ANCOVA Using G*Power

For *a priori* power, we can determine the total sample size needed for the main effects and/or interactions given an estimated effect size *f*, alpha level, desired power, numerator degrees of freedom (i.e., number of categories of our independent variable and/or interaction, depending on which *a priori* power we are interested in and depending on the number of independent variables), number of groups (i.e., the number of categories of the independent variable *in the case of only one independent variable* OR the product of the number of levels of the independent variables *in the case of multiple independent variables*), and the number of covariates. We follow Cohen's (1988) conventions for effect size (i.e.,

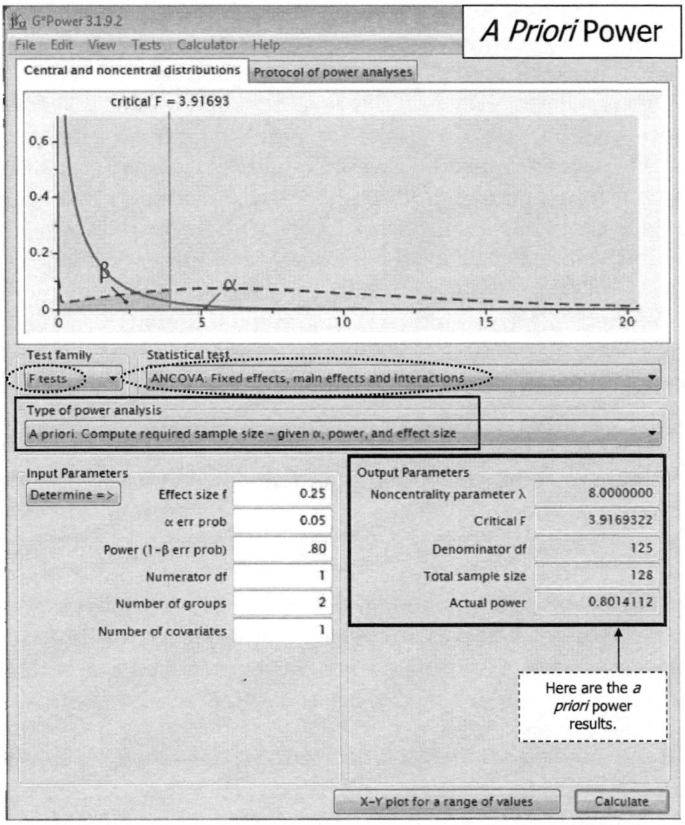

FIGURE 4.34
A priori power

small f = .10; moderate f = .25; large f = .40). In this example, had we estimated a moderate effect f of .25, alpha of .05, desired power of .80, numerator degrees of freedom of one (two categories in our independent variable, thus 2 − 1 = 1), number of groups of two (i.e., there is only one independent variable and there were two categories), and one covariate, we would need a total sample size of 128.

4.6 Research Question Template and Example Write-Up

Finally we come to an example paragraph of the results for the physical performance example. Recall that our graduate research assistant, Addie, was assisting Dr. Waung, the university's director of the Exercise Physiology and Wellness Institute, with an experimental study to determine if there was a mean difference in self-rated physical performance based on caffeine use in an attempt to facilitate improved athletic performance. Twelve athletes ingested either caffeinated (treatment) or decaffeinated (control) beverage prior to physical activity. Prior to random assignment to sections, participants were also measured on mental fatigue. After random assignment, participants completed a 2000-meter self-paced jog and were then asked to self-rate their physical performance. She was looking to see if there was a mean difference in physical performance based on the treatment group (two categories: ingesting caffeinated beverage or decaffeinated beverage prior to the self-paced jog) while controlling for mental fatigue. Her research question was: *Is there a mean difference in self-rated physical performance based on caffeine ingestion, controlling for mental fatigue?* Addie then generated an ANCOVA as the test of inference. A template for writing a research question for ANCOVA is presented below. This is illustrated assuming a one-factor (i.e., one independent variable) model, but it can easily be extended to two or more factors. As we noted in previous chapters, it is important to be sure the reader understands the levels or groups of the independent variables. This may be done parenthetically in the actual research question, as an operational definition, or specified within the methods section. In this example, parenthetically we could have stated the following: *Is there a mean difference in self-rated physical performance based on caffeine ingestion (caffeinated beverage or decaffeinated beverage), controlling for mental fatigue?*

Is there a mean difference in [dependent variable] based on [independent variable], controlling for [covariate]?

It may be helpful to preface the results of the ANCOVA with information on an examination of the extent to which the assumptions were met (recall that we tested several assumptions: (a) independence of observations; (b) homogeneity of variance; (c) normality; (d) linearity; (e) independence of the covariate and the independent variable; and (f) homogeneity of regression slopes.

An analysis of covariance (ANCOVA) was conducted to determine if the mean physical performance differed based on caffeine ingestion (caffeinated or decaffeinated beverage), while controlling for mental fatigue. The assumptions of ANCOVA, including independence, homogeneity of variance, normality, linearity, independence of the covariate and independent variable, and homogeneity of regression slopes were examined.

Independence of observations was met by random assignment of athletes to group. This assumption was also confirmed by review of a scatterplot of residuals against the levels of the independent variable. A random display of points around zero provided further evidence that the assumption of independence was met.

According to Levene's test, the **homogeneity of variance** assumption was not satisfied [F (1, 10) = 6.768, p = .026]. However, heterogeneity is less problematic with a balanced design *and* when the assumption of normality holds, as is the case with this study.

The assumption of **normality** was tested and met via examination of the residuals. Review of the Shapiro-Wilk test for normality (overall SW = .965, df = 12, p = .854; treatment SW = .902, df = 6, p = .383; control SW = .911, df = 6, p = .443)) and skewness (overall −.237; treatment = −.076; control = −1.296) and kurtosis (overall, −1.024; treatment = −2.303; control = 2.015) statistics generally suggested that normality was a reasonable assumption (though kurtosis was a bit high). Additional tests, including D'Agostino's test for skewness (z = −.390, p = .697) and the Bonett-Seier test for Geary's kurtosis (overall, z = −1.051, p = .293; treatment, z = −1.949, p = .0513; control, z = .268, p = .788) suggested evidence of normality. The boxplots by group suggested a relatively normal distributional shape (with no outliers) of the residuals. The histograms by group suggested some non-normality, but that was expected given the small sample size. The Q-Q plots suggested normality was reasonable. In general, there is evidence that normality has been met.

Linearity of the dependent variable with the covariate was examined with scatterplots, both overall and by group of the independent variable. Overall, the scatterplot of the dependent variable with the covariate suggested a positive linear relationship. This same pattern was present for the scatterplot of the dependent variable with the covariate when disaggregated by the categories of the independent variables.

Independence of the covariate and independent variable was met by random assignment of athletes to treatment group. This assumption was also confirmed by an independent t test which examined the mean difference on the covariate (i.e., mental fatigue) by independent variable (i.e., caffeinated or decaffeinated beverage ingestion). The results were not statistically significant, $t(10)$, = 1.604, p = .140, which further confirms evidence of independence of the covariate and independent variable. There was not a mean difference in physical performance based on whether or not the athlete ingested caffeine prior to the self-paced jog.

Homogeneity of regression slopes was suggested by similar regression lines evidenced in the scatterplots of the dependent variable and covariates by group (reported earlier as evidence for linearity). This assumption was confirmed by a nonstatistically significant interaction of aptitude by group, F (1, 8) = .000, p = 1.000.

Here is an APA-style example paragraph of results for the ANCOVA (remember that this will be prefaced by the previous paragraph reporting the extent to which the ANCOVA assumptions were met).

The results of the ANCOVA suggest a statistically significant effect of the covariate, mental fatigue, on the dependent variable, physical performance ($F_{fatigue}$ = 21.961; df = 1,9; p = .001). More importantly, there is a statistically significant effect for treatment group (F_{group} = 11.372; df = 1,9; p = .008), with a large effect size and strong power (ω^2 = .465, observed power = .850). The effect size suggests that about 47% of the variance in physical performance can be accounted for by treatment group when controlling for mental fatigue. Follow-up tests were conducted to evaluate the pairwise differences among the adjusted means of physical performance by treatment group. [Had there been more than two groups, including a statement similar to this would be needed: Follow-up tests were conducted to evaluate the pairwise differences among the adjusted means of physical performance by treatment group.]

The *unadjusted* group physical performance mean (i.e., prior to controlling for mental fatigue) was larger for the caffeinated group (M = 4.00, SD = 2.10) as compared to the decaffeinated group (M = 3.50, SD = 1.87) by only .50. However, the *adjusted mean* for the caffeinated group (M = 4.814, SE = .423) as compared to the decaffeinated group (M = 2.686, SE = .423) was larger by 2.128. Thus, the use of the covariate resulted in a large significant difference between the treatment groups. In summary, athletes assigned to the caffeinated group outperformed athletes in the decaffeinated group on physical performance when controlling for mental fatigue.

If our independent variable had more than two groups, we would have needed to evaluate and report the results of a post hoc multiple comparison procedure when generating SPSS (recall that we asked for Bonferroni post hoc results). Even though we only had two groups, let's see how we might have presented our results:

The Bonferroni adjustment was applied to alpha to control for the risk of increased Type I error across all pairwise comparisons. Pairwise comparisons revealed that athletes in the treatment group (adjusted M = 4.814, SE = .423) had statistically greater physical performance than athletes in the control group (adjusted M = 2.686, SE = .423) (p < .008).

The following provides a template for how these results may have been written, had our analyses required them.

Follow-up tests were conducted to evaluate the pairwise differences among the adjusted means of [dependent variable] based on [independent variable]. The [post hoc procedure selected, e.g., Bonferroni] was applied to control for the risk of increased Type I error across all pairwise comparisons. Pairwise comparisons revealed [report specific results, including means and standard deviations here].

4.7 Additional Resources

This chapter has provided a preview into conducting ANCOVA. However, there are a number of areas that space limitations prevent us from delving into. For those of you who are interested in learning more about ANCOVA, or if you find yourself in a sticky situation in your analyses, you may wish to look into the following, among many other excellent resources.

- For more in-depth coverage of ANCOVA models, see Huitema (2011), Maxwell et al. (2018), and Rutherford (2011).
- The use of ANCOVA when the design contains comparisons of participants sampled from different populations (i.e., classification designs where participants are classified into two or more mutually exclusive groups according to criteria (e.g., age, gender) and this classification is then used as a between-subjects factor in subsequent analysis) (Schneider et al., 2015).
- Alternative models that can be considered for ANCOVA when interested in examining pretest-posttest effects, such as using a change, gain, or difference score and using residual scores (Kisbu-Sakarya, MacKinnon, & Aiken, 2013; Maxwell et al., 2018).

Problems

Conceptual Problems

1. Oscar wants to determine whether adults who can elect to work at home differ in their work engagement as compared to adults who are required to work in an office setting. Oscar randomly assigns 10 employees to be assigned to work in an office setting or be provided the option to work at home. After six months, Oscar measures adults on their work engagement. Is ANCOVA appropriate given this scenario?
2. Joe wants to determine whether the time to run the Magic Mountain Marathon (ratio level variable) differs, on average, for nonprofessional athletes who complete a 12-week endurance training program as compared to those who complete a 4-week endurance training program. Joe randomly assigns nonprofessional athletes to one of the two training programs. In conducting this experiment, Joe also wants to control for the number of prior marathons in which the participant has run. Is ANCOVA appropriate given this scenario?
3. Tami has generated an ANCOVA. In testing the assumptions, she reviews a scatterplot of the residuals for each category of the independent variable. For which assumption is Tami likely reviewing evidence?
   a. Homogeneity of regression slopes
   b. Homogeneity of variance
   c. Independence of observations
   d. Independence of the covariate and the independent variable
   e. Linearity
4. Wesley has generated an ANCOVA. In his model, there is one independent variable which has three categories (type of phone: Blackberry, iPhone, and Droid) and one covariate (amount of time spent on desktop or laptop computer). In testing the assumptions, he reviews a one-way ANOVA, the dependent variable being amount of time spent on desktop or laptop computer and the independent variable being type of phone. For which assumption is Wesley likely reviewing evidence?
   a. Homogeneity of regression slopes
   b. Homogeneity of variance

c. Independence of observations
d. Independence of the covariate and the independent variable
e. Linearity

5. If the correlation between the covariate X and the dependent variable Y differs markedly in the two treatment groups, it seems likely that
   a. The assumption of normality is suspect.
   b. The assumption of homogeneity of slopes is suspect.
   c. A nonlinear relation exists between X and Y.
   d. The adjusted means for Y differ significantly.

6. If for both the treatment and control groups the correlation between the covariate X and the dependent variable Y is substantial but negative, the error variation for ANCOVA as compared to that for ANOVA is
   a. Less.
   b. About the same.
   c. Greater.
   d. Unpredictably different.

7. An experiment was conducted to compare three different instructional strategies. Fifteen subjects were included in each group. The same test was administered prior to and after the treatments. If both pretest and IQ are used as covariates, what are the degrees of freedom for the error term?
   a. 2
   b. 40
   c. 41
   d. 42

8. The effect of a training program concerned with educating heart attack patients to the benefits of moderate exercise was examined. A group of recent heart attack patients was randomly divided into two groups; one group received the training program and the other did not. The dependent variable was the amount of time taken to jog three laps, with the weight of the patient after the program used as a covariate. Examination of the data after the study revealed that the covariate means of the two groups differed. Which of the following assumptions is most clearly violated?
   a. Linearity
   b. Homogeneity of slopes
   c. Independence of the treatment and the covariate
   d. Normality

9. In ANCOVA, the covariate is a variable which should have a
   a. Low positive correlation with the dependent variable
   b. High positive correlation with the independent variable
   c. High positive correlation with the dependent variable
   d. Zero correlation with the dependent variable

10. In ANCOVA, how will the correlation of zero between the covariate and the dependent variable appear?

a. Unequal group means on the dependent variable

b. Unequal group means on the covariate

c. Regression of the dependent variable on the covariate with $b_w = 0$.

d. Regression of the dependent variable on the covariate with $b_w = 1$

11. Which of the following is not a necessary requirement for using ANCOVA?

a. Covariate scores are not affected by the treatment.

b. There is a linear relationship between the covariate and the dependent variable.

c. The covariate variable is the same measure as the dependent variable.

d. Regression slopes for the groups are similar.

12. Which of the following is the most desirable situation to use ANCOVA?

a. The slope of the regression line equals zero.

b. The variance of the dependent variable for a specific covariate score is relatively large.

c. The correlation between the covariate and the dependent variable is −.95.

d. The correlation between the covariate and the dependent variable is .60.

13. A group of students was randomly assigned to one of three instructional strategies. Data from the study indicated an interaction between slope and treatment group. It seems likely that

a. The assumption of normality is suspect.

b. The assumption of homogeneity of slopes is suspect.

c. A nonlinear relation exists between *X* and *Y*.

d. The covariate is not independent of the treatment.

14. If the mean on the dependent variable GPA (*Y*) for persons of middle social class (*X*) is higher than for persons of lower and higher social classes, one would expect that

a. The relationship between X and Y is curvilinear.

b. The covariate X contains substantial measurement error.

c. GPA is not normally distributed.

d. Social class is not related to GPA.

15. If both the covariate and the dependent variable are assessed after the treatment has been concluded, and if both are affected by the treatment, the use of ANCOVA for these data would likely result in

a. An inflated *F* ratio for the treatment effect

b. An exaggerated difference in the adjusted means

c. An underestimate of the treatment effect

d. An inflated value of the slope b_w

16. When the covariate correlates +.5 with the dependent variable, I assert that the adjusted MS_{with} from the ANCOVA will be less than the MS_{with} from the ANOVA. Am I correct?

17. For each of two groups, the correlation between the covariate and the dependent variable is substantial, but negative in direction. I assert that the error variance for ANCOVA, as compared to that for ANOVA, is greater. Am I correct?

18. True or false? In ANCOVA, X is known as a factor.

19. A study was conducted to compare six types of diets. Twelve subjects were included in each group. Their weights were taken prior to and after treatment. If pre-weight is used as a covariate, what are the degrees of freedom for the error term?

a. 5

b. 65

c. 66

d. 71

20. A researcher conducts both a one-factor ANOVA and a one-factor ANCOVA on the same data. In comparing the adjusted group means to the unadjusted group means, they find that for each group, the adjusted mean is equal to the unadjusted mean. I assert that the researcher must have made a computational error. Am I correct?

21. The correlation between the covariate and the dependent variable is zero. I assert that ANCOVA is still preferred over ANOVA. Am I correct?

22. If there is a nonlinear relationship between the covariate X and the dependent variable Y, then it is very likely that

a. There will be less reduction in SS_{with}.

b. The group effects will be biased.

c. The correlation between X and Y will be smaller in magnitude.

d. All of the above.

23. Which of the following assumptions is not shared between ANCOVA and ANOVA?

a. Homogeneity of variance

b. Independence

c. Linearity

d. Normality

24. The assumption of normality in ANCOVA is concerned with the distributional shape of which one of the following?

a. Covariate

b. Dependent variable

c. Independent variable

d. Residuals

25. The regression of the dependent variable on the covariate is assumed to be which one of the following?

a. Independent

b. Homogenous

c. Linear

d. Multicollinear

26. If units have been randomly assigned to conditions within the independent variable, which one of the following assumptions have likely been met?
    a. Homogeneity of variance
    b. Independence of the covariate and independent variable
    c. Linearity
    d. Normality
27. True or false? In ANCOVA, the independent variable is statistically adjusted to remove the effects of that part of uncontrolled variation in the covariate.

Answers to Conceptual Problems

1. **No** (there is no covariate mentioned for which to control)
3. **c** (evidence of meeting the assumption of independence can be examined by a scatterplot of residuals by group or category of the independent variable; a random display of points suggests the assumption is met)
5. **b** (see discussion on homogeneity of regression slopes)
7. **b** (14 *df* per group, 3 groups, 42 *df* – 2 *df* for covariates = 40)
9. **c** (want covariate having a high correlation with the dependent variable)
11. **c** (the covariate and dependent variable need not be the same measure; could be pretest and posttest, but it does not have to be)
13. **b** (an interaction indicates that the regression lines are not parallel across the groups)
15. **c** (a post hoc covariate typically results in an underestimate of the treatment effect, due to confounding or interference of the covariate)
17. **No** (if the correlation is substantial, then error variance will be reduced in ANCOVA regardless of its sign)
19. **b** (11 *df* per group, 6 groups, 66 *df* – 1 *df* for covariate = 65)
21. No (there will be no adjustment due to the covariate and one *df* will be lost from the error term)
23. **c** (linearity is an assumption applicable to ANCOVA but not ANOVA)
25. **c** (regression of the dependent variable on the covariate is assumed to be linear in ANCOVA)
27. **False** (in ANCOVA, the *dependent variable*, not the independent variable, is statistically adjusted to remove the effects of that part of uncontrolled variation in the covariate)

Computational Problems

1. Consider the analysis of covariance situation where the dependent variable Y is the posttest of an achievement test and the covariate X is the pretest of the same test. Given the data that follow, where there are three groups, (a) calculate the adjusted Y values assuming that $b_w = 1.00$, and (b) determine what effects the adjustment had on the posttest results.

Group	X	$\bar{X}$	Y	$\bar{Y}$
	40		120	
1	50	50	125	125
	60		130	
	70		140	
2	75	75	150	150
	80		160	
	90		160	
3	100	100	175	175
	110		190	

2. Malani wants to determine whether children whose preschool classroom has a window differ in their receptive vocabulary as compared to children whose classroom does not have a window. At the beginning of the school year, Malani randomly assigns 10 children at Sunshine Raindrop Preschool to one of two different classrooms: one classroom which has a window that looks out onto a grassy area or another classroom that has no windows. At the end of the school year, Malani measures children on their receptive vocabulary. Below are two independent random samples (classroom with and without window) of paired values on the covariate (X; receptive vocabulary measured at beginning of school year) and the dependent variable essay score (Y; receptive vocabulary measured at the end of the school year). Conduct an analysis of variance on Y, an analysis of covariance on Y using X as a covariate, and compare the results ($\alpha = .05$). Determine the unadjusted and adjusted means.

Classroom With Window		Classroom Without Window	
X	Y	X	Y
80	105	80	95
75	100	85	100
85	105	90	105
70	100	85	100
90	110	95	105

3. Below are four independent random samples (different methods of instruction) of paired values on the covariate IQ (X) and the dependent variable essay score (Y). Conduct an analysis of variance on Y, an analysis of covariance on Y using X as a covariate, and compare the results ($\alpha = .05$). Determine the unadjusted and adjusted means.

Group 1		Group 2		Group 3		Group 4	
X	Y	X	Y	X	Y	X	Y
94	14	80	38	92	55	94	24
96	19	84	34	96	53	94	37
98	17	90	43	99	55	98	22
100	38	97	43	101	52	100	43

Group 1		Group 2		Group 3		Group 4	
X	*Y*	*X*	*Y*	*X*	*Y*	*X*	*Y*
102	40	97	61	102	35	103	49
105	26	112	63	104	46	104	24
109	41	115	93	107	57	104	41
110	28	118	74	110	55	108	26
111	36	120	76	111	42	113	70
130	66	120	79	118	81	115	63

4. A communications researcher wants to know which of five versions of commercials for a new television show is most effective in terms of viewing likelihood. Each commercial is viewed by six students. A one-factor ANCOVA was used to analyze these data where the covariate was amount of television previously viewed per week. Complete the ANCOVA summary table below ($\alpha = .05$).

Source	*SS*	*df*	*MS*	*F*	Critical Value	Decision
Between adjusted	96	–	–	–	–	–
Within adjusted	192	–	–			
Covariate	–	–	–	–	–	–
Total	328	–				

5. Dr. Lee wants to determine whether prescribed workouts improve quality of movement for pre-professional athletes. College athletes were randomly assigned to either a specific prescribed movement plan (treatment) or instructed to choose their own workout (comparison group). Quality of movement was measured prior to random assignment and after 6 weeks of participation in the study. The data appear below, with athletes randomly assigned to treatment group and measured on the covariate (*X*; baseline quality of movement) and the dependent variable (*Y*; quality of movement at 6 weeks post treatment). Compute ANCOVA on *Y* using *X* as a covariate, and determine the results to the null hypothesis that the adjusted means are all equal ($\alpha = .05$). Should there be a statistically significant group effect, indicate which group had the higher mean quality of movement.

Group	Baseline (X)	Post (Y)
Prescribed movement plan	30	61
Prescribed movement plan	35	57
Prescribed movement plan	40	58
Prescribed movement plan	36	53
Prescribed movement plan	38	56
Comparison	42	49
Comparison	34	46
Comparison	39	45
Comparison	43	48
Comparison	41	47

Answers to Computational Problems

1. The adjusted group means are all equal to 150; this resulted because the adjustment moved the mean for Group 1 up to 150 and the mean for Group 3 down to 150.
3. ANOVA results: $SS_{betw} = 4{,}763.275$, $SS_{with} = 9{,}636.7$, $df_{betw} = 3$, $df_{with} = 36$, $MS_{betw} = 1{,}587.758$, $MS_{with} = 267.686$, $F = 5.931$, critical value approximately 2.88 (reject H_0).

   Unadjusted means in order: 32.5, 60.4, 53.1, 39.9.

   ANCOVA results: $SS_{betw} = 5{,}402.046$, $SS_{with} = 3{,}880.115$, $df_{betw} = 3$, $df_{with} = 35$, $MS_{betw} = 1{,}800.682$, $MS_{with} = 110.8604$, $F = 16.24$, critical value approximately 2.88 (reject H_0), $SS_{cov} = 5{,}117.815$, $F_{cov} = 46.164$, critical value approximately 4.12 (reject H_0).

   Adjusted means in order: 30.7617, 61.2544, 53.1295, 40.7544.
5. The results of the ANCOVA suggest a statistically significant effect of the covariate, baseline quality of movement, on the dependent variable post-treatment quality of movement ($F_{baseline} = 12.469$, $p = .010$). Additionally, there is a statistically significant effect for workout plan ($F_{workoutplan} = 27.792$, $p = .001$) with prescribed workouts producing greater average quality of movement ($M_{prescribed} = 56.870$, $SE_{prescribed} = 1.215$; $M_{comparison} = 47.130$, $SE_{comparison} = 1.215$).

Interpretive Problems

1. Using the same data you selected for the first interpretative problem for Chapter 1, select an appropriate covariate and then generate a one-factor ANCOVA (including testing the assumptions of both the ANOVA and ANCOVA). Compare and contrast the results of the ANOVA and ANCOVA. Which method would you select and why?
2. Using the same data you selected for the second interpretative problem for Chapter 1, select an appropriate covariate and then generate a one-factor ANCOVA (including testing the assumptions of both the ANOVA and ANCOVA). Compare and contrast the results of the ANOVA and ANCOVA. Which method would you select and why?
3. Using the same data you selected for the third interpretative problem for Chapter 1, select an appropriate covariate and then generate a one-factor ANCOVA (including testing the assumptions of both the ANOVA and ANCOVA). Compare and contrast the results of the ANOVA and ANCOVA. Which method would you select and why?

5

Random- and Mixed-Effects Analysis of Variance Models

Chapter Outline

Key Concepts

1. Fixed-, random-, and mixed-effects models
2. Repeated measures models
3. Compound symmetry/sphericity assumption
4. Friedman repeated measures test based on ranks
5. Split-plot or mixed designs (i.e., both between- and within-subjects factors)

In this chapter we continue our discussion of the analysis of variance (ANOVA) by considering models in which there is a random-effects factor, previously introduced in Chapter 1. These models include the one-factor and factorial designs, as well as repeated measures designs. As becomes evident, repeated measures designs are used when there is at least one factor where each individual is exposed to all levels of that factor. This factor is referred to as a **repeated factor**, for obvious reasons. This chapter is mostly concerned with one- and two-factor random-effects models, the two-factor mixed-effects model, and one- and two-factor repeated measures designs.

It should be noted that effect size measures, power, and confidence intervals can be determined in the same fashion for the models in this chapter as for previously described ANOVA models. The standard effect size measures already described are applicable (i.e.,

ω^2 and η^2), although the intraclass correlation coefficient, ρ_I, can be utilized for random effects (similarly interpreted). For additional discussion of these issues in the context of this chapter, see Cohen (1988), Fidler and Thompson (2001), Keppel and Wickens (2004), Murphy, Myors, and Wolach (2009), Wilcox (2003), and Wilcox (1996)

Many of the concepts used in this chapter are the same as those covered in Chapters 1 through 4. In addition, the following new concepts are addressed: random- and mixed-effects factors, repeated measures factors, the compound symmetry/sphericity assumption, and mixed designs. Our objectives are that by the end of this chapter, you will be able to (a) understand the characteristics and concepts underlying random- and mixed-effects ANOVA models, (b) determine and interpret the results of random- and mixed-effects ANOVA models, and (c) understand and evaluate the assumptions of random- and mixed-effects ANOVA models.

5.1 The One-Factor Random-Effects Model

This section describes the distinguishing characteristics of the one-factor random-effects ANOVA model, the linear model, the ANOVA summary table and expected mean squares, assumptions and their violation, and multiple comparison procedures.

5.1.1 Characteristics of the Model

The characteristics of the one-factor *fixed-effects* ANOVA model have already been covered in Chapter 1. These characteristics include (a) one factor (or independent variable) with two or more levels, (b) all levels of the factor of interest are included in the design (i.e., a fixed-effects factor), (c) subjects are randomly assigned to one level of the factor, and (d) the dependent variable is measured at least at the interval level. Thus, the overall design is a fixed-effects model, where there is one factor and the individuals respond to only one level of the factor. If individuals respond to more than one level of the factor, then this is a repeated measures design, as shown later in this chapter.

The characteristics of the one-factor *random-effects* ANOVA model are the same with one obvious exception. This has to do with the selection of the levels of the factor. In the fixed-effects case, researchers select all of the levels of interest because they are interested only in making generalizations (or inferences) about those particular levels. Thus, in replications of this design, each replicate would use precisely the same levels. Considering analyses that are conducted on individuals, examples of factors that are typically fixed include socioeconomic status, sex, specific types of drug treatment, age group, weight, or marital status.

In the random-effects case, researchers randomly select levels from the population of levels because they are interested in making generalizations (or inferences) about the entire population of levels, not merely those that have been sampled. Thus in replications of this design, each replicate need not have the same levels included. The concept of random selection of factor levels from the population of levels is the same as the random selection of subjects from the population. Here the researcher is making an inference from the sampled levels to the population of levels, instead of making an inference from the sample of individuals to the population of individuals. In a random-effects design, then, a random sample of factor levels is selected in the same way as a random sample of individuals is selected.

For instance, a researcher interested in instructor effectiveness may have randomly sampled instructors from one discipline (i.e., the independent variable) from the population of instructors in a university system. Generalizations can then be made about all instructors in that university system that could have been sampled. Other examples of factors that are typically random include *randomly selected* preexisting groups such as classrooms, organizations, buildings, observers or raters, or time (seconds, minutes, hours, days, weeks, etc.). It should be noted that in many settings, the random selection of groups or units such as organizations, schools, or classrooms is not often possible, as those decisions are not under the researcher's control. Here we would need to consider such factors as fixed rather than random effects.

5.1.2 The ANOVA Model

The one-factor ANOVA random-effects model is written in terms of population parameters as

$$Y_{ij} = \mu + a_j + \varepsilon_{ij}$$

where Y_{ij} is the observed score on the dependent variable for individual i in level j of factor A, μ is the overall or grand population mean, a_j is the random effect for level j of factor A, and ε_{ij} is the random residual error for individual i in level j. The residual error can be due to individual differences, measurement error, and/or other factors not under investigation. Note that we use a_j to designate the random effects to differentiate them from α_j in the fixed-effects model.

Because the random-effects model consists of only a sample of the effects from the population, the sum of the sampled effects is not necessarily zero. For instance, we may select a sample having only positive effects (e.g., all very effective instructors). If the entire population of effects were examined, then the sum of those effects would indeed be zero.

For the one-factor random-effects ANOVA model, the hypotheses for testing the effect of factor A are written in terms of equality of the variances among the means of the random levels, as follows (i.e., the means for each level are about the same and thus the variability among those means is about zero). It should be noted that the sign for the alternative hypothesis is "greater than," reflecting the fact that the variance cannot be negative.

$$H_0 : \sigma_a^2 = 0$$

$$H_1 : \sigma_a^2 > 0$$

Recall for the one-factor fixed-effects ANOVA model that the hypotheses for testing the effect of factor A are written in terms of equality of the means of the groups (as presented here):

$$H_0 : \mu_{.1} = \mu_{.2} = \ldots = \mu_{.J}$$

$$H_1 : not\ all\ the\ \mu_{.J}\ are\ equal$$

This reflects the difference in the inferences made in the random- and fixed-effects models. In the fixed-effects case, the null hypothesis is about specific population means; in the random-effects case, the null hypothesis is about variation among the entire population of means. As becomes evident, the difference in the models is reflected in the multiple comparison procedures.

5.1.3 ANOVA Summary Table and Expected Mean Squares

Here there are very few differences between the one-factor random-effects and one-factor fixed-effects models. The sources of variation are still A (or between), within, and total. The sums of squares, degrees of freedom, mean squares, *F* test statistic, and critical value are determined in the same way as in the fixed-effects case. Obviously then, the ANOVA summary table looks the same as well. Using the example from Chapter 1, assuming the model is now a random-effects model, we obtain a test statistic $F = 6.8177$, which is again significant at the .05 level.

As in Chapters 1 and 3, the formation of a proper *F* ratio is related to the expected mean squares. If H_0 is actually *true*, then the *expected mean squares* are as follows:

$$E(MS_A) = \sigma_\varepsilon^2$$

$$E(MS_{with}) = \sigma_\varepsilon^2$$

and thus the ratio of expected mean squares is:

$$\frac{E(MS_A)}{E(MS_{with})} = 1$$

where the expected value of *F* is $E(F) = df_{with} / (df_{with} - 2)$, and σ_ε^2 is the population variance of the residual errors.

If H_0 is actually *false*, then the expected mean squares are as follows:

$$E(MS_A) = \sigma_\varepsilon^2 + n\sigma_a^2$$

$$E(MS_{with}) = \sigma_\varepsilon^2$$

and thus the ratio of the expected mean squares is as follows:

$$\frac{E(MS_A)}{E(MS_{with})} > 1$$

where $E(F) > df_{with} / (df_{with} - 2)$ and σ_a^2 is the population variance of the levels of factor A. Thus the important part of $E(MS_A)$ is the magnitude of the second term $n\sigma_a^2$.

As in previous ANOVA models, the proper *F* ratio should be formed as follows:

F = (*systematic variability* + *error variability*) / *error variability*

For the one-factor random-effects model, the only appropriate *F* ratio is MS_A / MS_{with} because it does serve to isolate the systematic variability (i.e., the variability between the levels or groups in factor A, the independent variable). That is, the within term must be utilized as the error term in the *F* ratio.

5.1.4 Assumptions and Violation of Assumptions

In Chapter 1 we described the assumptions for the one-factor fixed-effects model. The assumptions are nearly the same for the one-factor random-effects model, and we need not devote much attention to them here. In short, the assumptions are again concerned with the distribution of the dependent variable scores, specifically that scores are random and independent, coming from normally distributed populations with equal population variances. The effect of assumption violations and how to deal with them have been thoroughly discussed in Chapter 1 (although see, for example, Wilcox, 2003, for alternative procedures when variances are unequal).

Additional assumptions must be made for the random-effects model. These assumptions deal with the effects for the levels of the independent variable, the a_j. First, here are a few words about the a_j. The random group effects a_j are computed, in the population, by the following:

$$a_j = \mu_{.j} - \mu_{..}$$

For example, a_3 represents the effect for being a member of Group 3. If the overall mean $\mu_{..}$ is 60 and the mean of Group 3 (i.e., $\mu_{.3}$) is 100, then the group effect would be

$$a_3 = \mu_{.3} - \mu_{..} = 100 - 60 = 40$$

In other words, the effect for being a member of Group 3 is an increase of 40 points over the overall mean.

The assumptions are that the a_j group effects are randomly and independently sampled from the normally distributed population of group effects, with a population mean of zero and a population variance of σ_a^2. Stated another way, there is a population of group effects out there from which we are taking a random sample. For example, with teacher as the factor of interest, we are interested in examining the effectiveness of teachers as measured by academic performance of students in their class. We take a random sample of teachers from the population of second-grade teachers. For these teachers we measure their effectiveness in the classroom via student performance and generate an effect for each teacher (i.e., the a_j). These effects indicate the extent to which a particular teacher is more or less effective than the population average of teachers. Their effects are known as random effects as the teachers are randomly selected. In selecting teachers, each teacher is selected independently of all other teachers to prevent a biased sample.

The effects of the violation of the assumptions about the a_j are the same as with the dependent variable scores. The F test is quite robust to nonnormality of the a_j terms, and unequal variances of the a_j terms. However, the F test is quite sensitive to nonindependence among the a_j terms, with no known solutions. A summary of the assumptions and the effects of their violation for the one-factor random-effects model is presented in Table 5.1.

TABLE 5.1

Assumptions and Effects of Violations: One-Factor Random-Effects Model

Assumption	Effect of Assumption Violation
Independence	• Increased likelihood of a Type I and/or Type II error in F • Affects standard errors of means and inferences about those means
Homogeneity of variance	• Bias in SS_{with}; increased likelihood of a Type I and/or Type II error • Small effect with equal or nearly equal n's; otherwise effect decreases as n increases
Normality	• Minimal effect with equal or nearly equal n's

5.1.5 Multiple Comparison Procedures

Let us think for a moment about the use of multiple comparison procedures for the random-effects model. In general, the researcher is not usually interested in making inferences about just the levels of A that were sampled. Thus, estimation of the a_j terms does not provide us with any information about the a_j terms that were not sampled. Also, the a_j terms cannot be summarized by their mean, as they do not necessarily sum to zero for the levels sampled, only for the population of levels.

5.2 The Two-Factor Random-Effects Model

In this section, we describe the distinguishing characteristics of the two-factor random-effects ANOVA model, the linear model, the ANOVA summary table and expected mean squares, assumptions of the model and their violation, and multiple comparison procedures.

5.2.1 Characteristics of the Model

The characteristics of the one-factor random-effects ANOVA model have already been covered in this chapter, and those of the two-factor fixed-effects model in Chapter 3. Here we extend and combine these characteristics to form the two-factor random-effects model. These characteristics include (a) two factors (or independent variables) each with two or more levels, (b) the levels of each of the factors are randomly sampled from the population of levels (i.e., two random-effects factors), (c) subjects are randomly assigned to one combination of the levels of the two factors, and (d) the dependent variable is measured at least at the interval level. Thus the overall design is a random-effects model, with two factors, and the individuals respond to only one combination of the levels of the two factors (note that this is not a popular model in education and the behavioral sciences; in factorial designs we typically see a random-effects factor with a fixed-effects factor). If individuals respond to more than one combination of the levels of the two factors, then this is a repeated measures design (discussed later in this chapter).

5.2.2 The ANOVA Model

The two-factor ANOVA random-effects model is written in terms of population parameters as

$$Y_{ijk} = \mu + a_j + b_k + (ab)_{jk} + \varepsilon_{ijk}$$

where Y_{ijk} is the observed score on the dependent variable for individual i in level j of factor A and level k of factor B (or in the jk cell), μ is the overall or grand population mean (i.e., regardless of cell designation), a_j is the random effect for level j of factor A (row effect), b_k is the random effect for level k of factor B (column effect), $(ab)_{jk}$ is the interaction random effect for the combination of level j of factor A and level k of factor B, and ε_{ijk} is the random residual error for individual i in cell jk. The residual error can be due to individual differences, measurement error, and/or other factors not under investigation. Note that we use a_j, b_k, and $(ab)_{jk}$ to designate the random effects to differentiate them from the α_j, β_k, and $(\alpha\beta)_{jk}$ in the fixed-effects model. Finally, there is no requirement that the sum of the main or interaction effects is equal to zero as only a sample of these effects are taken from the population of effects.

There are three sets of hypotheses, one for each of the two main effects and one for the interaction effect. The null and alternative hypotheses, respectively, for testing the main effect of factor A (i.e., independent variable A) follows. The null hypothesis tests whether the variance among the means for the random effect of independent variable A is equal to zero (i.e., the means for each level of factor A are about the same; thus, the variability among those means is about zero). It should be noted that the sign for the alternative hypothesis is "greater than," reflecting the fact that the variance cannot be negative.

$$H_{01}: \sigma_a^2 = 0$$

$$H_{11}: \sigma_a^2 > 0$$

The hypotheses for testing the main effect of factor B (i.e., independent variable B) similarly test whether the variance among the means for the random effect of independent variable B is equal to zero (i.e., the means for each level of factor B are about the same and thus the variability among those means is about zero). It should be noted that the sign for the alternative hypothesis is "greater than," reflecting the fact that the variance cannot be negative.

$$H_{02}: \sigma_b^2 = 0$$

$$H_{12}: \sigma_b^2 > 0$$

Finally, the hypotheses for testing the interaction effect are presented next. In this case, the null hypothesis tests whether the variance among the means for the interaction of the random effects of factors A and B is equal to zero (i.e., the means for each AB cell are about the same and thus the variability among those means is about zero). It should be noted that the sign for the alternative hypothesis is "greater than," reflecting the fact that the variance cannot be negative.

$$H_{03}: \sigma_{ab}^2 = 0$$

$$H_{13}: \sigma_{ab}^2 > 0$$

These hypotheses again reflect the difference in the inferences made in the random- and fixed-effects models. In the fixed-effects case, the null hypotheses are about means, whereas in the random-effects case, the null hypotheses are about *variation* among the means.

5.2.3 ANOVA Summary Table and Expected Mean Squares

Here there are very few differences between the two-factor fixed-effects and random-effects models. The sources of variation are still A, B, AB, within, and total. The sums of squares, degrees of freedom, and mean squares are determined the same as in the fixed-effects case.

However, the F test statistics are different due to the expected mean squares, as are the critical values used. The F test statistics are formed for the test of factor A (i.e., the main effect for independent variable A) as follows:

$$F = \frac{MS_A}{MS_{AB}}$$

for the test of factor B (i.e., the main effect for independent variable B) as presented here:

$$F = \frac{MS_B}{MS_{AB}}$$

and for the test of the AB interaction as indicated:

$$F = \frac{MS_{AB}}{MS_{with}}$$

Recall that in the fixed-effects model, the MS_{with} was used as the error term for all three hypotheses. However, in the random-effects model, the MS_{with} is used as the error term *only* for the test of the interaction. The MS_{AB} is used as the error term for the tests of both main effects. The critical values used are those based on the degrees of freedom for the numerator and denominator of each hypothesis tested. Thus, using the example from Chapter 3, assuming that the model is now a random-effects model, we obtain the following as our test statistic for the test of factor A (i.e., the main effect for independent variable A):

$$F_A = \frac{MS_A}{MS_{AB}} = \frac{246.1979}{7.2813} = 33.8124$$

for the test of factor B, the test statistic is computed as follows:

$$F_B = \frac{MS_B}{MS_{AB}} = \frac{712.5313}{7.2813} = 97.8577$$

and for the test of the AB interaction, we find the following:

$$F_{AB} = \frac{MS_{AB}}{MS_{with}} = \frac{7.2813}{11.5313} = 0.6314$$

The critical value for the test of factor A is found in the F table of Appendix Table A.4 as ${}_{\alpha}F_{J-1,(J-1)(K-1)}$, which for the example is ${}_{.05}F_{3,3} = 9.28$, and is significant at the .05 level. The critical value for the test of factor B is found in the F table as ${}_{\alpha}F_{K-1,(J-1)(K-1)}$, which for the example is ${}_{.05}F_{1,3} = 10.13$, and is significant at the .05 level. The critical value for the test of the interaction is found in the F table as ${}_{\alpha}F_{(J-1)(K-1),N-JK}$, which for the example is ${}_{.05}F_{3,24} = 3.01$, and is not significant at the .05 level. It just so happens for the example data that the results for the random- and fixed-effects models are the same. This will not always be the case.

The formation of the proper F ratios is again related to the expected mean squares. Recall that our hypotheses for the two-factor random-effects model are based on variation among the means of the random effects (rather than the means as seen in the fixed-effects case). If

H_0 is actually *true* (i.e., there is no variation among the means of the random effects), then the *expected mean squares* are all equals, as noted as follows:

$$E(MS_A) = \sigma_\varepsilon^2$$

$$E(MS_B) = \sigma_\varepsilon^2$$

$$E(MS_{AB}) = \sigma_\varepsilon^2$$

$$E(MS_{with}) = \sigma_\varepsilon^2$$

where σ_ε^2 is the population variance of the residual errors.

If H_0 is actually *false* (i.e., there *is* variation among the means of the random effects), then the expected mean squares are as follows:

$$E(MS_A) = \sigma_\varepsilon^2 + n\sigma_{ab}^2 + Kn\sigma_a^2$$

$$E(MS_B) = \sigma_\varepsilon^2 + n\sigma_{ab}^2 + Jn\sigma_b^2$$

$$E(MS_{AB}) = \sigma_\varepsilon^2 + n\sigma_{ab}^2$$

$$E(MS_{with}) = \sigma_\varepsilon^2$$

where σ_a^2, σ_b^2, and σ_{ab}^2. are the population variances of A, B and AB, respectively.

As in previous ANOVA models, the proper *F* ratio should be formed as follows:

$$F = (systematic\ variability + error\ variability) / error\ variability$$

For the two-factor random-effects model, the appropriate error term for the main effects is MS_{AB} and the appropriate error term for the interaction effect is MS_{with}.

5.2.4 Assumptions and Violation of Assumptions

Previously we described the assumptions for the one-factor random-effects model. The assumptions are nearly the same for the two-factor random-effects model, and we need not devote much attention to them here. As before, the assumptions are concerned with the distribution of the dependent variable scores, and of the random-effects [sampled levels of the independent variables, the a_j, b_k, and their interaction $(ab)_{jk}$]. However, there are a few new wrinkles. Little is known about the effect of unequal variances (i.e., heterogeneity) or dependence (i.e., violation of the assumption of independence) for this random-effects model, although we expect the effects to be the same as for the fixed-effects model. For violation of the normality assumption, effects are known to be substantial. A summary of the assumptions and the effects of their violation for the two-factor random-effects model is presented in Table 5.2.

5.2.5 Multiple Comparison Procedures

The story of multiple comparisons for the two-factor random-effects model is the same as that for the one-factor random-effects model. In general, the researcher is not usually interested in making inferences about just the levels of A, B, or AB that were sampled, and

TABLE 5.2

Assumptions and Effects of Violations: Two-Factor Random-Effects Model

Assumption	Effect of Assumption Violation
Independence	Little is known about the effects of dependence; however, based on the fixed-effects model, we might expect the following: • Increased likelihood of a Type I and/or Type II error in F • Affects standard errors of means and inferences about those means
Homogeneity of variance	Little is known about the effects of heteroscedasticity; however, based on the fixed-effects model, we might expect the following: • Bias in SS_{with} • Increased likelihood of a Type I and/or Type II error • Small effect with equal or nearly equal n's • Otherwise effect decreases as n increases
Normality	• Minimal effect with equal or nearly equal n's • Otherwise substantial effects

thus performing multiple comparison procedures in a two-factor random-effects model is a moot point. Thus, estimation of the a_j, b_k, or $(ab)_{jk}$ terms do not provide us with any information about the a_j, b_k, or $(ab)_{jk}$ terms that were not sampled. Also, the a_j, b_k, or $(ab)_{jk}$ terms cannot be summarized by their means, as they will not necessarily sum to zero for the levels sampled, only for the population of levels.

5.3 The Two-Factor Mixed-Effects Model

This section describes the distinguishing characteristics of the two-factor *mixed-effects* ANOVA model, the linear model, the ANOVA summary table and expected mean squares, assumptions of the model and their violation, and multiple comparison procedures.

5.3.1 Characteristics of the Model

The characteristics of the two-factor random-effects ANOVA model have already been covered in the preceding section, and those of the two-factor fixed-effects model in Chapter 3. Here we combine these characteristics to form the two-factor mixed-effects model. These characteristics include (a) two factors (or independent variables) each with two or more levels, (b) the levels for one of the factors are randomly sampled from the population of levels (i.e., the random-effects factor) and all of the levels of interest for the second factor are included in the design (i.e., the fixed-effects factor), (c) subjects are randomly selected and assigned to one combination of the levels of the two factors, and (d) the dependent variable is measured at least at the interval level. Thus, the overall design is a mixed-effects model, with one fixed-effects factor and one random-effects factor, and individuals respond to only one combination of the levels of the two factors. If individuals respond to more than one combination, then this is a repeated measures design.

5.3.2 The ANOVA Model

There are actually two variations of the two-factor mixed-effects model, one where factor A is fixed and factor B is random, and the other where factor A is random and factor B is fixed. The labeling of a factor as A or B is arbitrary, so we consider only the former variation where A is fixed and B is random. For the latter variation, merely switch the labels of the factors. The two-factor ANOVA mixed-effects model is written in terms of population parameters as

$$Y_{ijk} = \mu + \alpha_j + b_k + (\alpha b)_{jk} + \varepsilon_{ijk}$$

where Y_{ijk} is the observed score on the dependent variable for individual i in level j of factor A and level k of factor B (or in the jk cell), μ is the overall or grand population mean (i.e., regardless of cell designation), α_j is the fixed effect for level j of factor A (row effect), b_k is the random effect for level k of factor B (column effect), $(\alpha b)_{jk}$ is the interaction mixed effect for the combination of level j of factor A and level k of factor B, and ε_{ijk} is the random residual error for individual i in cell jk. The residual error can be due to individual differences, measurement error, and/or other factors not under investigation. Note that we use b_k and $(\alpha b)_{jk}$ to designate the random and mixed effects respectively to differentiate them from β_k and $(\alpha\beta)_{jk}$ in the fixed-effects model.

As shown in Figure 5.1, due to the nature of the mixed-effects model, only some of the columns are randomly selected for inclusion in the design. Each cell of the design will include row (α), column (b), and interaction (αb) effects. With an equal n's model, if we sum these effects for a given column, then the effects will sum to zero. However, if we sum these effects for a given row, then the effects will not sum to zero, as some columns were not sampled.

The null and alternative hypotheses, respectively, for testing the effect of factor A are presented below. These hypotheses reflect testing the equality of means of the levels of independent variable A (the fixed-effect).

$$H_{01} : \mu_{.1.} = \mu_{.2.} = \ldots \mu_{.J.}$$

$$H_{11} : not\ all\ the\ \mu_{.j.}\ are\ equal$$

The hypotheses for testing the effect of factor B, the random effect, follow. The null hypothesis tests whether the variance among the means for the random effect of independent

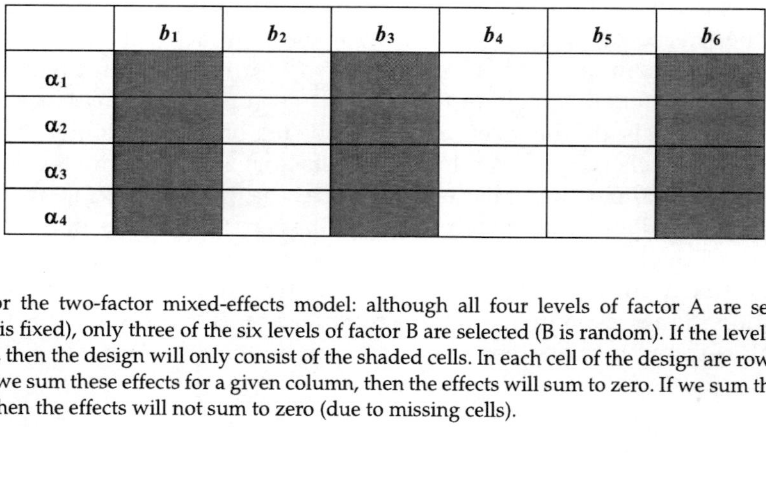

FIGURE 5.1
Conditions for the two-factor mixed-effects model: although all four levels of factor A are selected by the researcher (A is fixed), only three of the six levels of factor B are selected (B is random). If the levels of B selected are 1, 3, and 6, then the design will only consist of the shaded cells. In each cell of the design are row, column, and cell effects. If we sum these effects for a given column, then the effects will sum to zero. If we sum these effects for a given row, then the effects will not sum to zero (due to missing cells).

variable B is equal to zero (i.e., the means for each level of factor B are about the same and thus the variability among those means is about zero). It should be noted that the sign for the alternative hypothesis is "greater than," reflecting the fact that the variance cannot be negative.

$$H_{02} : \sigma_b^2 = 0$$

$$H_{12} : \sigma_b^2 > 0$$

Finally, the hypotheses for testing the interaction effect are presented next. In this case, the null hypothesis tests whether the variance among the means for the interaction of the random effects of factors A and B is equal to zero (i.e., the means for each AB cell are about the same and thus the variability among those means is about zero). It should be noted that the sign for the alternative hypothesis is "greater than," reflecting the fact that the variance cannot be negative.

$$H_{03} : \sigma_{\alpha b}^2 = 0$$

$$H_{13} : \sigma_{\alpha b}^2 > 0$$

These hypotheses reflect the difference in the inferences made in the mixed-effects model. Here we see that the hypotheses about the fixed-effect A (i.e., the main effect for independent variable A) are about *means*, whereas the hypotheses involving the random-effect B (i.e., the main effect of B and the interaction effect AB) are about *variation among the means* as these involve a random effect.

5.3.3 ANOVA Summary Table and Expected Mean Squares

There are very few differences between the two-factor fixed-effects, random-effects, and mixed-effects models. The sources of variation for the mixed-effects model are again A (the fixed effect), B (the random effect), AB (the interaction effect), within, and total. The sums of squares, degrees of freedom, and mean squares are determined the same as in the fixed-effects case. However, the *F* test statistics are different in each of these models, as well as the critical values used. The *F* test statistics are formed for the test of factor A, the fixed effect, as seen here:

$$F_A = \frac{MS_A}{MS_{AB}}$$

for the test of factor B, the random effect, is computed as follows:

$$F_B = \frac{MS_B}{MS_{with}}$$

and for the test of the AB interaction, the mixed effect, as indicated here:

$$F_{AB} = \frac{MS_{AB}}{MS_{with}}$$

Recall that in the fixed-effects model, the MS_{with} is used as the error term for all three hypotheses. However, in the random-effects model, the MS_{with} is used as the error term only for the test of the interaction, and the MS_{AB} is used as the error term for the tests of both main effects. Finally, in the mixed-effects model, the MS_{with} is used as the error term for the test of factor B (the random effect) and the interaction (i.e., AB), whereas the MS_{AB} is used as the error term for the test of factor A (the fixed effect). The critical values used are those based on the degrees of freedom for the numerator and denominator of each hypothesis tested.

Thus, using the example from Chapter 3, let us assume the model is now a mixed-effects model where factor A, the fixed effect, is the type of sport in which the athlete participates (four categories). Factor B, the random effect, is selection status (two randomly chosen categories from levels such as selected as starter, selected as second string, etc.). We obtain as our test statistic for the test of factor A, the fixed effect of type of sport, as follows:

$$F_A = \frac{MS_A}{MS_{AB}} = \frac{246.1979}{7.2813} = 33.8124$$

for the test of factor B, the random effect of selection status, the test statistic is computed as:

$$F_B = \frac{MS_B}{MS_{with}} = \frac{712.5313}{11.5313} = 61.7911$$

and for the test of the AB (fixed by random effect, type of sport by selection status) interaction, we find a test statistic as follows:

$$F_{AB} = \frac{MS_{AB}}{MS_{with}} = \frac{7.2813}{11.5313} = 0.6314$$

The critical value for the test of factor A (the fixed effect, type of sport) is found in the F table as ${}_{\alpha}F_{J-1,(J-1)(K-1)}$, which for the example is ${}_{.05}F_{3,3} = 9.28$, and is statistically significant at the .05 level. The critical value for the test of factor B (the random-effect, selection status) is found in the F table as ${}_{\alpha}F_{K-1,N-JK}$, which for the example is ${}_{.05}F_{1,24} = 4.26$, and is significant at the .05 level. The critical value for the test of the interaction between type of sport and selection status is found in the F table as ${}_{\alpha}F_{(J-1)(K-1),N-JK}$, which for the example is ${}_{.05}F_{3,24} = 3.01$, and is not significant at the .05 level. It just so happens for the example data that the results for the mixed-, random-, and fixed-effects models are the same. This is not always the case.

The formation of the proper F ratio is again related to the expected mean squares. If H_0 is actually *true* (i.e., the variance among the means is zero), then the *expected mean squares* are as follows:

$$E(MS_A) = \sigma_\varepsilon^2$$

$$E(MS_B) = \sigma_\varepsilon^2$$

$$E(MS_{AB}) = \sigma_\varepsilon^2$$

$$E(MS_{with}) = \sigma_\varepsilon^2$$

where σ_ε^2 is the population variance of the residual errors.

If H_0 is actually *false* (the variance among the means is *not* equal to zero), then the expected mean squares are as follows:

$$E(MS_A) = \sigma_\varepsilon^2 + n\sigma_{\alpha b}^2 + Kn\left[\sum_{J=1}^{J} \alpha_j^2 / (J-1)\right]$$

$$E(MS_B) = \sigma_\varepsilon^2 + Jn\sigma_b^2$$

$$E(MS_{AB}) = \sigma_\varepsilon^2 + n\sigma_{\alpha b}^2$$

$$E(MS_{with}) = \sigma_\varepsilon^2$$

where all terms have been previously defined.

As in previous ANOVA models, the proper F ratio should be formed as follows:

$$F = (systematic\ variability + error\ variability) / error\ variability$$

For the two-factor mixed-effects model, MS_{AB} must be used as the error term for the test of A, and MS_{with} must be used as the error term for the test of B and for the interaction test.

5.3.4 Assumptions and Violation of Assumptions

Previously we described the assumptions for the two-factor random-effects model. The assumptions are nearly the same for the two-factor mixed-effects model, and we need not devote much attention to them here. As before, the assumptions are concerned with the distribution of the dependent variable scores and of the random effects. However, note that not much is known about the effects of dependence or heteroscedasticity for random effects, although we expect the effects are the same as for the fixed-effects case. A summary of the assumptions and the effects of their violation for the two-factor mixed-effects model are presented in Table 5.3.

TABLE 5.3

Assumptions and Effects of Violations: Two-Factor Mixed-Effects Model

Assumption	Effect of Assumption Violation
Independence	Little is known about the effects of dependence; however, based on the fixed-effects model, we might expect the following: • Increased likelihood of a Type I and/or Type II error in F • Affects standard errors of means and inferences about those means
Homogeneity of variance	Little is known about the effects of heteroscedasticity; however, based on the fixed-effects model, we might expect the following: • Bias in SS_{with} • Increased likelihood of a Type I and/or Type II error • Small effect with equal or nearly equal n's • Otherwise effect decreases as n increases
Normality	• Minimal effect with equal or nearly equal n's • Otherwise substantial effects

5.3.5 Multiple Comparison Procedures

For multiple comparisons in the two-factor mixed-effects model, the researcher is not usually interested in making inferences about just the levels of the random-effect factor (i.e., B) or the interaction (i.e., AB) that were randomly sampled. Thus, estimation of the b_k or $(\alpha b)_{jk}$ terms does not provide us with any information about the b_k or $(\alpha b)_{jk}$ terms not sampled. Also, the b_k or $(\alpha b)_{jk}$ terms cannot be summarized by their means as they will not necessarily sum to zero for the levels sampled, only for the population of levels. However, inferences about the fixed-factor A can be made in the same way they were made for the two-factor fixed-effects model. We have already used the example data to look at some multiple comparison procedures in Chapter 3.

This concludes our discussion of random- and mixed-effects models for the one- and two-factor designs. For three-factor designs, see Keppel and Wickens (2004). In the major statistical software, the analysis of random effects can be treated as follows: in SAS PROC GLM, use the RANDOM statement to designate random effects; in SPSS GLM, random effects can also be designated, either in the point-and-click mode (by using the "Random Factor(s)" box) or in the syntax mode to designate random effects.

5.4 The One-Factor Repeated Measures Design

In this section, we describe the distinguishing characteristics of the one-factor repeated measures ANOVA model, the layout of the data, the linear model, assumptions of the model and their violation, the ANOVA summary table and expected mean squares, multiple comparison procedures, alternative ANOVA procedures, and an example.

5.4.1 Characteristics of the Model

The one-factor repeated measures model is the logical extension to the dependent t test. Although in the dependent t test there are only two measurements for each subject (e.g., the same individuals measured prior to an intervention and then again after an intervention), in the one-factor repeated measures model two *or more* measurements can be examined. The characteristics of the one-factor repeated measures ANOVA model are somewhat similar to the one-factor fixed-effects model, yet there are a number of obvious exceptions. The first unique characteristic has to do with the fact that each subject responds to each level of factor A. This is in contrast to the nonrepeated case where each subject is exposed to only one level of factor A. This design is often referred to as a **within-subjects design**, as each subject responds to each level of factor A. Thus, subjects serve as their own controls such that individual differences are taken into account. This was not the case in any of the previously discussed ANOVA models. As a result, subjects' scores are not independent across the levels of factor A. Compare this design to the one-factor fixed-effects model where total variation was decomposed into variation due to A (or between) and due to the residual (or within). In the one-factor repeated measures design, residual variation is further decomposed into variation due to subjects and variation due to the interaction between A and subjects. The reduction in the residual sum of squares yields a more powerful design as well as more precision in estimating the effects of A, and thus is more economical in that fewer subjects are necessary than in previously discussed models (Murphy et al., 2009).

The one-factor repeated measures design is also a mixed model. The subjects factor is a random effect, whereas the A factor is almost always a fixed effect. For example, if time is the fixed effect, then the researcher can examine phenomena over time. Finally, the one-factor repeated measures design is similar in some ways to the two-factor mixed-effects design except with one subject per cell. In other words, the one-factor repeated measures design is really a special case of the two-factor mixed-effects design with $n = 1$ per cell. Unequal n's can happen only when subjects miss the administration of one or more levels of factor A.

On the down side, the repeated measures design includes some risk of carry-over effects from one level of A to another because each subject responds to all levels of A. Remember that the repeated factor must be the same measure (or equated) at each measurement occasion. As examples of the carry-over effect, subjects' performance may be altered due to fatigue (decreased performance), practice (increased performance), or sensitization (increased performance) effects. These effects may be minimized by (a) counterbalancing the order of administration of the levels of A so that each subject does not receive the same order of the levels of A (this can also minimize problems with the compound symmetry assumption; see subsequent discussion), (b) allowing some time to pass between the administration of the levels of A, or (c) matching or blocking similar subjects with the assumption of subjects within a block being randomly assigned to a level of A. This last method is a type of randomized block design (see Chapter 6).

5.4.2 The Layout of the Data

The layout of the data for the one-factor repeated measures model is shown in Table 5.4. Here we see the columns designated as the levels of factor A and the rows as the subject. Thus, the columns or "levels" of factor A represent the different measurements. An example is measuring children on reading performance before, immediately after, and six months after they participate in a reading intervention. Row, column, and overall means are also shown in Table 5.4, although the subject means are seldom of any utility (and thus are not reported in research studies). Here you see that the layout of the data looks the same as the two-factor model, although there is only one observation per cell.

TABLE 5.4

Layout for the One-Factor Repeated Measures ANOVA

	Level of Factor A (Repeated Factor)				
Level of Factor S	**1**	**2**	**...**	**J**	**Row Mean**
1	Y_{11}	Y_{12}	...	Y_{1J}	$\bar{Y}_{1.}$
2	Y_{21}	Y_{22}	...	Y_{2J}	$\bar{Y}_{2.}$
...	...	...	...	...	...
n	Y_{n1}	Y_{n2}		Y_{nJ}	$\bar{Y}_{n.}$
Column Mean	$\bar{Y}_{.1}$	$\bar{Y}_{.2}$	...	$\bar{Y}_{.J}$	$\bar{Y}_{..}$

5.4.3 The ANOVA Model

The one-factor repeated measures ANOVA model is written in terms of population parameters as

$$Y_{ij} = \mu + \alpha_j + s_i + (s\alpha)_{ij} + \varepsilon_{ij}$$

where Y_{ij} is the observed score on the dependent variable for individual i responding to level j of factor A, μ is the overall or grand population mean, α_j is the fixed effect for level j of factor A, s_i is the random effect for subject i of the subject factor, $(s\alpha)_{ij}$ is the interaction between subject i and level j, and ε_{ij} is the random residual error for individual i in level j. The residual error can be due to measurement error and/or other factors not under investigation. From the model you can see this is similar to the two-factor model only with one observation per cell. Also, the fixed effect is denoted by α and the random effect by s; thus we have a mixed-effects model. Lastly, for the equal n's model the effects for α and $s\alpha$ sum to zero for each subject (or row).

The hypotheses for testing the effect of factor A are as follows. The null hypothesis indicates that the means for each measurement are the same.

$$H_{01}: \mu_{.1} = \mu_{.2} = \ldots \mu_{.J}$$

$$H_{11}: \textit{not all the } \mu_{.j} \textit{ are equal}$$

The hypotheses are written in terms of means because factor A is a fixed effect (i.e., all sampled cases have been measured).

5.4.4 Assumptions and Violation of Assumptions

Previously we described the assumptions for the two-factor mixed-effects model. The assumptions are nearly the same for the one-factor repeated measures model (since it is similar to the two-factor mixed-effects model) and are again mainly concerned with the distribution of the dependent variable scores and of the random effects.

A new assumption is known as **compound symmetry** and states that the covariances between the scores of the subjects across the levels of the repeated factor A are constant. In other words, the covariances for all pairs of levels of the fixed factor are the same across the population of random effects (i.e., the subjects). The analysis of variance is not particularly robust to a violation of this assumption. In particular, the assumption is often violated when factor A is time, as the relationship between adjacent levels of A is stronger than when the levels are farther apart. For example, consider the previous illustration of children measured in reading performance before, after, and six months after intervention. The means of the pre- and immediate post-reading performance will likely be more similar than the means of the pre- and six months post-reading performance. If the assumption is violated, three alternative procedures are available. The first is to limit the levels of factor A (i.e., the repeated measures factor) either to those that meet the assumption, or to limit the number of repeated measures to two (in which case there would be only one covariance and thus nothing to assume). The second and more plausible alternative is to use adjusted F tests. These are reported shortly. The third is to use multivariate analysis of variance (MANOVA), which makes no compound symmetry assumption, but is slightly less powerful. For readers interested in MANOVA, a number of excellent multivariate textbooks can be referred to (e.g., Hahs-Vaughn, 2016).

TABLE 5.5

Assumptions and Effects of Violations: One-Factor Repeated Measures Model

Assumption	Effect of Assumption Violation
Independence	Little is known about the effects of dependence; however, based on the fixed-effects model, we might expect the following: • Increased likelihood of a Type I and/or Type II error in *F* • Affects standard errors of means and inferences about those means
Homogeneity of variance	Little is known about the effects of heteroscedasticity; however, based on the fixed-effects model, we might expect the following: • Bias in SS_{SA} • Increased likelihood of a Type I and/or Type II error • Small effect with equal or nearly equal *n*'s • Otherwise effect decreases as *n* increases
Normality	• Minimal effect with equal or nearly equal *n*'s • Otherwise substantial effects
Sphericity	• *F* not particularly robust • Consider usual *F* test, Geisser-Greenhouse conservative *F* test, and adjusted (Huynh-Feldt) *F* test, if necessary

Huynh and Feldt (1970) showed that the compound symmetry assumption is a sufficient but not necessary condition for the validity of the *F* test. Thus, the *F* test may also be valid under less stringent conditions. The necessary and sufficient condition for the validity of the *F* test is known as **sphericity**. This assumes that the variance of the difference scores for each pair of factor levels is the same (e.g., with $J = 3$ levels, the variance of the difference score between levels 1 and 2 is the same as the variance of the difference score between levels 1 and 3, which is the same as the variance of the difference score between levels 2 and 3; thus another type of homogeneity of variance assumption). Further discussion of sphericity is beyond the scope of this text (see, for example, Keppel & Wickens, 2004; Kirk, 2014; Myers, Lorch, & Well, 2010). A summary of the assumptions and the effects of their violation for the one-factor repeated measures design is presented in Table 5.5.

5.4.5 ANOVA Summary Table and Expected Mean Squares

The sources of variation for this model are similar to those for the two-factor model, except that there is no within-cell variation. The ANOVA summary table is shown in Table 5.6, where we see the following sources of variation: A (i.e., the repeated measure), subjects (denoted by S), the SA interaction, and total. The test of subject differences is of no real

TABLE 5.6

One-Factor Repeated Measures ANOVA Summary Table

Source	*SS*	*df*	*MS*	*F*
A	SS_A	$J - 1$	MS_A	MS_A / MS_{SA}
S	SS_S	$n - 1$	MS_S	
SA	SS_{SA}	$(J - 1)(n - 1)$	MS_{SA}	
Total	SS_{total}	$N - 1$		

interest. Quite naturally, we expect there to be variation among the subjects. From the table, we see that although three mean square terms can be computed, only one F ratio results for the test of factor A; thus, the subjects effect cannot be tested anyway as there is no appropriate error term. This is subsequently shown through the expected mean squares.

Next we need to consider the sums of squares for the one-factor repeated measures model. If we take the total sum of squares and decompose it, we have the following:

$$SS_{total} = SS_A + SS_B + SS_{SA}$$

These three terms can then be computed by statistical software. The degrees of freedom, mean squares, and F ratio are determined as shown in Table 5.6.

The formation of the proper F ratio is again related to the expected mean squares. If H_0 is actually *true* (in other words, the means are the same for each of the measures), then the *expected mean squares* are as follows:

$$E(MS_A) = \sigma_\varepsilon^2$$

$$E(MS_S) = \sigma_\varepsilon^2$$

$$E(MS_{SA}) = \sigma_\varepsilon^2$$

where σ_ε^2 is the population variance of the residual errors.

If H_0 is actually *false* (i.e., the means are not the same for each of the measures), then the expected mean squares are as follows:

$$E(MS_A) = \sigma_\varepsilon^2 + \sigma_{s\alpha}^2 + n\sum_{j=1}^{J}\alpha_j^2/(J-1)$$

$$E(MS_S) = \sigma_\varepsilon^2 + J\sigma_s^2$$

$$E(MS_{SA}) = \sigma_\varepsilon^2 + \sigma_{s\alpha}^2$$

where σ_s^2 and $\sigma_{s\alpha}^2$ represent variability due to subjects and to the interaction of factor A and subjects, respectively, and other terms are as before.

As in previous ANOVA models, the proper F ratio should be formed as follows:

F = (*systematic variability* + *error variability*) / *error variability*

For the one-factor repeated measures model, MS_{SA} must be used as the error term for the test of A and there is no appropriate error term for the test of S or the test of SA (although that is fine as we are not really interested in those tests anyway since they refer to the individual cases).

As noted earlier in the discussion of assumptions for this model, the F test is not very robust to violation of the compound symmetry assumption. This assumption is often violated in education and the behavioral sciences; consequently, statisticians have spent considerable time studying this problem. Research suggests that the following sequential procedure be used in the test of factor A. First, do the usual F test that is quite liberal in terms of rejecting H_0 too often. If H_0 is not rejected, then stop. If H_0 is rejected, then continue with step 2, which is to use the Geisser and Greenhouse (1958) conservative F test. For the model being considered here, the degrees of freedom for the F critical value are adjusted to be 1 and $n - 1$. If H_0 is rejected, then stop. This would indicate that both the liberal and conservative tests reached the same conclusion to reject H_0. If H_0 is not rejected, then the two tests did not reach the same conclusion, and a further test (a tie-breaker) should be undertaken. Thus, in step 3, an adjusted F test is conducted. The adjustment is known as Box's (Box, 1954) correction, usually referred to as the Huynh and Feldt (1970) procedure. Here the numerator degrees of freedom are $(J-1)\varepsilon$, and the denominator degrees of freedom are $(J-1)(n-1)\varepsilon$, where ε is a correction factor (not to be confused with the residual term ε). The correction factor is quite complex and is not shown here (see, for example, Keppel & Wickens, 2004; Myers et al., 2010). Most major statistical software conducts the Geisser-Greenhouse and Huynh-Feldt tests. The Huynh-Feldt test is recommended due to greater power (Keppel & Wickens, 2004; Myers et al., 2010); thus, when available, you can simply use the Huynh-Feldt procedure rather than the previously recommended sequence.

5.4.6 Multiple Comparison Procedures

If the null hypothesis for repeated factor (i.e., factor A) is rejected and there are more than two levels of the factor, then the researcher may be interested in which means or combinations of means are different (in other words, which measurement means differ from one another). This could be assessed, as we have seen in previous chapters, by the use of some multiple comparison procedure (MCP). In general, most of the MCPs outlined in Chapter 2 can be used in the one-factor repeated measures model (see additional discussion in Keppel & Wickens, 2004; Mickey, Dunn, & Clark, 2004).

It has been shown that these MCPs are seriously affected by a violation of the compound symmetry assumption. In this situation, two alternatives are recommended. The first alternative is, rather than using the same error term for each contrast (i.e., MS_{SA}), to use a separate error term for each contrast tested. Then many of the MCPs previously covered in Chapter 2 can be used. This complicates matters considerably (Keppel & Wickens, 2004; Kirk, 2013). A second alternative, recommended by Maxwell (1980) and Wilcox (1987), involves the use of multiple dependent t tests where the α level is adjusted much like the Bonferroni procedure. Maxwell concluded that this procedure is better than many of the other MCPs. For other similar procedures, see Hochberg and Tamhane (1987).

5.4.7 Alternative ANOVA Procedures

There are several alternative procedures to the one-factor repeated measures ANOVA model. These include the Friedman (1937) test, as well as others, such as the Agresti and Pendergast (1986) test. The Friedman test, like the Kruskal-Wallis test, is a nonparametric procedure based on ranks. However, the Kruskal-Wallis test cannot be used in a repeated measures model as it assumes that the individual scores are independent. This is obviously not the case in the one-factor repeated measures model where each individual is exposed to all levels of factor A.

Let us outline how the Friedman test is conducted. First, scores are ranked within subject. For instance, if there are $J = 4$ levels of factor A, then the scores for each subject would be ranked from 1 to 4. From this, one can compute a mean ranking for each level of factor A. The null hypothesis essentially becomes a test of whether the mean rankings for the levels of A are equal. The test statistic is a χ^2 statistic. In the case of tied ranks, either the available ranks can be averaged, or a correction factor can be used as done with the Kruskal-Wallis test (see Chapter 1). The test statistic is compared to the critical value of ${}_{\alpha}\chi^2_{J-1}$ (see Appendix Table A.3). The null hypothesis that the mean rankings are the same for the levels of factor A will be rejected if the test statistic exceeds the critical value.

You may also recall from the Kruskal-Wallis test the problem with small n's in terms of the test statistic not being precisely distributed as χ^2. The same problem exists with the Friedman test when $J < 6$ and $n < 6$, so we suggest you consult the table of critical values in (Marascuilo & McSweeney, 1977, Table A22). The Friedman test, like the Kruskal-Wallis test, assumes that the population distributions have the same shape (although not necessarily normal) and variability, and that the dependent measure is continuous. For a discussion of other alternative nonparametric procedures, see Agresti and Pendergast (1986), Myers and Well (1995), and Wilcox (1987, 1996, 2003). For information on more advanced within subjects ANOVA models, see Cotton (1998), Keppel and Wickens (2004), and Myers et al. (2010).

Various multiple comparison procedures (MCPs) can be used for the Friedman test. For the most part, these MCPs are analogs to their parametric equivalents. In the case of planned (or *a priori*) pairwise comparisons, one may use multiple matched-pair Wilcoxon tests (i.e., a form of the Kruskal-Wallis test for two groups) in a Bonferroni form (i.e., taking the number of contrasts into account through an adjustment of the α level; for example, if there are six contrasts with an alpha of .05, the adjusted alpha would be .05/6 or .008). For post hoc comparisons, numerous parametric analogs are available. For additional discussion on MCPs for this model, see Marascuilo and McSweeney (1977).

5.4.8 An Example

Let us consider an example to illustrate the procedures used for this model. The data are shown in Table 5.7 where there are eight dancers, each of whom has been evaluated by four ballet instructors (who will be referred to as "raters") on ballet technique. First, let us take a look at the results for the parametric ANOVA model, as shown in Table 5.8. The F test statistic is compared to the usual F test critical value of ${}_{.05}F_{3,21} = 3.07$, which is significant. For the Geisser-Greenhouse conservative procedure, the test statistic is compared to the critical value of ${}_{.05}F_{1,7} = 5.59$, which is also significant. The two procedures both yield a statistically significant result; thus we need not be concerned with a violation of the compound symmetry assumption. As an example MCP, the Bonferroni procedure determined that all pairs of raters are significantly different from one another, except for Rater 1 versus Rater 2.

TABLE 5.7

Data for the Ballet Technique Example One-Factor Design: Raw Scores and Rank Scores on the Ballet Technique Task by Subject and Instructor

	Rater 1		Rater 2		Rater 3		Rater 4	
Subject	Raw	Rank	Raw	Rank	Raw	Rank	Raw	Rank
1	3	1	4	2	7	3	8	4
2	6	2	5	1	8	3	9	4
3	3	1	4	2	7	3	9	4
4	3	1	4	2	6	3	8	4
5	1	1	2	2	5	3	10	4
6	2	1	3	2	6	3	10	4
7	2	1	4	2	5	3	9	4
8	2	1	3	2	6	3	10	4

TABLE 5.8

One-Factor Repeated Measures ANOVA Summary Table for the Ballet Technique Example

Source	*SS*	*df*	*MS*	*F*
Within subjects:				
Rater (A)	198.125	3	66.042	73.477*
Error (SA)	18.875	21	.899	
Between subjects:				
Error (S)	14.875	7	2.125	
Total	231.875	31		

$^{*}{}_{.05}F_{3,21} = 3.07$

Finally, let us take a look at the Friedman test. The test statistic is $\chi^2 = 22.9500$. This test statistic is compared to the critical value ${}_{.05}\chi^2_3 = 7.8147$, which is significant. Thus the conclusions for the parametric ANOVA and nonparametric Friedman tests are the same here. This will not always be the case, particularly when ANOVA assumptions are violated.

5.5 The Two-Factor Split-Plot or Mixed Design

Through the previous chapters, we have learned about many statistical procedures as our talented set of graduate students have assisted others and conducted studies of their own. What is in store for the group now?

For the past few chapters, we have followed Addie, Challie, Oso, and Ott, an extraordinarily talented group of graduate students working in a research lab, as they have successfully examined various questions. Knowing the success this group of students have achieved thus far, their faculty advisor feels confident that Oso can assist another faculty member at the university. Oso is working with Dr. Kilauea, the coordinator of the dance program. Dr. Kilauea has conducted an experiment in which eight ballet dancers were randomly assigned to one of two dance instructors. Each dancer was then assessed on ballet technique (e.g., body alignment, hip placement, feet placement) by four raters. Dr. Kilauea wants to know the following: if there is a mean difference in ballet technique based on instructor; if there is a mean difference in ballet technique based on rater; and if there is a mean difference in ballet technique based on the rater by instructor interaction. The research questions presented to Dr. Kilauea from Oso includes the following:

- *Is there a mean difference in ballet technique based on instructor?*
- *Is there a mean difference in ballet technique based on rater?*
- *Is there a mean difference in ballet technique based on rater by instructor?*

With one between-subjects independent variable (i.e., instructor) and one within-subjects factor (i.e., rating on ballet technique), Oso determines that a two-factor split-plot ANOVA is the best statistical procedure to use to answer Dr. Kilauea's question. His next task is to assist Dr. Kilauea in analyzing the data.

In this section, we describe the distinguishing characteristics of the two-factor split-plot or mixed ANOVA design, the layout of the data, the linear model, assumptions and their violation, the ANOVA summary table and expected mean squares, multiple comparison procedures, and an example.

5.5.1 Characteristics of the Model

The characteristics of the two-factor split-plot or mixed ANOVA design are a combination of the characteristics of the one-factor repeated measures and the two-factor fixed-effects models. It is unique because there are two factors, only one of which is repeated. For this reason the design is often called a **mixed design**. Thus, one of the factors is a *between-subjects* factor, the other is a *within-subjects* factor, and the result is known as a **split-plot design** (from agricultural research). Each subject then responds to every level of the repeated factor, but to only one level of the nonrepeated factor. Subjects then serve as their own controls for the repeated factor, but not for the nonrepeated factor. The other characteristics carry over from the one-factor repeated measures model and the two-factor model.

5.5.2 The Layout of the Data

The layout of the data for the two-factor split-plot or mixed design is shown in Table 5.9. Here we see the rows designated as the levels of factor A, the between-subjects or nonrepeated factor, and the columns as the levels of factor B, the within-subjects or repeated factor. Within each factor level combination or cell are the subjects. Notice that the same subjects appear at all levels of factor B (the within-subjects factor, the repeated measure), but only at one level of factor A (the between-subjects factor). Row, column, cell, and overall means are also shown. Here you see that the layout of the data looks much the same as the two-factor model.

TABLE 5.9

Layout for the Two-Factor Split-Plot or Mixed ANOVA

Level of Factor A (Nonrepeated Factor)	Level of Factor B (Repeated Factor) 1	2	...	K	Row Mean
1	Y_{111}	Y_{112}	...	Y_{11K}	
	.	.	...	.	
	.	.	...	.	$\bar{Y}_{.1.}$
	.	.	...	.	
	Y_{n11}	Y_{n12}	...	Y_{n1K}	
	—	—		—	
	$\bar{Y}_{.11}$	$\bar{Y}_{.12}$		$\bar{Y}_{.1K}$	

TABLE 5.9 (continued)

Layout for the Two-Factor Split-Plot or Mixed ANOVA

Level of Factor A (Nonrepeated Factor)	Level of Factor B (Repeated Factor) 1	2	...	K	Row Mean
2	Y_{121}	Y_{122}	...	Y_{12K}	
	.	.	...	.	
	.	.	...	.	$\bar{Y}_{.2.}$
	.	.	...	.	
	Y_{n21}	Y_{n22}	...	Y_{n2K}	
	$\bar{Y}_{.21}$	$\bar{Y}_{.22}$	...	$\bar{Y}_{.2K}$	
.	.	.	...	.	.
.	.	.	...	.	.
.	.	.	...	.	.
J	Y_{1J1}	Y_{1J2}	...	Y_{1JK}	
	.	.	...	.	
	.	.	...	.	$\bar{Y}_{.J.}$
	.	.	...	.	
	Y_{nJ1}	Y_{nJ2}	...	Y_{nJK}	
	$\bar{Y}_{.J1}$	$\bar{Y}_{.J2}$	...	$\bar{Y}_{.JK}$	
Column Mean	$\bar{Y}_{..1}$	$\bar{Y}_{..2}$		$\bar{Y}_{..K}$	$\bar{Y}_{...}$

Note: Each subject is measured at all levels of factor B, but at only one level of factor A.

5.5.3 The ANOVA Model

The two-factor split-plot model can be written in terms of population parameters as follows:

$$Y_{ijk} = \mu + \alpha_j + s_{i(j)} + \beta_k + (\alpha\beta)_{jk} + (\beta s)_{ki(j)} + \varepsilon_{ijk}$$

where Y_{ijk} is the observed score on the dependent variable for individual *i* in level *j* of factor A (the between-subjects factor) and level *k* of factor B (i.e., the *jk* cell, the within-subjects factor or repeated measure), μ is the overall or grand population mean (i.e., regardless of cell designation), a_j is the effect for level *j* of factor A (row effect for the nonrepeated factor), $s_{i(j)}$ is the effect of subject *i* that is nested within level *j* of factor A (i.e., "*i*(*j*)" denotes that *i* is nested within *j*), b_k is the effect for level *k* of factor B (column effect for the repeated factor), $(\alpha\beta)_{jk}$ is the interaction effect for the combination of level *j* of factor A and level *k* of factor B, $(\beta s)_{ki(j)}$ is the interaction effect for the combination of level *k* of factor B (the within-subjects factor, the repeated measure) and subject *i* that is nested within level *j* of factor A (the between-subjects factor), and ε_{ijk} is the random residual error for individual *i* in cell *jk*.

We use the terminology "subjects are nested within factor A" to indicate that a particular subject s_i is exposed to only one level of factor A (the between-subjects factor), level *j*. This observation is then denoted in the subjects effect by $s_{i(j)}$ and in the interaction effect by $(\beta s)_{ki(j)}$. This is due to the fact that not all possible combinations of subject with the levels of factor A are included in the model. A more extended discussion of designs with nested factors is given in Chapter 6. The residual error can be due to individual differences, measurement error, and/or other factors not under investigation. We assume for now that A and B are fixed-effects factors and that S is a random-effects factor.

It should be mentioned that for the equal n's model, the sum of the row effects, the sum of the column effects, and the sum of the interaction effects are all equal to zero, both across rows and across columns. This implies, for example, that if there are any nonzero row effects, then the row effects will balance out around zero with some positive and some negative effects.

The hypotheses to be tested here are exactly the same as in the nonrepeated two-factor ANOVA model (see Chapter 3). For the two-factor ANOVA model, there are three sets of hypotheses, one for each of the main effects, and one for the interaction effect. The null and alternative hypotheses, respectively, for testing the main effect of factor A (between subjects factor) are as follows:

$$H_{01} : \mu_{.1.} = \mu_{.2.} = \ldots \mu_{.J.}$$

$$H_{11} : not\ all\ the\ \mu_{.j.}\ are\ equal$$

The hypotheses for testing the main effect of factor B (within-subjects factor, i.e., the repeated measure) are noted as:

$$H_{02} : \mu_{..1} = \mu_{..2} = \ldots \mu_{..K}$$

$$H_{12} : not\ all\ the\ \mu_{..K}\ are\ equal$$

Finally, the hypotheses for testing the interaction effect (i.e., the between- by within-factors effect) are as follows:

$$H_{03} : \left(\mu_{.jk} - \mu_{.j.} - \mu_{..k} + \mu\right) = 0\ for\ all\ j\ and\ k$$

$$H_{13} : not\ all\ the \left(\mu_{.jk} - \mu_{.j.} - \mu_{..k} + \mu\right) = 0$$

If one of the null hypotheses is rejected, then the researcher may want to consider a multiple comparison procedure so as to determine which means or combination of means are significantly different (discussed later in this chapter).

5.5.4 Assumptions and Violation of Assumptions

Previously we described the assumptions for the different two-factor models and the one-factor repeated measures model. The assumptions for the two-factor split-plot or mixed design are actually a combination of these two sets of assumptions.

The assumptions can be divided into two sets of assumptions, one for the between-subjects factor and one for the within-subjects (or repeated measures) factor. For the between-subjects factor, we have the usual assumptions of population scores being random, independent, and normally distributed with equal variances. For the within-subjects factor (i.e., the repeated measure), the assumption is the already familiar compound symmetry assumption. For this design, the assumption involves the population covariances for all pairs of the levels of the within-subjects factor (i.e., k and k') being equal, at each level of the between-subjects factor (for all levels j). To deal with this assumption, we look at alternative F tests in the next section. A summary of the assumptions and the effects of their violation for the two-factor split-plot or mixed design are presented in Table 5.10.

TABLE 5.10

Assumptions and Effects of Violations: Two-Factor Split-Plot or Mixed Model

Assumption	Effect of Assumption Violation
Independence	• Increased likelihood of a Type I and/or Type II error in F • Affects standard errors of means and inferences about those means
Homogeneity of variance	• Bias in error terms • Increased likelihood of a Type I and/or Type II error • Small effect with equal or nearly equal n's • Otherwise effect decreases as n increases
Normality	• Minimal effect with equal or nearly equal n's • Otherwise substantial effects
Sphericity	• F not particularly robust • Consider usual F test, Geisser-Greenhouse conservative F test, and adjusted (Huynh-Feldt) F test, if necessary

5.5.5 ANOVA Summary Table and Expected Mean Squares

The ANOVA summary table is shown in Table 5.11, where we see the following sources of variation: A, S, B, AB, BS, and total. The table is divided into within-subjects sources and between-subjects sources. The between-subjects sources are A and S, where S will be used as the error term for the test of factor A. The within-subjects sources are B, AB, and BS, where BS will be used as the error term for the test of factor B and of the AB interaction. This will become clear when we examine the expected mean squares shortly.

Next we need to consider the sums of squares for the two-factor mixed design. Taking the total sum of squares and decomposing it yields the following:

$$SS_{total} = SS_A + SS_S + SS_B + SS_{AB} + SS_{BS}$$

We leave the computation of these five terms for statistical software. The degrees of freedom, mean squares, and F ratios are computed as shown in Table 5.11.

TABLE 5.11

Two-Factor Split-Plot or Mixed Model ANOVA Summary Table

Source	*SS*	*df*	*MS*	*F*
Between subjects:				
A	SS_A	$J - 1$	MS_A	MS_A / MS_A
S	SS_S	$J(n - 1)$	MS_S	
Within subjects:				
B	SS_B	$K - 1$	MS_B	MS_B / MS_{BS}
AB	SS_{AB}	$(J - 1)(K - 1)$	MS_{AB}	MS_{AB} / MS_{BS}
BS	SS_{BS}	$(K - 1)J(n - 1)$	MS_{BS}	
Total	SS_{total}	$N - 1$		

The formation of the proper F ratio is again related to the expected mean squares. If H_0 is actually *true* (i.e., the means are really equal), then the *expected mean squares* are as follows:

$$E(MS_A) = \sigma_\varepsilon^2$$

$$E(MS_S) = \sigma_\varepsilon^2$$

$$E(MS_B) = \sigma_\varepsilon^2$$

$$E(MS_{AB}) = \sigma_\varepsilon^2$$

$$E(MS_{BS}) = \sigma_\varepsilon^2$$

where σ_ε^2 is the population variance of the residual errors.

If H_0 is actually *false* (i.e., the means are really not equal), then the expected mean squares are as follows:

$$E(MS_A) = \sigma_\varepsilon^2 + K\sigma_s^2 + nK\left[\sum_{j=1}^{J}\alpha_j^2 / (J-1)\right]$$

$$E(MS_S) = \sigma_\varepsilon^2 + K\sigma_s^2$$

$$E(MS_B) = \sigma_\varepsilon^2 + \sigma_{\beta s}^2 + nJ\left[\sum_{k=1}^{K}\beta_k^2 / (K-1)\right]$$

$$E(MS_{AB}) = \sigma_\varepsilon^2 + \sigma_{\beta s}^2 + n\left[\sum_{j=1}^{J}\sum_{k=1}^{K}(\alpha\beta)_{jk}^2 / (J-1)(K-1)\right]$$

$$E(MS_{BS}) = \sigma_\varepsilon^2 + \sigma_{\beta s}^2$$

where $\sigma_{\beta s}^2$ represents variability due to the interaction of factor B (the within-subjects or repeated measures factor) and subjects, and the other terms are as before.

As in previous ANOVA models, the proper F ratio should be formed as follows:

F = (*systematic variability* + *error variability*) / *error variability*

For the two-factor split-plot design, the error term for the proper test of factor A (the between-subjects factor) is the S term, whereas the error term for the proper tests of factor B (the within-subjects or repeated measures factor) and the AB interaction is the BS interaction. For models where factors A and B are not both fixed-effects factors, see Keppel and Wickens (2004).

As the compound symmetry assumption is often violated, we again suggest the following sequential procedure to test for B (the repeated measure) and for AB (the within-by between-subjects factor interaction). First, do the usual F test, which is quite liberal in terms of rejecting H_0 too often. If H_0 is not rejected, then stop. If H_0 is rejected, then continue with Step 2, which is to use the Geisser-Greenhouse (1958) conservative F test. For the model under consideration here, the degrees of freedom for the F critical values are adjusted to be 1 and $J(n - 1)$ for the test of B, and $J - 1$ and $J(n - 1)$ for the test of the AB interaction. There is no conservative test necessary for factor A, the between-subjects nonrepeated factor, as the assumption does not apply; thus, the usual test is all that is necessary for the test of A. If H_0 for B and/or AB is rejected, then stop. This would indicate that both the liberal and conservative tests reached the same conclusion to reject H_0. If H_0 is not rejected, then the two tests did not yield the same conclusion, and an adjusted F test is conducted. The adjustment is known as Box's (1954) correction [or the Huynh and Feldt (1970) procedure]. Most major statistical software conducts the Geisser-Greenhouse and Huynh-Feldt tests.

5.5.6 Multiple Comparison Procedures

Consider the situation where the null hypothesis for any of the three hypotheses is rejected (i.e., for A, B, and/or AB). If there is more than one degree of freedom in the numerator for any of these hypotheses, then the researcher may be interested in which means or combinations of means are different. This could be assessed again by the use of some multiple comparison procedure (MCP). Thus, the procedures outlined in Chapter 3 (i.e., for main effects, and for simple and complex interaction contrasts) for the regular two-factor ANOVA model can be adapted to this model.

However, it has been shown that the MCPs involving the repeated factor are seriously affected by a violation of the compound symmetry assumption. In this situation, two alternatives are recommended. The first alternative is, rather than using the same error term for each contrast involving the repeated factor (i.e., MS_B or MS_{AB}), to use a separate error term for each contrast tested. Then many of the MCPs previously covered in Chapter 2 can be used. This complicates matters considerably (Keppel & Wickens, 2004; Kirk, 2014). The second and simpler alternative is suggested by Shavelson (1996). He recommended that the appropriate error terms be used in MCPs involving the main effects, but for interaction contrasts both error terms be pooled (or added) together (this procedure is conservative, yet simpler than the first alternative).

5.5.7 An Example

Consider now an example problem to illustrate the two-factor mixed design. Here we expand on the example presented earlier in this chapter by adding a second factor to the model. The data are shown in Table 5.12 where there are eight dancers, each of whom has been evaluated by four raters on ballet technique (rater is the *within-subjects factor* as each individual has been evaluated by four raters). Ratings on ballet technique can range from 1 (lowest rating) to 10 (highest rating). Each dancer was also randomly assigned to one of two ballet instructors. Thus, factor A is the between-subjects factor. In this illustration, factor A represents the dance instructors, where we see that four subjects are randomly assigned to level 1 of factor A (i.e., ballet instructor 1) and the remaining four to level 2 of factor A (i.e., ballet instructor 2). Thus, factor B (i.e., rater) is repeated (the within-subjects factor) and factor A (i.e., ballet instructor) is not repeated (the between-subjects factor). The ANOVA summary table is shown in Table 5.13.

TABLE 5.12

Data for the Ballet Technique Example Two-Factor Design: Raw Scores on the Ballet Technique Task by Instructor and Rater

Factor A (Nonrepeated Factor)		Factor B (Repeated Factor)			
Ballet Instructor	**Subject**	**Rater 1**	**Rater 2**	**Rater 3**	**Rater 4**
1	1	3	4	7	8
	2	6	5	8	9
	3	3	4	7	9
	4	3	4	6	8
2	5	1	2	5	10
	6	2	3	6	10
	7	2	4	5	9
	8	2	3	6	10

TABLE 5.13

Two-Factor Split-Plot ANOVA Summary Table for the Ballet Technique Example

Source	SS	df	MS	F
Between subjects:				
Instructor (A)	6.125	1	6.125	4.200**
Error (S)	8.750	6	1.458	
Within subjects:				
Rater (B)	198.125	3	66.042	190.200*
Instructor x Rater	12.625	3	4.208	12.120*
Error (BS)	6.250	18	0.347	
Total	231.875	31		

$^{*}\,{}_{.05}F_{3,18} = 3.16$

$^{**}\,{}_{.05}F_{1,6} = 5.99$

The test statistics are compared to the following usual F test critical values: for factor A (the between-subjects factor that tests mean differences based on instructor), ${}_{.05}F_{1,6} = 5.99$, which is not statistically significant; for factor B (the within-subjects factor that tests mean differences based on repeated ratings), ${}_{.05}F_{3,18} = 3.16$, which is significant; and for AB, ${}_{.05}F_{3,18} = 3.16$, which is also statistically significant. For the Geisser-Greenhouse conservative procedure, the test statistics are compared to the following critical values: for factor A (i.e., between-subjects factor, ballet instructor) no conservative procedure is necessary; for factor B (i.e., within-subjects factor or the repeated measure), ${}_{.05}F_{1,6} = 5.99$, which is

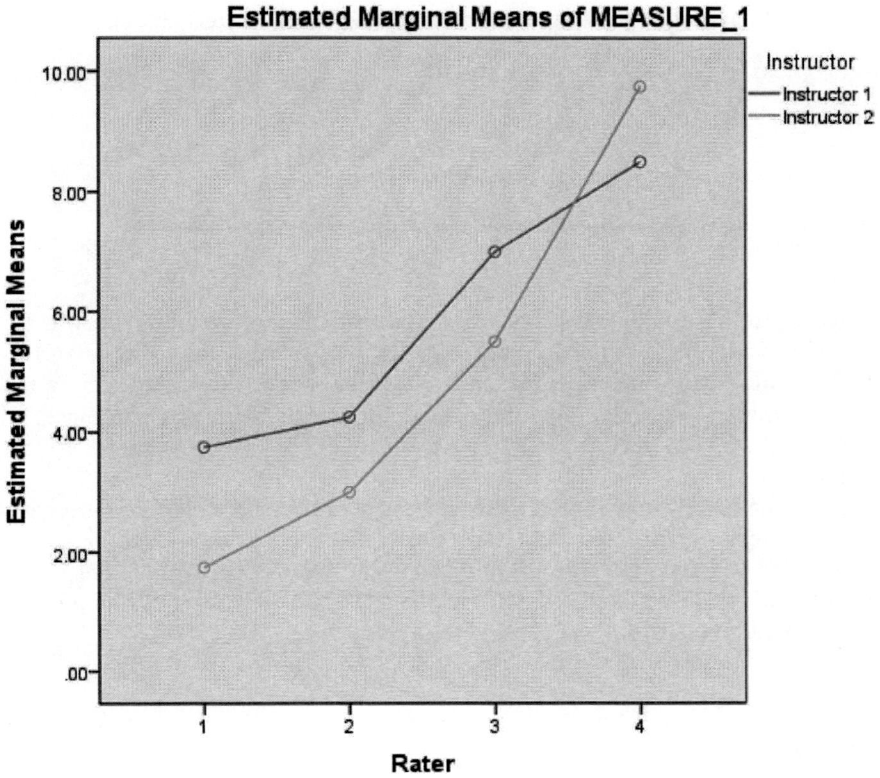

FIGURE 5.2
Profile plot for example data.

also significant; and for the interaction AB (ballet instructor by rater), ${}_{.05}F_{1,6} = 5.99$, which is also significant. The usual and Geisser-Greenhouse procedures both yield a statistically significant result for factor B (rater) and for the interaction AB (ballet instructor by rater); thus we need not be concerned with a violation of the sphericity assumption. A profile plot of the interaction is shown in Figure 5.2.

There is a significant AB (i.e., instructor by rater) interaction, so we should follow this up with simple interaction contrasts, each involving only four cell means. As an example of a MCP, consider the following contrast:

$$\psi' = \frac{(\bar{Y}_{.11} - \bar{Y}_{.21}) - (\bar{Y}_{.14} - \bar{Y}_{.24})}{4} = \frac{(3.75 - 1.75) - (8.50 - 9.75)}{4} = 0.8125$$

with a standard error computed as follows:

$$se_{\psi'} = \sqrt{MS_{BS}\left(\frac{\sum_{j=1}^{J}\sum_{k=1}^{K} c_{jk}^2}{n_{jk}}\right)} = \sqrt{0.3472\left(\frac{1/16 + 1/16 + 1/16 + 1/16}{4}\right)} = 0.1473$$

Using the Scheffé procedure, we formulate the following as the test statistic:

$$t = \frac{\psi'}{se_{\psi'}} = \frac{0.8125}{0.1473} = 5.5160$$

This is compared with the critical value presented here:

$$\sqrt{(J-1)(K-1)\,{}_{\alpha}F_{(J-1)(K-1),(K-1)J(n-1)}} = \sqrt{3\left({}_{.05}F_{3,18}\right)} = \sqrt{3(3.16)} = 3.0790$$

Thus, we may conclude that the tetrad interaction difference between the first and second levels of factor A (ballet instructor) and the first and fourth levels of factor B (rater, the repeated measure) is significant. In other words, Rater 1 finds better ballet technique among the dancers of ballet instructor 1 than ballet instructor 2, whereas Rater 4 finds better ballet technique among the dancers of ballet instructor 2 than ballet instructor 1.

Although we have only considered the basic repeated measures designs here, more complex repeated measures designs also exist. For further information, see any number of excellent sources (Cotton, 1998; Glass & Hopkins, 1996; Keppel & Wickens, 2004; Kirk, 2014; Myers et al., 2010) as well as alternative ANOVA procedures described by Wilcox (2003) and McCulloch (2005).

5.6 Computing ANOVA Models Using SPSS

Next we consider SPSS for the models presented in this chapter.

5.6.1 One-Factor Random-Effects ANOVA

To conduct a one-factor random-effects ANOVA analysis, there are only two differences from the one-factor fixed-effects ANOVA (Chapter 1). Otherwise, the form of the data and the conduct of the analyses are exactly the same. In terms of the form of the data, one column or variable indicates the levels or categories of the independent variable (i.e., the random factor), and the second is for the dependent variable. Each row then represents one individual, indicating the level or group that individual is a member of (1, 2, 3, or 4 in our example; recall that for the one-factor random-effects ANOVA, these categories are randomly selected from the population of categories), and their score on the dependent variable. Thus, we wind up with two long columns of group values and scores as shown in the screenshot in Figure 5.3. The data used to illustrate has measured customer satisfaction based on the location visited by the customer. The independent variable, location, is a random factor rather than fixed as the locations were randomly selected from the population of all locations. The dependent variable is a measure of customer satisfaction with their experience.

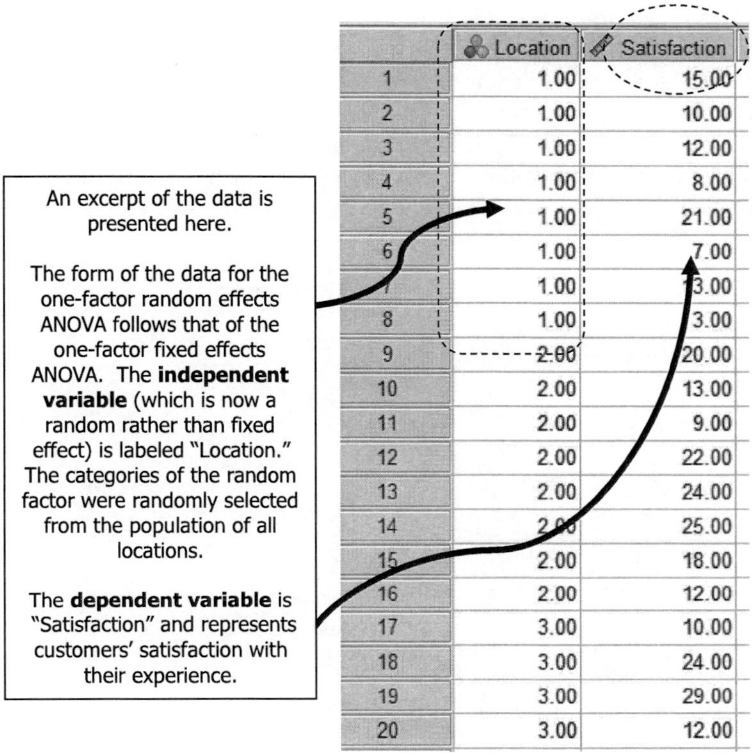

FIGURE 5.3
One-factor random-effects ANOVA data.

Step 1. To conduct a one-factor random-effects ANOVA, go to "Analyze" in the top pull-down menu, then select "General Linear Model," and then select "Univariate." Following the screenshot for Step 1 (shown in Figure 5.4) produces the Univariate dialog box.

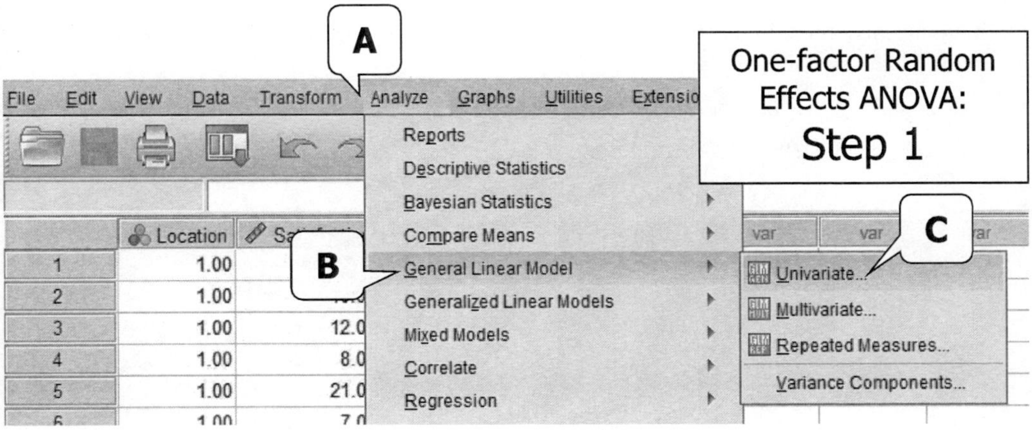

FIGURE 5.4
One-factor random-effects ANOVA: Step 1.

Step 2. Click the dependent variable (e.g., satisfaction) and move it into the "Dependent Variable" box by clicking the arrow button. Click the independent variable (e.g., location; this is the random-effects factor) and move it into the "Random Factors" box by clicking the arrow button. On this Univariate dialog screen, you will notice that while the "Post hoc" option button is active, clicking on "Post hoc" will produce a dialog box with no active options as we are now dealing with a random factor rather than a fixed factor. Post hoc multiple comparison procedures are available only from the "Options" screen, as we will see in the following screenshots.

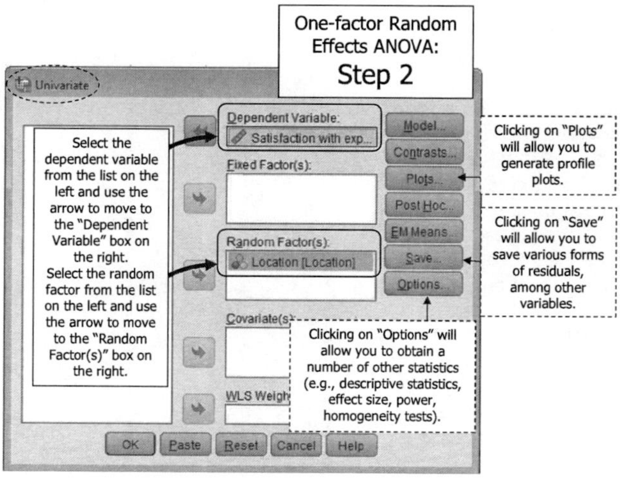

FIGURE 5.5
One-factor random-effects ANOVA: Step 2.

Step 3. Clicking on "EM Means" provides the option to display marginal and overall means. Click on "Continue" to return to the original dialog box. *Note that if you are interested in a multiple comparison procedure for testing mean differences of the random effect, post hoc MCPs are available only from this screen.* To select a post hoc procedure, click on "Compare main effects" and use the toggle menu to reveal the Tukey LSD, Bonferroni, and Sidak procedures. However, we have already mentioned that MCPs are not generally of interest for this model.

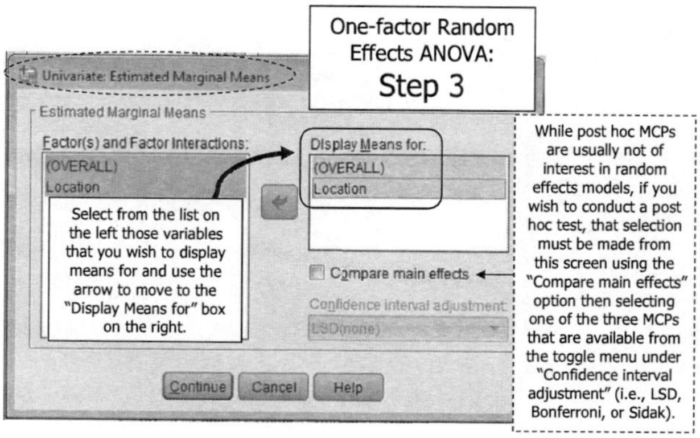

FIGURE 5.6
One-factor random-effects ANOVA: Step 3.

Step 4. Clicking on "Options" provides the option to select such information as "Descriptive statistics," "Estimates of effect size," "Observed power," and "Homogeneity tests" (i.e., Levene's test for equal variances). Click on "Continue" to return to the original dialog box.

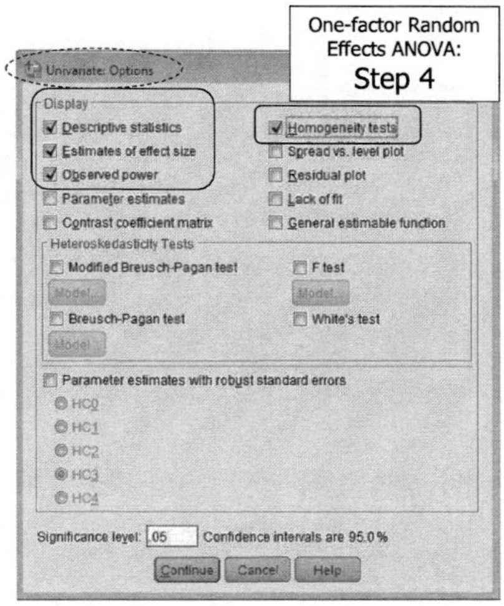

FIGURE 5.7
One-factor random-effects ANOVA: Step 4.

Step 5. From the Univariate dialog box, click on "Plots" to obtain a profile plot of means. Click the random factor (e.g., "Location") and move it into the "Horizontal Axis" box by clicking the arrow button (see screenshot for Step 5a, Figure 5.8). Then click on "Add" to move the variable into the "Plots" box at the bottom of the dialog box (see screenshot for Step 5b, Figure 5.9). Click on "Continue" to return to the original dialog box.

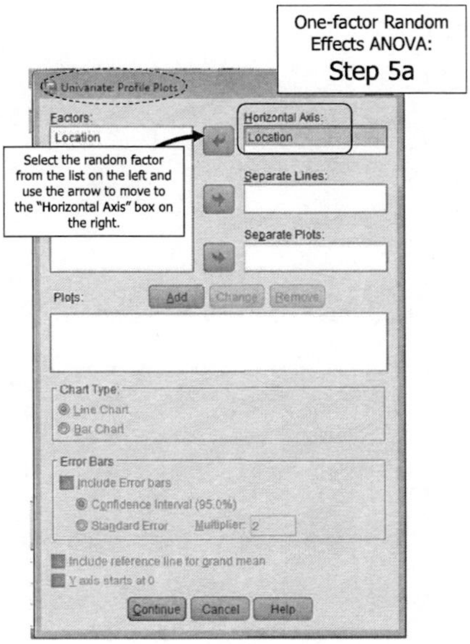

FIGURE 5.8
One-factor random-effects ANOVA: Step 5a.

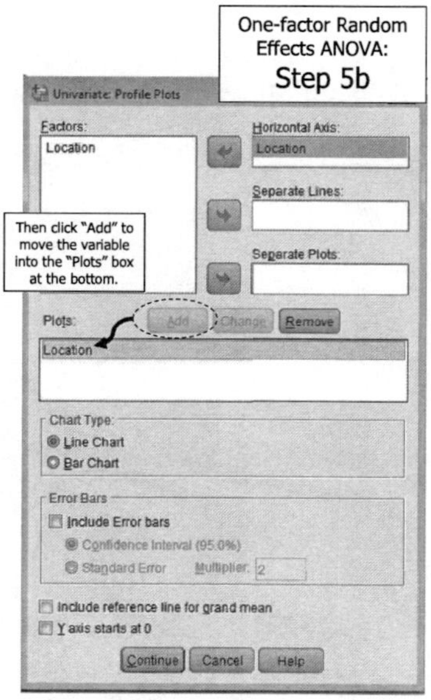

FIGURE 5.9
One-factor random-effects ANOVA: Step 5b.

Step 6. From the Univariate dialog box (see the screenshot for Step 2, Figure 5.5), click on "Save" to select those elements that you want to save. In our case, we want to save the unstandardized residuals which will be used later to examine the extent to which normality and independence are met. Thus, place a checkmark in the box next to "Unstandardized." Click "Continue" to return to the main Univariate dialog box. From the Univariate dialog box, click on "OK" to return to generate the output.

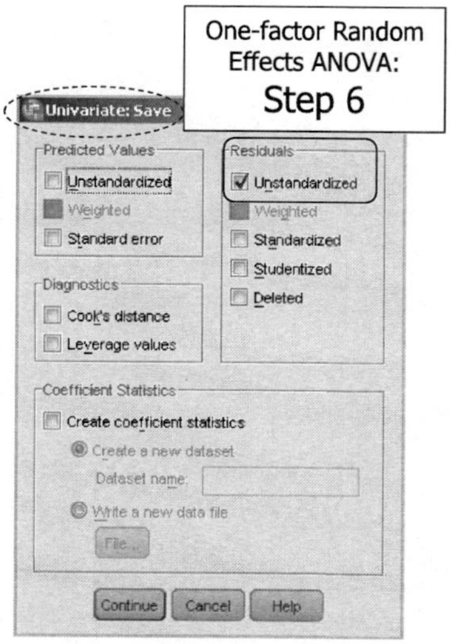

FIGURE 5.10
One-factor random-effects ANOVA: Step 6.

5.6.2 Two-Factor Random-Effects ANOVA

To run a two-factor random-effects ANOVA model, there are the same two differences from the two-factor fixed-effects ANOVA (covered in Chapter 3). First, on the GLM screen (shown in the screenshot in Figure 5.11), click both factor names into the "Random Factor(s)" box rather than the "Fixed Factor(s)" box. Second, the same situation exists with MCPs: if you are interested in a multiple comparison procedure for the random factors, post hoc MCPs are available only from the "EM Means" screen. However, we have already mentioned that MCPs are not generally of interest for this model. For brevity, the subsequent screenshots are not presented.

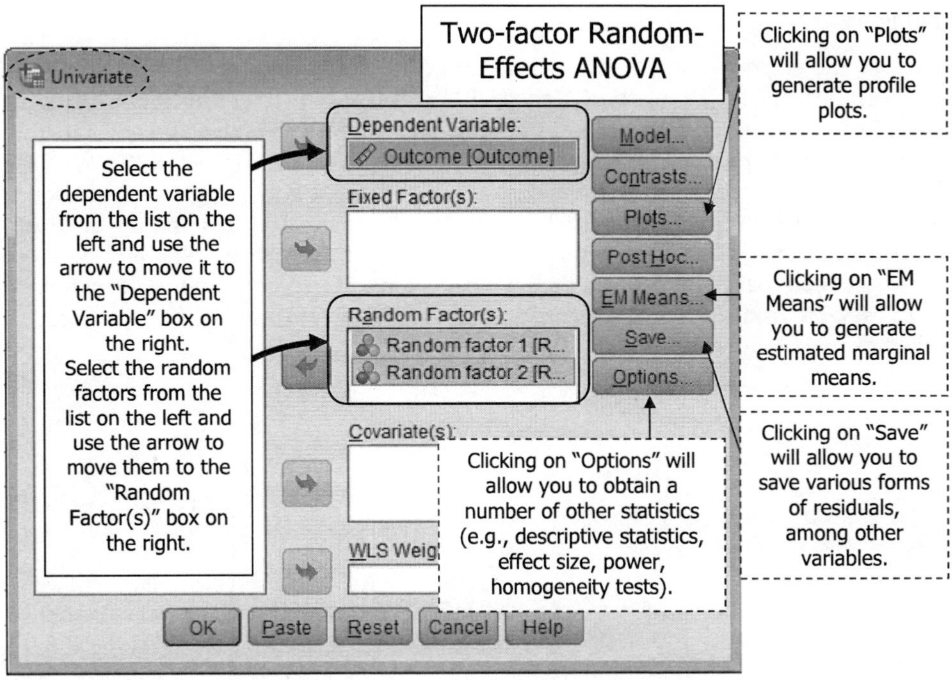

FIGURE 5.11
Two-factor random-effects ANOVA.

5.6.3 Two-Factor Mixed-Effects ANOVA

To conduct a two-factor mixed-effects ANOVA, there are three differences from the two-factor fixed-effects ANOVA when using SPSS to analyze the model. The first is that both a random and a fixed effect factor must be defined (see the screenshot for Step 2, Figure 5.12). The second difference is that post hoc MCPs for the fixed-effects factor are available from either the "Post Hoc" or "EM Means" screens, while for the random-effects factor they are available only from the "EM Means" screen. The third difference is related to the output provided by SPSS. Unfortunately, the *F* statistic for any main effect that is random in a mixed-effects model is computed incorrectly in SPSS because the wrong error term is used when implementing the SPSS point-and-click mode. As described in Lomax and Surman (2007) and extended by Li and Lomax (2011), you need to either (a) compute the *F* statistics by hand from the *MS* values (which are correct), (b) use SPSS syntax where the user indicates the proper error terms, or (c) use a different software package (e.g., SAS, where the user also provides the proper error terms). These options are not presented here. Rather, readers are referred to the appropriate references. For the purpose of this illustration, we will use the satisfaction data. The dependent variable is satisfaction with the customer experience. Assignment to an intervention or comparison group will be a fixed factor, and the location will be a random factor.

Step 1. To conduct a one-factor fixed-effects ANOVA, go to "Analyze" in the top pulldown menu, then select "General Linear Model," and then select "Univariate." Following screenshot one for the one-factor random-effects ANOVA presented previously produces the Univariate dialog box.

Step 2. Per the screenshot that follows (Figure 5.12), click the dependent variable (e.g., satisfaction) and move it into the "Dependent Variable" box by clicking the arrow button. Click the fixed factor (e.g., treatment) and move it into the "Fixed Factors" box by clicking the arrow button. Click the random factor (e.g., location) and move it into the "Random Factors" box by clicking the arrow button. Next, click on "EM Means." Please note that post hoc MCPs for the fixed-effects factor (in this case, treatment) are available from either the Post Hoc or EM Means screens, while for the random-effects factor they are available only from the EM Means screen. Because these steps have been presented in previous screenshots (e.g., Chapter 2 for MCPs and the one-factor random-effects previously shown in this chapter), they are not repeated here.

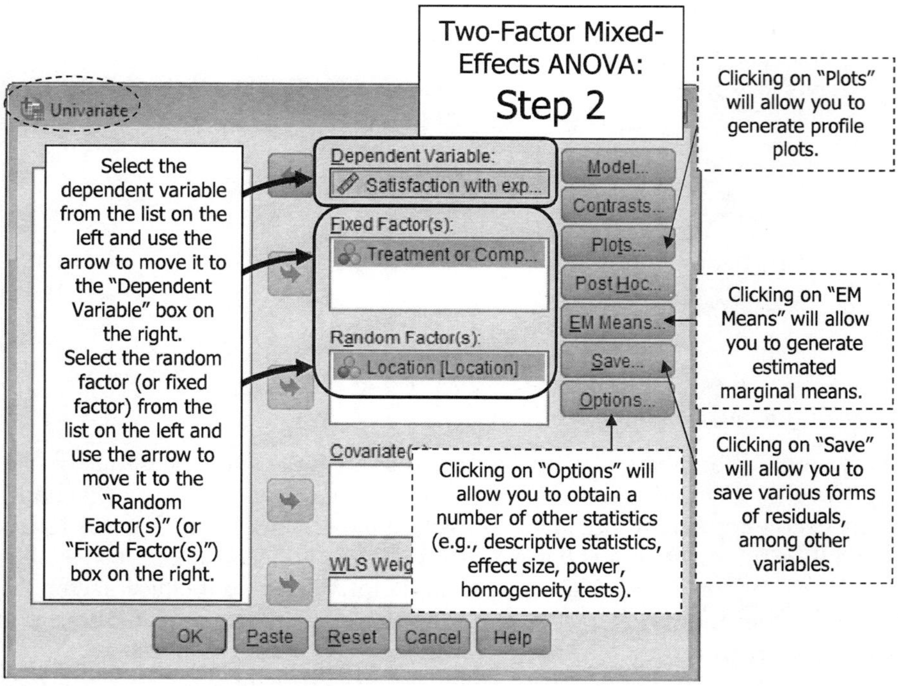

FIGURE 5.12
Two-factor mixed-effects ANOVA: Step 2.

5.6.4 One-Factor Repeated Measures ANOVA

In order to run a one-factor repeated measures ANOVA model, the data have to be in the form suggested by the following screenshot. Each row represents one dancer in our sample. All of the scores for each subject must be in one row of the dataset, and each level of the repeated factor is a separate variable (represented by the columns). For example, if there are four raters who assess each dancer's ballet technique, there will be variables for each rater (e.g., Rater1 through Rater4; example dataset on the website). In this illustration, we have both raw scores and ranked data for each of the four raters. When using ANOVA for repeated measures, we will apply the raw scores. The ranked scores will be of value only when computing the nonparametric version of ANOVA (i.e., the Friedman test) which will be covered later in this chapter.

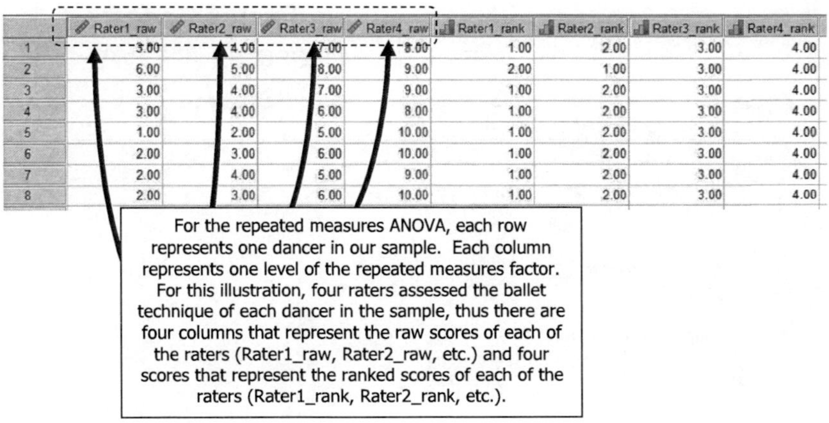

FIGURE 5.13
Repeated measures ANOVA data.

Step 1. To conduct a one-factor repeated measures ANOVA, go to "Analyze" in the top pull-down menu, then select "General Linear Model," and then select "Repeated Measures." Following the screenshot for Step 1 (Figure 5.14) produces the Repeated Measures dialog box.

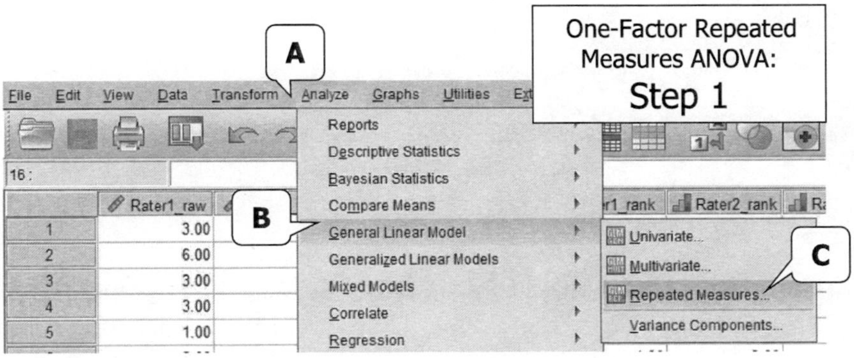

FIGURE 5.14
One-factor repeated measures ANOVA: Step 1.

Step 2. The "Repeated Measures Define Factor(s)" dialog box will appear (see Figure 5.15). In the box under "Within-Subject Factor Name," enter the name you wish to call the repeated factor. For this illustration, we will label the repeated measure "Rating." It is necessary to define a name for the repeated factor as there is no single variable representing this factor (recall that the columns in the dataset represent the repeated measures); in the dataset there is one variable for each level of the factor (in other words, one variable for each different rater or measurement). Again, in our example, there are four levels of rater (i.e., four raters) and thus four variables. Thus we name the within-subjects factor "Rating." The "Number of Levels" indicates the number of measurements of the repeated measure. In this example, there were four raters, and thus the "number of levels" of the factor is 4.

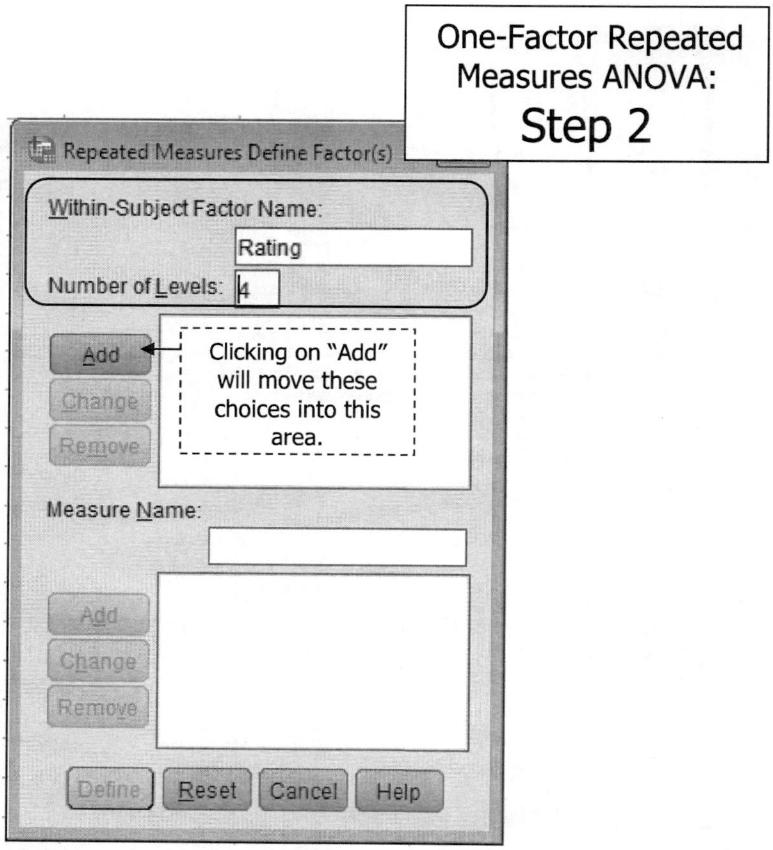

FIGURE 5.15
One-factor repeated measures ANOVA: Step 2.

Step 3. After we have defined the "Within-Subject Factor Name" and the "Number of Levels," then click on ADD to move this information into the middle box. In the screenshot for Step 3 (Figure 5.16), we see our newly defined repeated measures factor (i.e., Rater) with "4" indicating that there are four levels: Rater(4). Finally, click on "Define" to open the main Repeated Measures dialog box.

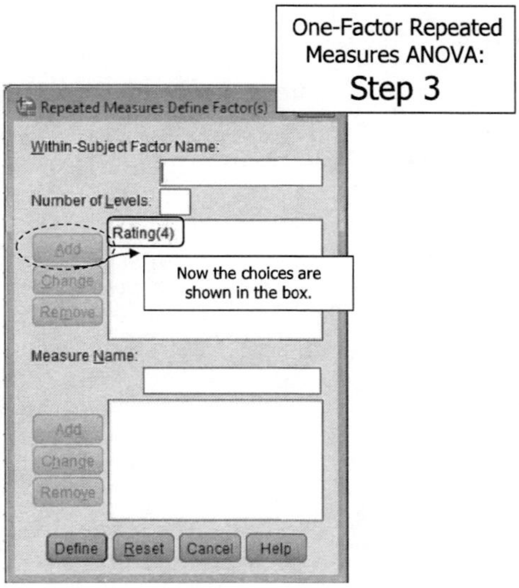

FIGURE 5.16
One-factor repeated measures ANOVA: Step 3.

Step 4a. From the Repeated Measures dialog box (see the screenshot for Step 4a, Figure 5.17), we see a heading called "Within-Subjects Variables" with the newly defined factor rater in parentheses. In this illustration, the values of 1 through 4 represent each one of the four raters that we just defined through the screenshot in Figure 5.16. Preceding each of the levels of the repeated factor are lines with question marks. This is the software's way of asking us to define which variable from the list on the left represents the first measurement (or the first rater in our illustration).

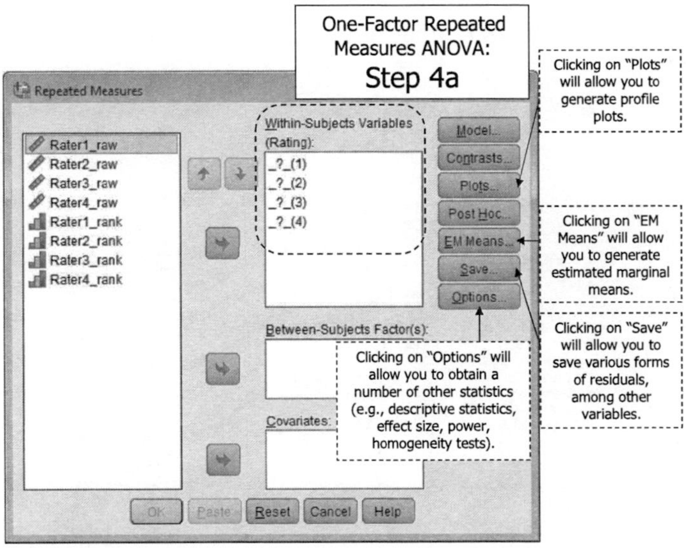

FIGURE 5.17
One-factor repeated measures ANOVA: Step 4a.

Step 4b. Move the appropriate variables from the variable list on the left into the "Within-Subjects Variables" box on the right. It is important to make sure that the first measurement is matched up with "1," the second measurement is matched with "2," and so forth so that the correct order of repeated measures is defined. This is especially critical when there is some temporal order to the repeated measures (e.g., pre-, post-, 3 months after post, etc.).

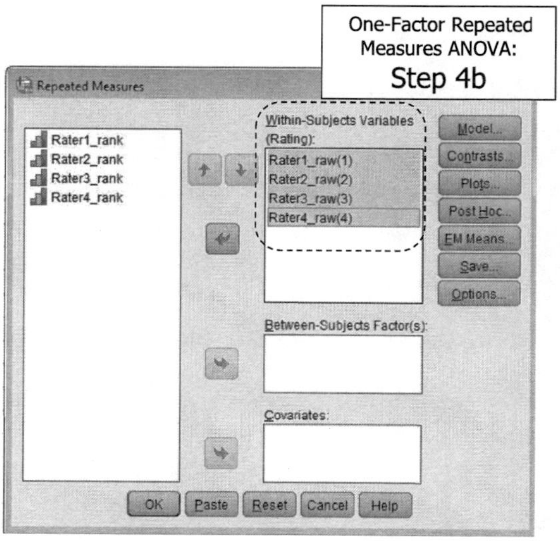

FIGURE 5.18
One-factor repeated measures ANOVA: Step 4b.

Step 5. From the Univariate dialog box (see the screenshot in Figure 5.17), clicking on "EM Means" will provide the option to compute marginal and overall means. For the one-factor repeated measures ANOVA, this dialog box is the proper place to obtain post hoc multiple comparison procedures including Tukey's LSD, Bonferroni, and Sidak procedures. Click on "Continue" to return to the original dialog box.

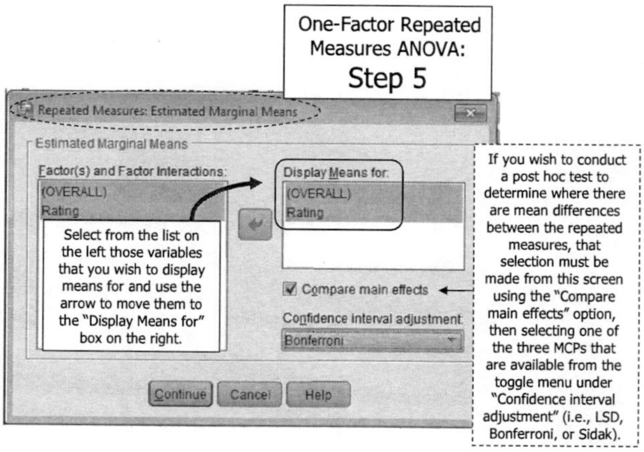

FIGURE 5.19
One-factor repeated measures ANOVA: Step 5.

Step 6. From the Univariate dialog box (see the screenshot in Figure 5.17), clicking on "Options" will provide the option to select such information as "Descriptive statistics," "Estimates of effect size," "Observed power," and "Homogeneity tests." Click on "Continue" to return to the original dialog box.

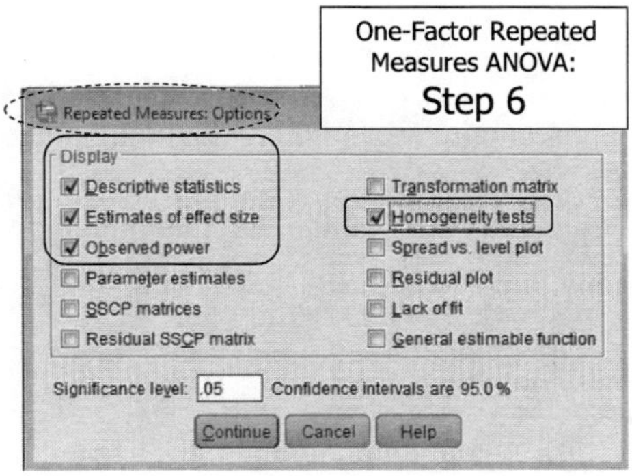

FIGURE 5.20
One-factor repeated measures ANOVA: Step 6.

Step 7. From the Univariate dialog box (see the screenshot in Figure 5.17), click on "Plots" to obtain a profile plot of means. Click the repeated measure factor (e.g., "Rater") and move it into the "Horizontal Axis" box by clicking the arrow button (see the screenshot for Step 7a in Figure 5.21). Then click on "Add" to move the variable into the "Plots" box at the bottom of the dialog box (see the screenshot for Step 7b in Figure 5.22). Click on "Continue" to return to the original dialog box.

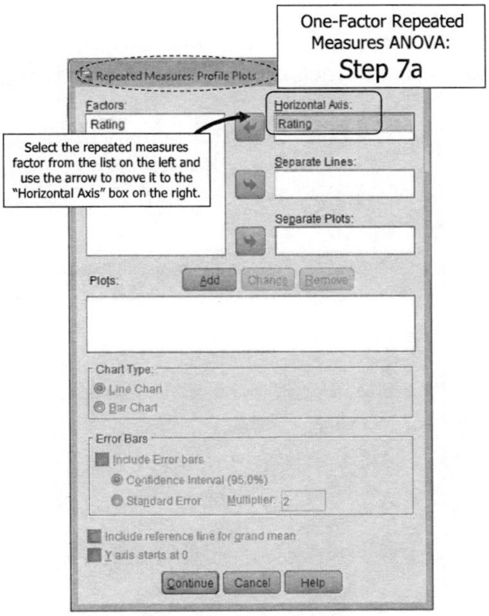

FIGURE 5.21
One-factor repeated measures ANOVA: Step 7a.

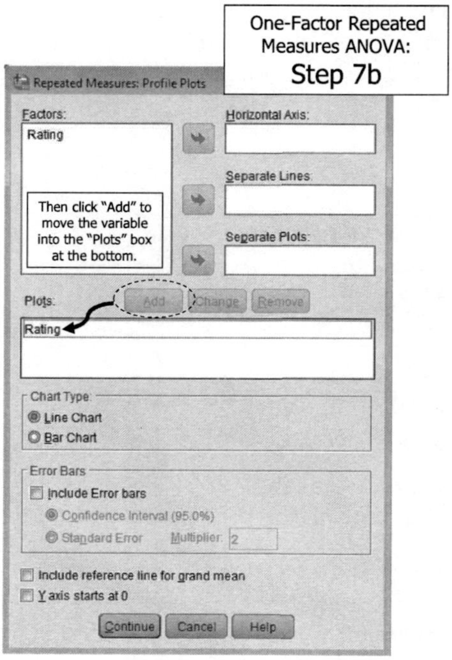

FIGURE 5.22
One-factor repeated measures ANOVA: Step 7b.

Step 8. From the Univariate dialog box (see the screenshot in Figure 5.17), click on "Save" to select those elements that you want to save (in our case, we want to save the unstandardized residuals which will be used later to examine the extent to which normality and independence are met). To do this, place a checkmark next to "Unstandardized." Click "Continue" to return to the main Univariate dialog box, and then click on "OK" to return to generate the output.

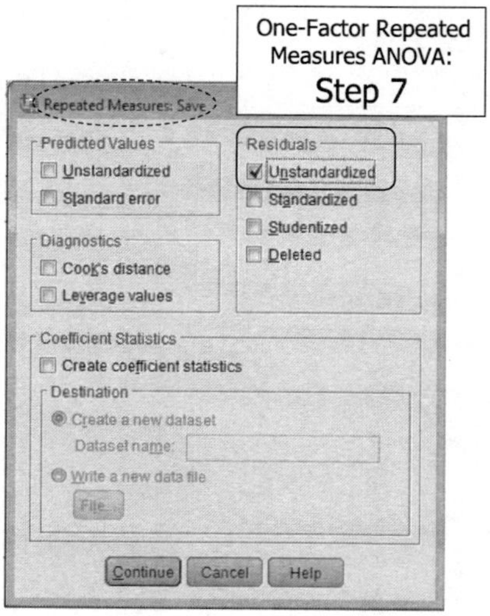

FIGURE 5.23
One-factor repeated measures ANOVA: Step 8.

Interpreting the output. Annotated results are presented in Table 5.14.

TABLE 5.14

One-Factor Repeated Measures ANOVA SPSS Results for the Ballet Technique Example

Within-Subjects Factors

Measure: MEASURE_1

Rating	Dependent Variable
1	Rater1_raw
2	Rater2_raw
3	Rater3_raw
4	Rater4_raw

Descriptive Statistics

	Mean	Std. Deviation	N
Rater1_raw	2.7500	1.48805	8
Rater2_raw	3.6250	.91613	8
Rater3_raw	6.2500	1.03510	8
Rater4_raw	9.1250	.83452	8

The table labeled "Descriptive Statistics" provides basic descriptive statistics (means, standard deviations, and sample sizes) for each rater of the repeated measure.

Multivariate Tests[a]

Effect		Value	F	Hypothesis df	Error df	Sig.	Partial Eta Squared	Noncent. Parameter	Observed Power[c]
Rating	Pillai's Trace	.967	48.650[b]	3.000	5.000	.000	.967	145.949	1.000
	Wilks' Lambda	.033	48.650[b]	3.000	5.000	.000	.967	145.949	1.000
	Hotelling's Trace	29.190	48.650[b]	3.000	5.000	.000	.967	145.949	1.000
	Roy's Largest Root	29.190	48.650[b]	3.000	5.000	.000	.967	145.949	1.000

a. Design: Intercept
Within-Subjects Design: Rating

b. Exact statistic

c. Computed using alpha = .05

The table labeled "Multivariate Tests" provides results for the multivariate test of mean differences between the repeated measures. Multivariate tests are provided when there are three or more levels of the within-subjects factor. These results are generally more conservative than the univariate results (in other words, you may be less likely to find statistically significant multivariate results as compared to univariate results.) *Note that the multivariate tests do not require meeting the assumption of sphericity.* Thus, if the assumption of sphericity is met, reporting univariate results is recommended.

If results for the multivariate tests are reported, of the four test results, Wilks' Lambda is recommended. In this example, all four multivariate criteria produce the same results—specifically that there is a statistically significant multivariate mean difference (as noted by p less than α.)

(continued)

TABLE 5.14 (continued)

One-Factor Repeated Measures ANOVA SPSS Results for the Ballet Technique Example

"Mauchly's Test of Sphericity" can be reviewed to determine if the assumption of sphericity is met. If the *p* value is larger than α (as in this illustration), we have met the assumption of sphericity.

"Epsilon" is a gauge of differences in the variances of the repeated measures and is used to adjust the degrees of freedom when sphericity is violated. The closer the epsilon values is to 1.0, the more homogenous are the variances. Complete heterogeneity of variances is specified by the "Lower-bound" and is computed as $1/(K-1)$ where K is the number of within subjects factors. For this example, with four raters, the lower bound is $1/(4-1)$ or .333.

Mauchly's Test of Sphericity[a]

Measure: MEASURE_1

Within Subjects Effect	Mauchly's W	Approx. Chi-Square	df	Sig.	Epsilon[b] Greenhouse-Geisser	Huynh-Feldt	Lower-bound
Rating	.155	10.679	5	.062	.476	.564	.333

Tests the null hypothesis that the error covariance matrix of the orthonormalized transformed dependent variables is proportional to an identity matrix.

a. Design: Intercept
Within-Subjects Design: Rating

b. May be used to adjust the degrees of freedom for the averaged tests of significance. Corrected tests are displayed in the Tests of Within-Subjects Effects table.

Since we met the assumption of sphericity, we use the results from the row labeled "sphericity assumed."

Rater *df* is computed as $(J-1) = 4-1 = 3$

Comparing *p* to α, we find a statistically significant difference in the mean ratings. This is an omnibus test. We will look at our MCP to determine which mean ratings differ.

Partial eta squared is one measure of effect size:

$$\eta^2_{partial} = \frac{SS_{betw}}{SS_{betw} + SS_{error}}$$

$$\eta^2_{partial} = \frac{198.125}{198.125 + 18.875} = .913$$

We can interpret this to say that approximately 91% of the variation in the rating is accounted for by the differences in the raters.

Tests of Within-Subjects Effects

Measure: MEASURE_1

Source		Type III Sum of Squares	df	Mean Square	F	Sig.	Partial Eta Squared	Noncent. Parameter	Observed Power[a]
Rating	Sphericity Assumed	198.125	3	66.042	73.477	.000	.913	220.430	1.000
	Greenhouse-Geisser	198.125	1.428	138.760	73.477	.000	.913	104.912	1.000
	Huynh-Feldt	198.125	1.691	117.163	73.477	.000	.913	124.250	1.000
	Lower-bound	198.125	1.000	198.125	73.477	.000	.913	73.477	1.000
Error (Rating)	Sphericity Assumed	18.875	21	.899					
	Greenhouse-Geisser	18.875	9.995	1.888					
	Huynh-Feldt	18.875	11.837	1.595					
	Lower-bound	18.875	7.000	2.696					

a. Computed using alpha = .05

Error sum of squares indicates how much variability is unexplained across the conditions of the repeated measures.

Error *df* is computed as $(J-1)(N-1) = (4-1)(8-1) = 21$

Had we violated the assumption of sphericity, we would have wanted to use a different set of results (e.g., Greenhouse-Geisser, Huynh-Feldt, Lower-bound). Notice that in all four sets of results, the sum of squares is the same value, however the *degrees of freedom differs* for each. The *F* ratio is computed the same for each (i.e., MS_{rater}/MS_{error}). Of the three results that can be used when sphericity is violated, the Lower-bound is the most conservative, followed by Greenhouse-Geisser (use when epsilon is ≤ .75 and then Huynh-Feldt (use when .75 < epsilon < 1.0).

Observed power tells whether our test is powerful enough to detect mean differences if they really exist. Power of 1.000 indicates maximum power, the probability of rejecting the null hypothesis if it is really false is 1.00.

TABLE 5.14 (continued)

One-Factor Repeated Measures ANOVA SPSS Results for the Ballet Technique Example

The output from the "Tests of Within-Subjects Contrasts" will not be used. Polynomial contrasts do not make sense for the rater factor.

Tests of Within-Subjects Contrasts

Measure: MEASURE_1

Source	Rating	Type III Sum of Squares	df	Mean Square	F	Sig.	Partial Eta Squared	Noncent. Parameter	Observed Power[a]
Rating	Linear	189.225	1	189.225	103.685	.000	.937	103.685	1.000
	Quadratic	8.000	1	8.000	18.667	.003	.727	18.667	.957
	Cubic	.900	1	.900	2.032	.197	.225	2.032	.235
Error (Rating)	Linear	12.775	7	1.825					
	Quadratic	3.000	7	.429					
	Cubic	3.100	7	.443					

a. Computed using alpha = .05

The output from the "Tests of Between-Subjects Effects" will not be used as there is no between-subjects factor.

Tests of Between-Subjects Effects

Measure: MEASURE_1

Transformed Variable: Average

Source	Type III Sum of Squares	df	Mean Square	F	Sig.	Partial Eta Squared	Noncent. Parameter	Observed Power[a]
Intercept	946.125	1	946.125	445.235	.000	.985	445.235	1.000
Error	14.875	7	2.125					

a. Computed using alpha = .05

Estimated Marginal Means

1. Grand Mean

Measure: MEASURE_1

Mean	Std. Error	95% Confidence Interval Lower Bound	95% Confidence Interval Upper Bound
5.438	.258	4.828	6.047

The "Grand Mean" (in this case, 5.438) represents the overall mean, regardless of the rater. The 95% CI represents the CI of the grand mean.

(continued)

TABLE 5.14 (continued)

One-Factor Repeated Measures ANOVA SPSS Results for the Ballet Technique Example

2. Rating

Estimates

Measure: MEASURE_1

Rating	Mean	Std. Error	95% Confidence Interval Lower Bound	95% Confidence Interval Upper Bound
1	2.750	.526	1.506	3.994
2	3.625	.324	2.859	4.391
3	6.250	.366	5.385	7.115
4	9.125	.295	8.427	9.823

The table labeled **"Rating"** provides descriptive statistics for each of the four raters. In addition to means, the *SE* and 95% CI of the means are reported.

"Mean Difference" is simply the difference between the means of the two raters being compared. For example, the mean difference of rater 1 and rater 2 is calculated as 2.750 – 3.625 = –.875.

Pairwise Comparisons

Measure: MEASURE_1

(I) Rating	(J) Rating	Mean Difference (I-J)	Std. Error	Sig.[b]	95% Confidence Interval for Difference[b] Lower Bound	95% Confidence Interval for Difference[b] Upper Bound
1	2	-.875	.295	.126	-1.948	.198
	3	-3.500*	.267	.000	-4.472	-2.528
	4	-6.375*	.706	.000	-8.940	-3.810
2	1	.875	.295	.126	-.198	1.948
	3	-2.625*	.263	.000	-3.581	-1.669
	4	-5.500*	.567	.000	-7.561	-3.439
3	1	3.500*	.267	.000	2.528	4.472
	2	2.625*	.263	.000	1.669	3.581
	4	-2.875*	.549	.007	-4.871	-.879
4	1	6.375*	.706	.000	3.810	8.940
	2	5.500*	.567	.000	3.439	7.561
	3	2.875*	.549	.007	.879	4.871

Based on estimated marginal means

*. The mean difference is significant at the .05 level.

b. Adjustment for multiple comparisons: Bonferroni.

"Sig." denotes the observed *p* value and provides the results of the Bonferroni post hoc procedure. There is a statistically significant mean difference between:

1. rater 1 and rater 3
2. rater 1 and rater 4
3. rater 2 and rater 3
4. rater 2 and rater 4
5. rater 3 and rater 4

The only groups for which there is not a statistically significant mean difference is between raters 1 and 2.

Note there are redundant results presented in the table. The comparison of raters 1 and 2 (presented in results for rater 1) is the same as the comparison of raters 2 and 1 (presented in results for rater 2) and so forth.

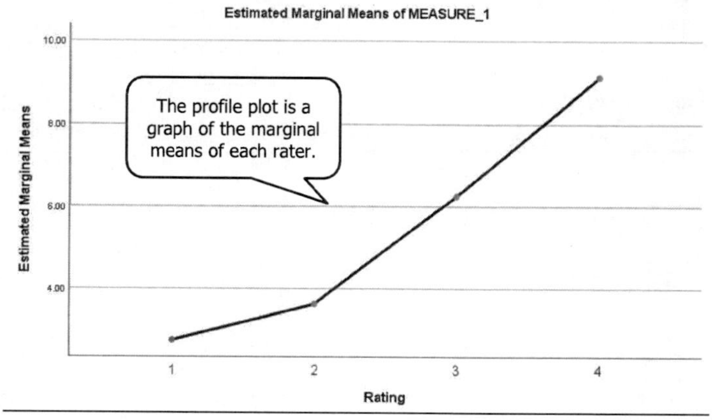

5.6.5 Friedman's Test: Nonparametric One-Factor Repeated Measures ANOVA

Step 1. The nonparametric version of the repeated measures ANOVA is Friedman's test. To compute Friedman's test, go to "Analyze" in the top pulldown menu, then select "Nonparametric Tests," then "Legacy Dialogs," and then finally "K Related Samples." Following the screenshot for Step 1 (shown in Figure 5.24) produces the "Tests for Several Related Samples" dialog box.

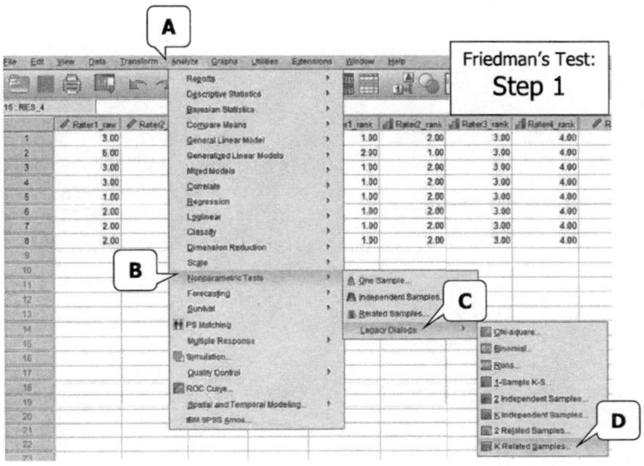

FIGURE 5.24
Friedman's test: Step 1.

Step 2. Recall that the Friedman test operates using ranked data, not continuous raw scores as with the repeated measures ANOVA; thus we will work with the ranked variables in our dataset for this test. From the Tests for Several Related Samples dialog box, click the variables representing the *ranked levels* of the repeated factor into the "Test Variables" box by using the arrow key in the middle of the dialog box. Under "Test Type" at the bottom left, check "Friedman." Then click on "OK" to return to generate the output.

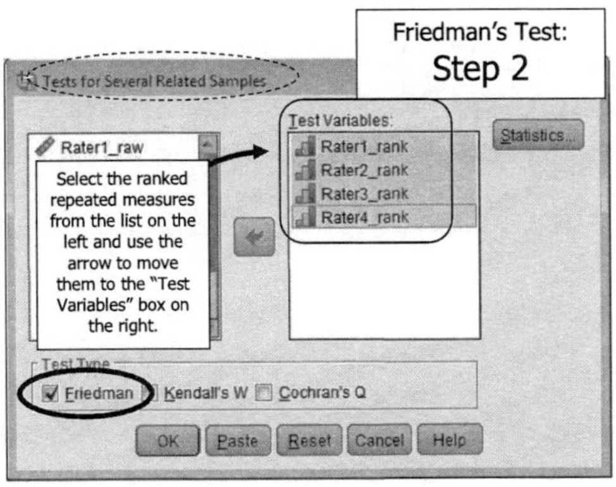

FIGURE 5.25
Friedman's test: Step 2.

Interpreting the output. Annotated results are presented in Table 5.15.

TABLE 5.15

Friedman's test SPSS results for the ballet technique example.

Ranks

	Mean Rank
Rater1_rank	1.13
Rater2_rank	1.88
Rater3_rank	3.00
Rater4_rank	4.00

The table labeled "Ranks" provides the average rank for each of the repeated measures levels.

Test Statistics[a]

N	8
Chi-Square	22.950
df	3
Asymp. Sig.	.000

a. Friedman Test

The table labeled "Test Statistics" provides the results for the hypothesis test of the difference in the mean ranks. Since p is less than α, this tells us there is a statistically significant difference in the mean ranks of the raters.

5.6.6 Two-Factor Split-Plot ANOVA

To conduct the two-factor split-plot ANOVA, the dataset must include variables for each level of the repeated factor (as in the one-factor repeated measures ANOVA), and another variable for the nonrepeated factor. Here our repeated measures or within-subjects factor is reflected in the raw scores of the four raters and the nonrepeated or between-subjects factor is the instructor.

	Instructor	Rater1_raw	Rater2_raw	Rater3_raw	Rater4_raw
1	1.00	3.00	4.00	7.00	8.00
2	1.00	6.00	5.00	8.00	9.00
3	1.00	3.00	4.00	7.00	9.00
4	1.00	3.00	4.00	6.00	8.00
5	2.00	1.00	2.00	5.00	10.00
6	2.00	2.00	3.00	6.00	10.00
7	2.00	2.00	4.00	5.00	9.00
8	2.00	2.00	3.00	6.00	10.00

The **nonrepeated or between-subjects factor** is labeled "Instructor" where each value represents the instructor to which the dancers were randomly assigned. Four dancers were randomly assigned to instructor 1 and four were randomly assigned to instructor 2.

The **repeated measures** or **within-subjects factor** is labeled "Rater" where there are four different raters, each reflected in the score they assigned to each of the eight dancers. (We will use the raw scores of the raters for the two-factor split-plot ANOVA.)

FIGURE 5.26

Two-factor split-plot ANOVA data.

Step 1. To conduct a two-factor split-plot ANOVA, go to "Analyze" in the top pulldown menu, then select "General Linear Model," and then select "Repeated Measures." This will produce the Repeated Measures dialog box. This step has been presented previously (see Figure 5.14 for the one-factor repeated measures design) and will not be reiterated here.

Step 2. The "Repeated Measures Define Factor(s)" dialog box will appear (see Figure 5.15 for the one-factor repeated measures design presented previously). In the box under "Within-Subjects Factor Name," enter the name you wish to call the repeated factor. For this example we label the repeated factor "Rating." It is necessary to define a name for the repeated factor as there is no single variable representing this factor (recall that the columns in the dataset represent the repeated measures); in the dataset there is one variable for each level of the factor (in other words, one variable for each different rater or measurement). Again, in our example, there are four levels of rater (i.e., four raters) and thus four variables. The "number of levels" indicates the number of measurements of the repeated factor. Here there were four raters, and thus the "number of levels" of the factor is 4.

Step 3. After defining the "Within-Subjects Factor Name" and the "number of levels," then click on ADD to move this information into the middle box. In Figure 5.16 for the one-factor repeated measures design presented previously, we see our newly defined repeated factor (i.e., Rating) with "4" indicating it was measured by four raters: Rating(4). Finally, click on "Define" to open the main Repeated Measures dialog box.

Step 4a. From the Repeated Measures dialog box (see Figures 5.17 and 5.18 for the one-factor repeated measures design presented previously), we see a heading called "Within-Subjects Variables" with the newly defined factor rater in parentheses. Here the values of 1 through 4 represent each one of the four raters. Preceding each of the levels of the repeated factor are lines with question marks. This is the software's way of asking us to define which variable represents the first measurement (or the first rater in our illustration).

Step 4b. Move the appropriate variables from the variable list on the left into the "Within-Subjects Variables" box on the right. It is important to make sure that the first measurement is matched up with "1," the second measurement is matched with "2," and so forth so that the correct order of repeated measures is defined.

Step 5. Once the "Within-Subjects Variables" are defined, the next step is to define the between-subjects or nonrepeated factor, as we see in the screenshot that follows (Figure 5.27). Move the appropriate variable from the variable list on the left into the "Between-Subjects Factor(s)" box on the right. From this, point, the options and selections work as we have seen when conducting other ANOVA models.

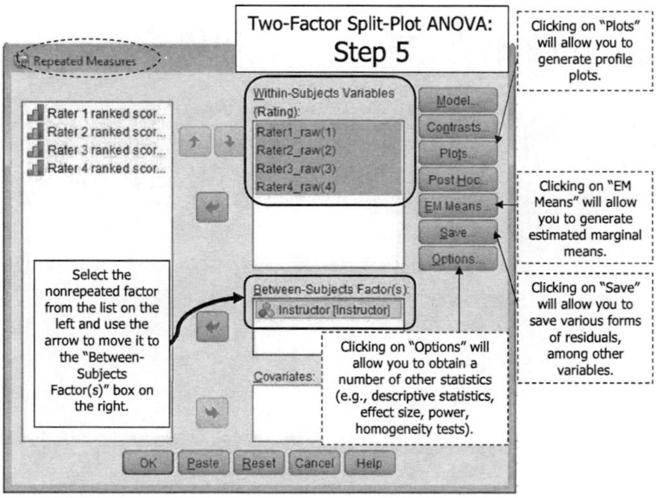

FIGURE 5.27
Two-factor split-plot ANOVA: Step 5.

Step 6. From the Repeated Measures dialog box, clicking on "EM Means" will provide the option to display overall and marginal means (see the screenshot in Figure 5.28). For the two-factor split-plot ANOVA, this dialog box is the proper place to obtain post hoc multiple comparison procedures for the *repeated measure.* Post hoc procedures include Tukey's LSD, Bonferroni, and Sidak procedures. Click on "Continue" to return to the original dialog box.

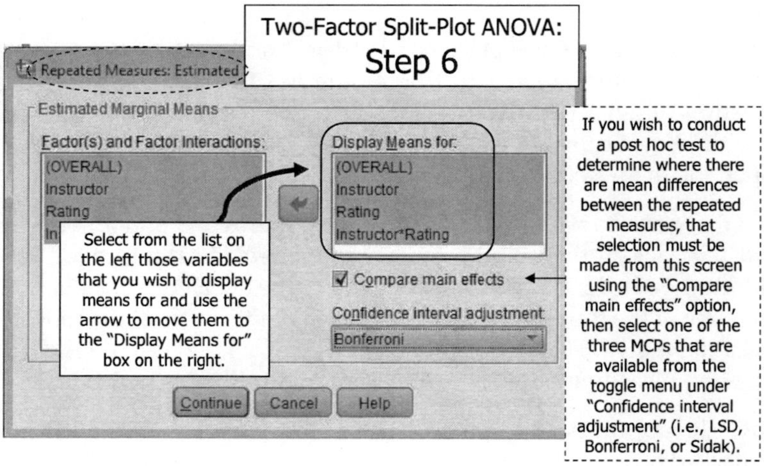

FIGURE 5.28
Two-factor split-plot ANOVA: Step 6.

Step 7. From the Repeated Measures dialog box, clicking on "Options" will provide the option to select such information as "Descriptive statistics," "Estimates of effect size," "Observed power," and "Homogeneity tests." Click on "Continue" to return to the original dialog box.

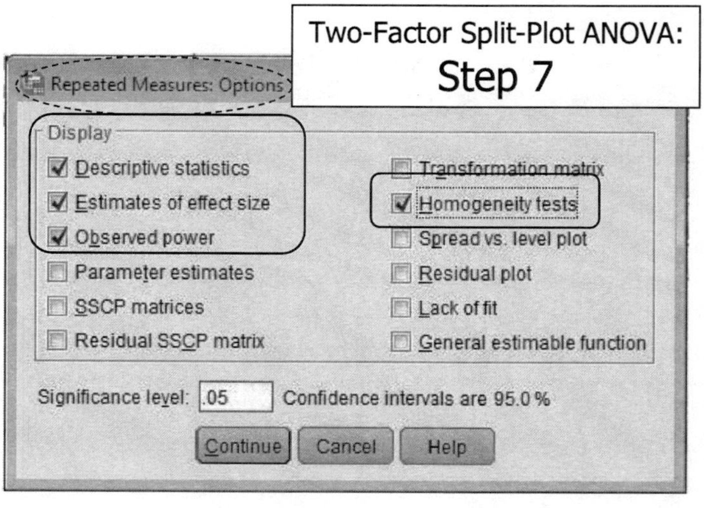

FIGURE 5.29
Two-factor split-plot ANOVA: Step 7.

Step 8. Click on the name of the nonrepeated or between-subjects factor in the "Factor(s)" list box in the top left and move it to the "Post Hoc Tests for" box in the top right by clicking on the arrow key. Check an appropriate MCP for your situation by placing a checkmark in the box next to the desired MCP. In this example, we select Tukey (see the screenshot for Step 8 shown in Figure 5.30). Click on "Continue" to return to the original dialog box.

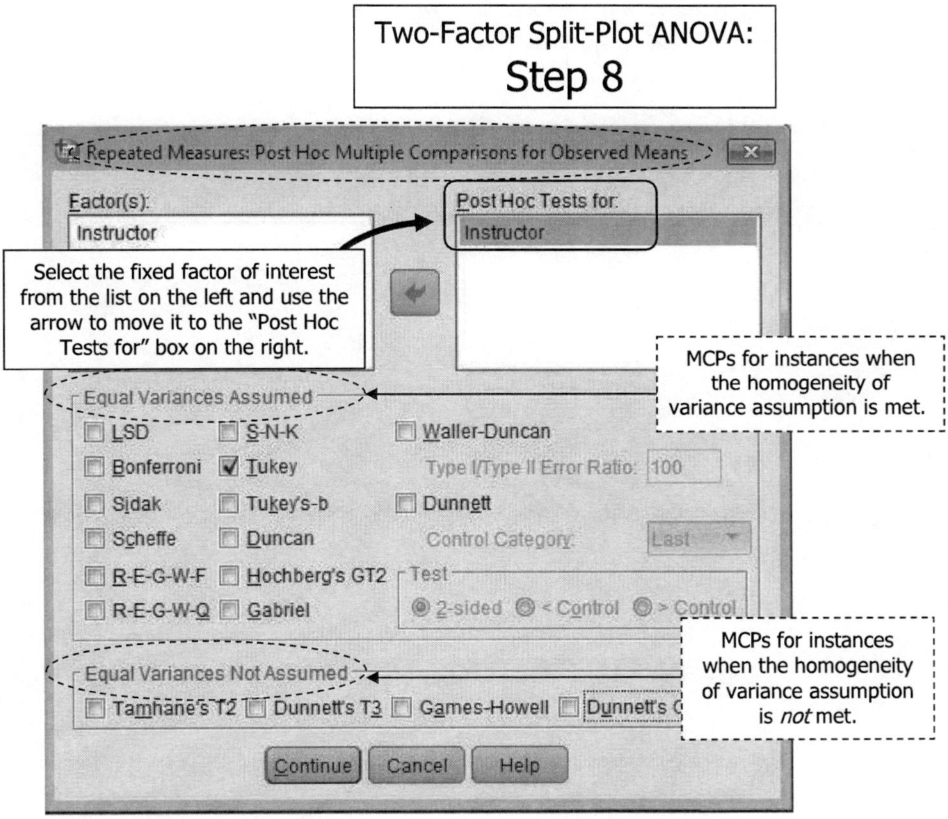

FIGURE 5.30
Two-factor split-plot ANOVA: Step 8.

Step 9. From the Repeated Measures dialog box, click on "Plots" to obtain a profile plot of means. Click the repeated measures factor (e.g., rating) and move it into the "Horizontal Axis" box by clicking the arrow button. Then click the nonrepeated factor (e.g., instructor) and move it into the "Separate Lines" box by clicking the arrow button. Then click on "Add" to move this into the "Plots" box at the bottom of the dialog box (see the screenshots for Steps 9a and 9b, Figures 5.31 and 5.32). Click on "Continue" to return to the original dialog box. (*Tip:* Placing the factor that has the most categories or levels on the horizontal axis of the profile plot will make for easier interpretation of the graph. In this case, there were four raters and two instructors, thus we placed "rater" on the horizontal axis. You can graph multiple plots, so trying different placement of factors on the axis or lines may produce a more desirable plot given your situation.)

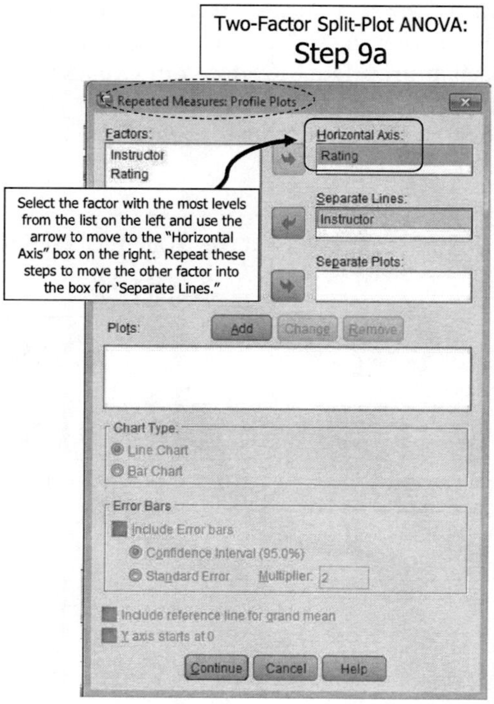

FIGURE 5.31
Two-factor split-plot ANOVA: Step 9a.

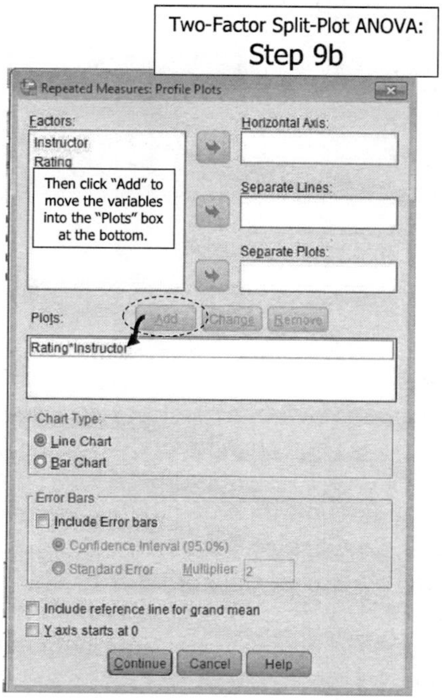

FIGURE 5.32
Two-factor split-plot ANOVA: Step 9b.

Step 10. From the Repeated Measures dialog box, click on "Save" to select those elements that you want to save (here we want to save the unstandardized residuals which will be used later to examine the extent to which normality and independence are met). To do this, place a checkmark next to "Unstandardized." Click "Continue" to return to the main Repeated Measures dialog box. From there, click on "OK" to generate the output.

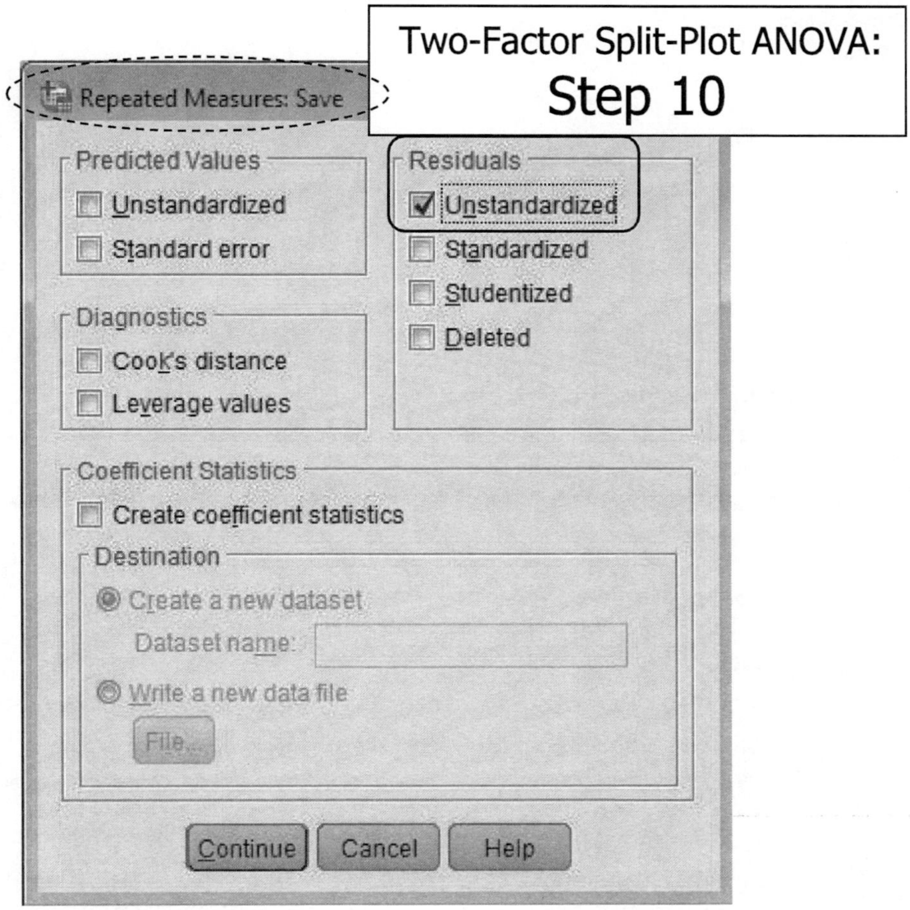

FIGURE 5.33
Two-factor split-plot ANOVA: Step 10.

Interpreting the output. Annotated results are presented in Table 5.16. Note that statistically significant interactions can be examined with simple effect, following the same steps as detailed in factorial ANOVA.

TABLE 5.16

Two-Factor Split-Plot ANOVA SPSS Results for the Ballet Technique Example

Within-Subjects Factors

Measure: MEASURE_1

Rating	Dependent Variable
1	Rater1_raw
2	Rater2_raw
3	Rater3_raw
4	Rater4_raw

The table labeled "Within-Subjects Factors" lists the variable names for levels of the repeated factor.

Between-Subjects Factors

		Value Label	N
Instructor	1.00	Instructor 1	4
	2.00	Instructor 2	4

The table labeled "Between-Subjects Factors" lists the names and sample sizes for the levels of the nonrepeated factor.

Descriptive Statistics

	Instructor	Mean	Std. Deviation	N
Rater 1 raw score	Instructor 1	3.7500	1.50000	4
	Instructor 2	1.7500	.50000	4
	Total	2.7500	1.48805	8
Rater 2 raw score	Instructor 1	4.2500	.50000	4
	Instructor 2	3.0000	.81650	4
	Total	3.6250	.91613	8
Rater 3 raw score	Instructor 1	7.0000	.81650	4
	Instructor 2	5.5000	.57735	4
	Total	6.2500	1.03510	8
Rater 4 raw score	Instructor 1	8.5000	.57735	4
	Instructor 2	9.7500	.50000	4
	Total	9.1250	.83452	8

The table labeled "Descriptive Statistics" lists the means, standard deviations, and sample sizes for each of the between subjects factors (i.e., instructors) by each of the repeated measures (i.e., raters).

TABLE 5.16 (continued)

Two-Factor Split-Plot ANOVA SPSS Results for the Ballet Technique Example

Multivariate Tests[a]

Effect		Value	F	Hypothesis df	Error df	Sig.	Partial Eta Squared	Noncent. Parameter	Observed Power[c]
Rating	Pillai's Trace	.983	74.892[b]	3.000	4.000	.001	.983	224.677	1.000
	Wilks' Lambda	.017	74.892[b]	3.000	4.000	.001	.983	224.677	1.000
	Hotelling's Trace	56.169	74.892[b]	3.000	4.000	.001	.983	224.677	1.000
	Roy's Largest Root	56.169	74.892[b]	3.000	4.000	.001	.983	224.677	1.000
Rating * Instructor	Pillai's Trace	.899	11.925[b]	3.000	4.000	.018	.899	35.774	.860
	Wilks' Lambda	.101	11.925[b]	3.000	4.000	.018	.899	35.774	.860
	Hotelling's Trace	8.944	11.925[b]	3.000	4.000	.018	.899	35.774	.860
	Roy's Largest Root	8.944	11.925[b]	3.000	4.000	.018	.899	35.774	.860

a. Design: Intercept + Instructor
 Within Subjects Design: Rating

b. Exact statistic

c. Computed using alpha = .05

The table labeled "Multivariate Tests" provides results for the multivariate test of mean differences for the repeated measures factor (i.e., "Rating"), and for the between- by within-subjects interaction (i.e., "Rating Instructor"). Multivariate tests are provided when there are three or more levels of the within-subjects factor. These results are generally more conservative than the univariate results (in other words, you may be less likely to find statistically significant multivariate results as compared to univariate results.). *Note that the multivariate tests do not require meeting the assumption of sphericity.* Thus if the assumption of sphericity is met, reporting univariate results is recommended.

If results for the multivariate tests are reported, of the four test criteria, Wilks' Lambda is recommended. In this example, all four multivariate criteria produce the same results—specifically that there is a statistically significant multivariate mean difference for the repeated measures factor and a statistically significant between- by within-subjects interaction (as noted by p less than α).

"Mauchly's Test of Sphericity" can be reviewed to determine if the assumption of sphericity is met. If the p value is larger than α (as in this illustration), we have met the assumption of sphericity.

'Epsilon' is a gauge of differences in the variances of the repeated measures. The closer the epsilon value is to 1.0, the more homogenous are the variances. Complete heterogeneity of variances is specified by the 'Lower-bound' and is computed as $1/(K-1)$ where K is the number of within subjects levels. For this example, with four raters, the lower bound is $1/(4-1)$ or .333.

Mauchly's Test of Sphericity[a]

Measure: MEASURE_1

Within Subjects Effect	Mauchly's W	Approx. Chi-Square	df	Sig.	Epsilon[b] Greenhouse-Geisser	Huynh-Feldt	Lower-bound
Rating	.429	4.001	5	.557	.706	1.000	.333

Tests the null hypothesis that the error covariance matrix of the orthonormalized transformed dependent variables is proportional to an identity matrix.

a. Design: Intercept + Instructor
 Within Subjects Design: Rating

b. May be used to adjust the degrees of freedom for the averaged tests of significance. Corrected tests are displayed in the Tests of Within-Subjects Effects table.

(*continued*)

TABLE 5.16 (continued)

Two-Factor Split-Plot ANOVA SPSS Results for the Ballet Technique Example

The table labeled "Tests of Within-Subjects Effects" provides results for the univariate test of mean differences for the within-subjects factor (i.e., "rating") and within-between subjects interaction (i.e., "rating*instructor").

Since we met the assumption of sphericity, we use the results from the row labeled "sphericity assumed."

Rating *df* is computed as $(K-1) = 4-1 = 3$

Comparing *p* to α, we find a statistically significant difference in the ratings (repeated factor) *and* a statistically significant rating by instructor interaction. These are omnibus tests. We will look at our MCPs to determine which raters differ and which ratings differ by instructor.

Partial eta squared is one measure of effect size:

$$\eta^2 = \frac{SS_{betw}}{SS_{betw} + SS_{error}}$$

$$\eta^2 = \frac{198.125}{198.125 + 6.250} = .969$$

We can interpret this to say that approximately 97% of the variation in the ratings is accounted for by the differences in the raters.

Tests of Within-Subjects Effects

Measure: MEASURE_1

Source		Type III Sum of Squares	df	Mean Square	F	Sig.	Partial Eta Squared	Noncent. Parameter	Observed Power[a]
Rating	Sphericity Assumed	198.125	3	66.042	190.200	.000	.969	570.600	1.000
	Greenhouse-Geisser	198.125	2.119	93.515	190.200	.000	.969	402.966	1.000
	Huynh-Feldt	198.125	3.000	66.042	190.200	.000	.969	570.600	1.000
	Lower-bound	198.125	1.000	198.125	190.200	.000	.969	190.200	1.000
Rating * Instructor	Sphericity Assumed	12.625	3	4.208	12.120	.000	.669	36.360	.998
	Greenhouse-Geisser	12.625	2.119	5.959	12.120	.001	.669	25.678	.983
	Huynh-Feldt	12.625	3.000	4.208	12.120	.000	.669	36.360	.998
	Lower-bound	12.625	1.000	12.625	12.120	.013	.669	12.120	.825
Error (Rating)	Sphericity Assumed	6.250	18	.347					
	Greenhouse-Geisser	6.250	12.712	.492					
	Huynh-Feldt	6.250	18.000	.347					
	Lower-bound	6.250	6.000	1.042					

a. Computed using alpha = .05

Error sum of squares indicates how much variability is unexplained across the conditions of the repeated measures.

Within*Between interaction *df* is computed as $(K-1)(J-1) = (4-1)(2-1) = 3$

Error *df* is computed as: $(J)(K-1)(n-1) = (2)(4-1)(4-1) = 18$

Had we violated the assumption of sphericity, we would have wanted to use a different set of results (e.g., Greenhouse-Geisser, Huynh-Feldt, Lower-bound). Notice that in all four sets of results, the sum of squares is the same value, however the *degrees of freedom differs* for each. The *F* ratio is computed the same for each. Of the three results that can be used when sphericity is violated, the lower-bound is the most conservative, followed by Greenhouse-Geisser and then Huynh-Feldt.

Observed power tells whether our test is powerful enough to detect mean differences if they really exist. Power of 1.000 indicates maximum power, the probability of rejecting the null hypothesis if it is really false is 1.00. Power of .998 is only slightly below maximum power of 1.00; this is extremely strong power.

TABLE 5.16 (continued)

Two-Factor Split-Plot ANOVA SPSS Results for the Ballet Technique Example

The output from the "Tests of Within-Subjects Contrasts" will not be used as polynomial contrasts do not make sense here.

Tests of Within-Subjects Contrasts

Measure: MEASURE_1

Source	Rating	Type III Sum of Squares	df	Mean Square	F	Sig.	Partial Eta Squared	Noncent. Parameter	Observed Power[a]
Rating	Linear	189.225	1	189.225	302.760	.000	.981	302.760	1.000
	Quadratic	8.000	1	8.000	48.000	.000	.889	48.000	1.000
	Cubic	.900	1	.900	3.600	.107	.375	3.600	.359
Rating * Instructor	Linear	9.025	1	9.025	14.440	.009	.706	14.440	.883
	Quadratic	2.000	1	2.000	12.000	.013	.667	12.000	.821
	Cubic	1.600	1	1.600	6.400	.045	.516	6.400	.563
Error (Rating)	Linear	3.750	6	.625					
	Quadratic	1.000	6	.167					
	Cubic	1.500	6	.250					

a. Computed using alpha = .05

Levene's Test of Equality of Error Variances[a]

		Levene Statistic	df1	df2	Sig.
Rater 1 raw score	Based on Mean	3.600	1	6	.107
	Based on Median	.400	1	6	.550
	Based on Median and with adjusted df	.400	1	3.659	.564
	Based on trimmed mean	2.704	1	6	.151
Rater 2 raw score	Based on Mean	.158	1	6	.705
	Based on Median	.429	1	6	.537
	Based on Median and with adjusted df	.429	1	5.880	.537
	Based on trimmed mean	.188	1	6	.680
Rater 3 raw score	Based on Mean	.000	1	6	1.000
	Based on Median	.000	1	6	1.000
	Based on Median and with adjusted df	.000	1	3.000	1.000
	Based on trimmed mean	.000	1	6	1.000
Rater 4 raw score	Based on Mean	1.000	1	6	.356
	Based on Median	1.000	1	6	.356
	Based on Median and with adjusted df	1.000	1	3.000	.391
	Based on trimmed mean	1.000	1	6	.356

Tests the null hypothesis that the error variance of the dependent variable is equal across groups.

a. Design: Intercept + Instructor
Within Subjects Design: Rating

The *F* test (and associated *p* values) for Levene's Test for Equality of Error Variances is reviewed to determine if equal variances can be assumed. In this case, we meet the assumption (as *p* is greater than α).

Note that *df1* is degrees of freedom for the numerator (calculated as $J - 1$) and *df2* are the degrees of freedom for the denominator (calculated as $N - J$).

SPSS computes Levene's test four different ways and reports the associated statistic and significance for each. All test the same null hypothesis (which is noted in the footnote of the table) and are thus interpreted the same way (i.e., as a test of equal population error variances across all cells). For this illustration, we will interpret the results "based on mean" and will use the corresponding *p* value to interpret the extent of meeting the assumption of homogeneity.

(*continued*)

TABLE 5.16 (continued)

Two-Factor Split-Plot ANOVA SPSS Results for the Ballet Technique Example

The table labeled "Tests of Between-Subjects Effects" provides results for the univariate test of mean differences for the between-subjects factor (i.e., "instructor").

Instructor *df* is computed as $(J-1) = 2-1 = 1$

Comparing p to α, we find a statistically significant difference in the mean ratings by instructor. These are omnibus tests. We look at MCPs to determine which mean ratings differ by instructor.

Partial eta squared is one measure of effect size:

$$\eta^2 = \frac{SS_{betw}}{SS_{betw} + SS_{error}}$$

$$\eta^2 = \frac{6.125}{6.125 + 8.750} = .412$$

We can interpret this to say that approximately 41% of the variation in the ratings is accounted for by the differences in the instructors.

Tests of Between-Subjects Effects

Measure: MEASURE_1

Transformed Variable: Average

Source	Type III Sum of Squares	df	Mean Square	F	Sig.	Partial Eta Squared	Noncent. Parameter	Observed Power[a]
Intercept	946.125	1	946.125	648.771	.000	.991	648.771	1.000
Instructor	6.125	1	6.125	4.200	.086	.412	4.200	.407
Error	8.750	6	1.458					

a. Computed using alpha = .05

Observed power tells whether our test is powerful enough to detect mean differences if they really exist. Power of .407 indicates low power; the probability of rejecting the null hypothesis if it is really false is about .41 or 41%.

Estimated Marginal Means

1. Grand Mean

Measure: MEASURE_1

Mean	Std. Error	95% Confidence Interval Lower Bound	95% Confidence Interval Upper Bound
5.438	.213	4.915	5.960

The "Grand Mean" (in this case, 5.438) represents the overall mean, regardless of the rater or instructor. The 95% CI represents the CI of the grand mean.

2. Instructor

Estimates

Measure: MEASURE_1

Instructor	Mean	Std. Error	95% Confidence Interval Lower Bound	95% Confidence Interval Upper Bound
Instructor 1	5.875	.302	5.136	6.614
Instructor 2	5.000	.302	4.261	5.739

The table for **"Instructor"** provides descriptive statistics for each of the levels of our between-subjects factor. In addition to means, the *SE* and 95% CI of the means are reported.

TABLE 5.16 (continued)

Two-Factor Split-Plot ANOVA SPSS Results for the Ballet Technique Example

"Mean Difference" is simply the difference between the means of the two categories of our between-subjects factor. For example, the mean difference of instructor 1 and instructor 2 is calculated as 5.875 – 5.000 = .875

Pairwise Comparisons

Measure: MEASURE_1

(I) Instructor	(J) Instructor	Mean Difference (I-J)	Std. Error	Sig.[a]	95% Confidence Interval for Difference[a] Lower Bound	Upper Bound
Instructor 1	Instructor 2	.875	.427	.086	-.170	1.920
Instructor 2	Instructor 1	-.875	.427	.086	-1.920	.170

Based on estimated marginal means

a. Adjustment for multiple comparisons: Bonferroni.

"Sig." denotes the observed *p* value and provides the results of the Bonferroni post hoc procedure. There is not a statistically significant mean difference in ratings between instructor 1 and 2.

Note there are redundant results presented in the table. The comparison of instructor 1 and 2 (presented in the first row) is the same as the comparison of instructor 2 and 1 (presented in the second row).

The contrast output from the "Univariate Tests" will not be used here.

Univariate Tests

Measure: MEASURE_1

	Sum of Squares	df	Mean Square	F	Sig.	Partial Eta Squared	Noncent. Parameter	Observed Power[a]
Contrast	1.531	1	1.531	4.200	.086	.412	4.200	.407
Error	2.188	6	.365					

The F tests the effect of Instructor. This test is based on the linearly independent pairwise comparisons among the estimated marginal means.

a. Computed using alpha = .05

3. Rating

Estimates

Measure: MEASURE_1

Rating	Mean	Std. Error	95% Confidence Interval Lower Bound	Upper Bound
1	2.750	.395	1.783	3.717
2	3.625	.239	3.039	4.211
3	6.250	.250	5.638	6.862
4	9.125	.191	8.658	9.592

The table labeled **"Rating"** provides descriptive statistics for the rating of each of the four raters. In addition to means, the *SE* and 95% CI of the means are reported.

(continued)

TABLE 5.16 (continued)

Two-Factor Split-Plot ANOVA SPSS Results for the Ballet Technique Example

"Mean Difference" is simply the difference between the means of the two raters being compared. For example, the mean difference of rater 1 and rater 2 is calculated as 2.750 – 3.625 = -.875.

Pairwise Comparisons

Measure: MEASURE_1

(I) Rating	(J) Rating	Mean Difference (I-J)	Std. Error	Sig.[b]	95% Confidence Interval for Difference[b] Lower Bound	Upper Bound
1	2	-.875	.280	.122	-1.955	.205
	3	-3.500*	.270	.000	-4.543	-2.457
	4	-6.375*	.375	.000	-7.824	-4.926
2	1	.875	.280	.122	-.205	1.955
	3	-2.625*	.280	.000	-3.705	-1.545
	4	-5.500*	.339	.000	-6.808	-4.192
3	1	3.500*	.270	.000	2.457	4.543
	2	2.625*	.280	.000	1.545	3.705
	4	-2.875*	.191	.000	-3.613	-2.137
4	1	6.375*	.375	.000	4.926	7.824
	2	5.500*	.339	.000	4.192	6.808
	3	2.875*	.191	.000	2.137	3.613

Based on estimated marginal means

*. The mean difference is significant at the .05 level.

b. Adjustment for multiple comparisons: Bonferroni.

"Sig." denotes the observed *p* value and provides the results of the Bonferroni post hoc procedure. There is a statistically significant mean difference in ratings of writing between:

1. rater 1 and rater 3
2. rater 1 and rater 4
3. rater 2 and rater 3
4. rater 2 and rater 4
5. rater 3 and rater 4

The only groups for which there is not a statistically significant mean difference is raters 1 and 2.

Note there are redundant results presented in the table. The comparison of rater 1 and 2 (presented in results for rater 1) is the same as the comparison of group 2 and 1 (presented in results for rater 2) and so forth.

Multivariate test results for "rating," which were presented earlier in the output (note that earlier output included results for rating and rating*instructor), are provided again. See earlier output for interpretations.

Multivariate Tests

	Value	F	Hypothesis df	Error df	Sig.	Partial Eta Squared	Noncent. Parameter	Observed Power[b]
Pillai's trace	.983	74.892[a]	3.000	4.000	.001	.983	224.677	1.000
Wilks' lambda	.017	74.892[a]	3.000	4.000	.001	.983	224.677	1.000
Hotelling's trace	56.169	74.892[a]	3.000	4.000	.001	.983	224.677	1.000
Roy's largest root	56.169	74.892[a]	3.000	4.000	.001	.983	224.677	1.000

Each F tests the multivariate effect of Rating. These tests are based on the linearly independent pairwise comparisons among the estimated marginal means.

a. Exact statistic

b. Computed using alpha = .05

TABLE 5.16 (continued)

Two-Factor Split-Plot ANOVA SPSS Results for the Ballet Technique Example

4. Instructor*Rating

Measure: MEASURE_1

Instructor	Rating	Mean	Std. Error	95% Confidence Interval Lower Bound	Upper Bound
Instructor 1	1	3.750	.559	2.382	5.118
	2	4.250	.339	3.422	5.078
	3	7.000	.354	6.135	7.865
	4	8.500	.270	7.839	9.161
Instructor 2	1	1.750	.559	.382	3.118
	2	3.000	.339	2.172	3.828
	3	5.500	.354	4.635	6.365
	4	9.750	.270	9.089	10.411

The table for **"Instructor*Rating"** provides descriptive statistics for each of the combinations of instructor by rater (or cell). In addition to means, the *SE* and 95% CI of the means are reported.

Profile Plots

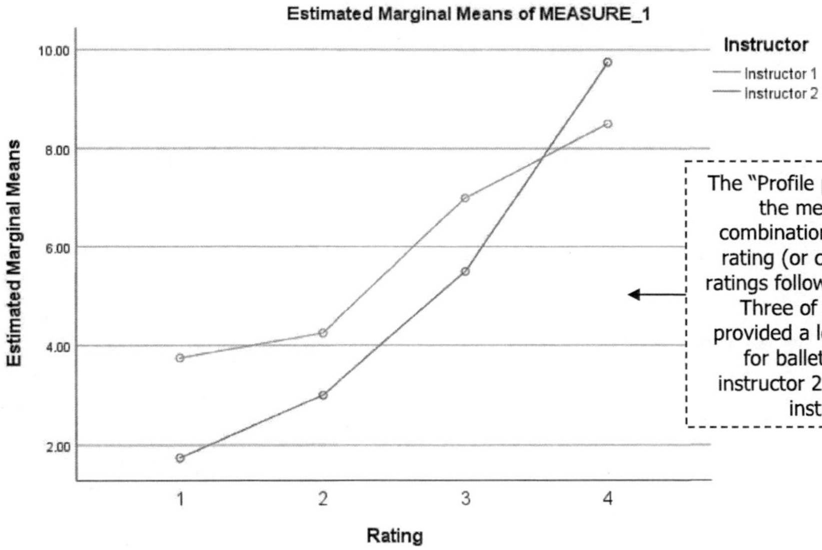

The "Profile plot" is a graph of the means for each combination of instructor by rating (or cell). We see the ratings follow a similar pattern. Three of the four raters provided a lower mean rating for ballet technique for instructor 2 (as compared to instructor 1).

5.7 Computing ANOVA Models Using R

5.7.1 The One-Factor Repeated Measures Design

Next we consider **R** for the one-factor repeated measures ANOVA model. Note that the scripts are provided within the blocks with additional annotation to assist in understanding how the command works. Should you want to write reminder notes and annotation to yourself as you write the commands in **R** (and we highly encourage doing so), remember that any text that follows a hashtag (i.e., #) is annotation only and not part of the **R** script. Thus, you can write annotations directly into **R** with hashtags. We encourage this practice so that when you call up the commands in the future, you'll understand what the various lines of code are doing. You may think you'll remember what you did. However, trust us. There is a good chance that you won't. Thus, consider it best practice when using **R** to annotate heavily!

```
getwd()
```

R is always pointed to a directory on your computer. The *get working directory* function can be used to determine to which directory **R** is pointed. We will assume that we need to change the working directory, and will use the next line of code to set the working directory to the desired path.

```
setwd("E:/FolderName")
```

We use the *setwd* function to establish the working directory. To set the working directory, change what is in quotation marks to your file location. Also, if you are copying the directory name from your properties, you will need to change the backslash (i.e., \) to a forward slash (i.e., /).

```
Ch5_repeat <- read.csv("Ch5_repeatraw.csv")
```

The *read.csv* function reads our data into **R**. What's to the left of the "<-" will be what the data will be called in **R**. In this example, we're calling the **R** dataframe "Ch5_repeat." What's to the right of the "<-" tells **R** to find this particular csv file. In this example, our file is called "Ch5_repeatraw.csv." Make sure the extension (i.e., .csv) is included in your script. Also note that the name of your file should be in quotation marks within the parentheses.

```
names(Ch5_repeat)
```

The *names* function will produce a list of variable names for each dataframe as follows. This is a good check to make sure your data have been read in correctly.

```
[1] "ID"    "R1_raw" "R2_raw" "R3_raw" "R4_raw"
```

```
View(Ch5_repeat)
```

The *View* function will let you view the dataset in spreadsheet format in RStudio.

```
summary(Ch5_repeat)
```

The *summary* function will produce basic descriptive statistics on all the variables in your dataframe. This is a great way to quickly check to see if the data have been read in correctly and to get a feel for your data, if you haven't already. The output from the summary statement for this dataframe looks like this:

```
       ID            R1_raw           R2_raw              R3_raw
Min.   :1.00   Min.   :1.00   Min.   :2.000   Min.   :5.00
1st Qu.:2.75   1st Qu.:2.00   1st Qu.:3.000   1st Qu.:5.75
Median :4.50   Median :2.50   Median :4.000   Median :6.00
Mean   :4.50   Mean   :2.75   Mean   :3.625   Mean   :6.25
3rd Qu.:6.25   3rd Qu.:3.00   3rd Qu.:4.000   3rd Qu.:7.00
Max.   :8.00   Max.   :6.00   Max.   :5.000   Max.   :8.00
    R4_raw
Min.   : 8.000
1st Qu.: 8.750
Median : 9.000
Mean   : 9.125
3rd Qu.:10.000
Max.   :10.000
```

FIGURE 5.34
Reading data into **R**.

5.7.2 Restructuring Data for the One-Factor Repeated Measures ANOVA Model

```
install.packages(reshape)
library(reshape2)
```

We will install the *reshape* package and load *reshape2* to convert our wide format data to long format.

```
Ch5long <-melt(Ch5_repeat, id.vars = c("ID"),
               measure.vars = ,
               variable.name = "rater",
               value.name = "ranking")
```

We are creating a new dataset named "Ch5long." The *melt* function will transform our "wide" format data to "long" format. In other words, there will be multiple rows of data for each measurement occasion but just one column for the repeated measure. Within parentheses, we see we are using the Ch5_repeat dataframe to do this. The ID variable we want to keep but not split apart so we include the *id.vars=c("ID")* command. If there were other nonrepeated measures, we would have listed those here as well. The last two lines tell us that *variable.name* defines the column heading for the Rater and *value.name* defines the column heading for the rank score.

```
names(Ch5long)
```

The *names* function will produce a list of variable names for our new dataframe as follows. This is a good check to make sure we have restructured the data correctly and retained the ID variable.

```
[1] "ID"    "rater"    "ranking"
```

```
View(Ch5long)
```

The *View* function will let you view the dataset in spreadsheet format in RStudio.

```
Ch5long$rater<-as.numeric(Ch5long$rater)
```

The *as.numeric* function will change the string to numeric values for the variable "rater" that is located in our Ch5long dataframe.

```
Ch5long<-within(Ch5long,
                {rater <-factor(rater)
                ID <- factor(ID)})
```

The *within* function will define *rater* and *ID* as factors within the *Ch5long* dataframe.

```
View(Ch5long)
```

The *View* function will let you view the dataset in spreadsheet format in RStudio.

FIGURE 5.35
Restructuring data for the one-factor repeated measures ANOVA model.

5.7.3 Generating the One-Factor Repeated Measures ANOVA Model

```
repeatedANOVA <- aov(ranking~rater+Error(ID), data=Ch5long)
```

The *aov* function will model the one-factor repeated measures ANOVA with "ranking" as the repeated measure, "rater" as the variable that defines who completed the ranking, and "ID" as the ID variable. The data come from the Ch5long dataframe, and we are creating an object called "repeatedANOVA" from the model.

```
summary(repeatedANOVA)
```

The *summary* function will provide output from our repeated measures ANOVA model.

```
Error: ID
          Df  Sum Sq  Mean Sq  F value    Pr(>F)
Residuals  7   14.88    2.125

Error: Within
          Df  Sum Sq  Mean Sq  F value    Pr(>F)
rater      3  198.13    66.04    73.48  2.66e-11 ***
Residuals 21   18.87     0.90
—
Signif. codes: 0 '***' 0.001 '**' 0.01 '*' 0.05 '.' 0.1 ' ' 1
```

```
resid <- proj(repeatedANOVA)
Ch5long$unstandardizedResiduals <- resid[[3]][, "Residuals"]
```

After running the repeated measures ANOVA, save the residuals using this script.

FIGURE 5.36
Generating the one-factor repeated measures ANOVA model.

5.7.4 Computing Friedman's Test in R: Nonparametric One-Factor Repeated Measures ANOVA

Next we consider **R** for Friedman's test, the nonparametric version of the one-factor repeated measures ANOVA model.

```
if (!require("devtools")) {
  install.packages("devtools")
}
devtools::install_github("b0rxa/scmamp")
library("scmamp")
```

The **R** package *devtools* will be used when generating our model. If this package is not installed, this command will install the pack and load what is required.

```
reprank <- read.csv (file="E:/FolderName/Ch5_repeatrank.csv", head=T, sep=",")
```

The *read.csv* function will read in our csv file, recognize the column headers as names (i.e., *head=T*) and recognize that it is a comma delimited file (i.e., *sep= ","*).

```
reprank
```

To see the data in our console, we type in its name and see the data displayed as follows.

FIGURE 5.37
Computing Friedman's test in **R**.

```
  R1_rank R2_rank R3_rank R4_rank
1       1       2       3       4
2       2       1       3       4
3       1       2       3       4
4       1       2       3       4
5       1       2       3       4
6       1       2       3       4
7       1       2       3       4
8       1       2       3       4
```

```
reprank.matrix <- data.matrix(reprank)
```

Next, using the *data.matrix* function, we take the dataframe "reprank" and convert it to a matrix labeled *reprank.matrix*.

```
reprank.matrix
```

To see the data in our console, we type in its name and see the data displayed as follows.

```
  R1_rank R2_rank R3_rank R4_rank
[1,]    1       2       3       4
[2,]    2       1       3       4
[3,]    1       2       3       4
[4,]    1       2       3       4
[5,]    1       2       3       4
[6,]    1       2       3       4
[7,]    1       2       3       4
[8,]    1       2       3       4
```

```
friedmanTest(reprank.matrix)
```

The *friedmanTest* function can be used to run Friedman's test on the matrix we just created, *reprank.matrix*. The results provided are as follows:

```
        Friedman's rank sum test
data:   reprank.matrix
Friedman's chi-squared = 22.95, df = 3, p-value = 4.136e-05
```

FIGURE 5.37 (continued)
Computing Friedman's test in **R**.

5.7.5 Computing the Two-Factor Split-Plot or Mixed Design in R

Next we consider **R** for the two-factor split-plot or mixed ANOVA model.

5.7.5.1 Reading Data Into R

```
getwd()
```

R is always pointed to a directory on your computer. The *get working directory* function can be used to determine to which directory **R** is pointed. We will assume that we need to change the working directory, and will use the next line of code to set the working directory to the desired path.

FIGURE 5.38
Reading data into **R**.

```
setwd("E:/FolderName")
```

We use the *setwd* function to establish the working directory. To set the working directory, change what is in quotation marks to your file location. Also, if you are copying the directory name from your properties, you will need to change the backslash (i.e., \) to a forward slash (i.e., /).

```
Ch5split <- read.csv("Ch5_splitplot.csv", header = TRUE)
```

The *read.csv* function reads our data into **R**. What's to the left of the "<-" will be what the data will be called in **R**. In this example, we're calling the **R** dataframe "Ch5_split." What's to the right of the "<-" tells **R** to find this particular csv file. In this example, our file is called "Ch5_splitplot.csv." Make sure the extension (i.e., .csv) is included in your script. Also note that the name of your file should be in quotation marks within the parentheses. We are reading in the first row of data as headers with "header = TRUE." Note that this data include the instructor variable and the raw rankings.

```
names(Ch5split)
```

The *names* function will produce a list of variable names for each dataframe as follows. This is a good check to make sure your data have been read in correctly.

```
[1] "ID"    "Instructor" "R1_raw"    "R2_raw"    "R3_raw"    "R4_raw"
```

```
View(Ch5split)
```

The *View* function will let you view the dataset in spreadsheet format in RStudio.

```
summary(Ch5split)
```

The *summary* function will produce basic descriptive statistics on all the variables in your dataframe. This is a great way to quickly check to see if the data have been read in correctly and to get a feel for your data, if you haven't already. The output from the summary statement for this dataframe looks like this:

```
       ID        Instructor         R1_raw          R2_raw            R3_raw
Min.   :1.00  Min.   :1.0   Min.   :1.00   Min.   :2.000   Min.   :5.00
1st Qu.:2.75  1st Qu.:1.0   1st Qu.:2.00   1st Qu.:3.000   1st Qu.:5.75
Median :4.50  Median :1.5   Median :2.50   Median :4.000   Median :6.00
Mean   :4.50  Mean   :1.5   Mean   :2.75   Mean   :3.625   Mean   :6.25
3rd Qu.:6.25  3rd Qu.:2.0   3rd Qu.:3.00   3rd Qu.:4.000   3rd Qu.:7.00
Max.   :8.00  Max.   :2.0   Max.   :6.00   Max.   :5.000   Max.   :8.00
     R4_raw
Min.   : 8.000
1st Qu.: 8.750
Median : 9.000
Mean   : 9.125
3rd Qu.:10.000
Max.   :10.000
```

```
install.packages(reshape)
library(reshape2)
```

We will install the *reshape* package and load *reshape2* to convert our wide format data to long format.

```
Ch5splitlong<-melt(Ch5split, id.vars = c("ID", "Instructor"),
                measure.vars = ,
                variable.name = "rater",
                value.name = "rating")
```

FIGURE 5.38 (continued)
Reading data into **R**.

We are creating a new dataset named "Ch5splitlong." The *melt* function will transform our "wide" format data to "long" format. In other words, there will be multiple rows of data for each measurement occasion but just one column for the repeated measure. Within parentheses, we see we are using the Ch5_split dataframe to do this. The ID and instructor variables we want to keep but not split apart so we include the *id.vars=c("ID", "Instructor")* command. If there were other nonrepeated measures, we would have listed those here as well. The last two lines tell us that *variable.name* defines the column heading for the rater and *value.name* defines the column heading for the rating.

```
names(Ch5splitlong)
```

The *names* function will produce a list of variable names for each variable in our dataframe as follows.

```
[1] "ID"    "Instructor" "rater"    "rating"
```

```
View(Ch5splitlong)
```

The *View* function will let you view the dataset in spreadsheet format in RStudio.

```
Ch5splitlong$rater<-as.numeric(Ch5splitlong$rater)
```

The *as.numeric* function will change the string to numeric values for the variable "rater" that is located in our "Ch5splitlong" dataframe.

```
Ch5splitlong$Instructor=factor(Ch5splitlong$Instructor)
Ch5splitlong$rater=factor(Ch5splitlong$rater)
Ch5splitlong$ID=factor(Ch5splitlong$ID)
```

The *factor* function will be used to define variables *Instructor, rater*, and *ID* as nominal within the dataframe Ch5splitlong.

FIGURE 5.38 (continued)
Reading data into **R**.

5.7.5.2 Generating the Two-Factor Split-Plot ANOVA

```
install.packages(reshape)
library(reshape2)
```

We will install the *reshape* package and load *reshape2* to convert our wide format data to long format.

```
Ch5splitlong<-melt(Ch5split, id.vars = c("ID", "Instructor"),
              measure.vars = ,
              variable.name = "rater",
              value.name = "rating")
```

We are creating a new dataset named "Ch5splitlong." The *melt* function will transform our "wide" format data to "long" format. In other words, there will be multiple rows of data for each measurement occasion but just one column for the repeated measure. Within parentheses, we see we are using the Ch5_split dataframe to do this. The *ID* and *Instructor* variables we want to keep but not split apart so we include the *id.vars=c("ID", "Instructor")* command. If there were other nonrepeated measures, we would have listed those here as well. The last two lines tell us that *variable.name* defines the column heading for the Rater and *value.name* defines the column heading for the rank score.

```
names(Ch5splitlong)
```

FIGURE 5.39
Two-factor split-plot ANOVA in **R**.

The *names* function will produce a list of variable names for our new dataframe as follows. This is a good check to make sure we have restructured the data correctly and retained the ID variable.

```
[1] "ID"    "Instructor" "rater"    "rating"
```

```
View(Ch5splitlong)
```

The *View* function will let you view the dataset in spreadsheet format in RStudio.

```
Ch5splitlong$raterF<-as.numeric(Ch5splitlong$rater)
```

The variable "rater" is currently in our dataframe as a string variable. The *as.numeric* function will change rater to numeric (i.e., define a numeric value for each string category). To the left of "<-" tells **R** to create a new variable in our dataframe called "raterF."

```
Ch5splitlong$InstructorF=factor(Ch5splitlong$Instructor)
Ch5splitlong$rater=factor(Ch5splitlong$rater)
Ch5splitlong$ID=factor(Ch5splitlong$ID)
```

The *factor* function defines "InstructorF," "rater," and "ID" as factors in our dataframe.

```
modelsplit <- aov(rating ~ Instructor*rater + Error(ID),
data = Ch5splitlong)
```

The *aov* function is used to generate the two-factor split-plot ANOVA model using the dataframe Ch5splitlong. The dependent variable is "rating," and the within-subjects factor is "rater." The repeated measures are nested within rater. We model error based on the ID.

```
summary(modelsplit)
```

The *summary* function will provide output from our split-plot ANOVA model.

```
Error: ID
           Df  Sum Sq  Mean Sq  F value  Pr(>F)
Instructor  1   6.125    6.125      4.2  0.0863 .
Residuals   6   8.750    1.458
—
Signif. codes: 0 '***' 0.001 '**' 0.01 '*' 0.05 '.' 0.1 ' ' 1
Error: Within
                  Df  Sum Sq  Mean Sq  F value     Pr(>F)
rater              3  198.13    66.04   190.20  8.13e-14 ***
Instructor:rater   3   12.62     4.21    12.12  0.000141 ***
Residuals         18    6.25     0.35
—
Signif. codes: 0 '***' 0.001 '**' 0.01 '*' 0.05 '.' 0.1 ' ' 1
```

Our output corresponds to the tests of within-subjects effects (i.e., rater, instructor*rater, and error) and between-subjects effects (i.e., instructor) that we found using SPSS.

```
Ch5splitlong$unstandardizedResiduals <- residuals(modelsplit)
```

We also want to save our unstandardized residuals to the dataframe. We use the *residuals* function to compute unstandardized residuals from our *modelsplit* model. To the left of "<-" we will save the residuals as a variable named "unstandardizedResiduals" in our dataframe, "Ch5splitlong$unstandardizedResiduals."

```
install.packages("ggplot2")
```

FIGURE 5.39 (continued)
Two-factor split-plot ANOVA in **R**.

We can graph the data using the *ggplot2* package. If this is not already installed, the *install.packages* function is used to install the package in **R**.

```
library(ggplot2)
```

The *library* function is used to call up the *ggplot2* package.

```
ggplot(Ch5splitlong, aes(y=rating, x=rater,
shape=Instructor, color=rating)) + geom_point()
```

The *ggplot* function can be used to create a plot from our dataframe, Ch5splitlong, with "rating" on the *Y* axis and "rater" on the *X* axis. We allow shapes to define the different instructors using the *shaper=Instructor* command.

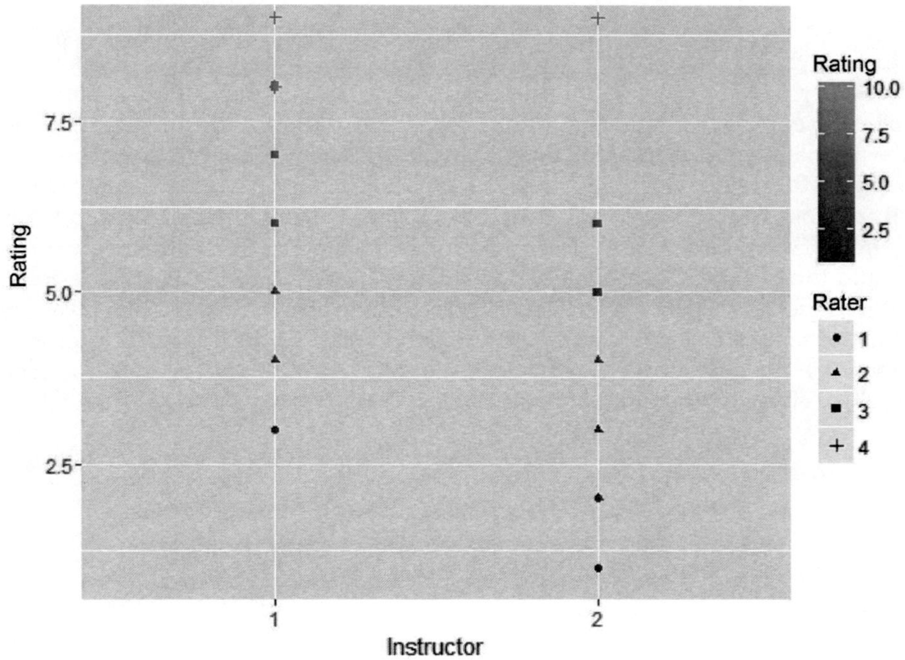

FIGURE 5.39 (continued)
Two-factor split-plot ANOVA in **R**.

5.8 Data Screening for the Two-Factor Split-Plot ANOVA

Now let's examine our data for the assumptions of the two-factor split-plot ANOVA.

5.8.1 Normality

We use the residuals (which we requested and created through the "Save" option when generating our two-factor split-plot ANOVA) to examine the extent to which normality was met.

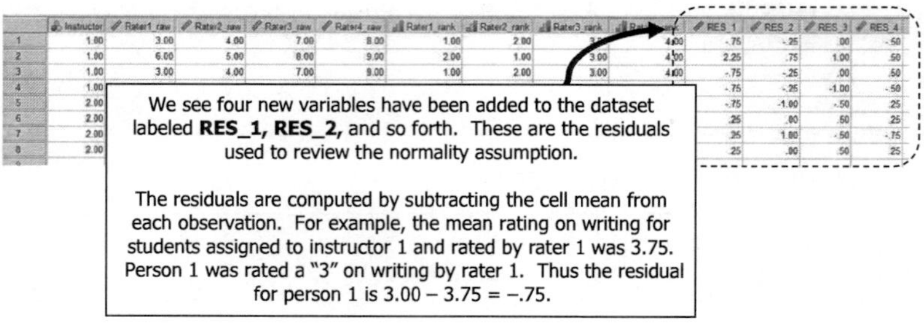

FIGURE 5.40
Normality data.

5.8.1.1 Generating Normality Evidence

As mentioned in previous chapters, understanding the distributional shape, specifically the extent to which normality is a reasonable assumption, is important. For the two-factor mixed design ANOVA, the distributional shape for the residuals should be a normal distribution. Because we have multiple residuals to reflect the multiple measurements, we need to examine normality for *each* residual. For brevity, we provide SPSS excerpts only for "RES_1" which reflects the residual for time 1; however we will narratively discuss all of the residuals.

As in previous chapters, we can again use "Explore" to examine the extent to which the assumption of normality is met. The steps for accessing Explore have already been presented, and thus we provide only a basic overview of the process. Click the residual and move it into the "Dependent List" box by clicking on the arrow button. The procedures for selecting normality statistics are as follows: click on "Plots" in the upper right corner. Place a checkmark in the boxes for "Normality plots with tests" and also for "Histogram." Then click "Continue" to return to the main Explore dialog box. Finally, click "OK" to generate the output.

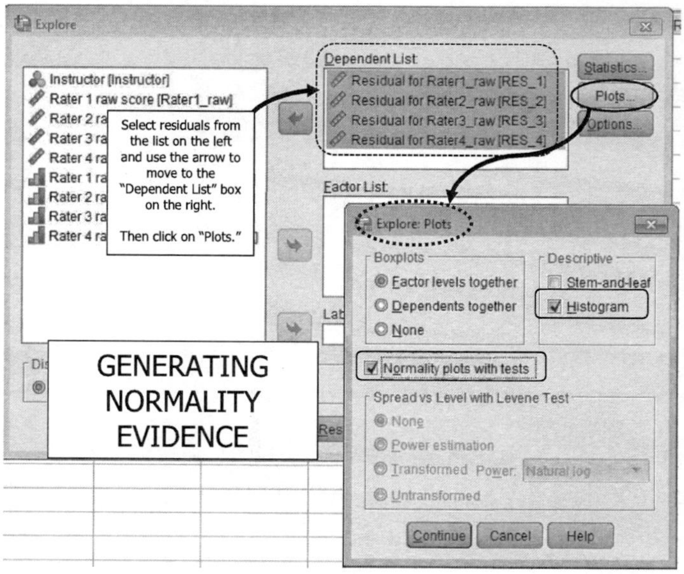

FIGURE 5.41
Generating normality evidence.

5.8.1.2 Interpreting Normality Evidence

We have already developed a good understanding of how to interpret some forms of evidence of normality including skewness and kurtosis, histograms, and boxplots. Next we see the output for this evidence. The skewness statistic of the residuals for rater 1 is 1.675 and kurtosis is 3.136—skewness being within the range of an absolute value of 2.0, suggesting evidence of normality, but some non-normality based on kurtosis. For the other three residuals, all skewness and kurtosis statistics (not shown here) are within an absolute value of 2.0, suggesting evidence of normality.

Descriptives

			Statistic	Std. Error
Residual for Rater1_raw	Mean		.0000	.36596
	95% Confidence Interval for Mean	Lower Bound	–.8654	
		Upper Bound	.8654	
	5% Trimmed Mean		–.0833	
	Median		–.2500	
	Variance		1.071	
	Std. Deviation		1.03510	
	Minimum		–.75	
	Maximum		2.25	
	Range		3.00	
	Interquartile Range		1.00	
	Skewness		1.675	.752
	Kurtosis		3.136	1.481

FIGURE 5.42
Normality evidence.

As suggested by the skewness statistic, the histogram of residuals is positively skewed, and the histogram also provides a visual display of the leptokurtic distribution.

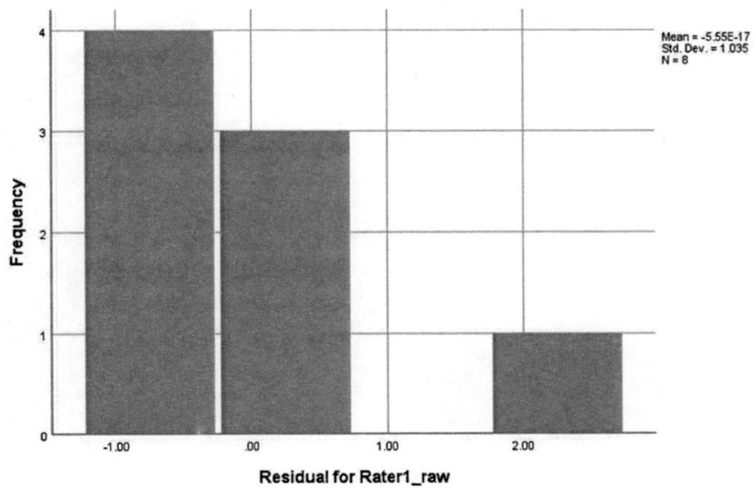

FIGURE 5.43
Histogram.

There are a few other statistics that can be used to gauge normality. The formal test of normality, the Shapiro-Wilk test (*SW*) (Shapiro & Wilk, 1965), provides evidence of the extent to which the sample distribution is statistically different from a normal distribution. The output for the Shapiro-Wilk test is presented in Figure 5.44 and suggests that our sample distributions for three of the four residuals (specifically residuals for raters 2, 3, and 4) are not statistically significantly different than what would be expected from a normal distribution, as those p values are greater than alpha. However, the distribution for the residual for rater 1 *is* statistically significantly different than a normal distribution ($SW = .745$, $df = 8$, $p = .007$).

Tests of Normality

	Kolmogorov-Smirnov[a]			Shapiro-Wilk		
	Statistic	df	Sig.	Statistic	df	Sig.
Residual for Rater1_raw	.280	8	.065	.745	8	.007
Residual for Rater2_raw	.250	8	.150	.913	8	.374
Residual for Rater3_raw	.152	8	.200*	.965	8	.857
Residual for Rater4_raw	.316	8	.018	.828	8	.057

*. This is a lower bound of the true significance.

a. Lilliefors Significance Correction

FIGURE 5.44
Shapiro-Wilk test of normality.

Quantile-quantile (Q-Q) plots are also often examined to determine evidence of normality. These graphs plot quantiles of the theoretical normal distribution against quantiles of the sample distribution. Points that fall on or close to the diagonal line suggest evidence of normality. The Q-Q plot of residuals for rater 1 shown below suggests some nonnormality (for brevity, the plots for the other raters are not shown).

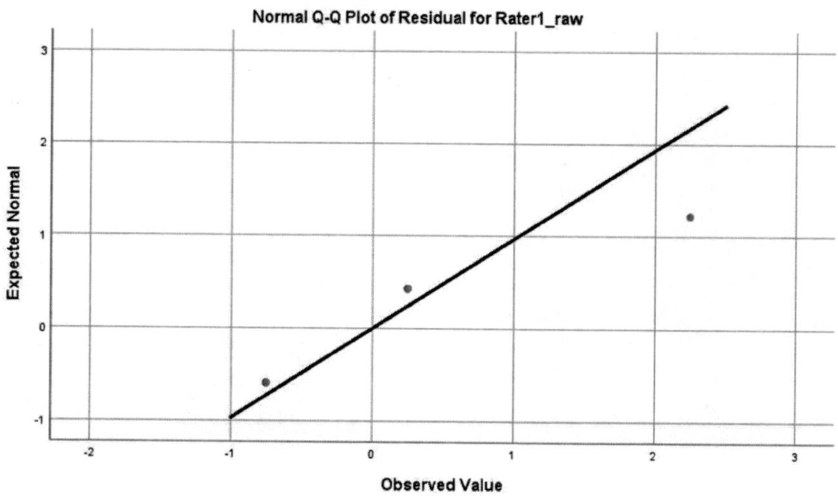

FIGURE 5.45
Normal Q-Q plot.

Examination of the boxplot for rater 1 (Figure 5.46) also suggests a nonnormal distributional shape of residuals with one outlier. For brevity, the boxplots for the remaining residuals are not presented but suggest normality.

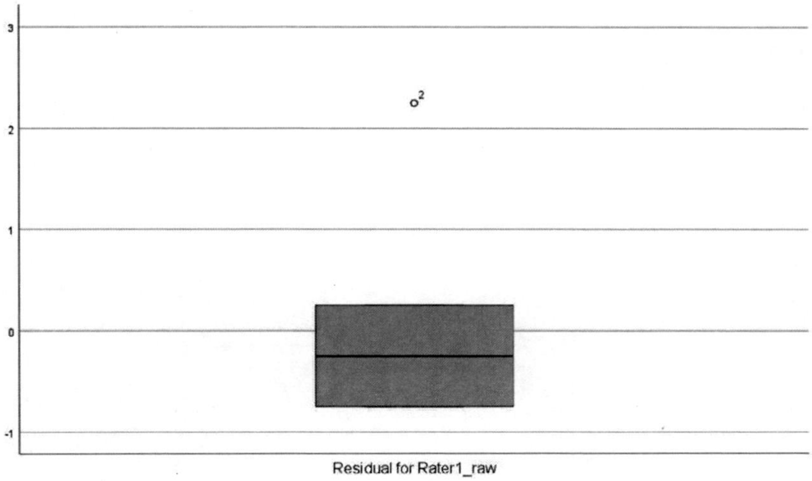

FIGURE 5.46
Boxplot.

For three of the four residuals (residuals for raters 2, 3, and 4), the forms of evidence we have examined—skewness and kurtosis statistics, the Shapiro-Wilk test, the Q-Q plot, and the boxplot—all suggest normality is a reasonable assumption. We can be reasonably assured we have met the assumption of normality for residuals for raters 2, 3, and 4. However, all forms of evidence suggest nonnormality for the residual for rater 1.

5.8.2 Independence

The only assumption we have not tested for yet is independence. As we discussed in reference to the one-way ANOVA, if subjects have been randomly assigned to conditions (in other words, the different levels of the between-subjects factor), the assumption of independence has been met. In this illustration, students were randomly assigned to instructor and thus the assumption of independence was met. However, we often use between-subjects factors that do not allow random assignment, such as preexisting characteristics (e.g., sex or education level). We can plot residuals against levels of our between-subjects factor using a scatterplot to get an idea of whether or not there are patterns in the data and thereby provide an indication of whether we have met this assumption. In this illustration, we only have one between-subjects factor. If there were multiple between-subjects factors, we would split the scatterplot by levels of one between-subjects factor and then generate a bivariate scatterplot for the other between-subjects factor by residual (as we did with factorial ANOVA). Remember that the residual was added to the dataset by saving it when we generated the two-factor split-plot ANOVA model.

Please note that some researchers do not believe that the assumption of independence can be tested. If there is not random assignment to groups, then these researchers believe this assumption has been violated—period. The plot that we generate will give us a general idea of patterns, however, in situations where random assignment was not performed.

5.8.2.1 Generating the Scatterplot

The general steps for generating a simple scatterplot through "Scatter/dot" have been presented in Chapter 10 of the previous volume and will not be reiterated here. From the "Simple Scatterplot" dialog screen, click the residual variable and move it into the "Y Axis" box by clicking on the arrow. Click the between-subjects factor (e.g., "Instructor") and move it into the "X Axis" box by clicking on the arrow. Then click "OK." Repeat these steps for each of the four residuals.

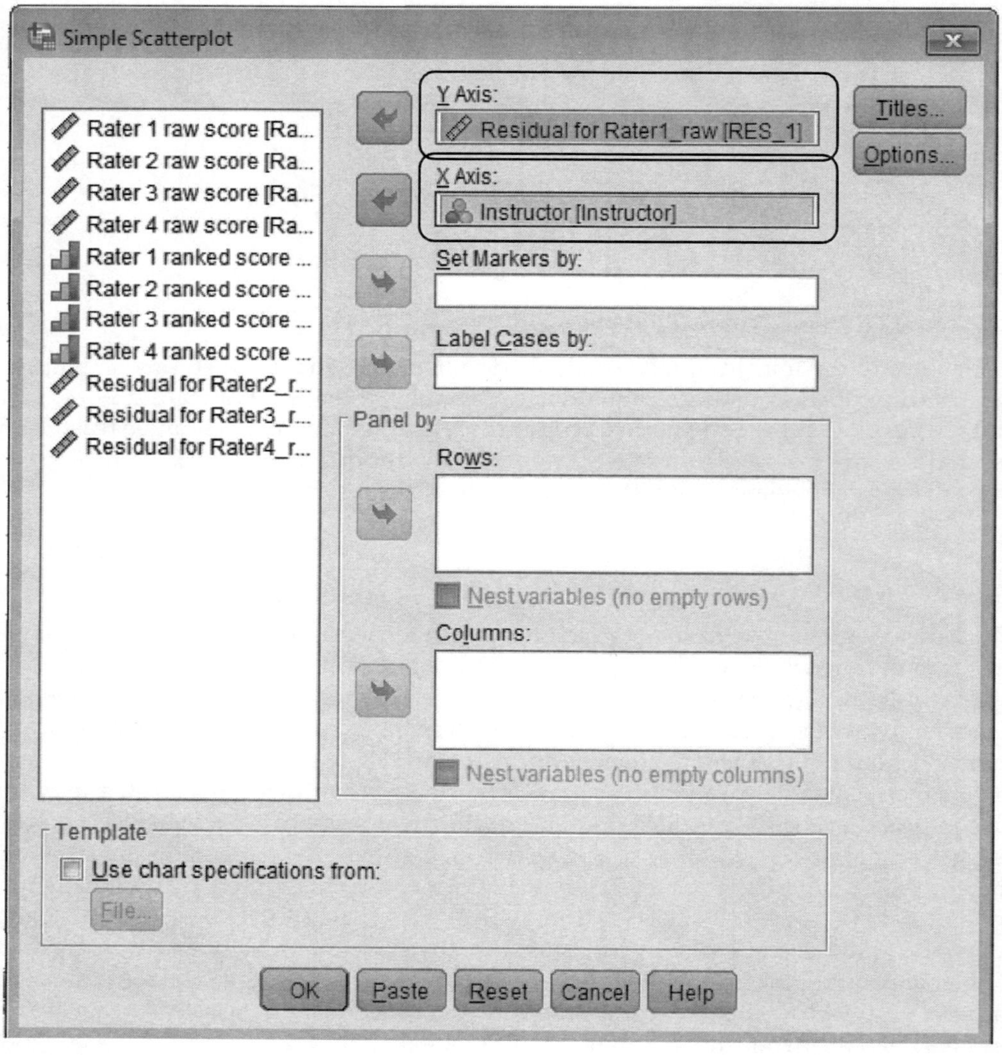

FIGURE 5.47
Generating scatterplot.

5.8.2.2 Interpreting Independence Evidence

In examining the scatterplots for evidence of independence, the points should fall relatively randomly above and below a horizontal line at zero. (You may recall in Chapter 1 that we added a reference line to the graph using Chart Editor. To add a reference line, double click on the graph in the output to activate the chart editor. Select "Options" in the top pulldown menu, then "Y axis reference line." This will bring up the "Properties" dialog box. Change the value of the position to be "0." Then click on "Apply" and "Close" to generate the graph with a horizontal line at zero.)

Here our scatterplot for each residual generally suggests evidence of independence with a relatively random display of residuals above and below the horizontal line at zero for each category of time (note that only the scatterplot of the residual for rater 3 by instructor is presented). If we had not met the assumption of independence through random assignment of cases to groups, this provides evidence that independence was a reasonable assumption.

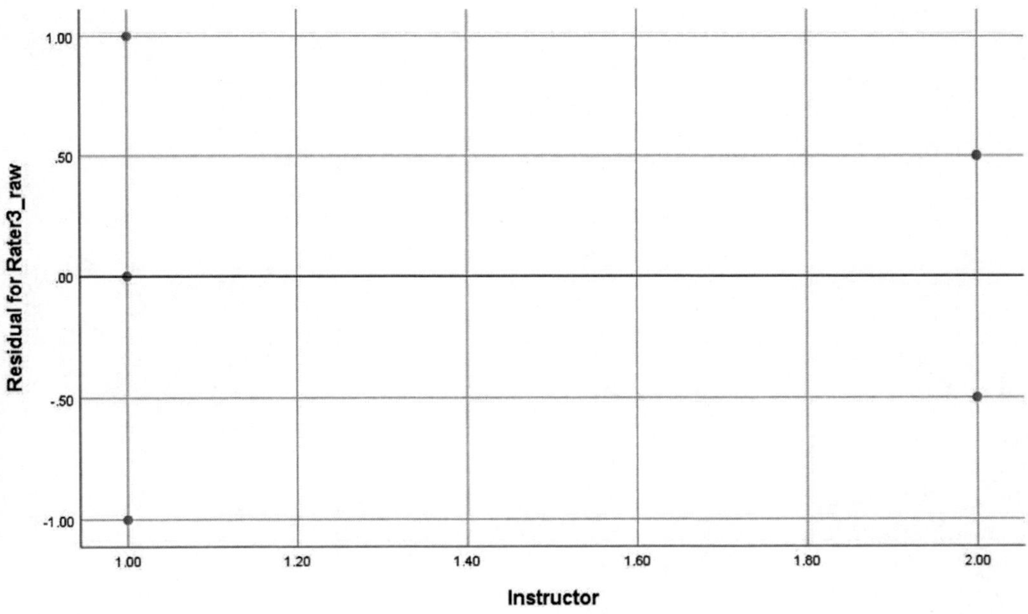

FIGURE 5.48
Scatterplot.

5.9 Power Using G*Power

5.9.1 Post Hoc Power for Two-factor Split-Plot ANOVA

Generating power analyses for a two-factor split-plot ANOVA models follow similarly to that for ANOVA, factorial ANOVA, and ANCOVA. In particular, if there is more than one independent variable, we must test for main effects and interactions separately. The first thing that must be done when using G*Power for computing post hoc power is to select the correct test family. In our case, we conducted a two-factor split-plot ANOVA. Because we have both between, within, and interaction terms, the type of statistical test selected depends on which part of the model power is to be estimated. In this illustration, let us first determine power for the within-between subjects interaction. To find this design, we select "Tests" in the top pulldown menu, then "Means," and then "ANOVA: Repeated measures, within-between interactions." Once that selection is made, the "Test family" automatically changes to "F tests." (Note that had we wanted to determine power for the between-subjects main effect, we would have selected "ANOVA: Repeated measures, between factors." For the within-subjects main effect, we would have selected "ANOVA: Repeated measures, within factors.")

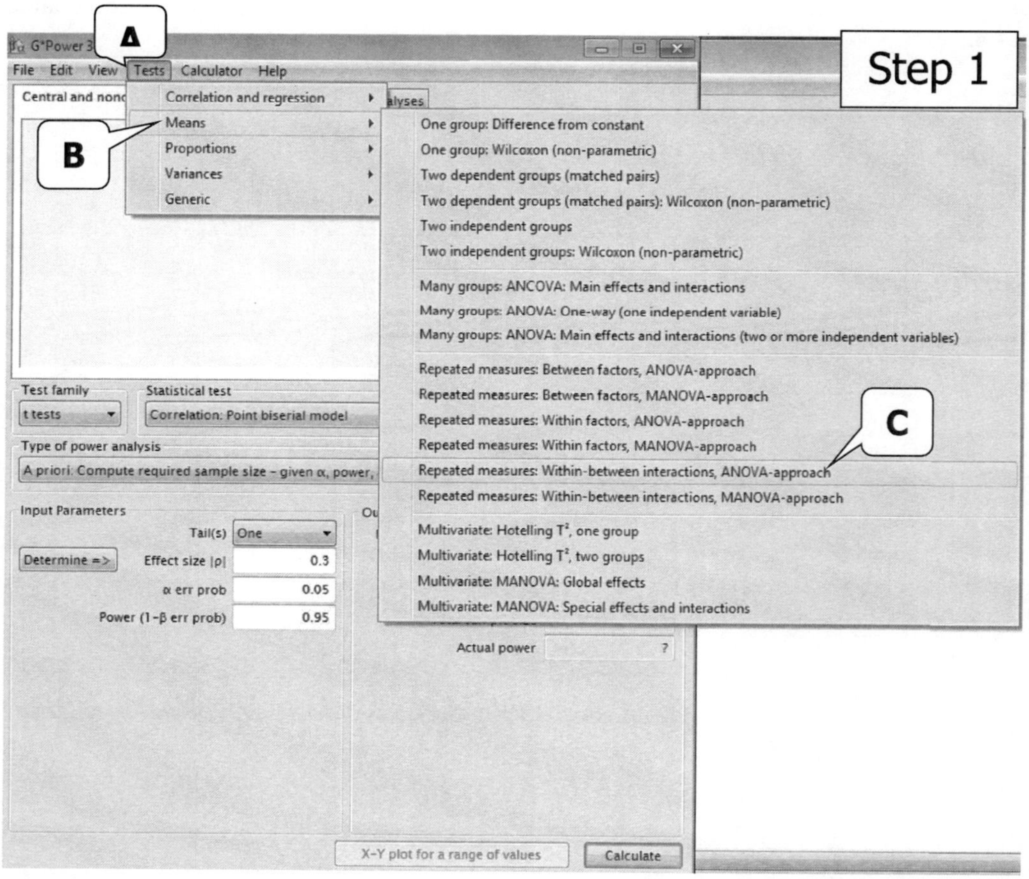

FIGURE 5.49
Post hoc power for two-factor split-plot ANOVA using G*Power.

The "Type of power analysis" desired needs to be selected. To compute post hoc power, select "Post hoc: Compute achieved power—given α, sample size, and effect size."

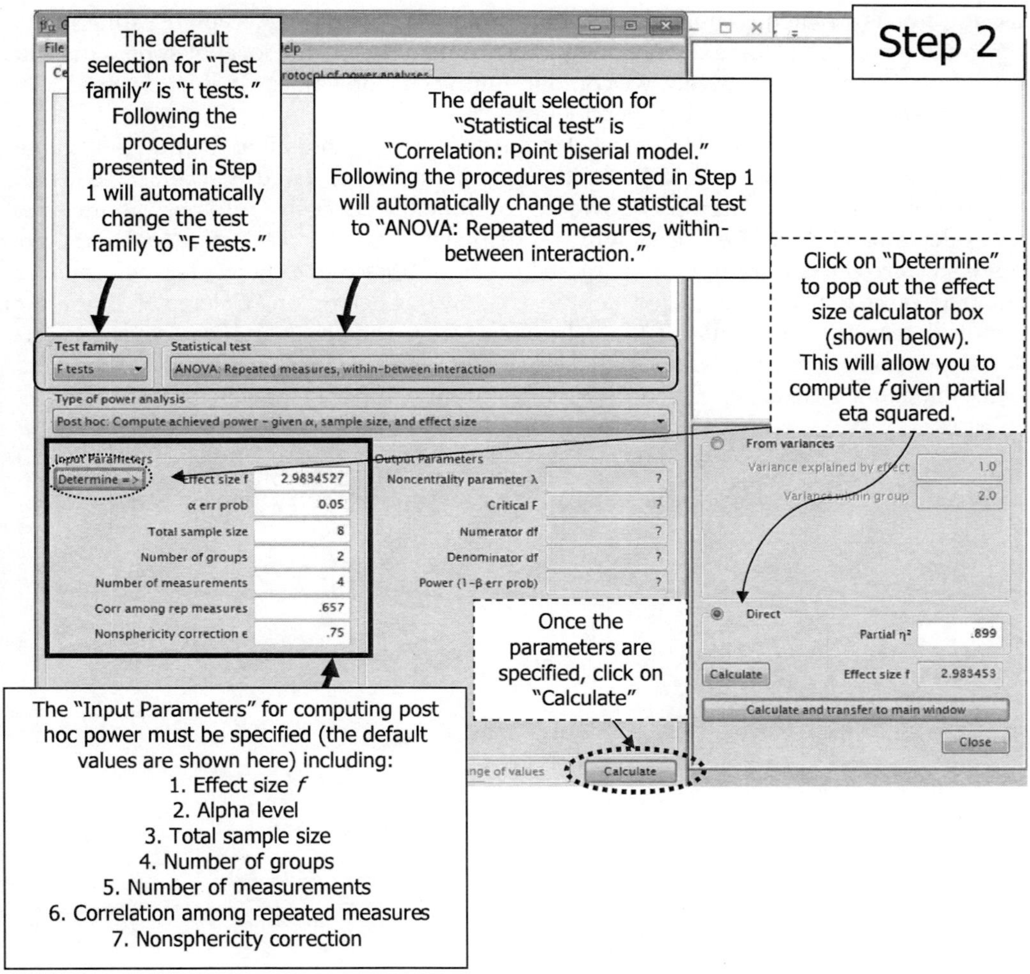

FIGURE 5.50
Post hoc power for two-factor Split-plot ANOVA: Step 2.

The "Input Parameters" must then be specified. We will compute the effect size f last, so we skip that for the moment. In our example, the alpha level we used was .05 and the total sample size was 8. The *number of groups*, in the case of a two-factor split-plot ANOVA with one nonrepeated factor having two categories, equals two. The next parameter is the number of measurements. This refers to the number of levels of the repeated factor, which in this illustration is four. Next, we have to input the correlation among repeated measures. We will estimate this parameter as the average correlation among all bivariate correlations of the repeated measures. For our raters, the Pearson correlation coefficients were: $r_{12} = .865$, $r_{13} = .881$, $r_{14} = -.431$, $r_{23} = .716$, $r_{24} = -.677$, and $r_{34} = -.372$ and thus the average correlation was .657 (in absolute value terms). The last parameter to define is the nonsphericity correction epsilon, ε. Epsilon ranges from 0 to 1, with 0 indicating the assumption is violated completely and 1 being perfect sphericity. Acceptable sphericity is approximately .75 or higher. One option is to input an acceptable level of sphericity; thus we input .75 here. Alternatively, we could input the epsilon values obtained for the usual, Geisser-Greenhouse, and Huynh-Feldt F tests.

We skipped filling in the first parameter, the effect size f, until all of the previous values were input. This is because SPSS provides only a partial eta squared effect size. We use the pop out effect size calculator in G*Power to compute the effect size f. To pop out the effect size calculator, click on "Determine" which is displayed under "Input Parameters." In the pop out effect size calculator, click on the radio button for "Direct" and then enter the partial eta squared value that was calculated in SPSS (i.e., .899). Clicking on "Calculate" in the pop out effect size calculator will calculate the effect size f. Then click on "Calculate and transfer to main window" to transfer the calculated effect size (i.e., 2.9834527) to the "Input Parameters." Once the parameters are specified, click on "Calculate" to find the power statistics.

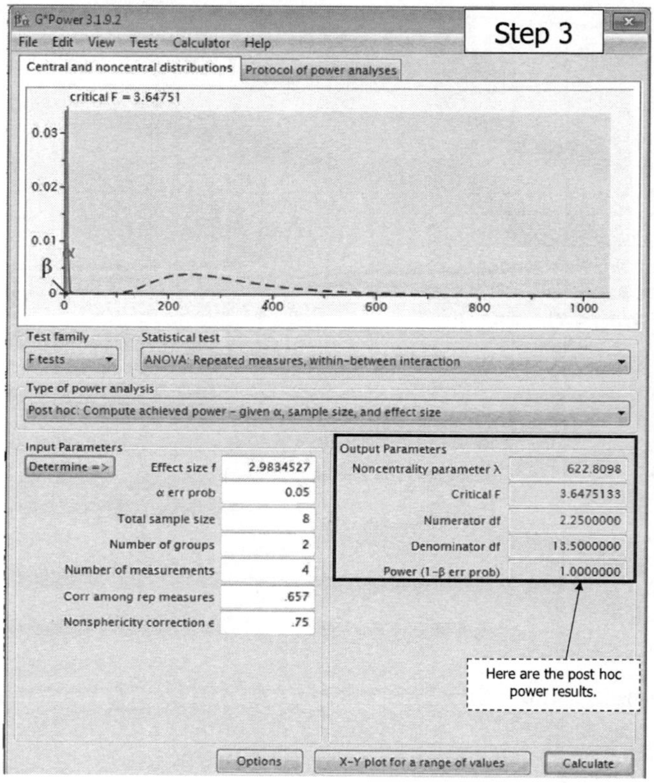

FIGURE 5.51
Post hoc power for two-factor Split-plot ANOVA: Step 3.

The "Output Parameters" provide the relevant statistics given the input just specified. In this example, we were interested in determining post hoc power for the within-between interaction in a two-factor split-plot ANOVA with a computed effect size f of 2.9834527, an alpha level of .05, total sample size of 8, two groups, four measurements, an average correlation among repeated measures of .657, and epsilon sphericity correction of .75. Based on those criteria, the post hoc power of our within-between interaction effect for this test was 1.000—the probability of rejecting the null hypothesis when it is really false (in this case, the probability that the means of the dependent variable would be equal for each level of the independent variable) was at the maximum (i.e., 100%) (sufficient power is often .80 or above). Note that this is the same value as that reported in SPSS. Keep in mind that conducting power analysis *a priori* is recommended so that you avoid a situation where, post hoc, you find that the sample size was not sufficient to reach the desired level of power (given the observed parameters).

5.9.2 *A Priori* Power for Two-Factor Split-Plot ANOVA

For *a priori* power, we can determine the total sample size needed for the main effects and/or interactions given an estimated effect size *f*, alpha level, desired power, number of groups (i.e., the number of categories of the independent *variable in the case of only one independent variable* OR the product of the number of levels of the independent variables *in the case of multiple independent variables*), number of measurements, correlation among repeated measures, and nonsphericity correction epsilon. We follow Cohen's (1988) convention for effect size (i.e., small $f = .10$; moderate $f = .25$; large $f = .40$). In this example, had we wanted to determine *a priori* power for a within-between interaction and had estimated a moderate effect *f* of .25, alpha of .05, desired power of .80, number of groups was two (i.e., we have only one independent variable and there were two categories), four measurements, a moderate correlation among repeated measures of .50, and a nonsphericity correction epsilon of .75, we would need a total sample size of 30 (i.e., 15 cases per group given two levels to our independent variable).

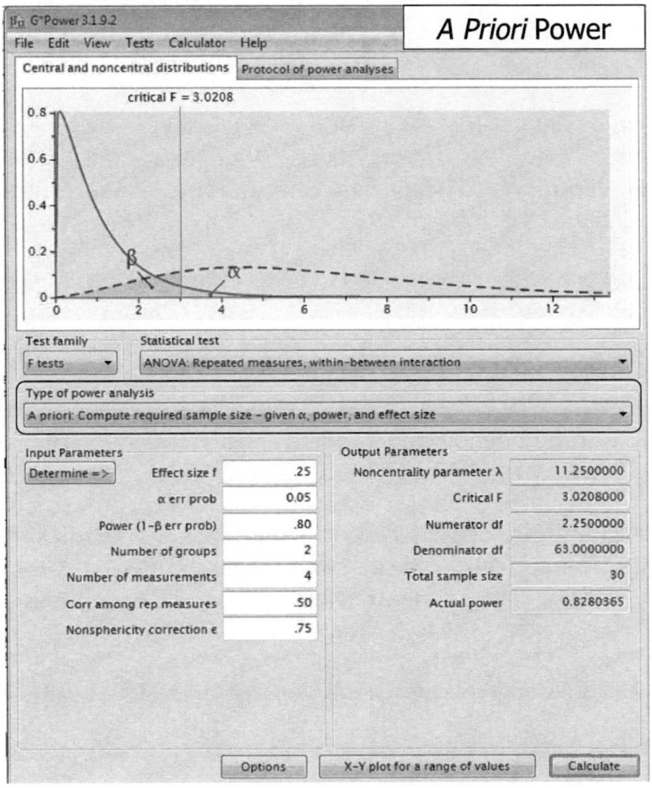

FIGURE 5.52
A priori power for two-factor split-plot ANOVA.

5.10 Research Question Template and Example Write-Up

Finally, here is an example paragraph just for the results of the two-factor split-plot design (feel free to write similar paragraphs for the other models in this chapter). Recall that our graduate research assistant, Oso Wyse, was assisting the coordinator of the dance program, Dr. Kilauea. Dr. Kilauea wanted to know if there is a mean difference in ballet technique based on instructor; if there is a mean difference in ballet technique based on rater; and if there is a mean difference in ballet technique based on rater by instructor. The research questions presented to Dr. Kilauea from Oso's work include the following:

- *Is there a mean difference in ballet technique based on instructor?*
- *Is there a mean difference in ballet technique based on rater?*
- *Is there a mean difference in ballet technique based on rater by instructor?*

Oso then assisted Dr. Kilauea in generating a two-factor split-plot ANOVA as the test of inference, and a template for writing the research questions for this design is presented in this section. As we noted in previous chapters, it is important to ensure the reader understands the levels or groups of the factor(s). This may be done parenthetically in the actual research question, as an operational definition, or specified within the methods section.

- Is there a mean difference in [dependent variable] based on [between-subjects factor]?
- Is there a mean difference in [dependent variable] based on [within-subjects factor]?
- Is there a mean difference in [dependent variable] based on [between-subjects factor] by [within-subjects factor]?

It may be helpful to preface the results of the two-factor split-plot ANOVA with information on an examination of the extent to which the assumptions were met (recall that we tested several assumptions). For the between-subjects factor (i.e., the nonrepeated factor), assumptions include: (a) independence of observations; (b) homogeneity of variance; and (c) normality. For the within-subjects factor (i.e., the repeated factor), we examine the assumption of sphericity.

A two-factor split-plot (one within-subjects factor and one between-subjects factor) analysis of variance (ANOVA) was conducted. The within-subjects factor was rating on ballet technique (four independent raters) and the between-subjects factor was dance instructor (two dance instructors). The null hypotheses tested include: (1) the mean ballet technique rating was equal for each of the four different raters; (2) the mean ballet technique rating for each dance instructor was equal; and (3) the mean ballet technique rating by rater given dance instructor were equal.

There were no missing data and no univariate outliers. The assumption of *sphericity* was met ($\chi^2 = 4.001$, Mauchly's $W = .429$, $df = 5$, $p = .557$); therefore, the results reported reflect univariate results. The sphericity assumption was further upheld in that the same results were obtained for the usual, Geisser-Greenhouse, and Huynh-Feldt F tests. The assumption of *homogeneity of variance* was met for the ballet technique rating of all raters [rater 1, $F(1, 6) = 3.600$, $p = .107$; rater 2, $F(1, 6) = .158$, $p = .705$; rater 3, $F(1, 6) = 0.000$, $p = 1.000$; and rater 4, $F(1, 6) = 1.000$, $p = .356$].

The assumption of *normality* was tested via examination of the residuals. Review of the Shapiro-Wilk test for normality (SW_{rater1} = .745, *df* = 8, *p* = .007; SW_{rater2} = .913, *df* = 8, *p* = .374; SW_{rater3} = .965, *df* = 8, *p* = .857; SW_{rater4} = .828, *df* = 8, *p* = .057), and skewness (rater 1 = 1.675; rater 2 = .290; rater 3 = .000; rater 4 = –.571) and kurtosis (rater 1 = 3.136; rater 2 = .272; rater 3 = –.700; rater 4 = –1.729) statistics suggest that normality was a reasonable assumption for raters 2, 3, and 4, but nonnormality was suggested for rater 1. The boxplot suggested a relatively normal distributional shape (with no outliers) of the residuals for raters 2 through 4. The boxplot of the residuals for rater 1 suggested nonnormality with one outlier. The Q-Q plots suggested normality was reasonable for the residuals of raters 2, 3, and 4, but suggested nonnormality for rater 1. Thus, while there was nonnormality suggested by the residuals for rater 1, the two-factor split-plot ANOVA is robust to violations of normality with equal sample sizes of groups as is evident in this design.

Random assignment of individuals to dance instructor helped ensure that the assumption of *independence* was met. Additionally, a scatterplot of residuals against the levels of the between-subjects factor was reviewed. A relatively random display of points around zero provided further evidence that the assumption of independence was met.

Here is an APA-style example paragraph of results for the two-factor split-plot ANOVA (remember that this will be prefaced by the previous paragraph reporting the extent to which the assumptions of the test were met).

The results for the univariate ANOVA indicate:

1. A statistically significant within-subjects main effect for rater (F_{rater} = 198.125, *df* = 3,18, *p* = .001) (rater 1, *M* = 2.750, *SE* = .395; rater 2, *M* = 3.625, *SE* = .239; rater 3, *M* = 6.250, *SE* = .250; rater 4, *M* = 9.125, *SE* = .191).
2. A statistically significant within-between subjects interaction effect between rater and dance instructor ($F_{\text{rater x instructor}}$ = 12.625, *df* = 3,18, *p* = .001) (for brevity, we have not included the means and standard errors here; however, you may want to include those in the narrative or in tabular form).
3. A nonstatistically significant between-subjects main effect for dance instructor ($F_{\text{instructor}}$ = 4.200, *df* = 1,6, *p* = .086) (dance instructor 1, *M* = 5.875, *SE* = .302; dance instructor 2, *M* = 5.000, *SE* = .302).

Effect sizes were rather large for the significant effects (partial $\eta^2_{rater} = .969$, power = 1.000; partial $\eta^2_{rater\ x\ instructor} = .669$, power = .998) with more than sufficient observed power, but less so for the nonsignificant effect (partial $\eta^2_{instructor} = .412$, power = .407) which had less than desired power.

The statistically significant *main effect for the within-subjects factor* suggests that there are mean differences in ballet technique rating by rater. The raters were quite inconsistent in that Bonferroni multiple comparison procedures revealed statistically significant differences among all pairs of raters except for rater 1 versus rater 2. The nonstatistically significant *main effect for the between-subjects factor* suggests that there are not differences, on average, in ballet technique rating per dance instructor. Most value in our findings is the interaction for the between-within factor (i.e., dance instructor by rater).

In examining confidence intervals of the *interaction for the between-within factor* (i.e., dance instructor by rater), nonoverlapping confidence intervals suggest statistically significant differences. We see that the patterns evident for the within-subjects factors echo here as well. For both dance instructor 1 and dance instructor 2, there are statistically significant differences among all pairs of raters except for rater 1 versus rater 2. Examining the statistically significant interaction using simple effects, we find for all raters, there is a statistically significant difference between instructor 1 and instructor 2 (rater 1, $p = .045$; rater 2, $p = .040$; . . .) . For instructor 1, there is a statistically significant difference between rater 1 and 4 ($p = .001$) and rater 1 and 4 ($p = .001$) . . . [report results of simple effects; for brevity only a few are included here!]. From the profile plot in Figure 5.2, we see that while rater 4 found the dancers of instructor 2 to have better ballet technique, the other raters liked the ballet technique by the dancers of instructor 1.

5.11 Additional Resources

This chapter has provided a preview into conducting a number of ANOVA models. However, there are a number of areas that space limitations prevent us from delving into. For those of you who are interested in learning more about ANOVA models, or if you find yourself in a sticky situation in your analyses, you may wish to look into the following, among many other excellent resources.

- For more in-depth coverage of ANOVA models, see Maxwell, Delaney, and Kelley (2018), Kirk (2014) and Keppel and Wickens (2004), among others.

Problems

Conceptual Problems

1. When an ANOVA design includes a random factor that is crossed with a fixed factor, the design illustrates which type of model?
   a. Fixed
   b. Mixed
   c. Random
   d. Crossed
2. The denominator of the F ratio used to test the interaction in a two-factor ANOVA is MS_{with} in which one of the following?
   a. Fixed-effects model
   b. Random-effects model
   c. Mixed-effects model
   d. All of the above

3. A course consists of five units, the order of presentation of which is varied (counterbalanced). A researcher used a 5 x 2 ANOVA design with order (five different randomly selected orders) and gender serving as factors. Which ANOVA model is illustrated by this design?
   a. Fixed-effects model
   b. Random-effects model
   c. Mixed-effects model
   d. Nested model
4. A researcher conducts a study where children are measured on frequency of sharing at three different times over the course of the academic year. Which ANOVA model is most appropriate for analysis of this data?
   a. One-factor random-effects model
   b. Two-factor random-effect model
   c. Two-factor mixed-effects model
   d. One-factor repeated measures design
   e. Two-factor split-plot design
5. A health care researcher wants to make generalizations about the number of patients served by after hour clinics in her region. She randomly samples clinics and collects data on the number of patients served. Which ANOVA model is most appropriate for analysis of this data?
   a. One-factor random-effects model
   b. Two-factor random-effect model
   c. Two-factor mixed-effects model
   d. One-factor repeated measures design
   e. Two-factor split-plot design
6. A preschool teacher randomly assigns children to classrooms—some with windows and some without windows. She wants to know if there is a mean difference in receptive vocabulary based on type of classroom (with and without windows) and whether this varies by classroom teacher. Which ANOVA model is most appropriate for analysis of this data?
   a. One-factor random-effects model
   b. Two-factor random-effect model
   c. Two-factor mixed-effects model
   d. One-factor repeated measures design
   e. Two-factor split-plot design
7. True or false? If a given set of data was analyzed with both a one-factor fixed-effects model and a one-factor random-effects model, the F ratio for the random-effects model will be greater than the F ratio for the fixed-effects model.
8. True or false? A repeated measures design is necessarily an example of the random-effects model.
9. Suppose researchers A and B perform a two-factor ANOVA on the same data, but that A assumes a fixed-effects model and B assumes a random-effects model. I assert

that if A finds the interaction significant at the .05 level, B will also find the interaction significant at the .05 level. Am I correct?

10. I assert that MS_{with} should always be used as the denominator for all F ratios in any two-factor analysis of variance. Am I correct?

11. I assert that in a one-factor repeated measures ANOVA and a two-factor split-plot ANOVA, the SS_{total} will be exactly the same when using the same data. Am I correct?

12. Football players are each exposed to all three different counterbalanced coaching strategies, one per month. This is an example of which type of model?

a. One-factor fixed-effects ANOVA model

b. One-factor repeated-measures ANOVA model

c. One-factor random-effects ANOVA model

d. One-factor fixed-effects ANCOVA model

13. A two-factor split-plot design involves which of the following?

a. Two repeated factors

b. Two nonrepeated factors

c. One repeated factor and one nonrepeated factor

d. Farmers splitting up their land into plots

14. The interaction between factors L and M can be assessed only if which one of the following occurs?

a. Both factors are crossed.

b. Both factors are random.

c. Both factors are fixed.

d. Factor L is a repeated factor.

15. True or false? A student factor is almost always random.

16. In a two-factor split-plot design, there are two interaction terms. Hypotheses can actually be tested for how many of those interactions?

a. 0

b. 1

c. 2

d. Cannot be determined

17. True or false? In a one-factor repeated measures ANOVA design, the F test is quite robust to violation of the sphericity assumption, and thus we never need to worry about it.

18. True or false? Assumptions for the two-factor split-plot ANOVA include consideration only for the between-subjects factors.

19. The assumption of sphericity is applicable to which ANOVA models? Select all that apply.

a. One-factor random effects

b. Two-factor random effects

c. Two-factor mixed effects

d. One-factor repeated measures
e. Two-factor split-plot

20. Which one of the following is a type of equal variance assumption?
a. Independence
b. Multicollinearity
c. Normality
d. Sphericity

Answers to Conceptual Problems

1. **b** (when there are both random and fixed factors, then the design is mixed)
3. **c** (gender is fixed, and order is random; thus it is a mixed-effects model)
5. **a** (clinics were randomly selected from the population; thus the one-factor random-effects model is appropriate)
7. **False** (the *F* ratio will be the same for both the one-factor random- and fixed-effects models)
9. **Yes** (the test of the interaction is exactly the same for both models yielding the same *F* ratio)
11. **Yes** (SS_{total} is the same for both models; the total amount of variation is the same, it is just divided up in different ways; review the example dataset in this chapter)
13. **c** (see definition of design)
15. **True** (rarely is one interested in particular students, thus students are usually random)
17. **False** (the *F* test is not very robust in this situation and we should be concerned about it).
19. **d** and **e** (the assumption of sphericity is applicable to the within subjects factor—i.e., repeated factor—so it is applicable to both the one-factor repeated measures and two-factor split-plot ANOVA designs)

Computational Problems

1. Complete the following ANOVA summary table for a two-factor model, where there are three levels of factor A (fixed method effect) and two levels of factor B (random teacher effect). Each cell of the design includes 4 students ($\alpha = .01$).

Source	*SS*	*df*	*MS*	*F*	Critical Value	Decision
A	3.64	–	–	–	–	–
B	0.57	–	–	–	–	–
AB	2.07	–	–	–	–	–
Within	–	–	–			
Total	8.18	–				

2. A researcher tested whether aerobics increased the fitness level of eight undergraduate students participating over a four-month period. Students were measured at the end of each month using a 10-point fitness measure (10 being most fit). The data are shown here. Conduct an ANOVA to determine the effectiveness of the program, using $\alpha = .05$. Use the Bonferroni method to detect exactly where the differences are among the time points (if they are different).

Subject	Time 1	Time 2	Time 3	Time 4
1	3	4	6	9
2	4	7	5	10
3	5	7	7	8
4	1	3	5	7
5	3	4	7	9
6	2	5	6	7
7	1	4	6	9
8	2	4	5	6

3. Using the same data as in Computational Problem #2, conduct a two-factor split-plot ANOVA, where the first four subjects participate in a step aerobics problem and the last four subjects participate in a spinning program ($\alpha = 05$).
4. To examine changes in teaching self-efficacy, 10 teachers were measured on their self-efficacy towards teaching at the beginning of their teaching career and at the end of their first and third years of teaching. The teaching self-efficacy scale ranged from 0 to 100, with higher scores reflecting greater teaching self-efficacy. The data are shown here. Conduct a one-factor repeated measures ANOVA to determine mean differences across time, using $\alpha = .05$. Use the Bonferroni method to detect if and/or where the differences are among the time points.

Subject	Beginning Year 1	End Year 1	End Year 3
1	35	50	45
2	50	75	82
3	42	51	56
4	70	72	71
5	65	50	81
6	92	42	69
7	80	82	88
8	78	76	79
9	85	60	83
10	64	71	89

5. Using the same data as in Computational Problem #4, conduct a two-factor split-plot ANOVA, where the first five subjects participate in a mentoring program and the last five subjects do not participate in a mentoring program ($\alpha = 05$).

6. You are a statistical consultant, and a researcher comes to you with the following partial SPSS output (sphericity assumed). In a two-factor split-plot ANOVA design, rater is the repeated (or within-subjects) factor, gender of the rater is the nonrepeated (or between-subjects) factor, and the dependent variable is history exam scores. (a) Are the effects significant (which you must determine, as significance is missing, using $\alpha = .05$)? (b) What are the implications of these results in terms of rating the history exam?

Tests of Within-Subjects Effects

Source	Type III *SS*	*df*	*MS*	*F*
RATER	298.38	3	99.46	30.47
RATER*GENDER	184.38	3	61.46	18.83
ERROR(RATER)	58.75	18	3.26	

Tests of Between-Subjects Effects

Source	Type III *SS*	*df*	*MS*	*F*
GENDER	153.13	1	153.13	20.76
ERROR	44.25	6	7.38	

7. To examine changes in stress, 10 patients with generalized anxiety disorder were measured on their subjective stress at baseline, after 6 weeks of participating in mindfulness meditation training, and after 12 weeks of participation. Self-reported stress ranged from 0 to 50, with higher scores reflecting greater stress. The data are shown here. Conduct a one-factor repeated measures ANOVA to determine mean differences across time, using $\alpha = .05$. Use the Bonferroni method to detect if and/or where the differences are among the time points.

Subject	Baseline	After 6 Weeks	After 12 Weeks
1	48	43	40
2	40	38	35
3	43	40	34
4	48	44	41
5	46	42	36
6	41	38	35
7	49	45	39
8	44	40	37
9	43	40	34
10	42	37	33

8. To examine changes in stress, 20 patients with generalized anxiety disorder were measured on their subjective stress at baseline and then randomly assigned to receive either mindfulness meditation intervention (1) or stress management education (0).

The patients were measured again on subjective stress after 6 weeks and after 12 weeks of participation in the study. Self-reported stress ranged from 0 to 50, with higher scores reflecting greater stress. The data are shown here. Conduct a two-factor split-plot ANOVA to determine mean differences across time and group, using $\alpha = .05$.

Subject	Intervention	Baseline	After 6 Weeks	After 12 Weeks
1	1	48	43	40
2	1	40	38	35
3	1	43	40	34
4	1	48	44	41
5	1	46	42	36
6	1	41	38	35
7	1	49	45	39
8	1	44	40	37
9	1	43	40	34
10	1	42	37	33
11	0	47	46	43
12	0	46	46	44
13	0	49	47	45
14	0	48	45	43
15	0	40	39	39
16	0	43	41	40
17	0	42	40	39
18	0	44	42	41
19	0	45	44	43
20	0	47	45	44

Answers to Computational Problems

1. $SS_{with} = 1.9$, $df_A = 2$, $df_B = 1$, $df_{AB} = 2$, $df_{with} = 18$, $df_{total} = 23$, $MS_A = 1.82$, $MS_B = .57$, $MS_{AB} = 1.035$, $MS_{with} = .1056$, $F_A = 1.7585$, $F_B = 5.3977$, $F_{AB} = 9.8011$, critical value for AB = 6.01 (reject H_0 for AB), critical value for B = 8.29 (fail to reject H_0 for B), critical value for A = 99 (fail to reject H_0 for A).
3. $SS_{\text{time}} = 126.094$, $SS_{\text{time x program}} = 2.594$, $SS_{\text{program}} = 3.781$, $MS_{\text{time}} = 42.031$, $MS_{\text{time x program}} = 0.865$, $MS_{\text{program}} = 3.781$, $F_{\text{time}} = 43.078$ ($p < .001$), $F_{\text{time x program}} = 0.886$ ($p > .05$), $F_{\text{program}} = 0.978$ ($p > .05$).
5. $SS_{\text{time}} = 691.467$, $SS_{\text{time x mentor}} = 550.400$, $SS_{\text{mentor}} = 1968.300$, $MS_{\text{time}} = 345.733$, $MS_{\text{time x mentor}} = 275.200$, $MS_{\text{mentor}} = 1968.300$, $F_{\text{time}} = 2.719$ ($p = .096$), $F_{\text{time x mentor}} = 2.164$ ($p = .147$), $F_{\text{mentor}} = 7.073$ ($p < .001$).
7. $SS_{\text{subjects}} = 206.833$, $SS_{\text{time}} = 320.600$, $SS_{\text{subjects x time}} = 18.067$, $MS_{\text{subjects}} = 22.981$, $MS_{\text{time}} = 160.300$, $MS_{\text{subjects x time}} = 1.004$, $F = 159.708$, $p < .000$ (reject H_0); with Bonferroni, all contrasts are statistically significant at alpha = .05.

Interpretive Problems

1. In Chapter 3, you built on the interpretive problem from Chapter 1 utilizing the survey1 dataset from the website. SPSS or **R** was used to conduct a two-factor fixed-effects ANOVA, including effect size, where political view is factor A ($J = 5$), gender is factor B ($K = 2$), and the dependent variable is the same one that you used for Interpretative problem #1 in Chapter 1. Now, in addition to the two-factor fixed-effects ANOVA, conduct both a random-effects and a mixed-effects design. Determine whether the nature of the factors makes any difference in the results.
2. In Chapter 3, you built on the interpretive problem from Chapter 1 utilizing the survey1 dataset from the website. SPSS or **R** was used to conduct a two-factor fixed-effects ANOVA, including effect size, where hair color is factor A (i.e., one independent variable) ($J = 5$), gender is factor B (a new factor, $K = 2$), and the dependent variable is an interval or ratio variable of your choice. Now, in addition to the two-factor fixed-effects ANOVA, conduct both a random-effects and a mixed-effects design. Determine whether the nature of the factors makes any difference in the results.

6

Hierarchical and Randomized Block Analysis of Variance Models

Chapter Outline

Key Concepts

1. Crossed designs and nested designs
2. Confounding
3. Randomized block designs
4. Methods of blocking

In the last several chapters our discussion has dealt with different analysis of variance (ANOVA) models. In this chapter we complete our discussion of the analysis of variance by considering models in which there are multiple factors, but where at least one of the factors is either a hierarchical (or nested) factor or a blocking factor. As we define these models, summarized in Box 6.1, we shall see that this results in a hierarchical (or nested) design and a blocking design, respectively.

BOX 6.1 Summary of Hierarchical and Randomized Block ANOVA Models

Model	Summary
Two-factor hierarchical ANOVA model	One factor is nested within another factor. • A two-factor nested design (or incomplete factorial design) of factor B being nested within factor A is one where the levels of factor B occur for only one level of factor A. **Nesting** is a particular type of confounding among the factors being investigated, where the AB interaction is part of the B effect (or is **confounded** with B) and therefore cannot be investigated. • Also known as a nested design, hierarchical design, or multilevel model
Two-factor randomized block design for $n = 1$	Two factors, each with at least two levels. One factor is known as the **treatment factor** (although this factor could also be an observable factor). The second factor is known as the **blocking factor,** which is a nuisance factor for which control is desired. • Each block represents the formation of a matched set of individuals; that is, matched on the blocking variable but not necessarily matched on any other nuisance variable. The purpose of the blocking factor is to reduce residual variation. • Each subject falls into only one block in the design and is subsequently randomly assigned to one level of the treatment factor within that block. There is only one subject for each treatment-block level combination. As a result, the model does not include an interaction term, and this is a distinguishing feature of this model. • Designs that include one or more blocking factors are known as **randomized block designs,** *matching designs*, or *treatment by block designs*.
Two-factor randomized block design for $n > 1$	For two-factor randomized block designs with more than one observation per cell, the characteristics are exactly the same as with the $n = 1$ model, with the obvious exception that when $n > 1$, an interaction term exists.
Friedman test	This is the nonparametric equivalent to the two-factor randomized block ANOVA model, and it is based on mean ranks.

In this chapter we are mostly concerned with the two-factor hierarchical (or nested) model and the two-factor randomized block model, although these models can be generalized to designs with more than two factors. Most of the concepts used in this chapter are the same as those covered in previous chapters. In addition, new concepts include crossed and nested factors, confounding, blocking factors, and methods of blocking. Our objectives are that by the end of this chapter, you will be able to (a) understand the characteristics and concepts underlying hierarchical and randomized block ANOVA models, (b) determine and interpret the results of hierarchical and randomized block ANOVA models, (c) understand and evaluate the assumptions of hierarchical and randomized block ANOVA models, and (d) compare different ANOVA models and select an appropriate model.

6.1 What Hierarchical and Randomized Block ANOVA Models Are and How They Work

Throughout the text, we have followed a savvy group of graduate students on statistical analysis adventures. In this chapter, we see one of those students, Challie Lenge, embarking on a new journey.

The quad of graduate students have enjoyed the complex statistical analyses that they have been tasked with and are looking forward to another challenging task. This time, Challie Lenge will be working with a psychology faculty member involved in a clinical trial through their institution's medical center. Dr. Mayfield has conducted an experiment in which hospice patients were randomly assigned to one of two interventions (massage therapy or music therapy) and one of four different interventionists. There were 24 hospice patients who participated; thus there were six patients in each intervention-interventionist combination. Each patient was assessed on quality of life at the conclusion of the study. Dr. Mayfield wants to know the following: if there is a mean difference in quality of life based on intervention (music therapy or massage therapy) and if there is a mean difference in quality of life between interventionist. Challie suggests the following research questions to Dr. Mayfield:

- *Is there a mean difference in quality of life based on intervention?*
- *Is there a mean difference in quality of life based on interventionist?*

With one between-subjects independent variable (i.e., intervention, either music therapy or massage therapy) and one hierarchical or nested factor (i.e., interventionist/clinician), Challie determines that a two-factor hierarchical ANOVA is the best statistical procedure to use to answer Dr. Mayfield's question. Her next task is to assist Dr. Mayfield in analyzing the data.

6.1.1 Characteristics of the Two-Factor Hierarchical Model

In this section, we describe the distinguishing characteristics of the two-factor hierarchical ANOVA model, the layout of the data, the linear model, the ANOVA summary table and expected mean squares, and multiple comparison procedures.

The characteristics of the two-factor fixed-, random-, and mixed-effects models have already been covered in earlier chapters. Here we consider a special form of the two-factor model where *one factor is nested within another factor*. The best introduction to this model is via an example. Suppose you are interested in which of several different interventions (e.g., music therapy, massage therapy, art therapy) results in the highest level of quality of life among hospice patients. Thus, quality of life is the dependent variable and type of intervention is one factor. A second factor is the interventionist or therapist (i.e., the person who performs the intervention, such as the massage therapist or music therapist). That is, you may also believe that some therapists are more effective than others, which results in different levels of quality of life. However, each therapist has only one caseload of patients and only one type of intervention in which they are trained. In other words, all combinations of the intervention and interventionist (aka therapist) factors are not possible. This design is known as a **nested design**, **hierarchical design**, or **multilevel model** because the interventionist factor is nested within the intervention factor. This is in contrast to a two-factor **crossed design**, where all possible combinations of the two factors are included. The two-factor designs described in Chapters 3 and 5 were all crossed designs.

Let us give a more precise definition of crossed and nested designs. A two-factor completely crossed design (or **complete factorial design**) is one where every level of factor A occurs in combination with every level of factor B. A two-factor nested design (or **incomplete factorial design**) of factor B being nested within factor A is one where the levels of factor B occur for only one level of factor A. We denote this particular nested design as **B(A)**, which is read as *factor B being nested within factor A* (in other references, you may see this written as B:A or as B | A). To return to our example, the therapist factor (factor B) is nested within the intervention factor (factor A), as each therapist utilizes only one type of intervention (e.g., music therapy or massage therapy). The outcome measured is quality of life. Thus, a researcher may select a nested design to examine the extent to which patient quality of life differs given that therapists are nested within intervention. The researcher is likely most interested in the treatment (e.g., type of intervention) but recognizes that the context (i.e., the person providing the intervention, i.e., the interventionist, therapist, or clinician) may contribute to differences in the outcome, and can model this statistically through a hierarchical ANOVA.

These models are shown graphically in Figure 6.1. In Figure 6.1a, a **completely crossed or complete factorial design** is shown where there are two levels of factor A and six levels of factor B. Thus, there are 12 possible factor combinations that would all be included in a completely crossed design. The shaded region indicates the combinations that might be included in a nested or incomplete factorial design where factor B (e.g., interventionist) is nested within factor A (e.g., intervention). Although the number of levels of each factor remains the same, factor B now has only three levels within each level of factor A. For A_1 we see only B_1, B_2, and B_3, whereas for A_2 we see only B_4, B_5, and B_6. Thus, only 6 of the possible 12 factor combinations are included in the nested design. For example, level 1 of factor B occurs only in combination with level 1 of factor A. In summary, Figure 6.1a shows that the nested or incomplete factorial design consists of only a portion of the completely crossed design (the shaded regions).

In Figure 6.1b, we see the **nested design** depicted in its more traditional form. Here you see that the six factor combinations not included are not even shown (e.g., A_1 with B_4). Other examples of the two-factor nested design are as follows: (a) student is nested within teacher (or classroom), (b) faculty member is nested within department, (c) individual is nested within neighborhood, (d) county is nested within state, (e) employee is nested within employer, (f) patient is nested within doctor, (g) chapter is nested within book.

	B_1	B_2	B_3	B_4	B_5	B_6
A_1						
A_2						

Part (a)

A_1			A_2		
B_1	B_2	B_3	B_4	B_5	B_6

Part (b)

(a) **The *completely crossed design*.** The shaded region indicates the cells that would be included in a nested design where factor B is nested within factor A. In the nested design, factor A has two levels and factor B has three levels within each level of factor A. You see that only 6 of the 12 possible cells are filled in the nested design.

(b) **The same nested design in *traditional form*.** The shaded region indicates the cells included in the nested design (i.e., the same six as shown in the first part).

FIGURE 6.1
Two-factor completely crossed versus nested designs.

Thus, with this design, one factor is nested within another factor, rather than the two factors being crossed. As is shown in more detail later in this chapter, the nesting characteristic has some interesting and distinct outcomes. For now, some brief mention should be made of these outcomes. **Nesting** is a particular type of confounding among the factors being investigated, where the AB interaction is part of the B effect (or is **confounded** with B) and therefore cannot be investigated. (Going back to the previous example, this means that the therapist by intervention interaction effect is confounded with the therapist main effect, and thus teasing apart those effects is not possible.) In the ANOVA model and the ANOVA summary table, there will not be an interaction term or source of variation. This is due to the fact that each level of factor B (the nested factor, such as the therapist) occurs in combination with only one level of factor A (the nonnested factor, such as the treatment). We cannot compare for a particular level of B (e.g., the interventionist) all levels of factor A (e.g., intervention), as a certain level of B only occurs with one level of A.

Confounding may occur for two reasons. First, the confounding may be intentional due to practical reasons, such as a reduction in the number of individuals to be observed. Fewer individuals would be necessary in a nested design, as compared to a crossed design, due to the fact that there are fewer cells in the model. Second, the confounding may be absolutely necessary because crossing may not be possible. For example, school is nested within school district because a particular school can be a member of only one school district. The nested factor (here factor B) may be a nuisance variable that the researcher wants to take into account in terms of explaining or predicting the dependent variable Y. An error commonly made is to ignore the nuisance variable B and go ahead with a one-factor design using only factor A. This design may result in a biased test of factor A such that the F ratio is inflated. Thus H_0 would be rejected more often that it should be, serving to increase the actual α level over that specified by the researcher and thereby increase the likelihood of a Type I error. The F test is then too liberal.

Let us make two further points about this first characteristic. First, in the one-factor ANOVA design discussed in Chapter 1, we have already seen nesting going on in a different way. Here subjects were nested within factor A because each subject only responded to one level of factor A. It was only when we got to repeated measures designs in Chapter 5 that individuals were allowed to respond to more than one level of a factor. For the repeated measures design, we actually had a completely crossed design of subjects by factor A. Second, Glass and Hopkins (1996) give a nice conceptual example of a nested design with teachers being nested within schools, where each school is like a nest having multiple eggs or teachers.

The remaining characteristics should be familiar. These include the following: (a) two factors (or independent variables) that are nominal or ordinal in scale, each with two or more levels; (b) the levels of each of the factors may be either randomly sampled from the population of levels or fixed by the researcher (i.e., the model may be fixed, mixed, or random); (c) subjects are randomly assigned to only one combination of the levels of the two factors; and (d) the dependent variable is measured at least at the interval level. If individuals respond to more than one combination of the levels of the two factors, then this is a repeated measures design (see Chapter 5).

For simplicity, we again assume the design is balanced. For the two-factor nested design, a design is balanced if (a) the number of observations within each factor combination (or cell) is the same (in other words, the sample size for each cell of the design is the same), and (b) the number of levels of the nested factor within each level of the other factor is the same. The first portion of this statement should be quite familiar from factorial designs, so no further explanation is necessary. The second portion of this statement is unique to this design and requires a brief explanation. As an example, say factor B is nested within factor A (i.e., the nonnested factor) and factor A has two levels. On the one hand, factor B may have the same number of levels for each level of factor A. This occurs if there are three levels of factor B under level 1 of factor A (i.e., A_1) and also three levels of factor B under level 2 of factor A (i.e., A_2). On the other hand, factor B may not have the same number of levels for each level of factor A. This occurs if there are three levels of factor B under A_1 and only two levels of factor B under A_2. If the design is unbalanced, you are encouraged to use a more modern hierarchical analytic approach that goes beyond least squares estimation and uses, for example, maximum likelihood estimation (Maxwell, Delaney, & Kelley, 2018). See the discussion, for example, in Kirk (2013) and Dunn and Clark (1987).

6.1.1.1 The Layout of the Data for the Two-Factor Hierarchical Model

The layout of the data for the two-factor nested design is shown in Table 6.1. To simplify matters, we have limited the number of levels of the factors to two levels of factor A (e.g., intervention or treatment group) and three levels of factor B (e.g., interventionist or therapist). This serves only as an example layout because many other possibilities obviously exist. Here we see the major set of columns designated as the levels of factor A, the nonnested factor (e.g., intervention), and for each level of A, the minor set of columns are the levels of factor B, the nested factor (e.g., interventionist). Within each factor level combination or cell are the subjects. Means are shown for each cell, for the levels of factor A, and overall. Note that the means for the levels of factor B need not be shown, as they are the same as the cell means. For instance, $\bar{Y}_{.11}$ is the same as $\bar{Y}_{..1}$ (not shown) as B_1 only occurs once. This is another result of the nesting.

6.1.1.2 The Two-Factor Hierarchical ANOVA Model

The nested factor is almost always random (Glass & Hopkins, 1996; Keppel, 1991; Mickey, Dunn, & Clark, 2004; Page, Braver, & MacKinnon, 2003). In other words, the levels of the

TABLE 6.1

Layout for the Two-Factor Nested Design

	A_1			A_2		
	B_1	B_2	B_3	B_4	B_5	B_6
	Y_{111}	Y_{112}	Y_{113}	Y_{124}	Y_{125}	Y_{126}
	.	.	.	.	.	.
	.	.	.	.	.	.
	.	.	.	.	.	.
	Y_{n11}	Y_{n12}	Y_{n13}	Y_{n24}	Y_{n25}	Y_{n26}
Cell means	$\bar{Y}_{.11}$	$\bar{Y}_{.12}$	$\bar{Y}_{.13}$	$\bar{Y}_{.24}$	$\bar{Y}_{.25}$	$\bar{Y}_{.26}$
A means		$\bar{Y}_{.1.}$			$\bar{Y}_{.2.}$	
Overall mean			$\bar{Y}_{...}$			

nested factor are a random sample of the population of levels. For example, in the case of teachers (or classrooms) nested within teaching pedagogy, it is often the case that a random sample of the teachers (or classrooms) is selected rather than specific teachers (which would be a fixed-effects factor). This can be extended to any number of examples where groups or clusters are nested within the factor of interest (e.g., intervention). Thus, the nested factor (i.e., the teacher factor) is a random factor. As a result, the two-factor nested ANOVA is often a mixed-effects model where the nonnested factor is fixed (i.e., all the levels of interest for the nonnested factor are included in the model) and the nested factor is random. The two-factor mixed-effects nested ANOVA model is written in terms of **population parameters**, as follows:

$$Y_{ijk} = \mu + \alpha_j + b_{k(j)} + \varepsilon_{ijk}$$

where Y_{ijk} is the observed score on the dependent variable for individual i in level j of factor A (where A is the nonnested factor) and level k of factor B (or in the jk cell) (where B is the nested factor), μ is the overall or grand population mean (i.e., regardless of cell designation), α_j is the fixed effect for level j of factor A, $b_{k(j)}$ is the random effect for level k of factor B, and e_{ijk} is the random residual error for individual i in cell jk. Notice that there is no interaction term in the model, and also that the effect for factor B is denoted by $b_{k(j)}$. *This tells us that the levels of factor B are nested within factor A*. The residual error can be due to individual differences, measurement error, and/or other factors not under investigation. We consider the fixed-, mixed-, and random-effects cases later in this chapter.

For the two-factor mixed-effects nested ANOVA model, there are only two sets of hypotheses, one for each of the main effects, because there is no interaction effect. The null and alternative hypotheses, respectively, for testing the effect of factor A (nonnested factor) are as follows. The null hypothesis for testing the effect of factor A is similar to what we have seen in previous chapters for fixed-effects factors and written as the means of the levels of factor A are the same.

$$H_{01}: \mu_{.1.} = \mu_{.2.} = \cdots = \mu_{.j.}$$
$$H_{11}: \text{not all the} \, \mu_{.j.} \text{ are equal}$$

The hypotheses for testing the effect of factor B, because this is a random-effects factor, are written as the *variation among the means*, and are presented as follows:

$$H_{02}: \sigma_b^2 = 0$$
$$H_{12}: \sigma_b^2 > 0$$

These hypotheses reflect the inferences made in the fixed-, mixed-, and random-effects models (as fully described in Chapter 5). For *fixed main effects*, the null hypotheses are about *means*, whereas for *random main effects*, the null hypotheses are about *variation among the means*. As we already know, the difference in the models is also reflected in the multiple comparison procedures. As before, we do need to pay particular attention to whether the model is fixed, mixed, or random. The assumptions about the two-factor nested model are exactly the same as with the two-factor crossed model (discussed in Chapters 3 and 5), and thus we need not provide any additional discussion other than to remind you of the assumptions regarding normality, homogeneity of variance, and independence (of observations within cells). In addition, procedures for determining power and confidence intervals are the same as with the two-factor crossed model.

6.1.1.3 ANOVA Summary Table and Expected Mean Squares for the Two-Factor Hierarchical Model

The computations of the two-factor mixed-effects nested model are somewhat similar to those of the two-factor mixed-effects crossed model. The main difference lies in the fact that there is no interaction term. The ANOVA summary table is shown in Table 6.2, where we see the following sources of variation: A, B(A), within cells, and total. There we see that only two F ratios can be formed, one for each of the two main effects, because no interaction term is estimated (recall that this is because not all possible combinations of A and B occur).

If we take the **total sum of squares** and decompose it, we have the following:

$$SS_{total} = SS_A + SS_{B(A)} + SS_{with}$$

We leave the computations involving these terms to the statistical software. The degrees of freedom, mean squares, and F ratios are determined as shown in Table 6.2, assuming a mixed-effects model. The critical value for the test of factor A is ${}_{\alpha}F_{J-1,J(K_{(j)}-1)}$ and for the

TABLE 6.2

Two-Factor Nested Design ANOVA Summary Table: Mixed Effects Model

Source	*SS*	*df*	*MS*	F
A	SS_A	$J-1$	MS_A	$MS_A/MS_{B(A)}$
B(A)	$SS_{B(A)}$	$J(K_{(j)}-1)$	$MS_{B(A)}$	$MS_{B(A)}/MS_{with}$
Within	SS_{with}	$JK_{(j)}(n-1)$	MS_{with}	
Total	SS_{total}	$N-1$		

test of factor B is ${}_{\alpha}F_{J\left(K_{(j)}-1\right),JK_{(j)}(n-1)}$. Let us explain something about the degrees of freedom. The degrees of freedom for B(A) are equal to $J(K_{(j)} - 1)$. This means that for a design with two levels of factor A (e.g., intervention, the nonnested factor) and three levels of factor B (e.g., interventionist, the nested factor) within each level of A (for a total of six levels of B), the degrees of freedom are equal to 2(3 – 1) = 4. This is not the same as the degrees of freedom for a completely crossed design where df_B would be 5 (i.e., 6 – 1 = 5). The degrees of freedom for within are equal to $JK_{(j)}(n - 1)$. For this same design with $n = 10$, then the degrees of freedom within are equal to (2)(3)(10 – 1) = 54 (i.e., 6 cells with 9 degrees of freedom per cell).

The appropriate **error terms** for each of the fixed-, random-, and mixed-effects models are described in the following two paragraphs. For the *fixed-effects model*, both F ratios use the within source as the error term. For the *random-effects model*, the appropriate error term for the test of A is $MS_{B(A)}$ and for the test of B is MS_{with}. For the *mixed-effects model where A is fixed and B is random*, the appropriate error term for the test of A is $MS_{B(A)}$ and for the test of B is MS_{with}. As already mentioned, this is the predominant model in the social sciences. Finally, for the *mixed-effects model where A is random and B is fixed*, both F ratios use the within source as the error term. These are now described by the expected mean squares.

The formation of the proper F ratios is again related to the **expected mean squares**. If H_0 is actually *true*, then the expected mean squares are as follows:

$$E(MS_A) = \sigma_\varepsilon^2$$

$$E\left(MS_{B(A)}\right) = \sigma_\varepsilon^2$$

$$E(MS_{with}) = \sigma_\varepsilon^2$$

If H_0 is actually *false*, then the expected mean squares for the *fixed-effects case* are as follows:

$$E(MS_A) = \sigma_\varepsilon^2 + nK_{(j)}\left[\frac{\sum_{j=1}^{J}\alpha_j^2}{J-1}\right]$$

$$E\left(MS_{B(A)}\right) = \sigma_\varepsilon^2 + n\left[\frac{\sum_{j=1}^{J}\sum_{k=1}^{K}\beta_{k(j)}^2}{J\left(K_{(j)}-1\right)}\right]$$

$$E(MS_{with}) = \sigma_\varepsilon^2$$

Thus, the appropriate F ratios both involve using the *within source* as the error term.

If H_0 is actually *false*, then the expected mean squares for the *random-effects case* are as follows:

$$E(MS_A) = \sigma_\varepsilon^2 + n\sigma_{b(a)}^2 + nK_{(j)}\sigma_a^2$$

$$E\left(MS_{B(A)}\right) = \sigma_\varepsilon^2 + n\sigma_{b(a)}^2$$

$$E(MS_{with}) = \sigma_\varepsilon^2$$

Thus, the appropriate error term for the test of A (i.e., the nonnested factor) is $MS_{B(A)}$ and the appropriate error term for the test of B (i.e., the nested factor) is MS_{with}.

If H_0 is actually *false*, then the expected mean squares for the *mixed-effects case where A is fixed and B is random* are as follows:

$$E(MS_A) = \sigma_\varepsilon^2 + n\sigma_{b(a)}^2 + nK_{(j)}\left(\frac{\sum_{j=1}^{J}\alpha_j^2}{J-1}\right)$$

$$E\left(MS_{B(A)}\right) = \sigma_\varepsilon^2 + n\sigma_{b(a)}^2$$

$$E(MS_{with}) = \sigma_\varepsilon^2$$

Thus, the appropriate error term for the test of A (nonnested) is $MS_{B(A)}$ and the appropriate error term for the test of B (nested) is MS_{with}.

Finally, if H_0 is actually *false*, then the expected mean squares for the *mixed-effects case where A is random and B is fixed* are as follows:

$$E(MS_A) = \sigma_\varepsilon^2 + nK_{(j)}\sigma_a^2$$

$$E\left(MS_{B(A)}\right) = \sigma_\varepsilon^2 + n\left[\frac{\sum_{j=1}^{J}\sum_{k=1}^{K}\beta_{k(j)}^2}{J\left(K_{(j)}-1\right)}\right]$$

$$E(MS_{with}) = \sigma_\varepsilon^2$$

Thus, the appropriate F ratios both involve using the *within source* as the error term.

6.1.1.4 Multiple Comparison Procedures for the Two-Factor Hierarchical Model

This section considers multiple comparison procedures (MCPs) for the two-factor nested design. First of all, the researcher is usually not interested in making inferences about random effects. Second, for MCPs based on the levels of factor A (the nonnested factor), there is nothing new to report. Third, for MCPs based on the levels of factor B (the nested factor), this is a different situation. The researcher is not usually as interested in MCPs about the nested factor as compared to the nonnested factor because inferences about the levels of factor B are not even generalizable across the levels of factor A, due to the nesting. If you are nonetheless interested in MCPs for factor B, by necessity you have to look within a level of A to formulate a contrast. Otherwise, MCPs are conducted as before. For more complex nested designs, see Myers (1979), Keppel and Wickens (2004), Kirk (2013), Mickey et al. (2004), or Myers, Lorch, and Well (2010).

6.1.1.5 An Example of the Two-Factor Hierarchical Model

Let us consider an example to illustrate the procedures in this section. The data are shown in Table 6.3. Factor A is approach to the teaching of reading (basal vs. whole language approaches), and factor B is teacher. Thus, there are two teachers using the basal approach and two different teachers using the whole language approach. The researcher is interested in the effects these factors have on student's reading comprehension in the first grade. Thus the dependent variable is a measure of reading comprehension. Six students are randomly assigned to each approach-teacher combination for small-group instruction. This particular example is a *mixed model*, where factor A (instructional method) is a fixed effect and factor B (teacher) is a random effect. This could easily translate to other examples. For example, factor A is a healthcare treatment and factor B is provider, with some doctors using one type of healthcare approach and the remaining doctors using a different approach. The outcome could be improvement in health (e.g., lower blood pressure). The results are shown in the ANOVA summary table of Table 6.4.

TABLE 6.3

Data for the Teaching Reading Example: Two-Factor Nested Design

	Reading Approaches			
	A_1 (Basal)		A_2 (Whole Language)	
	Teacher B_1	Teacher B_2	Teacher B_3	Teacher B_4
	1	1	7	8
	1	3	8	9
	2	3	8	11
	4	4	10	13
	4	6	12	14
	5	6	15	15
Cell means	2.8333	3.8333	10.0000	11.6667
A means	3.3333		10.8333	
Overall mean	7.0833			

TABLE 6.4

Two-Factor Nested Design ANOVA Summary Table: Teaching Reading Example

Source	*SS*	*df*	*MS*	*F*
A	337.5000	1	337.5000	59.5585*
B(A)	11.3333	2	5.6667	0.9524**
Within	119.0000	20	5.9500	
Total	467.8333	23		

* ${}_{.05}F_{1,2} = 18.51$
** ${}_{.05}F_{2,20} = 3.49$

From Appendix Table A.4, the critical value for the test of factor A is ${}_{\alpha}F_{J-1,J(K_{(j)}-1)} = {}_{.05}F_{1,2} = 18.51$, and the critical value for the test of factor B is ${}_{\alpha}F_{J(K_{(j)}-1),JK_{(j)}(n-1)} = {}_{.05}F_{2,20} = 3.49$. Thus there is a statistically significant difference between the two approaches to reading instruction at the .05 level of significance, and there is no significant difference between the teachers. When we look at the means for the levels of factor A, we see that the mean comprehension score for the whole language approach ($\bar{Y}_{.2.} = 10.8333$) is greater than the mean for the basal approach ($\bar{Y}_{.1.} = 3.3333$). Because there were only two levels of the reading approach tested (whole language and basal), no post hoc multiple comparisons are really necessary. Rather, the mean reading comprehension scores for each approach can be merely examined to determine which mean was statistically significantly larger.

6.1.2 Characteristics of the Two-Factor Randomized Block Model for $n = 1$

In this section, we describe the distinguishing characteristics of the two-factor randomized block ANOVA model for one observation per cell, the layout of the data, the linear model, assumptions and their violation, the ANOVA summary table and expected mean squares, multiple comparison procedures, and methods of block formation.

The characteristics of the two-factor randomized block ANOVA model are quite similar to those of the regular two-factor ANOVA model, as well as sharing a few characteristics with the one-factor repeated measures ANOVA design. There is one obvious exception, which has to do with the nature of the factors being used. Here there will be two factors, each with at least two levels. One factor is known as the **treatment factor** and is referred to here as factor A (a treatment factor is technically what we have been considering in Chapters 1 through 5; although as we'll soon discuss, this factor does not have to truly be a "treatment" but can be an observable attribute). The second factor is known as the **blocking factor** and is referred to here as factor B. A blocking factor is a new concept and requires some discussion.

Take an ordinary one-factor ANOVA design, where the single factor is a treatment factor (e.g., method of exercising) and the researcher is interested in its effect on some dependent variable (e.g., percentage of body fat). Despite individuals being randomly assigned to a treatment group, the groups may be different due to a nuisance variable operating in a nonrandom way. For instance, Group 1 may consist of mostly older adults and Group 2 may consist of mostly younger adults. Thus, it is likely that Group 2 will be favored over Group 1 because age, the nuisance variable, has not been properly balanced out across the groups by the randomization process.

One way to deal with this problem is to control the effect of the nuisance variable by incorporating it into the design of the study. Including the blocking or nuisance variable as a factor in the design should result in a reduction in residual variation (due to some additional portion of individual differences being explained) and an increase in power (Glass & Hopkins, 1996; Keppel & Wickens, 2004). The blocking factor is selected based on the strength of its relationship to the dependent variable, where an unrelated blocking variable would not reduce residual variation. It would be reasonable to expect, then, that variability among individuals within a block (e.g., within younger adults) should be less than variability among individuals between blocks (e.g., between younger and older adults). *Thus, each block represents the formation of a matched set of individuals, that is, matched on the blocking variable, but not necessarily matched on any other nuisance variable.* Using our example, we expect that in general, adults within a particular age block (i.e., the older or younger blocks) will be more similar in terms of variables related to body fat than adults across blocks.

Let us consider several examples of blocking factors. Some blocking factors are naturally occurring blocks such as siblings, friends, neighbors, plots of land, and time. Other blocking factors are not naturally occurring, but can be formulated by the researcher. Examples of this type include grade point average, age, weight, aptitude test scores, intelligence test scores, socioeconomic status, and school or district size. Note that the examples of blocking factors here represent a variety of measurement scales (categorical as well as continuous). Later we will discuss how to deal with the blocking factor based on its measurement scale in the discussion of method of block formation.

Let us make some summary statements about characteristics of blocking designs. First, designs that include one or more blocking factors are known as **randomized block designs**, also known as *matching designs* or *treatment by block designs. The researcher's main interest is in the treatment factor*. The purpose of the blocking factor is to reduce residual variation. Thus, the researcher is not as much interested in the test of the blocking factor (possibly not at all) as compared to the treatment factor. Thus, there is at least one blocking factor and one treatment factor, each with two or more levels. Second, each subject falls into only one block in the design and is subsequently randomly assigned to one level of the treatment factor within that block. Thus subjects within a block serve as their own controls such that some portion of their individual differences is taken into account. As a result, the scores of subjects are not independent within a particular block. Third, for purposes of this section, we assume there is only one subject for each treatment-block level combination. *As a result, the model does not include an interaction term*, and this is a distinguishing feature of this model. Later in this chapter, we consider the multiple observations case, where there is an interaction term in the model. Finally, the dependent variable is measured at least at the interval level.

6.1.2.1 The Layout of the Data for the Two-Factor Randomized Block Design for *n* = 1

The layout of the data for the two-factor randomized block model is shown in Table 6.5. Here we see the columns designated as the levels of the blocking factor B and the rows as the levels of the treatment factor A. Row, block, and overall means are also shown. Here you see that the layout of the data looks the same as the two-factor model, but with a single observation per cell.

TABLE 6.5

Layout for the Two-Factor Randomized Block Design

	Level of Factor B				
Level of Factor A	**1**	**2**	**. . .**	**K**	**Row Mean**
1	Y_{11}	Y_{12}	. . .	Y_{1k}	$\bar{Y}_{1.}$
2	Y_{21}	Y_{22}	. . .	Y_{2k}	$\bar{Y}_{2.}$
.	.	.	. . .	.	.
.	.	.	. . .	.	.
.	.	.	. . .	.	.
J	Y_{J1}	Y_{J2}		Y_{JK}	$\bar{Y}_{J.}$
Block mean	$\bar{Y}_{.1}$	$\bar{Y}_{.2}$	. . .	$\bar{Y}_{.K}$	$\bar{Y}_{..}$ *(overall mean)*

6.1.2.2 The Two-Factor Randomized Block Design for n = 1 ANOVA Model

The two-factor fixed-effects randomized block ANOVA model is written in terms of *population parameters* as follows:

$$Y_{jk} = \mu + \alpha_j + \beta_k + \varepsilon_{jk}$$

where Y_{jk} is the observed score on the dependent variable for the individual responding to level j of factor A and level k of block B, μ is the overall or grand population mean, α_j is the fixed effect for level j of factor A, β_k is the fixed effect for level k of the block B, and ε_{jk} is the random residual error for the individual in cell jk. The residual error can be due to measurement error, individual differences, and/or other factors not under investigation. You can see this is similar to the two-factor fully crossed model with one observation per cell (i.e., $i = 1$ making the i subscript unnecessary), and with no interaction term included. Also, the effects are denoted by α and β given we have a fixed-effects model. Note that the row and column effects both sum to zero in the fixed-effects model.

The hypotheses for testing the effect of factor A are as follows, where the null indicates that the means of the levels of factor A are equal:

$$H_{01}: \mu_{1.} = \mu_{2.} = \ldots = \mu_{J.}$$

$$H_{11}: \text{not all the } \mu_{j.} \text{ are equal}$$

For testing the effect of factor B (the blocking factor), the hypotheses are presented here, where the null hypothesis is that the means of the levels of the blocking factor are equal.

$$H_{02}: \mu_{.1} = \mu_{.2} = \ldots = \mu_{.K}$$

$$H_{12}: \text{not all the } \mu_{.k} \text{ are equal}$$

The factors are both fixed, so the hypotheses are written in terms of means.

6.1.2.3 ANOVA Summary Table and Expected Mean Squares

The sources of variation for this model are similar to those of the regular two-factor model, except that there is no interaction term. The ANOVA summary table is shown in Table 6.6, where we see the following sources of variation: A (treatments), B (blocks), residual, and total. The test of block differences is usually of no real interest. In general, we expect there to be differences between the blocks. From the table, we see that two F ratios can be formed.

TABLE 6.6

Two-Factor Randomized Block Design ANOVA Summary Table

Source	*SS*	*df*	*MS*	*F*
A	SS_A	$J-1$	MS_A	MS_A/MS_{res}
B	SS_B	$K-1$	MS_B	MS_B/MS_{res}
Residual	SS_{res}	$(J-1)(K-1)$	MS_{res}	
Total	SS_{total}	$N-1$		

If we take the total sum of squares and decompose it, we have the following equation:

$$SS_{total} = SS_A + SS_B + SS_{res}$$

The remaining computations are determined by the statistical software. The degrees of freedom, mean squares, and F ratios are also shown in Table 6.6.

Earlier in our discussion of the two-factor randomized block design, we mentioned that the F test is not very robust to violation of the sphericity assumption. We again recommend the following sequential procedure be used in the test of factor A. First, perform the usual F test, which is quite liberal in terms of rejecting H_0 too often, where the degrees of freedom are $J - 1$ and $(J - 1)(K - 1)$. If H_0 is not rejected, then stop. If H_0 is rejected, then continue with step 2, which is to use the Geisser and Greenhouse (1958) conservative F test. For the model we are considering here, the degrees of freedom for the F critical value are adjusted to be 1 and $K - 1$. If H_0 is rejected, then stop. This would indicate that both the liberal and conservative tests reached the same conclusion, that is, to reject H_0. If H_0 is not rejected, then the two tests did not reach the same conclusion, and a further test should be undertaken. Thus, in step 3, an adjusted F test is conducted. The adjustment is known as Box's (1954) correction (the Huynh and Feldt (1970) procedure). Here the degrees of freedom are equal to $(J - 1)(\varepsilon)$ and $(J - 1)(K - 1)(\varepsilon)$, where ε is the correction factor (e.g., Kirk, 2013). It is now fairly standard for the major statistical software to conduct the Geisser-Greenhouse and Huynh-Feldt tests.

Based on the expected mean squares (not shown here for simplicity), the residual is the proper error term for the fixed-, random-, and mixed-effects models. *Thus, MS_{res} is the proper error term for every version of this model.* One may also be interested in an assessment of the effect size for the treatment factor A; note that the effect size of the blocking factor B is usually not of interest, and further discussion on effect size is provided later in the chapter. Finally, the procedures for determining confidence intervals and power are the same as in previous models.

6.1.2.4 Multiple Comparison Procedures

If the null hypothesis for either the A (treatment) or B (blocking) factor is rejected and there are more than two levels of the factor for which statistical significance was found, then the researcher may be interested in which means or combinations of means are different. This could be assessed, as put forth in previous chapters, by the use of some multiple comparison procedure (MCP). In general, the use of MCPs outlined in Chapter 2 is unchanged as long as the sphericity assumption is met. If the assumption is not met, then MS_{res} is not the appropriate error term, and the alternatives recommended in Chapter 5 should be considered (e.g., Boik, 1981; Kirk, 2013; Maxwell, 1980).

6.1.2.5 Methods of Block Formation

There are different methods available for the formation of blocks depending on the nature of the blocking variable. As we see, the methods have to do with whether the blocking factor is an ordinal or an interval/ratio variable, and whether the blocking factor is a fixed or a random effect. This discussion borrows heavily from the work of Pingel (1969) in defining five such methods. The first method is the **predefined value blocking method**, *where the blocking factor is an ordinal variable.* Here the researcher specifies K different population values of the blocking variable. For each of these values (i.e., a fixed effect), individuals are randomly assigned to the levels of the treatment factor. Thus, individuals within a block have the same value on the blocking variable. For example, if class rank is the blocking variable, the levels might be the top third, middle third, and bottom third of the class.

The second method is the **predefined range blocking method**, *where the blocking factor is an interval or ratio variable*. Here the researcher specifies K mutually exclusive ranges in the population distribution of the blocking variable, where the probability of obtaining a value of the blocking variable in each range may be specified as $1/K$. For each of these ranges (i.e., a fixed effect), individuals are randomly assigned to the levels of the treatment factor. Thus, individuals within a block are in the same range on the blocking variable. For example, if a score that ranges from 0 to 100 is the blocking variable, the levels might be 0–32, 33–66, and 67–100.

The third method is the **sampled value blocking method**, *where the blocking variable is an ordinal variable*. Here the researcher randomly samples K population values of the blocking variable (i.e., a random effect). For each of these values, individuals are randomly assigned to the levels of the treatment factor. Thus individuals within a block have the same value on the blocking variable. For example, if class rank is again the blocking variable, only this time measured in tenths, the researcher might randomly select 3 levels from the population of 10 levels.

The fourth method is the **sampled range blocking method**, *where the blocking variable is an interval or ratio variable*. Here the researcher randomly samples N individuals from the population, such that $N = JK$, where K is the number of blocks desired (i.e., a fixed effect) and J is the number of treatment groups. These individuals are ranked according to their values on the blocking variable from 1 to N. The first block consists of those individuals ranked from 1 to J, the second block of those ranked from $J + 1$ to $2J$, and so on. Finally, individuals within a block are randomly assigned to the J treatment groups. For example, consider a placement exam score as the blocking variable, where there are $J = 4$ treatment groups, $K = 10$ blocks, and thus $N = JK = 40$ individuals. The top four ranked individuals on the placement exam would constitute the first block, and they would be randomly assigned to the four groups. The next four ranked individuals would constitute the second block, and so on.

The fifth method is the **post hoc blocking method**. Here the researcher has already designed the study and collected the data, without the benefit of a blocking variable. After the fact, a blocking variable is identified and incorporated into the analysis. It is possible to implement any of the four preceding procedures on a post hoc basis.

Based on the research of Pingel (1969), some statements can be made about the precision of these blocking methods in terms of a reduction in residual variability as well as better estimation of the treatment effect. In general, for an ordinal blocking variable, the predefined value blocking method is more precise than the sampled value blocking method. Likewise, for an interval or ratio blocking variable, the predefined range blocking method is more precise than the sampled range blocking method. Finally, the post hoc blocking method is the least precise of the methods discussed. For discussion of selecting an optimal number of blocks, we suggest you consider Feldt (1958; highly recommended), as well as Keppel and Wickens (2004) and Myers et al. (2010). These researchers make the following recommendations about the optimal number of blocks (where r_{XY} is the correlation between the blocking factor X, in a randomized block design, and the dependent variable Y):

- if $r_{XY} = .2$, then use five blocks;
- if $r_{XY} = .4$, then use four blocks;

- if r_{XY} = .6, then use three blocks; and
- if r_{XY} = .8, then use two blocks.

6.1.2.6 An Example

Let us consider an example to illustrate the procedures in this section. The data are shown in Table 6.7. The blocking factor is age (i.e., 20, 30, 40, and 50 years of age), the treatment factor is number of workouts per week (i.e., 1, 2, 3, and 4), and the dependent variable is amount of weight lost during the first month. Presume we have a fixed-effects model. Table 6.8 contains the resultant ANOVA summary table.

The test statistics are both compared to the usual *F* test critical value of ${}_{.05}F_{3,9} = 3.86$ (from Appendix Table A.4), so that both main effects tests are statistically significant. The Geisser-Greenhouse conservative procedure is necessary for the test of factor A; here the test statistic is compared to the critical value of ${}_{.05}F_{1,3} = 10.13,$ which is also significant. The two procedures both yield a statistically significant result, so we need not be concerned with a violation of the sphericity assumption for the test of A. In summary, the effects of amount of exercise undertaken and age on amount of weight lost are both statistically significant at the .05 level of significance.

Next we need to test the *additivity assumption* using Tukey's (1949) test of additivity. The *F* test statistic is equal to 0.1010, which is compared to the critical value of ${}_{.05}F_{1,8} = 5.32$ from Appendix Table A.4. The test is nonsignificant, so the model is additive and the assumption has been met.

TABLE 6.7

Data for the Exercise Example: Two-Factor Randomized Block Design

	Age				
Exercise Program	**20**	**30**	**40**	**50**	**Row Means**
1/week	3	2	1	0	1.5000
2/week	6	5	4	2	4.2500
3/week	10	8	7	6	7.7500
4/week	9	7	8	7	7.7500
Block means	7.0000	5.5000	5.0000	3.7500	5.3125 *(overall mean)*

TABLE 6.8

Two-Factor Randomized Block Design ANOVA Summary Table: Exercise Example

Source	*SS*	*df*	*MS*	*F*
A	21.6875	3	7.2292	18.2648*
B	110.1875	3	36.7292	92.7974*
Residual	3.5625	9	0.3958	
Total	135.4375	15		

* ${}_{.05}F_{3,9} = 3.86$

As an example of a MCP, the Tukey HSD procedure is used to test for the equivalence of exercising once a week (j = 1) and four times a week (j = 4), where the contrast is written as $\bar{Y}_{4.} - \bar{Y}_{1.}$. The mean amount of weight lost for these groups are 1.5000 for the once a week program and 7.7500 for the four times a week program. The standard error is computed as:

$$s_{\psi'} = \sqrt{\frac{MS_{res}}{J}} = \sqrt{\frac{0.3958}{4}} = 0.3146$$

and the studentized range statistic is as follows:

$$q = \frac{\bar{Y}_{4.} - \bar{Y}_{1.}}{s_{\psi'}} = \frac{7.75 - 1.50}{0.3146} = 19.8665$$

The critical value is ${}_{\alpha}q_{9,4} = 4.415$ (from Appendix Table A.9). The test statistic exceeds the critical value; thus we conclude that the mean amount of weight lost for groups 1 (exercise once per week) and 4 (exercise four times per week) are statistically significantly different at the .05 level (i.e., more frequent exercise helps one to lose more weight).

6.1.3 Characteristics of the Two-Factor Randomized Block Design for $n > 1$

For two-factor randomized block designs with more than one observation per cell, there is little that we have not already covered. First, the characteristics are exactly the same as with the n = 1 model, with the obvious exception that when $n > 1$, an interaction term exists. Second, the layout of the data, the model, the ANOVA summary table, and the multiple comparison procedures are the same as in the regular two-factor model. Third, the assumptions are the same as with the n = 1 model, except the assumption of additivity is not necessary because an interaction term exists. The sphericity assumption is required for those tests using MS_{AB} as the error term. We do not mean to minimize the importance of this popular model; however, there really is no additional information to provide beyond what we have already presented. For a discussion of other randomized block designs, see Kirk (2014).

6.1.4 Characteristics of the Friedman Test

There is a nonparametric equivalent to the two-factor randomized block ANOVA model. The test was developed by Friedman (1937) and is based on mean ranks. For the case of n = 1, the procedure is precisely the same as the Friedman test for the one-factor repeated measures model (see Chapter 5). For the case of $n > 1$, the procedure is slightly different. First, all of the scores within each block are ranked for that block. For instance, if there are J = 4 levels of factor A and n = 10 individuals per cell, then each block's scores would be ranked from 1 to 40 (i.e., nJ). From this, a mean ranking can be determined for each level of factor A. The null hypothesis tests whether the mean rankings for each of the levels of A are equal. The test statistic is a χ^2, which is compared to the critical value of ${}_{\alpha}\chi^2_{J-1}$ (see Appendix Table A.3), where the null hypothesis is rejected if the test statistic exceeds the critical value.

In the case of tied ranks, either the available ranks can be averaged, or a correction factor can be used (see Chapter 5). You may also recall the problem with small n's in terms of the test statistic not being precisely distributed as a χ^2. For situations where $J < 6$ and $n < 6$, consult the table of critical values in Marascuilo and McSweeney (1977, Table A-22, p. 521). The Friedman test assumes that the population distributions have the same shape (although not necessarily normal) and the same variability, and that the dependent measure is continuous. For alternative nonparametric procedures, see the discussion in Chapter 5.

Various multiple comparison procedures (MCPs) can be used for the nonparametric two-factor randomized block model. For the most part, these MCPs are analogs to their parametric equivalents. In the case of planned pairwise comparisons, one may use multiple matched-pair Wilcoxon tests in a Bonferroni form (i.e., taking the number of contrasts into account by splitting up the α level). Due to the nature of planned comparisons, these are more powerful than the Friedman test. For post hoc comparisons, two example MCPs are the Tukey HSD analog for pairwise contrasts, and the Scheffé analog for complex contrasts. For additional discussion about the use of MCPs for this model, see Marascuilo and McSweeney (1977). For an example of the Friedman test, return to Chapter 5. Finally, note that MCPs are not usually conducted on the blocking factor as they are rarely of interest to the applied researcher.

6.1.5 Comparison of Various ANOVA Models

How do some of the ANOVA models we have considered compare in terms of power and precision? Recall again that **power** is defined as the probability of rejecting H_0 when H_0 is false, and **precision** is defined as a measure of our ability to obtain good estimates of the treatment effects. The classic literature on this topic revolves around the correlation between the dependent variable Y and the concomitant variable X (i.e., r_{XY}), where the concomitant variable can be either a covariate or a blocking factor. First, let us compare the one-factor ANOVA and one-factor ANCOVA models. If r_{XY}, the correlation between the covariate X and the dependent variable Y, is not statistically significantly different from zero, then the amount of unexplained variation will be the same in the two models. Thus, no statistical adjustment will be made on the group means. In this situation, the ANOVA model is more powerful, as we lose one degree of freedom for each covariate used in the ANCOVA model. If r_{XY} *is* significantly different from zero, then the amount of unexplained variation will be smaller in the ANCOVA model as compared to the ANOVA model. Here the ANCOVA model is more powerful and is more precise as compared to the ANOVA model. Second, compare the one-factor ANOVA and two-factor randomized block designs. If $2 < r_{XY} < 2, < .4$, the correlation between the blocking factor X and the dependent variable Y, is not statistically significantly different from zero, then the blocking factor will not account for much variability in the dependent variable. One rule of thumb states that if $r_{XY} < 2$, then ignore the concomitant variable (whether it is a covariate or a blocking factor), and use the one-factor analysis of variance. Otherwise, take the concomitant variable into account somehow, either as a covariate or blocking factor.

How should we take the concomitant variable into account if it correlates with the dependent variable at *greater* than .20 (i.e., $r_{XY} > 20$)? The two best possibilities are the analysis of covariance design (ANCOVA, Chapter 4) and the randomized block ANOVA design (discussed in this chapter). That is, the concomitant variable can be used either as a covariate through a statistical form of control (i.e., ANCOVA), or as a blocking factor through an experimental design form of control (i.e., randomized block ANOVA). As suggested by the classic work of Feldt (1958), if $.2 < r_{XY} < .4$, then use the concomitant variable as a blocking factor in a randomized block design as it is the most powerful and precise design. If $r_{XY} > .6$, then use the concomitant variable as a covariate in an ANCOVA design as it is the most powerful and precise design. If $.4 < r_{XY} < .6$, then the randomized block and ANCOVA designs are about equal in terms of power and precision.

However, Maxwell, Delaney, and Dill (1984) showed that the correlation between the covariate and dependent variable should not be the ultimate criterion in deciding whether to use an ANCOVA or a randomized block design. These designs differ in the following two ways: (a) whether the concomitant variable is treated as continuous (ANCOVA) or categorical (randomized block), and (b) whether individuals are assigned to groups based on the concomitant variable (randomized blocks) or without regard to the concomitant variable (ANCOVA). Thus the Feldt (1958) comparison of these particular models is not a fair one in that the models differ in these two ways. The ANCOVA model makes full use of the information contained in the concomitant variable, whereas in the randomized block model, some information is lost due to the categorization. In examining nine different models, Maxwell and colleagues suggest that r_{XY} should not be the sole factor in the choice of a design (given that r_{XY} is at least .3), but that two other factors be considered. The first factor is whether scores on the concomitant variable are available prior to the assignment of individuals to groups. If so, power will be increased by assigning individuals to groups based on the concomitant variable (i.e., blocking). The second factor is whether X (the concomitant variable) and Y (the dependent variable) are linearly related. If so, the use of ANCOVA with a continuous concomitant variable is more powerful because linearity is an assumption of the model (Keppel & Wickens, 2004; Myers et al., 2010). If not, either the concomitant variable should be used as a blocking variable, or some sort of nonlinear ANCOVA model should be used.

There are a few other decision criteria you may want to consider in choosing between the randomized block and ANCOVA designs. First, in some situations, blocking may be difficult to carry out. For instance, we may not be able to find enough homogeneous individuals to constitute a block. If the blocks formed are not very homogeneous, this defeats the whole purpose of blocking. Second, the interaction of the independent variable and the concomitant variable may be an important effect to study. In this case, use the randomized block design with multiple individuals per cell. If the interaction is significant, this violates the assumption of homogeneity of regression slopes in the analysis of covariance design, but does not violate any assumption in the randomized block design with $n > 1$. Third, it should be obvious by now that the assumptions of the ANCOVA design are much more restrictive than in the randomized block design. Thus when important assumptions are likely to be seriously violated, the randomized block design is preferable.

There are other alternative designs for incorporating the concomitant variable as a pretest, such as an analysis of variance on gain (the difference between posttest and pretest), or a mixed (split-plot) design where the pretest and posttest measures are treated as the levels of a repeated factor. Based on the research of Huck and McLean (1975) and Jennings (1988), the ANCOVA model is generally preferred over these other two models. For further discussion see Reichardt (1979), Huitema (2011), or Kirk (2013).

6.1.6 Sample Size

6.1.6.1 Hierarchical ANOVA Model Sample Size

Sample size is often a difficult question to answer with single-level analyses, and the question of sufficient sample size becomes even more complex to answer with multilevel models. In general, in multilevel models (i.e., hierarchical models), *the sample size at the highest level is primarily of most concern because the sample size at that level is always smaller than at the lowest level* (Maas & Hox, 2005). In a two-level model, such as a two-factor hierarchical ANOVA, the "highest-level" sample size would be the sample size at the group or cluster level (i.e., nested factor). The following discussion of sample size is in the context of multilevel modeling in general and goes a bit beyond what has been covered in this chapter as most current research on hierarchical models has been in the context of estimation methods such as full maximum likelihood or restricted maximum likelihood. We'll proceed regardless, as this may help framing how to think about sample size in a hierarchical design. Additionally, you may be using hierarchical ANOVA with a maximum likelihood estimation, and in those cases, this is completely applicable. A few guidelines exist for minimum group sample size, including more than 10 groups (Snijders & Bosker, 1999), assuming restricted maximum likelihood is the estimation method, and a minimum of 30 groups (Kreft & de Leeuw, 1998). Sample size of the number of cases within groups (i.e., at the lowest level in a multilevel model) is less of a concern, and groups with even just one observation in them should be retained. While those groups will not contribute to the within-group variances, they will contribute to the between-group variance and overall average.

In addition to considering the sample sizes at each level, the proportion of variation in the outcome between groups, i.e., **intraclass correlation coefficient (ICC)**, as well as the estimation method, i.e., **full maximum likelihood (FML)** or **restricted maximum likelihood (RML)**, are also considerations. Simulation research has been conducted that has conditioned on estimation methods (FML, RML), number of groups (30, 50, 100), size of groups (5, 30, 50), and ICC (.1, .2, .3) (Maas & Hox, 2005). In all conditions, regression coefficients and variance components are unbiased. However, the standard errors of the variances at level 2 are underestimated when the number of groups is fewer than 100; however, the bias is, "in practice, probably acceptable" (Maas & Hox, 2005, p. 91). Conditions were also tested with only 10 groups based on work by Snijders and Bosker (1999). While the regression coefficients and level 1 variance components were unbiased, the level 2 variance components were overestimated, and the standard errors were unacceptably underestimated, suggesting that 10 groups at level 2 is insufficient for estimating MLM (Maas & Hox, 2005). Optimal Design (Spybrook, Raudenbush, Liu, Congdon, & Martinez, 2006) is a freely accessible online program designed to estimate power and sample size in group randomized designs and can be used *a priori* or post hoc. Even if you are not in a situation where randomization of groups will be or has been done, Optimal Design may provide the best available information for estimating sample size in a hierarchical design.

6.1.6.2 Randomized Block ANOVA Sample Size

In general, randomized block designs have more power that completely randomized designs of equal size (Festing, 2014). In terms of sample size, there are no magic numbers that can be suggested. Rather, we encourage you to determine sample size based on power tables or software (e.g., G*Power).

6.1.7 Power

A discussion of power has been intertwined throughout this chapter. As noted previously, recall that procedures for determining power in the hierarchical ANOVA model are the same as with the two-factor crossed model. Depending on your situation, the use of software such as Optimal Design may be appropriate.

6.1.8 Effect Size

Traditional effect sizes considered in ANOVA include omega squared (ω^2), eta squared (η^2), and partial eta squared $\left(\eta_p^2\right)$. *Omega squared* is interpreted as the proportion of the variation of the dependent variable that is attributed to variation in the independent variable. *Eta squared* is interpreted as the proportion of total variability in the dependent variable that is accounted for by variation in each main effect, interaction, and error in the model. *Partial eta squared* is interpreted as the proportion of total variability in the dependent variable attributed to a factor and that is not explained by other factors in the model. However, as pointed out by Olejnik and Algina (2003), Cohen's effect sizes are based on "unrestricted populations" (p. 446); in other words, designs that do *not* include controls or blocking variables. Thus, effect sizes that work well in simpler ANOVA models (e.g., omega squared and eta squared) do not work well with more complex ANOVA models such as nested and randomized block designs. *Failing to consider the design (e.g., nested random effects) when calculating the effect size can result in biased estimates.* Wampold and Serlin (2000) found that ignoring the nested model can lead not only to inflated Type I error rates but grossly overstated effects. For example, when 30% of the variance in the outcome was due to the cluster, a moderate effect was produced when the actual treatment effect was zero (Wampold & Serlin, 2000).

To address this and other shortcomings of effect size in more complicated ANOVA models, researchers are encouraged to consider other, more appropriate, effect size indices. For example, Olejnik and Algina (2003) proposed generalized eta squared and generalized omega squared effect size statistics that take into account research design features.

6.1.8.1 Hierarchical ANOVA Effect Size

For the hierarchical ANOVA model with a random nested factor, the **overall omega squared ($\hat{\omega}^2$)** is the appropriate effect size measure (Olejnik & Algina, 2000). This effect size represents the proportion of total variance of the dependent variable accounted for by the respective factor. In this case, both factors may be random, or one factor is fixed while the other is random. Applying Cohen's (1988) conventions for interpretation, a small overall omega squared hat is .01, medium is .06, and large is .14. Maxwell et al. (2018) provide the following formula:

$$\hat{\omega}_A^2 = \frac{\sum \hat{\alpha}_j^2 / a}{\left(\sum \hat{\alpha}_j^2 / a\right) + \hat{\alpha}_\beta^2 + \hat{\alpha}_\varepsilon^2}$$

Where

$$\frac{\sum \hat{\alpha}_j^2}{a} = \left(\frac{(a-1)}{a}\right)\left(\frac{\left(MS_A - MS_{B(A)}\right)}{bn}\right)$$

$$\hat{\alpha}_\beta^2 = \frac{MS_{B(A)} - MS_{with}}{n}$$

$$\hat{\alpha}_\varepsilon^2 = MS_{with}$$

a = number of levels of factor A
b = number of levels per nest (not the total number of levels of factor B)

Partial omega squared, $\hat{\omega}^2_{partial}$, is for assessing partial variance; e.g., other factors in the design are controlled by excluding them from the computation (proportion of total variability in the dependent variable attributed to a factor and that is not explained by other factors in the model). Generally, for the same effect, *a proportion of partial variance effect size will be larger than the proportion of total variance effect size* (Olejnik & Algina, 2000). Applying Cohen's (1988) conventions for interpretation, a small partial omega squared hat is .01, medium is .06, and large is .14.

$$\hat{\omega}^2_{A.partial} = \frac{MS_A - MS_{AB}}{MS_A + (n)(K)(MS_{error}) - MS_{AB}}$$

$$\hat{\omega}^2_{B.partial} = \frac{MS_{AB} - MS_{AB}}{MS_B + (n)(J)(MS_{error}) - MS_{AB}}$$

where
J = number of levels in factor A
K = number of levels in factor B

Maxwell et al. (2018) provide the following formula for the effect of the nonnested factor, where $\sum \hat{\alpha}_j^2 / a$ and $\hat{\sigma}^2_\varepsilon$ were defined previously.

$$\hat{\omega}^2_{A,partial} = \frac{\sum \hat{\alpha}_j^2 / a}{\left(\sum \hat{\alpha}_j^2 / a\right) + \hat{\sigma}^2_\varepsilon}$$

Partial intraclass correlation coefficient, $\hat{\rho}^2_{I:B(A),partial}$, for assessing the effect of the nested factor, can be computed as follows (Maxwell et al., 2018):

$$\hat{\rho}^2_{I:B(A),partial} = \frac{\hat{\sigma}^2_\beta}{\hat{\sigma}^2_\beta + \hat{\sigma}^2_\varepsilon}$$

where

$$\hat{\sigma}^2_\beta = \frac{MS_{B(A)} - MS_{with}}{n}$$

$$\hat{\sigma}^2_\varepsilon = MS_{with}$$

If we follow conventions for ICC in general that are presented by Hox, Moerbeek, and van de Schoot (2017) and apply these to partial ICC, a small effect is .05, moderate is .10, and large is .15. In cases where higher ICCs are reasonable based on a prior information, small is .10, medium is .20, and large is .30. *We caution readers on applying conventions for interpreting the size of the effect, regardless of which effect size is interpreted.* As noted by Hox et al. (2017), what is small versus moderate versus large very much depends on the context.

Thus, we encourage readers to review related literature to compare and make interpretations of the size of the effect rather than apply effect size conventions.

For the two-factor nested ANOVA example presented in the illustration throughout the test, we find the following overall $\hat{\omega}_A^2$ of .70:

$$\hat{\omega}_A^2 = \frac{\sum \hat{\alpha}_j^2 / a}{\left(\sum \hat{\alpha}_j^2 / a\right) + \hat{\sigma}_\beta^2 + \hat{\sigma}_\varepsilon^2} = \frac{13.83}{13.83 + (-.012) + 5.95} = \frac{13.83}{19.77} = .70$$

where

$$\frac{\sum \hat{\alpha}_j^2}{a} = \left(\frac{(a-1)}{a}\right)\left(\frac{(MS_A - MS_{B(A)})}{bn}\right) = \left(\frac{2-1}{2}\right)\left(\frac{337.50 - 5.667}{(2)(24)}\right) = .5\left(\frac{331.833}{48}\right) = 13.83$$

$$\hat{\sigma}_\beta^2 = \frac{MS_{B(A)} - MS_{with}}{n} = \frac{5.667 - 5.95}{24} = \frac{-.283}{24} = -.012$$

$$\hat{\sigma}_\varepsilon^2 = MS_{with} = 5.95$$

a = number of levels of factor A = 2
b = number of levels per nest (not the total number of levels of factor B) = 2

And, the effect of level A (i.e., nonnested factor, in this case the intervention) can be computed as follows:

$$\hat{\omega}_{A,partial}^2 = \frac{\sum \hat{\alpha}_j^2 / a}{\left(\sum \hat{\alpha}_j^2 / a\right) + \hat{\sigma}_\varepsilon^2} = \frac{13.83}{13.83 + 5.95} = \frac{13.83}{19.78} = .70$$

6.1.8.2 Two-Factor Randomized Block Effect Size

For the two-factor randomized block, the overall omega squared is the appropriate effect size measure and can be calculated as follows (Olejnik & Algina, 2000):

$$\hat{\omega}_A^2 = \frac{J\,(MS_A - MS_{AB})}{SS_{total} + MS_A + MS_B - MS_{AB}}$$

$$\hat{\omega}_B^2 = \frac{K\,(MS_B - MS_{AB})}{SS_{total} + MS_A + MS_B - MS_{AB}}$$

$$\hat{\omega}_{AB}^2 = \frac{JK\,(MS_{AB} - MS_{error})}{SS_{total} + MS_A + MS_B - MS_{AB}}$$

where
J = number of levels in factor A
K = number of levels in factor B

TABLE 6.9

Effect Sizes and Interpretations

Effect Size	Interpretation
Overall omega squared ($\hat{\omega}^2$)	Proportion of the variation of the dependent variable that is attributed to variation in the factor (i.e., independent variable) • Small effect, $\omega_A^2 = .01$ • Medium effect, $\omega_A^2 = .06$ • Large effect, $\omega_A^2 = .14$
Partial omega squared for level A (nonnested factor) $\left(\hat{\omega}^2_{partial}\right)$	Proportion of total variability in the dependent variable attributed to the nonnested factor that is not explained by other variables in the model • Small effect, $\omega^2_{A,partial} = .01$ • Medium effect, $\omega^2_{A,partial} = .06$ • Large effect, $\omega^2_{A,partial} = .14$
Partial intraclass correlation coefficient for the effect of level B (nested factor) $\left(\hat{\rho}^2_{I:B(A),\,partial}\right)$	Proportion of variation in the dependent variable due to the random factor of B nested within A; conventions based on Hox et al. (2017) with values in parentheses denoting cases where higher ICCs are reasonable based on *a priori* information • Small effect $\hat{\rho}^2_{I:B(A),partial} = .05(.10)$ • Medium effect $\hat{\rho}^2_{I:B(A),partial} = .10(.20)$ • Large effect $\hat{\rho}^2_{I:B(A),partial} = .15(.30)$

Hox, J. J., Moerbeek, M., & van de Schoot, R. (2017). *Multilevel analysis: Techniques and applications* (3rd ed.). New York, NY: Routledge.

Partial omega squared is for assessing partial variance, e.g., other factors in the design are controlled by excluding them from the computation. Generally, for the same effect, *a proportion of partial variance effect size will be larger than the proportion of total variance effect size* (Olejnik & Algina, 2000). Applying Cohen's (1988) conventions for interpretation, a small partial omega squared hat is .01, medium is .06, and large is .14. Partial omega squared can be computed as follows:

$$\hat{\omega}^2_{A.partial} = \frac{MS_A - MS_{AB}}{MS_A + (n)(K)(MS_{error}) - MS_{AB}}$$

$$\hat{\omega}^2_{B.partial} = \frac{MS_B - MS_{AB}}{MS_B + (n)(J)(MS_{error}) - MS_{AB}}$$

where

J = number of levels in factor A

K = number of levels in factor B

6.1.9 Assumptions

6.1.9.1 Assumptions of Hierarchical Models

As noted previously, the assumptions of the two-factor hierarchical model are the same as with the two-factor crossed model (Chapters 3 and 5). These include normality, homogeneity of variance, and independence of observations within cells.

6.1.9.2 Assumptions of the Two-Factor Randomized Block ANOVA

In Chapter 5 we described the assumptions for the one-factor repeated measures ANOVA model. The assumptions are nearly the same for the two-factor randomized block model, and we need not devote much attention to them here. As before, the assumptions are mainly concerned with independence, normality, and homogeneity of variance. As these have been presented previously, we will not devote additional time on them here.

Another assumption is **compound symmetry** and is necessary because the observations within a block are not independent. *The assumption states that the population covariances for all pairs of the levels of the treatment factor A (i.e., j and j') are equal*. The analysis of variance is not particularly robust to a violation of this assumption. If the assumption is violated, three alternative procedures are available. The first is to limit the levels of factor A, either to those that meet the assumption, or to two levels (in which case there is only one covariance). The second, and more plausible, alternative is to use adjusted *F* tests. These are reported shortly. The third is to use multivariate analysis of variance, which has no compound symmetry assumption but is slightly less powerful. This method is beyond the scope of this text, but you may refer to Hahs-Vaughn (2016).

Huynh and Feldt (1970) showed that the compound symmetry assumption is a sufficient but unnecessary condition for the test of treatment factor A to be *F* distributed. Thus the *F* test may also be valid under less stringent conditions. The necessary and sufficient condition for the validity of the F test of A is known as **sphericity**. The assumption of sphericity is met when the variance of the difference scores for each pair of factor levels is the same. Further discussion of sphericity is beyond the scope of this text (e.g., Keppel & Wickens, 2004; Kirk, 2013), although we have previously discussed sphericity for repeated measures designs in Chapter 5.

A final assumption purports that there is no interaction between the treatment and blocking factors. This is obviously an assumption of the model because no interaction term is included. Such a model is often referred to as an *additive model*, and thus this assumption is referred to as the **assumption of additivity**. As was mentioned previously, in this model the interaction is confounded with the error term. Violation of the additivity assumption results in the test of factor A to be negatively biased; thus there is an increased probability of committing a Type II error. As a result, if H_0 is rejected, then we are confident that H_0 is really false. If H_0 is not rejected, then our interpretation is ambiguous as H_0 may or may not be really true (due to an increased probability of a Type II error). Here you would not know whether H_0 was true or not, as there might really be a difference, but the test may not be powerful enough to detect it. Also, the power of the test of factor A is reduced by a violation of the additivity assumption. The assumption may be tested by Tukey's (1949) test of additivity (see Kirk, 2013; Timm, 2002), which generates an *F* test statistic that is compared to the critical value of ${}_{\alpha}F_{1,[(J-1)(K-1)-1]}$. If the test is not statistically significant, then the model is additive and the assumption has been met. If the test *is* significant, then the model is *not* additive and the assumption has *not* been met. A summary of the assumptions and the effects of their violation for this model are presented in Table 6.10.

TABLE 6.10

Assumptions and Effects of Violations: Two-Factor Randomized Block ANOVA

Assumption	Effect of Assumption Violation
Independence	• Increased likelihood of a Type I and/or Type II error in *F* • Affects standard errors of means and inferences about those means
Homogeneity of variance	• Small effect with equal or nearly equal *n*'s • Otherwise effect decreases as *n* increases
Normality	• Minimal effect with equal or nearly equal *n*'s
Sphericity	• Fairly serious effect
No interaction between treatment and blocks	• Increased likelihood of a Type II error for the test of factor A and thus reduced power

6.2 Mathematical Introduction Snapshot

Let's summarize some of the mathematics that underlie the models we've covered. The *two-factor mixed-effects nested ANOVA* model is written in terms of population parameters as follows:

$$Y_{ijk} = \mu + \alpha_j + b_{k(j)} + \varepsilon_{ijk}$$

where Y_{ijk} is the observed score on the dependent variable for individual i in level j of factor A and level k of factor B (or in the jk cell), μ is the overall or grand population mean (i.e., regardless of cell designation), α_j is the fixed effect for level j of factor A, $b_{k(j)}$ is the random effect for level k of factor B, and ε_{ijk} is the random residual error for individual i in cell jk. The distinguishing feature of this model is the lack of an interaction term.

The *two-factor fixed-effects randomized block ANOVA* model is written in terms of population parameters as follows:

$$Y_{jk} = \mu + \alpha_j + \beta_k + \varepsilon_{jk}$$

where Y_{jk} is the observed score on the dependent variable for the individual responding to level j of factor A and level k of block B, μ is the overall or grand population mean, α_j is the fixed effect for level j of factor A, β_k is the fixed effect for level k of the block B, and ε_{jk} is the random residual error for the individual in cell jk. This is similar to the two-factor fully crossed model with one observation per cell (i.e., $i = 1$ making the i subscript unnecessary), and with no interaction term included. Also, the effects are denoted by α and β given we have a fixed-effects model.

6.3 Computing Hierarchical and Randomized Block ANOVA Models Using SPSS

In this section we examine SPSS for the models presented in this chapter. We begin with the two-factor hierarchical ANOVA and then follow with the two-factor randomized block ANOVA.

6.3.1 Computing the Two-Factor Hierarchical ANOVA Using SPSS

To conduct a two-factor hierarchical (or nested) ANOVA, there are a few differences from other ANOVA models we have considered in this text. We will illustrate computation of the model that follows the point-and-click method, as we have done in previous chapters, and will be using the "twofactor_nested.sav" data. It is important to note that the most recent versions of SPSS offer increasing ability to generate multilevel models using more modern analytic procedures (i.e., going beyond least squares estimation), and readers interested in more complex regression models using SPSS are referred to Heck, Tabata, and Thomas (2014). For this illustration, we will walk through the GLM steps as we have with previous ANOVA models.

In terms of the form of the data, one column or variable indicates the levels or categories of the independent variable (i.e., the fixed factor), one column indicates the levels of the nested factor, and the one variable represents the outcome or the dependent variable. Each row represents one individual, indicating the level or group of the nonnested factor (massage therapy or music therapy, in our example), the level or group of the nested factor (interventionist, therapist, or clinician 1, 2, 3, or 4), and their score on the dependent variable. Thus we have three columns which represent the nonnested factor (factor A), the nested factor (factor B), and the outcome value or dependent variable, as shown in Figure 6.2.

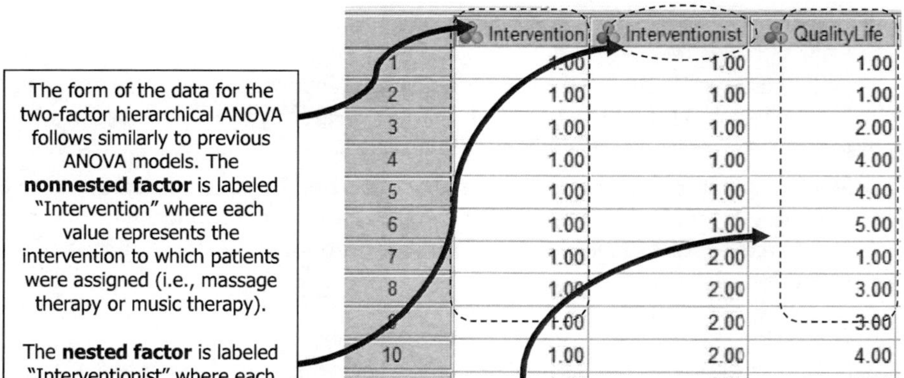

	Intervention	Interventionist	QualityLife
1	1.00	1.00	1.00
2	1.00	1.00	1.00
3	1.00	1.00	2.00
4	1.00	1.00	4.00
5	1.00	1.00	4.00
6	1.00	1.00	5.00
7	1.00	2.00	1.00
8	1.00	2.00	3.00
9	1.00	2.00	3.00
10	1.00	2.00	4.00
11	1.00	2.00	6.00
12	1.00	2.00	6.00
13	2.00	3.00	7.00
14	2.00	3.00	8.00
15	2.00	3.00	8.00
16	2.00	3.00	10.00
17	2.00	3.00	12.00
18	2.00	3.00	15.00
19	2.00	4.00	8.00
20	2.00	4.00	9.00

FIGURE 6.2
Data for the two-factor hierarchical ANOVA.

Step 1. To conduct a two-factor hierarchical ANOVA, go to "Analyze" in the top pulldown menu, then select "General Linear Model," and then select "Univariate." Following the screenshot for Step 1 (shown in Figure 6.3) produces the Univariate dialog box.

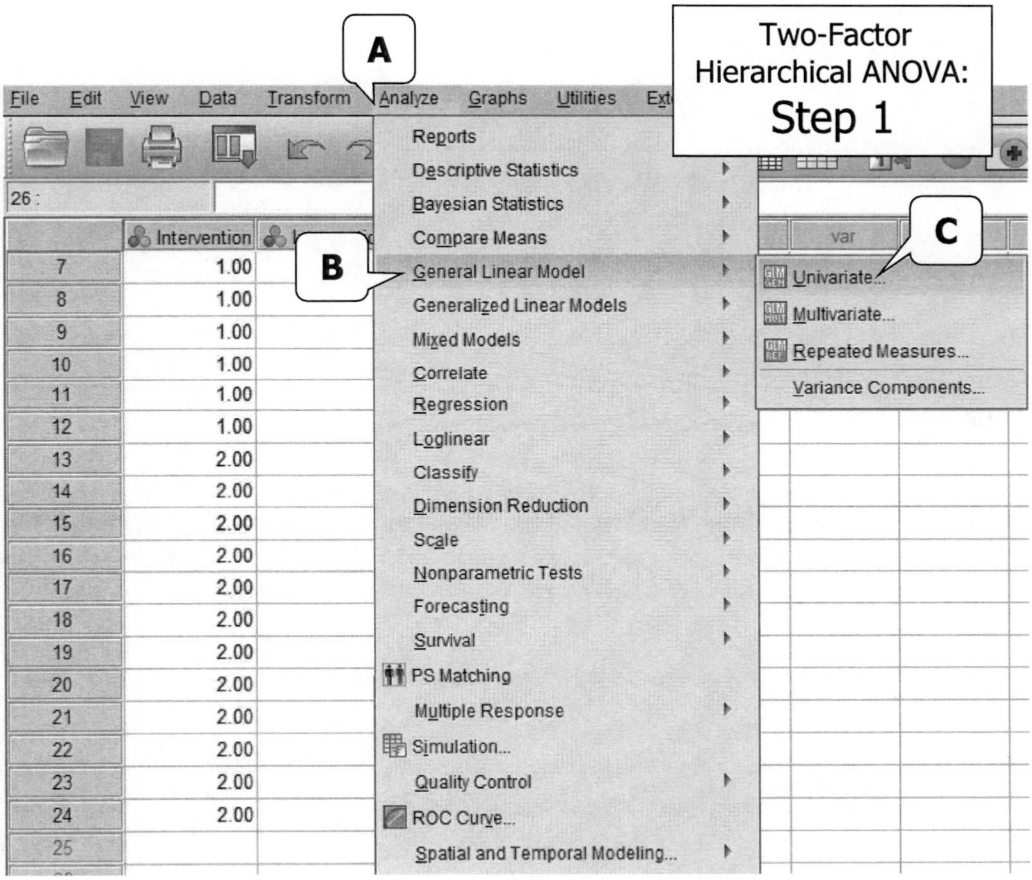

FIGURE 6.3
Two-factor hierarchical ANOVA: Step 1.

Step 2. Click the dependent variable (e.g., Quality of Life score) and move it into the "Dependent Variable" box by clicking the arrow button. Click the nonnested factor (e.g., intervention; this is a fixed-effects factor) and move it into the "Fixed Factor(s)" box by clicking the arrow button. Click the nested variable (e.g., interventionist; this is a random-effects factor) and move it into the "Random Factor(s)" box by clicking the arrow button.

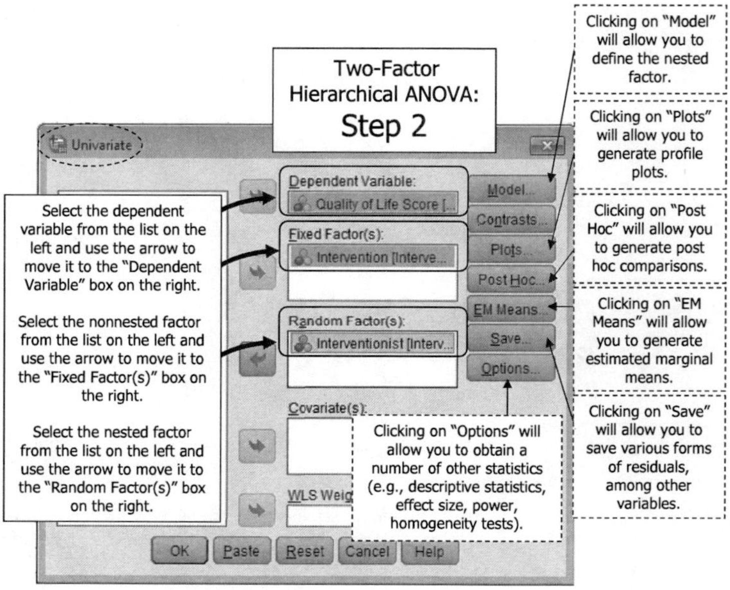

FIGURE 6.4
Two-factor hierarchical ANOVA: Step 2.

Step 3a. From the main Univariate dialog box (see the screenshot in Figure 6.4), click on "Model" to enact the Univariate Model dialog box. From the Univariate Model dialog box, click the "Build terms" radio button located in the top toolbar under "Specify Model" (see the screenshot for Step 3a in Figure 6.5). We will now define a *main effect* for intervention (see

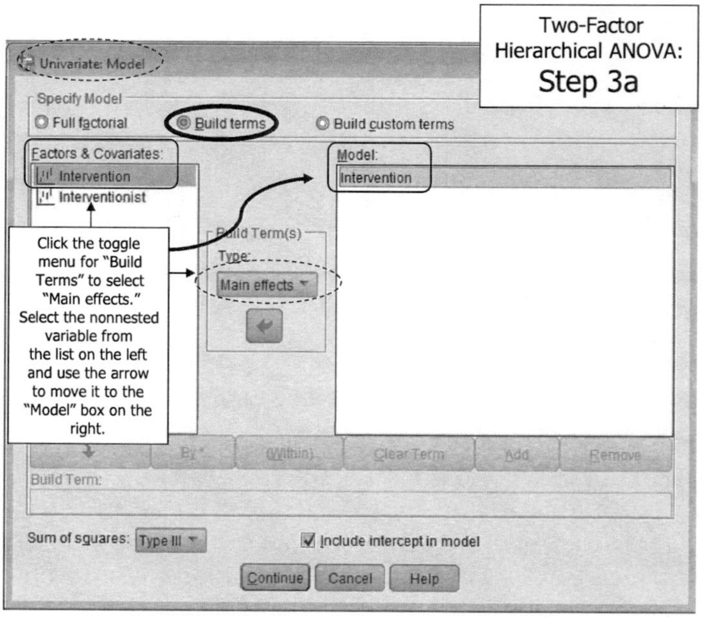

FIGURE 6.5
Two-factor hierarchical ANOVA: Step 3a.

Figure 6.5). To do this, click the "Build terms" toggle menu in the center of the page and select "Main effects." Click the nonnested factor (in this illustration, "Intervention") from the Factors & Covariates list on the left and move to the "Model" box on the right by clicking the arrow.

Step 3b. We will now define an *interaction effect* for intervention by interventionist (see the screenshot for Step 3b in Figure 6.6). To do this, click the "Build terms" toggle menu in the center of the page and select "Interaction." Click both the nonnested factor (e.g., "Intervention") and nested factor (e.g., "Interventionist") from the Factors & Covariates list on the left and move them to the "Model" box on the right by clicking the arrow. The interaction term is necessary to trick SPSS into computing the main effect of B(A) for the nested factor (which SPSS calls "intervention*interventionist," but is actually "interventionist"), and thus generate the proper ANOVA summary table. Thus the model should *not* include a main effect term for "Interventionist."

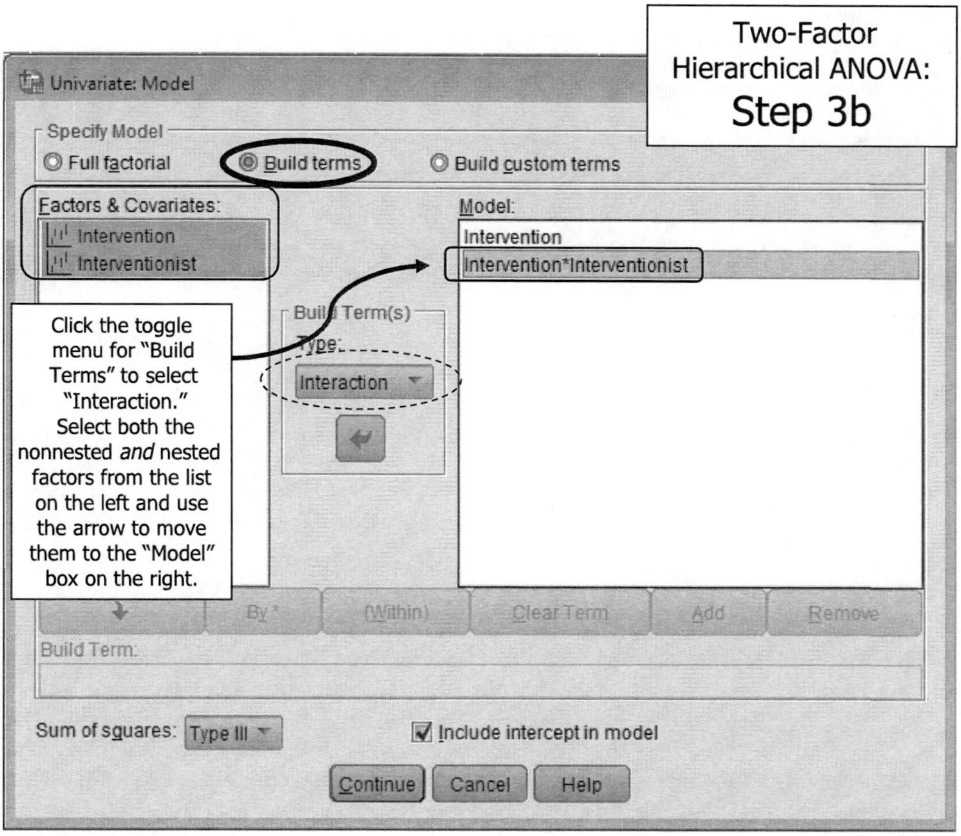

FIGURE 6.6
Two-factor hierarchical ANOVA: Step 3b.

Step 4. From the Univariate dialog box (see Figure 6.4), clicking on "Plots" will provide the option to graph various profile plots. From the Univariate Profile Plots dialog box, click on the name of the nonnested factor in the "Factor(s)" list box in the top left and move it to the "Horizontal Axis" box in the top right by clicking on the arrow key. Then click on the name of the nested factor in the Factor(s) list box in the top left and move it to the "Separate Lines" box in the right by clicking on the arrow key. Next, click on "Add" to create the command to generate the plot in the dialog box in the middle. Select the radio button for "Line Chart" under Chart Type. Click on "Continue" to return to the original dialog box.

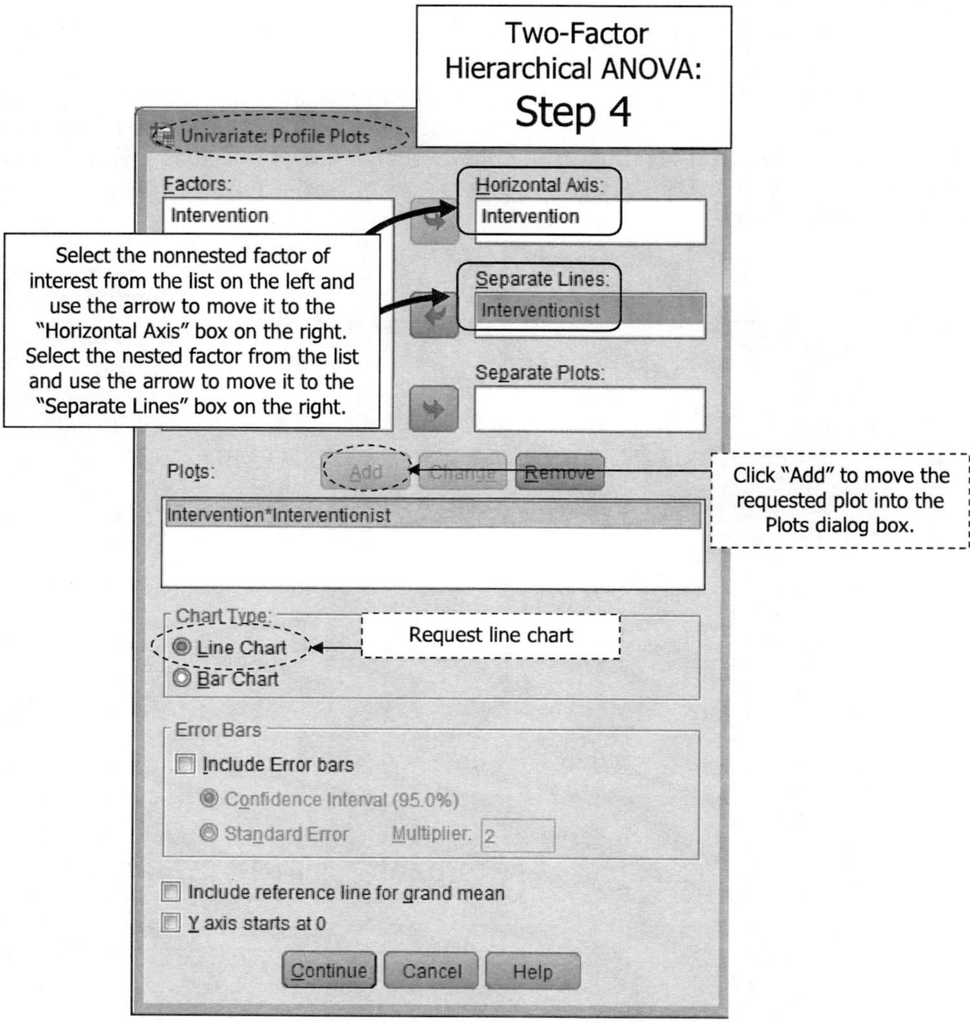

FIGURE 6.7
Two-factor hierarchical ANOVA: Step 4.

Step 5. From the Univariate dialog box (see Figure 6.4), clicking on "Post hoc" will provide the option to select post hoc multiple comparison procedures for the nonnested factor. From the Univariate Post Hoc Multiple Comparisons for Observed Means dialog box, click on the name of the nonnested factor in the Factor(s) list box in the top left and move it to the "Post Hoc Tests for" box in the top right by clicking on the arrow key. Check an appropriate MCP for your situation by placing a checkmark in the box next to the desired MCP. In this example, we select Tukey. Click on "Continue" to return to the original dialog box. (*Note:* Because we only have two treatments, this step is unnecessary as we can simply compare the means of the groups. The steps have been provided so that you can see the process in the event you have three or more groups in your own research.)

It is important to note that Li and Lomax (2011) found that the standard errors of the MCPs for the nonnested factor in SPSS point-and-click mode are not correct. More specifically, SPSS point-and-click uses MS_{with} as the error term in computing the MCP standard error rather than $MS_{B(A)}$ as the error term. There is no way to generate the correct results solely with SPSS point-and-click, unless hand computations using the correct error term are utilized or other software programs (e.g., SPSS syntax) are also involved.

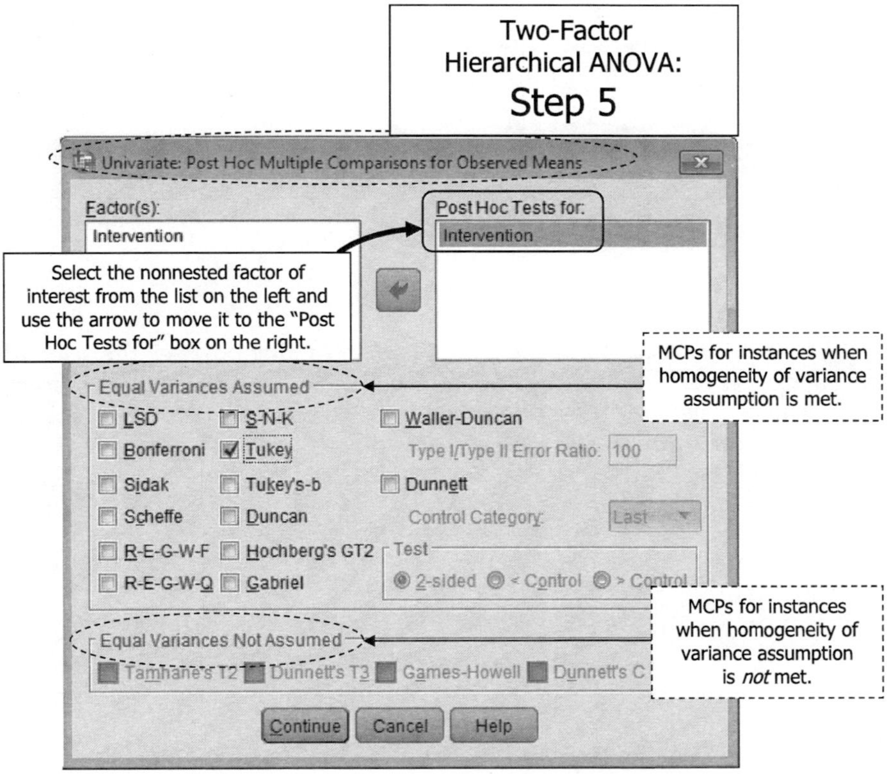

FIGURE 6.8
Two-factor hierarchical ANOVA: Step 5.

Step 6. From the Univariate dialog box (see Figure 6.4), clicking on "EM Means" will provide the option to select estimated marginal means. From the Univariate Estimated Marginal Means dialog box, click on "(OVERALL)" and the names of the nonnested and nested factors in the "Factor(s) and Factor Interactions" list box in the top left and move it to the "Display Means for" box in the top right by clicking on the arrow key. *Note that if you are interested in a multiple comparison procedure for the nested factor (although generally not of interest for this model), post hoc MCPs are available only from this screen.* To select a post hoc procedure, click on "Compare main effects" and use the toggle menu to reveal the Tukey LSD, Bonferroni, and Sidak procedures. For illustration purposes, we'll select Bonferroni. However, we have already mentioned that MCPs are not generally of interest for the nested factor. Click on "Continue" to return to the original dialog box.

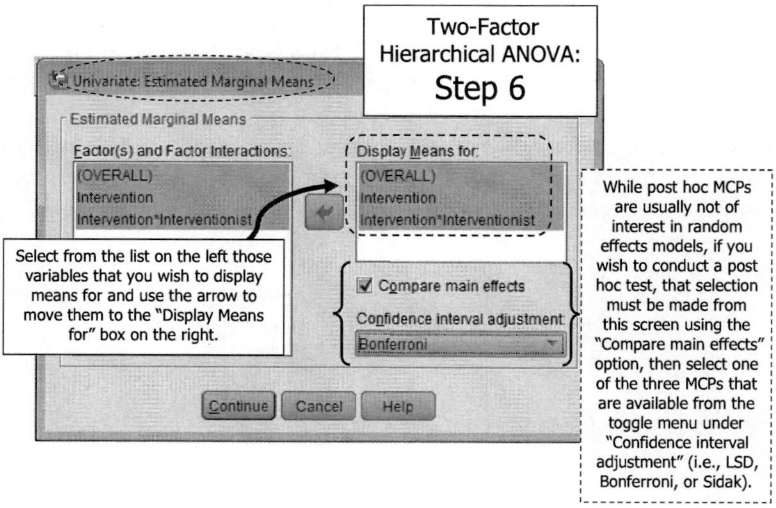

FIGURE 6.9
Two-factor hierarchical ANOVA: Step 6.

Step 7. Clicking on "Options" from the main Univariate dialog box (see the screenshot for Step 2 in Figure 6.4) will provide the option to select such information as "Descriptive statistics," "Estimates of effect size," "Observed power," and "Homogeneity tests" (i.e., Levene's test). Click on "Continue" to return to the original dialog box.

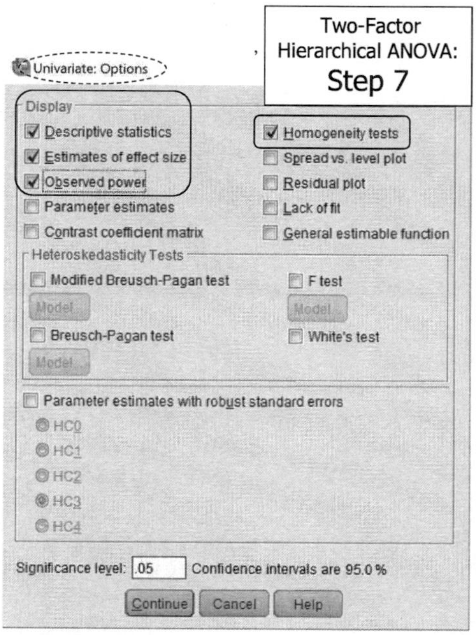

FIGURE 6.10
Two-factor Hierarchical ANOVA: Step 7.

Step 8. From the Univariate dialog box (see the screenshot for Step 2 in Figure 6.4), click on "Save" to select those elements you want to save. Here we want to save the unstandardized residuals to be used to examine the extent to which normality and independence are met. Thus, place a checkmark in the box next to "Unstandardized." Click "Continue" to return to the main Univariate dialog box. From there, click on "OK" to generate the output.

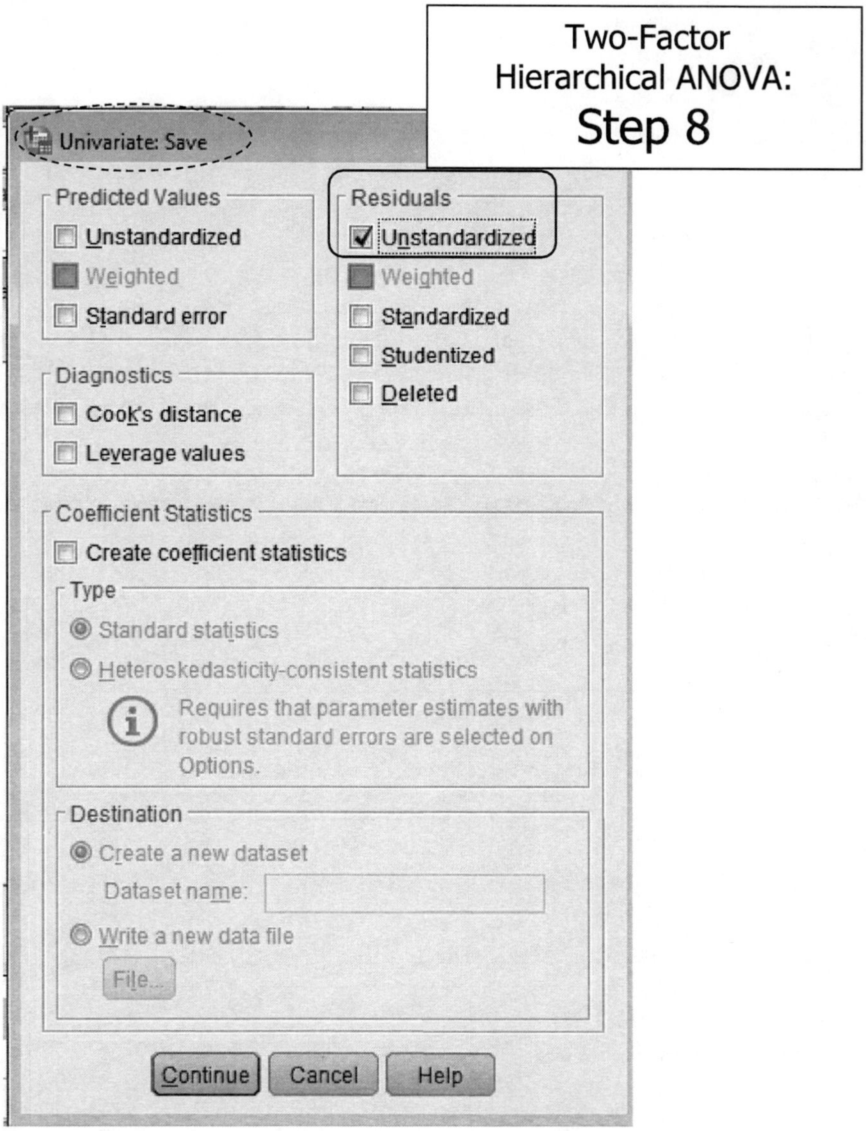

FIGURE 6.11
Two-factor hierarchical ANOVA: Step 8.

Interpreting the Output. Annotated results are presented in Table 6.11.

TABLE 6.11

Two-Factor Hierarchical ANOVA SPSS Results for the Quality of Life Intervention Example

Between-Subjects Factors

		Value Label	N
Intervention	1.00	Massage Therapy	12
	2.00	Music Therapy	12
Interventionist	1.00	Clinician B1	6
	2.00	Clinician B2	6
	3.00	Clinician B3	6
	4.00	Clinician B4	6

The table labeled "Between-Subjects Factors" lists the variable names and sample sizes for the non-nested factor (i.e., 'Intervention') and the nested factor (i.e., 'Interventionist').

Descriptive Statistics

Dependent Variable: Quality of Life Score

Intervention	Interventionist	Mean	Std. Deviation	N
Massage Therapy	Clinician B1	2.8333	1.72240	6
	Clinician B2	3.8333	1.94079	6
	Total	3.3333	1.82574	12
Music Therapy	Clinician B3	10.0000	3.03315	6
	Clinician B4	11.6667	2.80476	6
	Total	10.8333	2.91807	12
Total	Clinician B1	2.8333	1.72240	6
	Clinician B2	3.8333	1.94079	6
	Clinician B3	10.0000	3.03315	6
	Clinician B4	11.6667	2.80476	6
	Total	7.0833	4.51005	24

The table labeled "Descriptive Statistics" provides basic descriptive statistics (means, standard deviations, and sample sizes) for each non-nested factor and nested factor combination (or cell).

Levene's Test of Equality of Error Variances[a,b]

		Levene Statistic	df1	df2	Sig.
Quality of Life Score	Based on Mean	1.042	3	20	.396
	Based on Median	.813	3	20	.502
	Based on Median and with adjusted df	.813	3	12.531	.510
	Based on trimmed mean	1.038	3	20	.397

Tests the null hypothesis that the error variance of the dependent variable is equal across groups.

a. Dependent variable: Quality of Life Score

b. Design: Intercept + Intervention + Intervention * Interventionist

The *F* test (and associated *p* value) for Levene's Test for Equality of Error Variances is reviewed to determine if equal variances can be assumed. We will review Levene's 'based on the mean.' In this case, we meet the assumption (as *p* is greater than α).

TABLE 6.11 (continued)

Two-Factor Hierarchical ANOVA SPSS Results for the Quality of Life Intervention Example

Comparing p to α, we find a statistically significant difference in intervention. This is an omnibus test. We will look at our MCPs to determine which mean ratings differ.

Partial eta squared is one measure of effect size:

$$\eta_p^2 = \frac{SS_{intervention}}{SS_{intervention} + SS_{intervention\ error}}$$

$$\eta_p^2 = \frac{337.50}{337.50 + 11.33} = .968$$

We can interpret this to say that approximately 97% of the variation in quality of life is explained by intervention that is not explained by the nesting of treatment within therapist.

Tests of Between-Subjects Effects

Dependent Variable: Quality of Life Score

Source		Type III Sum of Squares	df	Mean Square	F	Sig.	Partial Eta Squared	Noncent. Parameter	Observed Power[c]
Intercept	Hypothesis	1204.167	1	1204.167	212.500	.005	.991	212.500	1.000
	Error	11.333	2	5.667[a]					
Intervention	Hypothesis	337.500	1	337.500	59.559	.016	.968	59.559	.948
	Error	11.333	2	5.667[a]					
Intervention * Interventionist	Hypothesis	11.333	2	5.667	.952	.403	.087	1.905	.192
	Error	119.000	20	5.950[b]					

a. MS(Intervention * Interventionist)

b. MS(Error)

c. Computed using alpha = .05

Observed power tells whether our test is powerful enough to detect mean differences if they really exist. Power of .948 is strong. The probability of rejecting the null hypothesis, if it is really false, is about 95%.

Estimated Marginal Means

1. Grand Mean

Dependent Variable: Quality of Life Score

Mean	Std. Error	95% Confidence Interval	
		Lower Bound	Upper Bound
7.083[a]	.498	6.045	8.122

a. Based on modified population marginal mean.

The **'Grand Mean'** (in this case, 7.083) represents the overall Quality of Life score, regardless of the intervention or interventionistThe 95% CI represents the CI of the grand mean.

2. Intervention

Estimates

Dependent Variable: Quality of Life Score

Intervention	Mean	Std. Error	95% Confidence Interval	
			Lower Bound	Upper Bound
Massage Therapy	3.333[a]	.704	1.864	4.802
Music Therapy	10.833[a]	.704	9.364	12.302

a. Based on modified population marginal mean.

The table for **"Intervention"** provides descriptive statistics for each of the interventions. In addition to means, the *SE* and 95% CI of the means are reported.

(continued)

TABLE 6.11 (continued)

Two-Factor Hierarchical ANOVA SPSS Results for the Quality of Life Intervention Example

Pairwise Comparisons

Dependent Variable: Quality of Life Score

(I) Intervention	(J) Intervention	Mean Difference (I-J)	Std. Error	Sig.[d]	95% Confidence Interval for Difference[d] Lower Bound	Upper Bound
Massage Therapy	Music Therapy	-7.500[*,b,c]	.996	.000	-9.577	-5.423
Music Therapy	Massage Therapy	7.500[*,b,c]	.996	.000	5.423	9.577

Based on estimated marginal means

*. The mean difference is significant at the .05 level.

b. An estimate of the modified population marginal mean (I).

c. An estimate of the modified population marginal mean (J).

d. Adjustment for multiple comparisons: Bonferroni.

'Mean difference' is simply the difference between the means of the two interventions. For example, the mean difference of massage therapy and music therapy is calculated as 3.333 − 10.833 = -7.500.

Univariate Tests

Dependent Variable: Quality of Life Score

	Sum of Squares	df	Mean Square	F	Sig.	Partial Eta Squared	Noncent. Parameter	Observed Power[a]
Contrast	337.500	1	337.500	56.723	.000	.739	56.723	1.000
Error	119.000	20	5.950					

The F tests the effect of Intervention. This test is based on the linearly independent pairwise comparisons among the estimated marginal means.

a. Computed using alpha = .05

The error term represents the within cells source of variation.

3. Intervention * Interventionist

Dependent Variable: Quality of Life Score

Intervention	Interventionist	Mean	Std. Error	95% Confidence Interval Lower Bound	Upper Bound
Massage Therapy	Clinician B1	2.833	.996	.756	4.911
	Clinician B2	3.833	.996	1.756	5.911
	Clinician B3	.[a]	.	.	.
	Clinician B4	.[a]	.	.	.
Music Therapy	Clinician B1	.[a]	.	.	.
	Clinician B2	.[a]	.	.	.
	Clinician B3	10.000	.996	7.923	12.077
	Clinician B4	11.667	.996	9.589	13.744

a. This level combination of factors is not observed, thus the corresponding population marginal mean is not estimable.

The table for **"Intervention* Interventionist"** provides descriptive statistics for each of the intervention-interventionist combinations. In addition to means, the *SE* and 95% CI of the means are reported.
Note the footnote in reference to the missing mean values. This is because this is not a completely crossed design (i.e., the clinicians provided only one intervention).

TABLE 6.11 (continued)

Two-Factor Hierarchical ANOVA SPSS Results for the Quality of Life Intervention Example

Profile Plots

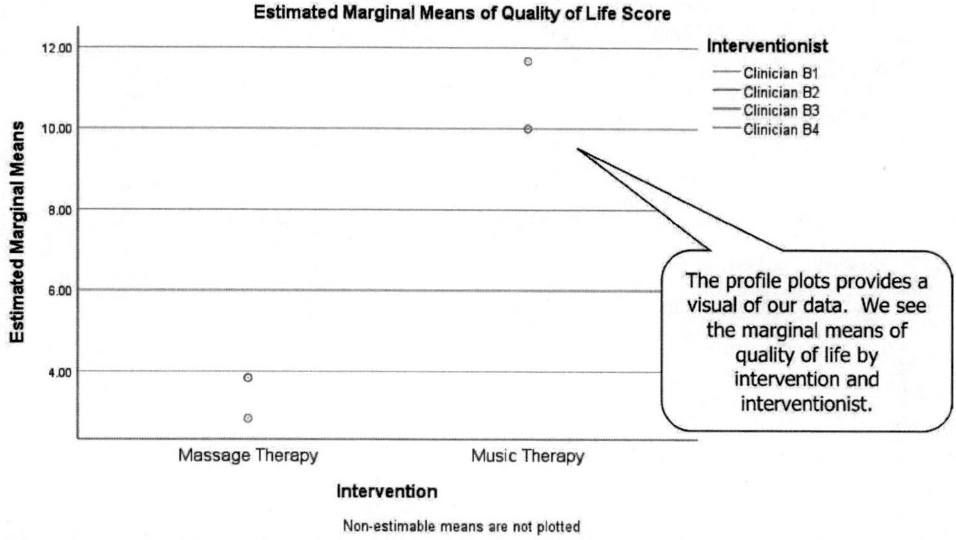

6.3.2 Computing the Two-Factor Fixed-Effects Randomized Block ANOVA for n = 1 Using SPSS

To run a two-factor fixed-effects randomized block ANOVA for n = 1, there a few differences from the regular two-factor fixed-effects ANOVA that we see later as we build the model in SPSS. Additionally, the test of additivity is not available in SPSS, nor are the adjusted F tests (i.e., the Geisser-Greenhouse and Huynh-Feldt procedures). All other ANOVA procedures that you are familiar with will operate as before.

In terms of the form of the data, it looks just as we saw with the two-factor fixed-effects ANOVA, with the exception that now we have one treatment factor and one blocking variable. The dataset must therefore consist of three variables or columns, one for the level of the treatment factor, one for the level of the blocking factor, and the third for the dependent variable. Each row still represents one individual, indicating the levels of the treatment and blocking factors to which the individual is a member, and their score on the dependent variable. As seen in the screenshot (Figure 6.12), for a two-factor fixed-effects randomized block ANOVA, the SPSS data take the form of two columns that represent the group values (i.e., the treatment and blocking factors) and one column that represents the scores on the dependent variable.

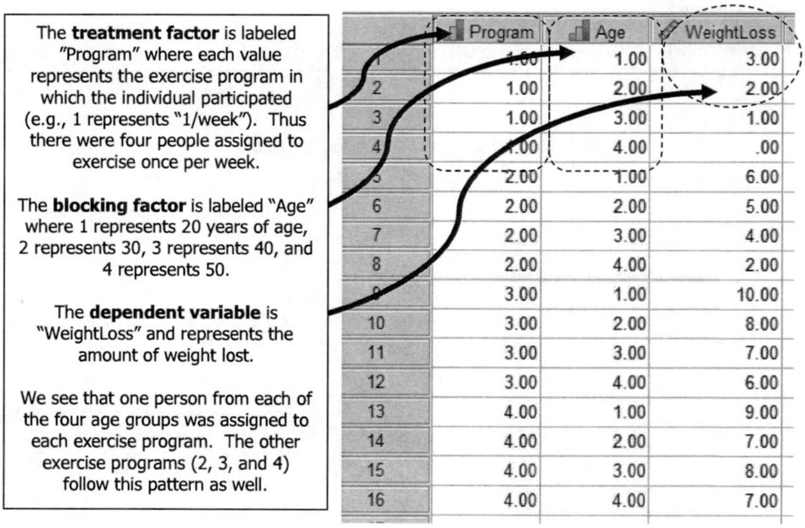

FIGURE 6.12
Two-factor fixed-effects randomized block ANOVA for $n = 1$ data.

Step 1. To conduct a two-factor randomized block ANOVA for $n = 1$, go to "Analyze" in the top pulldown menu, then select "General Linear Model," and then select "Univariate." Following the screenshot for Step 1 (Figure 6.13) produces the Univariate dialog box.

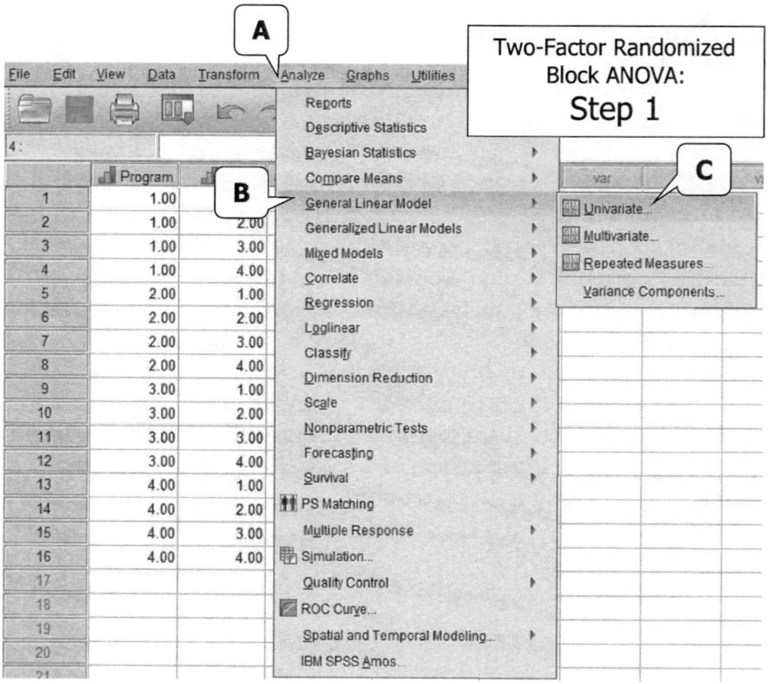

FIGURE 6.13
Two-factor fixed-effects randomized block ANOVA for $n = 1$: Step 1.

Step 2. Click the dependent variable (e.g., weight loss) and move it into the "Dependent Variable" box by clicking the arrow button. Click the treatment factor and the blocking factor and move them into the "Fixed Factor(s)" box by clicking the arrow button.

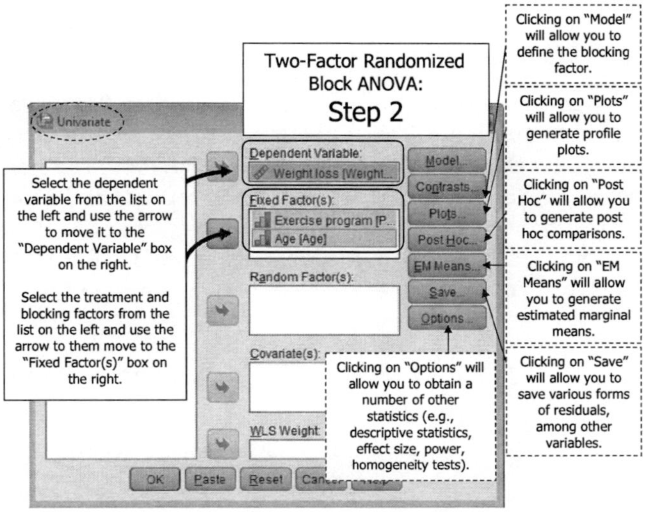

FIGURE 6.14
Two-factor fixed-effects randomized block ANOVA for $n = 1$: Step 2.

Step 3. From the main Univariate dialog box (see the screenshot for Step 2, Figure 6.14), click on "Model" to enact the Univariate Model dialog box. From the Univariate Model dialog box, click the "Custom" radio button (see the screenshot for Step 3 in Figure 6.15). We will now

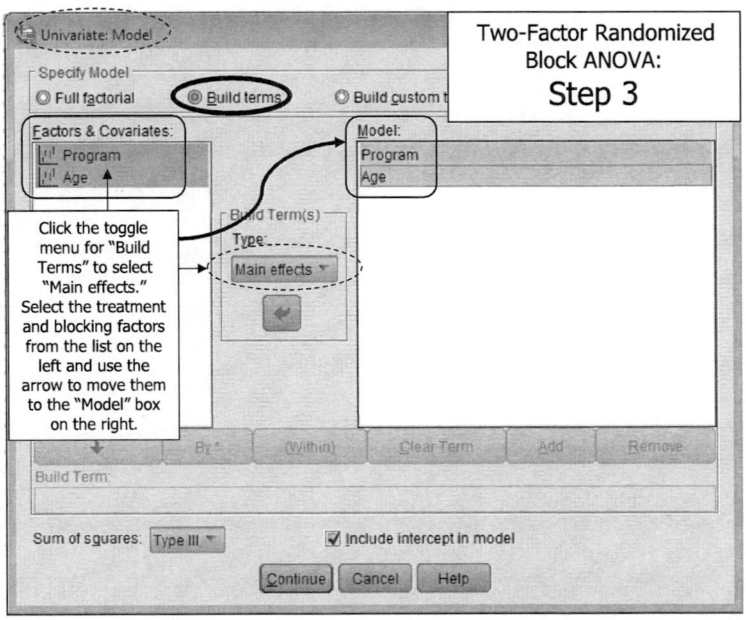

FIGURE 6.15
Two-factor fixed-effects randomized block ANOVA for $n = 1$: Step 3.

define the effects necessary for this model, a main effect for both exercise program and for age. We will *not* define an interaction. To do this, click the "Build terms" toggle menu in the center of the page and select "Main effect." Click the treatment factor (i.e., "Program") and the blocking factor (i.e., "Age") from the Factors & Covariates list on the left and move them to the "Model" box on the right by clicking the arrow. Thus. the model should *not* include an interaction effect for "Program*Age."

Step 4. From the Univariate dialog box (see screenshot for Step 2, Figure 6.14), clicking on "Post hoc" will provide the option to select post hoc MCPs for both factors. From the Post Hoc Multiple Comparisons for Observed Means dialog box, click on the name of the factors (i.e., "Program" and "Age") in the "Factor(s)" list box in the top left and move to the "Post Hoc Tests for" box in the top right by clicking on the arrow key. Check an appropriate MCP for your situation by placing a checkmark in the box next to the desired MCP. In this example, we select Tukey. Click on "Continue" to return to the original dialog box.

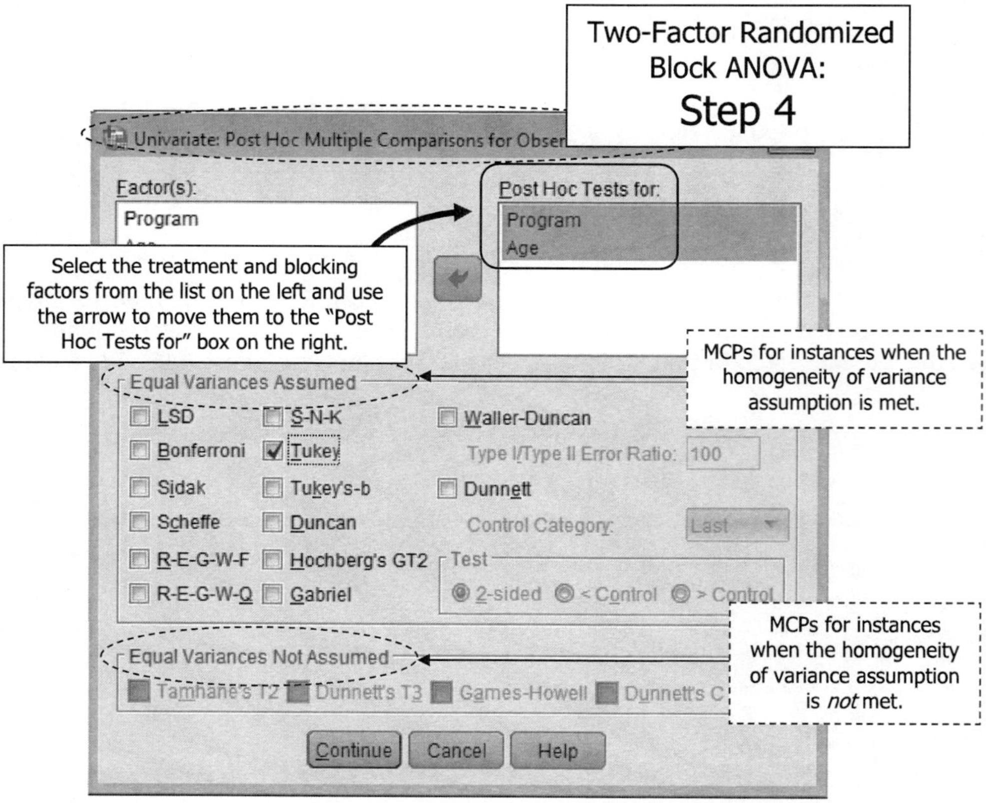

FIGURE 6.16
Two-factor fixed-effects randomized block ANOVA for $n = 1$: Step 4.

Step 5. Clicking on "Options" from the main Univariate dialog box (see the screenshot for Step 2, Figure 6.14) will provide the option to select such information as "Descriptive statistics," "Estimates of effect size," and "Observed power." Notice that for the two-factor fixed-effects randomized block ANOVA for $n = 1$, we do not select "Homogeneity tests" as the results for Levene's cannot be generated from the design we have specified—recall that there is only one individual per age group in each exercise program. Thus, there is no within-cell variation to calculate. This is not an issue with randomized block designs with $n > 1$. Click on "Continue" to return to the original dialog box.

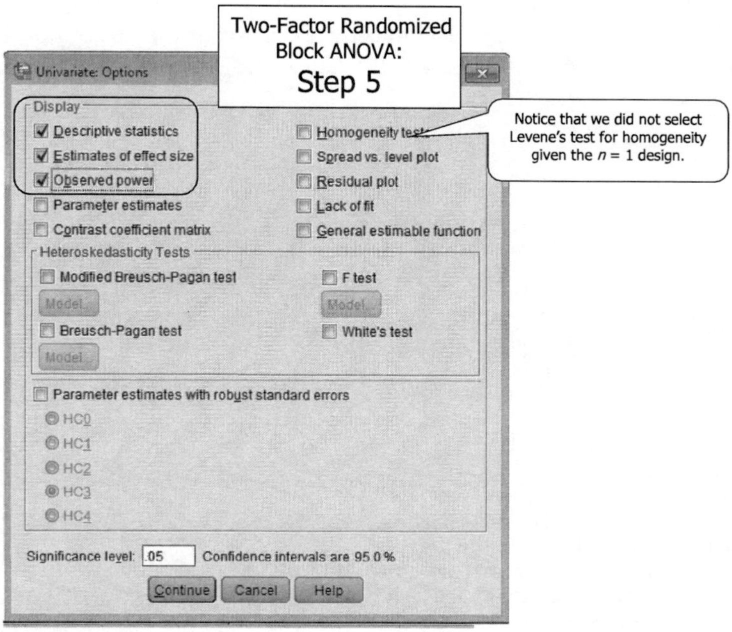

FIGURE 6.17
Two-factor fixed-effects randomized block ANOVA for $n = 1$: Step 5.

Step 6. From the Univariate dialog box (see the screenshot for Step 2, Figure 6.14), clicking on "EM Means" will provide the option to select estimated marginal means. From the Univariate Estimated Marginal Means dialog box, click on "(OVERALL)" and the names of the non-nested and nested factors in the "Factor(s) and Factor Interactions" list box in the top left and move it to the "Display Means for" box in the top right by clicking on the arrow key. Click on "Continue" to return to the original dialog box.

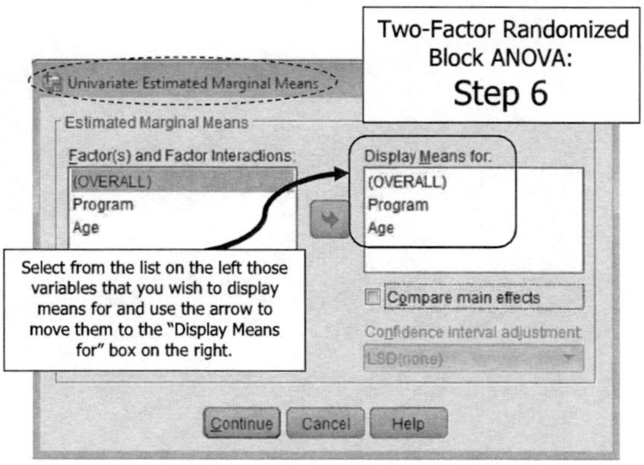

FIGURE 6.18
Two-factor fixed-effects randomized block ANOVA for $n = 1$: Step 6.

Step 7. From the Univariate dialog box, click on "Plots" to obtain a profile plot of means. Click the treatment factor (e.g., "Program") and move it into the "Horizontal Axis" box by clicking the arrow button. Click the blocking factor (e.g., "Age") and move it into the "Separate Lines" box by clicking the arrow button (see the screenshot for Step 6a in Figure 6.19). Then click on "Add" to move this arrangement into the Plots box at the bottom of the dialog box (see the screenshot for Step 6b in Figure 6.20). Click on "Continue" to return to the original dialog box.

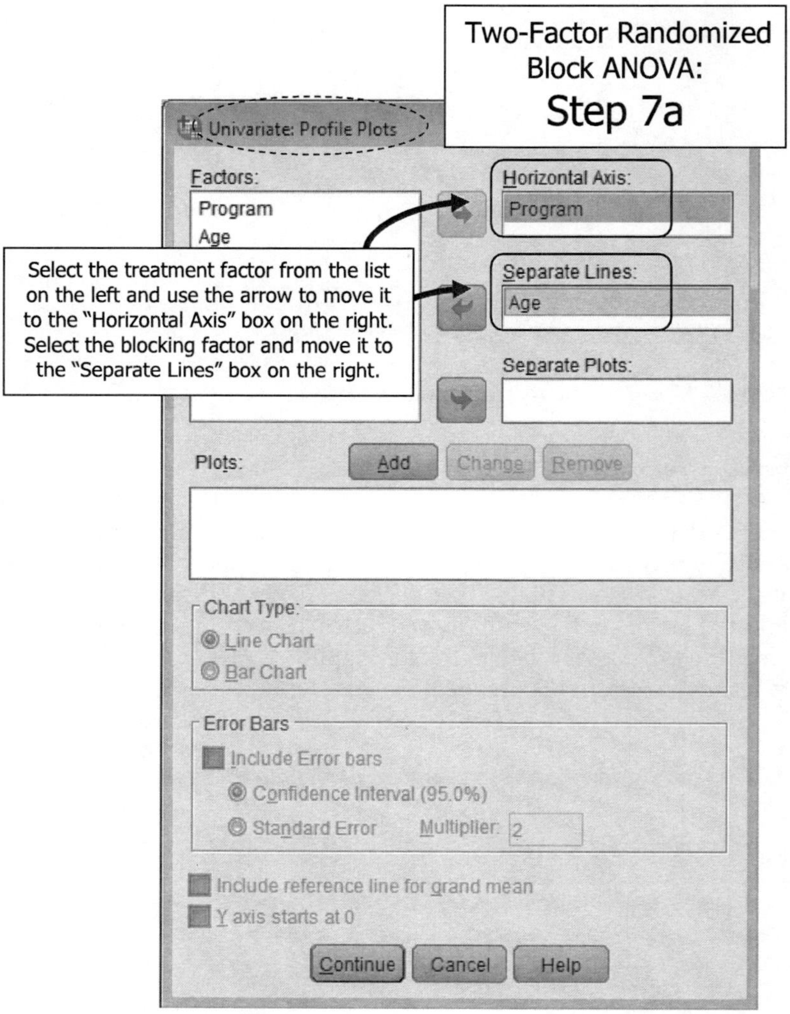

FIGURE 6.19
Two-factor fixed-effects randomized block ANOVA for $n = 1$: Step 7a.

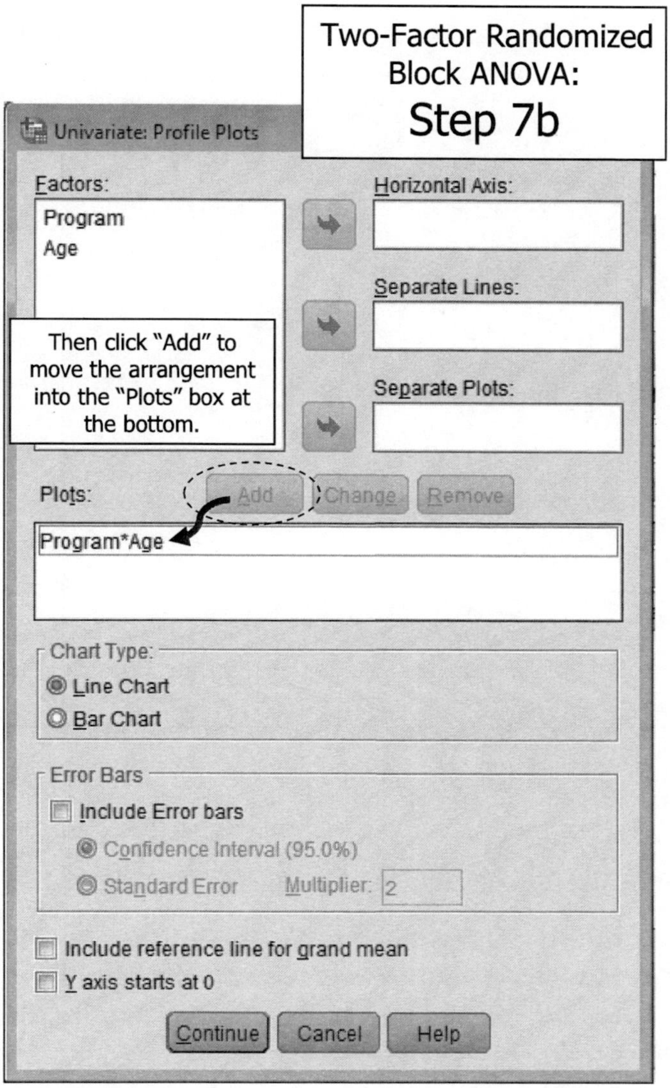

FIGURE 6.20 Two-factor fixed-effects randomized block ANOVA for $n = 1$: Step 7b.

Step 8. From the Univariate dialog box (see the screenshot for Step 2, Figure 6.14), click on "Save" to select those elements you want to save. Here we save the unstandardized residuals to use later to examine the extent to which normality and independence are met. Thus, place a checkmark in the box next to "Unstandardized." Click "Continue" to return to the main Univariate dialog box. From there, click on "OK" to return and generate the output.

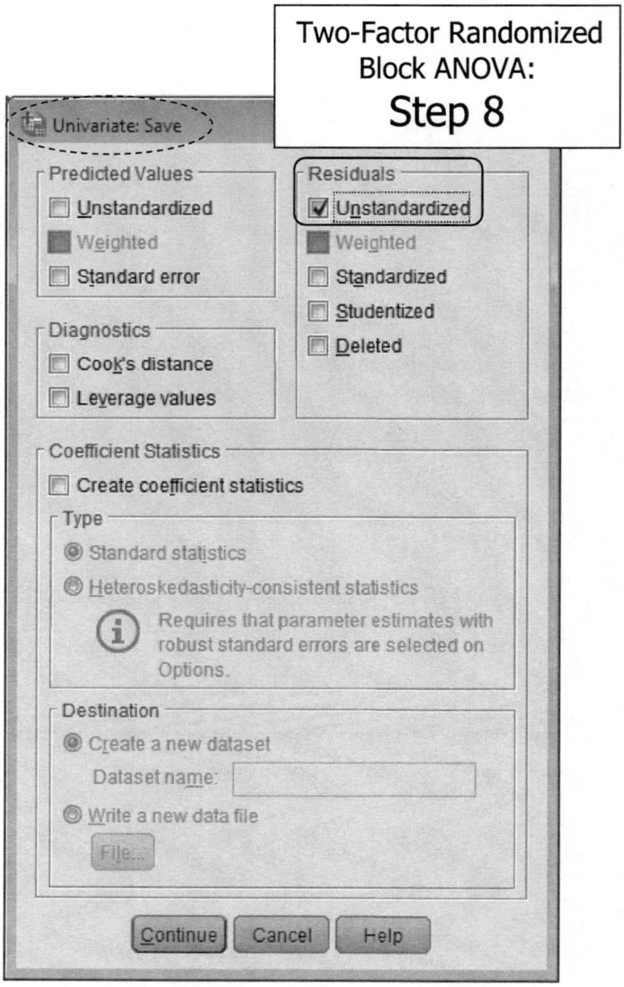

FIGURE 6.21
Two-factor fixed-effects randomized block ANOVA for $n = 1$: Step 8.

6.3.2.1 Interpreting the Output

Annotated results are presented in Table 6.12.

6.3.3 Computing the Two-Factor Fixed-Effects Randomized Block ANOVA for $n > 1$ Using SPSS

To run a two-factor randomized block ANOVA for $n > 1$, the procedures are exactly the same as with the regular two-factor ANOVA. However, the adjusted F tests are not available.

TABLE 6.12

Two-Factor Randomized Block ANOVA for $n = 1$ SPSS Results for the Exercise Program Example

Between-Subjects Factors

		Value Label	N
Exercise program	1.00	1/week	4
	2.00	2/week	4
	3.00	3/week	4
	4.00	4/week	4
Age	1.00	20 years old	4
	2.00	30 years old	4
	3.00	40 years old	4
	4.00	50 years old	4

The table labeled "Between-Subjects Factors" lists the variable names and sample sizes for the levels of treatment factor (i.e., "exercise program") and the blocking factor (i.e., "age").

Descriptive Statistics

Dependent Variable: Weight loss

Exercise program	Age	Mean	Std. Deviation	N
1/week	20 years old	3.0000	.	1
	30 years old	2.0000	.	1
	40 years old	1.0000	.	1
	50 years old	.0000	.	1
	Total	1.5000	1.29099	4
2/week	20 years old	6.0000	.	1
	30 years old	5.0000	.	1
	40 years old	4.0000	.	1
	50 years old	2.0000	.	1
	Total	4.2500	1.70783	4
3/week	20 years old	10.0000	.	1
	30 years old	8.0000	.	1
	40 years old	7.0000	.	1
	50 years old	6.0000	.	1
	Total	7.7500	1.70783	4
4/week	20 years old	9.0000	.	1
	30 years old	7.0000	.	1
	40 years old	8.0000	.	1
	50 years old	7.0000	.	1
	Total	7.7500	.95743	4
Total	20 years old	7.0000	3.16228	4
	30 years old	5.5000	2.64575	4
	40 years old	5.0000	3.16228	4
	50 years old	3.7500	3.30404	4
	Total	5.3125	3.00486	16

The table labeled "Descriptive Statistics" provides basic descriptive statistics (means, standard deviations, and sample sizes) for each treatment factor-blocking factor combination.

Because there was only one individual per age group in each exercise program, there is no within-cell variation to calculate (and thus missing values for the standard deviation).

(continued)

TABLE 6.12 (continued)

Two-Factor Randomized Block ANOVA for $n = 1$ SPSS Results for the Exercise Program Example

Comparing *p* to α, we find a statistically significant difference in weight loss based for both exercise program and age group. These are omnibus tests. We will look at *post hoc* tests to determine which exercise programs and age groups statistically differ on weight loss.

Partial eta squared is one measure of effect size calculated as:

$$\eta^2 = \frac{SS_{program}}{SS_{program} + SS_{error}}$$

$$\eta^2 = \frac{110.187}{110.187 + 3.563} = .969$$

Tests of Between-Subjects Effects

Dependent Variable: Weight loss

Source	Type III Sum of Squares	df	Mean Square	F	Sig.	Partial Eta Squared	Noncent. Parameter	Observed Power[b]
Corrected Model	131.875[a]	6	21.979	55.526	.000	.974	333.158	1.000
Intercept	451.563	1	451.563	1140.789	.000	.992	1140.789	1.000
Program	110.187	3	36.729	92.789	.000	.969	278.368	1.000
Age	21.688	3	7.229	18.263	.000	.859	54.789	.999
Error	3.563	9	.396					
Total	587.000	16						
Corrected Total	135.438	15						

a. R Squared = .974 (Adjusted R Squared = .956)

b. Computed using alpha = .05

Observed power tells whether our test is powerful enough to detect mean differences if they really exist. Power of 1.00 indicates maximum power; the probability of rejecting the null hypothesis, if it is really false, is about 100%.

Estimated Marginal Means

1. Grand Mean

Dependent Variable: Weight loss

Mean	Std. Error	95% Confidence Interval Lower Bound	Upper Bound
5.313	.157	4.957	5.668

The "Grand Mean" (in this case, 5.313) represents the overall mean, regardless of the exercise program or age. The 95% CI represents the CI of the grand mean.

2. Exercise program

Dependent Variable: Weight loss

Exercise program	Mean	Std. Error	95% Confidence Interval Lower Bound	Upper Bound
1/week	1.500	.315	.788	2.212
2/week	4.250	.315	3.538	4.962
3/week	7.750	.315	7.038	8.462
4/week	7.750	.315	7.038	8.462

The table for **"Exercise program"** provides descriptive statistics for each of the programs. In addition to means, the *SE* and 95% CI of the means are reported.

TABLE 6.12 (continued)

Two-Factor Randomized Block ANOVA for $n = 1$ SPSS Results for the Exercise Program Example

3. Age

Dependent Variable: Weight loss

Age	Mean	Std. Error	95% Confidence Interval Lower Bound	Upper Bound
20 years old	7.000	.315	6.288	7.712
30 years old	5.500	.315	4.788	6.212
40 years old	5.000	.315	4.288	5.712
50 years old	3.750	.315	3.038	4.462

The table for **"Age"** provides descriptive statistics for each of the age groups. In addition to means, the *SE* and 95% CI of the means are reported.

Post Hoc Tests

Exercise program

"Mean Difference" is simply the difference between the means of the categories of our program factor. For example, the mean difference of exercising once per week and exercising twice per week is calculated as 1.500 – 4.250 = -2.750.

Multiple Comparisons

Dependent Variable: Weight loss

Tukey HSD

(I) Exercise program	(J) Exercise program	Mean Difference (I-J)	Std. Error	Sig.	95% Confidence Interval Lower Bound	Upper Bound
1/week	2/week	-2.7500*	.44488	.001	-4.1388	-1.3612
	3/week	-6.2500*	.44488	.000	-7.6388	-4.8612
	4/week	-6.2500*	.44488	.000	-7.6388	-4.8612
2/week	1/week	2.7500*	.44488	.001	1.3612	4.1388
	3/week	-3.5000*	.44488	.000	-4.8888	-2.1112
	4/week	-3.5000*	.44488	.000	-4.8888	-2.1112
3/week	1/week	6.2500*	.44488	.000	4.8612	7.6388
	2/week	3.5000*	.44488	.000	2.1112	4.8888
	4/week	.0000	.44488	1.000	-1.3888	1.3888
4/week	1/week	6.2500*	.44488	.000	4.8612	7.6388
	2/week	3.5000*	.44488	.000	2.1112	4.8888
	3/week	.0000	.44488	1.000	-1.3888	1.3888

Based on observed means.

The error term is Mean Square(Error) = .396.

*. The mean difference is significant at the .05 level.

"Sig." denotes the observed p value and provides the results of the Tukey post hoc procedure. There is a statistically significant mean difference in weight loss for all exercise programs except for exercising 3 vs. 4 times per week ($p = 1.000$). Note there are redundant results presented in the table. The comparison of exercising 1/week vs. 2/week (row 1) is the same as the comparison of 2/week vs. 1/week (row 4).

(continued)

TABLE 6.12 (continued)

Two-Factor Randomized Block ANOVA for $n = 1$ SPSS Results for the Exercise Program Example

Homogeneous Subsets

Weight loss

Tukey HSD[a,b]

Exercise program	N	Subset 1	Subset 2	Subset 3
1/week	4	1.5000		
2/week	4		4.2500	
3/week	4			7.7500
4/week	4			7.7500
Sig.		1.000	1.000	1.000

Means for groups in homogeneous subsets are displayed.

Based on observed means.

The error term is Mean Square(Error) = .396.

a. Uses Harmonic Mean Sample Size = 4.000.

b. Alpha = .05.

"Homogenous Subsets" provides a visual representation of the MCP. For each subset, the means that are printed are homogeneous, or not significantly different. For example, in subset 1 the mean weight loss for exercising once per week (regardless of age group) is 1.50. This is statistically significantly different than the mean weight loss for exercising 2, 3, or 4 times per week (as reflected by empty cells in row 1). Similar interpretations are made for contrasts involving exercising 2, 3, and 4 times per week.

Age

"Mean Difference" is simply the difference between the means of the age groups (i.e., the blocking factor). For example, the mean weight loss difference of 20-year-olds to 30-year-olds is calculated as 7.000 − 5.500 = 1.5000.

Dependent Variable: Weight loss

Tukey HSD

(I) Age	(J) Age	Mean Difference (I-J)	Std. Error	Sig.	95% Confidence Interval Lower Bound	Upper Bound
20 years old	30 years old	1.5000*	.44488	.034	.1112	2.8888
	40 years old	2.0000*	.44488	.007	.6112	3.3888
	50 years old	3.2500*	.44488	.000	1.8612	4.6388
30 years old	20 years old	-1.5000*	.44488	.034	-2.8888	-.1112
	40 years old	.5000	.44488	.685	-.8888	1.8888
	50 years old	1.7500*	.44488	.015	.3612	3.1388
40 years old	20 years old	-2.0000*	.44488	.007	-3.3888	-.6112
	30 years old	-.5000	.44488	.685	-1.8888	.8888
	50 years old	1.2500	.44488	.080	-.1388	2.6388
50 years old	20 years old	-3.2500*	.44488	.000	-4.6388	-1.8612
	30 years old	-1.7500*	.44488	.015	-3.1388	-.3612
	40 years old	-1.2500	.44488	.080	-2.6388	.1388

Based on observed means.

The error term is Mean Square(Error) = .396.

*. The mean difference is significant at the .05 level.

"Sig." denotes the observed p value and provides the results of the Tukey post hoc procedure. There is a statistically significant mean difference in weight loss for:

- 20-and 30-year-olds ($p = .034$)
- 20-and 40-year-olds ($p = .007$)
- 20-and 50-year-olds ($p < .001$)
- 30-and 50-year-olds ($p = .015$)

Note there are redundant results presented in the table. The comparison of 20-year-olds to 30-year-olds is the same as the comparison of 30-year-olds to 20-year-olds, and so forth.

TABLE 6.12 (continued)

Two-Factor Randomized Block ANOVA for $n = 1$ SPSS Results for the Exercise Program Example

Homogeneous Subsets

Weight loss

Tukey HSD[a,b]

Age	N	Subset 1	Subset 2	Subset 3
50 years old	4	3.7500		
40 years old	4	5.0000	5.0000	
30 years old	4		5.5000	
20 years old	4			7.0000
Sig.		.080	.685	1.000

Means for groups in homogeneous subsets are displayed.

Based on observed means.

The error term is Mean Square(Error) = .396.

a. Uses Harmonic Mean Sample Size = 4.000.

b. Alpha = .05.

"Homogenous Subsets" provides a visual representation of the MCP. For each subset, the means that are printed are homogeneous, or not significantly different. For example, in subset 1 the mean weight loss for 50 year olds (regardless of exercise program) is 3.750. This is statistically significantly different than the mean weight loss for individuals in the 30 and 20 year old age groups (as they are not printed in subset 1).

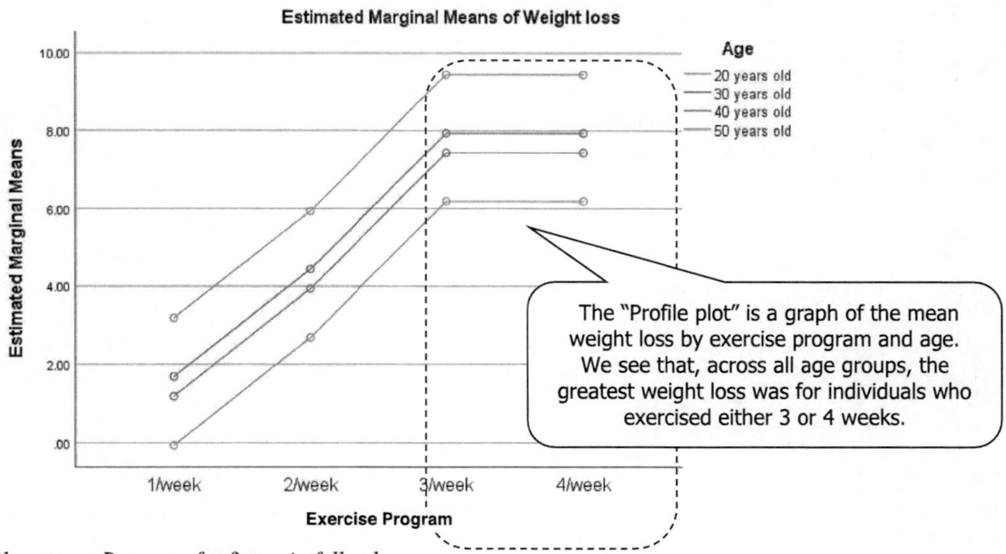

Please see eResource for figure in full color.

6.3.4 Computing the Friedman Test Using SPSS

Lastly, the Friedman test can be run as previously described in Chapter 5.

6.4 Computing Hierarchical and Randomized Block Analysis of Variance Models Using R

6.4.1 Two-Factor Hierarchical ANOVA in R

Next we consider **R** for the two-factor hierarchical ANOVA model. The commands are provided within the blocks with additional annotation to assist in understanding how the command works. Should you want to write reminder notes and annotation to yourself as you write the commands in **R** (and we highly encourage doing so), remember that any text that follows a hashtag (i.e., #) is annotation only and not part of the **R** code. Thus, you can write annotations directly into **R** with hashtags. We encourage this practice so that when you call up the commands in the future, you'll understand what the various lines of code are doing. You may think you'll remember what you did. However, trust us. There is a good chance that you won't. Thus, consider it best practice when using **R** to annotate heavily!

6.4.1.1 Reading Data Into R

```
getwd()
```

R is always pointed to a directory on your computer. The *get working directory* function can be used to determine to which directory **R** is pointed. We will assume that we need to change the working directory, and will use the next line of code to set the working directory to the desired path.

```
setwd("E:/FolderName")
```

We use the *setwd* function to establish the working directory. To set the working directory, change what is in quotation marks to your file location. Also, if you are copying the directory name from your properties, you will need to change the backslash (i.e., \) to a forward slash (i.e., /).

```
Ch6_nested <- read.csv("Ch6_nested.csv")
```

The *read.csv* function reads our data into **R**. What's to the left of the "<-" will be what the data will be called in **R**. In this example, we're calling the **R** dataframe "Ch6_nested." What's to the right of the "<-" tells **R** to find this particular csv file. In this example, our file is called "Ch6_nested.csv." Make sure the extension (i.e., .csv) is included in your script. Also note that the name of your file should be in quotation marks within the parentheses.

```
names(Ch6_nested)
```

The *names* function will produce a list of variable names for each dataframe as follows. This is a good check to make sure your data have been read in correctly.

```
[1] "Tx"  "Intvnst"  "Quality"
```

```
View(Ch6_nested)
```

FIGURE 6.22
Reading data into **R**.

The *View* function will let you view the dataset in spreadsheet format in RStudio.

```
Ch6_nested$TxF <- factor(Ch6_nested$Tx,
                    labels = c("massage therapy", "music therapy"))
Ch6_nested$IntvnstF <- factor(Ch6_nested$Intvnst,
                    labels = c(1,2,3,4))
```

This script will create a new variable in our dataframe named "TxF." We use the *factor* function to define the variable "Tx" as categorical with the two groups defined here (i.e., *massage therapy, music therapy*). We do this similarly for the "Intvnst" variable. What is to the left of "<-" in the script creates two new variables in our dataframe named "TxF" and "IntvnstF." We could have also done this by just renaming the variables rather than creating new ones.

```
summary(Ch6_nested)
```

The *summary* function will produce basic descriptive statistics on all the variables in your dataframe. This is a great way to quickly check to see if the data have been read in correctly and to get a feel for your data, if you haven't already. The output from the summary statement for this dataframe looks like this. Because we defined IntvnstF and TxF as factors, we are provided only the frequencies for each category in those variables.

```
       Tx        Intvnst          Quality       IntvnstF              TxF
Min.   :1.0   Min.   :1.00   Min.   : 1.000   1:6      massage therapy:12
1st Qu.:1.0   1st Qu.:1.75   1st Qu.: 3.750   2:6      music therapy  :12
Median :1.5   Median :2.50   Median : 6.500   3:6
Mean   :1.5   Mean   :2.50   Mean   : 7.083   4:6
3rd Qu.:2.0   3rd Qu.:3.25   3rd Qu.:10.250
Max.   :2.0   Max.   :4.00   Max.   :15.000
```

FIGURE 6.22 (continued)
Reading data into **R**.

6.4.1.2 Generating the Two-Factor Nested ANOVA Model

```
install.packages("nlme")
library(nlme)
```

The *install.packages* and *library* functions will be used to, respectively, install the *nlme* package and load it into our library. This packages is used for linear and nonlinear mixed effects modeling.

```
Ch6nest = aov(Quality ~ TxF +Error(IntvnstF), Ch6_nested)
```

The *aov* function will be used to define our model. We will create an object from the results called "Ch6nest." The dependent variable is "Quality" and we include one nonnested factor, *TxF*, and one random effect for the interventionist, defined as *Error(IntvnstF)*. The dataframe is Ch6_nested.

```
summary(Ch6nest)
```

The *summary* function will output the results of our model, *Ch6nest*.

FIGURE 6.23
Generating the two-factor nested ANOVA.

```
Error: IntvnstF
          Df Sum Sq Mean Sq F value Pr(>F)
TxF        1  337.5 337.5     59.56 0.0164 *
Residuals  2   11.3   5.7
---

Signif. codes: 0 '***' 0.001 '**' 0.01 '*' 0.05 '.' 0.1 ' ' 1

Error: Within
          Df Sum Sq Mean Sq F value Pr(>F)
Residuals 20    119    5.95
```

FIGURE 6.23 (continued)
Generating the two-factor nested ANOVA.

6.4.1.3 Generating a Post Hoc Test

```
install.packages("TukeyC")
library(TukeyC)
```

The *install.packages* and *library* functions will be used to, respectively, install the *TukeyC* package and load it into our library. This package is used for post hoc analyses.

```
tuk = TukeyC(Ch6_nested,
             model = 'Quality ~ TxF + Error(IntvnstF)',
             error = 'IntvnstF',
             which = 'TxF',
             fl1=1,
             sig.level = 0.05)
```

The *TukeyC* function will be used to generate the post hoc test. The dataframe is Ch6_nested. We define our model, error, and predictor for which we are examining the post hoc results (in this example, *TxF*).

```
summary(tuk)
```

The *summary* function will output the results.

```
Groups of means at sig.level = 0.05
                 Means G1 G2
music therapy    10.83 a
massage therapy   3.33    b

Matrix of the difference of means above diagonal and
respective p-values of the Tukey test below diagonal values
                music therapy massage therapy
music therapy           0.000             7.5
massage therapy         0.016             0.0
```

FIGURE 6.24
Generating a post hoc test.

6.4.2 Two-Factor Fixed-Effects Randomized Block ANOVA in R

Next we consider **R** for the two-factor fixed-effects randomized block ANOVA model. The commands are provided within the blocks with additional annotation to assist in understanding how the command works. As noted previously, should you want to write reminder notes and annotation to yourself as you write the commands in **R**, any text that follows a hashtag (i.e., #) is annotation only and not part of the **R** code.

6.4.2.1 Reading Data Into R

```
getwd()
```

R is always pointed to a directory on your computer. The *get working directory* function can be used to determine to which directory **R** is pointed. We will assume that we need to change the working directory, and will use the next line of code to set the working directory to the desired path.

```
setwd("E:/FolderName")
```

We use the *setwd* function to establish the working directory. To set the working directory, change what is in quotation marks to your file location. Also, if you are copying the directory name from your properties, you will need to change the backslash (i.e., \) to a forward slash (i.e., /).

```
Ch6_block <- read.csv("Ch6_block.csv")
```

The *read.csv* function reads our data into **R**. What's to the left of the "<-" will be what the data will be called in **R**. In this example, we're calling the **R** dataframe "Ch6_block." What's to the right of the "<-" tells **R** to find this particular csv file. In this example, our file is called "Ch6_block.csv." Make sure the extension (i.e., .csv) is included in your script. Also note that the name of your file should be in quotation marks within the parentheses.

```
names(Ch6_block)
```

The *names* function will produce a list of variable names for each dataframe as follows. This is a good check to make sure your data have been read in correctly.

```
[1] "Program" "Age" "WtLoss"
```

```
View(Ch6_block)
```

The *View* function will let you view the dataset in spreadsheet format in RStudio.

```
Ch6_block$ProgramF <- factor(Ch6_block$Program,
                             labels=c("1/week", "2/week","3/week","4/week"))
```

The *factor* command is used to define the categorical variables. What is to the left of "<-" is creating a new variable in our dataframe (i.e., Ch6_block), named "ProgramF." To the right of "<-" is defining the variable "Program" in the dataframe as a categorical variable with four categories with the labels defined here.

```
Ch6_block$Age <-ordered(Ch6_block$Age,
                        labels=c("20", "30", "40", "50"))
```

The command to the left of "<-" is writing over the variable in our dataframe (i.e., Ch6_block), named "Age," and defining it as an ordinal variable. To the right of "<-" is defining the variable "Age" in the dataframe with four categories which have labels of 20, 30, 40, and 50.

FIGURE 6.25
Reading data into **R**.

```
summary(Ch6_block)
```

The *summary* command will produce basic descriptive statistics on all the variables in your dataframe. This is a great way to quickly check to see if the data have been read in correctly and get a feel for your data, if you haven't already. The output from the summary statement for this dataframe looks like this.

```
   Program       Age      WtLoss           ProgramF
Min.   :1.00   20:4   Min.   : 0.000   1/week:4
1st Qu.:1.75   30:4   1st Qu.: 2.750   2/week:4
Median :2.50   40:4   Median : 6.000   3/week:4
Mean   :2.50   50:4   Mean   : 5.312   4/week:4
3rd Qu.:3.25          3rd Qu.: 7.250
Max.   :4.00          Max.   : 10.000
```

FIGURE 6.25 (continued)
Reading data into **R**.

6.4.2.2 Generating the Two-Factor Fixed-Effects Randomized Block ANOVA

```
BlockModel <- aov(WtLoss ~ ProgramF + Age, Ch6_block)
```

The *aov* function is used to generate our model randomized block ANOVA model. We create an object from those results called "BlockModel." "WtLoss" is our dependent variable. "ProgramF" is the treatment (i.e., independent variable), and "Age" is the blocking factor. We are using data from the dataframe Ch6_block.

```
summary(BlockModel)
```

Because we created an object from our model, we run the *summary* function to output the results.

```
          Df Sum Sq Mean Sq F value    Pr(>F)
ProgramF   3 110.19   36.73   92.79 4.35e-07 ***
Age        3  21.69    7.23   18.26 0.000362 ***
Residuals  9   3.56    0.40
—
Signif. codes: 0 '***' 0.001 '**' 0.01 '*' 0.05 '.' 0.1 ' ' 1
```

```
Ch6_block$unstandardizedResiduals <- resid(BlockModel)
```

What is to the left of "<-" tells **R** to save a new variable in our dataframe (i.e., "Ch6_block$unstandardized Residuals") that is called "unstandardizedResiduals." What is to the right of "<-" tells **R** to created unstandardized residuals using the results from the object "BlockModel."

FIGURE 6.26
Generating the two-factor fixed-effects randomized block ANOVA.

6.5 Data Screening

6.5.1 Examining Assumptions for the Two-Factor Hierarchical ANOVA

The assumptions for the two-factor hierarchical ANOVA that we will examine include normality homogeneity of variance, and independence of observations within cells.

6.5.1.1 Normality

We will use the residuals (which were requested and created through the "Save" option) to examine the extent to which normality was met.

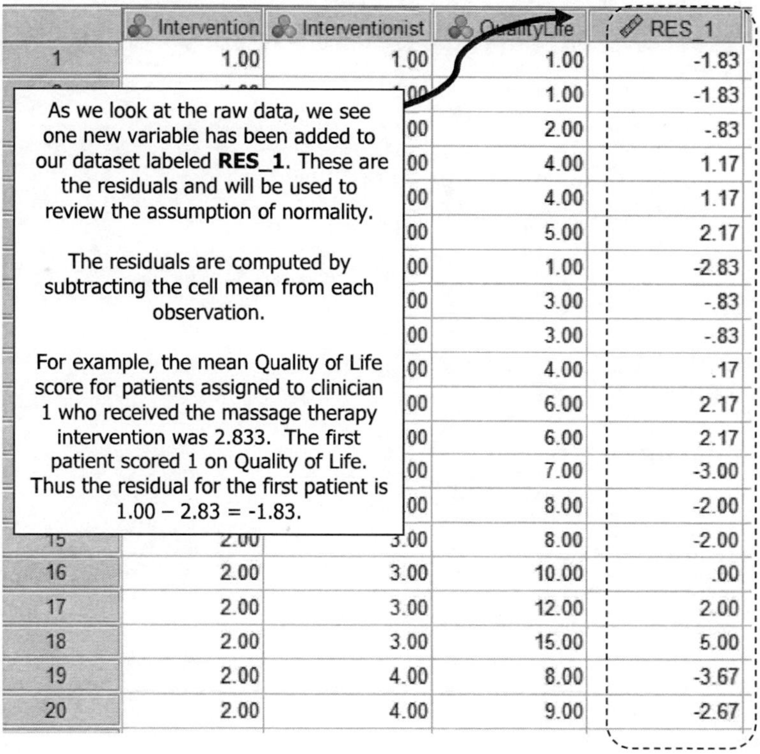

Working in **R,** we can save the unstandardized residuals using the following command. The *proj* function will create a matrix giving projections of the data given the terms of the model. That matrix will be used to compute residuals. We created a new variable in our dataframe called *unstandardizedResiduals* (i.e., *Ch6_nested$unstandardizedResiduals*).

```
Ch6nest.pr <- proj(Ch6nest)
Ch6_nested$unstandardizedResiduals <- Ch6nest.pr[['Within']][,'Residuals']
```

FIGURE 6.27
Two-factor hierarchical ANOVA residuals.

As described in earlier ANOVA chapters, understanding the distributional shape, specifically whether normality is a reasonable assumption, is important. For the two-factor hierarchical ANOVA, the residuals should be normally distributed.

As in previous chapters, we use "Explore" to examine whether the assumption of normality is met. The general steps for accessing Explore have been presented in previous chapters and will not be repeated here. Click the residual and move it into the "Dependent List" box by clicking on the arrow button. The procedures for selecting normality statistics remain the same what we have done previously. Click on "Plots" in the upper right corner. Place a checkmark in the boxes for "Normality plots with tests" and also for "Histogram." Then click "Continue" to return to the main Explore dialog box, and click "OK" to generate the output.

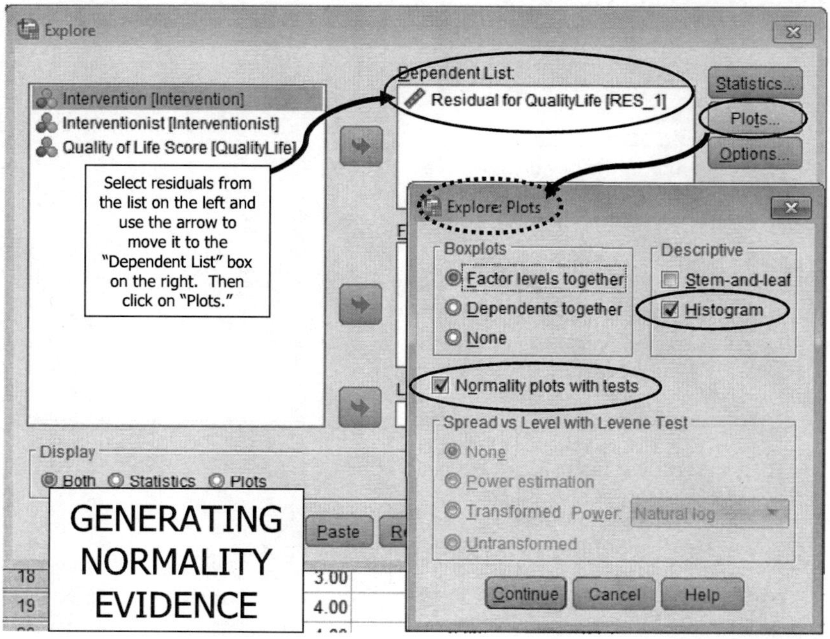

Working in **R**, we can generate various normality statistics as well.

```
install.packages("pastecs")
```

The *install.packages* command will install the *pastecs* package which we will use to generate various forms of normality evidence.

```
library(pastecs)
```

The *library* function will load the *pastecs* package.

```
stat.desc(Ch6_nested$unstandardizedResiduals,
          norm = TRUE)
```

The *stat.desc* will generate normality indices on the variable "unstandardizedResiduals" in the dataframe Ch6_nested as follows. The *norm=TRUE* command will produce Shapiro-Wilk results (*SW*), which are displayed as *normtest.W* (which is the *SW* statistic value) and *normtest.p* (which is the observed probability value). Here, we see $SW = .960$ and the related $p = .44$.

We see skew (.249) and kurtosis (−.976) along with $SW = .960$, $p = .442$ for the "unstandardizedResidual" variable. All indicate the assumption of normality has been met. As we know, we can divide the skew and kurtosis values by their standard errors to get a standardized value that can be used to determine if the skew and/or kurtosis is statistically different from zero. Since this output provides "2SE," we would simply divide this value by 2 to arrive at the standard error.

FIGURE 6.28
Generating normality evidence.

Note: You may have noticed that the skewness and kurtosis value that we've just generated differs from what we found in SPSS, which was skew = .284 and kurtosis = −.693 This is because there are different ways to calculate skewness and kurtosis. Let's use another package in **R** to calculate these statistics with different algorithms.

```
     nbr.val      nbr.null        nbr.na           min
2.400000e+01  0.000000e+00  0.000000e+00 -3.666667e+00

         max         range           sum        median
5.000000e+00  8.666667e+00  7.216450e-16 -3.333333e-01

        mean       SE.mean  CI.mean.0.95           var
3.008661e-17  4.643056e-01  9.604894e-01  5.173913e+00

     std.dev      coef.var      skewness      skew.2SE
2.274624e+00  7.560253e+16  2.494060e-01  2.640553e-01

    kurtosis      kurt.2SE    normtest.W    normtest.p
-9.764138e-01 -5.319450e-01 9.601899e-01  4.422514e-01
```

```
install.packages("e1071")
```

The *install.packages* function will install the **e1071** package which we will use to generate skewness and kurtosis.

```
library(e1071)
```

The *library* function will load the **e1071** package.

```
skewness(Ch6_block$unstandardizedResiduals, type=3)
skewness(Ch6_block$unstandardizedResiduals, type=2)
skewness(Ch6_block$unstandardizedResiduals, type=1)
```

The *skewness* function will generate skewness statistics on the variable(s) we specify. The "type=" script defines how skewness is calculated. Specifying "type=2" will use the algorithm that is used by SPSS. Readers interested in learning more, including the algorithms for each of the three methods, are encouraged to review Joanes and Gill (1998). We see that using "type=2," our skew is .284, the same value as generated using SPSS.

```
# skewness(Ch6_nested$unstandardizedResiduals, type=3)
[1] 0.249406

# skewness(Ch6_nested$unstandardizedResiduals, type=2)
[1] 0.2839088

# skewness(Ch6_nested$unstandardizedResiduals, type=1)
[1] 0.2658471
```

```
kurtosis(Ch6_block$unstandardizedResiduals, type=3)
kurtosis(Ch6_block$unstandardizedResiduals, type=2)
kurtosis(Ch6_block$unstandardizedResiduals, type=1)
```

The *kurtosis* function will generate kurtosis statistics on the variable(s) we specify. The "type=" script defines how kurtosis is calculated. Specifying "type=2" will use the algorithm that is used by SPSS. Readers interested in learning more, including the algorithms for each of the three methods, are encouraged to review Joanes and Gill (1998). We see that using "type=2," our kurtosis is −.693, the same value as generated using SPSS.

FIGURE 6.28 (continued)
Generating normality evidence.

```
# kurtosis(Ch6_nested$unstandardizedResiduals, type=3)
[1] -0.9764138

# kurtosis(Ch6_nested$unstandardizedResiduals, type=2)
[1] -0.6927686

# kurtosis(Ch6_nested$unstandardizedResiduals, type=1)
[1] -0.7966245
```

```
shapiro.test(Ch6_nested$unstandardizedResiduals)
```

Had we wanted to generate only the Shapiro-Wilk test, the *shapiro.test* function could be used.

```
        Shapiro-Wilk normality test

data: Ch6_nested$unstandardizedResiduals
W = 0.96019, p-value = 0.4423
```

Working in **R**, another way to test for normality is **D'Agostino's test** for skewness and the **Bonett-Seier test** for **Geary's kurtosis**.

```
install.packages("moments")
library(moments)
```

To conduct D'Agostino's test, we first have to install the *moments* package and then load it into our library. The null hypothesis for this test is that skewness equals zero. Thus, a statistically significant D'Agostino's test would indicate that there is statistically significant skewness.

```
agostino.test(Ch6_nested$unstandardizedResiduals)
```

The function *agostino.test* is generated using the variable "unstandardizedResiduals" from our Ch6_nested dataframe. The results suggest evidence of normality as $p = .526$, greater than alpha.

```
        D'Agostino skewness test

data:  Ch6_nested$unstandardizedResiduals
skew = 0.26585, z = 0.63406, p-value = 0.526
alternative hypothesis: data have a skewness
```

```
bonett.test((Ch6_nested$unstandardizedResiduals))
```

The *bonett.test* function, generated using the variable "unstandardizedResiduals" from our Ch6_nested dataframe, performs the Bonett-Seier test for Geary's kurtosis for data that are normally distributed. The null hypothesis states that data should have a Geary's kurtosis value equal to $\sqrt{2/\pi} = .7979$. The results suggest evidence of normality as $p = .147$, greater than alpha.

```
        Bonett-Seier test for Geary kurtosis

data: (Ch6_nested$unstandardizedResiduals)
tau = 1.9167, z = -1.4508, p-value = 0.1468
alternative hypothesis: kurtosis is not equal to sqrt(2/pi)
```

FIGURE 6.28 (continued)
Generating normality evidence.

6.5.1.1.1 Interpreting Normality Evidence

By this point, we have had a substantial amount of practice in interpreting quite a range of normality statistics and interpret them again in reference to the hierarchical ANOVA model assumption of normality. The skewness statistic of the residuals is .284 and kurtosis is −.693—both being within the range of what would be considered normal (i.e., an absolute value of 2.0), suggesting some evidence of normality. Working in **R** (see Figure 6.28), D'Agostino's test (D'Agostino, 1970) can be used to examine the null hypothesis that skewness equals zero. Thus, a statistically significant D'Agostino's test would indicate that there is statistically significant skewness. For kurtosis, we can use the Bonett-Seier test for Geary's kurtosis (Bonett & Seier, 2002) for data that are normally distributed. The null hypothesis states that data should have a Geary's kurtosis value equal to $\sqrt{2/\pi} = .7979$. Thus, a statistically significant Bonett-Seier test for Geary's kurtosis would indicate that there is statistically significant kurtosis. Thus, with these tests, as with Kolmogorov-Smirnov and Shapiro-Wilk, we do *not* want to find statistically significant results—which is exactly what was found in this illustration.

As suggested by the skewness statistic, the histogram of residuals is slightly positively skewed, and the histogram also provides a visual display of the slightly platykurtic distribution.

Descriptives

			Statistic	Std. Error
Residual for QualityLife	Mean		.0000	.46431
	95% Confidence Interval for Mean	Lower Bound	-.9605	
		Upper Bound	.9605	
	5% Trimmed Mean		-.0648	
	Median		-.3333	
	Variance		5.174	
	Std. Deviation		2.27462	
	Minimum		-3.67	
	Maximum		5.00	
	Range		8.67	
	Interquartile Range		4.08	
	Skewness		.284	.472
	Kurtosis		-.693	.918

FIGURE 6.29
Normality evidence.

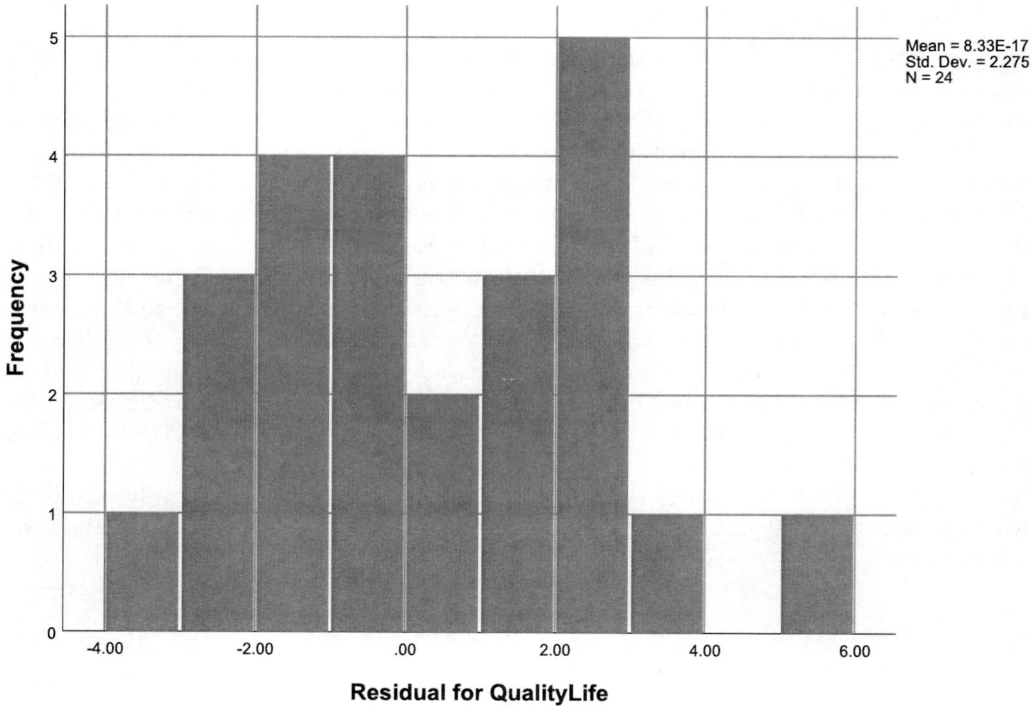

Working in **R**, we can generate a histogram using the *ggplot2* package.

```
install.packages("ggplot2")
```

The *install.packages* function will install the *ggplot2* package which we can use to create various graphs and plots. Remember, if this package has previously been installed, there is no need to install again.

```
library(ggplot2)
```

The *library* function will load the *ggplot2* package.

```
qplot(Ch6_nested$unstandardizedResiduals,
      geom="histogram",
      binwidth=0.5,
      main = "Histogram of Unstandardized Residuals",
      xlab = "Unstandardized Residual", ylab = "Count",
      fill=I("gray"),
      col=I("white"))
```

Using the *qplot* command, we create a histogram (i.e., geom = "histogram") from our dataframe (i.e., Ch6_nested) using the variable "unstandardizedResiduals." We can add a few commands to change the width of the bars (i.e., *binwidth=0.5*), color of the bars (i.e., *fill=I("gray")*), and outline of the bars (i.e., *col=I("white")*). We can also add a title (i.e., *main = "Histogram of Unstandardized Residuals"*) and change the *X* and *Y* axes (*xlab = "Unstandardized Residual"*, *ylab = "Count"*).

FIGURE 6.30
Histogram.

There are a few other statistics that can be used to gauge normality. The formal test of normality, the Shapiro-Wilk test (*SW*) (Shapiro & Wilk, 1965), provides evidence of the extent to which our sample distribution is statistically different from a normal distribution. The output for the Shapiro-Wilk test is presented in Figure 6.31 and suggests that our sample distribution for the residual is not statistically significantly different than what would be expected from a normal distribution as the p value is greater than α.

Tests of Normality

	Kolmogorov-Smirnov[a]			Shapiro-Wilk		
	Statistic	df	Sig.	Statistic	df	**Sig.**
Residual for QualityLife	.123	24	.200*	.960	24	**.442**

*. This is a lower bound of the true significance.

a. Lilliefors Significance

Working in **R**, we used the *stat.desc* function from the *pastecs* package to generate *SW* earlier, along with many other statistics.

```
shapiro.test(Ch6_nested$unstandardizedResiduals)
```

Had we wanted to generate only the Shapiro-Wilk test, the *shapiro.test* function could be used.

```
     Shapiro-Wilk normality test

data:  Ch6_nested$unstandardizedResiduals
W = 0.96019, p-value = 0.4423
```

FIGURE 6.31
Shapiro-Wilk test of normality.

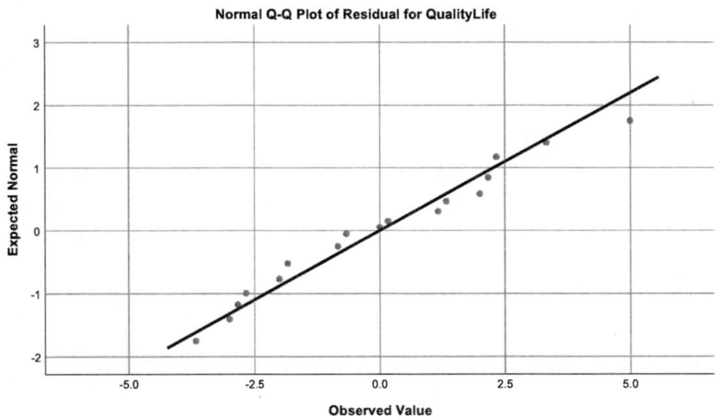

Working in **R,** we can use the *qplot* command to create a Q-Q plot of unstandardized residuals.

```
qplot(sample=unstandardizedResiduals,
      data = Ch6_nested)
```

FIGURE 6.32
Normal Q-Q plot.

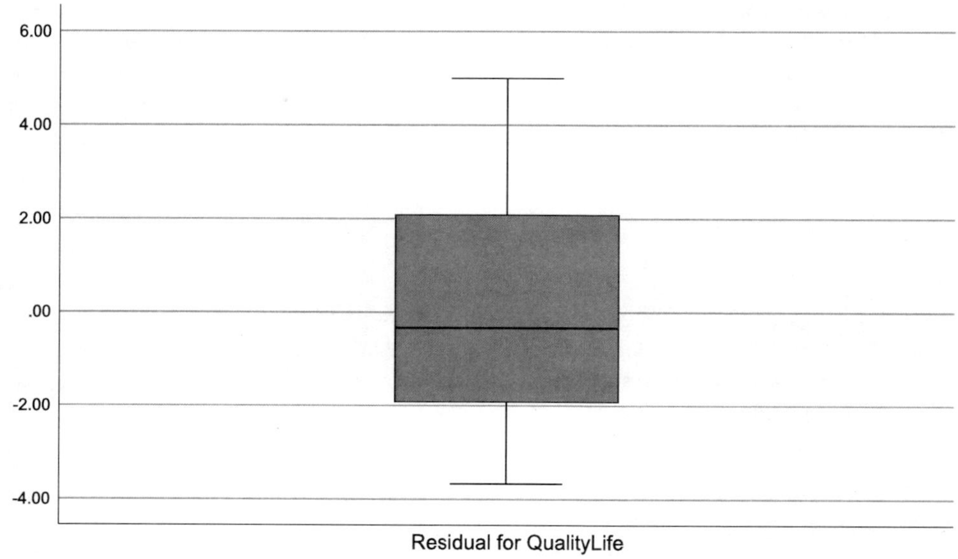

Working in **R**, we can generate a boxplot for unstandardized residuals using the *boxplot* function. To label the *Y* axis, we include the *ylab* command.

```
boxplot(Ch6_nested$unstandardizedResiduals,
        ylab="unstandardized residual")
```

FIGURE 6.33
Residual boxplot.

Quantile-quantile (Q-Q) plots are also often examined to determine evidence of normality, where quantiles of the theoretical normal distribution are plotted against quantiles of the sample distribution. Points that fall on or close to the diagonal line suggest evidence of normality. The Q-Q plot of residuals shown below suggests relative normality.

Examination of the boxplot in Figure 6.33 also suggests a relatively normal distributional shape of residuals with no outliers.

Considering the forms of evidence we have examined, skewness and kurtosis statistics, the Shapiro-Wilk test, histogram, the Q-Q plot, and the boxplot, all suggest normality is a reasonable assumption. We can be reasonably assured we have met the assumption of normality.

6.5.1.2 Independence

Another assumption for which to test is independence. As we have seen this tested in other designs, we do not consider it further here.

6.5.1.3 Homogeneity of Variance

Homogeneity is the assumption of equal variances. In SPSS, Levene's test is used to examine this assumption. The results are provided in Table 6.11 and suggest that the variance of the error term is constant across groups in our model. In other words, we have met the homogeneity of variance assumption.

6.5.2 Examining Assumptions for the Two-Factor Fixed-Effects Randomized Block ANOVA for $n = 1$

The assumptions for the two-factor randomized block ANOVA that we will examine include normality, independence, and homoscedasticity (or homogeneity of variance).

6.5.2.1 Normality

We use the residuals (which were requested and created through the "Save" option when generating our model) to examine the extent to which normality was met. As shown in previous ANOVA chapters, understanding the distributional shape, specifically the extent to which normality is a reasonable assumption, is important. For the two-factor randomized block ANOVA, the residuals should be a normal distribution. Because the steps for generating normality evidence were presented previously in the chapter for the two-factor hierarchical ANOVA model, they will not be reiterated here.

6.5.2.1.1 Interpreting Normality Evidence

By this point, we have had a substantial amount of practice in interpreting quite a range of normality statistics. Here we interpret them again, only now in reference to the two-factor randomized block ANOVA model. The skewness statistic of the residuals is −.154 and kurtosis is −.496—both being within the range of what would be considered normal (i.e., an absolute value of 2.0), suggesting some evidence of normality.

Descriptives

			Statistic	Std. Error
Residual for WeightLoss	Mean		.0000	.12183
	95% Confidence Interval for Mean	Lower Bound	-.2597	
		Upper Bound	.2597	
	5% Trimmed Mean		.0069	
	Median		.0625	
	Variance		.238	
	Std. Deviation		.48734	
	Minimum		-.94	
	Maximum		.81	
	Range		1.75	
	Interquartile Range		.87	
	Skewness		-.154	.564
	Kurtosis		-.496	1.091

FIGURE 6.34
Two-factor randomized block ANOVA normality evidence.

Working in **R**, we can generate various normality statistics as well.

```
install.packages("pastecs")
```

The *install.packages* function will install the *pastecs* package which we will use to generate various forms of normality evidence.

```
library(pastecs)
```

The *library* function will load the *pastecs* package.

```
stat.desc(Ch6_block$unstandardizedResiduals,
norm = TRUE
```

The *stat.desc* will generate normality indices on the variable "unstandardizedResiduals" in the dataframe Ch6_block as follows. The *norm=TRUE* command will produce Shapiro-Wilk results (*SW*). We see skew (−.127) and kurtosis (−.985) along with $SW = .965$, $p = .757$ for the "unstandardizedResidual" variable. All indicate the assumption of normality has been met. As we know, we can divide the skew and kurtosis values by their standard errors to get a standardized value that can be used to determine if the skew and/or kurtosis is statistically different from zero. Since this output provides "2SE," we would simply divide this value by 2 to arrive at the standard error.

Note: You may have noticed that the skewness and kurtosis value that we've just generated differs from what we found in SPSS, which was skew = .284 and kurtosis = −.693 This is because there are different ways to calculate skewness and kurtosis. Let's use another package in **R** to calculate these statistics with different algorithms.

```
      nbr.val        nbr.null          nbr.na             min
 1.600000e+01    0.000000e+00    0.000000e+00   -9.375000e-01

          max           range             sum          median
 8.125000e-01    1.750000e+00   -5.551115e-17    6.250000e-02

         mean         SE.mean    CI.mean.0.95             var
-3.469447e-18    1.218349e-01    2.596850e-01    2.375000e-01

      std.dev        coef.var        skewness        skew.2SE
 4.873397e-01   -1.404661e+17   -1.265598e-01   -1.121373e-01

     kurtosis        kurt.2SE      normtest.W      normtest.p
-9.849269e-01   -4.514808e-01    9.652256e-01    7.566056e-01
```

```
install.packages("e1071")
```

The *install.packages* function will install the **e1071** package which we will use to generate skewness and kurtosis.

```
library(e1071)
```

The *library* function will load the **e1071** package.

```
skewness(Ch6_block$unstandardizedResiduals, type=3)
skewness(Ch6_block$unstandardizedResiduals, type=2)
skewness(Ch6_block$unstandardizedResiduals, type=1)
```

FIGURE 6.34 (continued)
Two-factor randomized block ANOVA normality evidence.

The *skewness* function will generate skewness statistics on the variable(s) we specify. The "type=" script defines how skewness is calculated. Specifying "type=2" will use the algorithm that is used by SPSS. Readers interested in learning more, including the algorithms for each of the three methods, are encouraged to review Joanes and Gill (1998). We see that using "type=2," our skew is −.154, the same value as generated using SPSS.

```
# skewness(Ch6_block$unstandardizedResiduals, type=3)
[1] -0.1265598

# skewness(Ch6_block$unstandardizedResiduals, type=2)
[1] -0.1542825

# skewness(Ch6_block$unstandardizedResiduals, type=1)
[1] -0.1394245
```

```
kurtosis(Ch6_block$unstandardizedResiduals, type=3)
kurtosis(Ch6_block$unstandardizedResiduals, type=2)
kurtosis(Ch6_block$unstandardizedResiduals, type=1)
```

The *kurtosis* function will generate kurtosis statistics on the variable(s) we specify. The "type=" script defines how kurtosis is calculated. Specifying "type=2" will use the algorithm that is used by SPSS. Readers interested in learning more, including the algorithms for each of the three methods, are encouraged to review Joanes and Gill (1998). We see that using "type=2," our kurtosis is −.496, the same value as generated using SPSS.

```
# kurtosis(Ch6_block$unstandardizedResiduals, type=3)
[1] -0.9849269

# kurtosis(Ch6_block$unstandardizedResiduals, type=2)
[1] -0.4964841

# kurtosis(Ch6_block$unstandardizedResiduals, type=1)
[1] -0.7072946
```

FIGURE 6.34 (continued)
Two-factor randomized block ANOVA normality evidence.

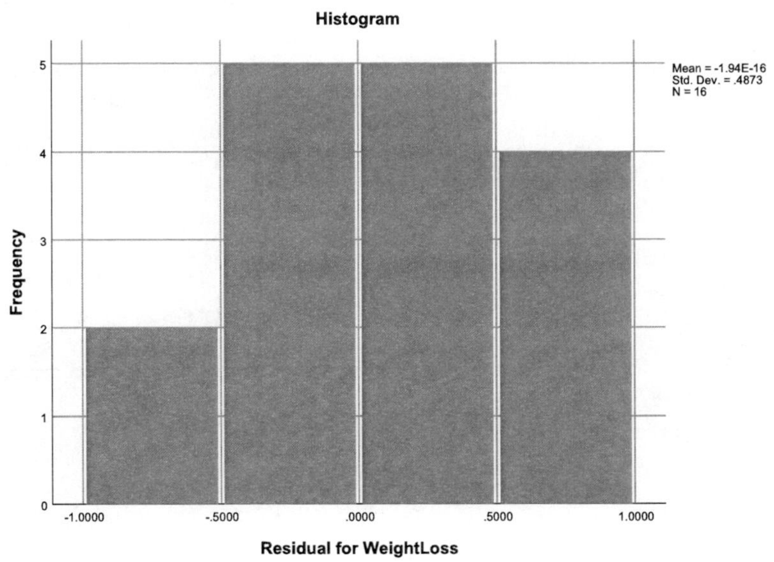

FIGURE 6.35
Histogram.

Working in **R**, we can generate a histogram using the *ggplot2* package.

```
install.packages("ggplot2")
```

The *install.packages* function will install the *ggplot2* package which we can use to create various graphs and plots. Remember, if this package has previously been installed, there is no need to install again.

```
library(ggplot2)
```

The *library* function will load the *ggplot2* package.

```
qplot(Ch6_block$unstandardizedResiduals,
      geom="histogram",
      binwidth=0.5,
      main = "Histogram of Unstandardized Residuals",
      xlab = "Unstandardized Residual", ylab = "Count",
      fill=I("gray"),
      col=I("white"))
```

Using the *qplot* function, we create a histogram (i.e., geom = "histogram") from our dataframe (i.e., Ch6_block) using the variable "unstandardizedResiduals." We can add a few commands to change the width of the bars (i.e., *binwidth=0.5*), color of the bars (i.e., *fill=I("gray")*), and outline of the bars (i.e., *col=I("white")*). We can also add a title (i.e., *main = "Histogram of Unstandardized Residuals"*) and change the *X* and *Y* axes (*xlab = "Unstandardized Residual", ylab = "Count"*).

FIGURE 6.35 (continued)
Histogram.

As suggested by the skewness statistic, the histogram of residuals is slightly negatively skewed, and the histogram also provides a visual display of the slightly platykurtic distribution.

There are a few other statistics that can be used to gauge normality. The formal test of normality, the Shapiro-Wilk test (*SW*) (Shapiro & Wilk, 1965), provides evidence of the extent to which our sample distribution is statistically different from a normal distribution. The output for the Shapiro-Wilk test is presented in Figure 6.36 and suggests that our sample distribution for the residuals is *not* statistically significantly different than what would be expected from a normal distribution as the p value is greater than α.

Tests of Normality

	Kolmogorov-Smirnov[a]			Shapiro-Wilk		
	Statistic	df	Sig.	Statistic	df	**Sig.**
Residual for WeightLoss	.136	16	.200*	.965	16	**.757**

*. This is a lower bound of the true significance.

a. Lilliefors Significance Correction

FIGURE 6.36
Shapiro-Wilk test of normality.

Quantile-quantile (Q-Q) plots are also often examined to determine evidence of normality where quantiles of the theoretical normal distribution are plotted against quantiles of the sample distribution. Points that fall on or close to the diagonal line suggest evidence of normality. The Q-Q plot of residuals shown below suggests relative normality.

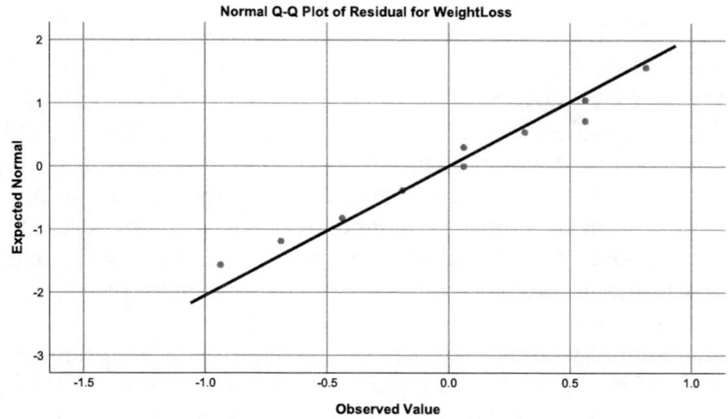

Working in **R,** we can use the *qplot* function to create a Q-Q plot of unstandardized residuals.

```
qplot(sample=unstandardizedResiduals,
      data = Ch6_block)
```

FIGURE 6.37
Q-Q plot.

Examination of the boxplot in Figure 6.38 also suggests a relatively normal distributional shape of residuals with no outliers.

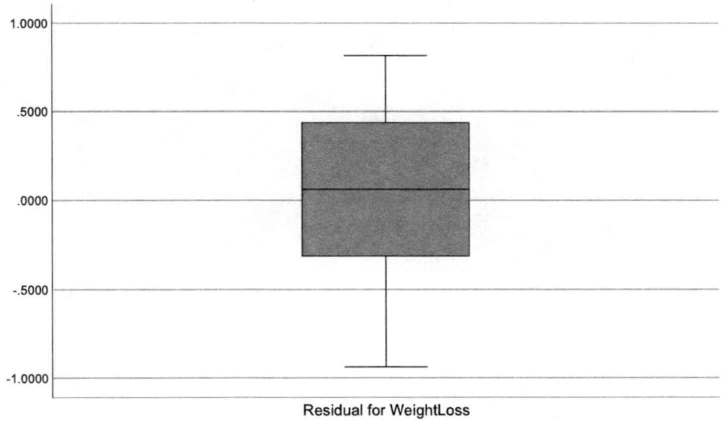

Working in **R,** we can generate a boxplot for unstandardized residuals using the *boxplot* function. To label the *Y* axis, we include the *ylab* command.

```
boxplot(Ch6_block$unstandardizedResiduals,
        ylab="unstandardized residual")
```

FIGURE 6.38
Residual boxplot.

Considering the forms of evidence we have examined, skewness and kurtosis statistics, the Shapiro-Wilk test, histogram, the Q-Q plot, and the boxplot, all suggest normality is a reasonable assumption. We can be reasonably assured we have met the assumption of normality.

6.5.2.2 Independence

The only assumption we have not tested for yet is independence. As we discussed in reference to the one-way ANOVA, if subjects have been randomly assigned to conditions (in other words, the different levels of the treatment factor in a two-factor randomized block ANOVA), the assumption of independence has likely been met. In our example, individuals were randomly assigned to an exercise program, and thus the assumption of independence was met. However, we often use independent variables that do not allow random assignment. We can plot residuals against levels of our treatment factor using a scatterplot to see whether or not there are patterns in the data and thereby provide an indication of whether we have met this assumption.

Please note that some researchers do not believe that the assumption of independence can be tested. If there is not random assignment to groups, then these researchers believe this assumption has been violated—period. The plot that we generate will give us a general idea of patterns, however, in situations where random assignment was not performed.

6.4.2.2.1 Generating the Scatterplot

The general steps for generating a simple scatterplot through "Scatter/dot" have been presented in Chapter 10 of the previous volume, and they will not be reiterated here. From the "Simple Scatterplot" dialog screen, click the residual variable and move it into the "Y Axis" box by clicking on the arrow. Click the independent variable that we wish to display (e.g., "Exercise Program") and move it into the "X Axis" box by clicking on the arrow. Then click "OK."

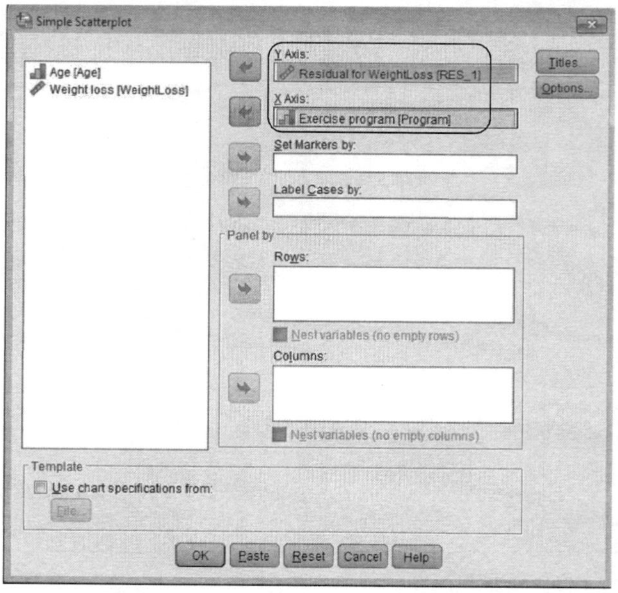

Working in **R**, we create a similar scatterplot using the following *plot* command, with the first variable listed displaying on the *X* axis (e.g., "Ch6_block$Program"), and the second variable displaying on the *Y* axis (i.e., "Ch6_block$unstandardized.residuals"). Additional commands are provided to label the axes (*xlab* and *ylab*) and title the graph (*main*). *Note:* We use the "Program" (not the "ProgramF") variable on the *X* axis. Had we generated the plot with "ProgramF," a scatterplot would have automatically been generated.

FIGURE 6.39
Generating a scatterplot.

```
plot(Ch6_block$Program,
     Ch6_block$unstandardizedResiduals,
     xlab = "program",
     ylab = "unstandardized residuals",
     main = "Scatterplot for independence")
```

Using the *plot* function, additional plots (one of which is the Q-Q plot) that can be used for diagnostic purposes are created.

```
plot(BlockModel)
```

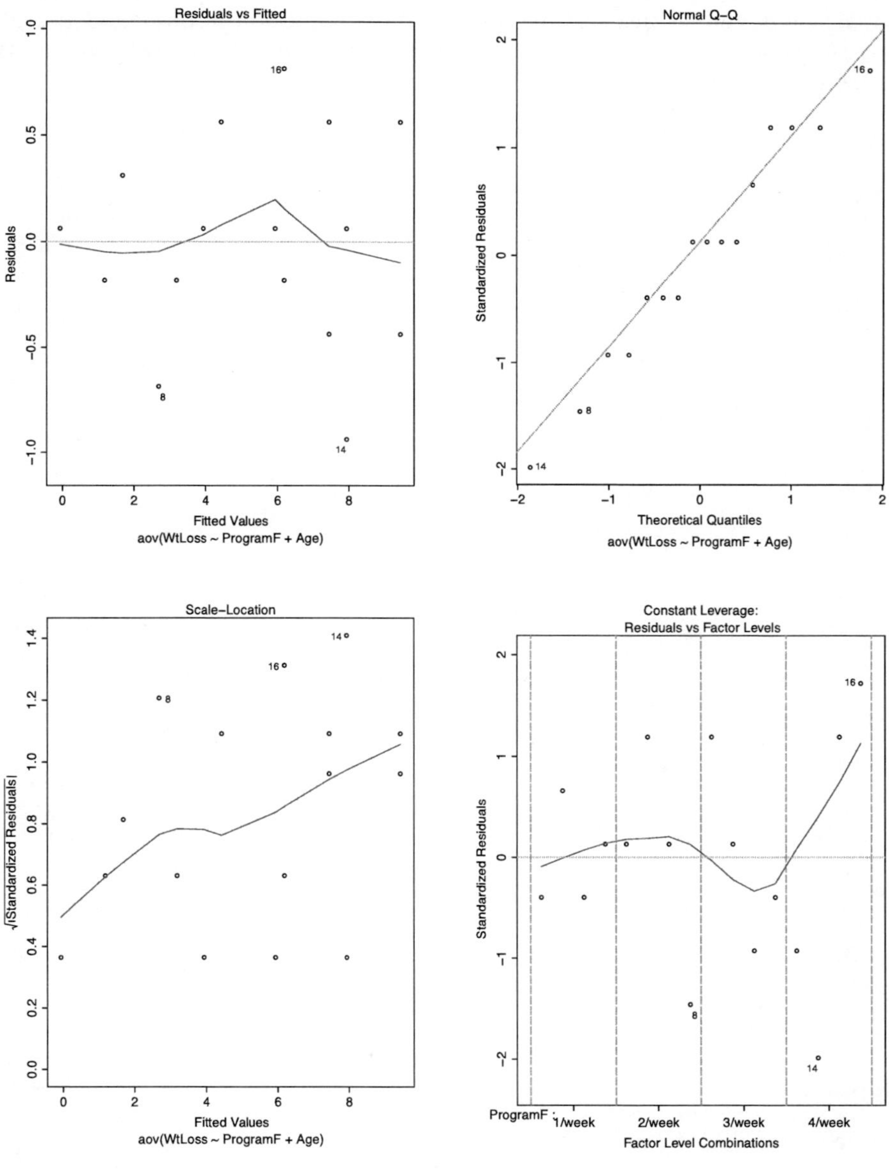

FIGURE 6.39 (continued)
Generating a scatterplot.

6.4.2.2.2 Interpreting Independence Evidence

In examining the scatterplot for evidence of independence, the points should be fall relatively randomly above and below a horizontal line at zero. (You may recall in Chapter 1 that we added a reference line to the graph using Chart Editor. To add a reference line, double click on the graph in the output to activate the chart editor. Select "Options" in the top pulldown menu, then "Y axis reference line." This will bring up the "Properties" dialog box. Change the value of the position to be "0." Then click on "Apply" and "Close" to generate the graph with a horizontal line at zero.)

In this example, our scatterplot for exercise program by residual generally suggests evidence of independence with a relatively random display of residuals above and below the horizontal line at zero. Thus, had we not met the assumption of independence through random assignment of cases to groups, this would have provided evidence that independence was a reasonable assumption.

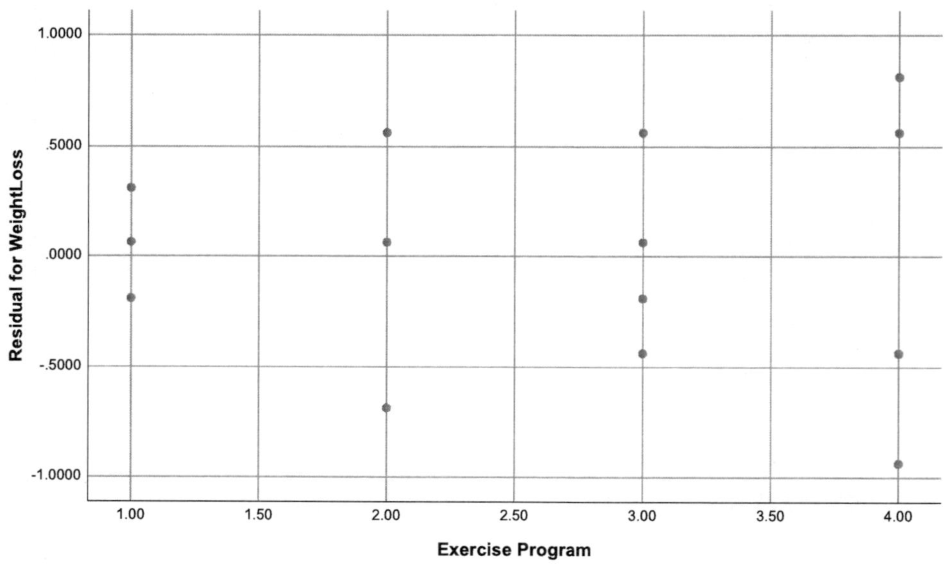

FIGURE 6.40
Scatterplot.

6.5.2.3 Homogeneity of Variance

Homogeneity of variance is the assumption that the variances of the groups are equal. Because of the design of our study, there is not an option for testing this.

6.6 Power Using G*Power

G*Power provides power calculations for the two-factor randomized block ANOVA model. In G*Power, just treat this design as if it were a regular two-factor ANOVA model.

6.7 Research Question Template and Example Write-Up

Finally, here is an example paragraph just for the results of the *two-factor hierarchical ANOVA* design (feel free to write a similar paragraph for the two-factor randomized block ANOVA example). Recall that our graduate research assistant, Challie Lenge, was assisting a psychology faculty member, Dr. Mayfield, in a clinical trial conducted through the institution's medical center. Dr. Mayfield wanted to know the following: if there is a mean difference in quality of life for hospice patients based on type of intervention (massage therapy or music therapy), and if there is a mean difference in quality of life based on interventionist or clinician assigned to provide the intervention. The research questions presented to Dr. Mayfield from Challie include the following:

- *Is there a mean difference in quality of life based on intervention?*
- *Is there a mean difference in quality of life based on the interventionist?*

Challie then assisted Dr. Mayfield in generating a two-factor hierarchical ANOVA as the test of inference, and a template for writing the research questions for this design is presented here. As we noted in previous chapters, it is important to ensure the reader understands the levels of the factor(s). This may be done parenthetically in the actual research question, as an operational definition, or specified within the methods section.

- Is there a mean difference in [dependent variable] based on [nonnested factor]?
- Is there a mean difference in [dependent variable] based on [nested factor]?

It may be helpful to preface the results of the two-factor hierarchical ANOVA with information on an examination of the extent to which the assumptions were met. The assumptions include: (a) homogeneity of variance, and (b) normality.

A two-factor hierarchical analysis of variance (ANOVA) was conducted. The dependent variable was quality of life. The nonnested factor was intervention (massage therapy or music therapy) and the nested factor was interventionist or clinician (four interventionists or clinicians). The null hypotheses tested included: (1) the mean quality of life was equal for each of the interventions, and (2) the mean quality of life for each interventionist was equal.

The data were screened for missingness and violation of assumptions prior to analysis. There were no missing data. The assumption of *homogeneity of variance* was met via Levene's test (F (3, 20) = 1.042, p = .396). The assumption of *normality* was tested via examination of the residuals. Review of the Shapiro-Wilk test (SW = .960, df = 24, p = .442) and skewness (.284) and kurtosis (–.693) statistics suggested that normality was a reasonable assumption. Additional tests, including D'Agostino's test for skewness (z = .634, p = .526) and the Bonett-Seier test for Geary's kurtosis (z = –1.451, p = .147) suggested evidence of normality. The boxplot displayed a relatively normal distributional shape (with no outliers) of the residuals. The Q-Q plot and histogram suggested normality was tenable.

Here is an APA-style example paragraph of results for the two-factor hierarchical ANOVA (remember that this will be prefaced by the previous paragraph reporting the extent to which the assumptions of the test were met).

The results for the two-factor hierarchical ANOVA indicate:

1. A statistically significant main effect for intervention ($F_{\text{intervention}} = 59.559$, $df = 1,2$, $p = .016$).
2. A nonstatistically significant nested effect for interventionist (i.e., clinician) ($F_{\text{interventionist}} = .952$, $df = 2, 20$, $p = .403$).

Overall effect size as measured by $\hat{\omega}_A^2$ was .70 with high observed power (.948). The partial effect for the effect of intervention was also large (partial $\hat{\omega}_{A,partial}^2 = .70$) but with lower power (.192). The results of this study provide evidence to suggest that quality of life is significantly higher for hospice patients who received music therapy as the intervention (M = 10.833, SE = .704) as compared to massage therapy (M = 3.333, SE = .704). The results also suggest that mean scores for quality of life are comparable for hospice patients regardless of the clinician who provided the intervention.

6.8 Additional Resources

This chapter has provided a preview into conducting hierarchical (nested) and randomized block ANOVA. However, there are a number of areas that space limitations prevent us from delving into. For those of you who are interested in learning more about ANOVA models, or if you find yourself in a sticky situation in your analyses, you may wish to look into the following, among many other excellent resources.

- For more in-depth coverage of ANOVA models, see Maxwell et al. (2018), Kirk (2014), and Keppel and Wickens (2004), among others.
- To learn more about multilevel models, in general, see Raudenbush and Bryk (2002), Hox et al. (2017), and Snijders and Bosker (2012), among many other excellent sources.

Problems

Conceptual Problems

1. A researcher wants to know if the number of professional development courses that a teacher completes differs based on the format that the professional development is offered (online, mixed mode, face-to-face). The researcher randomly samples 100 teachers employed in the district. Believing that years of teaching experience may be a concomitant variable, the researcher ranks the teachers on years of experience and places them in categories that represent five-year intervals. The researcher then randomly selects four years of experience blocks. The teachers within those blocks are then randomly assigned to professional development format. Which of the following methods of blocking is employed here?

a. Predefined value blocking
b. Predefined range blocking
c. Sampled value blocking
d. Sampled range blocking

2. To study the effectiveness of three spelling methods, 45 subjects are randomly selected from the fourth graders in a particular elementary school. Based on the order of their IQ scores, subjects are grouped into IQ groups (low = 75–99, average = 100–115, high = 116–130), 15 in each group. Subjects in each group are randomly assigned to one of the three methods of spelling, five each. Which of the following methods of blocking is employed here?
   a. Predefined value blocking
   b. Predefined range blocking
   c. Sampled value blocking
   d. Sampled range blocking
3. A researcher is examining preschoolers' knowledge of number identification. Fifty preschoolers are grouped based on socioeconomic status (low, moderate, high). Within each SES group, students are randomly assigned to one of two treatment groups: one which incorporates numbers through individual, small group, and whole group work with manipulatives, music, and art; and a second which incorporates numbers through whole group study only. Which of the following methods of blocking is employed here?
   a. Predefined value blocking
   b. Predefined range blocking
   c. Sampled value blocking
   d. Sampled range blocking
4. If three teachers employ method A and three other teachers employ method B, then which one of the following is suggested?
   a. Teachers are nested within method.
   b. Teachers are crossed with methods.
   c. Methods are nested within teacher.
   d. Cannot be determined.
5. The interaction of factors A and B can be assessed only if which one of the following occurs?
   a. Both factors are fixed.
   b. Both factors are random.
   c. Factor A is nested within factor B.
   d. Factors A and B are crossed.
6. In a two-factor design, factor A is nested within factor B for which one of the following?
   a. At each level of A, each level of B appears.
   b. At each level of A, unique levels of B appear.
   c. At each level of B, unique levels of A appear.
   d. Cannot be determined.

7. Five teachers use an experimental method of teaching statistics, and five other teachers use the traditional method. If factor M is method of teaching, and factor T is teacher, this design can be denoted by which one of the following?
   a. T(M)
   b. T x M
   c. M x T
   d. M(T)
8. True or false? If factor C is nested within factors A and B, this is denoted as AB(C).
9. True or false? A design in which all levels of each factor are found in combination with each level of every other factor is necessarily a nested design.
10. True or false? To determine if counseling method E is uniformly superior to method C for the population of counselors, from which random samples are taken to conduct a study, one needs a nested design with a mixed model.
11. I assert that the predefined value method of block formation is more effective than the sampled value method in reducing unexplained variability. Am I correct?
12. For the interaction to be tested in a two-factor randomized block design, it is required that which one of the following occurs?
   a. Both factors be fixed
   b. Both factors be random
   c. $n = 1$
   d. $n > 1$
13. Five medical professors use a computer-based method of teaching, and five other medical professors use a lecture-based method of teaching. A researcher is interested in student outcomes for those enrolled in classes taught by these instructional methods. This is an example of which type of design?
   a. Completely crossed design
   b. Repeated measures design
   c. Hierarchical design
   d. Randomized block design
14. In a randomized block study, the correlation between the blocking factor and the dependent variable is .35. I assert that the residual variation will be smaller when using the blocking variable than without. Am I correct?
15. A researcher is interested in examining the number of suspensions of high school students based on random assignment participation in a series of self-awareness workshops. The researcher believes that age may be a concomitant variable. Applying a two-factor randomized block ANOVA design to the data, is age an appropriate blocking factor?
16. In a two-factor hierarchical design with two levels of factor A and three levels of factor B nested within each level of A, how many F ratios can be tested?
   a. 1
   b. 2
   c. 3
   d. Cannot be determined

17. If the correlation between the concomitant variable and dependent variable is −.80, which of the following designs is recommended?
    a. ANCOVA
    b. One-factor ANOVA
    c. Randomized block ANOVA
    d. All of the above
18. True or false? IQ must be used as a treatment factor.
19. Which of the following blocking methods best estimates the treatment effects?
    a. Predefined value blocking
    b. Post hoc predefined value blocking
    c. Sampled value blocking
    d. Sampled range blocking
20. The assumption of normality for the two-factor hierarchical ANOVA is concerned with which of the following?
    a. Dependent variable
    b. Independent variable
    c. Nested factor
    d. Residual
21. True or false? The assumption of normality for the two-factor hierarchical ANOVA differs from the two-factor crossed model.
22. The assumptions for the two-factor randomized block ANOVA model are nearly identical to which one of the following?
    a. Dependent *t* test
    b. Factorial ANOVA
    c. One-factor repeated measures ANOVA
    d. Two-factor hierarchical ANOVA
23. Why is the assumption of compound symmetry with two-factor randomized block ANOVA necessary?
    a. Observations within a block are not independent.
    b. The distribution of residuals cannot be assumed normal.
    c. The means of the levels of the blocking factor are equal.
    d. The population covariances for all pairs of the dependent variable are equal.
24. An interaction between the treatment and blocking factors in a two-factor randomized block ANOVA results in which one of the following?
    a. Compound symmetry
    b. Multicollinearity
    c. Rejection of the null hypothesis
    d. Violation of the additivity assumption
25. In a two-factor randomized block ANOVA, a multiple comparison procedure is needed in which of the following situations?
    a. When the null hypothesis for the treatment is rejected and it has more than two levels.

b. When the null hypothesis for the blocking factor is rejected and it has more than two levels.
c. Both a and b only.
d. Either a or b.

Answers to Conceptual Problems

1. **d** (teachers are ranked according to a ratio blocking variable; a random sample of blocks are drawn; then teachers within the blocks are assigned to treatment)
3. **a** (children are randomly assigned to treatment based on ordinal SES value)
5. **d** (interactions occur only among factors that are crossed)
7. **a** (this is the notation for teachers nested within methods; see also Problem 2)
9. **False** (cannot be a nested design; must be a crossed design)
11. **Yes** (see the discussion on the types of blocking)
13. **c** (physician is nested within method)
15. **Yes** (age is an appropriate blocking factor here)
17. **a** (use of a covariate is best for large correlations)
19. **a** (see the summary of the blocking methods)
21. **False** (assumptions for the two-factor nested model and assumptions for the two-factor crossed model are the same)
23. **a** (the assumption of compound symmetry with two-factor randomized block ANOVA is needed because the observations within a block are not independent)
25. **d** (a multiple comparison procedure is needed for two-factor randomized block ANOVA when either or both of the following occur: the null hypothesis for the treatment is rejected and it has more than two levels; *or* when the null hypothesis for the blocking factor is rejected and it has more than two levels)

Computational Problems

1. An experiment was conducted to compare three types of behavior modification (1, 2, and 3) using age as a blocking variable (4-, 6-, and 8-year-old children). The mean scores on the dependent variable, number of instances of disruptive behavior, are listed here for each cell. The intention of the treatments is to minimize the number of disruptions.

Type of Behavior Modification	Age		
	4 years	6 years	8 years
1	20	40	40
2	50	30	20
3	50	40	30

Use these cell means to graph the interaction between type of behavior modification and age.

a. Is there an interaction between type of behavior modification and age?

b. What kind of recommendation would you make to teachers?

2. An experiment was conducted to compare four different preschool curricula that were adopted in four different classrooms. Reading readiness proficiency was used as a blocking variable (below proficient, at proficient, above proficient). The mean scores on the dependent variable, letter recognition, are listed here for each cell. The intention of the treatment (i.e., the curriculum) is to increase letter recognition.

	Reading Readiness Proficiency		
Curriculum	**Below**	**At**	**Above**
1	12	20	22
2	20	24	18
3	16	16	20
4	15	18	25

Use these cell means to graph the interaction between curriculum and reading readiness proficiency.

a. Is there an interaction between type of curriculum and reading readiness proficiency?

b. What kind of recommendation would you make to teachers?

3. An experimenter tested three sales pitches (subtle, moderate, pushy) on morning versus afternoon shoppers. Thus, shopping time of day (morning or afternoon) is a blocking variable. The dependent measure was the number of sales during a 2-week period. There were five subjects in each cell. Complete the ANOVA summary table below, assuming a fixed-effects model, where $\alpha = .50$.

Source	*SS*	*df*	*MS*	*F*	Critical Value	Decision
Sales pitch (A)	200	–	–	–	–	–
Time of day (B)	100	–	–	–	–	–
Interaction (AB)	20	–	–	–	–	–
Within	240	–	–			
Total	–	–				

4. An experiment was conducted to determine if there was a mean difference in weight for women based on type of aerobics exercise program participated (low impact vs. high impact). Body mass index (BMI) was used as a blocking variable to represent below, at, or above recommended BMI. The data are shown below. Conduct a two-factor randomized block ANOVA ($\alpha = .50$) and Bonferroni MCPs using SPSS to determine the results of the study.

Subject	Exercise Program	BMI	Weight
1	1	1	100
2	1	2	135
3	1	3	200
4	1	1	95
5	1	2	140
6	1	3	180
7	2	1	120
8	2	2	152
9	2	3	176
10	2	1	128
11	2	2	142
12	2	3	220

5. A mathematics professor wants to know which of three approaches to teaching calculus resulted in the best test performance (section 1, 2, or 3). Scores on a placement exam were used as a blocking variable (block 1: 200–400; block 2: 401–600; block 3: 601–800). The data are shown below. Conduct a two-factor randomized block ANOVA ($\alpha = .50$) and Bonferroni MCPs using SPSS to determine the results of the study.

Subject	Section	Placement Exam	Test Score
1	1	1	90
2	1	2	93
3	1	3	100
4	2	1	88
5	2	2	90
6	2	3	97
7	3	1	79
8	3	2	85
9	3	3	92

6. A restaurant owner (who owns multiple franchise locations) wants to know which of three recipes for a signature dish (mild, medium, spicy) resulted in the best sales, blocking on section of town in which the restaurant is located (section 1, 2, or 3). The data are shown below. Conduct a two-factor randomized block ANOVA ($a = .50$) and Bonferroni MCPs using SPSS to determine the results of the study.

Subject	Section of Town	Recipe	Sales
1	1	1	90
2	1	2	93
3	1	3	100

4	2	1	88
5	2	2	90
6	2	3	97
7	3	1	79
8	3	2	85
9	3	3	92

7. A restaurant owner wants to know which of three recipes for a signature dish (mild =1, medium = 2, spicy = 3) resulted in the best sales, with recipe nested within chef (chef 1, 2, or 3). The data are shown below. Conduct a two-factor hierarchical ANOVA ($\alpha = .50$) and Bonferroni MCPs using SPSS to determine the results of the study.

Chef	Recipe	Sales
1.00	1.00	45.00
1.00	2.00	52.00
1.00	3.00	59.00
2.00	1.00	38.00
2.00	2.00	41.00
2.00	3.00	52.00
3.00	1.00	40.00
3.00	2.00	50.00
3.00	3.00	62.00
1.00	1.00	45.00
1.00	2.00	48.00
1.00	3.00	60.00
2.00	1.00	38.00
2.00	2.00	41.00
2.00	3.00	50.00
3.00	1.00	43.00
3.00	2.00	55.00
3.00	3.00	65.00

Answers to Computational Problems

1. a. Yes

   b. At age 4, type 1 is most effective; at age 6, type 2 is most effective; and at age 8, type 2 is most effective.

3. $SS_{total} = 560$, $df_A = 2$, $df_B = 1$, $df_{AB} = 2$, $df_{with} = 24$, $df_{total} = 29$, $MS_A = 100$, $MS_B = 100$, $MS_{AB} = 10$, $MS_{with} = 10$, $F_A = 10$, $F_B = 10$, $F_{AB} = 1$, critical value for B = 4.26 (reject H_0 for B), critical value for A and AB = 3.40 (reject H_0 for A and fail to reject H_0 for AB).

5. $F_{section} = 44.385$, $p = .002$; $F_{placement} = 61.000$, $p = .001$; thus reject H_0 for both effects. Bonferroni results: all but sections 1 and 2 are different, and all blocks are statistically different.

7. $F_{recipe} = 6.961$, $p < .001$; thus reject H_0 for the main effect of recipe. Bonferroni results: all flavors (i.e., mild, medium, and spicy) of recipes are statistically different.

Interpretive Problems

1. The following is the first one-factor ANOVA interpretive problem you developed in Chapter 1: *Using the survey1 dataset, which is accessible from the website, use SPSS or **R** to conduct a one-factor fixed-effects ANOVA, where political view is the grouping variable (i.e., independent variable) ($J = 5$) and the dependent variable is an interval or ratio variable of your choice. Also compute effect size and test for assumptions. Then write an APA-style paragraph describing the results.*

   Take the one-factor ANOVA interpretive problem you developed in Chapter 1. What are some reasonable blocking variables to consider? Which type of blocking would be best in your situation? Select this blocking variable from the same dataset and conduct a two-factor randomized block ANOVA. Compare these results with the one-factor ANOVA results (without the blocking factor) to determine how useful the blocking variable was in terms of reducing residual variability.

2. The following is the second one-factor ANOVA interpretive problem you developed in Chapter 1: *Using the survey1 dataset, which is accessible from the website, use SPSS or **R** to conduct a one-factor fixed-effects ANOVA, where hair color is the grouping variable (i.e., independent variable) ($J = 5$) and the dependent variable is an interval or ratio variable of your choice. Also compute effect size and test for assumptions. Then write an APA-style paragraph describing the results.*

   Take this one-factor ANOVA interpretive problem you developed in Chapter 1. What are some reasonable blocking variables to consider? Which type of blocking would be best in your situation? Select this blocking variable from the same dataset and conduct a two-factor randomized ANOVA. Compare these results with the one-factor ANOVA results (without the blocking factor) to determine how useful the blocking variable was in terms of reducing residual variability.

3. The following is the third one-factor ANOVA interpretive problem you developed in Chapter 1: *Using the IPEDS2017 dataset, which is accessible from the website, use SPSS or **R** to conduct a one-factor fixed-effects ANOVA. Select an appropriate independent variable (e.g., land grant institution, LANDGRNT) and appropriate dependent variable (e.g., total dormitory capacity, ROOMCAP). Also compute effect size and test for assumptions. Then write an APA-style paragraph describing the results.*

   Take this one-factor ANOVA interpretive problem you developed in Chapter 1. What are some reasonable blocking variables to consider? Which type of blocking would be best in your situation? Select this blocking variable from the same dataset and conduct a two-factor randomized ANOVA. Compare these results with the one-factor ANOVA results (without the blocking factor) to determine how useful the blocking variable was in terms of reducing residual variability.

7

Simple Linear Regression

Chapter Outline

Key Concepts

1. Slope and intercept of a straight line
2. Regression model
3. Prediction errors/residuals
4. Standardized and unstandardized regression coefficients
5. Proportion of variation accounted for; coefficient of determination

In one or more previous statistics classes that you've had, you've likely considered various bivariate measures of association. Specifically, topics of scatterplot, covariance, types of correlation coefficients, and their resulting inferential tests would have been covered. Bivariate measures of association are concerned with addressing the question of the extent to which two variables are associated or related. In this chapter we extend our discussion of two variables to address the question of the extent to which one variable can be used to predict or explain another variable.

Beginning in Chapter 1 we examined various analysis of variance (ANOVA) models. It should be mentioned again that ANOVA and regression are both forms of the same general linear model (GLM), where the relationship between one or more independent variables and one dependent variable is evaluated. The major difference between the two procedures is that in ANOVA, the independent variables are discrete variables (i.e., nominal or ordinal), while in regression, the independent variables are continuous variables (i.e., interval or ratio; however, we will see later how we can apply dichotomous variables in regression models). Otherwise there is considerable overlap of these two procedures in terms of concepts and their implementation. Note that a continuous variable can be transformed into a discrete variable. For example, the GRE-Quantitative exam is a continuous variable scaled from 130 to 170. It could be made into a discrete variable, such as low (130–139), average (140–159), and high (160–170).

When considering the relationship between two variables (say X and Y), the researcher usually determines some measure of relationship between those variables, such as a correlation coefficient (e.g., r_{XY}, the Pearson product-moment correlation coefficient). Another way of looking at how two variables may be related is through regression analysis, in terms of prediction or explanation. That is, we evaluate the ability of one variable to predict or explain a second variable. Here we adopt the usual notation where X is defined as the **independent** or **predictor variable**, and Y as the **dependent** or **criterion variable**.

For example, an admissions officer might want to use a placement exam score to predict graduate-level grade point averages (GPA) to make admissions decisions for a sample of applicants to a university or college. The research question of interest is how well does the placement exam (the independent or predictor variable) predict or explain performance in graduate school (the dependent or criterion variable)? This is an example of simple linear regression where only a single predictor variable is included in the analysis. The utility of the placement exam in predicting GPA requires that these variables have a correlation different from zero. Otherwise the placement exam will not be very useful in predicting GPA. For education and the behavioral sciences, the use of a single predictor does not usually result in reasonable prediction or explanation. Thus, Chapter 8 considers the case of multiple predictor variables through multiple linear regression analysis.

In this chapter, we consider the concepts of slope, intercept, regression model, unstandardized and standardized regression coefficients, residuals, proportion of variation accounted for, tests of significance, and statistical assumptions. Our objectives are that by the end of this chapter, you will be able to (a) understand the concepts underlying simple linear regression, (b) determine and interpret the results of simple linear regression, and (c) understand and evaluate the assumptions of simple linear regression.

7.1 What Simple Linear Regression Is and How It Works

In this chapter, we find Ott Lier stretching his statistical skills.

Ott Lier, along with the additional graduate research assistants working in the statistics and research lab, has continued to expand his palette of statistical skills and has been brought into a project with the Human Resources Department of a large employer in their area. Ott will be working with Dr. Randall, the director of Human Resources. Dr. Randall wants to know if work optimism can be used to predict employment success. If this is possible, Dr. Randall anticipates changes to their onboarding and training of employees to hopefully increase employment success. Ott suggests the following research question to Dr. Randall: *Can employment success be predicted from work optimism?* Ott determines that a simple linear regression is the best statistical procedure to use to answer Dr. Randall's question. His next task is to assist Dr. Randall in analyzing the data.

Let us consider the basic concepts involved in simple linear regression. Many years ago when you had algebra, you learned about an equation used to describe a straight line:

$$Y = bX + a$$

Here the predictor variable X is used to predict the criterion variable Y. The **slope** of the line is denoted by b and indicates the number of Y units the line changes for a one-unit change in X. You may find it easier to think about the slope as measuring tilt or steepness. The Y-intercept is denoted by a and is the point at which the line intersects or crosses the Y axis. To be more specific, a is the value of Y when X is equal to zero. Hereafter we use the term **intercept** rather than Y-intercept to keep it simple.

Consider the plot of the straight line $Y = 0.5X + 1.0$ as shown in Figure 7.1. Here we see that the line clearly intersects the Y axis at $Y = 1.0$; thus the intercept is equal to one. The

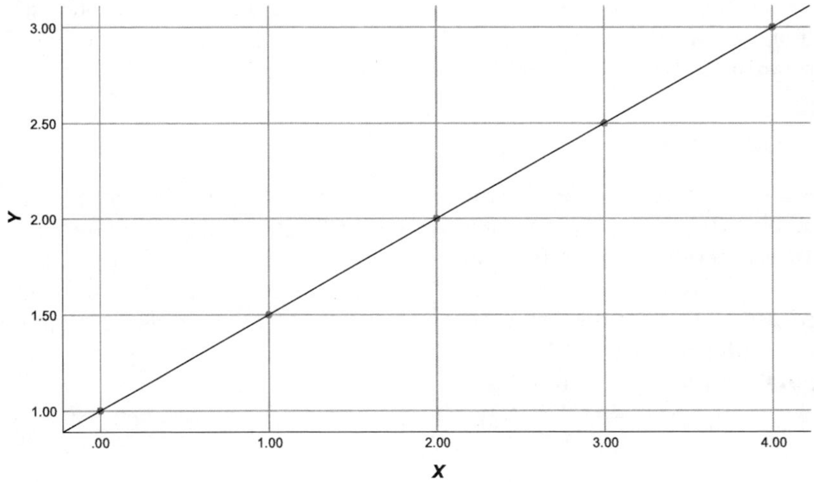

FIGURE 7.1
Plot of line: $Y = 0.5\,X + 1.0$.

slope of a line is defined, more specifically, as the change in Y (numerator) divided by the change in X (denominator).

$$b = \frac{\Delta Y}{\Delta X} = \frac{Y_2 - Y_1}{X_2 - X_1}$$

For instance, take two points shown in Figure 7.1, (X_1, Y_1) and (X_2, Y_2), that fall on the straight line with coordinates (0, 1) and (4, 3), respectively. We compute the slope for those two points to be $(3 - 1)/(4 - 0) = 0.5$. If we were to select any other two points that fall on the straight line, then the slope for those two points would also be equal to 0.5. That is, regardless of the two points on the line that we select, the slope will always be the same, constant value of 0.5. This is true because we only need two points to define a particular straight line. That is, with the points (0, 1) and (4, 3) we can draw only one straight line that passes through both of those points, and that line has a slope of 0.5 and an intercept of 1.0.

Let us take the concepts of slope, intercept, and straight line and apply them in the context of correlation so that we can study the relationship between the variables X and Y. If the slope of the line is a positive value (e.g., Figure 7.1), as X increases, Y also increases, then the correlation will be positive. If the slope of the line is zero, such that the line is parallel or horizontal to the X axis, as X increases Y remains constant, then the correlation will be zero. If the slope of the line is a negative value, as X increases Y decreases (i.e., the line decreases from left to right), then the correlation will be negative. Thus the sign of the slope corresponds to the sign of the correlation.

7.1.1 Characteristics

7.1.1.1 The Population Simple Linear Regression Model

Let us take these concepts and apply them to simple linear regression. Consider the situation where we have the entire population of individual's scores on both variables X (the independent variable, such as work optimism) and Y (the dependent variable, such as employment success). We define the linear regression model as the equation for a straight line. This yields an equation for the regression of Y, *the criterion*, given X, *the predictor*, often stated as the **regression of Y on X**, although more easily understood as Y being predicted by X.

The **population regression model** for Y being predicted by X is as follows:

$$Y_i = \beta_{YX} X_i + \alpha_{YX} + \varepsilon_i$$

where Y is the criterion variable, X is the predictor variable, β_{YX} is the population slope for Y predicted by X, α_{YX} is the population intercept for Y predicted by X, ε_i are the population residuals or errors of prediction (the part of Y_i not predicted from X_i), and i represents an index for a particular case (an individual or object; in other words, the unit of analysis that has been measured). The index i can take on values from 1 to N, where N is the size of the population, written as $i = 1, \ldots, N$.

The **population prediction model** is

$$Y'_i = \beta_{YX} X_i + \alpha_{YX}$$

where Y'_i is the predicted value of Y for a specific value of X. That is, Y_i is the *actual or observed score* obtained by individual i, while Y'_i is the *predicted score* based on their X score for that same individual (in other words, you are using the value of X to predict what Y will be). Thus, we see that the population prediction error is defined as follows:

$$\varepsilon_i = Y_i - Y_i'$$

There is only one difference between the regression and prediction models. The regression model explicitly includes prediction error as ε_i, whereas the prediction model includes prediction error implicitly as part of the predicted score Y_i' (i.e., there is some error in the predicted values).

Consider for a moment a practical application of the difference between the regression and prediction models. Frequently a researcher will develop a regression model for a population where X and Y are both known, and then use the prediction model to actually predict Y when only X is known (i.e., Y will not be known until later). Using the employment example, the human resources officer first develops a regression model for a population of employees currently employed at the organization so as to have a current measure of work optimism. This yields the slope and intercept. Then the prediction model is used to predict future employment success and to help make training and onboarding decisions for future populations of incoming employees based on their work optimism.

A simple method for determining the population slope (β_{YX}) and intercept (α_{YX}) is computed as follows:

$$\beta_{YX} = \rho_{XY}\left(\frac{\sigma_Y}{\sigma_X}\right)$$

and

$$\alpha_{YX} = \mu_Y - \beta_{YX}\mu_X$$

where σ_Y and σ_X are the population standard deviations for Y and X respectively, ρ_{XY} is the population correlation between X and Y (simply the Pearson correlation coefficient, rho), and μ_Y and μ_X are the population means for Y and X respectively. Note that the previously used mathematical method for determining the slope and intercept of a straight line is not appropriate in regression analysis with real data.

7.1.1.2 The Sample Simple Linear Regression Model

Our discussion of the sample simple linear regression model begins with coverage of the unstandardized and standardized models. This is followed by prediction errors, least squares criterion, coefficient tests, significance tests, and confidence intervals.

7.1.1.2.1 Unstandardized Regression Model

Let us return to the real world of sample statistics and consider the sample simple linear regression model. As usual, Greek letters refer to population parameters and English letters refer to sample statistics. The **sample regression model** for predicting Y from X is computed as:

$$Y_i = b_{YX}X_i + a_{YX} + e_i$$

where Y and X are as before (i.e., the dependent and independent variables respectively), b_{YX} is the sample slope for Y predicted by X, a_{YX} is the sample intercept for Y predicted by X, e_i are sample residuals or errors of prediction (the part of Y_i not predictable from X_i), and i represents an index for a case (an individual or object). The index i can take on values from 1 to n, where n is the size of the sample, and is written as $i = 1, \ldots, n$.

The **sample prediction model** is computed as follows:

$$Y_i' = b_{YX}X_i + a_{YX}$$

where Y_i' is the predicted value of Y for a specific value of X. We define the sample prediction error as the difference between the *actual score* obtained by individual i (i.e., Yi) and the *predicted score* based on the X score for that individual (i.e., Y_i'). In other words, the residual is that part of Y that is *not* predicted by X. The goal of the prediction model is to include an independent variable X that minimizes the residual; this means that the independent variable does a nice job of predicting the outcome. Computationally, the residual (or error) is computed as:

$$e_i = Y_i - Y_i'$$

The difference between the regression and prediction models is the same as previously discussed, except now we are dealing with a sample rather than a population.

The sample slope (b_{YX}) and intercept (a_{YX}) can be determined by

$$b_{YX} = r_{YX}\left(\frac{s_Y}{s_X}\right)$$

and

$$a_{YX} = \bar{Y} - b_{YX}\bar{X}$$

where s_Y and s_X are the sample standard deviations for Y and X respectively, r_{YX} is the sample correlation between X and Y (again the Pearson correlation coefficient, rho), and $\bar{Y}$ and $\bar{X}$ are the sample means for Y and X, respectively. The **sample slope (b_{YX})** is referred to alternately as (a) the expected or predicted change in Y for a one-unit change in X, and (b) the unstandardized or raw regression coefficient. The **sample intercept (a_{YX})** is referred to alternately as (a) the point at which the regression line intersects (or crosses) the Y axis, and (b) the value of Y when X is zero.

Consider now the analysis of a realistic example to be followed throughout this chapter. Let us use work optimism (a continuous score) to predict employment success (also a continuous score). The work optimism scale has a possible range of 20 to 80 points, and the employment success scale has a possible range of 0 to 50 points. Given the sample of 10 employees shown in Table 7.1, let us work through a simple linear regression analysis. The observation numbers ($i = 1, \ldots, 10$), and values for the work optimism score (the independent variable, X) and employment success core (the dependent variable, Y) variables are given in the first three columns of the table, respectively. The other columns are discussed as we go along.

TABLE 7.1

Employment example regression data.

Employee	Work Optimism (X)	Employment Success (Y)	Residual (e)	Predicted Employment Success (Y′)
1	37	32	3.7125	28.2875
2	45	36	3.5125	32.4875
3	43	27	−4.4375	31.4375
4	50	34	−1.1125	35.1125
5	65	45	2.0125	42.9875
6	72	49	2.3375	46.6625
7	61	42	1.1125	40.8875
8	57	38	−0.7875	38.7875
9	48	30	−4.0625	34.0625
10	77	47	−2.2875	49.2875

The sample statistics for the work optimism score (the independent variable) are $\bar{X} = 55.5$ and $s_X = 13.1339$, for the employment success core (the dependent variable) are $\bar{Y} = 38$ and $s_Y = 7.5130$, and the correlation r_{YX} is 0.9177. The sample slope (b_{YX}) and intercept (a_{YX}) are computed as follows:

$$b_{YX} = r_{YX}\left(\frac{s_Y}{s_X}\right) = 0.9177\left(\frac{7.5130}{13.1339}\right) = 0.5250$$

and

$$a_{YX} = \bar{Y} - b_{YX}\bar{X} = 38 - 0.5250(55.5) = 8.8625$$

Let us interpret the slope and intercept values. A **slope** of 0.5250 means that if your score on work optimism is increased by one point, then your predicted score on employment success (i.e., the dependent variable) will be increased by 0.5250 points or about one-half of one point. An **intercept** of 8.8625 means that if your score on work optimism is zero, then your score on employment success is 8.8625. The sample simple linear regression model, given these values, becomes

$$Y_i = b_{YX}X_i + a_{YX} + e_i = 0.5250X_i + 8.8625 + e_i$$

If your score on work optimism is 63, then your **predicted score** on employment success is the following:

$$Y_i' = 0.5250(63) + 8.8625 = 41.9375$$

Thus, based on the prediction model developed, your predicted score on employment success is approximately 42; however, as becomes evident, predictions are generally not perfect.

7.1.1.2.2 Standardized Regression Model

Up until now, the computations in simple linear regression have involved the use of raw scores. For this reason, we call this the *unstandardized regression model*. The slope estimate is an unstandardized or raw regression slope because it is the predicted change in Y raw score units for a one raw score unit change in X. We can also express regression in **standard z score units** for both X and Y as follows:

$$z(X_i) = \frac{X_i - \bar{X}}{s_X}$$

and

$$z(Y_i) = \frac{Y_i - \bar{Y}}{s_Y}$$

In both cases, the numerator is the difference between the observed score and the mean, and the denominator is the standard deviation (and dividing by the standard deviation standardizes the value). The means and variances of both standardized variables (i.e., z_X and z_Y) are 0 and 1, respectively.

The **sample standardized linear prediction model** becomes the following where $z(Y_i')$ is the standardized predicted value of Y:

$$z(Y_i') = (b_{YX}^*)(z(X_i)) = (r_{YX})(z(X_i))$$

Thus the **standardized regression slope, b_{YX}^***, sometimes referred to as a **beta weight**, is equal to r_{XY}, i.e., the simple bivariate correlation between X and Y. No intercept term is necessary in the prediction model as the mean of the z scores for both X and Y is zero (i.e., $a_{YX}^* = \bar{z}_Y - b_{YX}^* \bar{z}_X = 0$). In summary, *the standardized slope is equal to the correlation coefficient* and *the standardized intercept is equal to zero*.

For our employment example, the sample standardized linear prediction model is

$$z(Y_i') = (.9177)(z(X_i))$$

The slope of .9177 would be interpreted as the expected increase in employment success in z score (i.e., standardized score) units for a one z score (i.e., standardized score) unit increase in the work optimism score. A one z score unit increase is also the same as a one standard deviation increase because the standard deviation of z is equal to one (recall that the mean of a standardized z score is 0 with a standard deviation of 1).

When should you consider use of the standardized versus unstandardized regression analyses? According to Pedhazur (1997), the standardized regression slope b^* is not very stable from sample to sample. For example, at Organization Q, the standardized regression slope b^* would vary across different employee types (or samples), whereas the unstandardized regression slope b would be much more consistent across employee types. Thus, in simple regression most researchers prefer the use of b. We see later that the standardized regression slope b^* has some utility in multiple regression analysis.

7.1.1.2.3 Prediction Errors

Previously we mentioned that perfect prediction of Y from X is extremely unlikely, only occurring with a perfect correlation between X and Y (i.e., $r_{YX'}$ also noted as $r_{XY'} = \pm 1.0$). When developing the regression model, the values of the outcome, Y, are known. Once the slope and intercept have been estimated, we can then use the prediction model to predict the outcome (Y) from the independent variable (X) when the values of Y are *unknown*. We have already defined the predicted values of Y as Y'. In other words, *a predicted value Y' can be computed by plugging the obtained value for X into the prediction model.* It can be shown that $Y_i = Y_i'$ for all i only when there is perfect prediction. However, this is extremely unlikely in reality, particularly in simple linear regression using a single predictor.

We can determine a value of Y' for each of the i cases (individuals or objects) from the prediction model. In comparing the actual Y values to the predicted Y values, we obtain the **residuals** as the difference between the observed (Y_i) and predicted values (Y_i'), computed as follows:

$$e_i = Y_i - Y_i'$$

for all $i = 1, \ldots, n$ individuals or objects in the sample. The residuals, e_i, are also known as **errors of estimate**, or **prediction errors**, and are that portion of Y_i that is not predictable from X_i. The residual terms are random values that are unique to each individual or object.

The residuals and predicted values for the employment example are shown in the last two columns of Table 7.1, respectively. Consider observation 2, where the observed work optimism score is 45 and the observed employment success core is 36. The predicted employment success score is 32.4875 and the residual is +3.5125. This indicates that person 2 had a higher observed employment success score than was predicted using the work optimism score as a predictor. We see that a **positive residual** indicates the observed criterion score is larger than the predicted criterion score, whereas a **negative residual** (such as in observation 3) indicates the observed criterion score is smaller than the predicted criterion score. For observation 3, the observed work optimism score is 43, the observed employment success score is 27, the predicted employment success score is 31.4375, and thus the residual is −4.4375. Person 2 scored higher on employment success than we predicted, and person 3 scored lower on employment success than we predicted.

The regression example is shown graphically in the **scatterplot** of Figure 7.2, where the straight diagonal line represents the regression line. *Individuals falling above the regression line have positive residuals* (e.g., observation 1) (in other words, the difference between the observed score, represented as a dot, is greater in value than the predicted value, which is represented by the regression line) *and individuals falling below the regression line have negative residuals* (e.g., observation 3) (in other words, the difference between the observed score, represented as a dot, is less in value than the predicted value, which is represented by the regression line). The residual is, very simply, *the vertical distance between the observed score (represented by the "dots" in the scatterplot (Figure 7.2) and the regression line.* In the residual column of Table 7.1 we see that one-half of the residuals are positive and one-half are negative, and in Figure 7.2 that one-half of the points fall above the regression line and one-half below the regression line. It can be shown that the mean of the residuals is always zero (i.e., $\bar{e} = 0$), as the sum of the residuals is always zero. This results from the fact that the mean of the observed criterion scores is equal to the mean of the predicted criterion scores (i.e., $\bar{Y} = \bar{Y}'$; 38 for the example data).

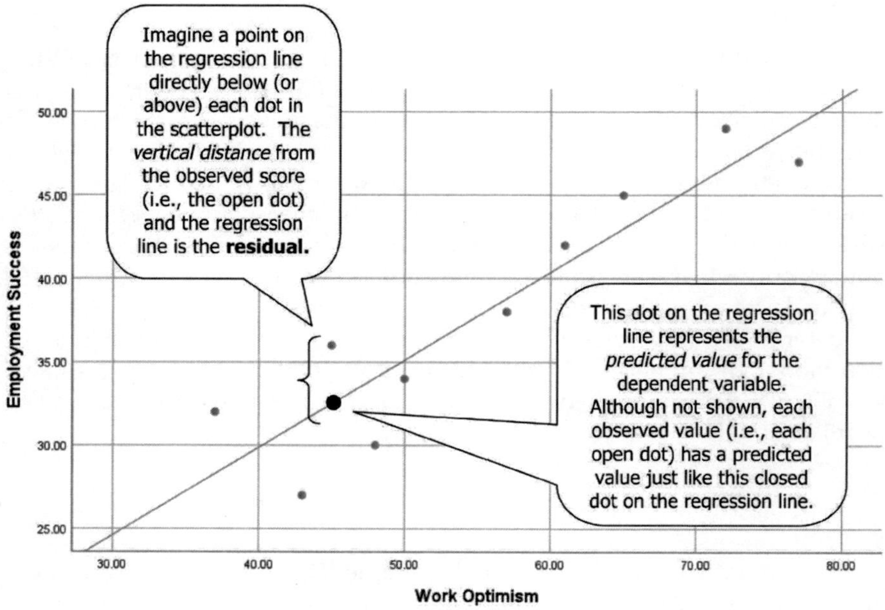

FIGURE 7.2
Scatterplot for employment example.

7.1.1.2.4 Least Squares Criterion

How was one particular method selected for determining of the slope and intercept? Obviously, some standard procedure has to be used. Thus, there are statistical criteria that help us decide which method to use in determining the slope and intercept. The criterion usually used in linear regression analysis (and in all general linear models, for that matter) is the **least squares criterion**. According to the least squares criterion, *the sum of the squared prediction errors or residuals is smallest*. That is, we want to find that regression line, defined by a particular slope and intercept, which results in the smallest sum of the squared residuals (recall that the residual is the difference between the observed and predicted values for the outcome). Since the residual is the vertical difference between the observed and predicted value, the regression line is simply the line that minimizes that vertical distance. Given the value that we place on the accuracy of prediction, this is the most logical choice of a method for estimating the slope and intercept.

In summary then, the least squares criterion gives us a particular slope and intercept, and thus a particular regression line, such that the sum of the squared residuals is smallest. We often refer to this particular method for determining the slope and intercept as **least squares estimation** or ordinary least squares (OLS), because b and a represent sample estimates of the population parameters b and α obtained using the least squares criterion.

7.1.1.2.5 Proportion of Predictable Variation (Coefficient of Determination)

How well is the criterion variable Y predicted by the predictor variable X? For our example, we want to know how well employment success scores are predicted by work optimism scores. Let us consider two possible situations with respect to this example. First, if the work optimism score is found to be a really good predictor of employment success scores, then instructors could use the work optimism score to individualize onboarding and training based on work optimism of the employee. They could, for example, provide enhanced onboarding to employees with low work optimism scores, or in general, adjust the level of training to fit the optimism of their employees. Second, if work optimism is not found to be a very good predictor of employment success scores, then human resources representatives would not find very much use for the work optimism score in terms of their preparation for employment. They could search for some other more useful predictor, such as employee engagement or career readiness. In other words, if a predictor is not found to be particularly useful in predicting the criterion variable, then other relevant predictors should be considered.

How do we determine the utility of a predictor variable? The simplest method involves partitioning the total sum of squares in Y, which we denote as SS_{total} (sometimes written as SS_Y). This process is much like partitioning the sum of squares in the analysis of variance.

In simple linear regression, we can partition SS_{total} as follows:

$$SS_{total} = SS_{reg} + SS_{res}$$

$$\sum_{i=1}^{n}\left(Y-\bar{Y}\right)^2 = \sum_{i=1}^{n}\left(Y'-\bar{Y}\right)^2 + \sum_{i=1}^{n}\left(Y-Y'\right)^2$$

where SS_{total} is the total sum of squares in Y, SS_{reg} is the sum of squares of the regression of Y predicted by X (sometimes written as $SS_{Y'}$) (and represented in the equation as $\sum_{(i=1)}^{n}\left(Y'-\bar{Y}\right)^2$), SS_{res} is the sum of squares of the residuals (and represented in the equation as $\sum_{i=1}^{n}\left(Y-Y'\right)^2$), and the sums are taken over all observations from $i = 1, \ldots, n$. Thus, SS_{total} represents the total variation in the observed Y scores, SS_{reg} the variation in Y predicted by X, and SS_{res} the variation in Y not predicted by X.

The equation for SS_{reg} uses information about the difference between the predicted value of Y and the mean of Y: $\sum_{(i=1)}^{n}\left(Y'-\bar{Y}\right)^2$. Thus, the SS_{reg} is essentially examining how much better the line of best fit (i.e., the predicted value of Y) is as compared to the mean of Y (recall that a slope of zero is a horizontal line, which is the mean of Y). The equation for SS_{res} uses information about the difference between the observed value of Y and the predicted value of Y: $\sum_{i=1}^{n}\left(Y-Y'\right)^2$. Thus, the SS_{res} *is providing an indication of how "off" or inaccurate the model is.* The closer SS_{res} is to zero, the better the model fit (as more variability of the dependent variable is being explained by the model; in other words, the independent variables are doing a good job of prediction when the SS_{res} is smaller). Since $r_{XY}^2 = \frac{SS_{reg}}{SS_{total}}$, we can write SS_{total}, SS_{reg}, and SS_{res} as follows:

$$SS_{total} = \frac{n\sum_{i=1}^{n}Y^2 - \left(\sum_{i=1}^{n}Y\right)^2}{n}$$

$$SS_{reg} = \left(r_{XY}^2\right)\left(SS_{total}\right)$$

$$SS_{res} = \left(1 - r_{XY}^2\right)\left(SS_{total}\right)$$

where r_{XY}^2 is the *squared sample correlation* between X and Y **(which, as we know, is the same as the squared sample correlation between Y and X, r_{YX}^2)**, commonly referred to as the **coefficient of determination**. The coefficient of determination in simple linear regression is not only the squared simple bivariate Pearson correlation between X and Y, but also $r_{XY}^2 = \frac{SS_{reg}}{SS_{total}}$ which tells us that it is the proportion of the total variation of the dependent variable (i.e., the denominator) that has been explained by the regression model (i.e., the numerator). Thus, the coefficient of determination can be used both as a measure of **effect size** (described in a later section) and as a **test of significance** (described in the next section).

With the sample data of predicting employment success scores from work optimism, let us determine the sums of squares. We can write SS_{total} as follows:

$$SS_{total} = \frac{n\sum_{i=1}^{n} Y^2 - \left(\sum_{i=1}^{n} Y\right)^2}{n} = \frac{10(14,948) - (380^2)}{10} = 508.00$$

We already know that $r_{XY} = .9177$, so by squaring it, we obtain $r_{XY}^2 = .8422$. Next, we can determine SS_{reg} and SS_{res} as follows:

$$SS_{reg} = (r_{XY}^2)(SS_{total}) = (.8422)(508.00) = 427.8376$$

$$SS_{res} = (1 - r_{XY}^2)(SS_{total}) = (1 - .8422)(508.00) = 80.1624$$

Given the squared correlation between X and Y ($r_{XY}^2 = .8422$), work optimism predicts approximately 84% of the variation in employment success, which is clearly a large effect size. Significance tests are discussed in the next section.

7.1.1.2.6 Significance Tests and Confidence Intervals

This section describes four procedures used in the simple linear regression context. The first two are tests of statistical significance that generally involve testing whether or not X is a significant predictor of Y. Then we consider two confidence interval techniques.

Test of Significance of r_{XY}^2. The first test is the test of the significance of r_{XY}^2 (*alternatively known as the test of the proportion of variation in Y predicted or explained by* X). It is important that r_{XY}^2 be different from zero in order to have reasonable prediction. The null and alternative hypotheses, respectively, are as follows where the null indicates that the correlation between X and Y will be zero:

$$H_0: \rho_{XY}^2 = 0$$

$$H_1: \rho_{XY}^2 > 0$$

This test is based on the following test statistic:

$$F = \frac{r^2/m}{(1 - r^2)/(n - m - 1)}$$

where F indicates that this is an F statistic, r_2 is the coefficient of determination, $1 - r^2$ is the proportion of variation in Y that is not predicted by X, m is the number of predictors (which in the case of simple linear regression is always 1), and n is the sample size. The F test statistic is compared to the F critical value, always a one-tailed test (given that a squared value cannot be negative) and at the designated level of significance α, with degrees of freedom equal to m (i.e., the number of independent variables) and $(n - m - 1)$, as taken from the F table in Appendix Table A.4. That is, the tabled critical value is ${}_{\alpha}F_{m,(n-m-1)}$.

For the employment example, we determine the test statistic to be the following:

$$F = \frac{r^2/m}{(1-r^2)/(n-m-1)}$$

$$F = \frac{.8422/1}{(1-.8422)/(10-1-1)} = 42.6971$$

From Appendix Table A.4, the critical value, at the .05 level of significance, with degrees of freedom of 1 (i.e., one predictor) and 8 (i.e., $n - m - 1 = 10 - 1 - 1 = 8$) is ${}_{.05}F_{1,8} = 5.32$. The test statistic exceeds the critical value; thus we reject H_0 and conclude that ρ^2_{XY} is not equal to zero at the .05 level of significance (i.e., work optimism does predict a significant proportion of the variation on employment success).

Test of Significance of b_{YX}. The second test is the test of the significance of the slope or regression coefficient, b_{YX}. In other words, *is the unstandardized regression coefficient statistically significantly different from zero*? This is actually the same as the test of b^*, the standardized regression coefficient, so we need not develop a separate test for the standardized regression coefficient. The null and alternative hypotheses, respectively, are as follows, where the null hypothesis states that the regression coefficient is equal to zero and the alternative states that it is not equal to zero.

$$H_0: \beta_{YX} = 0$$
$$H_1: \beta_{YX} \neq 0$$

To test whether the regression coefficient is equal to zero, we need a standard error for the slope b. However, first we need to develop some new concepts. The first new concept is the **variance error of estimate**. Although this is the correct term, it is easier to consider this as the **variance of the residuals**. The variance error of estimate, or variance of the residuals, is defined as follows:

$$s^2_{res} = \frac{\sum e_i^2}{df_{res}} = \frac{SS_{res}}{df_{res}} = MS_{res}$$

where the summation is taken from $i = 1, \ldots, n$ and $df_{res} = (n - m - 1)$ (or $n - 2$ if there is only a single predictor). Two degrees of freedom are lost because we have to estimate the population slope and intercept, β and α, from the sample data. The variance error of estimate indicates the amount of variation among the residuals. *If there are some extremely large residuals, this will result in a relatively large value of s^2_{res}, indicating poor prediction overall. If the residuals are generally small, this will result in a comparatively small value of s^2_{res}, indicating good prediction overall.*

The next new concept is the **standard error of estimate** (sometimes known as the **root mean square error**). *The standard error of estimate is simply the positive square root of the variance error of estimate, and thus is the standard deviation of the residuals or errors of estimate.* We denote the standard error of estimate as s_{res}.

The final new concept is the **standard error of *b***. We denote the standard error of b as s_b and define it as

$$s_b = \frac{s_{res}}{\sqrt{\frac{n\sum X^2 - \left(\sum X\right)^2}{n}}} = \frac{s_{res}}{\sqrt{SS_X}}$$

where the summation is taken over $i = 1, \ldots, n$. We want s_b to be small to reject H_0, so we need s_{res} to be small and SS_X to be large. In other words, we want there to be a large spread of scores in X. *If the variability in X is small, it is difficult for X to be a significant predictor of Y.*

Now we can put these concepts together into a **test statistic** to test the significance of the slope b. As in many significance tests, the test statistic is formed by the ratio of a parameter estimate divided by its respective standard error. A ratio of the parameter estimate of the slope b to its standard error s_b is formed as follows:

$$t = \frac{b}{s_b}$$

The test statistic t is compared to the critical values of t (in Appendix Table A.2), a two-tailed test for a nondirectional H_1, at the designated level of significance α, and with degrees of freedom of $(n - m - 1)$. That is, the tabled critical values are $\pm\ _{(\alpha/2)}t_{(n-m-1)}$ for a two-tailed test.

In addition, all other things being equal (i.e., same data, same degrees of freedom, same level of significance), both of these significance tests (i.e., the test of significance of the squared bivariate correlation between X and Y and the test of significance of the slope) will yield the exact same result. That is, if X is a significant predictor of Y, then H_0 will be rejected in both tests. If X *is not* a significant predictor of Y, then H_0 will not be rejected for either test. *In simple linear regression, each of these tests is a method for testing the same general hypothesis and logically should lead the researcher to the exact same conclusion.* Thus, there is no need to implement both tests.

We can also form a **confidence interval around the slope b**. As in most confidence interval procedures, it follows the form of the sample estimate plus or minus the tabled critical value multiplied by the standard error. The confidence interval (CI) around b is formed as follows:

$$CI(b) = b \pm {}_{(\alpha/2)}t_{(n-m-1)}(s_b)$$

Recall that the null hypothesis was written as $H_0\!: \beta_{YX} = 0$. Therefore, *if the confidence interval contains zero, then β is not significantly different from zero at the specified α level.* This is interpreted to mean that in $(1 - \alpha)\%$ of the sample confidence intervals that would be formed from multiple samples, β will be included. This procedure assumes homogeneity of variance (discussed later in this chapter); for alternative procedures see Wilcox (1996, 2003).

Now we can determine the second test statistic for the employment example. We specify $H_0 : \beta_{YX} = 0$. (i.e., the null hypothesis is that the slope is equal to zero; visually, a slope of zero is a horizontal line) and conduct a two-tailed test. First the variance error of estimate is as follows:

$$s^2_{res} = \frac{\sum e_i^2}{df_{res}} = \frac{SS_{res}}{df_{res}} = MS_{res}$$

$$s^2_{res} = \frac{80.1578}{8} = 10.0197$$

The standard error of estimate, S_{res}, is $\sqrt{10.0197} = 3.1654$. Next, the standard error of b is computed as:

$$s_b = \frac{s_{res}}{\sqrt{\dfrac{n\sum X^2 - \left(\sum X\right)^2}{n}}} = \frac{s_{res}}{\sqrt{SS_X}} = \frac{3.1654}{\sqrt{1552.50}} = .0803$$

Finally, we determine the test statistic to be as follows:

$$t = \frac{b}{s_b} = \frac{.5250}{.0803} = 6.5380$$

To evaluate the null hypothesis, we compare this test statistic to its critical values $\pm_{(.025)}t_{(8)} = \pm 2.306$. The test statistic exceeds the critical value, so H_0 is rejected in favor of H_1 (recall that we're not "accepting" the alternative hypothesis, simply finding evidence to support the alternative hypothesis). We conclude that the slope is indeed significantly different from zero, at the .05 level of significance.

Finally let us determine the confidence interval for the slope b as follows:

$$CI\,(b) = b \pm_{(\alpha/2)} t_{(n-m-1)}\,(s_b) = b \pm_{.025} t_8\,(s_b)$$

$$CI\,(b) = 0.5250 \pm (2.306)(0.0803) = (0.3398, 0.7102)$$

The interval does not contain zero, the value specified in H_0; thus we conclude that the slope β is significantly different from zero, at the .05 level of significance.

Confidence Interval for the Predicted Mean Value of Y. The third procedure is to develop a confidence interval for the predicted mean value of Y, denoted by $\bar{Y}_0'$, for a specific value of X_0. Alternatively, $\bar{Y}_0'$ is referred to as the **conditional mean of *Y* given X_0** (more about conditional distributions in the next section). In other words, for a particular predictor score X_0, how confident can we be in the predicted mean for Y?

The standard error of $\bar{Y}_0'$ is as follows:

$$s\left(\bar{Y}_0'\right) = s_{res}\sqrt{\left(\frac{1}{n}\right) + \left[\frac{\left(X_0 - \bar{X}\right)^2}{SS_X}\right]}$$

In looking at this equation, the further X_0 is from $\bar{X}$, the larger the standard error. Thus, the standard error depends on the particular value of X_0 selected. In other words, we expect to make our best predictions at the center of the distribution of X scores, and to make our poorest predictions for extreme values of X. Thus, the closer the value of the predictor is to the center of the distribution of the X scores, the better the prediction will be.

A confidence interval around $\bar{Y}_0'$ is formed as follows:

$$CI\left(\bar{Y}_0'\right) = \bar{Y}_0' \pm_{(\alpha/2)} t_{(n-2)}\left[s\left(\bar{Y}_0'\right)\right]$$

Our interpretation is that in (1 – α)% of the sample confidence intervals that would be formed from multiple samples, the population mean value of Y for a given value of X will be included.

Let us consider an example of this confidence interval procedure with the employment data. If we take a work optimism score of 50, the predicted score on employment success is 35.1125. A confidence interval for the predicted mean value of 35.1125 is as follows:

$$s\left(\bar{Y}_0'\right) = s_{res}\sqrt{\left(\frac{1}{n}\right) + \left[\frac{\left(X_0 - \bar{X}\right)^2}{SS_X}\right]}$$

$$s\left(\bar{Y}_0'\right) = 3.1654\sqrt{\left(\frac{1}{10}\right) + \left[\frac{\left(50 - 55\right)^2}{1552.50}\right]} = 1.0786$$

$$CI\left(\bar{Y}_0'\right) = \bar{Y}_0' \pm_{(\alpha/2)} t_{(n-2)}\left[s\left(\bar{Y}_0'\right)\right] = \bar{Y}_0' \pm_{.025} t_8\left[s\left(\bar{Y}_0'\right)\right]$$

$$CI\left(\bar{Y}_0'\right) = 35.1125 \pm \left(2.306\right)\left(1.0786\right) = \left(32.6252, 37.5998\right)$$

In Figure 7.3 the confidence interval around $\bar{Y}_0'$ given X_0 is plotted as the pair of curved lines closest to the regression line. Here we see graphically that the width of the confidence interval increases the further we move from X (where $X = 55.5000$).

Prediction Interval for Individual Values of Y. The fourth and final procedure is to develop a prediction interval for an individual predicted value of $\bar{Y}_0'$ at a specific individual value of X_0. That is, the predictor score for a particular individual is known, but the

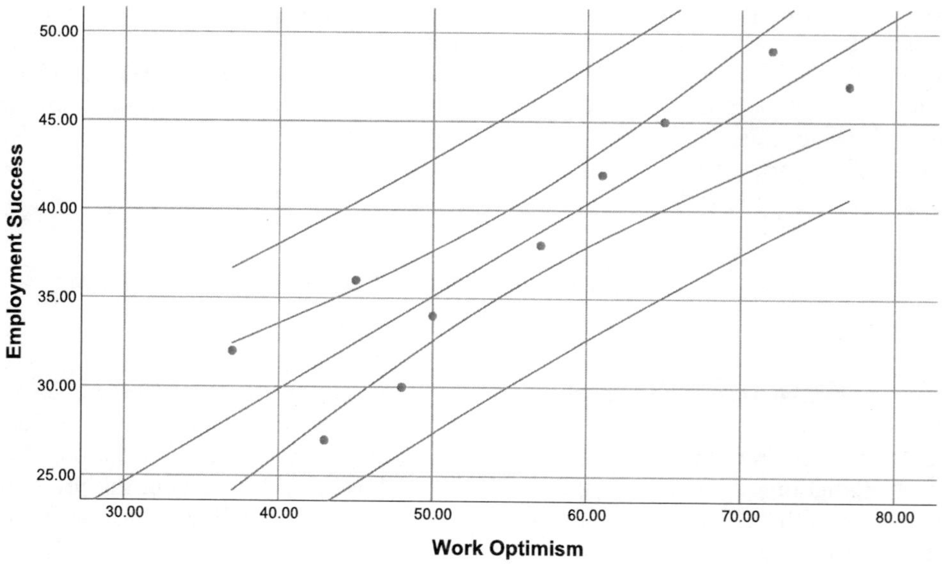

FIGURE 7.3
Confidence intervals for the employment example. Curved lines closest to the regression line represent the 95% CI; lines furthest from the regression line represent the 95% predicted interval (PI)

criterion score for that individual has not yet been observed. This is in contrast to the confidence interval just discussed where the individual Y scores have already been observed. Thus, the *confidence interval deals with the mean of the predicted values, while the prediction interval deals with an individual predicted value not yet observed.*

The standard error of $\bar{Y}_0'$ is as follows:

$$s\left(Y_0'\right) = s_{res}\sqrt{1+\left(\frac{1}{n}\right)+\left[\frac{\left(X_0-\bar{X}\right)^2}{SS_X}\right]}$$

The standard error of $\bar{Y}_0'$ is similar to the standard error of $\bar{Y}_0'$ with the addition of 1 to the equation. Thus the standard error of $\bar{Y}_0'$ will always be greater than the standard error of $\bar{Y}_0'$ as there is greater uncertainty about individual values than about the mean. The further X_0 is from $\bar{X}$, the larger the standard error. Thus the standard error again depends on the particular value of X, where we have more confidence in predictions for values of X close to $\bar{X}$.

The **prediction interval (PI)** around $\bar{Y}_0'$ is formed as follows:

$$PI\left(\bar{Y}_0'\right) = \bar{Y}_0' \pm_{(\alpha/2)} t_{(n-2)}\left[s\left(\bar{Y}_0'\right)\right]$$

Our interpretation of the prediction interval is that in (1 – α)% of the sample prediction intervals that would be formed from multiple samples, the new observation Y_0 for a given value of X will be included.

Consider an example of this prediction interval procedure with the employment data. If we take a work optimism score of 50, the predicted score on employment success is 35.1125. A prediction interval for the predicted individual value of 35.1125 is as follows:

$$s\left(Y_0'\right) = s_{res}\sqrt{1+\left(\frac{1}{n}\right)+\left[\frac{\left(X_0-\bar{X}\right)^2}{SS_X}\right]} = 3.1654\sqrt{1+\left(\frac{1}{10}\right)+\left[\frac{(50-55)^2}{1552.50}\right]} = 3.3441$$

$$PI\left(Y_0'\right) = Y_0' \pm_{(\alpha/2)} t_{(n-2)}\left[s\left(Y_0'\right)\right] = Y_0' \pm_{.025} t_8\left[s\left(Y_0'\right)\right]$$

$$PI\left(Y_0'\right) = 35.1125 \pm (2.306)(3.3441) = (27.4010, 42.8240)$$

In Figure 7.3, the prediction interval around Y_0' given X_0 is plotted as the pair of curved lines furthest from the regression line. Here we see graphically that the *prediction interval is always wider than its corresponding confidence interval.*

7.1.2 Sample Size

A widely heard convention for sample size in regression is that a researcher needs at least 10 cases for every independent variable in the model. In the case of simple linear regression, that would suggest a sample size of 10 provides sufficient power. In some cases, this may be sufficient, but in other cases, this may be quite insufficient. Rather than suggest there are general guidelines that work for "guesstimating" sample size, we recommend performing power analyses to estimate sample size given known or anticipated parameters. Should you choose to throw caution to the wind and decide not to systematically explore sample size as a function of power, we will reiterate Darlington and Hayes (2017, p. 521) in that "larger is generally better."

7.1.3 Power

With simple linear regression, we have only one predictor and one dependent variable. As we will later see in the illustration using G*Power, power in simple linear regression is a function of directionality of the test (i.e., one- or two-tailed), size of the population effect (i.e., effect size), level of significance, slope specified in the null hypothesis (usually 0), and the standard deviation of both the predictor and outcome. To determine sample size for a desired level of power, we suggest that you consult power tables (e.g., Cohen, 1988) or power software such as G*Power (note that Liu includes syntax for using **R**, SAS, and SPSS) (Erdfelder, Faul, & Buchner, 1996; Faul, Erdfelder, Buchner, & Lang, 2009; Faul, Erdfelder, Lang, & Buchner, 2007; Liu, 2014). If you're interested in learning more about power, we encourage you to consult any of a number of excellent resources (e.g., Aberson, 2010; Cohen, 1988; Liu, 2014; Murphy, Myors, & Wolach, 2014).

7.1.4 Effect Size

There are multiple effect size indices that can be considered in simple linear regression. We will discuss the coefficient of determination and f^2.

7.1.4.1 Coefficient of Determination

Recall that r_{XY}^2 is the squared sample correlation between X and Y, i.e., the *coefficient of determination*, introduced earlier. The coefficient of determination in simple linear regression is not only the squared simple bivariate Pearson correlation between X and Y, but also $r_{XY}^2 = \frac{SS_{reg}}{SS_{total}}$ which tells us that it is the proportion of the total variation of the dependent variable (i.e., the denominator) that has been explained by the regression model (i.e., the numerator). In other words, r_{XY}^2 indicates the proportion of total variation in the dependent variable Y that is predicted from the set of predictor variables. There is no objective gold standard as to how large the coefficient of determination needs to be in order to say a meaningful proportion of variation has been predicted. The coefficient is determined not just by the quality of the predictor variable included in the model, but also by the quality of relevant predictor variables not included in the model, as well as by the amount of total variation in the dependent variable Y. According to the subjective standards of Cohen (1988), small effect size for the coefficient of determination is defined as $r_{XY}^2 = .01$, a medium effect size as $r_{XY}^2 = .09$, and a large effect size as $r_{XY}^2 = .25$. Interpretation of effect size can be made based on a comparison to similar studies; what is considered a "small" effect using Cohen's convention may actually be quite large in comparison to other related studies that have been conducted. In lieu of a comparison to other studies, such as in those cases where there are no or minimal related studies, then Cohen's subjective standards may be appropriate. For additional information on effect size measures in regression, we suggest you consider Steiger and Fouladi (1992), Mendoza and Stafford (2001), and Smithson (2001) (which also includes some discussion of power).

7.1.4.2 f²

The coefficient of determination (i.e., the squared multiple correlation coefficient), r_{XY}^2, can also be used to compute a globalized f^2, sometimes referred to as **Cohen's f^2** (Cohen, 1988), which is $f^2 = \left(r_{XY}^2\right)/\left(1 - r_{XY}^2\right)$. *Note that Cohen's f^2 is a ratio of two proportions.* More specifically, it is the ratio of (1) the proportion of variation in the dependent variables uniquely explained by the independent variable to (2) the proportion of variation in the dependent variable unexplained by *any* variable in the model (Darlington & Hayes, 2017). In simple linear regression, with only one predictor, this ratio of proportions is expressed very simply as the globalized f^2. Note that Cohen's f^2 is *not* a proportion itself, because while it cannot be smaller than zero, it has no upper bound (Darlington & Hayes, 2017). Thus, *f^2 can be greater than one, and thus interpretations of f^2 cannot follow similarly as correlation coefficients.*

In many instances, *a localized effect* is of interest. In other words, the proportion of variation in the outcome is uniquely explained by one variable in the model. As we'll see in the multiple regression chapter, the computation for the localized effect differs given multiple predictors. With a single predictor in simple linear regression, we are concerned only with the globalized f^2.

7.1.4.3 Confidence Intervals for Effect Size

Confidence intervals (CI) can be computed for correlations, and thus in the case of simple linear regression, these CI are also the CI for the regression model and the CI for the effect size. *Larger CI suggest lower precision, and smaller CI reflect higher precision.* An excellent online calculator for computing all types of effect sizes and their confidence intervals is provided by Dr. David B. Wilson and is available through the Campbell Collaboration (see https://campbellcollaboration.org/research-resources/effect-size-calculator.html). Although designed for use when conducting meta-analyses, the online calculator comes in handy whenever an effect size and its CI are desired.

Let's take an example with our employment data that will be used later. Correlating work optimism and employment success, we find a Pearson correlation of .918 (which will also be the model R in our simple linear regression). Using Campbell's effect size calculator for a correlation, along with the sample size, we find the 95% CI of (.6833, .9808) (see Figure 7.4). Because the confidence interval does not contain 0, our null value (i.e., reflecting no relationship), this

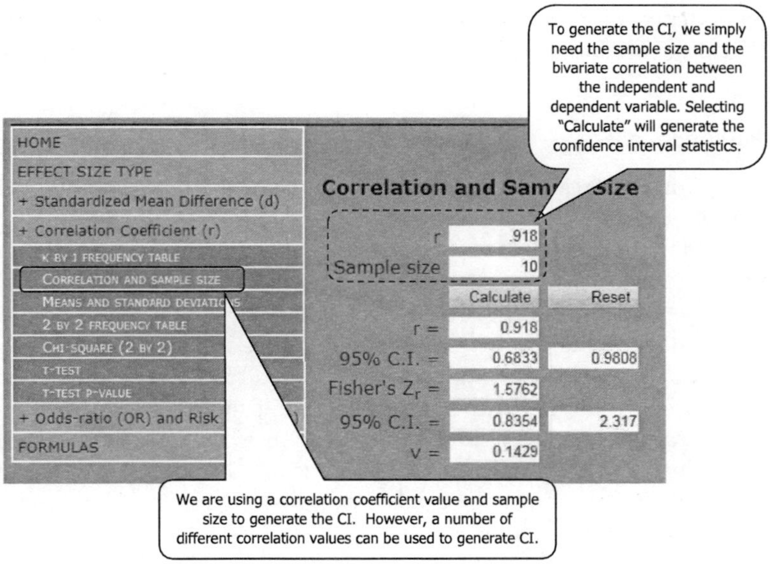

FIGURE 7.4
Computing correlation CI using the Campbell Collaboration Online Calculator.

TABLE 7.2

Effect sizes and interpretations.

Effect Size	Interpretation
r^2_{XY}	• Squared simple bivariate Pearson correlation between X and Y • Proportion of the total variation of the dependent variable (i.e., the denominator) that has been explained by the regression model (i.e., the numerator) • Cohen's standards: o $r^2_{XY} = .01$, small o $r^2_{XY} = .09$, medium o $r^2_{XY} = .25$, large
Cohen's f^2	• Ratio of (1) the proportion of variation in the dependent variables uniquely explained by the independent variable to (2) the proportion of variation in the dependent variable unexplained by *any* variable in the model • Cohen's standards: o $f^2 = .02$, small o $f^2 = .15$, medium o $f^2 = .35$, large

may provide evidence to suggest a statistically significant relationship between the independent variable and the outcome.

For additional information on effect size measures in regression, we suggest you consider Darlington and Hayes (2017), Steiger and Fouladi (1992), Mendoza and Stafford (2001), and Smithson (2001, which also includes some discussion of power).

7.1.5 Assumptions

In this section, we consider the following assumptions involved in simple linear regression: (a) independence; (b) homogeneity; (c) normality; (d) linearity; and (e) fixed X. Some discussion is also devoted to the effects of assumption violations and how to detect them.

7.1.5.1 Independence

The first assumption is concerned with independence of the observations. We should be familiar with this assumption from previous chapters (e.g., ANOVA). In regression analysis, another way to think about this assumption is that the errors in prediction or the residuals (i.e., e_i) are assumed to be random and independent. That is, there is no systematic pattern about the errors, and each error is independent of the other errors. An example of a systematic pattern would be where for small values of X the residuals tended to be small, whereas for large values of X the residuals tended to be large. Thus, there would be a relationship between the independent variable X and the residual e. Dependent errors occur when the error for one individual depends on or is related to the error for another individual as a result of some predictor not being included in the model. For our employment example, students similar in age might have similar residuals because age was not included as a predictor in the model.

Note that there are several different types of residuals. The e_i are known as **raw residuals** for the same reason that X_i and Y_i are called raw scores, all being in their original scale. The raw residuals are on the same raw score scale as Y, but with a mean of zero and a variance of s_{res}^2. Some researchers dislike raw residuals as their scale depends on the scale of Y, and therefore they must temper their interpretation of the residual values. Several different types of **standardized residuals** have been developed, including the original form of standardized residual e_i/s_{res}. These values are measured along the z score scale with a mean of 0 and a variance of 1, and approximately 95% of the values are within ±2 units of zero. Later in our illustration of SPSS, we will use **studentized residuals** for diagnostic checks. Studentized residuals, a type of standardized residual, are more sensitive to detecting outliers. Some researchers prefer these or other variants of standardized residuals over raw residuals because they find it easier to detect large residuals. However, if you really think about it, one can easily look at the middle 95% of the raw residuals by just considering the range of ±2 standard errors (i.e., $\pm 2s_{res}$) around zero. Readers interested in learning more about other types of standardized residuals are referred to a number of excellent resources (e.g., Atkinson, 1987; Cook & Weisberg, 1982; Dunn & Clark, 1987; Kleinbaum, Kupper, Muller, & Nizam, 1998; Weisberg, 2014).

The simplest procedure for assessing the assumption of independence is to examine a scatterplot (Y versus X) or a residual plot (e.g., e versus X). *If the independence assumption is satisfied, there should be a random display of points. If the assumption is violated, the plot will display some type of pattern*. For example, the negative residuals tend to cluster together, and positive residuals tend to cluster together. As we know from ANOVA, violation of the independence assumption generally occurs in the following three situations: (a) when the observations are collected over time [the independent variable is a measure of time; consider using the Durban-Watson test (Durbin & Watson, 1950, 1951, 1971)]; (b) observations are made within blocks, such that the observations within a particular block are more similar than observations in different blocks; or (c) when observation involves replication. Lack of independence affects the estimated standard errors, being under- or overestimated. For serious violations one could consider using generalized or weighted least squares as the method of estimation.

7.1.5.2 Homoscedasticity

The second assumption is **homogeneity of variance** or **homoscedasticity**, which should also be a familiar assumption (e.g., ANOVA). When discussed in the context of ANOVA, this assumption is usually referred to as homogeneity of variance; in the context of regression, it is usually referred to as homoscedasticity. This assumption must be reframed a bit in the regression context by examining the concept of a **conditional distribution**. In regression analysis, a conditional distribution is defined as the distribution of Y for a particular value of X. For instance, in the employment example, we could consider the conditional distribution of employment success scores when work optimism = 50; in other words, what the distribution of Y looks like for $X = 50$. *We call this a conditional distribution because it represents the distribution of Y conditional on a particular value of X (sometimes denoted as $Y|X$, read as Y given X)*. Alternatively we could examine the conditional distribution of the prediction errors, that is, the distribution of the prediction errors conditional on a particular value of X (i.e., $e|X$, read as e given X). Thus, the homogeneity or homoscedasticity assumption is that the conditional distributions have a constant variance for all values of X.

In a plot of the Y scores or the residuals versus X, the consistency of the variance of the conditional distributions can be examined. A common violation of this assumption occurs when the conditional residual variance increases as X increases. Here the residual plot is cone- or fan-shaped where the cone opens toward the right. An example of this violation would be where weight is predicted by age, as weight is more easily predicted for young children than it is for adults. Thus, residuals would tend to be larger for adults than for children.

If the homogeneity assumption is violated, estimates of the standard errors are larger, and although the regression coefficients remain unbiased, the validity of the significance tests is affected. In fact, with larger standard errors, it is more difficult to reject H_0, therefore resulting in a larger number of Type II errors. Minor violations of this assumption will have a small net effect; more serious violations occur when the variances are greatly different. In addition, nonconstant variances may also result in the conditional distributions being nonnormal in shape.

If the homogeneity assumption is seriously violated, the simplest solution is to use some sort of transformation, known as **variance stabilizing transformations** (e.g., Weisberg, 2014). Commonly used transformations are the log or square root of Y (e.g., Kleinbaum et al., 1998). These transformations can also often improve on the nonnormality of the conditional distributions. However, this complicates things in terms of dealing with transformed variables rather than the original variables. A better solution is to use generalized or weighted least squares (Weisberg, 2014). A third solution is to use a form of robust estimation (e.g., Carroll & Ruppert, 1982; Kleinbaum et al., 1998; Wilcox, 2003).

7.1.5.3 Normality

The third assumption of **normality** should also be a familiar one. In regression, the normality assumption is that the conditional distributions of either Y or the prediction errors (i.e., residuals) are normal in shape. That is, *for all values of X, the scores on Y or the prediction errors are normally distributed.* Oftentimes nonnormal distributions are largely a function of one or a few extreme observations, known as **outliers**, and thus we will begin our discussion here. Extreme values (i.e., outliers) may cause nonnormality and seriously affect the regression results. The regression estimates are quite sensitive to outlying observations such that the precision of the estimates is affected, particularly the slope. Also, the coefficient of determination can be affected. In general, the regression line will be pulled toward the outlier, because the least squares principle always attempts to find the line that best fits all of the points.

There are a number of different recommendations for crudely detecting outliers from a residual plot or scatterplot. A commonly used convention is to define an outlier as an observation that is more than two or three standard errors from the mean (i.e., a large distance from the mean). The outlier observation may be a result of (a) a simple recording or data entry error, (b) an error in observation, (c) an improperly functioning instrument, (d) inappropriate use of administration instructions, or (e) a true outlier. If the outlier is the result of an error, correct the error if possible and redo the regression analysis. If the error cannot be corrected, then the observation could be deleted. If the outlier represents an accurate observation, then this observation may contain important theoretical information, and one would be more hesitant to delete it (or perhaps seek out similar observations).

A simple procedure to use for single case outliers (i.e., in situations where there is just one outlier) is to perform *two* regression analyses, both with and without the outlier being included. A comparison of the regression results will provide some indication of the effects of the outlier. Other methods for detecting and dealing with outliers are available, but are not described here (e.g., Barnett & Lewis, 1994; Beckman & Cook, 1983; Cook, 1977, 2000; David & Daryl, 1978; Hawkins, 1980; Kleinbaum et al., 1998; Mickey, Dunn, & Clark, 2004; Pedhazur, 1997; Rousseeuw & Leroy, 1987; Wilcox, 2003).

Beyond examination and treatment for outliers, how does one go about detecting violation of the normality assumption? There are two commonly used procedures. *The simplest procedure involves checking for symmetry in a histogram, frequency distribution, boxplot, or skewness and kurtosis statistics*. Although **nonzero kurtosis** (i.e., a distribution that is either flat, platykurtic, or has a sharp peak, leptokurtic) will have minimal effect on the regression estimates, **nonzero skewness** (i.e., a distribution that is not symmetric with either a positive or negative skew) will have much greater impact on these estimates. Thus, finding asymmetrical distributions is a must. There are different conventions for determining how extreme skewness can be and still retain a relatively normal distribution. One simple guideline is that skewness values within ±2.0 are considered relatively normal, with more liberal researchers applying a ±3.0 guideline, and more conservative researchers using ±1.0. Another recommendation for determining how extreme a skewness value must be for the distribution to be considered nonnormal is as follows: Skewness values outside the range of plus or minus two standard errors of skewness suggest a distribution that is nonnormal. Applying this suggestion to a hypothetical example, if the standard error of skewness is .85, then any value of skewness outside of −2(.85) to +2(.85), or −1.7 to +1.7, would be considered nonnormal. It is important to note that this second recommendation is sensitive to small sample sizes and should only be considered as a general guide. For the employment example, the skewness value for the raw residuals is −0.2692. Based on the simple guideline, and the most stringent convention that skewness values within ±1.0 are considered relatively normal, there is evidence of normality in this illustration.

Another useful graphical technique is the normal probability plot (or Q-Q plot). With normally distributed data or residuals, the points on the normal probability plot will fall along a straight diagonal line, whereas nonnormal data will not. There is a difficulty with this plot because there is no criterion with which to judge deviation from linearity. A normal probability plot of the raw residuals for the employment example is shown in Figure 7.5. Together, the skewness and normal probability plot results indicate that the normality assumption is satisfied. It is recommended that skewness and/or the normal probability plot be considered at a minimum.

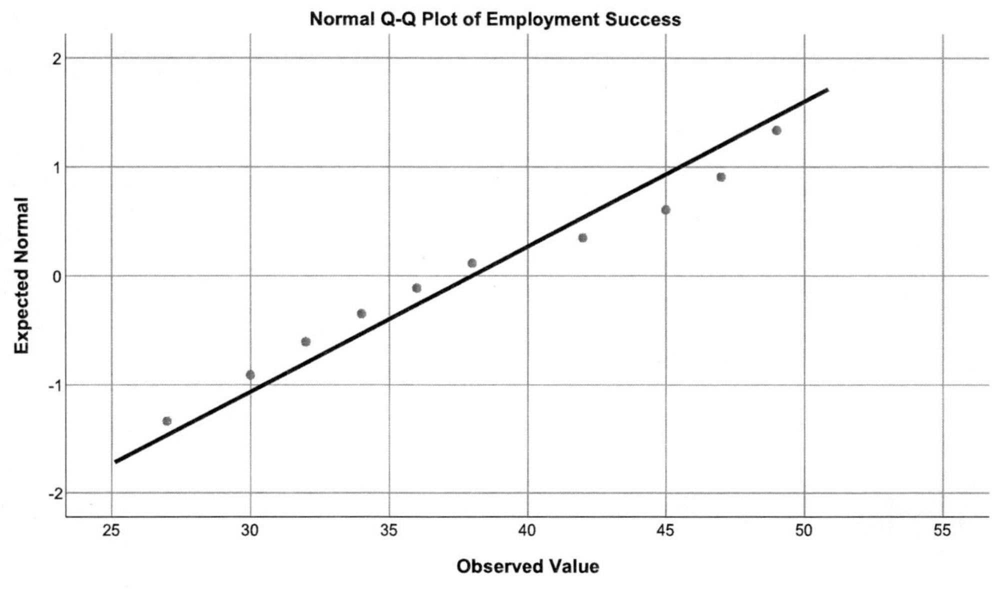

FIGURE 7.5
Normal probability plot for employment example.

Several statistical procedures are available for the detection of nonnormality (e.g., Andrews, 1971; Belsley, Kuh, & Welsch, 1980; Ruppert & Carroll, 1980; Wu, 1985). As we learned in previous chapters, the Kolmogorov-Smirnov (K-S) (Chakravart, Laha, & Roy, 1967) with Lilliefor's significance (Lilliefors, 1967), and the Shapiro-Wilk (*SW*) (Shapiro & Wilk, 1965) are tests that provide evidence of the extent to which our sample distribution is statistically different from a normal distribution.

In cases of nonnormality, various transformations are available to transform a nonnormal distribution into a normal distribution. The most commonly used transformations to correct for nonnormality in regression analysis are to transform the dependent variable using the log (to correct for positive skew) or the square root (to correct for positive or negative skew). However, again there is the problem of dealing with transformed variables measured along some other scale than that of the original variables.

7.1.5.4 Linearity

The fourth assumption is **linearity**. This assumption simply indicates that there is a linear relationship between X and Y, which is also assumed for most types of correlations. Consider the scatterplot and regression line in Figure 7.6 where X and Y are not linearly related. Here X and Y form a perfect curvilinear relationship as all of the points fall precisely on a curve. However, fitting a straight line to these points will result in a slope of zero as indicated by the solid horizontal line, not useful at all for predicting Y from X (as the predicted score for all cases will be the mean of Y). For example, age and performance are not linearly related.

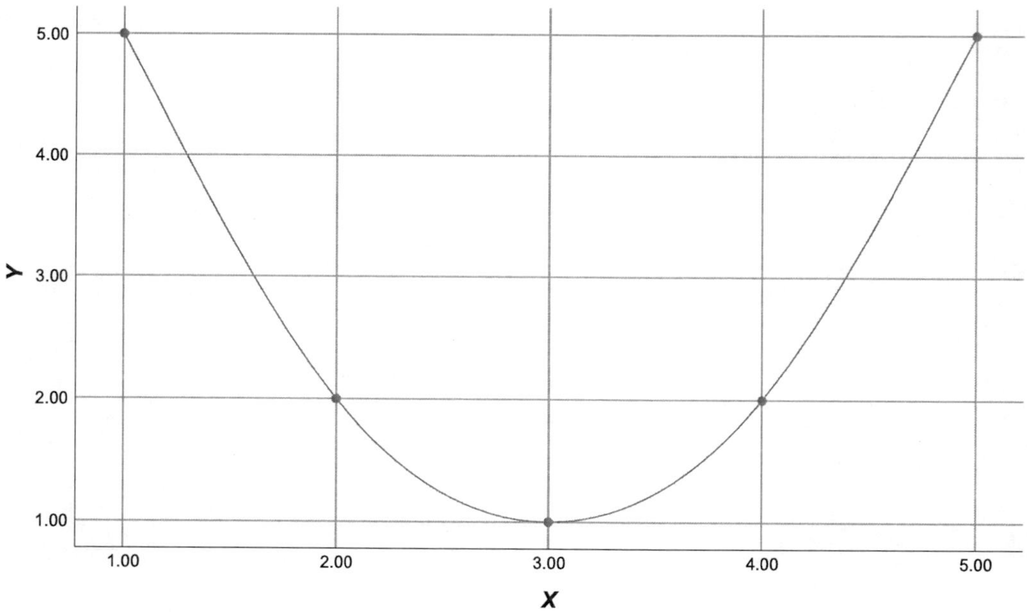

FIGURE 7.6
Nonlinear regression example.

If the relationship between X and Y is linear, then the sample slope and intercept will be unbiased estimators of the population slope and intercept, respectively. The linearity assumption is important because, regardless of the value of X_i, we always expect Y_i to increase by b_{XY} units for a one-unit increase in X_i. If a nonlinear relationship exists, this means that the expected increase in *Yi* depends on the value of X_i. Strictly speaking, linearity in a model refers to there being linearity in the parameters of the model (i.e., slope β and intercept α).

Detecting violation of the linearity assumption can often be done by looking at the scatterplot of *Y* versus *X*. If the linearity assumption is met, we expect to see no systematic pattern of points. While this plot is often satisfactory in simple linear regression, less obvious violations are more easily detected in a residual plot. If the linearity assumption is met, we expect to see a horizontal band of residuals mainly contained within $\pm 2s_{res}$ or $\pm 3s_{res}$ (or standard errors) across the values of *X*. If the assumption is violated, we expect to see a systematic pattern between *e* and *X*. Therefore, we recommend you examine both the scatterplot and the residual plot. A residual plot for the employment example is shown in Figure 7.7. Even with a very small sample, we see a fairly random display of residuals, and therefore feel fairly confident that the linearity assumption has been satisfied.

A hypothesis test for linearity can also be conducted in which a linear relationship is compared to a quadratic or cubic relationship. We will illustrate this later using SPSS.

If a serious violation of the linearity assumption has been detected, how should we deal with it? There are two alternative procedures that the researcher can utilize, **transformations** or **nonlinear models**. The first option is to transform either one or both of the variables to achieve linearity. That is, the researcher selects a transformation that subsequently results in a linear relationship between the transformed variables. Then the

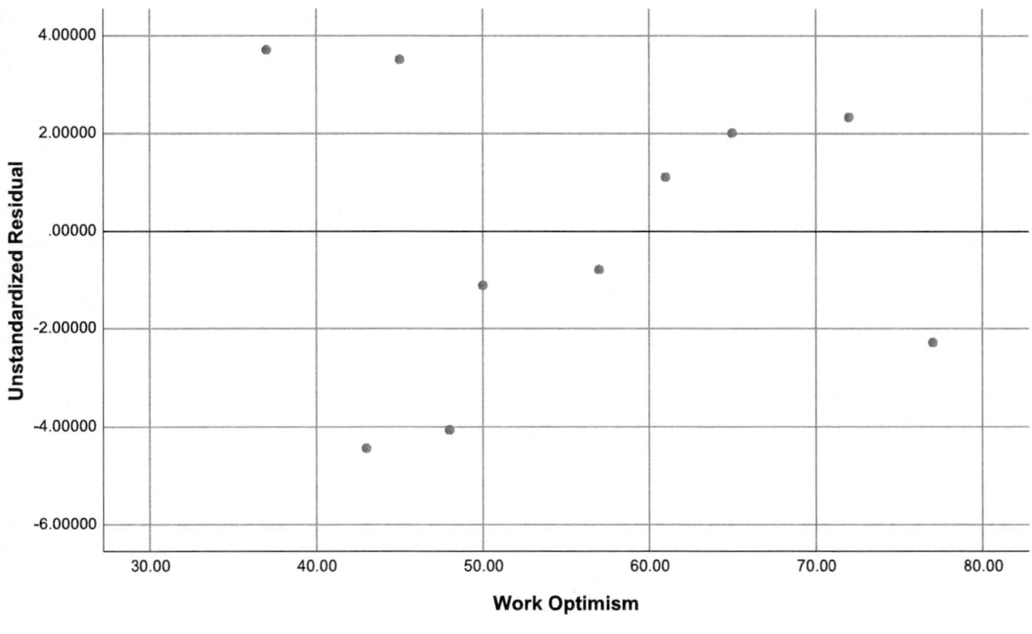

FIGURE 7.7
Residual plot for employment example.

method of least squares can be used to perform a linear regression analysis on the transformed variables. However, when dealing with transformed variables measured along a different scale, results need to be described in terms of the transformed rather than the original variables. A better option is to use a nonlinear model to examine the relationship between the variables in their original scale (see Wilcox, 1996, 2003; also discussed in Chapter 8).

7.1.5.5 Fixed X

The fifth and final assumption is that the values of X are **fixed**. That is, X is a fixed variable rather than a random variable. This results in the regression model being valid only for those particular values of X that were actually observed and used in the analysis. Thus, the same values of X would be used in replications or repeated samples. You may recall a similar concept in the fixed-effects analysis of variance models previously considered.

Strictly speaking, the regression model and its parameter estimates are valid only for those values of X actually sampled. The use of a prediction model, based on one sample of individuals, to predict Y for another sample of individuals may also be suspect. Depending on the circumstances, the new sample of individuals may actually call for a different set of parameter estimates. Two obvious situations that come to mind are the **extrapolation** and **interpolation** of values of X. In general we may not want to make predictions about individuals having X scores (i.e., scores on the independent variable) that are outside of the range of values used in developing the prediction model; this is defined as *extrapolating* beyond the sample predictor data and is more problematic than interpolation. We cannot assume that the function defined by the prediction model is the same outside of the values of X that were initially sampled. The prediction errors for the new nonsampled X values would be expected to be larger than those for the sampled X values because there are no supportive prediction data for the former.

On the other hand, we are not quite as concerned in making predictions about individuals having X scores within the range of values used in developing the prediction model; this is defined as *interpolating* within the range of the sample predictor data. We would feel somewhat more comfortable in assuming that the function defined by the prediction model is the same for other new values of X within the range of those initially sampled. For the most part, the fixed X assumption is satisfied if the new observations behave like those in the prediction sample. In the interpolation situation, we expect the prediction errors to be somewhat smaller as compared to the extrapolation situation because there are at least some similar supportive prediction data for the former. It has been shown that when other assumptions are met, regression analysis performs just as well when X is a random variable (e.g., Glass & Hopkins, 1996; Myers & Well, 1995; Pedhazur, 1997). There is no corresponding assumption about the nature of Y.

In our employment example, we have more confidence in our prediction for a work optimism value of 52 (which did not occur in the sample, but falls within the range of sampled values), than in a value of 20 (which also did not occur, and is much smaller than the smallest value sampled, 37). In fact, this is precisely the rationale underlying the prediction interval previously developed, where the width of the interval increased as an individual's score on the predictor (X_i) moved away from the predictor mean ($\bar{X}$).

A summary of the assumptions and the effects of their violation for simple linear regression is presented in Table 7.3.

TABLE 7.3

Assumptions and violation of assumptions: simple linear regression.

Assumption	Effect of Assumption Violation
Independence	• Influences standard errors of the model
Homogeneity	• Bias in s^2_{res} • May inflate standard errors and thus increase likelihood of a Type II error • May result in nonnormal conditional distributions
Normality	• Less precise slope, intercept, and R^2
Linearity	• Bias in slope and intercept • Expected change in *Y* is not a constant and depends on value of *X* • Reduced magnitude of coefficient of determination
Values of *X* fixed	• Extrapolating beyond the range of *X*: prediction errors larger, may also bias slope and intercept • Interpolating within the range of *X*: smaller effects than in extrapolation; if other assumptions met, negligible effect

7.1.5.6 Summary

The simplest procedure for assessing assumptions, and thus perhaps where you want to begin (but not end!) your examination of assumptions, is via plots of residuals. Take the employment problem as an example. Although sample size is quite small in terms of looking at conditional distributions, it would appear that all of our assumptions have been satisfied. All of the residuals are within two standard errors of zero, and there does not seem to be any systematic pattern in the residuals. The distribution of the residuals is nearly symmetrical, and the normal probability plot looks good. The scatterplot also strongly suggests a linear relationship.

7.2 Mathematical Introduction Snapshot

To summarize the mathematics that underlie simple linear regression, we first examined the *population regression model* for *Y* being predicted by *X*, which was as follows:

$$Y_i = \beta_{YX} X_i + \alpha_{YX} + \varepsilon_i$$

where *Y* is the criterion variable, *X* is the predictor variable, β_{YX} is the population slope for *Y* predicted by *X*, α_{YX} is the population intercept for *Y* predicted by *X*, ε_i are the population residuals or errors of prediction (the part of Y_i not predicted from X_i), and *i* represents an index for a particular case (an individual or object; in other words, the unit of analysis that has been measured). The index *i* can take on values from 1 to *N*, where *N* is the size of the population, written as $i = 1, \ldots, N$.

The *population prediction model* is as follows:

$$Y_i' = \beta_{YX} X_i + \alpha_{YX}$$

where Y_i' is the predicted value of Y for a specific value of X. That is, Y_i is the *actual or observed score* obtained by individual i, while Y_i' is the *predicted score* based on their X score for that same individual (in other words, you are using the value of X to predict what Y will be).

The *sample regression model* for predicting Y from X is computed as:

$$Y_i = b_{YX}X_i + a_{YX} + e_i$$

where Y and X are as before (i.e., the dependent and independent variables respectively), b_{YX} is the sample slope for Y predicted by X, a_{YX} is the sample intercept for Y predicted by X, e_i are sample residuals or errors of prediction (the part of Y_i not predictable from X_i), and i represents an index for a case (an individual or object). The index i can take on values from 1 to n, where n is the size of the sample, and is written as $I = 1, \ldots, n$.

The *sample prediction model* is computed as follows:

$$Y_i' = b_{YX}X_i + a_{YX}$$

where Y_i' is the predicted value of Y for a specific value of X. We define the sample prediction error as the difference between the *actual score* obtained by individual i (i.e., Y_i) and the *predicted score* based on the X score for that individual (i.e., Y_i'). The sample slope (b_{YX}) and intercept (a_{YX}) can be determined by

$$b_{YX} = r_{YX}\left(\frac{s_Y}{s_X}\right)$$

and

$$a_{YX} = \bar{Y} - b_{YX}\bar{X}$$

where s_Y and s_X are the sample standard deviations for Y and X respectively, r_{YX} is the sample correlation between X and Y (again the Pearson correlation coefficient, rho), and $\bar{Y}$ and $\bar{X}$ are the sample means for Y and X, respectively.

One method for determining the utility of a predictor variable is by partitioning the total sum of squares in Y, which we denote as SS_{total} (also SS_Y):

$$SS_{total} = SS_{reg} + SS_{res} = \sum_{i=1}^{n}\left(Y - \bar{Y}\right)^2 = \sum_{i=1}^{n}\left(Y' - \bar{Y}\right)^2 + \sum_{i=1}^{n}\left(Y - Y'\right)^2$$

where SS_{total} is the total variation in the observed Y scores Y, SS_{reg} is the sum of squares of the regression of Y predicted by X (i.e., variation in Y predicted by X, also $SS_{Y'}$ (and represented in the equation as $\sum_{i=1}^{n}\left(Y' - \bar{Y}\right)^2$, SS_{res} is the sum of squares of the residuals (i.e., the variation in Y not predicted by X; and represented in the equation as $\sum_{i=1}^{n}\left(Y - Y'\right)^2$) and the sums are taken over all observations from $i = 1, \ldots n$. Since $r_{XY}^2 = \frac{SS_{reg}}{SS_{total}}$, we can write SS_{total}, SS_{reg}, and SS_{res} as follows:

$$SS_{total} = \frac{n\sum_{i=1}^{n}Y^2 - \left(\sum_{i=1}^{n}Y\right)^2}{n}$$

$$SS_{reg} = \left(r_{XY}^2\right)\left(SS_{total}\right)$$

$$SS_{res} = \left(1 - r_{XY}^2\right)\left(SS_{total}\right)$$

where r_{XY}^2 is the squared sample correlation between X and Y, i.e., the *coefficient of determination*. The coefficient of determination in simple linear regression is not only the squared simple bivariate Pearson correlation between X and Y, but also $r_{XY}^2 = \frac{SS_{reg}}{SS_{total}}$, which tells us that it is the proportion of the total variation of the dependent variable (i.e., the denominator) that has been explained by the regression model (i.e., the numerator) and thus is a valuable effect size index.

7.3 Computing Simple Linear Regression Using SPSS

Next we consider SPSS for the simple linear regression model. Before we conduct the analysis, let us review the data. With one independent variable and one dependent variable, the dataset must consist of two variables or columns, *one for the independent variable* and *one for the dependent variable*. Each row still represents one individual or unit that has been measured, with the value of the independent variable for that particular case and their score on the dependent variable. In the screenshot in Figure 7.8, we see the SPSS dataset is in the form of two columns representing one independent variable (work optimism) and one dependent variable (employment success).

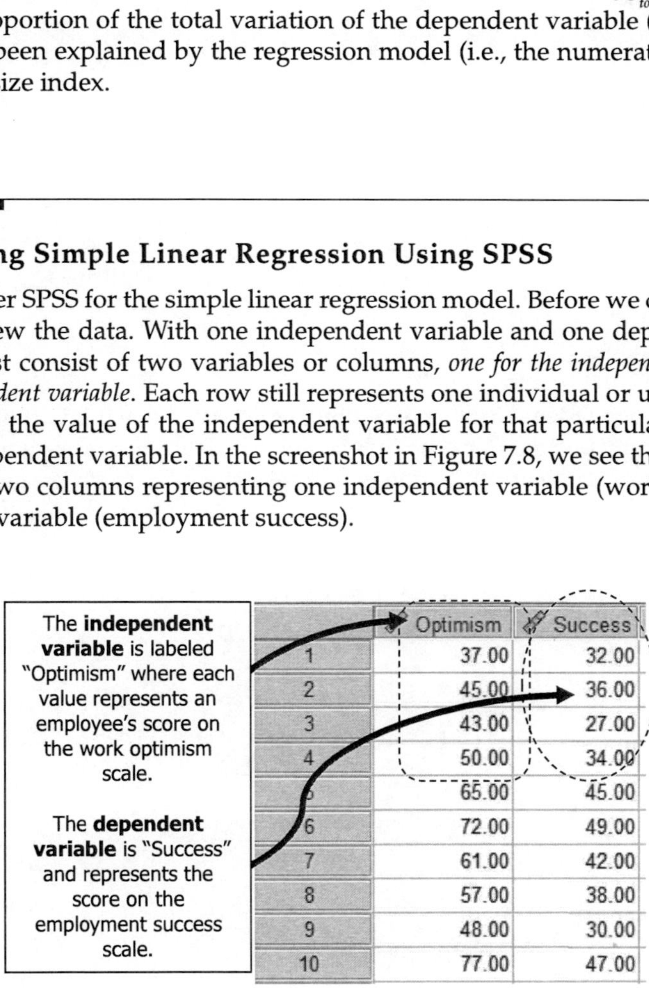

FIGURE 7.8
Data for the simple linear regression model.

Step 1. To conduct a simple linear regression, go to "Analyze" in the top pulldown menu, then select "Regression," and then select "Linear." Following the screenshot for Step 1 (Figure 7.9) produces the "Linear Regression" dialog box.

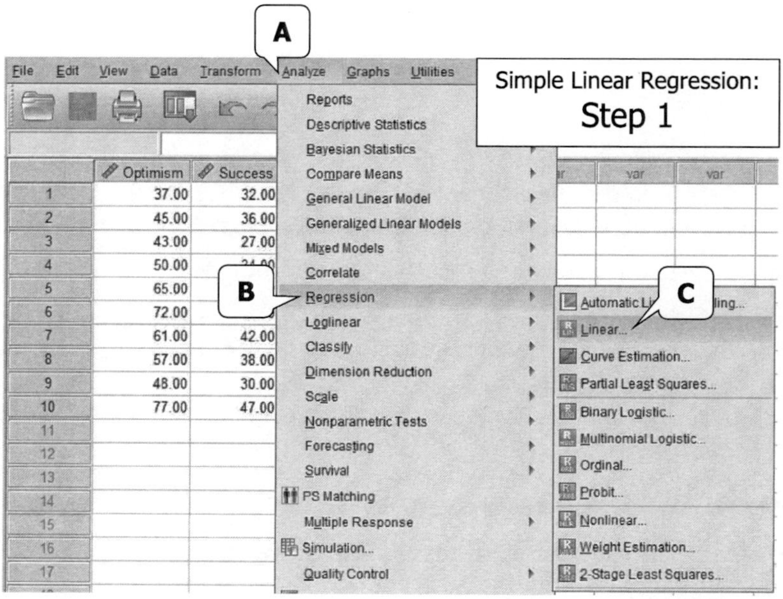

FIGURE 7.9
Conducting simple linear regression: Step 1.

Step 2. Click the dependent variable (e.g., "Success") and move it into the "Dependent" box by clicking the arrow button. Click the independent variable and move it into the "Independent(s)" box by clicking the arrow button (see the screenshot for Step 2, Figure 7.10).

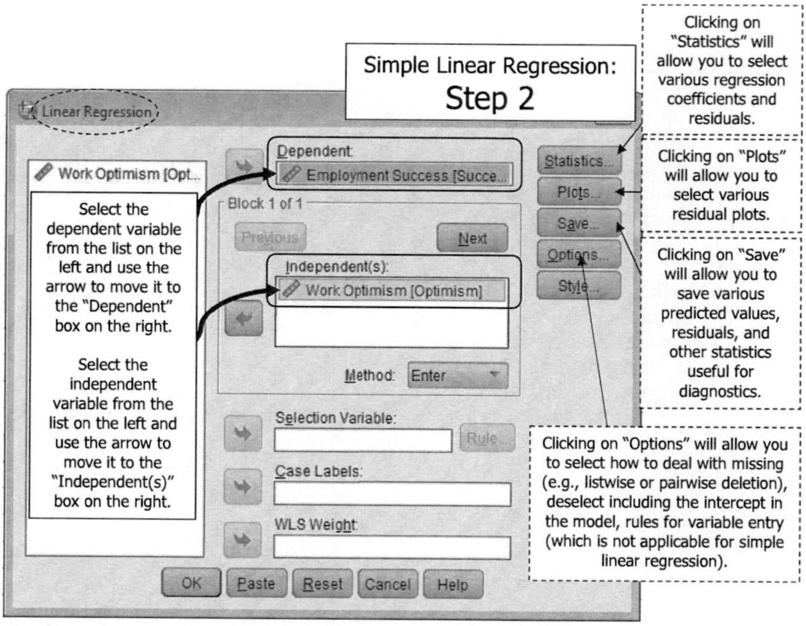

FIGURE 7.10
Conducting simple linear regression: Step 2.

Step 3. From the Linear Regression dialog box (see Figure 7.10), clicking on "Statistics" will provide the option to select various regression coefficients and residuals. From the Statistics dialog box (see the screenshot for Step 3, Figure 7.11), place a checkmark in the box next to the following: (1) estimates; (2) confidence intervals; (3) model fit; (4) descriptives; (5) Durbin-Watson; and (6) casewise diagnostics. Click on "Continue" to return to the original dialog box.

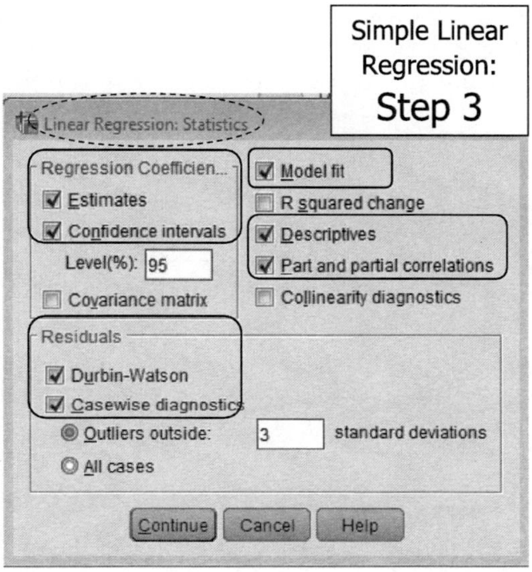

FIGURE 7.11
Conducting simple linear regression: Step 3.

Step 4. From the Linear Regression dialog box (see Figure 7.10), clicking on "Plots" will provide the option to select various residual plots. From the Plots dialog box, place a checkmark in the box next to the following: (1) histogram; (2) normal probability plot. Click on "Continue" to return to the original dialog box.

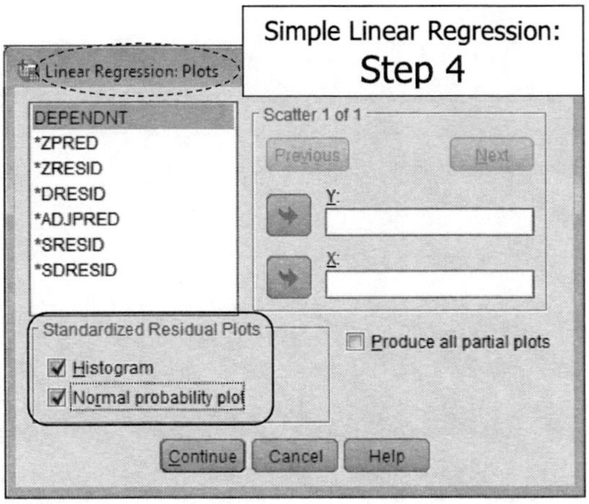

FIGURE 7.12
Conducting simple linear regression: Step 4.

Step 5. From the Linear Regression dialog box (see Figure 7.10), clicking on "Save" will provide the option to save various predicted values, residuals, and statistics that can be used for diagnostic examination. From the Save dialog box under "Predicted Values," place a checkmark in the box next to "Unstandardized." Under the heading "Residuals," place checkmarks in the boxes next to "Unstandardized" and "Studentized." Under the heading "Distances," place checkmarks in the boxes next to "Mahalanobis" and "Cook's." Under the heading "Influence Statistics," place checkmarks in the boxes next to "DfBeta(s)" and "Standardized DfBeta(s)." Click on "Continue" to return to the original dialog box. From the Linear Regression dialog box, click on "OK" to return to generate the output.

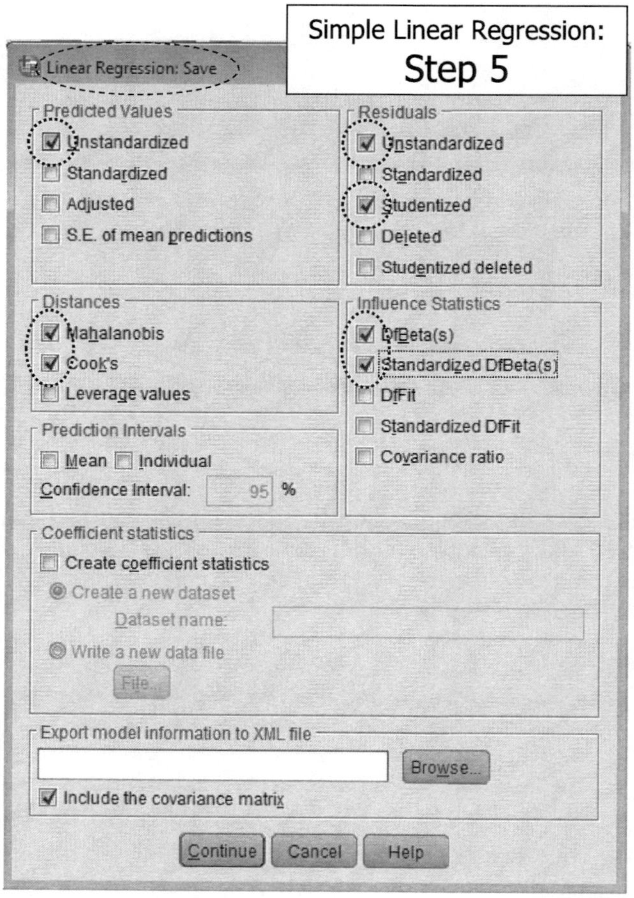

FIGURE 7.13
Conducting simple linear regression: Step 5.

Interpreting the output. Annotated results are presented in Table 7.4. In Chapters 8 and 9 we see other regression modules in SPSS which allow you to consider, for example, generalized or weighted least squares regression, nonlinear regression, and logistic regression. Additional information on regression analysis in SPSS is provided in texts such as Darlington and Hayes (2017).

TABLE 7.4

Selected SPSS results for the employment example.

Descriptive Statistics

	Mean	Std. Deviation	N
Employment Success	38.0000	7.51295	10
Work Optimism	55.5000	13.13393	10

The table labeled "Descriptive Statistics" provides basic descriptive statistics (means, standard deviations, and sample sizes) for the independent and dependent variables.

Correlations

		Employment Success	Work Optimism
Pearson Correlation	Employment Success	1.000	.918
	Work Optimism	.918	1.000
Sig. (1-tailed)	Employment Success	.	.000
	Work Optimism	.000	.
N	Employment Success	10	10
	Work Optimism	10	10

The table labeled "Correlations" provides the correlation coefficient value ($r = .918$), p value (<.001), and sample size ($N = 10$) for the simple bivariate Pearson correlation between the independent and dependent variables.
There is a statistically significant bivariate correlation between GRE-Q and midterm exam score.

Variables Entered/Removed[a]

Model	Variables Entered	Variables Removed	Method
1	Work Optimism[b]	.	Enter

a. Dependent Variable: Employment Success

b. All requested variables entered.

"Variables Entered/Removed" lists the independent variables included in the model and the method by which they were entered (i.e., "Enter"). With a single predictor, there is only one way for variables to enter the model. However, we will talk further about this in multiple linear regression.

"Adjusted R square" is an estimate of how well the model would fit other data from the same population and is calculated as:

$$R^2_{adj} = 1 - \left(1 - R^2\right)\left(\frac{n-1}{n-m-1}\right)$$

If an additional independent variable were entered in the model, an increase in R^2_{adj} indicates the new variable is adding value to the model. Negative R^2_{adj} values can occur and indicate the model fits the data VERY poorly

Model Summary[b]

					Change Statistics					
Model	R	R Square	Adjusted R Square	Std. Error of the Estimate	R Square Change	F Change	df1	df2	Sig. F Change	Durbin-Watson
1	.918[a]	.842	.822	3.16540	.842	42.700	1	8	.000	1.287

a. Predictors: (Constant), Work Optimism

b. Dependent Variable: Employment Success

R in simple linear regression is the simple bivariate Pearson correlation between X and Y.

R^2 in simple linear regression is the squared simple bivariate Pearson correlation between X and Y. It represents the proportion of variance in the dependent variable that is explained by the independent variable.

Durbin-Watson is a test of independence of the residuals. Ranging from 0 to 4, values of 2 indicate uncorrelated errors. Values less than 1 or greater than 3 indicate a likely assumption violation.

(*continued*)

TABLE 7.4 (continued)

Selected SPSS results for the employment example.

Total sum of squares is partitioned into *SS* regression and *SS* residual. When the regression *SS* equals zero, this indicates that the independent variable has provided no information in terms of explaining the dependent variable.

The *F* statistic is computed as

$$F = \frac{MS_{regression}}{MS_{residual}} = \frac{427.842}{10.020}$$

The *p* value (.000) indicates we reject the null hypothesis. The prediction equation provides a better fit to the data than estimating the predicted value of *Y* to be equal to the mean of *Y*.

ANOVA[a]

Model		Sum of Squares	df	Mean Square	F	Sig.
1	Regression	427.842	1	427.842	42.700	.000[b]
	Residual	80.158	8	10.020		
	Total	508.000	9			

a. Dependent Variable: Employment Success

b. Predictors: (Constant), Work Optimism

The **"Constant"** is the **intercept** and tells us that if work optimism (the independent variable) was zero, the employment success score (the dependent variable) would be 8.865. The estimate for **"work optimism"** is the **slope** and tells us that for a one point increase in work optimism, the employment success score will increase by about ½ of one point.

The test statistic, *t*, is calculated as the unstandardized coefficient divided by its standard error. Thus for the slope, the test statistic is:

$$t = \frac{b}{SE_b} = \frac{.525}{.080} = 6.535$$

The *p* value for the intercept (the "constant") ($p = .088$) indicates that the intercept is *not* statistically significantly different from zero (this finding is usually of less interest than the slope). The *p* value for work optimism (the independent variable) ($p = .000$) indicates that the slope is statistically significantly different from zero.

Coefficients[a]

Model		Unstandardized Coefficients B	Unstandardized Coefficients Std. Error	Standardized Coefficients Beta	t	Sig.	95.0% Confidence Interval for B Lower Bound	95.0% Confidence Interval for B Upper Bound	Correlations Zero-order	Correlations Partial	Correlations Part	Collinearity Statistics Tolerance	Collinearity Statistics VIF
1	(Constant)	8.865	4.570		1.940	.088	-1.673	19.402					
	Work Optimism	.525	.080	.918	6.535	.000	.340	.710	.918	.918	.918	1.000	1.00

a. Dependent Variable: Employment Success

In simple linear regression, the zero-order (i.e., bivariate), partial, and part correlation coefficients are all the same since there is just one independent variable.

"Residuals statistics" and related graphs (histogram and Q-Q plot, not shown here) will be examined in our discussion of assumptions.

Residuals Statistics[a]

	Minimum	Maximum	Mean	Std. Deviation	N
Predicted Value	28.2882	49.2866	38.0000	6.89478	10
Std. Predicted Value	-1.409	1.637	.000	1.000	10
Standard Error of Predicted Value	1.008	1.996	1.380	.333	10
Adjusted Predicted Value	26.5379	50.7968	37.9612	7.24166	10
Residual	-4.43800	3.71176	.00000	2.98436	10
Std. Residual	-1.402	1.173	.000	.943	10
Stud. Residual	-1.568	1.422	.006	1.071	10
Deleted Residual	-5.55197	5.46209	.03876	3.87616	10
Stud. Deleted Residual	-1.763	1.539	-.009	1.135	10
Mahal. Distance	.013	2.680	.900	.893	10
Cook's Distance	.004	.477	.159	.157	10
Centered Leverage Value	.001	.298	.100	.099	10

a. Dependent Variable: Employment Success

7.4 Computing Simple Linear Regression Using R

Next we consider **R** for the simple linear regression model. The commands are provided within the blocks with additional annotation to assist in understanding how the command works. Should you want to write reminder notes and annotation to yourself as you write the commands in **R** (and we highly encourage doing so), remember that any text that follows a hashtag (i.e., #) is annotation only and not part of the **R** code. Thus, you can write annotations directly into **R** with hashtags. We encourage this practice so that when you call up the commands in the future, you'll understand what the various lines of code are doing. You may think you'll remember what you did. However, trust us. There is a good chance that you won't. Thus, consider it best practice when using **R** to annotate heavily!

7.4.1 Reading Data Into R

```
getwd()
```

R is always directed to a directory on your computer. To find out to which directory it's pointed, run the *get working directory* command. We will assume that we need to change the working directory, and will use the next line of code to set the working directory to the desired path.

```
setwd("E:/FolderName")
```

This command will set your working directory to a specific folder that you name. Change what is in parentheses to your file location. Also, if you are copying the directory name, it will copy in slashes. You will need to change the backslash (i.e., \) to a forward slash (i.e., /) in the **R** command. Also note that you need the name of your folder enclosed in quotation marks.

```
Ch7_EmpSuccess <- read.csv("EmpSuccess.csv")
```

This command reads your data into **R**. To the left of "<-" will be what you want to call the dataframe in **R**. In this example, we're calling this **R** dataframe "Ch7_EmpSuccess." What's to the right of "<-" tells **R** to find this particular csv file. In this example, our file is called "EmpSuccess.csv." Make sure the extension (i.e., .csv) is there. Also note that you need the name of the file enclosed in quotations.

```
names(Ch7_EmpSuccess)
```

This command will produce a list of variable names for the dataframe that is noted in parentheses. For this illustration, our variable names are as follows. This is a good check to make sure your data have been read in correctly.

```
[1] "Optimism" "Success"
```

```
View(Ch7_EmpSuccess)
```

This command will let you view the dataset in spreadsheet format in RStudio.

```
summary(Ch7_EmpSuccess)
```

The *summary* command will produce basic descriptive statistics on all the variables in your dataframe. This is a great way to quickly check to see if the data have been read in correctly and get a feel for your data, if you haven't already. The output from the summary statement for this dataframe looks like this.

FIGURE 7.14
Reading Data into **R**

```
   Optimism          Success
Min.   :37.00   Min.   :27.00
1st Qu.:45.75   1st Qu.:32.50
Median :53.50   Median :37.00
Mean   :55.50   Mean   :38.00
3rd Qu.:64.00   3rd Qu.:44.25
Max.   :77.00   Max.   :49.00
```

FIGURE 7.14 (continued)
Reading Data into **R**

7.4.2 Generating the Simple Linear Regression Model

```
EmpSuccess <- lm(formula = Success ~ Optimism,
data = Ch7_EmpSuccess)
```

The *lm* command is the code to run the multiple linear regression model. In this example, we're creating an object named "EmpSuccess." The formula defines our dependent variable as "Success," and it is predicted by "Optimism." The data come from "Ch7_EmpSuccess."

```
summary(EmpSuccess)
```

Run the *summary* command to see the results from the multiple linear regression model displayed in the RStudio console. If you don't run the summary line of code, since we created an object from our model, there won't be any results output!

```
Residuals:
    Min       1Q  Median      3Q     Max
-4.4380  -1.9932  0.1626  2.2568  3.7118

Coefficients:
            Estimate  Std. Error  t value  Pr(>|t|)
(Intercept) 8.86473      4.56965    1.940  0.088358 .
Optimism    0.52496      0.08034    6.535  0.000181 ***
---
Signif. codes: 0 '***'  0.001 '**'  0.01 '*' 0.05 '.' 0.1 ' ' 1

Residual standard error: 3.165 on 8 degrees of freedom
Multiple R-squared: 0.8422, Adjusted R-squared: 0.8225
F-statistic: 42.7 on 1 and 8 DF, p-value: 0.0001814
```

```
anova(EmpSuccess)
```

This command generates the ANOVA summary table from the multiple regression model, i.e., the object we created called "EmpSuccess."

```
Analysis of Variance Table

Response: Success
          Df  Sum Sq  Mean Sq  F value     Pr(>F)
Optimism   1  427.84   427.84     42.7  0.0001814 ***
Residuals  8   80.16    10.02
---
Signif. codes: 0 '***'  0.001 '**' 0.01 '*' 0.05 '.' 0.1 ' ' 1
```

FIGURE 7.15
Generating the simple linear regression model and ANOVA summary table.

Comparing our output from **R** to SPSS, we see that, with the exception of small rounding error, the results for the coefficients are the same. There is additional output from **R** that we don't receive from SPSS.

7.4.3 Generating Correlation Coefficients

```
install.packages("Hmisc")
```

This command will install a package, *Hmisc*, that will allow us to generate the correlation matrix and related *p* values.

```
library("Hmisc")
```

We need to install the package only once in **R**. However, we need to load the package to our library each time we use it. Thus, after installing the package, we load the package with the *library* command.

```
cor(Ch7_EmpSuccess)
```

This command will generate a correlation table using all the variables in our dataframe. The default matrix is Pearson.

```
          Optimism   Success
Optimism 1.0000000 0.9177195
Success  0.9177195 1.0000000
```

FIGURE 7.16
Generating correlation coefficients.

7.4.4 Generating Confidence Intervals of Coefficient Estimates

```
confint(EmpSuccess, level =.95)
```

Because we created an object from our model (i.e., EmpSuccess), we can easily request additional stats. With the *confint* command, we can obtain confidence intervals for the coefficient estimates. With the *level* command, we set the confidence interval to 95% (i.e., the complement of our alpha level). The lower confidence interval is displayed as 2.5%, and the upper confidence interval is displayed as 97.5%.

```
                 2.5 %     97.5 %
(Intercept) -1.6728948 19.4023634
Optimism     0.3397038  0.7102157
```

FIGURE 7.17
Generating confidence intervals of coefficient estimates

7.5 Data Screening

As you may recall, there were a number of assumptions associated with simple linear regression. These included the following: (a) independence; (b) homogeneity of variance; (c) linearity; and (d) normality. Although fixed values of X are assumed, this is not an assumption that can be tested, but is instead related to the use of the results (i.e., extrapolation and interpolation).

Before we begin to examine assumptions, let us review the values that we requested to be saved to our datafile (see the dataset screenshot in Figure 7.18).

1. PRE_1 values are the **unstandardized predicted values** (i.e., Y_i').
2. RES_1 values are the **unstandardized residuals**, simply the difference between the observed and predicted values. For person 1, for example, the observed value for employment success (i.e., the dependent variable) was 32 and the predicted value was 28.28824. Thus, the unstandardized residual is simply 32–28.28824 or 3.71176.
3. SRE_1 values are the **studentized residuals**, a type of standardized residual that is more sensitive to outliers as compared to standardized residuals. Studentized residuals are computed as the unstandardized residual divided by an estimate of the standard deviation with that case removed. As a guideline, studentized residuals with an absolute value greater than 3 are considered outliers (Stevens, 1984). Studentized residuals, as compared to standardized residuals, are more sensitive for detecting outlying cases.
4. MAH_1 values are **Mahalanobis distance values**, which can be helpful in detecting outliers. These values can be reviewed to determine cases that are exerting leverage. Barnett and Lewis (1994) produced a table of critical values for evaluating Mahalanobis distance. Squared Mahalanobis distances divided by the number of variables (D^2/df) which are greater than 2.5 (for small samples) or 3 to 4 (for large samples) are suggestive of outliers (Hair et al, 2006). Later, we will follow another convention for examining these values using the chi-square distribution.
5. COO_1 values are **Cook's distance values** and provide an indication of influence of individual cases. As a guideline, Cook's values greater than 1.0 suggest that case is potentially problematic.
6. DFB0_1 and DFB1_1 values are **unstandardized DfBeta values** for the intercept and slope, respectively. These values provide estimates of the intercept and slope when the case is removed.
7. SDB0_1 and SDB1_1 values are **standardized DfBeta values** for the intercept and slope, respectively, and are easier to interpret as compared to their unstandardized counterparts. Standardized DfBeta values greater than an absolute value of two suggest that the case may be exerting undue influence on the parameters of the model (i.e., the slope and intercept).

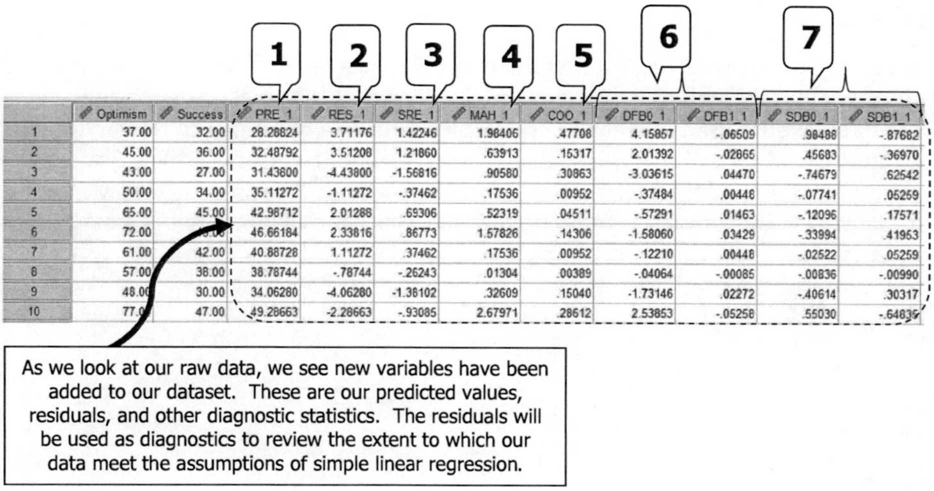

	Optimism	Success	PRE_1	RES_1	SRE_1	MAH_1	COO_1	DFB0_1	DFB1_1	SDB0_1	SDB1_1
1	37.00	32.00	28.28824	3.71176	1.42246	1.98406	.47708	4.15857	-.06509	.98488	-.87682
2	45.00	36.00	32.48792	3.51208	1.21860	.63913	.15317	2.01392	-.02865	.45683	-.36970
3	43.00	27.00	31.43800	-4.43800	-1.56816	.90580	.30863	-3.03615	.04470	-.74679	.62542
4	50.00	34.00	35.11272	-1.11272	-.37462	.17536	.00952	-.37484	.00448	-.07741	.05259
5	65.00	45.00	42.98712	2.01288	.69306	.52319	.04511	-.57291	.01463	-.12096	.17571
6	72.00	[illegible]	46.66184	2.33816	.86773	1.57826	.14306	-1.58060	.03429	-.33994	.41953
7	61.00	42.00	40.88728	1.11272	.37462	.17536	.00952	-.12210	.00448	-.02522	.05259
8	57.00	38.00	38.78744	-.78744	-.26243	.01304	.00389	-.04064	-.00085	-.00836	-.00990
9	48.0	30.00	34.06280	-4.06280	-1.36102	.32609	.15040	-1.73146	.02272	-.40614	.30317
10	77.0	47.00	49.28663	-2.28663	-.93085	2.67971	.28612	2.53853	-.05258	.55030	-.64836

Working in **R**, we can include the following commands to produce similar additional variables in our dataframe.

```
Ch7_EmpSuccess$unstandardizedPredicted <- predict(EmpSuccess)
```

What is to the left of "<-" tells **R** to save a new variable in our dataframe (i.e., Ch7_EmpSuccess) that is called "unstandardizedPredicted." What is to the right of "<-" tells **R** to created unstandardized predicted values using the simple linear regression results from the object EmpSuccess.

```
Ch7_EmpSuccess$unstandardizedResiduals <- resid(EmpSuccess)
```

Similarly, this command saves unstandardized residuals, using the simple linear regression results from the object EmpSuccess, into our dataframe.

```
Ch7_EmpSuccess$studentized.residuals <- rstudent(EmpSuccess)
```

Similarly, this command saves studentized residuals, using the simple linear regression results from the object EmpSuccess, into our dataframe.

```
Ch7_EmpSuccess$cook <- cooks.distance(EmpSuccess)
```

Similarly, this command saved Cook's distance, an influence statistic, using the simple linear regression results from the object EmpSuccess.

```
Ch7_EmpSuccess$dfbeta <- dfbeta(EmpSuccess)
```

Similarly, this command saves dfbeta values, using the simple linear regression results from the object EmpSuccess.

FIGURE 7.18
Saved variables.

7.5.1 Independence

We now plot the studentized residuals (which were requested and created through the "Save" option) against the values of X to examine the extent to which independence was met. You are likely well versed in creating scatterplots, but as a reminder, in the top toolbar in SPSS, go to "Graphs" then "Legacy Dialog" then "Scatter/Dot." From the "Simple Scatterplot" dialog screen, click the studentized residual variable and move it into the "Y Axis" box by clicking on the arrow. Click the independent variable X and move it into the "X Axis" box by clicking on the arrow. Then click "OK."

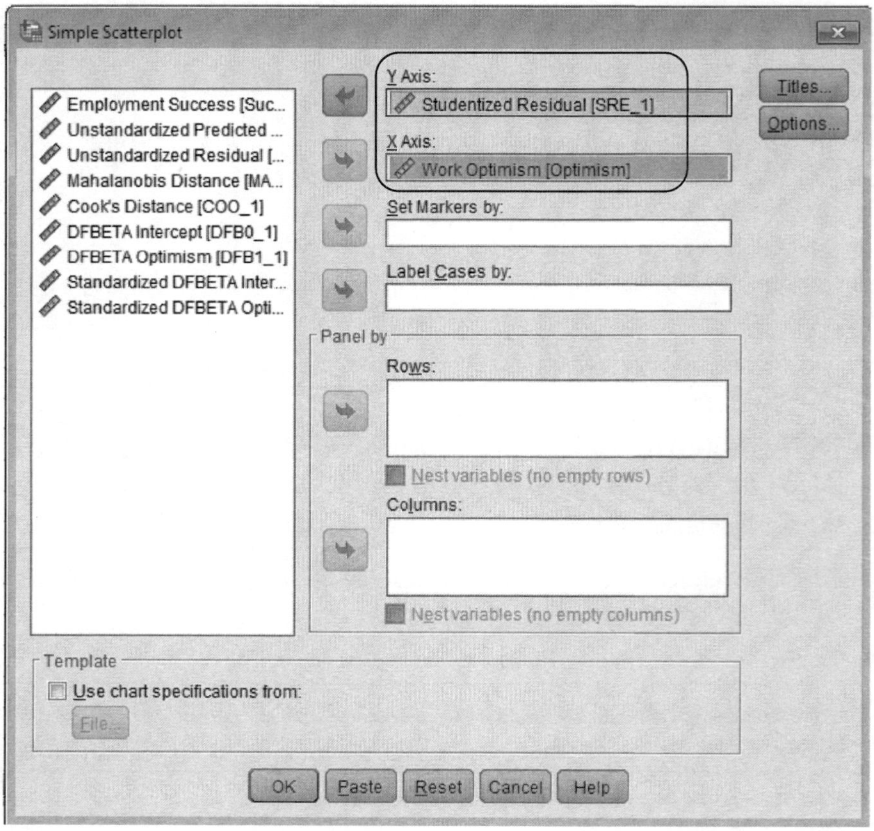

Working in **R**, we create a similar scatterplot using the following *plot* command, with the first variable listed displaying on the X axis (e.g., "Ch7_Empsuccess$Optimism"), and the second variable displaying on the Y axis (i.e., "Ch7_Empsuccess$studentized.residuals"). Additional commands are provided to label the axes (*xlab* and *ylab*) and title the graph (*main*).

```
plot(Ch7_EmpSuccess$Optimism,
     Ch7_EmpSuccess$studentized.residuals,
     xlab = "work optimism",
     ylab = "studentized residuals",
     main = "Scatterplot for independence")
```

FIGURE 7.19
Plotting to examine independence.

Interpreting independence evidence. If the assumption of independence is met, the points should fall randomly within a band of −2.0 to +2.0. Here we have evidence of independence, especially given the small sample size, as all points are within an absolute value of 2.0 and fall relatively randomly.

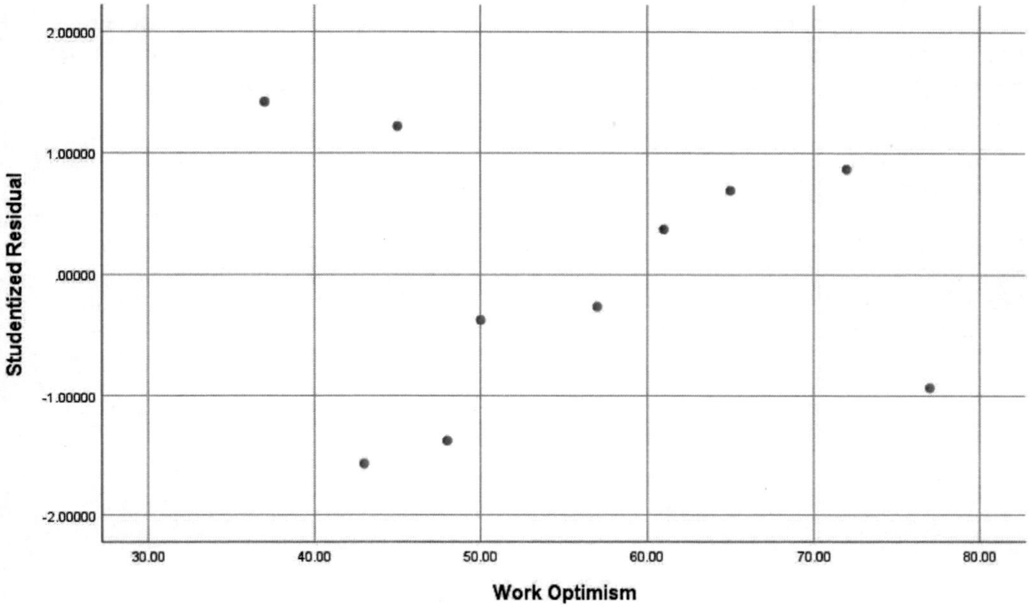

FIGURE 7.20
Scatterplot for examining the assumption of independence.

7.5.2 Homoscedasticity

We can use the same plot of studentized residuals against *X* values (used earlier for independence) to examine the extent to which homogeneity was met. Recall that homogeneity is when the dependent variable has the same variance for all values of the independent variable. Evidence of meeting the assumption of homogeneity is a plot where the spread of residuals appears fairly constant over the range of *X* values (i.e., a random display of points). If the spread of the residuals increases or decreases across the plot from left to right, this may indicate that the assumption of homogeneity has been violated. Here we have evidence of homogeneity.

There are a number of additional plots that are helpful diagnostics. We can also examine homogeneity of variance, or homoscedasticity, by looking at the spread of residuals over the range of predicted values, referred to as the *residual to fitted (or predicted value) plot*—again, looking for there to be a fairly constant spread. In other words, we're looking for a relatively random display of points. If the display of residuals increases or decreases across the plot, then there may be an indication that the assumption of homoscedasticity has been violated.

The *scale-location plot* provides evidence of the extent to which the residuals are spread equally across all values of the predictor. A random display of points suggests evidence of homoscedasticity.

The *residual vs. leverage plot* can be reviewed for influential cases, which would be evident by outlying values at the upper or lower right corners. Cases outside the dashed lines are suggestive of influential cases. In this example, there are cases that are close, but are not beyond the dashed lines, which suggests evidence that there are no outlying cases.

Working in **R**, we can also generate the *nonconstant error variance test* to determine if there is homogeneity of variance. The null hypothesis of this test is constant error variance, and the alternative hypothesis is that the error variance changes with the level of the fitted values, or with the linear combination of independent variables. A nonstatistically significant test suggests we have met the assumption of homoscedasticity, as we see here.

```
plot(EmpSuccess)
```

Using the *plot* command with our simple regression model, *EmpSuccess*, we can generate various *diagnostic plots*, including residuals to fitted values, Q-Q, scale-location, and residual vs. leverage. The *residual to fitted (or predicted value) plot* should have points that appear randomly around zero to provide evidence of meeting linearity and homoscedasticity. We see three cases (labeled 1, 3, and 9) that may be suggestive of outliers.

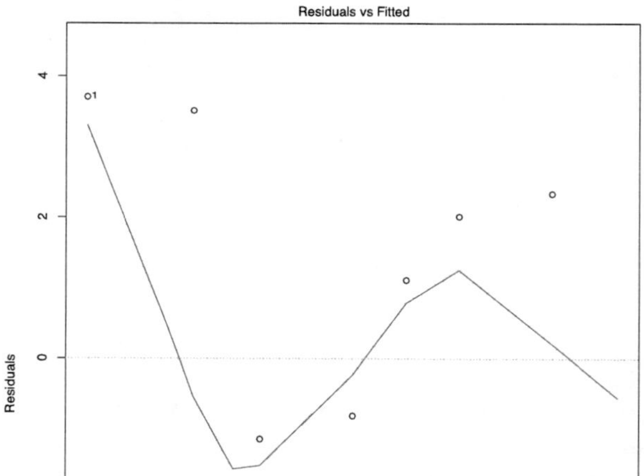

The *scale-location plot* provides evidence of the extent to which the residuals are spread equally across all values of the predictor. A random display of points suggests evidence of homoscedasticity.

FIGURE 7.21
Residual plots and nonconstant error variance test in **R**.

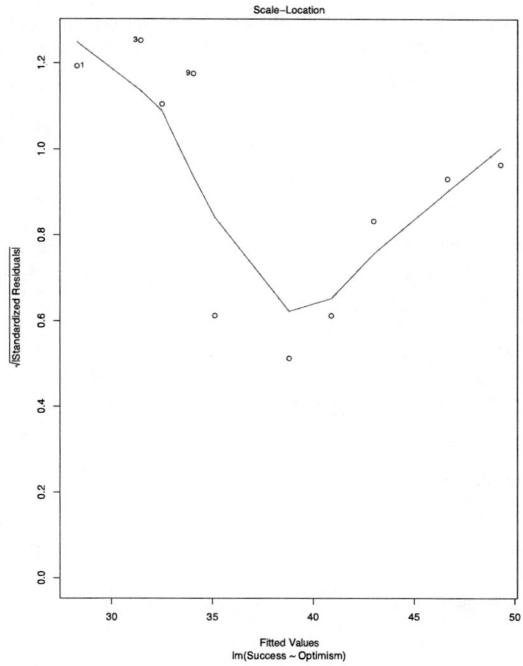

The *residual vs. leverage plot* can be reviewed for influential cases, which would be evident by outlying values at the upper or lower right corners. Cases outside the dashed lines are suggestive of influential cases. In this example, there are cases that are close, but are not beyond the dashed lines, which suggests evidence that there are no outlying cases.

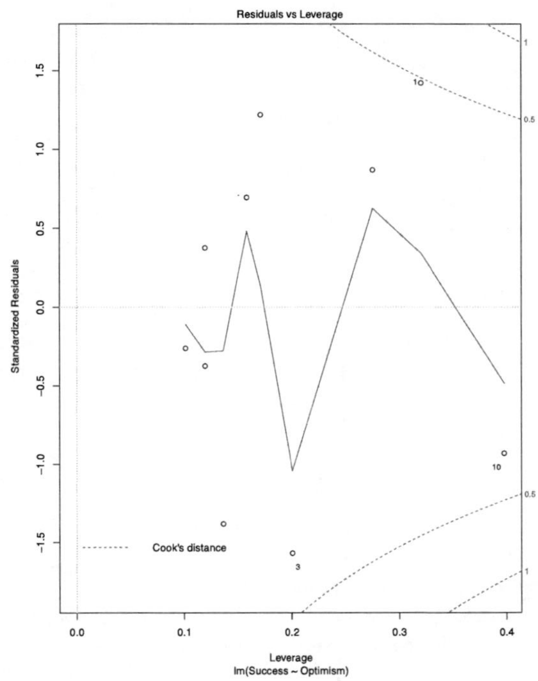

FIGURE 7.21 (continued)
Residual plots and nonconstant error variance test in **R**.

The nonconstant error variance test, the *ncvTest* function, is part of the *car* package, so we first install *car* using the *install.packages* function and then load it into our library using the *library* function.

We use our multiple linear regression object (i.e., EmpSuccess) with the *ncvTest* function to conduct the nonconstant error variance test.

```
install.packages("car")
library(car)
ncvTest(EmpSuccess)
```

The results produce a chi-squared test. Based on the *p* value (.251), our test is not statistically significant, which indicates we have met the assumption of homoscedasticity.

```
Nonconstant Variance Score Test
Variance formula: ~ fitted.values
Chisquare = 1.315736 Df = 1 p = 0.2513587
```

FIGURE 7.21 (continued)
Residual plots and nonconstant error variance test in **R**.

7.5.3 Linearity

Since we have only one independent variable, a simple bivariate scatterplot of the dependent variable (on the Y axis) and the independent variable (on the X axis) will provide a visual indication of the extent to which linearity is reasonable. As those steps have been presented previously in the discussion of independence, they will not be repeated here. For this scatterplot, there is a general positive linear relationship between the variables.

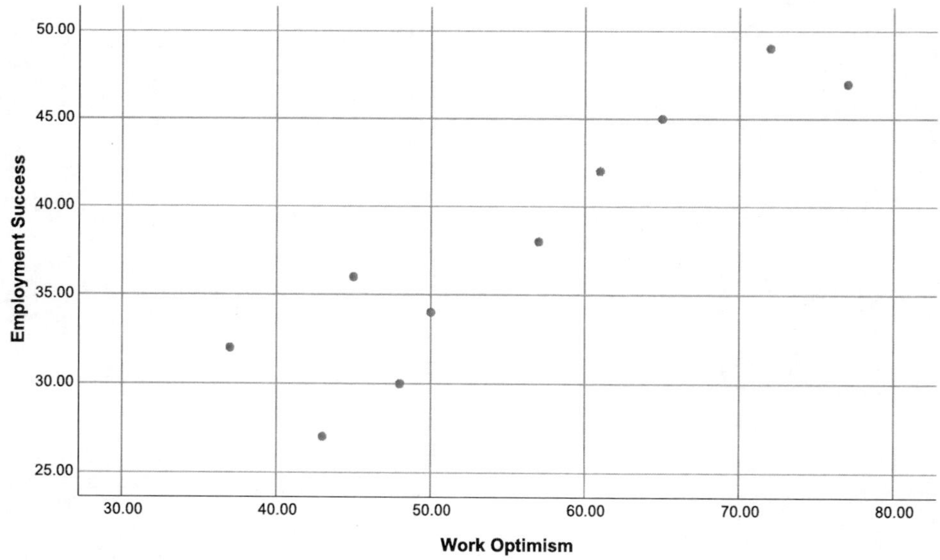

Working in **R**, we create a similar scatterplot using the following *plot* command, with the first variable listed displaying on the X axis (e.g., "Ch7_Empsuccess$Optimism"), and the second variable displaying on the Y axis (i.e., "Ch7_Empsuccess$Success"). Additional commands are provided to label the axes (*xlab* and *ylab*) and title the graph (*main*).

FIGURE 7.22
Scatterplot to Examine Linearity

```
plot(Ch7_EmpSuccess$Optimism,
     Ch7_EmpSuccess$Success,
    xlab = "work optimism",
    ylab = "employment success",
     main = "Scatterplot for Linearity")
```

FIGURE 7.22 (continued)
Scatterplot to Examine Linearity

Additionally, the plot of studentized residuals against *X* values (used earlier for independence) can be used to examine the extent to which linearity was met. We highly recommend examining this residual plot as it is more sensitive to detecting independence violations. Here a random display of points within an absolute value of 2 or 3 suggests further evident of linearity.

7.5.3.1 Hypothesis Tests to Examine Linearity Using SPSS

Another way to test for linearity is to conduct a hypothesis test using curve estimation to determine if there is a statistically significant linear (versus quadratic or cubic) relationship.

Step 1. To conduct curve estimation, go to "Analyze" in the top pulldown, then select "Regression," and then select the "Curve estimation" procedure.

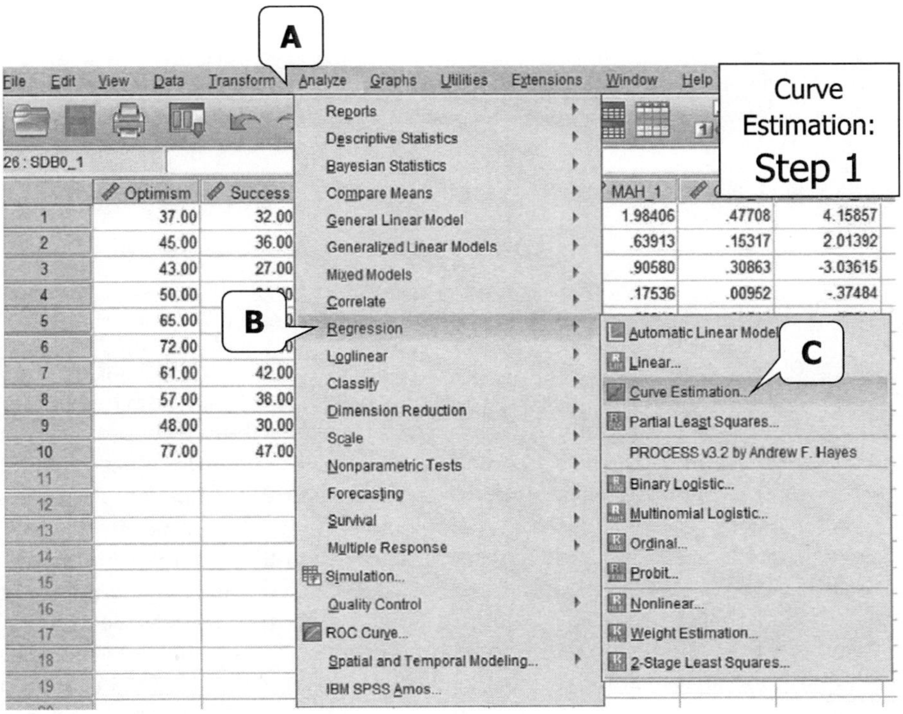

FIGURE 7.23
Hypothesis test for linearity: Step 1.

Step 2. Move the dependent variable to the "Dependent(s)" box, and the independent variable to the "Independent Variable" box. Under "Models," select "Linear," "Quadratic," and "Cubic."

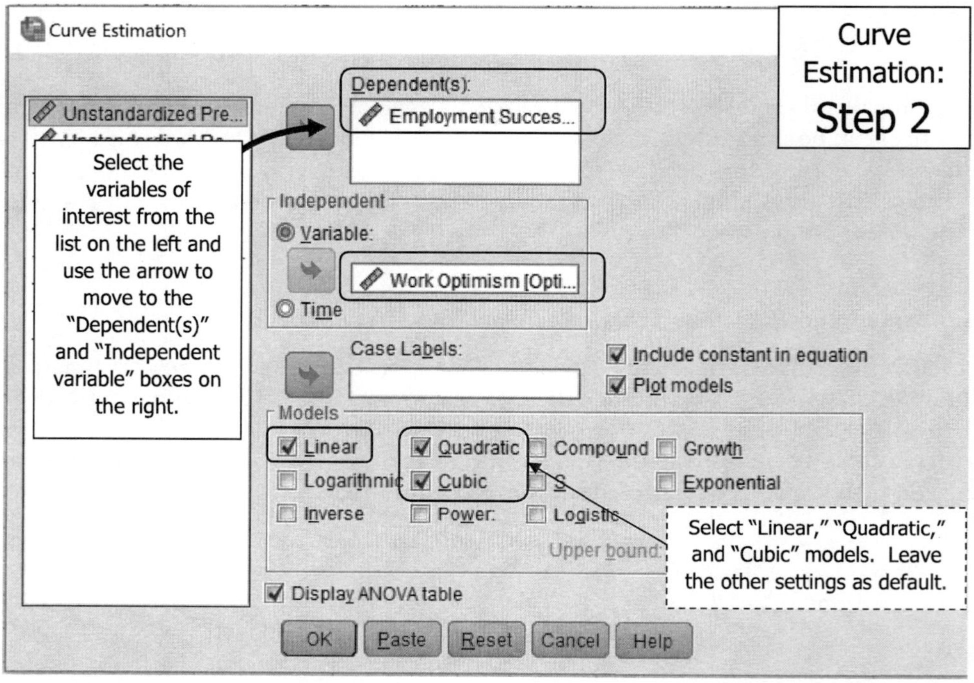

FIGURE 7.24
Hypothesis test for linearity: Step 2.

7.5.3.1.1 Interpreting Hypothesis Tests to Examine Linearity

For purposes of examining linearity, we are only concerned with the output for the "coefficients" (see Figure 7.25). Each coefficient hypothesis test is estimating whether the standardized coefficient is statistically different from zero. Finding statistical significance for the coefficient in the *linear model* provides evidence to suggest a linear relationship between the variables. For this illustration, we find a statistically significant linear relationship between work optimism and employment success, $t = 6.535$, $p < .001$.

For the quadratic model, we see we have parameter estimates for work optimism as well as "work optimism ** 2" where the latter term indicates that the independent variable has been squared (i.e., this is the quadratic term). Thus, in the quadratic model, the squared term is of interest. A statistically significant quadratic term indicates that the quadratic trend (i.e., quadratic relationship) is statistically significant beyond the linear relationship. In this illustration, we find a nonstatistically significant quadratic relationship, $t = .334$, $p = .748$, which provides evidence to suggest there is *not* a quadratic relationship between our variables.

Next, we examine the results of the cubic model. We find that a new term called "work optimism ** 3" has been estimated in this model. This term represented the cubic term. This model is estimating the extent to which there is a cubic trend, above and beyond the linear and quadratic relationships. A statistically significant cubic term suggests evidence that there is a cubic relationship between the variables. In this illustration, we find a non-statistically significant relationship, $t = -.511$, $p = .625$. Thus, we have evidence to suggest that there is not a cubic relationship between our variables.

Linear

Coefficients

	Unstandardized Coefficients B	Unstandardized Coefficients Std. Error	Standardized Coefficients Beta	t	Sig.
Work Optimism	.525	.080	.918	6.535	**.000**
(Constant)	8.865	4.570		1.940	.088

Quadratic

Coefficients

	Unstandardized Coefficients B	Unstandardized Coefficients Std. Error	Standardized Coefficients Beta	t	Sig.
Work Optimism	.234	.875	.409	.267	.797
Work Optimism ** 2	**.003**	**.008**	**.511**	**.334**	**.748**
(Constant)	16.797	24.224		.693	.510

Cubic

Coefficients

	Unstandardized Coefficients B	Unstandardized Coefficients Std. Error	Standardized Coefficients Beta	t	Sig.
Work Optimism ** 2	.009	.008	1.749	1.072	.319
Work Optimism ** 3	**-4.632E-5**	**.000**	**-.834**	**-.511**	**.625**
(Constant)	19.011	8.646		2.199	.064

FIGURE 7.25
Hypothesis test for linearity: results.

Looking at all three models, linear, quadratic, and cubic, we have evidence to suggest linearity between our variables given the nonstatistically significant nonlinear quadratic and cubic trends.

7.5.4 Normality

7.5.4.1 Generating Normality Evidence

Understanding the distributional shape, specifically the extent to which normality is a reasonable assumption, is important in simple linear regression just as it was in ANOVA models. We again examine residuals for normality, following the same steps as with the previous ANOVA designs. We also use various diagnostics to examine our data for influential cases. Let us begin by examining the unstandardized residuals for normality. For simple linear regression, the distributional shape of the unstandardized residuals should be a normal distribution. Because the steps for generating normality evidence were presented previously in the chapters for ANOVA models, they will not be provided here.

7.5.4.2 Interpreting Normality Evidence

By now we have had a substantial amount of practice in interpreting quite a range of normality statistics. We interpret them again in reference to the assumption of normality for the unstandardized residuals in simple linear regression. The skewness statistic of the residuals is −.269 and kurtosis is −1.369—both being within the range of what would be expected from a normal distribution (an absolute value of 2.0), suggesting some evidence of normality.

Descriptives

			Statistic	Std. Error
Unstandardized Residual	Mean		.0000000	.94373849
	95% Confidence Interval for Mean	Lower Bound	-2.1348848	
		Upper Bound	2.1348848	
	5% Trimmed Mean		.0403471	
	Median		.1626409	
	Variance		8.906	
	Std. Deviation		2.98436314	
	Minimum		-4.43800	
	Maximum		3.71176	
	Range		8.14976	
	Interquartile Range		5.36232	
	Skewness		-.269	.687
	Kurtosis		-1.369	1.334

Working in **R**, we can generate various normality statistics as well.

```
install.packages("pastecs")
```

This command will install the *pastecs* package which we will use to generate various forms of normality evidence.

```
library(pastecs)
```

FIGURE 7.26
Normality evidence.

This command will load the *pastecs* package.

```
stat.desc(Ch7_EmpSuccess$unstandardizedResiduals,
norm = TRUE)
```

This command will generate normality indices on the variable "unstandardizedResiduals" in the dataframe Ch7_EmpSuccess as follows. Should you want to generate normality indices on different residuals (e.g., studentized), just switch out the residual variable name in the *stat.desc* function. The *norm=TRUE* command will produce Shapiro-Wilk results (*SW*).

Looking at the results, we see skew (–.194) and kurtosis (–1.64)along with $SW = .927$, $p = .416$ for the "time" variable. All indicate the assumption of normality has been met. As we know, we can divide the skew and kurtosis values by their standard errors to get a standardized value that can be used to determine if the skew and/or kurtosis is statistically different from zero. Since this output provides "2SE," we would simply divide this value by 2 to arrive at the standard error.

Note: You may have noticed that the skewness and kurtosis value that we've just generated differs from what we found in SPSS, which was skew = –.269 and kurtosis = –.1.369. This is because there are different ways to calculate skewness and kurtosis. Let's use another package in **R** to calculate these statistics with different algorithms.

```
      nbr.val      nbr.null        nbr.na           min
 1.000000e+01  0.000000e+00  0.000000e+00 -4.438003e+00

          max         range           sum        median
 3.711755e+00  8.149758e+00  1.332268e-15  1.626409e-01

         mean       SE.mean  CI.mean.0.95           var
 1.333135e-16  9.437385e-01  2.134885e+00  8.906423e+00

      std.dev      coef.var      skewness      skew.2SE
 2.984363e+00  2.238605e+16 -1.938327e-01 -1.410630e-01

     kurtosis      kurt.2SE    normtest.W    normtest.p
-1.639214e+00 -6.142834e-01  9.267260e-01  4.164719e-01
```

```
install.packages("e1071")
```

This command will install the e1071 package which we will use to generate skewness and kurtosis.

```
library(e1071)
```

This command will load the e1071 package.

```
skewness(Ch7_EmpSuccess$unstandardizedResiduals, type=3)
skewness(Ch7_EmpSuccess$unstandardizedResiduals, type=2)
skewness(Ch7_EmpSuccess$unstandardizedResiduals, type=1)
```

The *skewness* function will generate skewness statistics on the variable(s) we specify. The "type=" script defines how skewness is calculated. Specifying "type=2" will use the algorithm that is used by SPSS. Readers interested in learning more, including the algorithms for each of the three methods, are encouraged to review Joanes and Gill (1998). We see that using "type=2," our skew is –.269, the same value as generated using SPSS.

```
# skewness(Ch7_EmpSuccess$unstandardizedResiduals, type=3)
[1] -0.1938327

# skewness(Ch7_EmpSuccess$unstandardizedResiduals, type=2)
[1] -0.2692121

# skewness(Ch7_EmpSuccess$unstandardizedResiduals, type=1)
[1] -0.2270196
```

FIGURE 7.26 (continued)
Normality evidence.

```
kurtosis(Ch7_EmpSuccess$unstandardizedResiduals, type=3)
kurtosis(Ch7_EmpSuccess$unstandardizedResiduals, type=2)
kurtosis(Ch7_EmpSuccess$unstandardizedResiduals, type=1)
```

The *kurtosis* function will generate kurtosis statistics on the variable(s) we specify. The "type=" script defines how kurtosis is calculated. Specifying "type=2" will use the algorithm that is used by SPSS. Readers interested in learning more, including the algorithms for each of the three methods, are encouraged to review Joanes and Gill (1998). We see that using "type=2," our kurtosis is −1.369, the same value as generated using SPSS.

```
# kurtosis(Ch7_EmpSuccess$unstandardizedResiduals, type=3)
[1] -1.639214

# kurtosis(Ch7_EmpSuccess$unstandardizedResiduals, type=2)
[1] -1.369316

# kurtosis(Ch7_EmpSuccess$unstandardizedResiduals, type=1)
[1] -1.320017
```

FIGURE 7.26 (continued)
Normality evidence.

While we have a very small sample size, the histogram reflects the skewness and kurtosis statistics.

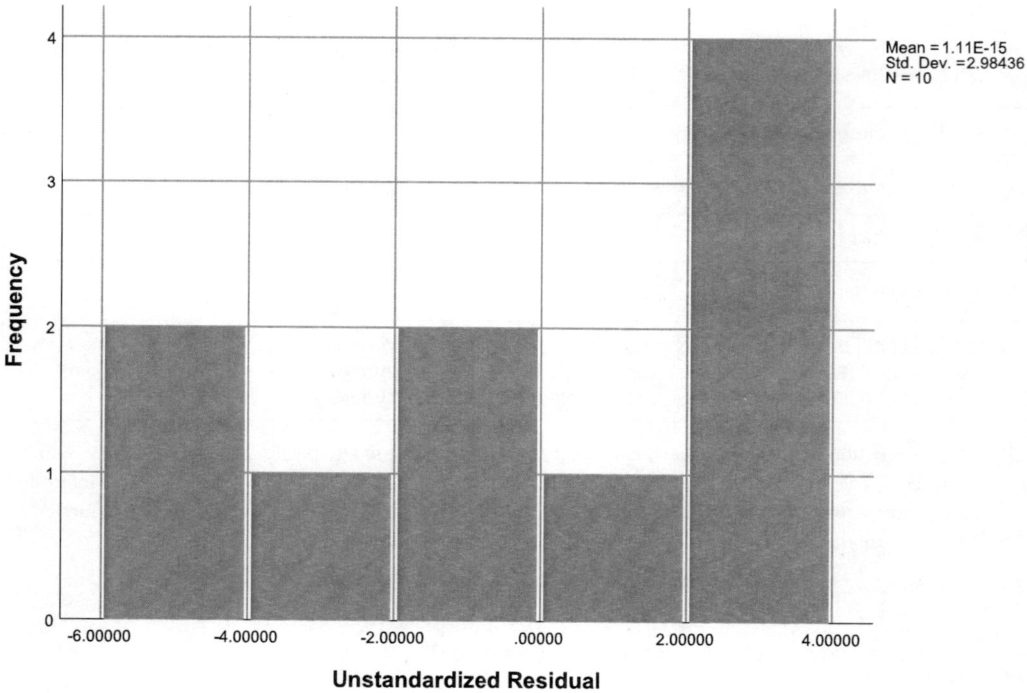

Working in **R**, we can generate a histogram using the *ggplot2* package.

```
install.packages("ggplot2")
```

The *install.packages* function will install the *ggplot2* package which we can use to create various graphs and plots.

```
library(ggplot2)
```

The *library* function will load the *ggplot2* package.

```
qplot(Ch7_EmpSuccess$unstandardizedResiduals,
      geom="histogram",
      binwidth=0.5,
      main = "Histogram of Unstandardized Residuals",
      xlab = "Unstandardized Residual", ylab = "Count",
      fill=I("gray"),
      col=I("white"))
```

Using the *gplot* function, we create a histogram (i.e., geom = "histogram") from our dataframe (i.e., Ch7_EmpSuccess) using the variable "unstandardizedResiduals." We can add a few commands to change the width of the bars (i.e., *binwidth=0.5*), color of the bars (i.e., *fill=I("gray")*), and outline of the bars (i.e., *col=I("white")*). We can also add a title (i.e., *main = "Histogram of Unstandardized Residuals"*) and change the *X* and *Y* axes (*xlab = "Unstandardized Residual", ylab = "Count"*).

FIGURE 7.27
Histogram of unstandardized residuals.

There are a few other statistics that can be used to gauge normality. The formal test of normality, the Shapiro-Wilk test (*SW*) (Shapiro & Wilk, 1965), provides evidence of the extent to which our sample distribution is statistically different from a normal distribution. The output for the Shapiro-Wilk test is presented in Figure 7.28 and suggests that our sample distribution for the residual is *not* statistically significantly different than what would be expected from a normal distribution as the p value is greater than α ($p = .416$).

Tests of Normality

	Kolmogorov-Smirnov[a]			Shapiro-Wilk		
	Statistic	df	Sig.	Statistic	df	Sig.
Unstandardized Residual	.150	10	.200*	.927	10	.416

* This is a lower bound of the true significance.

[a] Lilliefors Significance Correction

FIGURE 7.28
Shapiro-Wilk test of normality.

Quantile-quantile (Q-Q) plots are also often examined to determine evidence of normality. Q-Q plots graph quantiles of the theoretical normal distribution against quantiles of the sample distribution. Points that fall on or close to the diagonal line suggest evidence of normality. The Q-Q plot of residuals shown below suggests relative normality.

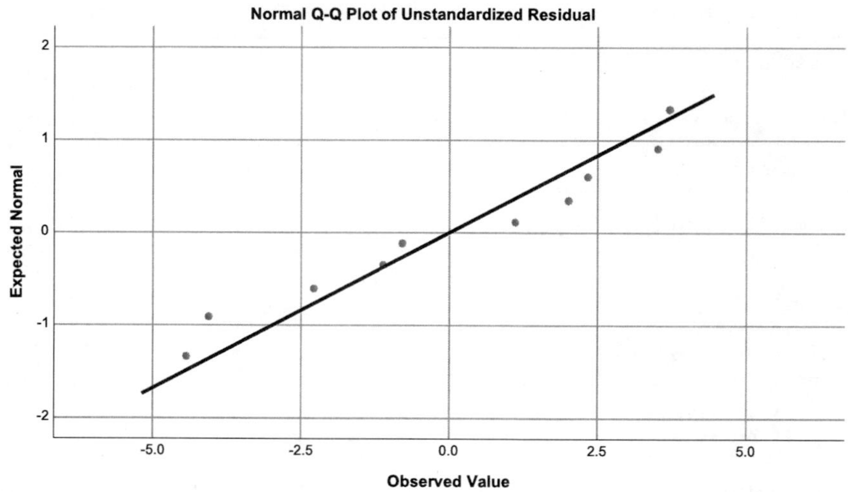

Working in **R**, we can use the *gplot* command to create a Q-Q plot of unstandardized residuals.

```
qplot(sample=unstandardizedResiduals,
      data = Ch7_EmpSuccess)
```

FIGURE 7.29
Normal Q-Q plot.

Examination of the boxplot shown in Figure 7.30 also suggests a relatively normal distributional shape of residuals with no outliers.

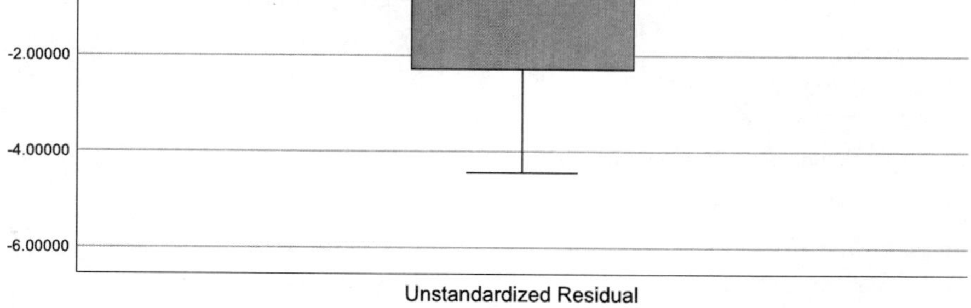

Working in **R**, we can generate a boxplot for unstandardized residuals using the *boxplot* function. To label the *Y* axis, we include the *ylab* command.

```
boxplot(Ch7_EmpSuccess$unstandardizedResiduals,
        ylab="unstandardized residual")
```

FIGURE 7.30
Boxplot of unstandardized residual.

Considering the forms of evidence we have examined, skewness and kurtosis statistics, the Shapiro-Wilk test, histogram, the Q-Q plot, and the boxplot, all suggest normality is a reasonable assumption. We can be reasonably assured we have met the assumption of normality of the residuals.

7.5.5 Screening Data for Influential Points

7.5.5.1 Casewise Diagnostics

Recall that we requested a number of statistics to help us in diagnostics and screening our data. One that we requested was for "Casewise diagnostics." If there were any cases with large values for the standardized residual (more than three standard deviations), there would have been information in our output to indicate the case number and values of the standardized residual, predicted value, and unstandardized residual. This information is useful for more closely examining case(s) with extreme standardized residuals.

7.5.5.2 Cook's Distance

Cook's distance provides an overall measure for the influence of individual cases. Values greater than one suggest that the case may be problematic in terms of undue influence on the model. In examining the residual statistics provided in the following table, we see that the maximum value for Cook's distance is .477, well under the point at which we should be concerned.

Residuals Statistics[a]

	Minimum	Maximum	Mean	Std. Deviation	N
Predicted Value	28.2882	49.2866	38.0000	6.89478	10
Std. Predicted Value	-1.409	1.637	.000	1.000	10
Standard Error of Predicted Value	1.008	1.996	1.380	.333	10
Adjusted Predicted Value	26.5379	50.7968	37.9612	7.24166	10
Residual	-4.43800	3.71176	.00000	2.98436	10
Std. Residual	-1.402	1.173	.000	.943	10
Stud. Residual	-1.568	1.422	.006	1.071	10
Deleted Residual	-5.55197	5.46209	.03876	3.87616	10
Stud. Deleted Residual	-1.763	1.539	-.009	1.135	10
Mahal. Distance	.013	2.680	.900	.893	10
Cook's Distance	.004	.477	.159	.157	10
Centered Leverage Value	.001	.298	.100	.099	10

a. Dependent Variable: Employment Success

Working in **R**, we can create a new variable in our dataframe (i.e., "Ch7_EmpSuccess$largeCook") that notes cases that have a Cook's distance that is greater than 1 using the following command:

```
Ch7_EmpSuccess$largeCook <- Ch7_EmpSuccess$cook > 1
```

We can then run the *sum* function to find out how many large Cook's values there are, and there are none.

```
sum(Ch7_EmpSuccess$largeCook)
```

```
[1] 0
```

FIGURE 7.31
Residual statistics.

7.5.5.3 Mahalanobis Distances

Mahalanobis distances are measures of the distance from each case to the mean of the independent variable for the remaining cases. We can use the value of Mahalanobis distance as a test statistic value using the chi-square distribution. With only one independent variable and one dependent variable, we have two degrees of freedom. Given an alpha level of .05, the chi-square critical value is 5.99. Thus any Mahalanobis distance greater than 5.99 suggests that case is an outlier. With a maximum distance of 2.680 (see the previous table), there is no evidence to suggest there are outliers in our data.

7.5.5.4 DfBeta

We also asked to save DfBeta values. These values provide another indication of the influence of cases. The DfBeta provides information on the change in the predicted value when the case is deleted from the model. For standardized DfBeta values, values greater than an absolute value of 2.0 should be examined more closely. Looking at the minimum (−.87682) and maximum (.62542) DfBeta values for the slope (i.e., Optimism), we do not have any cases that suggest undue influence.

Descriptive Statistics

	N	Minimum	Maximum	Mean	Std. Deviation
DFBETA Optimism	10	-.06509	.04470	-.0021866	.03608593
Standardized DFBETA Optimism	10	-.87682	.62542	-.0275752	.47302980
Valid N (listwise)	10				

Working in **R**, we can request DfBetas from our simple regression model, i.e., *EmpSuccess*, using the following command, and we will name this object "Ch7dfbeta":

```
Ch7dfbeta <- dfbetas(EmpSuccess)
```

Next, we want to define the range within which there may be influence. Values outside the range of an absolute value of 2 may be influential points. We define the range of our object (i.e., "Ch7dfbeta") to be < −2 and > 2. We will create an object from this called "Ch7dfbetasummary."

```
Ch7dfbetasummary <- Ch7dfbeta < -2 | Ch7dfbeta > 2
```

Now, all we need to do is run the *sum* function to see how many DfBeta values are outside this range, and we see there are none.

```
sum(Ch7dfbetasummary)
```

```
[1] 0
```

FIGURE 7.32
Interpreting DfBeta values for influential points.

7.6 Power Using G*Power

A priori and post hoc power could again be determined using the specialized software described previously in this text (e.g., G*Power); alternatively, you can consult *a priori* power tables (e.g., Cohen, 1988). As an illustration, we use G*Power to compute the post hoc power of our test.

7.6.1 Post Hoc Power

The first thing that must be done when using G*Power to compute post hoc power is to select the correct test family. Here we conducted simple linear regression. To find regression, select "Tests" in the top pulldown menu, then "Correlation and regression," and then "Linear bivariate regression: One group, size of slope." Once that selection is made, the "Test family" automatically changes to "t tests."

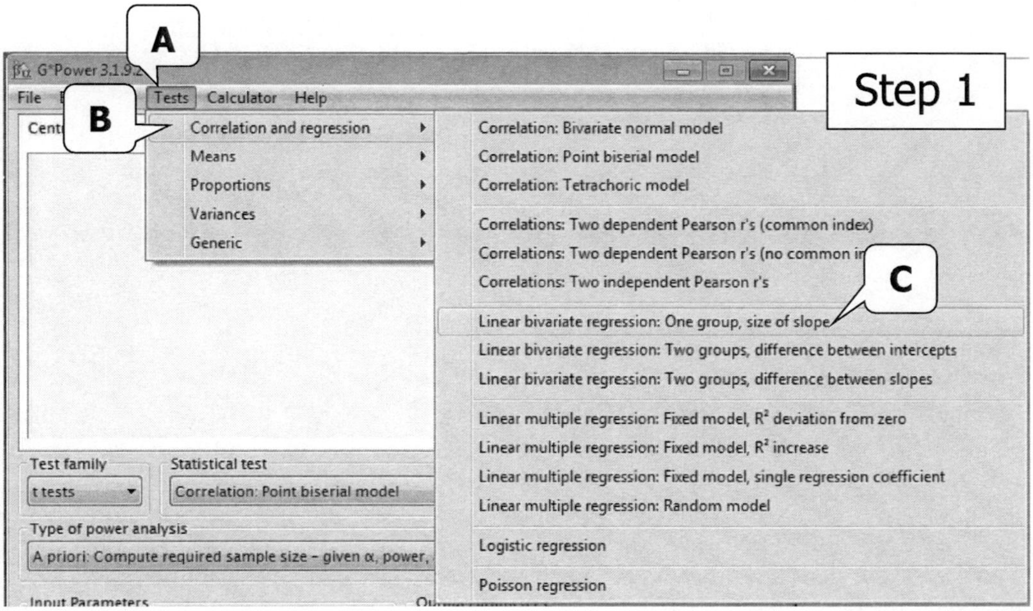

FIGURE 7.33
Post hoc power for simple linear regression: Step 1.

The "Type of power analysis" desired then needs to be selected. To compute post hoc power, select "Post hoc: Compute achieved power—given α, sample size, and effect size."

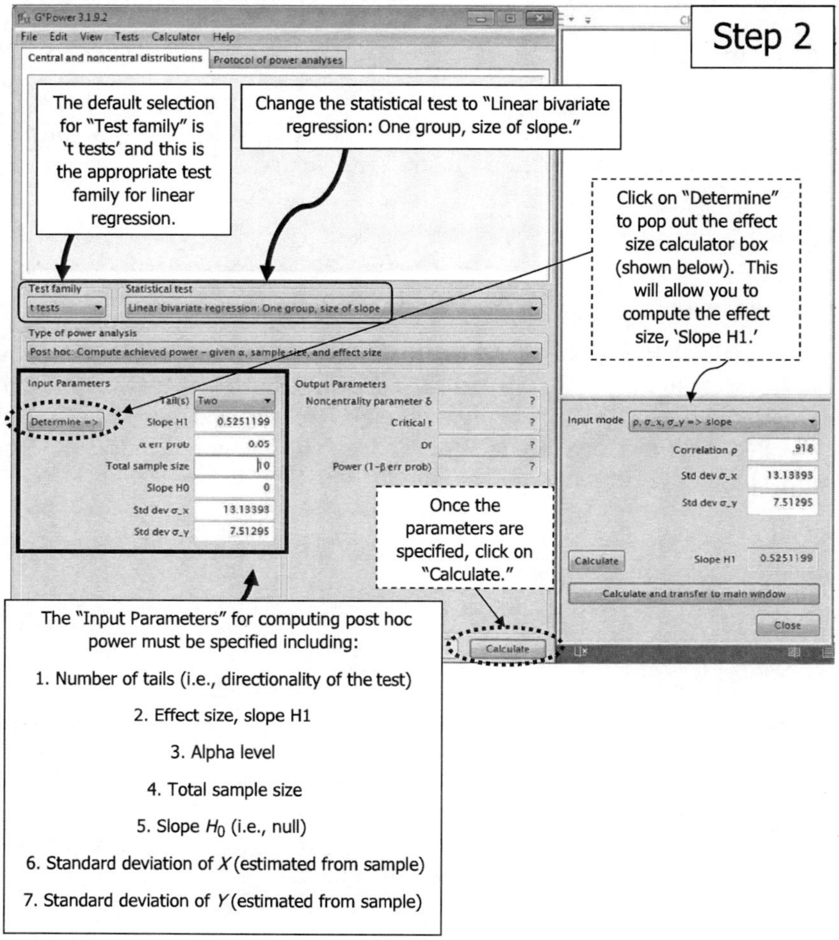

FIGURE 7.34
Computing post hoc power: Step 2.

The "Input Parameters" must then be specified. In our example, we conducted a two-tailed test. We will compute the effect size, *Slope H1*, last, so we skip that for the moment. The alpha level we used was .05 and the total sample size was 10. The *Slope H0* is the slope specified in the null hypothesis—thus a value of zero. The last two parameters to be specified are for the standard deviation of *X*, the independent variable, and the standard deviation of *Y*, the dependent variable.

We skipped filling in the second parameter, the effect size, *Slope H1*, for a reason. We will use the pop out effect size calculator in G*Power to compute the effect size *Slope H1*. To pop out the effect size calculator, click on "Determine" displayed under "Input Parameters." In the pop out effect size calculator, click the toggle menu to select $\rho, \sigma x, \sigma y$ => *slope*. Input the values for the correlation coefficient of *X* and *Y*, the standard deviation of *X*, and the standard deviation of *Y*. Click on "Calculate" in the pop out effect size calculator to compute the effect size *Slope H1*. Then click on "Calculate and transfer to main window" to transfer the calculated effect size (i.e., 0.5251199) to the "Input Parameters." Once the parameters are specified, click on "Calculate" to find the power statistics.

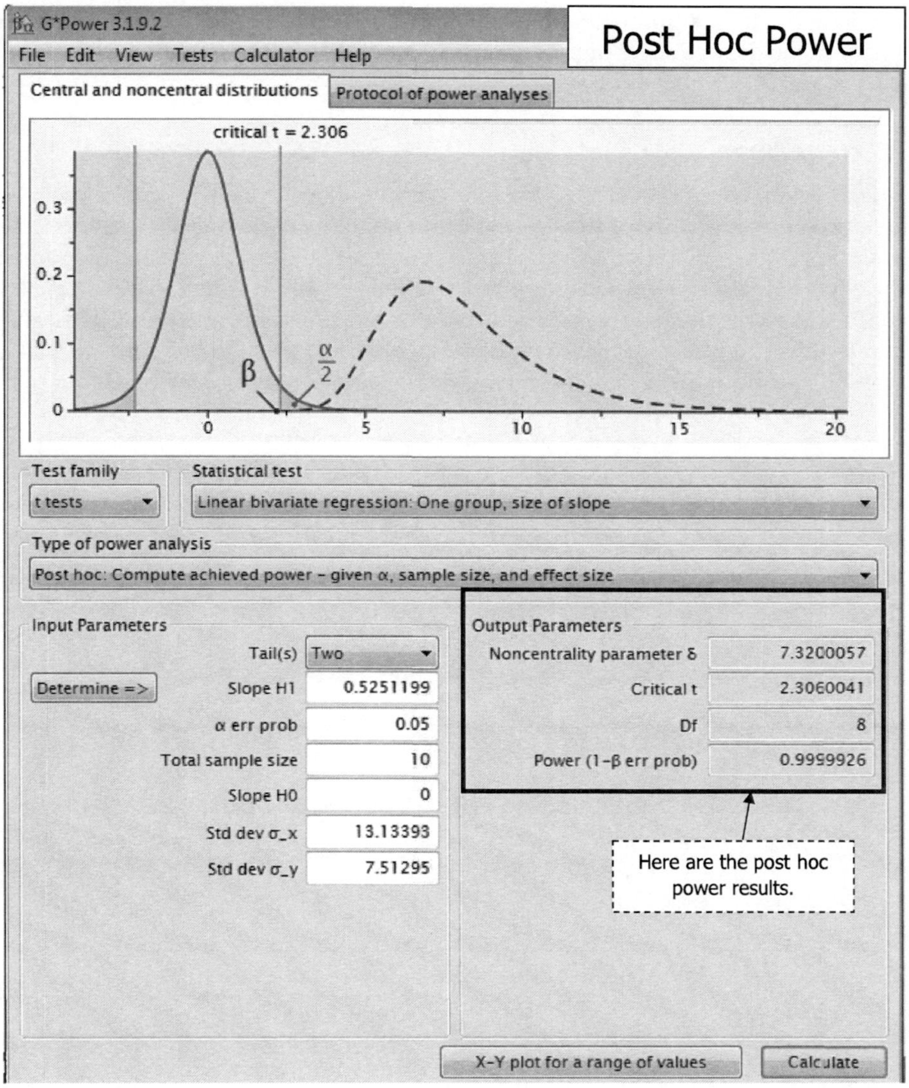

FIGURE 7.35
Post hoc power results.

The "Output Parameters" provide the relevant statistics given the input just specified. Here we were interested in determining post hoc power for simple linear regression with a two-tailed test, a computed effect size *Slope H1* of 0.5251199, an alpha level of .05, total sample size of 10, a hypothesized null slope of zero, a standard deviation of X (i.e., work optimism) of 13.13393, and a standard deviation of Y (i.e., employment success) of 7.51295. Based on those criteria, the post hoc power for the simple linear regression was .9999926. In other words, for these conditions the post hoc power of our simple linear regression was nearly 1.00—the probability of rejecting the null hypothesis when it is really false (in this case, the probability that the slope is zero) was around the maximum (i.e., 1.00; sufficient power is often .80 or above). Keep in mind that conducting power analysis *a priori* is recommended so that you avoid a situation where, post hoc, you find that the sample size was not sufficient to reach the desired level of power (given the observed parameters).

7.6.2 *A Priori* Power

For *a priori* power, we can determine the total sample size needed for simple linear regression given the directionality of the test, an estimated effect size *Slope H1*, α level, desired power, slope for the null hypothesis (i.e., zero), and the standard deviations of X and Y. We follow Cohen's (1988) conventions for effect size (i.e., small $r = .10$; moderate $r = .30$; large $r = .50$). In this example, had we wanted to determine *a priori* power and had estimated a moderate effect r of .30, α of .05, desired power of .80, null slope of zero, and standard deviation of 5 for both the X and Y, we would need a total sample size of 82.

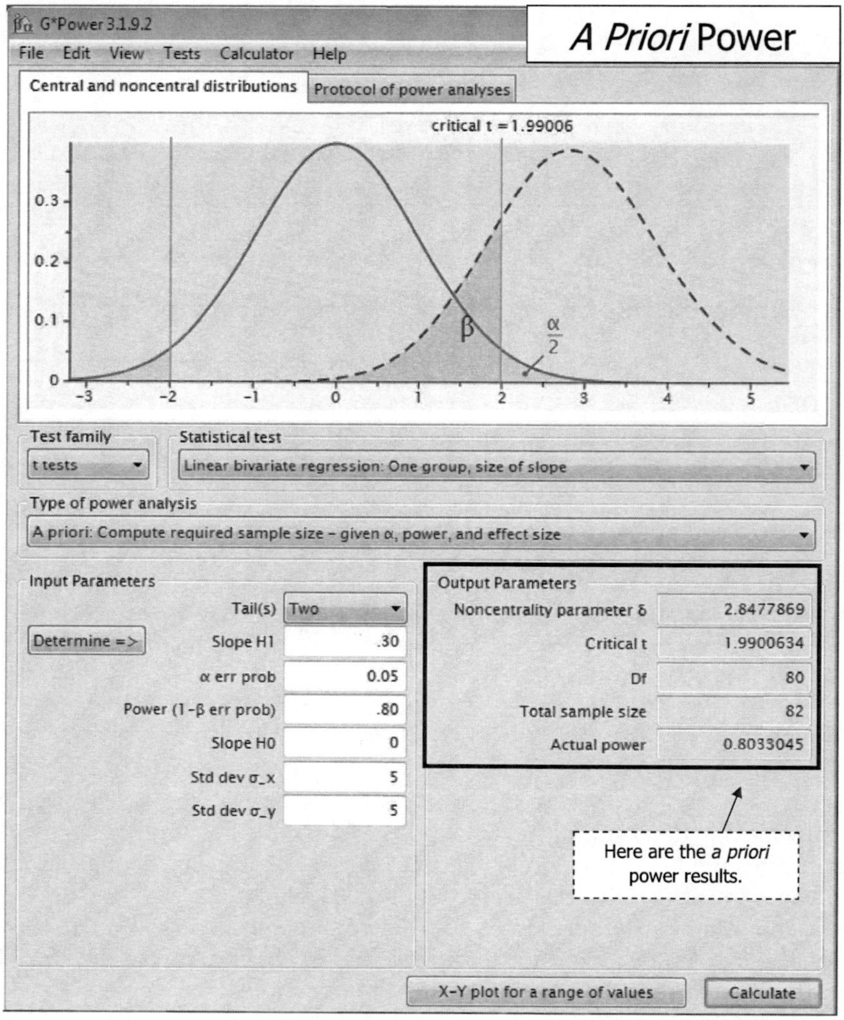

FIGURE 7.36
A priori power results.

7.7 Research Question Template and Example Write-Up

Finally, here is an example paragraph for the results of the simple linear regression analysis. Recall that our graduate research assistant, Ott Lier, was assisting the Human Resources director, Dr. Randall. Dr. Randall wanted to know if employment success could be predicted by work optimism. The research question presented to Dr. Randall from Ott included the following: *To what extent can employment success be predicted from work optimism?*

Ott then assisted Dr. Randall in generating a simple linear regression model as the test of inference. A template for writing the research question for this design is presented below.

- To what extent can [dependent variable] be predicted from [independent variable]?

It may be helpful to preface the results of the simple linear regression with information on an examination of the extent to which the assumptions were met. The assumptions include: (a) independence; (b) homogeneity of variance; (c) normality; (d) linearity; and (e) fixed values of X.

A simple linear regression analysis was conducted to determine if employment success (dependent variable) could be predicted from work optimism (independent variable). The null hypothesis tested was that the regression coefficient (i.e., the slope) was equal to zero. The data were screened for missingness and violation of assumptions prior to analysis. There was no missing data.

Linearity. The scatterplot of the independent variable (work optimism) and the dependent variable (employment success) indicates that the assumption of linearity is reasonable—as work optimism increases, employment success scores generally increase as well. With a random display of points falling within an absolute value of 2, a scatterplot of unstandardized residuals against values of the independent variable provided further evidence of linearity.

Normality. The assumption of normality was tested via examination of the unstandardized residuals. Review of the Shapiro-Wilk test for normality ($SW = .927$, $df = 10$, $p = .416$) and skewness (−.269) and kurtosis (−1.369) statistics suggested that normality was a reasonable assumption. The boxplot suggested a relatively normal distributional shape (with no outliers) of the residuals. The Q-Q plot and histogram suggested normality was reasonable.

Independence. A relatively random display of points in the scatterplot of studentized residuals against values of the independent variable provided evidence of independence. The Durbin-Watson statistic was computed to evaluate independence of errors and was 1.287, which is considered acceptable. This suggests that the assumption of independent errors has been met.

Homoscedasticity. A relatively random display of points, where the spread of residuals appears fairly constant over the range of values of the independent variable (in the scatterplot of studentized residuals against values of the independent variable) provided evidence of homogeneity of variance. The results of the nonconstant error variance test also provide evidence of homoscedasticity, $\chi^2 = 1.32$, $df = 1$, $p = .25$.

Here is an APA-style example paragraph of results for the simple linear regression analysis (remember that this will be prefaced by the previous paragraph reporting the extent to which the assumptions of the test were met).

The results of the simple linear regression suggest that a significant proportion of the total variation in employment success was predicted by work optimism. In other words, an employee's work optimism is a good predictor of their employment success, F (1, 8) = 42.700, $p < .001$. Additionally, we find: (a) the unstandardized slope (.525) and standardized slope (.918) are statistically significantly different from zero ($t = 6.535$, $df = 8$, $p < .001$); with every one point increase in work optimism, employment success will increase by approximately ½ of one point; (b) the confidence interval around the unstandardized slope does not include zero (.340, .710) further confirming that work optimism is a statistically significant predictor of employment success; and (c) the intercept (or average employment success score when work optimism is zero) was 8.865. Multiple R^2 indicates that approximately 84% of the variation in employment success was predicted by work optimism scores. According to Cohen (1988), this suggests a large effect.

7.8 Additional Resources

This chapter has provided a preview into conducting simple linear regression analysis. However, there are a number of areas that space limitations prevent us from delving into. For those of you who are interested in learning more about simple linear regression, or if you find yourself in a sticky situation in your analyses, you may wish to look into the following, among many other excellent resources.

- A comprehensive overview of regression, including managing irregularities, among other topics (Darlington & Hayes, 2017).
- A comprehensive treatment of the mathematics of regression (Olive, 2017).
- Comprehensive coverage of regression, including extensions of the linear model (e.g., boosting linear regression, Bayesian linear models), among other topics (Fahrmeir, Kneib, Lang, & Marx, 2013).

Problems

Conceptual Problems

1. A regression intercept represents which one of the following?
   a. The slope of the line
   b. The amount of change in Y given a one-unit change in X
   c. The value of Y when X is equal to zero
   d. The strength of the relationship between X and Y

2. The regression line for predicting final exam grades in history from midterm scores in the same course is found to be $Y' = .61X + 3.12$. If the value of X increases from 74 to 75, the value of Y will do which one of the following?
   a. Increase by .61 points
   b. Increase by 1.00 points
   c. Increase by 3.12 points
   d. Decrease by .61 points
3. The regression line for predicting salary of principals from cumulative GPA in graduate school is found to be $Y' = 35000X + 37000$. What does the value of 37000 represent?
   a. Average cumulative GPA
   b. The criterion value
   c. The mean salary of principals when cumulative GPA is zero
   d. The standardized regression coefficient given an intercept of zero
4. The regression line for predicting salary of principals from cumulative GPA in graduate school is found to be $Y' = 35000X + 37000$. What does the value of 35000 represent?
   a. The amount of change in Y given a one-unit change in X
   b. The correlation between X and Y
   c. The intercept value
   d. The value of Y when X is equal to zero
5. You are given that $\mu_X = 14$, $\sigma_X^2 = 36$, $\mu_Y = 14$, $\sigma_X^2 = 49$, and $Y = 14$ is the prediction equation for predicting Y from X. Which of the following is the variance of the predicted values of Y'?
   a. 0
   b. 14
   c. 36
   d. 49
6. In regression analysis, the prediction of Y is *most* accurate for which of the following correlations between X and Y?
   a. −.90
   b. −.30
   c. +.20
   d. +.80
7. If the relationship between two variables is linear, then which one of the following is correct?
   a. All of the points must fall on a curved line.
   b. The relationship is best represented by a curved line.
   c. All of the points must fall on a straight line.
   d. The relationship is best represented by a straight line.
8. If both X and Y are measured on a z score scale, the regression line will have a slope of which one of the following?
   a. 0.00
   b. +1 or −1

c. r_{XY}

d. s_Y/s_X

9. If the simple linear regression equation for predicting Y from X is $Y' = 25$, then the correlation between X and Y is which one of the following?

a. 0.00

b. 0.25

c. 0.50

d. 1.00

10. Which one of the following is correct for the unstandardized regression slope?

a. It may never be negative.

b. It may never be greater than +1.00.

c. It may never be greater than the correlation coefficient r_{XY}.

d. None of the above.

11. If two individuals have the same score on the predictor, their residual scores will be which one of the following?

a. Be necessarily equal

b. Depend only on their observed scores on Y

c. Depend only on their predicted scores on Y

d. Depend only on the number of individuals that have the same predicted score

12. If $r_{XY} = .6$, the proportion of variation in Y that is *not* predictable from X is which one of the following?

a. .36

b. .40

c. .60

d. .64

13. Homogeneity assumes which one of the following?

a. The range of Y is the same as the range of X.

b. The X and Y distributions have the same mean values.

c. The variability of the X and the Y distributions is the same.

d. The conditional variability of Y is the same for all values of X.

14. Which one of the following is suggested to examine the extent to which homogeneity of variance has been met?

a. Scatterplot of Mahalanobis distances against standardized residuals

b. Scatterplot of studentized residuals against unstandardized predicted values

c. Simple bivariate correlation between X and Y

d. Shapiro-Wilk test results for the unstandardized residuals

15. Which one of the following is suggested to examine the extent to which normality has been met?

a. Scatterplot of Mahalanobis distances against standardized residuals

b. Scatterplot of studentized residuals against unstandardized predicted values

c. Simple bivariate correlation between X and Y

d. Shapiro-Wilk test results for the unstandardized residuals

16. The linear regression slope b_{YX} represents which one of the following?
    a. Amount of change in X expected from a one-unit change in Y
    b. Amount of change in Y expected from a one-unit change in X
    c. Correlation between X and Y
    d. Error of estimate of Y from X
17. True or false? If the correlation between X and Y is zero, then the best prediction of Y that can be made is the mean of Y.
18. True or false? If X and Y are highly nonlinear, linear regression is more useful than the situation where X and Y are highly linear.
19. True or false? If the pretest (X) and the posttest (Y) are positively correlated, and your friend receives a pretest score below the mean, then the regression equation would predict that your friend would have a posttest score that is above the mean.
20. Two variables are linearly related so that given X, Y can be predicted without error. I assert that r_{XY} must be equal to either +1.0 or −1.0. Am I correct?
21. I assert that the simple regression model is structured so that at least two of the actual data points will necessarily fall on the regression line. Am I correct?
22. Which one of the following is *not* a metric that can be used as a measure of effect in simple linear regression?
    a. Coefficient of determination
    b. Mahalanobis distance
    c. r_{XY}^2
    d. Squared sample correlation between X and Y
23. What prevents Cohen's f^2 from being interpreted with the same conventions as correlation coefficients?
    a. It can be smaller than zero.
    b. It cannot be greater than one.
    c. It has no upper bound.
    d. It is a proportion.
24. The assumption of independence in simple linear regression deals with which one of the following?
    a. Bivariate correlation coefficient
    b. Independent variable
    c. Dependent variable
    d. Residuals

Answers to Conceptual Problems

1. **c** (see definition of intercept; a and b refer to the slope and d to the correlation)
3. **c** (the intercept is 37000, which represents average salary when cumulative GPA is zero)
5. **a** (the predicted value is a constant mean value of 14 regardless of X, thus the variance of the predicted values is 0)
7. **d** (linear relationships are best represented by a straight line, although all of the points need not fall on the line)

9. **a** (if the slope = 0, then the correlation = 0)
11. **b** (with the same predictor score, they will have the same residual score; whether the residuals are the same will depend only on the observed *Y*)
13. **d** (see definition of homogeneity)
15. **d** (various pieces of evidence for normality can be assessed, including formal tests such as the Shapiro-Wilk test)
17. **True** (the value of *Y* is irrelevant when the correlation = 0, so the mean of *Y* is the best prediction)
19. **False** (if the variables are positively correlated, then the slope would be positive and a low score on the pretest would predict a low score on the posttest)
21. **No** (the regression equation may generate any number of points on the regression line).
23. **c** (Cohen's f^2 is a ratio of two proportions but itself is *not* a proportion; thus, it has no upper bound and cannot be interpreted with the same conventions as correlation coefficients)

Computational Problems

1. You are given the following pairs of scores on *X* (number of hours studied) and *Y* (quiz score).

X	Y
4	5
4	6
3	4
7	8
2	4

   a. Find the linear regression model for predicting *Y* from *X*.
   b. Use the prediction model obtained to predict the value of *Y* for a new person who has a value of 6 for *X*.

2. You are given the following pairs of scores on *X* (preschool social skills) and *Y* (receptive vocabulary at the end of kindergarten).

X	Y
25	60
30	45
42	56
45	58
36	42
50	38
38	35
47	45
32	47
28	57
31	56

a. Find the linear regression model for predicting Y from X.
b. Use the prediction model obtained to predict the value of Y for a new child who has a value of 48 for X.

3. The prediction equation for predicting Y (pain indicator) from X (drug dosage) is $Y = 2.5X + 18$. What is the observed mean for Y if $\mu_X = 40$ and $\sigma_X^2 = 81$?

4. You are given the following pairs of scores on X (# of years working) and Y (# of raises).

X	Y
2	2
2	1
1	1
1	1
3	5
4	4
5	7
5	6
7	7
6	8
4	3
3	3
6	6
6	6
8	10
9	9
10	6
9	6
4	9
4	10

Perform the following computations using $\alpha = .05$.

a. The regression equation of Y predicted by X
b. Test of the significance of X as a predictor
c. Plot Y versus X
d. Compute the residuals
e. Plot residuals versus X

5. The prediction equation for predicting Y (customer satisfaction) from X (customer experience) is $Y = 5X + 8$. What is the observed mean for Y if $\mu_X = 25$ and $\sigma_X^2 = 16$?

Answers to Computational Problems

1. a. b (slope) = .8571, a (intercept) = 1.9716; b. Y= (outcome) = 7.1142

3. Given $Y = 2.5X + 18$ and knowing that $\mu_X = 40$ and $\sigma_X^2 = 81$. Then plug in the values to the equation for the intercept: $\alpha_{YX} = \mu_Y - \beta_{YX}\mu_X$. This leads to: $18 = \mu_Y - (2.5)(40)$. Thus, $\mu_Y - 18 + 100 = 118$.

5. The prediction equation for predicting Y (customer satisfaction) from X (customer experience) is $Y = 5X + 8$. We know that $\mu_X = 25$ and $\sigma_X^2 = 16$. Then plug in the values to the equation for the intercept: $\alpha_{YX} = \mu_Y - \beta_{YX}\mu_X$. This leads to: $8 = \mu_Y - (5)(25)$. Thus, $\mu_Y = 8 + 125 = 133$.

Interpretive Problems

1. With the survey1 data accessible from the website, your task is to use SPSS or **R** to find a suitable single predictor of current GPA. In other words, select several potential predictors that seem reasonable, and conduct a simple linear regression analysis for each of those predictors individually. Which of those is the best predictor of current GPA? What is the interpretation of the effect size? Write up the results following APA style.
2. With the survey1 data accessible from the website, your task is to use SPSS or **R** to find a suitable single predictor of the number of hours exercised per week. In other words, select several potential predictors that seem reasonable, and conduct a simple linear regression analysis for each of those predictors individually. Which of those is the best predictor of the number of hours of exercise? What is the interpretation of the effect size? Write up the results following APA style.
3. Using the volcano data (Ch7_volcano) accessible from the website, compute a simple linear regression using SPSS or **R** with number of injuries as the dependent variable and volcano elevation as the independent variable. Interpret the findings, including a measure of effect size. Write up the results following APA style.

8

Multiple Linear Regression

Chapter Outline

Key Concepts

1. Partial and semipartial (part) correlations
2. Standardized and unstandardized regression coefficients
3. Coefficient of multiple determination and multiple correlation

Modeling prediction is one of the most common methods of quantitative analysis. This leads us to multiple regression analysis, where we are able to model two or more predictors to predict or explain the criterion variable. Here we adopt the usual notation where the X's are defined as the *independent* or *predictor variables*, and Y as the *dependent* or *criterion variable*.

For example, an admissions officer might use Graduate Record Exam (GRE) scores to predict graduate-level grade point averages (GPA) to make admissions decisions for a sample of applicants to your favorite local university or college. The admissions office may decide that including only one variable omits a number of other factors that relate to GPA. Other potentially useful predictors might be undergraduate GPA, ratings of recommendation letters, scored writing samples, and/or an evaluation from a personal interview. The research question of interest would now be, *how well do the GRE, undergraduate GPA, recommendation ratings, writing sample scores, and/or interview scores (the independent or predictor variables) predict performance in graduate school (the dependent or criterion variable)*? This is an example of a situation where multiple regression analysis using multiple predictor variables might be the method of choice.

This chapter considers the concepts of partial, semipartial, and multiple correlations, standardized and unstandardized regression coefficients, and the coefficient of multiple determination, as well as introduces a number of other types of regression models. Our objectives are that by the end of this chapter, you will be able to (a) determine and interpret the results of partial and semipartial correlations, (b) understand the concepts underlying multiple linear regression, (c) determine and interpret the results of multiple linear regression, (d) understand and evaluate the assumptions of multiple linear regression, and (e) have a basic understanding of other types of regression models.

8.1 What Multiple Linear Regression Is and How It Works

The group of graduate students in the statistics lab have developed into quite a group of statistics gurus, their skills being sought from across the university campus and beyond. Today, we find Addie Venture taking the lead on an existing on-campus project.

Dr. Golly, the assistant dean in the Graduate Student Services office, seeks advice from Addie Venture on a special project. Dr. Golly is interested in estimating the extent to which graduate grade point average can be predicted by scores on the overall Graduate Record Exam (GRE-total) and undergraduate grade point average. From her recent statistical trek in regression, Addie knows that questions delving into relationships and prediction with continuous outcomes and multiple predictors can be examined using multiple regression. Addie suggests the following research question to Dr. Golly: *Can graduate grade point average be predicted by scores on the overall Graduate Record Exam (GRE-total) and undergraduate grade point average?* Addie determines that a multiple linear regression is the appropriate statistical procedure to use to answer Dr. Golly's question. Excited for the first project of the semester, Addie then proceeds to assist Dr. Golly in analyzing the data and interpreting the results.

8.1.1 Characteristics

Prior to a discussion of regression analysis, we need to consider two related concepts in correlational analysis, partial and semipartial correlations. Multiple regression analysis involves the use of two or more predictor variables, which can be either or both continuous or categorical, and one criterion variable that is continuous in scale (we will work with binary outcomes in the proceeding chapter); thus, there are at a minimum three variables involved in the analysis. If we think about these variables in the context of the Pearson correlation, we have a problem, because this correlation can be used to relate only two variables at a time. How do we incorporate additional variables into a correlational analysis? The answer is through partial and semipartial correlations, and later in this chapter, multiple correlations.

8.1.1.1 Partial Correlation

First we discuss the concept of **partial correlation**. The simplest situation consists of three variables, which we label X_1, X_2, and X_3. Here an example of a partial correlation would be the correlation between X_1 and X_2 where X_3 is *held constant* (i.e., controlled or partialed out). That is, *the influence of X_3 is removed from both X_1 and X_2 (both have been adjusted for X_3)*. Thus, the partial correlation here represents the linear relationship between X_1 and X_2 independent of the linear influence of X_3. This particular partial correlation is denoted by $\boldsymbol{r_{12.3}}$, where the X's are not shown for simplicity and the dot indicates that the variables preceding it are to be correlated and the variable(s) following it are to be partialed out. We compute $r_{12.3}$ as follows:

$$r_{12.3} = \frac{r_{12-}r_{13}r_{23}}{\sqrt{\left(1-r_{13}^2\right)\left(1-r_{23}^2\right)}}$$

Let us take an example of a situation where a partial correlation might be computed. Say a researcher is interested in the relationship between height (X_1) and weight (X_2). The sample consists of individuals ranging in age (X_3) from 6 months to 65 years. The sample correlations are for: height (X_1) and weight (X_2), $r_{12} = .7$; height (X_1) and age (X_3), $r_{13} = .1$; and weight (X_2) and age (X_3), $r_{23} = .6$. We compute the correlation between height and weight, controlling for age, $r_{12.3}$, as follows:

$$r_{12.3} = \frac{r_{12-}r_{13}r_{23}}{\sqrt{\left(1-r_{13}^2\right)\left(1-r_{23}^2\right)}} = \frac{.7-(.1)(.6)}{\sqrt{(1-.01)(1-.36)}} = .8040s$$

We see here that the bivariate correlation between height and weight, ignoring age ($r_{12} = .7$), is smaller than the partial correlation between height and weight controlling for age ($r_{12.3} = .8040$). *That is, the relationship between height and weight is stronger when age is held constant (i.e., for a particular age) than it is across all ages.* Although we often talk about holding a particular variable constant, in reality, variables such as age cannot be held constant artificially.

Holding age constant would be an **experimental control**—controlling for the effects of age by collecting height and weight data from everyone who has the same age. It is important to note that this is not the same as achieving **statistical control**—controlling for the effects of age by correlating the residuals of a regression to predict height from age with the residuals from a regression to predict weight from age.

Some rather interesting partial correlation results can occur in particular situations. At one extreme, if both the correlation between height (X_1) and age (X_3), r_{13}, and weight (X_2) and age (X_3), r_{23}, equal zero, then the correlation between height (X_1) and weight (X_2) will equal the partial correlation between height and weight controlling for age, $r_{12} = r_{12.3}$. That is, *if the variable being partialed out is uncorrelated with each of the other two variables, then the partialing process will logically not have any effect.*

At the other extreme, *if either r_{13} or r_{23} equals 1, then $r_{12.3}$ cannot be calculated as the denominator is equal to zero* (in other words, at least one of the terms in the denominator is equal to zero which results in the product of the two terms in the denominator equaling zero and thus a denominator of zero—and you cannot divide by zero). Thus, in this situation (where either r_{13} or r_{23} is perfectly correlated at 1.0), the partial correlation (i.e., $r_{12.3}$, partial correlation between height and weight controlling for age) is not defined. Later in this chapter, we refer to this as *perfect collinearity*, which is a serious problem.

In between these extremes, it is possible for the partial correlation to be greater than or less than its corresponding bivariate correlation (including a change in sign), and even for the partial correlation to be equal to zero when its bivariate correlation is not. For significance tests of partial and semipartial correlations, we refer you to your favorite statistical software.

8.1.1.2 Semipartial (Part) Correlation

Next, the concept of **semipartial correlation** (also called a **part correlation**) is discussed. The simplest situation consists again of three variables, which we label X_1, X_2, and X_3. Here an example of a semipartial correlation would be the correlation between X_1 and X_2 where X_3 is removed from X_2 only. That is, the influence of X_3 is removed from X_2 only. Thus the semipartial correlation here represents the linear relationship between X_1 and X_2 after that portion of X_2 that can be linearly predicted from X_3 has been removed from X_2. This particular semipartial correlation is denoted by $r_{1(2.3)}$, where the X's are not shown for simplicity, and within the parentheses the dot indicates that the variable(s) following it are to be removed from the variable preceding it. Another use of the semipartial correlation is when we want to examine the predictive power in the prediction of Y from X_1 after removing X_2 from the prediction. A method for computing $r_{1(2.3)}$ is as follows:

$$r_{1(2.3)} = \frac{r_{12} - r_{13}r_{23}}{\sqrt{\left(1 - r_{23}^2\right)}}$$

Let us take an example of a situation where a semipartial correlation might be computed. Say a researcher is interested in the relationship between GPA (X_1) and GRE scores (X_2). The researcher would like to remove the influence of intelligence (IQ: X_3) from GRE scores, but not from GPA. The simple bivariate correlation between GPA and GRE is $r_{12} = .5$; between GPA and IQ is $r_{13} = .3$; and between GRE and IQ is $r_{23} = .7$. We compute the semipartial correlation that removes the influence of intelligence (IQ: X_3) from GRE scores (X_2), but not from GPA (X_1) (i.e., $r_{1(2.3)}$) as follows:

$$r_{1(2.3)} = \frac{r_{12} - r_{13}r_{23}}{\sqrt{\left(1 - r_{23}^2\right)}} = \frac{.5 - (.3)(.7)}{\sqrt{1 - .49}} = .4061$$

Thus, the bivariate correlation between GPA (X_1) and GRE scores (X_2) ignoring IQ (X_3) (r_{12} = .50) is larger than the semipartial correlation between GPA and GRE controlling for IQ in GRE ($r_{1(2 \cdot 3)}$ = .4061). As was the case with partial correlations, various values of a semipartial correlation can be obtained depending on the combination of the bivariate correlations. For more information on partial and semipartial correlations, see Glass and Hopkins (1996), Hays (1988), and Pedhazur (1997).

Now that we have considered the correlational relationships among two or more variables (i.e., partial and semipartial correlations), let us move on to an examination of the multiple regression model where there are two or more predictor variables.

Let us take the concepts we have learned in this and the previous chapter and place them into the context of multiple linear regression. For purposes of brevity, we do not consider the population situation because the sample situation is invoked 99.44% of the time. In this section we discuss the unstandardized and standardized multiple regression models, the coefficient of multiple determination, multiple correlation, tests of significance, and statistical assumptions.

8.1.1.3 Unstandardized Regression Model

The sample multiple linear regression model for predicting Y from m predictors $X_{1, 2, \ldots,}$ m is

$$Y_i = b_1 X_{1i} + b_2 X_{2i} + \ldots + b_m X_{mi} + a + e_i$$

where Y is the **criterion variable** (also known as the dependent variable); the X_k's are the **predictor (or independent) variables** where $k = 1, \ldots, m$; b_k is the **sample partial slope** of the regression line for Y as predicted by X_k, a is the **sample intercept** of the regression line for Y as predicted by the set of X_k's; e_i are the **residuals or errors of prediction** (the part of Y not predictable from the X_k's); and i represents an index for an individual or object. The index i can take on values from 1 to n where n is the size of the sample (i.e., $i = 1, \ldots, n$). The term **partial slope** is used because it represents the slope of Y for a particular X_k in which we have partialed out the influence of the other X_k's, much as we did with the partial correlation.

The **sample prediction model** is

$$Y_i' = b_1 X_{1i} + b_2 X_{2i} + \ldots + b_m X_{mi} + a$$

Where Y_i' is the predicted value of Y for specific values of the X_k's, and the other terms are as before. There is only one difference between the regression and prediction models. The regression model explicitly includes prediction error as e_i, whereas the prediction model includes prediction error implicitly as part of the predicted score Y_i' (i.e., there is some error in the predicted values). The goal of the prediction model is to include an independent variable X that minimizes the residual; this means that the independent variable does a nice job of predicting the outcome. We can compute residuals, the e_i, for each of the i individuals or objects by comparing the actual Y values with the predicted Y values as

$$e_i = Y_i - Y_i'$$

for all $i = 1, \ldots, n$ individuals or objects in the sample.

Determining the sample partial slopes and the intercept in the multiple predictor case is rather complicated. To keep it simple, we use a two-predictor model for illustrative purposes. Generally we rely on statistical software for implementing multiple regression analysis. For the two-predictor case, the sample partial slopes (b_1 and b_2) and the intercept (a) can be determined as follows:

$$b_1 = \frac{\left(r_{Y1} - r_{Y2}r_{12}\right)s_Y}{\left(1 - r_{12}^2\right)s_1}$$

$$b_2 = \frac{\left(r_{Y2} - r_{Y1}r_{12}\right)s_Y}{\left(1 - r_{12}^2\right)s_2}$$

$$a = \bar{Y} - b_1\bar{X}_1 - b_2\bar{X}_2$$

The sample partial slope b_1 is referred to alternately as (a) the expected or predicted change in Y for a one-unit change in X_1 with X_2 held constant (or for individuals with the same score on X_2), and (b) the unstandardized or raw regression coefficient for X_1. Similar statements may be made for b_2. Note the similarity of the partial slope equation to the semipartial correlation. The sample intercept is referred to as the value of the dependent variable Y when the values of the independent variables X_1 and X_2 are both zero.

An alternative method for computing the sample partial slopes that involves the use of a partial correlation is as follows:

$$b_1 = r_{Y1.2}\left(\frac{s_Y\sqrt{1 - r_{Y2}^2}}{s_1\sqrt{1 - r_{12}^2}}\right)$$

$$b_2 = r_{Y2.1}\left(\frac{s_Y\sqrt{1 - r_{Y1}^2}}{s_2\sqrt{1 - r_{12}^2}}\right)$$

What statistical criterion is used to arrive at the particular values for the partial slopes and intercept? The criterion usually used in multiple linear regression analysis [and in all general linear models (GLM) for that matter] is the **least squares criterion**. *The least squares criterion arrives at those values for the partial slopes and intercept such that the sum of the squared prediction errors or residuals is smallest.* That is, we want to find that regression model, defined by a particular set of partial slopes and an intercept, which has the smallest sum of the squared residuals. We often refer to this particular method for calculating the slope and intercept as **least squares estimation,** because a and the b_k's represent sample estimates of the population parameters a and the β_k's, which are obtained using the least squares criterion. Recall from simple linear regression that the residual is simply the vertical distance from the observed value of Y to the predicted value of Y, and the line of best fit minimizes this distance. This concept still applies to multiple linear regression with the exception that we are now in a three-dimensional (or more) plane given there are multiple independent variables.

8.1.1.4 Standardized Regression Model

Up until this point in the chapter, everything in multiple linear regression analysis has involved the use of raw scores. For this reason we referred to the model as the

unstandardized regression model. Often we may want to express the regression in terms of standard z score units rather than in raw score units. The means and variances of the standardized variables (e.g., z_1, z_2, z_Y) are 0 and 1, respectively. The **sample standardized linear prediction model** becomes the following:

$$z(Y_i') = b_1^* z_{1i} + b_2^* z_{2i} + \ldots + b_m^* z_{mi}$$

where $\boldsymbol{b_k^*}$ represents a **sample standardized partial slope** (sometimes called **beta weights**) and the other terms are as before. As was the case in simple linear regression, no intercept term is necessary in the standardized prediction model as the mean of the z scores for all variables is 0. (Recall that the intercept is the value of the dependent variable when the scores on the independent variables are all zero. Thus, *in a standardized prediction model, the dependent variable will equal zero when the values of the independent variables are equal to their means—i.e., zero)*. The sample standardized partial slopes are, in general, computed by the following equation:

$$b_k^* = b_k \left(\frac{s_k}{s_Y} \right)$$

For the two-predictor case, the standardized partial slopes can be calculated by

$$b_1^* = b_1 \left(\frac{s_1}{s_Y} \right)$$

$$b_1^* = \frac{r_{Y1} - (r_{Y2} r_{12})}{(1 - r_{12}^2)}$$

and

$$b_2^* = b_2 \left(\frac{s_2}{s_Y} \right)$$

or

$$b_2^* = \frac{r_{Y2} - (r_{Y2} r_{12})}{(1 - r_{12}^2)}$$

If the two predictors are *uncorrelated* (i.e., $r_{12} = 0$), then the standardized partial slopes are equal to the simple bivariate correlations between the dependent variable and the independent variables (i.e., $b_1^* = r_{Y1}$ and $b_2^* = r_{Y2}$) because the rest of the equation goes away, as we see here. In the latter "mathematical introduction snapshot," we provide an illustration of this using the example data in the chapter.

$$b_1^* = \frac{r_{Y1} - (r_{Y2} r_{12})}{(1 - r_{12}^2)} = \frac{r_{Y1} - r_{Y2}(0)}{(1 - 0)} = r_{Y1}$$

When would you want to use the standardized versus unstandardized regression analyses? According to Pedhazur (1997), b_k^* is sample specific and is not very stable across different samples due to the variance of X_k changing (as the variance of X_k increases, the value of b_k^* also increases, all else being equal). For example, the example we will review later with data from Ivy-Covered University, b_k^* would vary across different graduating classes (or samples) while b_k would be much more consistent across classes. Thus most researchers prefer the use of b_k to compare the influence of a particular predictor variable across different samples and/or populations. Pedhazur also states that the b_k^* is of "limited value" (p. 321), but could be reported along with the b_k. As Pedhazur and others have reported, the b_k^* can be deceptive in determining the relative importance of the predictors as they are affected by the variances and covariances of both the included predictors and the predictors not included in the model. Thus we recommend the b_k for general purpose use.

8.1.1.5 Coefficient of Multiple Determination and Multiple Correlation

An obvious question now is, how well is the criterion variable predicted or explained by the set of predictor variables? For our example, we are interested in how well graduate grade point averages (the dependent variable) are predicted by GRE total scores and undergraduate grade point averages. In other words, *what is the utility of the set of predictor variables?*

The simplest method involves the partitioning of the familiar total sum of squares in Y, which we denote as SS_{total}. In multiple linear regression analysis, we can write SS_{total} as follows:

$$SS_{total} = \frac{\left[n\sum Y_i^2 - \left(\sum Y_i\right)^2\right]}{n}$$

$$\text{or } SS_{total} = (n-1)s_Y^2$$

where we sum over Y from $i = 1, \ldots, n$. Next we can conceptually partition SS_{total} as

$$SS_{total} = SS_{reg} + SS_{res}$$

$$\sum\left(Y_i - \bar{Y}\right)^2 = \sum\left(Y_i' - \bar{Y}\right)^2 + \sum\left(Y_i - Y_i'\right)^2$$

where SS_{reg} is the regression sum of squares due to the prediction of Y from the X_k's (often written as $SS_{Y'}$), and SS_{res} is the sum of squares due to the residuals.

Before we consider computation of SS_{reg} and SS_{res}, let us look at the **coefficient of multiple determination**. Recall the coefficient of determination that is applicable to simple linear regression, r_{XY}^2. We now consider the multiple predictor version of r_{XY}^2, here denoted as, $R_{Y.1,\ldots,m}^2$ which we will shorthand as R^2. The subscript tells us that Y is the criterion (or dependent) variable and that $X_{1,\ldots,}m$ are the predictor (or independent) variables (with m representing the total number of independent variables). The simplest procedure for computing R^2 is as follows:

$$R_{Y.1,\ldots,m}^2 = b_1^* r_{Y1} + b_2^* r_{Y2} + \ldots + b_m^* r_{Ym}$$

The coefficient of multiple determination tells us *the proportion of total variation in the dependent variable Y that is predicted from the set of predictor variables* (i.e., $X_{1,\ldots,}m$'s). Often we see the coefficient in terms of SS as follows:

$$R^2_{Y.1,\ldots,m} = \frac{SS_{reg}}{SS_{total}}$$

Thus, one method for computing the sums of squares regression and residual, SS_{reg} and SS_{res}, is from the coefficient of multiple determination, R^2 (an index that can also be used a measure of effect size) as follows:

$$SS_{reg} = R^2\left(SS_{total}\right)$$

$$SS_{res} = \left(1-R^2\right)\left(SS_{total}\right) = SS_{total} - SS_{reg}$$

Note also that $R_{Y.1,\ldots,m}$ is referred to as the **multiple correlation coefficient** so as not to confuse it with a simple bivariate correlation coefficient. In the latter "mathematical introduction snapshot," we provide an illustration using the example data in the chapter.

It should be noted that R^2 is sensitive to sample size and to the number of predictor variables. As sample size and/or the number of predictor variables increase, R^2 will increase as well. R is a biased estimate of the population multiple correlation due to sampling error in the bivariate correlations and in the standard deviations of X and Y. Because R systematically overestimates the population multiple correlation, an adjusted coefficient of multiple determination has been devised. The adjusted R^2 (denoted as R^2_{adj}) is calculated as follows:

$$R^2_{adj} = 1-\left(1-R^2\right)\left(\frac{n-1}{n-m-1}\right)$$

Thus, R^2_{adj} adjusts for sample size and for the number of predictors in the model; this allows us to compare models fitted to the same set of data with different numbers of predictors or with different samples of data. The difference between the squared multiple correlation (aka coefficient of multiple determination), R^2, and the adjusted squared multiple correlation (aka adjusted coefficient of multiple determination), R^2_{adj}, is called **shrinkage**.

When n is small relative to m, the amount of bias can be large as R^2 can be expected to be large by chance alone. In this case the adjustment will be quite large, as it should be. In addition, with small samples, the regression coefficients (i.e., the b_k's) may not be very good estimates of the population values. When n is large relative to m, bias will be minimized and generalizations are likely to be better about the population values.

For the example data, we determine the adjusted multiple coefficient of determination R^2_{adj} to be as follows:

$$R^2_{adj} = 1-\left(1-R^2\right)\left(\frac{n-1}{n-m-1}\right) = 1-\left(1-.9089\right)\left(\frac{11-1}{11-2-1}\right) = .8861$$

In this case, the adjusted multiple coefficient of determination indicates a very small adjustment in comparison to R^2.

8.1.1.6 Significance Tests

Here we describe two procedures used in multiple linear regression analysis. These involve testing the significance of the overall regression model and of each individual partial slope (or regression coefficient).

8.1.1.6.1 Test of Significance of the Overall Regression Model

The first test is the test of significance of the overall regression model, or alternatively the *test of significance of the coefficient of multiple determination*. This is a test of all of the b_k's simultaneously, an examination of overall model fit of the independent variables in aggregate. The null and alternative hypotheses, respectively, are as follows:

$$H_0\colon \beta_1 = \beta_2 = \ldots = \beta_k = 0$$

$$H_1\colon \textit{not all the } \beta_k = 0$$

If H_0 is rejected, then one or more of the individual regression coefficients (i.e., the b_k) is statistically significantly different from zero (if the assumptions are satisfied, as discussed later). If H_0 is not rejected, then none of the individual regression coefficients will be significantly different from zero.

The test is based on the following test statistic:

$$F = \frac{R^2/m}{(1-R^2)(n-m-1)}$$

where F indicates that this is an *F statistic*, m is the *number of predictors* or independent variables, and n is the *sample size*. The F test statistic is compared to the F critical value, always a one-tailed test (by default, this value can never be negative given the terms in the equation, so this will always be a *directional test*) and at the designated level of significance, with *degrees of freedom* being m and $(n - m - 1)$, as taken from the F table (see Appendix). That is, the tabled critical value is ${}_{\alpha}F_{m,(n-m-1)}$. The test statistic can also be written in equivalent form as

$$F = \frac{SS_{reg}/df_{reg}}{SS_{res}/df_{res}} = \frac{MS_{reg}}{MS_{res}}$$

Where the degrees of freedom regression equals the number of independent variables, $df_{reg} = m$, and degrees of freedom residual equals the difference between the sample size, number of independent variables, and one, $df_{res} = (n - m - 1)$.

8.1.1.6.2 Test of Significance of b_k

The second test is the test of the statistical significance of each individual partial slope or regression coefficient, b_k. *That is, are the individual unstandardized regression coefficients statistically significantly different from zero?* This is actually the same as the test of b^*_k, so we need not develop a separate test for b^*_k. The null and alternative hypotheses, respectively, are as follows:

$$H_0 : \beta_k = 0$$
$$H_1 : \beta_k \neq 0$$

where β_k is the **population partial slope** for X_k.

In multiple regression it is necessary to compute a standard error for each regression coefficient b_k. The **variance error of estimate, s^2_{res}**, is similarly defined for multiple linear regression and computed as:

$$s^2_{res} = \frac{SS_{res}}{df_{res}} = MS_{res}$$

where $df_{res} = (n - m - 1)$. Degrees of freedom are lost as we have to estimate the population partial slopes and intercept, the β_k's and α, respectively, from the sample data. *The variance error of estimate indicates the amount of variation among the residuals.* The standard error of estimate is simply the positive square root of the variance error of estimate and is the standard deviation of the residuals or errors of estimate. We call it the **standard error of estimate**, denoted as s_{res}.

Finally, we need to compute a **standard error** for each b_k. Denote the standard error of b_k as $s(b_k)$ and define it as follows:

$$s(b_k) = \frac{s_{res}}{\sqrt{(n-1)s_k^2(1-R_k^2)}}$$

where s_k^2 is the **sample variance** for predictor X_k, and R_k^2 is the **squared multiple correlation** between X_k and the remaining X_k's. R_k^2 represents the overlap between that predictor (X_k) and the remaining predictors. In the case of two predictors, the squared multiple correlation R_k^2 is equal to the simple bivariate correlation between the two independent variables r_{12}^2.

The test statistic, t, for testing the significance of the regression coefficients, b_k's, is as follows:

$$t = \frac{b_k}{s(b_k)}$$

The test statistic t is compared to the critical values of t, a two-tailed test for a nondirectional H_1, at the designated level of significance, and with degrees of freedom $(n - m - 1)$, as taken from the t table in Appendix Table A.2. Thus, the tabled critical values are $\pm_{(\alpha/2)}t_{(n-m-1)}$ for a two-tailed test.

We can also form a *confidence interval* around b_k as follows:

$$\text{CI}(b_k) = b_k \pm {}_{(\alpha/2)}t_{(n-m-1)}\, s(b_k)$$

Recall that the null hypothesis tested is $H_0 = \beta_k = 0$. Therefore, if the confidence interval contains zero, then the regression coefficient b_k is not statistically significantly different from zero at the specified α level. This is interpreted to mean that in $(1 - \alpha)$% of the sample confidence intervals that would be formed from multiple samples, β_k will be included. In the latter "mathematical introduction snapshot," we provide an illustration using the example data in the chapter.

8.1.1.6.3 Other Tests

One can also form confidence intervals for the predicted mean of Y and the prediction intervals for individual values of Y.

8.1.1.7 Methods of Entering Predictors

There are many different ways in which predictor variables can be entered in a regression model, none of which are necessarily right or wrong—although we highly discourage the use of a few. We will begin with what is likely the most common method of entering independent variables, and that is simultaneous regression. There are other methods of entering the independent variables where the predictor variables are entered (or selected) systematically; here the set of predictors has not been selected *a priori*. This class of models is referred to as **sequential regression** (also known as **variable selection procedures**). This section introduces a brief description of the following sequential regression procedures: backward elimination, forward selection, stepwise selection, all possible subsets regression, and hierarchical regression.

8.1.1.7.1 Simultaneous Regression

The multiple predictor model which we have considered thus far can be viewed as **simultaneous regression**. That is, *all of the predictors to be used are entered (or selected) simultaneously*, such that all of the regression parameters are estimated simultaneously; here the set of predictors has been selected *a priori*. In computing these regression models, we have used the default setting in SPSS of the method of entry as "Enter," which enters the set of independent variables in aggregate.

8.1.1.7.2 Backward Elimination

First consider the backward elimination procedure. Here variables are eliminated from the model based on their **minimal contribution** to the prediction of the criterion variable. In the first stage of the analysis, all potential predictors are included in the model. In the second stage, that predictor is deleted from the model that makes the smallest contribution to the prediction of the dependent variable. This can be done by eliminating that variable having the smallest t or F statistic such that it is making the smallest contribution to R^2_{adj}. In subsequent stages, that predictor is deleted that makes the next smallest contribution to the prediction of the outcome Y. The analysis continues until each of the remaining predictors in the model is a significant predictor of Y. This could be determined by comparing the t or F statistics for each predictor to the critical value, at a preselected level of significance. Some computer programs use as a stopping rule the maximum F-to-remove criterion, where the procedure is stopped when all of the selected predictors' F values are greater than the specified F criterion. Another stopping rule is where the researcher stops at a predetermined number of predictors (see Hocking, 1976; Thompson, 1978). In SPSS, this is the **backward** method of entering predictors.

8.1.1.7.3 Forward Selection

In the forward selection procedure, variables are added or selected into the model based on their **maximal contribution** to the prediction of the criterion variable. Initially, none of the potential predictors are included in the model. In the first stage, the predictor is added to the model that makes the largest contribution to the prediction of the dependent variable. This can be done by selecting that variable having the largest t or F statistic such that it is making the largest contribution to R^2_{adj}. In subsequent stages, the predictor is selected that makes the next largest contribution to the prediction of Y. The analysis continues until each of the selected predictors in the model is a significant predictor of the outcome Y, whereas none of the unselected predictors is a significant predictor. This could be determined by comparing the t or F statistics for each predictor to the critical value, at a preselected level of significance. Some computer programs use as a stopping rule the minimum F-to-enter criterion, where the procedure is stopped when all of the unselected predictors' F values are less than the specified F criterion. For the same set of data and at the same level of significance, the backward elimination and forward selection procedures may not necessarily result in the exact same final model due to the differences in how variables are selected. In SPSS, this is the **forward** method of entering predictors.

8.1.1.7.4 Stepwise Selection

The stepwise selection procedure is a modification of the forward selection procedure with one important difference. *Predictors that have been selected into the model can at a later step be deleted from the model*; thus, the modification conceptually involves a backward elimination mechanism. This situation can occur for a predictor when a significant contribution at an earlier step later becomes a nonsignificant contribution given the set of other predictors in the model. Thus a predictor loses its significance due to new predictors being added to the model.

The stepwise selection procedure is as follows. Initially, none of the potential predictors are included in the model. In the first step, that predictor is added to the model that makes the largest contribution to the explanation of the dependent variable. This can be done by selecting that variable having the largest t or F statistic such that it is making the largest contribution to R^2_{adj}. In subsequent stages, the predictor is selected that makes the next largest contribution to the prediction of Y. Those predictors that have entered at earlier stages are also checked to see if their contribution remains significant. If not, then that predictor is eliminated from the model. The analysis continues until each of the predictors remaining in the model is a significant predictor of Y, while none of the other predictors is a significant predictor. This could be determined by comparing the t or F statistics for each predictor to the critical value, at a specified level of significance. Some computer programs use as stopping rules the minimum F-to-enter and maximum F-to-remove criteria, where the F-to-enter value selected is usually equal to or slightly greater than the F-to-remove value selected (to prevent a predictor from continuously being entered and removed). For the same set of data and at the same level of significance, the backward elimination, forward selection, and stepwise selection procedures may not necessarily result in the exact same final model, due to differences in how variables are selected. In SPSS, this is the **stepwise** method of entering predictors.

8.1.1.7.5 All Possible Subsets Regression

Another sequential regression procedure is known as all possible subsets regression. Let us say, for example, that there are five potential predictors. In this procedure, all possible one-, two-, three-, and four-variable models are analyzed (with five predictors, there is only a single five-predictor model). Thus there will be 5 one-predictor models, 10 two-predictor models, 10 three-predictor models, and 5 four-predictor models. The best k predictor model can be selected as the model that yields the largest R^2_{adj}. For example, the best three-predictor model would be that model of the 10 estimated that yields the largest R^2_{adj}. With today's powerful computers, this procedure is easier and more cost efficient than in the past. However, the researcher is not advised to consider this procedure, or for that matter, any of the other sequential regression procedures, when the number of potential predictors is large. Here the researcher is allowing number crunching to take precedence over thoughtful analysis. Also, the number of models will be equal to $2m$, so that for 10 predictors there are 1,024 possible subsets. Obviously examining that number of models is not a thoughtful analysis.

8.1.1.7.6 Hierarchical Regression

In hierarchical regression, *the researcher specifies a priori a sequence for the individual predictor variables* (not to be confused with hierarchical linear models, which is a regression approach for analyzing nested data collected at multiple levels, such as child, classroom, and school). The analysis proceeds in a forward selection, backward elimination, or stepwise selection mode according to a researcher-specified, *theoretically based sequence*, rather than an unspecified statistically based sequence. This variable selection method is different from those previously discussed in that the *researcher determines the order of entry from a careful consideration of the available theory research*, instead of the software dictating the sequence.

A type of hierarchical regression is known as **setwise regression** (also called **blockwise, chunkwise**, or **forced stepwise regression**). Here the researcher specifies *a priori* a sequence for sets of predictor variables. This procedure is similar to hierarchical regression in that the researcher determines the order of entry of the predictors. The difference is that the setwise method uses sets of predictor variables at each stage rather than one individual predictor variable at a time. The sets of variables are determined by the researcher so that variables within a set share some common theoretical ground (e.g., home background variables in one set and aptitude variables in another set). Variables within a set are selected according to one of the sequential regression procedures. The variables selected for a particular set are then entered in the specified theoretically based sequence. In SPSS, this is conducted by entering predictors in **blocks** and selecting their desired method of entering variables in each block (e.g., simultaneously, forward, backward, stepwise).

8.1.1.7.7 Commentary on Sequential Regression Procedures

Let us make some comments and recommendations about the sequential regression procedures, which are summarized in Box 8.1. First, numerous statisticians have noted problems with stepwise methods (i.e., backward elimination, forward selection, and stepwise selection) (e.g., Derksen & Keselman, 1992; Huberty, 1989; Mickey, Dunn, & Clark, 2004; Miller, 1984, 1990; Wilcox, 2003). These problems include the following: (a) selecting noise rather than important predictors; (b) highly inflated R^2 and R^2_{adj} values; (c) confidence intervals for partial slopes that are too narrow; (d) p values that are not trustworthy; (e) important predictors being barely edged out of the model, making it possible to miss the true model; and (f) potentially heavy capitalization on chance given the number of models analyzed.

Second, theoretically based regression models have become the norm in many disciples (and the stepwise methods of entry are driven by mathematics of the models rather than theory). Thus hierarchical regression either has or will dominate the landscape of the sequential regression procedures. Thus, we strongly encourage you to consider more extended discussions of hierarchical regression (Cohen, Cohen, West, & Aiken, 2003; Pedhazur, 1997; Tabachnick & Fidell, 2013, 2019).

If you are working in an area of inquiry where research evidence is scarce or nonexistent, then you are conducting exploratory research. Thus, you are probably trying to simply identify the key variables. Here hierarchical regression is not appropriate, as a theoretically based sequence cannot be developed as there is no theory to guide its development. Here we recommend the use of all possible subsets regression (Kleinbaum, Kupper, Muller, & Nizam, 1998). For additional information on the sequential regression procedures, see Cohen and Cohen (1983), Weisberg (1985), Miller (1990), Pedhazur (1997), and Kleinbaum et al. (1998).

8.1.1.8 Nonlinear Relationships

Here we discuss how to deal with nonlinearity. We formally introduce several multiple regression models for when the criterion variable does not have a linear relationship with the predictor variables.

First consider polynomial regression models. In polynomial models, powers of the predictor variables (e.g., squared, cubed) are used. In general, a sample polynomial regression model that includes one quadratic term is as follows:

$$Y = b_1 X_1 + b_2 X^2 + \ldots + b_m X^m + a + e$$

where the independent variable X is taken from the first power through the mth power, and the i subscript for observations has been deleted to simplify matters. If the model consists only of X taken to the first power, then this is a **simple linear regression model** (or **first-degree polynomial**; this is a straight line and what we have studied to this point). A **second-degree polynomial** includes X taken to the second power (or **quadratic model**; this is a curve with one bend in it rather than a straight line). A **third-degree polynomial** includes X taken to the third power (or **cubic model**; this is a curve with two bends in it).

A polynomial model with multiple predictors can also be utilized. An example of a second-degree polynomial model with two predictors (X_1 and X_2) is illustrated in the following equation:

$$Y = b_1 X_1 + b_2 X_1^2 + b_3 X_2 + b_4 X_2^2 + a + e$$

It is important to note that when whenever a higher-order polynomial is included in a model (e.g., quadratic, cubic, and more), the first-order polynomial must also be included in the model. In other words, it is not appropriate to include a quadratic term X^2 without also including the first-order polynomial X. For more information on polynomial regression models, see Weisberg (1985), Bates and Watts (1988), Seber and Wild (1989), Pedhazur (1997), and Kleinbaum et al. (1998). Alternatively, one might transform the criterion variable and/or the predictor variables to obtain a more linear form, as previously discussed.

8.1.1.9 Interactions

Another type of model involves the use of an interaction term, a term with which you may be familiar from factorial ANOVA. These can be implemented in any type of regression model. We can write a simple two-predictor interaction-type model as follows:

$$Y = b_1 X_1 + b_2 X_1 + b_3 X_1 X_2 + a + e$$

where X_1X_2 represents the interaction of predictor variables 1 and 2. An interaction can be defined as occurring when the relationship between Y and X_1 depends on the level of X_2. In other words, X_2 is a **moderator variable**. For example, suppose one were to use years of education and age to predict political attitude. The relationship between education and attitude might be moderated by age. In other words, the relationship between education and attitude may be different for older versus younger individuals. If age were a moderator, we would expect there to be an interaction between age and education in a regression model. Note that if the predictors are very highly correlated, collinearity is likely. Moderation is covered in more detail in the final chapter. For more information on interaction models, see Cohen and Cohen (1983), Berry and Feldman (1985), Kleinbaum et al. (1998), Weinberg and Abramowitz (2002), and Meyers, Gamst, and Guarino (2006).

8.1.1.10 Categorical Predictors

So far we have only considered continuous predictors—independent variables that are interval or ratio in scale. There may be times, however, that you wish to use a categorical predictor—an independent variable that is nominal or ordinal in scale. For example, gender, grade level (e.g., freshman, sophomore, junior, senior), highest education earned (less than high school, high school graduate, etc.) are all categorical variables that may be very interesting and theoretically appropriate to include in either a simple or multiple regression model. Given their scale (i.e., nominal or ordinal), however, we must recode the values prior to analysis so that they are on a scale of zero and one. This is called "dummy coding" as this type of recoding makes the model work. For example, males might be coded as zero and females coded as one. When there are more than two categories to the categorical predictor, multiple dummy coded variables must be created—*specifically one minus the number of levels or categories of the categorical variable*. Thus, in the case of grade level where there are four categories (freshman, sophomore, junior, senior), three of the four categories would be dummy coded and included in the regression model as predictors. The category that is "left out" is the *reference category*, or that category to which all other levels are compared. The easiest way to understand this is perhaps to examine the data. In the screenshot in Figure 8.1, the first column represents grade level where 1 = freshman, 2 = sophomore,

	Grade	GPA
1	1.00	2.50
2	1.00	2.20
3	1.00	2.70
4	2.00	3.50
5	2.00	3.40
6	2.00	3.60
7	3.00	3.30
8	3.00	3.60
9	3.00	3.50
10	4.00	3.00
11	4.00	2.90
12	4.00	3.90

FIGURE 8.1
Grade level (categorical variable).

3 = junior, and 4 = senior. Dummy coding the grade levels will result in additional columns being created.

Dummy coding the grade levels will result in additional columns being created. To easily do this in SPSS, go to "Transform," then "Create Dummy Variables" (see Figure 8.2).

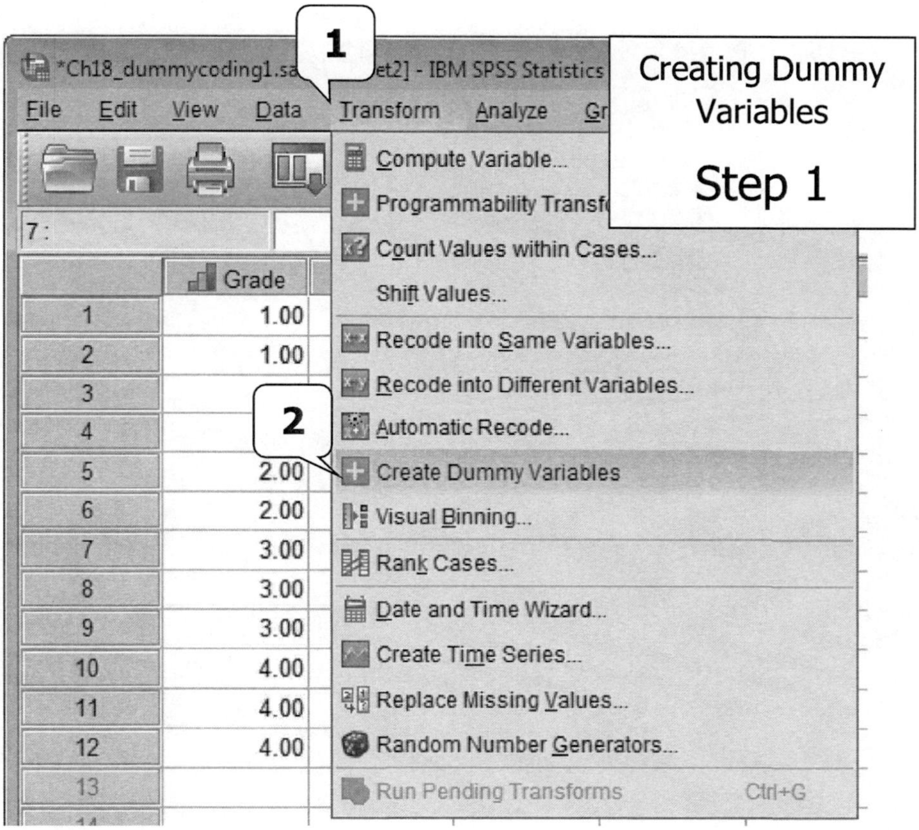

FIGURE 8.2
Creating dummy variables: Step 1.

From the Create Dummy Variables dialog box (see Figure 8.3), click on the categorical variable for which you want to create dummy values and click the arrow to move into the "Create Dummy Variables for" box. Place a check in the box to "Create main-effect dummies" and assign a "Root Name." The root name is essentially a prefix to the new dummy variables that will be created. If you have correctly specified the measurement scale of your variables as either nominal or ordinal in SPSS, then nothing else is needed. However, if your variables are not correctly specified, or if you have a scale (i.e., interval or ratio) variable that you want to create dummy variables for, then you'll need to select the radio button for "Create dummies for all variables." After your selections are made, click "OK." (Note that you are *highly encouraged* to make sure the measurement scales of your variables in your datafile are correctly defined! This is just good data cleaning practice!)

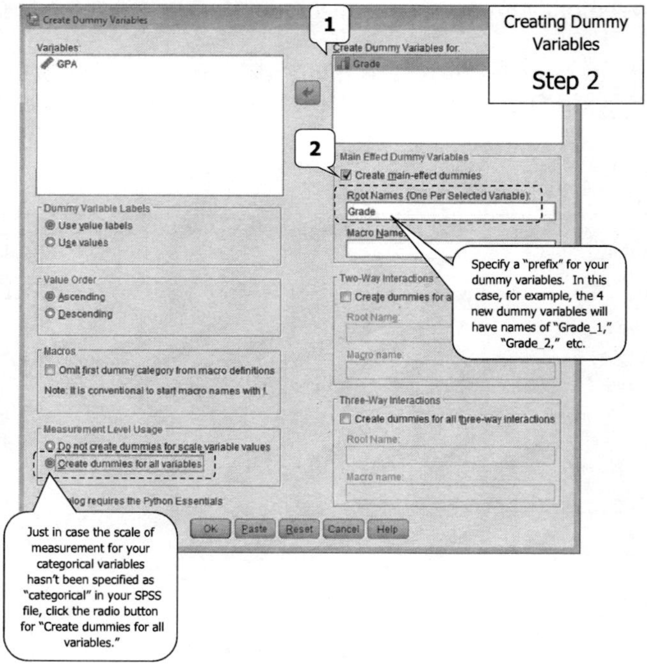

FIGURE 8.3
Creating dummy variables: Step 2.

Going back to your datafile, you'll see your new variables—one for each category of the variable.

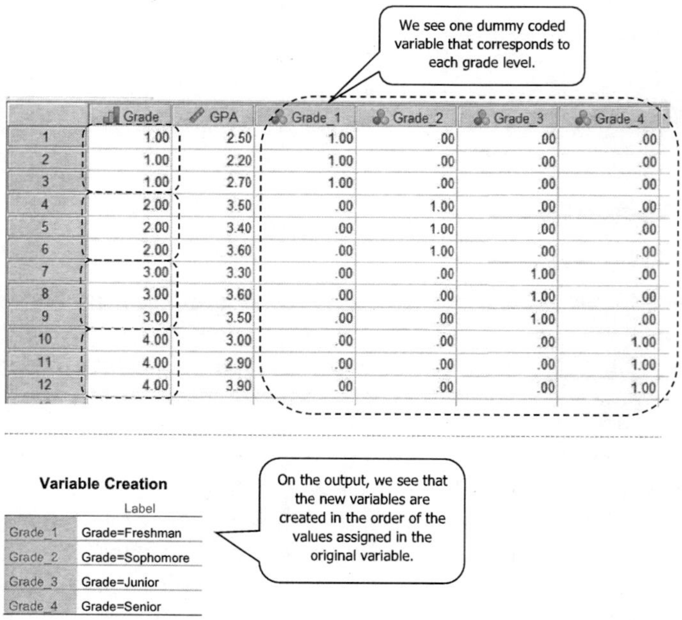

	Grade	GPA	Grade_1	Grade_2	Grade_3	Grade_4
1	1.00	2.50	1.00	.00	.00	.00
2	1.00	2.20	1.00	.00	.00	.00
3	1.00	2.70	1.00	.00	.00	.00
4	2.00	3.50	.00	1.00	.00	.00
5	2.00	3.40	.00	1.00	.00	.00
6	2.00	3.60	.00	1.00	.00	.00
7	3.00	3.30	.00	.00	1.00	.00
8	3.00	3.60	.00	.00	1.00	.00
9	3.00	3.50	.00	.00	1.00	.00
10	4.00	3.00	.00	.00	.00	1.00
11	4.00	2.90	.00	.00	.00	1.00
12	4.00	3.90	.00	.00	.00	1.00

Variable Creation

	Label
Grade_1	Grade=Freshman
Grade_2	Grade=Sophomore
Grade_3	Grade=Junior
Grade_4	Grade=Senior

FIGURE 8.4
Newly created dummy variables.

In terms of generating the analysis and the point-and-click use of SPSS to compute the regression model, nothing changes. The steps are the same regardless of whether the predictors are continuous or categorical. Now let us discuss *why* dummy coding works in this situation. The point biserial correlation is appropriate when one variable is dichotomous and the other variable is interval or ratio. The point biserial correlation is a variant of the Pearson product-moment correlation, and we can use the Pearson as a variant of the point biserial. Thus while we will not have a linear relationship between a continuous outcome and a binary variable, the mathematics that underlie the model will hold.

Consider the example output for predicting grade point average (GPA) based on grade level, where "senior" is the reference category. We see that the intercept (i.e., "constant") is statistically significant as is "freshman." The interpretation of the intercept remains the same regardless of the scale of the predictors. The intercept represents grade point average (the dependent variable) when all the predictors are zero. In this case, this means that grade point average is 3.267 for *seniors* (the reference category). The only statistically significant predictor is "freshman." This is interpreted to say that mean GPA decreases by .800 points for freshmen *as compared to seniors*. The nonstatistically significant regression coefficients for "sophomore" and "junior" indicate that mean GPA is similar for these grade levels as compared to seniors. The interpretation for dummy variable predictors is always in reference to the category that was "left out." In this case, that was "seniors."

Coefficients[a]

Model		Unstandardized Coefficients		Standardized Coefficients		
		B	Std. Error	Beta	t	Sig.
1	(Constant)	3.267	.183		17.892	.000
	Grade=Freshman	-.800	.258	-.704	-3.098	.015
	Grade=Sophomore	.233	.258	.205	.904	.393
	Grade=Junior	.200	.258	.176	.775	.461

a. Dependent Variable: GPA

FIGURE 8.5
Regression model with dummy variables.

It is important to note that even though "sophomore" and "junior" were not statistically significant, they should be retained in the model as they represent (along with "freshman") a group. Dropping one or more dummy coded indicator variables that represent a group will change the reference category. For example, if "sophomore" and "junior" were dropped from the model, the interpretation would then become the mean GPA for freshmen *as compared to all other grade levels*. Thus, careful thought needs to be put into dropping one or more indicators that are part of a set.

8.1.2 Sample Size

We will start and end with the same recommendation: Estimate sample size using power software or tables and consult current advances related to estimating sample size based on simulation research. With that said, you don't have to look far to find sample size guidelines and recommendations. For example, you have likely read in other textbooks or had a colleague recommend that there be at least 10 cases for every independent variable in the multiple regression model. This is an inappropriate way to estimate sample size. A fair body of research has examined minimum sample size in the context of multiple linear regression, and some consensus that sample size considerations differ depending on the goal of your research—either testing a hypothesis test estimating a parameter (Algina & Olejnik, 2000; Maxwell, 2000)—with larger sample sizes needed for estimation of a prediction equation (e.g., Pedhazur, 1997), as compared to sample sizes needed for testing a hypothesis related to the multiple correlation coefficient (Maxwell, 2000). Using simulation research, recent research suggests that the squared multiple correlation coefficient does have a relationship with overall sample size and the ratio of the sample size to predictors. Simulation research to examine the sample size needed to ensure the sample regression equation performed similarly to the population regression equation, Knofszynski (2008) examined more than 23 million simulated samples and found that the need for a large sample size increases as the squared multiple correlation coefficient diminishes (Knofszynski, 2008). More specifically, as the squared multiple correlation coefficient nears zero, there is a quicker increase in the need for a larger sample size, and this pattern is constant across varying numbers of predictors; however, the sample size does not dramatically increase as the number of predictors increases. For example, with a squared multiple correlation coefficient of .10 and three predictors, a sample size of 1800 is needed to achieve "excellent prediction level" (Knofszynski, 2008, p. 438). In comparison, with a square multiple correlation coefficient of .50, again with three predictors, a sample size of 220 is needed to achieve "excellent" (Knofszynski, 2008). When R^2 is .90, a sample size of only 15 or 7 is needed to achieve "excellent" or "good" prediction, respectively (Knofszynski, 2008). As we know, sample size and power are inextricably intertwined, and attempting to separate the two is futile. The best recommendation is to estimate sample size using power software and to consult current advances based on simulation research such as Knofszynski (2008).

8.1.3 Power

With a large number of predictors, power is reduced, and there is an increased likelihood of a Type I error across the total number of significance tests (i.e., one for each predictor and overall). In multiple regression, power is a function of sample size, the number of predictors, the level of significance, and the size of the population effect (i.e., for a given predictor, or overall). With multiple regression, there are several estimates of power that may be of interest to researchers including power for the *group of all predictors* (**R^2 model**), power for *one group of predictors as compared to another group of predictors* (**R^2 change**), and power for a *single predictor within the model* (i.e., **regression coefficient**) (Aberson, 2010). Because adding predictors increases the value of R^2, it is easy to fall into the trap of mindlessly including additional predictors (Murphy, Myors, & Wolach, 2014). However, powering for the group of all predictors (R^2 model) is affected by both the number of predictors and the amount of variance explained. *Increasing the number of predictors can decrease power as they use degrees of freedom for testing the null hypothesis, or may increase power by increasing the model* R^2 (Aberson, 2010). The ideal situation occurs when there is a parsimonious number of predictors that explain a large proportion of the variance (i.e., fewer predictors that explain a lot of variance is more powerful than a lot of predictors that explain the same amount of variance) (Aberson, 2010). As we will learn later in the "Assumptions" section, the more that predictors correlate with each other (i.e., multicollinearity), the less unique variance is explained (i.e., the value of R^2 change is determined by unique variation explained of the predictors; it is desirable to have predictors that explain substantial variance in the outcome over and above the other predictors), and thus power is reduced in the presence of multicollinearity (Aberson, 2010).

To determine how large a sample you need relative to the estimate of power in which you're interested, we suggest that you consult power tables (e.g., Cohen, 1988) or power software (were Liu includes syntax for using **R**, SAS, and SPSS) (Erdfelder, Faul, & Buchner, 1996; Faul, Erdfelder, Buchner, & Lang, 2009; Faul, Erdfelder, Lang, & Buchner, 2007; Liu, 2014). We will later illustrate how to use G*Power for estimating power. If you're interested in learning more about power, we encourage you to consult any of a number of excellent resources (e.g., Aberson, 2010; Cohen, 1988; Liu, 2014; Murphy et al., 2014).

8.1.4 Effect Size

There are multiple effect size indices that can be considered in the context of multiple linear regression. We will discuss effect size in the form of R^2 (i.e., multiple R squared or the coefficient of determination), partial R^2, f^2, and partial f^2.

8.1.4.1 Coefficient of Multiple Determination, R^2

One effect size in multiple linear regression is the *coefficient of multiple determination* or *multiple correlation coefficient*, introduced previously. The coefficient of multiple determination indicates the proportion of total variation in the dependent variable Y that is predicted from the set of predictor variables. There is no objective gold standard as to how large the coefficient of determination needs to be in order to say a meaningful proportion of variation has been predicted. The coefficient is determined not just by the quality of the predictor variables included in the model, but also by the quality of relevant predictor variables not included in the model, as well as by the amount of total variation in the dependent variable Y. According to the subjective standard of Cohen (1988), a small effect size is defined as $R^2 = .02$, a medium effect size as $R^2 = .13$, and a large effect size as $R^2 = .26$.

8.1.4.2 Multiple Partial R^2

The **multiple partial R^2**, symbolized by Cohen (1988) as $R^2_{YB.A}$, is computed as:

$$R^2_{YB.A} = \frac{R^2_{Y.A,B} - R^2_{Y.A}}{1 - R^2_{Y.A}} = \frac{R^2_{Y.(B.A)}}{1 - R^2_{Y.A}}$$

$R^2_{YB.A}$ is the proportion of that part of the total variation in the dependent variable, Y, uniquely explained by predictor, B, removing the influence of the set of predictors, A. In other words, A is partialed from both Y and B; A is held constant or statistically controlled. $R^2_{Y.A,B}$ represents the proportion of variance in Y accounted for by predictors A and B, and $R^2_{Y.A}$ represents the proportion of variance in Y accounted for by predictor A (Cohen, 1988). The numerator, therefore, represents the proportion of variation in Y that is uniquely accounted for by predictor, B, and can be conceived as a squared multiple semipartial (i.e., part) correlation. In the case of only two predictors, the multiple partial R^2 equates to the squared term of our earlier discussion of partial correlations. According to the subjective standard of Cohen (1988), a small effect size is defined as $R^2 = .02$, a medium effect size as $R^2 = .13$, and a large effect size as $R^2 = .26$.

8.1.4.3 f^2

The squared multiple correlation coefficient can also be used to compute f^2, which is computed as:

$$f^2 = \frac{R^2}{\left(1 - R^2\right)}$$

Interpreting f^2, it is the ratio of (1) the proportion of variation in the dependent variable uniquely explained by the independent variables to (2) the proportion of variation in the dependent variable unexplained by *any* variable in the model. The numerator, therefore, reflects the unique proportion of variance in Y for which the predictors account. The denominator reflects the proportion of variance in Y unaccounted for by the model. According to Cohen's (1988) conventions, a small effect size is defined as $f^2 = .02$, a medium effect size as $f^2 = .15$, and a large effect size as $f^2 = .35$.

8.1.4.4 Partial f^2

Similar to the computation of f^2, we can use the squared multiple partial correlations to compute $f^2_{partial}$ (Cohen, 1988). It is computed as:

$$f^2_{partial} = \frac{R^2_{YB.A}}{1 - R^2_{YB.A}}$$

Interpreting $f^2_{partial}$, $R^2_{YB.A}$ is the proportion of the total variation in the dependent variable, Y, uniquely explained by predictor(s), B, removing the influence of the set of predictors, A (i.e., the contribution of B over and above what is accounted for by A). In other words, A is partialed from both Y and B; A is held constant or statistically controlled. Thus, we can interpret $f^2_{partial}$ as the proportion of Y accounted for by predictor(s) B when the set of predictors, A, are held constant. According to the subjective standard of Cohen (1988), a small effect size is defined as $f^2_{partial} = .02$, a medium effect size as $f^2_{partial} = .13$, and a large effect size as $f^2_{partial} = .26$.

8.1.4.5 Additional Effect Size Considerations

For partial effects, standardized slopes or beta weights have been commonly reported as measures of effect size as they represent the number of standard deviation units the outcome variable will change for a one-unit standard deviation increase in the respective predictor variable. However, we discourage this practice. In simple linear regression, we saw that the standardized slope equals the Pearson correlation coefficient. With multiple linear regression, this is not the case as there are multiple independent variables. In multiple linear regression, the beta weight is influenced by the extent of overlap between the respective independent variable and the remaining independent variables (i.e., collinearity). The larger the overlap, the larger the beta weight. Thus, using the beta weight as a measure of effect size can be problematic, particularly if there is quite a bit of overlap between the independent variables.

TABLE 8.1

Effect sizes and interpretations.

Effect Size	Interpretation
R^2	• Coefficient of multiple determination or squared multiple correlation coefficient • Proportion of total variation in the dependent variable *Y* that is predicted from the set of predictor variables • Cohen's conventions: ∘ $R^2 = .02$, small ∘ $R^2 = .13$, medium ∘ $R^2 = .26$, large
$f^2 = \frac{R^2}{1-R^2}$	• Ratio of: (1) the proportion of variation in the dependent variable uniquely explained by the independent variables to (2) the proportion of variation in the dependent variable unexplained by *any* variable in the model • Cohen's standards: ∘ $f^2 = .02$, small ∘ $f^2 = .15$, medium ∘ $f^2 = .35$, large
$R^2_{YB.A} = \frac{R^2_{Y.A,B} - R^2_{Y.A}}{1 - R^2_{Y.A}}$	• Multiple partial R^2 • Proportion of variation in *Y* that is accounted for by predictor(s), *B*, when holding the set of independent variables *A* constant; i.e., *A* is partialed out from both *Y* and *B* • Cohen's conventions: ∘ $R^2 = .02$, small ∘ $R^2 = .13$, medium ∘ $R^2 = .26$, large
$f^2_{partial} = \frac{R^2_{YB.A}}{1 - R^2_{YB.A}}$	• Multiple partial f^2 • Proportion of variation in *Y* that is accounted for by predictor(s), *B*, when holding the set of independent variables *A* constant; i.e., *A* is partialed out from both *Y* and *B* • Cohen's conventions: ∘ $R^2 = .02$, small ∘ $R^2 = .13$, medium ∘ $R^2 = .26$, large

Table 8.1 provides a summary of multiple linear regression effect size indices and guidelines for interpretation. For additional information on effect size measures in regression, we suggest you consider Steiger and Fouladi (1992), Mendoza and Stafford (2001), and Smithson (2001, which also includes some discussion of power).

Confidence intervals (CI) can computed for R^2, and these CI reflect precision of the estimated R^2. *Larger CI suggest lower precision, and smaller CI reflect higher precision.* A R^2 CI that includes the null value (i.e., $R^2 = 0$) may provide evidence to suggest a nonstatistically significant relationship between the set of independent variables and the outcome. We can also use the online effect size calculator by Uanhoro (2017) to compute confidence intervals. There are four values that must be input: R^2, the confidence interval (i.e., complement of alpha), and the numerator (i.e., effect) and denominator (i.e., error) degrees of freedom. Inputting these values, we are provided the confidence interval of .5927733, .9690771.

8.1.5 Assumptions

For the most part, the assumptions of multiple linear regression analysis are the same as that with simple linear regression. The assumptions are concerned with:

(a) independence, (b) homoscedasticity, (c) normality, (d) linearity, (e) fixed X, and (f) noncollinearity. When the first four assumptions are met, the coefficients produced by ordinary least squares regression will be the best linear unbiased estimators (BLUE) (according to the Gauss-Markov theorem, ordinary least squares estimators will be BLUE in that they are unbiased, linear, and have the smallest variation of all estimators that are linear and unbiased). In other words, the smallest mean square error for the estimators will be produced (Meuleman, Loosveldt, & Emonds, 2013). This section also mentions those techniques appropriate for evaluating each assumption. Readers who are interested in expanded coverage of diagnostics related to assumptions are encouraged to review the chapter by Meuleman et al. (2013).

8.1.5.1 Independence

The first assumption is concerned with **independence** of the observations. The simplest procedure for assessing independence is to examine residual plots of e versus the predicted values of the dependent variable Y' and of e versus each independent variable X_k (alternatively, one can look at plots of observed values of the dependent variable Y versus predicted values of the dependent variable Y' and of observed values of the dependent variable Y versus each independent variable X_k). If the independence assumption is satisfied, the residuals should fall into a random display of points. If the assumption is violated, the residuals will fall into some sort of pattern. Lack of independence affects the estimated standard errors of the model. For serious violations, one could consider generalized or weighted least squares as the method of estimation (e.g., Myers, 1986; Weisberg, 1985), or some type of transformation. The residual plots shown in Figure 8.6 do not suggest any independence problems for the graduate grade point average (GGPA) example, where Figure 8.6a represents the residual e versus the predicted value of the dependent variable Y', Figure 8.6b represents e versus GRETOT, and Figure 8.6c represents e versus undergraduate grade point average (UGPA).

8.1.5.2 Homoscedasticity

The second assumption is **homoscedasticity**, where the conditional distributions have the same constant variance for all values of X. In the residual plots, the consistency of the variance of the conditional distributions may be examined.

The **nonconstant error variance test** can also be used to examine this assumption. The null hypothesis for this test is that there is constant error variance; the alternative hypothesis is that the error variance changes at levels of the fitted values (i.e., with the linear combination of the independent variables). A nonstatistically significant nonconstant error variance test suggests the assumption of homoscedasticity has been met.

If the homoscedasticity assumption is violated, estimates of the standard errors are larger, and the conditional distributions may also be nonnormal. Solutions to violation of this assumption include variance stabilizing transformations (such as the square root or log of Y), generalized or weighted least squares (Myers, 1986; Weisberg, 1985), or robust regression (Kleinbaum et al., 1998; Myers, 1986; Wilcox, 1996, 2003; Wu, 1985). Due to the small sample size, homoscedasticity cannot really be assessed for the example data.

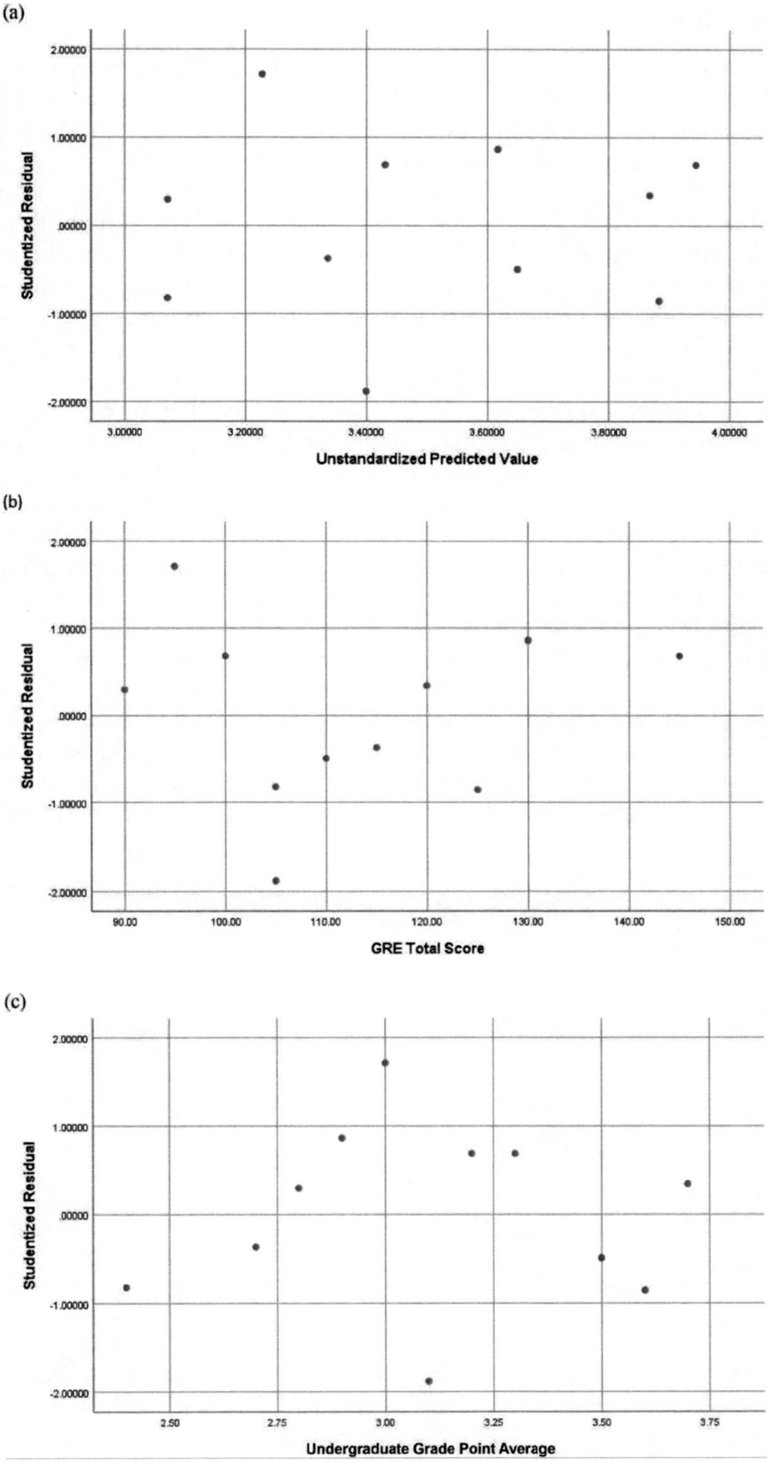

FIGURE 8.6
Residual plots for GRE-GPA example: (a), (b), (c).

8.1.5.3 Normality

The third assumption is that the conditional distributions of the scores on Y, or the prediction errors, are **normal** in shape. Violations of normality can lead to imprecision in the partial slopes and in the coefficient of determination. The following can be used to examine data for normality: frequency distributions, normal probability (Q-Q) plots, and skewness statistics. The simplest procedure involves checking for symmetry in a histogram, frequency distribution, boxplot, or skewness and kurtosis statistics. Although **nonzero kurtosis** (i.e., a distribution that is either flat, platykurtic, or has a sharp peak, leptokurtic) will have minimal effect on the regression estimates, **nonzero skewness** (i.e., a distribution that is not symmetric with either a positive or negative skew) will have much greater impact on these estimates. Thus, finding asymmetrical distributions is a must. One convention is to be concerned if the skewness value is larger than 2.0 in magnitude.

Another useful graphical technique is the **normal probability plot** (or Q-Q plot). With normally distributed residuals, the points on the normal probability plot will fall along a straight diagonal line, whereas nonnormal data will not. As is often the case with visual representations of data, there is a difficulty in evaluating this plot because there is no criterion with which to judge deviation from linearity. It is recommended that skewness and/or the normal probability plot be considered at a minimum when determining normality evidence. For the example data, the normal probability plot is shown in Figure 8.7, and even with a small sample looks good.

What causes nonnormality? Among other reasons, a violation of the normality assumption may be the result of **outliers**. Various conventions are used to crudely detect outliers from a residual plot or scatterplot. The simplest outlier detection procedure and a commonly used rule is to define an outlier as *an observation more than two or three standard errors from the mean (i.e., a large distance from the mean)*. The outlier observation may be a result of (a) a simple recording or data entry error, (b) an error in observation, (c) an improperly

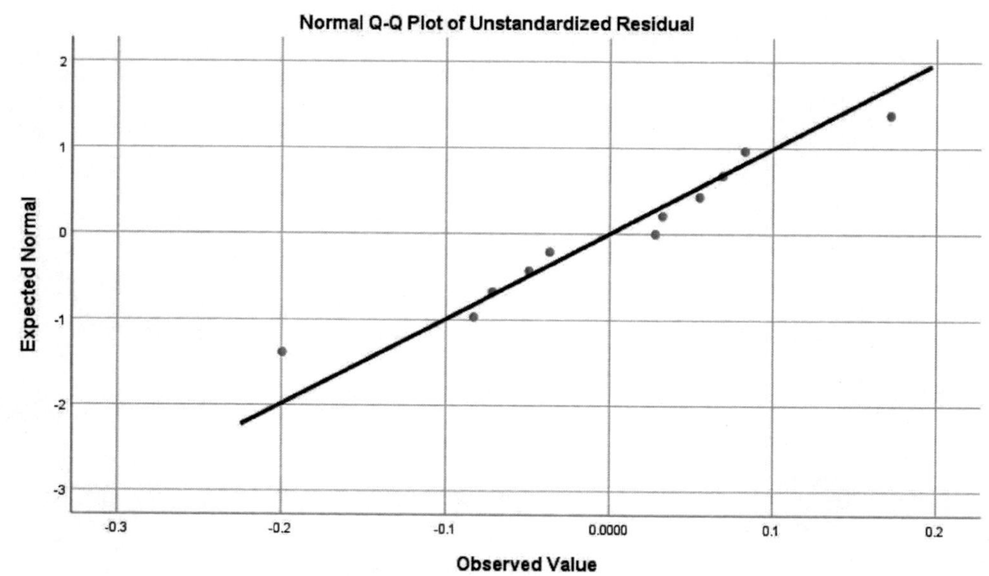

FIGURE 8.7
Q-Q plot.

functioning instrument, (d) inappropriate use of administration instructions, or (e) a true outlier. If the outlier results from an error, try to correct the error, and then redo the regression analysis. If the error cannot be corrected, then deleting the observation is possible. If the outlier represents an accurate observation, however, then this observation may contain important theoretical information, and one would be more hesitant to delete it (or perhaps seek out similar observations). Thus, implementing a different approach for dealing with the outlier is needed. A simple procedure to use for single case outliers (i.e., just one outlier) is to perform two regression analyses, one with the outlier being included and one without. Comparing the results of these analyses will provide some indication of the effects of the outlier. Other methods include robust regression (Kleinbaum et al., 1998; Wilcox, 1996, 2003) and nonparametric regression (Miller, 1997; Rousseeuw & Leroy, 1987; Wu, 1985).

What happens if you find other types of nonnormality (i.e., beyond outliers)? Transformations can be used to normalize the data. The most commonly used transformations to correct for nonnormality in regression analysis are to transform the dependent variable using the log (to correct for positive skew) or the square root (to correct for positive or negative skew). However, again there is the challenge of interpreting transformed variables measured along a scale other than that of the original variables.

8.1.5.4 Linearity

The fourth assumption is **linearity**, that there is a linear relationship between the observed scores on the dependent variable Y and the values of the independent variables, X_k's. If satisfied, then the sample partial slopes and intercept are unbiased estimators of the population partial slopes and intercept, respectively. The linearity assumption is important because regardless of the value of X_k, we always expect Y to increase by b_k units for a one-unit increase in X_k, controlling for the other X_k's. If a nonlinear relationship exists, this means that the expected increase in Y depends on the value of X_k; that is, the expected increase is not a constant value. Strictly speaking, *linearity in a model refers to there being linearity in the parameters of the model* (i.e., α and the β_k's).

Violation of the linearity assumption can be detected through residual plots. The residuals should be located within a band of +2 s_{res} (or standard errors), indicating no systematic pattern of points. Residual plots for the GGPA example were shown previously in Figure 8.1. Even with a very small sample, we see a fairly random pattern of residuals, and therefore feel fairly confident that the linearity assumption has been satisfied. Note also that there are other types of residual plots developed especially for multiple regression analysis, such as the added variable and partial residual plots (Larsen & McCleary, 1972; Mansfield & Conerly, 1987; Weisberg, 1985). Procedures to deal with nonlinearity include transformations (of one or more of the X_k's and/or of Y) and other regression models (discussed later in this chapter).

8.1.5.5 Fixed X

The fifth assumption is that the values of X_k are **fixed**, where the independent variables X_k are fixed variables rather than random variables. This results in the regression model being valid only for those particular values of X_k that were actually observed and used in the analysis. Thus, the same values of X_k would be used in replications or repeated samples.

Strictly speaking, the regression model and its parameter estimates are valid only for those values of X_k actually sampled. The use of a prediction model developed to predict the dependent variable Y, based on one sample of individuals, may be suspect for another sample of individuals. Depending on the circumstances, the new sample of individuals may actually call for a different set of parameter estimates. Generally we may not want to make predictions about individuals having combinations of X_k scores outside of the range of values used in developing the prediction model; this is defined as *extrapolating* beyond the sample predictor data. On the other hand, we may not be quite as concerned in making predictions about individuals having combinations of X_k scores within the range of values used in developing the prediction model; this is defined as *interpolating* within the range of the sample predictor data.

It has been shown that when other assumptions are met, regression analysis performs just as well when X is a random variable (e.g., Glass & Hopkins, 1996; Myers & Well, 1995; Pedhazur, 1997; Wonnacott & Wonnacott, 1981). There is no such assumption about Y.

8.1.5.6 Noncollinearity

Considering simple and multiple linear regression, the final assumption is unique to multiple linear regression analysis (as compared to simple linear regression), but will be quite common throughout multivariate procedures that we cover. A violation of this assumption is known as collinearity, where there is a very strong linear relationship between two or more of the predictors.

Although multicollinearity does not impact overall model fit or model predictions (Kutner, Nachtscheim, & Neter, 2005), the presence of severe collinearity is problematic in several respects. First, it will lead to instability of the regression coefficients across samples, where the estimates will bounce around quite a bit in terms of magnitude and even occasionally result in changes in sign (perhaps opposite of expectation). This occurs because the standard errors of the regression coefficients become larger, thus making it more difficult to achieve statistical significance. Another result that may occur involves an overall regression that is significant, but none of the individual predictors are significant. Collinearity will also restrict the utility and generalizability of the estimated regression model. While the potential impact of multicollinearity can be severe, it is not uncommon for authors to fail to report diagnostics for identifying multicollinearity (e.g., nearly 99.9% of clinical and epidemiological studies from 2004 to 2013 did *not* explicitly address examination of this assumption) (Vatcheva, Lee, McCormick, & Rahbar, 2016). *Don't be one of those authors!*

Recall from earlier in the chapter the notion of partial regression coefficients, where the other predictors were held constant. In the presence of severe collinearity, the other predictors cannot really be held constant because they are so highly intercorrelated. Collinearity may be indicated when there are large changes in estimated coefficients due to (a) a variable being added or deleted, and/or (b) an observation being added or deleted (Chatterjee & Price, 1977). **Singularity** is a special case of multicollinearity; it is perfect multicollinearity and occurs when two or more items/variables perfectly predict and are therefore perfectly redundant. This can occur, for example, when a composite variable as well as its component variables are used as predictors in the same model (e.g., including GRETOT, GRE-Quantitative, and GRE-Verbal as predictors).

How do we detect violations of this assumption? The simplest procedure is to conduct a series of special regression analyses, one for each X, where that predictor is predicted by all of the remaining X'_s (i.e., the criterion variable is not involved). If any of the resultant R_k^2 values are close to one (greater than .9 is a good guidelines), then there may be a collinearity problem. However, the large R^2 value may also be due to small sample size; thus more data would be useful in this type of situation. For the example data, $R^2_{12} = .091$, and therefore collinearity is not a concern.

Also, if the number of predictors is greater than or equal to n, then perfect collinearity is a possibility. Another statistical method for detecting collinearity is to compute a variance inflation factor (VIF) for each predictor, which is equal to $1/(1 - R^2_k)$. The VIF is defined as the inflation that occurs for each regression coefficient above the ideal situation of uncorrelated predictors. Many suggest that the largest VIF should be less than 10 in order to satisfy this assumption (Myers, 1990; Stevens, 2009; Wetherill, 1986).

There are several possible methods for dealing with a collinearity problem. First, one can remove one or more of the correlated predictors. Second, ridge regression techniques, and ridge-related techniques (e.g., partial ridge regression, which have been shown to be superior to other existing methods), can be used (Chandrasekhar, Bagyalakshmi, Srinivasan, & Gallo, 2016; Hoerl & Kennard, 1970a, 1970b; Marquardt & Snee, 1975; Singh, 2010; Wetherill, 1986). Third, principal component scores resulting from principal component analysis can be utilized rather than raw scores on each variable (Kleinbaum et al., 1998; Myers, 1986; Weisberg, 1985; Wetherill, 1986). Fourth, transformations of the variables can be used to remove or reduce the extent of the problem. The final solution, and probably our last choice, is to use simple linear regression, as collinearity cannot exist with a single predictor.

8.1.5.7 Summary of Assumptions

For the GGPA example, although sample size is quite small in terms of looking at conditional distributions, it would appear that all of our assumptions have been satisfied. All of the residuals are within two standard errors of zero, and there does not seem to be any systematic pattern in the residuals. The distribution of the residuals is nearly symmetric and the normal probability plot looks good. A summary of the assumptions and the effects of their violation for multiple linear regression analysis is presented in Table 8.2.

Table 8.2 Assumptions and violation of assumptions: multiple linear regression analysis.

Assumption	Effect of Assumption Violation
Independence	• Influences standard errors of the model
Homogeneity	• Bias in s^2_{res} • May inflate standard errors and thus increase likelihood of a Type II error • May result in nonnormal conditional distributions
Normality	• Less precise slopes, intercept, and R^2
Linearity	• Bias in slope and intercept • Expected change in Y is not a constant and depends on value of X
Fixed X values	• Extrapolating beyond the range of X combinations: prediction errors larger, may also bias slopes and intercept • Interpolating within the range of X combinations: smaller effects than above; if other assumptions met, negligible effect
Noncollinearity of X's	• Regression coefficients can be quite unstable across samples (as standard errors are larger) • R^2 may be significant, yet none of the predictors are significant • Restricted generalizability of the model

8.2 Mathematical Introduction Snapshot

Throughout the chapter we have woven some of the mathematics of multiple linear regression. Now, let's consider the analysis illustrated using the data with which Addie, our graduate researcher, is working. We use the GRE Quantitative + Verbal Total (GRETOT) and undergraduate grade point average (UGPA) to predict graduate grade point average (GGPA). GRETOT has a possible range of 40 to 160 points, sand GPA is defined as having a possible range of 0.00 to 4.00 points. Given the sample of 11 statistics students as shown in Table 8.3, let us work through a multiple linear regression analysis.

As sample statistics, we compute for GRETOT (X_1 or subscript 1) that the mean is $\bar{X}_1 = 112.7273$ and the variance is $s_1^2 = 266.8182$, for UGPA (X_2 or subscript 2) that the mean is $\bar{X}_2 = 3.1091$ and the variance is $s_2^2 = 0.1609$, and for GGPA (*Y*), a mean of $\bar{Y} = 3.5000$ and variance of $s_Y^2 = 0.1100$. In addition, we compute the bivariate correlation between the dependent variable (graduate GPA) and GRE total, $r_{Y1} = .7845$; between the dependent variable (graduate GPA) and undergraduate GPA, $r_{Y2} = .7516$; and between GRE total and undergraduate GPA, $r_{12} = .3011$. The sample partial slopes (b_1 and b_2) and intercept (*a*) are determined as follows:

$$b_1 = \frac{(r_{Y1} - r_{Y2}r_{12})s_Y}{(1 - r_{12}^2)s_1} = \frac{[.7845 - (.7516)(.3011)](.3317)}{(1 - .3011^2)(16.3346)} = .0125$$

$$b_2 = \frac{(r_{Y2} - r_{Y1}r_{12})s_Y}{(1 - r_{12}^2)s_2} = \frac{[.7516 - (.7845)(.3011)](.3317)}{(1 - .3011^2)(.4011)} = .4687$$

$$a = \bar{Y} - b_1\bar{X}_1 - b_2\bar{X}_2$$

$$a = 3.500 - (.0125)(112.7273) - (.4687)(3.1091) = .6337$$

Let us interpret the partial slope and intercept values. A partial slope of .0125 for GRETOT would mean that if your score on the GRETOT was increased by 1 point, then your graduate grade point average would be increased by .0125 points, controlling for undergraduate grade point average. Likewise, a partial slope of .4687 for UGPA would mean that if your

Table 8.3 GRE-GPA example data

Student	GRE-Total (X_1)	Undergraduate GPA (X_2)	Graduate GPA (*Y*)
1	145	3.2	4.0
2	120	3.7	3.9
3	125	3.6	3.8
4	130	2.9	3.7
5	110	3.5	3.6
6	100	3.3	3.5
7	95	3.0	3.4
8	115	2.7	3.3
9	105	3.1	3.2
10	90	2.8	3.1
11	105	2.4	3.0

undergraduate grade point average was increased by 1 point, then your graduate grade point average would be increased by .4687 points, controlling for GRETOT. An intercept of .6337 would mean that if your scores on the GRETOT and UGPA were both 0, then your graduate grade point average would be .6337. However, it is impossible to obtain a GRETOT score of 0 because 40 points is the minimum score possible. In a similar way, an undergraduate student could not obtain a UGPA of 0 and be admitted to graduate school. This is not to say that the regression equation is incorrect, but just to point out how the interpretation of "GRETOT and UGPA were both 0" is a bit meaningless in context.

To put all of this together then, the sample multiple linear regression model is

$$Y_1 = b_1 X_{1i} + b_2 X_{2i} + a + e_i$$
$$Y_1 = .0125\left(X_{1i}\right) + .4687\left(X_{2i}\right) + .6337 + e_i$$

In other words, if your score on the GRETOT was 130 and your UGPA was 3.5, then your predicted score on the GGPA would be computed as:

$$Y_i' = .0125(130) + .4687(3.50) + .6337 = 3.8992$$

Based on the prediction equation, we predict your GGPA to be around 3.9; however, predictions are usually somewhat less than perfect, even with two predictors.

For the GGPA example, we compute the overall F test statistic as the following:

$$F = \frac{R^2/m}{\left(1-R^2\right)\left(n-m-1\right)}$$
$$F = \frac{.9089/2}{(1-.9089)(11-2-1)} = 39.9078$$

or as

$$F = \frac{SS_{reg}/df_{reg}}{SS_{res}/df_{res}} = \frac{MS_{reg}}{MS_{res}} = \frac{.9998/2}{.1002/8} = 39.9122$$

The critical value, at the .05 level of significance, is ${}_{.05}F_{2,8} = 4.46$. The test statistic exceeds the critical value, so we reject H_0 and conclude that all of the partial slopes are not equal to zero at the .05 level of significance (the two F test statistics differ slightly due to rounding error).

For our graduate grade point average example, the standardized partial slopes are equal to

$$b_1^* = b_1\left(\frac{s_1}{s_Y}\right) = (.0125)\left(\frac{16.3346}{.3317}\right) = .6156$$

and

$$b_2^* = b_2\left(\frac{s_2}{s_Y}\right) = (.4687)\left(\frac{.4011}{.3317}\right) = .5668$$

The prediction model is then as follows:

$$z\left(Y_i'\right) = .6156 z_{1i} + .5668 z_{2i}$$

The standardized partial slope of .6156 for GRETOT would be interpreted as the expected increase in GGPA in *z* score units for a one *z* score unit increase in the GRETOT, controlling for UGPA. A similar statement may be made for the standardized partial slope of UGPA. The b_k^2 can also be interpreted as the expected standard deviation change in the dependent variable *Y* associated with a one standard deviation change in the independent variable X_k when the other X_k's are held constant.

With the example of predicting GGPA from GRETOT and UGPA, let us examine the partitioning of the total sum of squares SS_{total} as follows:

$$SS_{total} = (n-1)(s_Y^2) = (10)(.1100) = 1.100$$

Next, we can determine the multiple correlation coefficient R^2 as

$$R^2_{Y.1,\ldots,m} = b_1^*(r_{Y1}) + b_2^*(r_{Y2}) + \ldots + b_m^*(r_{Ym})$$
$$R^2_{Y.1,\ldots,m} = (.6156)(.7845) + (.5668)(.7516) = .9089$$

We can also partition SS_{total} into SS_{reg} and SS_{res}, where

$$SS_{reg} = (R^2)(SS_{total}) = (.9089)(1.1000) = .9998$$
$$SS_{res} = (1-R^2)(SS_{total}) = (1-.9089)(1.1000) = .1002$$

Finally, let us summarize these results for the example data. We found that the coefficient of multiple determination (R^2) was equal to .9089. Thus, the GRE total score and the undergraduate grade point average predicts around 91% of the variation in the graduate grade point average. This would be quite satisfactory for the college admissions officer in that there is little variation left to be explained, although this result is quite unlikely in actual research in education and the behavioral sciences. Obviously there is a large effect size here.

Let us compute the second test statistic for the GGPA example. We specify the null hypothesis to be $\beta_k = 0$ (i.e., the slope is zero) and conduct two-tailed tests. First the variance error of estimate is

$$s^2_{res} = \frac{SS_{res}}{df_{res}} = \frac{.1022}{8} = .0125$$

The standard error of estimate, s_{res}, is .1118. Next, the standard errors of the b_k are found to be

$$s(b_1) = \frac{s_{res}}{\sqrt{(n-1)(s_1^2)(1-r_{12}^2)}} = \frac{.1118}{\sqrt{(10)(266.8182)(1-.3011^2)}} = .0023$$

$$s(b_2) = \frac{s_{res}}{\sqrt{(n-1)(s_2^2)(1-r_{12}^2)}} = \frac{.1118}{\sqrt{(10)(.1609)(1-.3011^2)}} = .0924$$

Finally, we find the *t* test statistics to be computed as follows:

$$t_1 = \frac{b_1}{s(b_1)} = \frac{.0125}{.0023} = 5.4348$$

$$t_2 = \frac{b_2}{s(b_2)} = \frac{.4687}{.0924} = 5.0725$$

To evaluate the null hypotheses, we compare these test statistics to the critical values of $\pm_{.025}t_8 = \pm 2.306$. Both test statistics exceed the critical value; consequently H_0 is rejected in favor of H_1 for both predictors. We conclude that both partial slopes are indeed statistically significantly different from zero at the .05 level of significance.

Finally, let us compute the confidence intervals for the b_k's as follows:

$$CI(b_1) = b_1 \pm_{(\alpha/2)} t_{(n-m-1)} \left[s(b_1)\right]$$

$$CI(b_1) = b_1 \pm_{.025} t_8 \left[s(b_1)\right] = .0125 \pm (2.306)(.0023) = (.0072, .0178)$$

$$CI(b_2) = b_2 \pm_{(\alpha/2)} t_{(n-m-1)} \left[s(b_2)\right]$$

$$CI(b_2) = b_2 \pm_{.025} t_8 \left[s(b_2)\right] = .4687 \pm (2.306)(.0924) = (.2556, .6818)$$

The intervals do not contain zero, the value specified in H_0; thus we again conclude that both b_k's are significantly different from zero at the .05 level of significance.

8.3 Computing Multiple Linear Regression Using SPSS

Next we consider SPSS for the multiple linear regression model using the Ch8.GGPA data. Before we conduct the analysis, let us review the data. With one dependent variable and two independent variables, the dataset must consist of three variables or columns, one for each independent variable and one for the dependent variable. Each row still represents one individual, indicating the value of the independent variables for that particular case and their score on the dependent variable. As seen in the screenshot in Figure 8.8, for a multiple linear regression analysis therefore, the SPSS data are in the form of three columns that represent the two independent variables (GRE total score and undergraduate GPA) and one dependent variable (graduate grade point average).

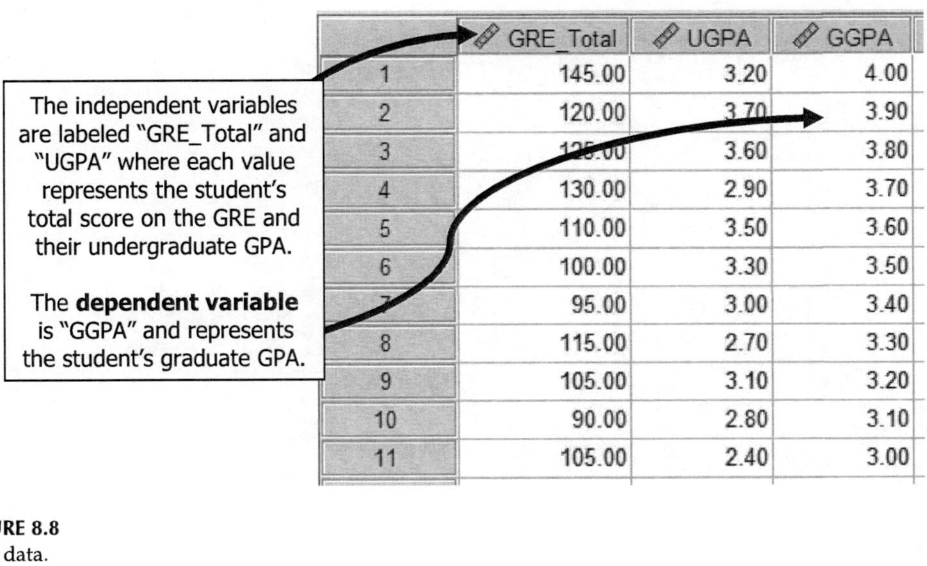

	GRE_Total	UGPA	GGPA
1	145.00	3.20	4.00
2	120.00	3.70	3.90
3	125.00	3.60	3.80
4	130.00	2.90	3.70
5	110.00	3.50	3.60
6	100.00	3.30	3.50
7	95.00	3.00	3.40
8	115.00	2.70	3.30
9	105.00	3.10	3.20
10	90.00	2.80	3.10
11	105.00	2.40	3.00

FIGURE 8.8
SPSS data.

Step 1. To conduct a simple linear regression, go to "Analyze" in the top pulldown menu, then select "Regression," and then select "Linear." Following the screenshot for Step 1 (Figure 8.9) produces the "Linear Regression" dialog box.

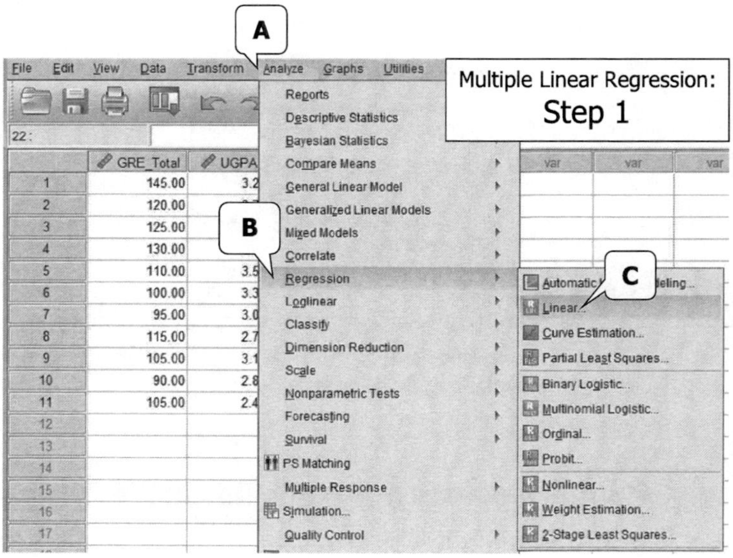

FIGURE 8.9
Multiple linear regression: Step 1.

Step 2. Click the dependent variable (e.g., "GGPA") and move it into the "Dependent" box by clicking the arrow button. Click the independent variables and move them into the "Independent(s)" box by clicking the arrow button (see the screenshot in Figure 8.10).

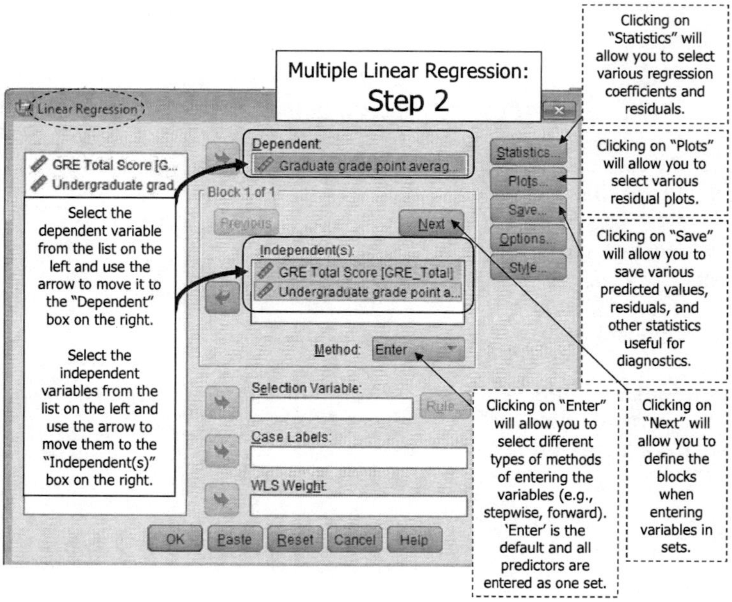

FIGURE 8.10
Multiple linear regression: Step 2.

Step 3. From the Linear Regression dialog box (see Figure 8.10), clicking on "Statistics" will provide the option to select various regression coefficients and residuals. From the Statistics dialog box (see the screenshot in Figure 8.11), place checkmarks in the box next to the following: (1) "Estimates"; (2) "Confidence intervals"; (3) "Model fit"; (4) "R squared change"; (5) "Descriptives'; (6) "Part and partial correlations"; (7) "Collinearity diagnostics"; (8) "Durbin-Watson"; and (9) "Casewise diagnostics." For this example we apply an alpha level of .05; thus, we will leave the default confidence interval percentage at 95. If we were using a different alpha, the confidence interval would be the complement of alpha (e.g., $\alpha = .01$, then CI = 1 – .01 = .99). We will also leave the default of "3 standard deviations" for defining outliers for the casewise diagnostics. Click on "Continue" to return to the original dialog box.

Let's quickly address the **Durbin-Watson test**. The Durbin-Watson test is a test of autocorrelation, and specifically whether adjacent residuals are correlated. It is a test that is usually conducted with time series data. The underlying principle is that correlated residuals should be more similar to their neighbors than other, random pairs of residuals. The Durbin-Watson test then examines the sum of squared differences between neighboring residuals to the sums of squared residuals. Because the examination is on adjacency, how the cases are ordered makes a difference.

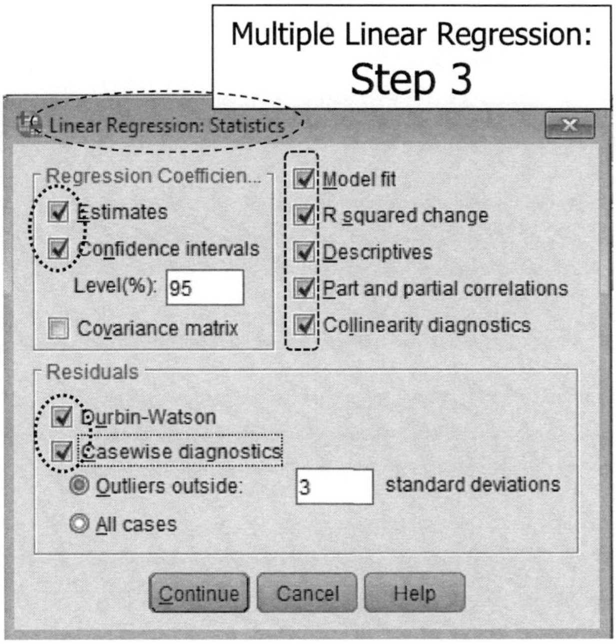

Working in **R**, we can generate the Durbin-Watson test using the following command, where 'GGPA_MultReg' is the object created when we generated our multiple linear regression model. In other words, you will need to compute the multiple linear regression model before you can produce the Durbin Watson test.

```
durbinWatsonTest(GGPA_MultReg)
```

FIGURE 8.11
Multiple linear regression: Step 3.

Step 4. From the Linear Regression dialog box (see Figure 8.10), clicking on "Plots" will provide the option to select various residual plots. From the Plots dialog box, place checkmarks in the boxes next to the following: (1) "Histogram"; (2) "Normal probability plot"; and (3) "Produce all partial plots." Click on "Continue" to return to the original dialog box. We can use this dialog box to generate various residual plots, which can be used to check assumptions. For this illustration, we'll save the residuals and plot them later as that provides greater flexibility in how we can work with them.

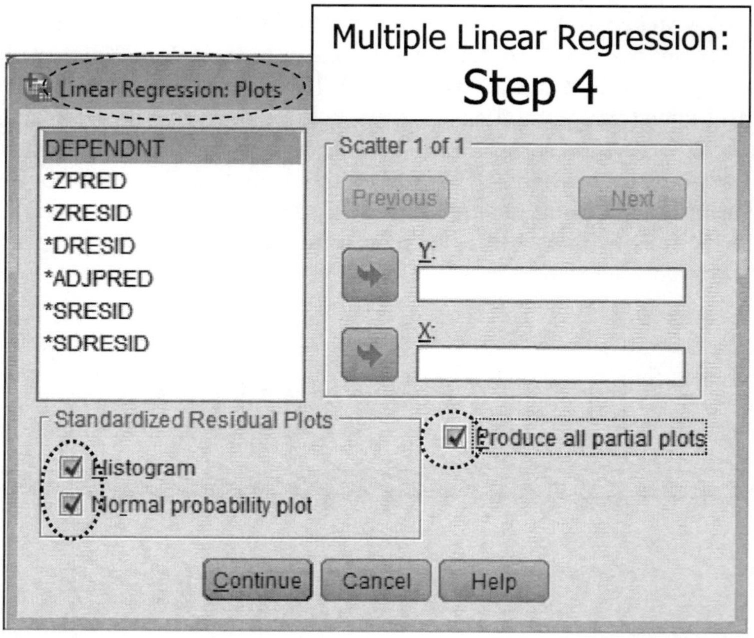

Working in **R**, to produce partial regression plots, install and load the 'car' package. Then generate the 'avPlots' command on the multiple linear regression model that was estimated (in this illustration, we called the regression model 'GGPA_MultReg').

```
install.packages("car")
library(car)
```

```
avPlots(GGPA_MultReg)
```

FIGURE 8.12
Multiple linear regression: Step 4.

Step 5. From the Linear Regression dialog box (see Figure 8.10), clicking on "Save" will provide the option to save various predicted values, residuals, and statistics that can be used for diagnostic examination. From the Save dialog box under the heading "Predicted Values," place a checkmark in the box next to "Unstandardized." Under the heading "Residuals," place checkmarks in the boxes next to "Unstandardized" and "Studentized." Under the heading of "Distances," place checkmarks in the boxes next to "Mahalanobis," "Cook's," and "Leverage values." Under the heading "Influence Statistics," place a checkmark in the box next to "Standardized DfBeta(s)." Click on "Continue" to return to the original dialog box. From the Linear Regression dialog box, click on "OK" to return and generate the output.

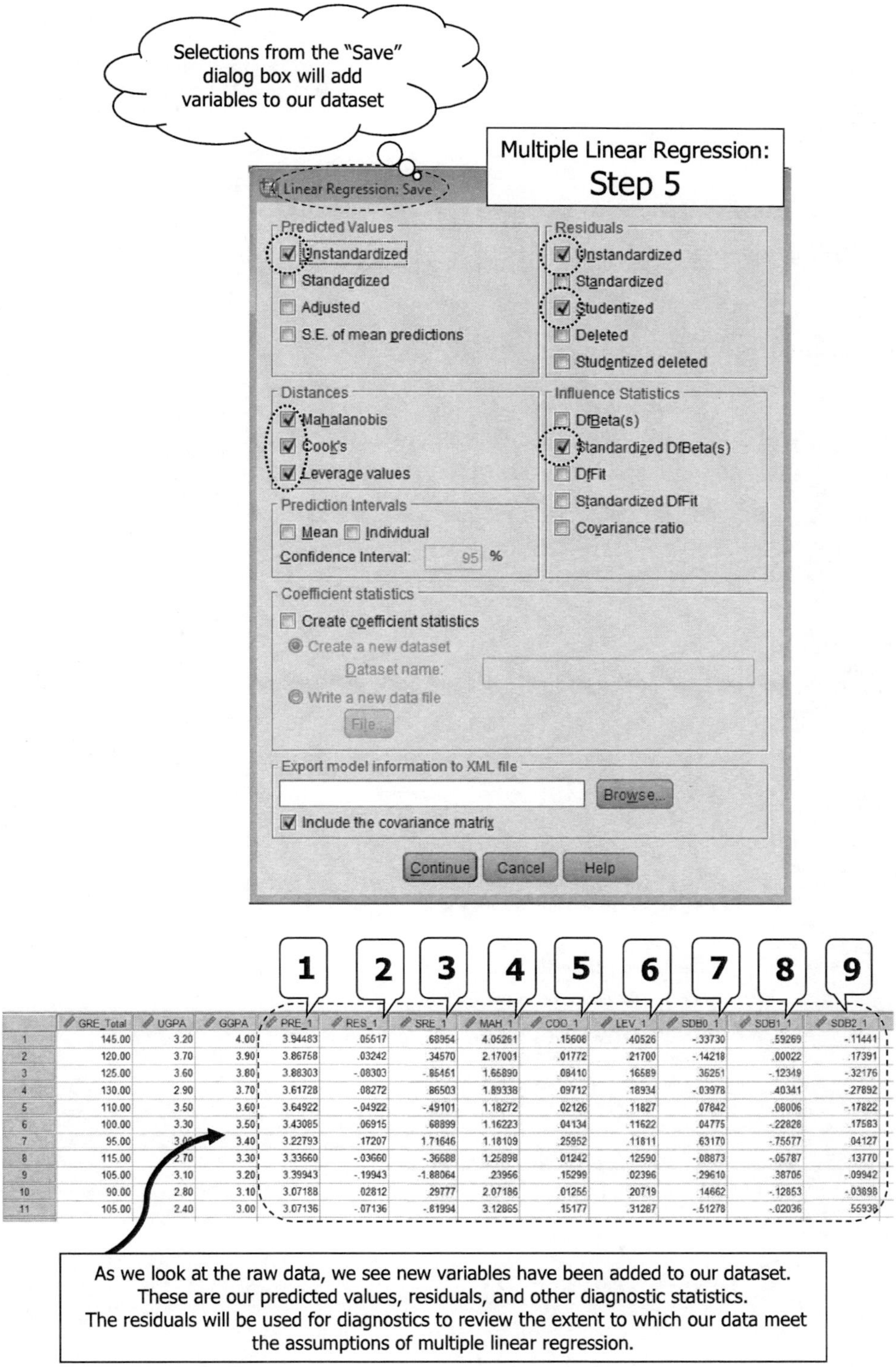

FIGURE 8.13
Multiple linear regression: Step 5.

Interpreting the output. Annotated results are shown in Table 8.4.

TABLE 8.4

SPSS Results for the Multiple Regression GRE-GPA Example

Descriptive Statistics

	Mean	Std. Deviation	N
Graduate grade point average	3.5000	.33166	11
GRE Total Score	112.7273	16.33457	11
Undergraduate grade point average	3.1091	.40113	11

The table labeled "Descriptive Statistics" provides basic descriptive statistics (means, standard deviations, and sample sizes) for the independent and dependent variables.

Correlations

		Graduate grade point average	GRE Total Score	Undergraduate grade point average
Pearson Correlation	Graduate grade point average	1.000	.784	.752
	GRE Total Score	.784	1.000	.301
	Undergraduate grade point average	.752	.301	1.000
Sig. (1-tailed)	Graduate grade point average	.	.002	.004
	GRE Total Score	.002	.	.184
	Undergraduate grade point average	.004	.184	.
N	Graduate grade point average	11	11	11
	GRE Total Score	11	11	11
	Undergraduate grade point average	11	11	11

The table labeled "Correlations" provides the: Pearson correlation coefficient values, *p* values, and sample size for the *bivariate Pearson correlation* between the independent and dependent variables.

The correlation between graduate GPA and GRE-Total (p = .002) and the correlation between graduate GPA and undergraduate GPA (p = .004) are statistically significant.

TABLE 8.4 (continued)

SPSS Results for the Multiple Regression GRE-GPA Example

Variables Entered/Removed[a]

Model	Variables Entered	Variables Removed	Method
1	Undergraduate grade point average, GRE Total Score[b]	.	Enter

a. Dependent Variable: Graduate grade point average

b. All requested variables entered.

"Variables Entered/Removed" lists the independent variables included in the model and the method they were entered (i.e., "Enter").

"Adjusted *R* square" adjusts for the number of independent variables and sample size. *Shrinkage* is the difference between R^2 and adjusted R^2. When sample size is small, given the number of independent variables, the difference between R^2 and adjusted R^2 will be large to compensate for a large amount of bias. If an additional independent variable were entered in the model, an increase in adjusted R^2 indicates the new variable is adding value to the model. Negative adjusted R^2 values can occur and indicate the model fits the data VERY poorly.

Change statistics are used when methods other than simultaneous entry (e.g., hierarchical, forward, backward) are used to enter the predictors in the model. In those cases, more than one row will be presented here. A *p* value less than α would indicate the additional variables are explaining additional variation.

Durbin-Watson is a test for independence of residuals. Ranging from 0 to 4, values of 2 indicate uncorrelated errors; values less than 1 or greater than 3 indicate a likely violation of this assumption.

Model Summary[b]

					Change Statistics					
Model	R	R Square	Adjusted R Square	Std. Error of the Estimate	R Square Change	F Change	df1	df2	Sig. F Change	Durbin-Watson
1	.953[a]	.908	.885	.11272	.908	39.291	2	8	.000	2.116

a. Predictors: (Constant), Undergraduate grade point average, GRE Total Score

b. Dependent Variable: Graduate grade point average

R is the **multiple correlation coefficient.**

R^2 is the **squared multiple correlation coefficient** (aka **coefficient of determination**). It represents the proportion of variance in the dependent variable that is explained by the independent variables.

Adjusted R^2 is interpreted as the percentage of variation in the dependent variable that is explained after adjusting for sample size and the number of predictors.

(continued)

TABLE 8.4 (continued)

SPSS Results for the Multiple Regression GRE-GPA Example

Working in **R**, the results for the Durbin-Watson test are as follows, where the p value (.904) indicates there is not statistically significant autocorrelation. This provides evidence that the assumption of independence has been met.

```
lag Autocorrelation D-W Statistic p-value
   1      -0.09800019       2.11595   0.904
 Alternative hypothesis: rho != 0
```

Total *SS* is partitioned into *SS* regression and *SS* residual. Regression sum of squares indicates variability explained by the regression model. Residual sum of squares indicates variability *not* explained by the regression model.

The *F* statistic tests the **overall regression model** (i.e., that the population multiple correlation coefficient is zero).

The *p* value (.000) indicates we reject the null hypothesis. The probability of finding a sample value of multiple R^2 of .908 or larger when the true population multiple correlation coefficient is zero is less than 1%.

ANOVA[a]

Model		Sum of Squares	df	Mean Square	F	Sig.
1	Regression	.998	2	.499	39.291	.000[b]
	Residual	.102	8	.013		
	Total	1.100	10			

a. Dependent Variable: Graduate grade point average

b. Predictors: (Constant), Undergraduate grade point average, GRE Total Score

TABLE 8.4 (continued)

SPSS Results for the Multiple Regression GRE-GPA Example

The **"constant"** is the **intercept** and the unstandardized coefficient tells us that when all the predictors were zero, graduate GPA (the dependent variable) would be .638. The "GRE-Total" and "UGPA" are the slopes. For every one point increase in GRE-Total, graduate GPA will increase by about 1/10 of one point (holding constant undergraduate GPA). For every one point increase in undergraduate GPA, graduate GPA will increase by about ½ of one point (holding constant GRE-Total).

The test statistic, *t*, is calculated as the *unstandardized coefficient divided by its standard error.* Thus the slope for *undergraduate GPA* is calculated as (difference due to rounding):

$$t \quad \frac{.469}{.093} = 5.043$$

The *p* value for the intercept (the "constant") (p = .087) indicates that the intercept is *not* statistically significantly different from zero (this finding is usually of less interest than the slopes). The *p* values for GRE-Total and undergraduate GPA (the independent variables) (p = .001) indicate that the slopes are statistically significantly different from zero.

Coefficients[a]

Model		Unstandardized Coefficients B	Unstandardized Coefficients Std. Error	Standardized Coefficients Beta	t	Sig.	95.0% Confidence Interval for B Lower Bound	95.0% Confidence Interval for B Upper Bound	Correlations Zero-order	Correlations Partial	Correlations Part	Collinearity Statistics Tolerance	Collinearity Statistics VIF
1	(Constant)	.638	.327		1.954	.087	-.115	1.391					
	GRE Total Score	.012	.002	.614	5.447	.001	.007	.018	.784	.887	.585	.909	1.100
	Undergrad GPA	.469	.093	.567	5.030	.001	.254	.684	.752	.872	.541	.909	1.100

a. Dependent Variable: Graduate grade point average

Zero-order correlations are the simple bivariate Pearson correlations between the dependent variable and the independent variables.

Collinearity statistics are reviewed under assumptions.

The **partial correlation** of .887 is the correlation between GRE-Total and graduate GPA (dependent variable) when the linear effect of undergraduate GPA has been removed from both GRE-Total and graduate GPA (i.e., "controlling" for or holding constant undergraduate GPA). Squaring this indicates that 78.7% of the variation in graduate GPA that is not explained by undergraduate GPA *is* explained by GRE-Total.

The **part correlation** of .585, when squared (i.e., .342) indicates that GRE-Total explains an additional 34% of the variance in graduate GPA over and above the variance in graduate GPA which is explained by undergraduate GPA.

(*continued*)

TABLE 8.4 (continued)

SPSS Results for the Multiple Regression GRE-GPA Example

"Collinearity diagnostics" will be examined in our discussion of assumptions.

Collinearity Diagnostics[a]

				Variance Proportions		
Model	Dimension	Eigenvalue	Condition Index	(Constant)	GRE Total Score	Undergraduate grade point average
1	1	2.981	1.000	.00	.00	.00
	2	.012	15.727	.03	.86	.40
	3	.007	20.537	.97	.13	.60

a. Dependent Variable: Graduate grade point average

"Residual statistics" and related graphs (histogram and Q-Q plot of standardized residuals, not presented here) will be examined in our discussion of assumptions.

Residuals Statistics[a]

	Minimum	Maximum	Mean	Std. Deviation	N
Predicted Value	3.0714	3.9448	3.5000	.31597	11
Std. Predicted Value	-1.357	1.408	.000	1.000	11
Standard Error of Predicted Value	.038	.079	.058	.011	11
Adjusted Predicted Value	3.0599	3.9117	3.4954	.30917	11
Residual	-.19943	.17207	.00000	.10082	11
Std. Residual	-1.769	1.527	.000	.894	11
Stud. Residual	-1.881	1.716	.017	1.008	11
Deleted Residual	-.22531	.21754	.00458	.12935	11
Stud. Deleted Residual	-2.355	2.020	.000	1.145	11
Mahal. Distance	.240	4.053	1.818	1.048	11
Cook's Distance	.012	.260	.092	.081	11
Centered Leverage Value	.024	.405	.182	.105	11

a. Dependent Variable: Graduate grade point average

TABLE 8.4 (continued)

SPSS Results for the Multiple Regression GRE-GPA Example

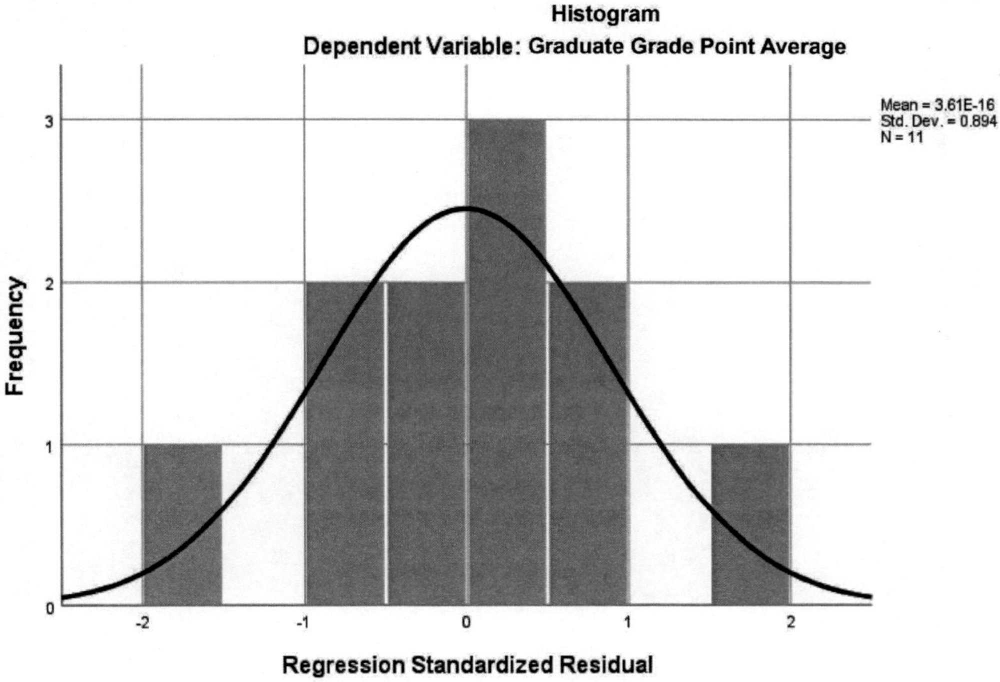

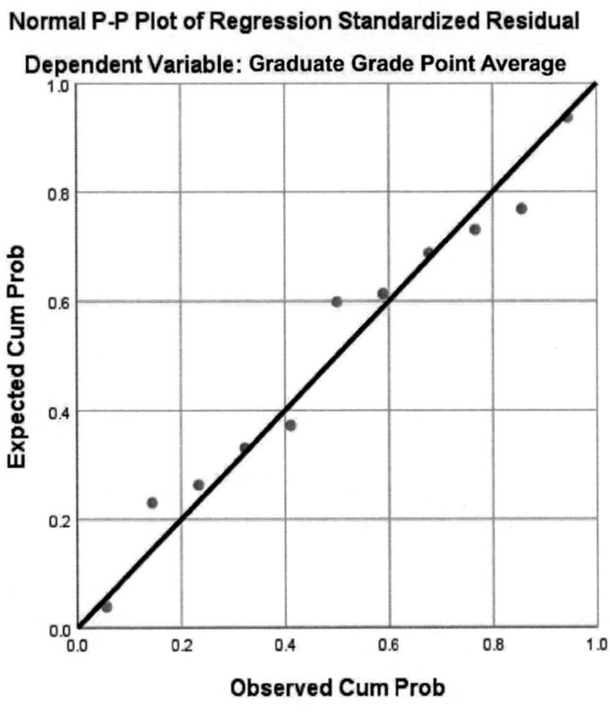

(*continued*)

TABLE 8.4 (continued)

SPSS Results for the Multiple Regression GRE-GPA Example

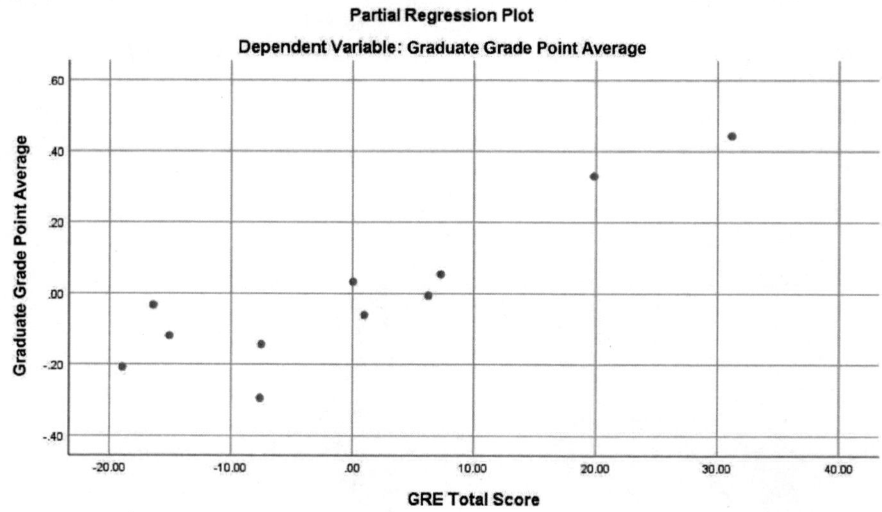

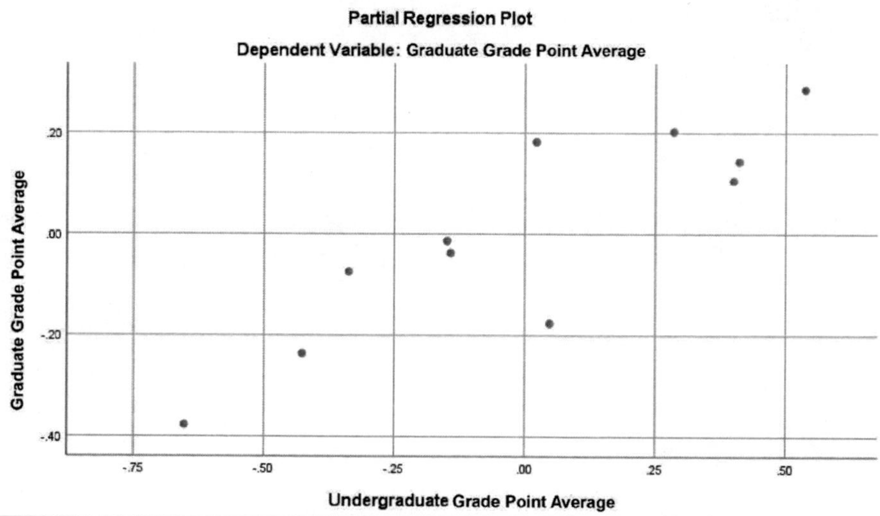

8.4 Computing Multiple Linear Regression Using R

Next we consider **R** for the multiple regression model. The commands are provided within the blocks with additional annotation to assist in understanding how the command works. Should you want to write reminder notes and annotation to yourself as you write the commands in **R** (and we highly encourage doing so), remember that any text that follows a hashtag (i.e., #) is annotation only and not part of the **R** code. Thus, you can write annotations directly into **R** with hashtags. We encourage this practice so that when you call up the commands in the future, you'll understand what the various lines of code are doing. You may think you'll remember what you did. However, trust us. There is a good chance that you won't. Thus, consider it best practice when using **R** to annotate heavily!

8.4.1 Reading Data Into R

In this illustration, we are pulling in data that are currently in a .csv file.

```
getwd()
```

R is always directed to a directory on your computer. To find out which directly it's pointed to, run the *get working directory* command. We will assume that we need to change the working directory, and will use the next line of code to set the working directory to the desired path.

```
setwd("E:/FolderName")
```

This command will set your working directory to a specific folder that you name. Change what is in parentheses to your file location. Also, if you are copying the directory name, it will copy in slashes. You will need to change the backslash (i.e., \) to a forward slash (i.e., /) in the **R** command. Also note that you need the name of your folder enclosed in quotation marks.

```
Ch8_GGPA <- read.csv("Ch8_GGPA.csv")
```

This command reads your data into **R**. To the left of "<-" will be what you want to call the dataframe in **R**. In this example, we're calling this **R** dataframe "Ch8_GGPA." What's to the right of "<-" tells **R** to find this particular csv file. In this example, our file is called "Ch8_GGPA.csv." Make sure the extension (i.e., .csv) is there. Also note that you need the name of the file enclosed in quotations.

```
names(Ch8_GGPA)
```

This command will produce a list of variable names for the dataframe that is noted in parentheses. For this illustration, our variable names are as follows. This is a good check to make sure your data have been read in correctly.

```
[1] "GRE_Total" "UGPA" "GGPA"
```

```
View(Ch8_GGPA)
```

This command will let you view the dataset in spreadsheet format in RStudio.

```
summary(Ch8_GGPA)
```

The *summary* command will produce basic descriptive statistics on all the variables in your dataframe. This is a great way to quickly check to see if the data have been read in correctly and get a feel for your data, if you haven't already. The output from the summary statement for this dataframe looks like this.

```
   GRE_Total          UGPA            GGPA
 Min.   : 90.0   Min.   :2.400   Min.   :3.00
 1st Qu.:102.5   1st Qu.:2.850   1st Qu.:3.25
 Median :110.0   Median :3.100   Median :3.50
 Mean   :112.7   Mean   :3.109   Mean   :3.50
 3rd Qu.:122.5   3rd Qu.:3.400   3rd Qu.:3.75
 Max.   :145.0   Max.   :3.700   Max.   :4.00
```

FIGURE 8.14
Reading data into **R**.

8.4.2 Generating the Multiple Regression Model and Saving Values

With these commands, we will generate the multiple regression model and save variables that can be used for data screening.

```
GGPA_MultReg <- lm(formula = GGPA ~ UGPA + GRE_Total,
                   data = Ch8_GGPA)
```

The *lm* command is the code to run the multiple linear regression model. In this example, we're naming our model (i.e., our object) "GGPA_MultReg." The formula defines our dependent variable as "GGPA" and it is predicted by "UGPA" and "GRE_Total." The data come from the Ch8_GGPA dataframe.

```
summary(GGPA_MultReg)
```

Run the *summary* command to see the results from the multiple regression model. The output includes a few residual statistics, coefficient estimates and related statistics, R^2, R^2_{adj}, and the overall F test. Note that if you don't run the summary line of code, since we created an object with our model, there won't be any results output from the *lm* command!

```
Residuals:
     Min       1Q   Median       3Q      Max
-0.19943 -0.06029  0.02812  0.06216  0.17207

Coefficients:
            Estimate Std. Error t value Pr(>|t|)
(Intercept) 0.637906   0.326537   1.954 0.086517 .
UGPA        0.468670   0.093181   5.030 0.001015 **
GRE_Total   0.012463   0.002288   5.447 0.000611 ***
---
Signif. codes:  0 '***' 0.001 '**' 0.01 '*' 0.05 '.' 0.1 ' ' 1

Residual standard error: 0.1127 on 8 degrees of freedom
Multiple R-squared:  0.9076,  Adjusted R-squared:  0.8845
F-statistic: 39.29 on 2 and 8 DF,  p-value: 7.289e-05
```

```
anova(GGPA_MultReg)
```

The *anova* command will generates the ANOVA summary table for the multiple regression model.

```
Analysis of Variance Table

Response: GGPA
          Df  Sum Sq Mean Sq F value    Pr(>F)
UGPA       1 0.62147 0.62147  48.916 0.0001133 ***
GRE_Total  1 0.37689 0.37689  29.665 0.0006112 ***
Residuals  8 0.10164 0.01270
---
Signif. codes:  0 '***' 0.001 '**' 0.01 '*' 0.05 '.' 0.1 ' ' 1
```

```
vcov(GGPA_MultReg)
```

The *vcov* command will generates the covariance matrix for the model parameters.

```
              (Intercept)          UGPA     GRE_Total
(Intercept)  0.1066264269 -1.975867e-02 -3.906768e-04
UGPA        -0.0197586732  8.682693e-03 -6.419572e-05
GRE_Total   -0.0003906768 -6.419572e-05  5.236241e-06
```

FIGURE 8.15
Generating multiple linear regression in **R** and saving variables.

Working in R, we can generate the Durbin-Watson test using the following command, where 'GGPA_MultReg' is the object created when we generated our multiple linear regression model. In other words, you will need to compute the multiple linear regression model before you can produce the Durbin Watson test.

```
durbinWatsonTest(GGPA_MultReg)
```

For diagnostic purposes, we can save additional variables to the dataframe which will be used later for checking assumptions.

```
Ch8_GGPA$unstandardizedPredicted <- predict(GGPA_MultReg)
```

This command saves unstandardized **predicted values**.

```
Ch8_GGPA$unstandardizedResiduals <- resid(GGPA_MultReg)
```

This command saves **unstandardized residuals**.

```
Ch8_GGPA$standardized.residuals <- rstandard(GGPA_MultReg)
```

This command saves **standardized residuals**. We expect standardized residuals to be within a range of −2.0 to +2.0.

```
Ch8_GGPA$studentized.residuals <- rstudent(GGPA_MultReg)
```

This command saves **studentized residuals**.

```
Ch8_GGPA$cook <- cooks.distance(GGPA_MultReg)
```

This command saves **Cook's distance**, an influence statistic.

```
Ch8_GGPA $leverage <- hatvalues(GGPA_MultReg)
```

This command saves the **leverage values**.

FIGURE 8.15 (continued)
Generating multiple linear regression in **R** and saving variables.

8.4.3 Generating Correlation Coefficients

```
install.packages("Hmisc")
```

There are many ways to generate correlations in **R**. For this illustration, we will use the *Hmisc* package. This command will install the *Hmisc* package that will allow us to generate the correlation matrix and related *p* values.

```
library("Hmisc")
```

This will load the package so we can use it.

```
cor(Ch8_GGPA)
```

This command will generate a simple correlation table. The default matrix is Pearson.

```
          GRE_Total      UGPA      GGPA
GRE_Total 1.0000000 0.3010711 0.7844854
UGPA      0.3010711 1.0000000 0.7516460
GGPA      0.7844854 0.7516460 1.0000000
```

FIGURE 8.16
Generating correlation coefficients in **R**.

```
res2 <- rcorr(as.matrix(Ch8_GGPA))
res2
```

This command will generate a matrix correlation table with all variables in our Ch8_GGPA dataframe and will also generate *p* values for the coefficients and sample size. We see the strongest correlation between GGPA and GRE_Total at $r = .78$.

```
          GRE_Total UGPA GGPA
GRE_Total      1.00 0.30 0.78
UGPA           0.30 1.00 0.75
GGPA           0.78 0.75 1.00

n= 11

P
          GRE_Total UGPA   GGPA
GRE_Total           0.3683 0.0042
UGPA      0.3683           0.0076
GGPA      0.0042    0.0076
```

FIGURE 8.16 (continued)
Generating correlation coefficients in **R**.

8.4.4 Generating Confidence Intervals of Coefficient Estimates

```
confint(GGPA_MultReg, level =.95)
```

Because we named our model as an object, we can easily request additional statistics on it. With the *confint* command, we can obtain confidence intervals for the coefficient estimates. With the *level* command, we set the confidence interval to 95% and thus are provided the lower confidence interval as 2.5% and upper confidence interval as 97.5%.

```
                   2.5 %      97.5 %
(Intercept) -0.115089982 1.39090147
UGPA         0.253794259 0.68354566
GRE_Total    0.007186535 0.01774012
```

FIGURE 8.17
Generating confidence intervals of coefficient estimates in **R**.

8.5 Data Screening

As you may recall, there were a number of assumptions associated with multiple linear regression. These included: (a) independence; (b) homoscedasticity; (c) linearity; (d) normality; and (e) noncollinearity. Although fixed values of *X* were discussed in assumptions, this is not an assumption that will be tested, but is instead related to the use of the results (i.e., extrapolation and interpolation). Before we begin to examine assumptions, let us review the values that we requested to be saved to our dataset (see Figure 8.13).

1. PRE_1 values are the unstandardized predicted values (i.e., Y_i').
2. RES_1 values are the unstandardized residuals, simply the difference between the observed and predicted values. For student 1, for example, the observed

value for the graduate GPA (i.e., the dependent variable) was 4 and the predicted value was 3.94483. Thus the unstandardized residual is simply 4 – 3.94483, or .05517.

3. SRE_1 values are the studentized residuals, a type of standardized residual that is more sensitive to outliers as compared to standardized residuals. Studentized residuals are computed as the unstandardized residual divided by an estimate of the standard deviation with that case removed. As a rule of thumb, studentized residuals with an absolute value greater than 3 are considered outliers (Stevens, 1984).
4. MAH_1 values are Mahalanobis distance values which measure how far that particular case is from the average of the independent variable and thus can be helpful in detecting outliers. These values can be reviewed to determine cases that are exerting leverage. Barnett and Lewis (1994) produced a table of critical values for evaluating Mahalanobis distance. Squared Mahalanobis distances divided by the number of variables $\left(D^2/df\right)$ which are greater than 2.5 (for small samples) or 3 to 4 (for large samples) are suggestive of outliers (Hair, Black, Babin, Anderson, & Tatham, 2006). Later, we follow another convention for examining these values using the chi-square distribution.
5. COO_1 values are Cook's distance values and provide an indication of influence of individual cases. As a general guidelines, Cook's values greater than one suggest that case is potentially problematic.
6. LEV_1 values are leverage values, a measure of distance from a respective case to the average of the predictor.
7. SDB0_1, **SDB1_1** and SDB2_1 values are standardized DfBeta values for the intercept and slopes, respectively, and are easier to interpret as compared to their unstandardized counterparts. Standardized DfBeta values greater than an absolute value of two suggest that the case may be exerting undue influence on the calculation of the parameters in the model (i.e., the slopes and intercept).

8.5.1 Independence

Here we will plot: (1) studentized residuals (which were requested and created through the "Save" option when generating our model) against unstandardized predicted values; and (2) studentized residuals against each independent variable to examine the extent to which independence was met. You are likely well versed in creating scatterplots, but as a reminder, in the top toolbar in SPSS, go to "Graphs," then "Legacy Dialog," then "Scatter/Dot." From the "Simple scatterplots" dialog screen, click the studentized residual and move it to the *Y* axis box by clicking the arrow. Similarly, move the unstandardized predicted value variable into the *X* axis box. Then click "OK" to generate the plot. Repeat these steps to plot the studentized residual to each independent variable. If the assumption of independence is met, the points should fall randomly within a band of –2.0 to +2.0, which is what we see in the graphs presented previously (see Figure 8.1).

Working in **R**, we create similar scatterplots using the following *plot* commands, with the first variable listed displaying on the *X* axis (e.g., "Ch8_GGPA$unstandardizedPredicted"), and the second variable displaying on the *Y* axis (i.e., "Ch8_GGPA$studentized.residuals"). Additional commands are provided to label the axes (*xlab* and *ylab*) and title the graph (*main*).

```
plot(Ch8_GGPA$unstandardizedPredicted,
     Ch8_GGPA$studentized.residuals,
     xlab = "unstandardized predicted values",
     ylab = "studentized residuals",
     main = "Scatterplot for independence")

plot(Ch8_GGPA$UGPA,
     Ch8_GGPA$studentized.residuals,
     xlab = "undergraduate GPA",
     ylab = "studentized residuals",
     main = "Scatterplot for independence")

plot(Ch8_GGPA$GRE_Total,
     Ch8_GGPA$studentized.residuals,
     xlab = "GRE Total",
     ylab = "studentized residuals",
     main = "Scatterplot for independence")
```

FIGURE 8.18
Generating plots in **R** for independence evidence.

8.5.2 Homoscedasticity

Recall that homogeneity of variance, or homoscedasticity, is evident when the spread of residuals is fairly constant over the range of unstandardized predicted values and observed values of the independent variables. In other words, we're looking for a relatively random display of points. If the display of residuals increases or decreases across the plot, then there may be an indication that the assumption of homoscedasticity has been violated. The plots used to examine independence (see Figure 8.1) can also be used for homoscedasticity: (1) studentized residuals against unstandardized predicted values and (2) studentized residuals against each independent variable to examine the extent to which independence was met.

Working in **R**, we can generate the *nonconstant error variance test* to determine if there is homogeneity of variance. The null hypothesis of this test is constant error variance, and the alternative hypothesis is that the error variance changes with the level of the fitted values, or with the linear combination of independent variables. A nonstatistically significant test suggests we have met the assumption, as we see here.

We use our multiple linear regression object (i.e., "GGPA_MultReg") with the *ncvTest* command to conduct the nonconstant error variance test. Note that this function runs from the package *car*; thus, make sure that *car* is installed and loaded in your library prior to running.

```
ncvTest(GGPA_MultReg)
```

The results produce a chi-squared test. Based on the *p* value (.463), our test is not statistically significant, which indicates we have met the assumption of homoscedasticity.

```
Non-constant Variance Score Test
Variance formula: ~ fitted.values
Chisquare = 0.5385073    Df = 1     p = 0.463052
```

FIGURE 8.19
Nonconstant error variance test in **R**.

8.5.3 Linearity

Since we have more than one independent variable, we have to take a different approach to examining linearity than what was done with simple linear regression. However, we can use the same information gleaned from our examination of independence and homoscedasticity for reviewing the assumption of linearity. As noted previously, the residuals should be located within a band of $\pm 2s_{res}$ (or standard errors), indicating no systematic pattern of points. Residual plots for the GGPA example are shown in Figure 8.1. Even with a very small sample, we see a fairly random pattern of residuals, and therefore feel fairly confident that the linearity assumption has been satisfied.

We can also review the partial regression plots that we asked for when generating the regression model (see Figure 8.12 for generating partial regression plots in **R**). A separate partial regression plot is provided for each independent variable, where we are looking for linearity (rather than some type of polynomial). Even with a small sample size, the partial regression plots suggest evidence of linearity.

8.5.4 Normality

Understanding the distributional shape, specifically the extent to which normality is a reasonable assumption, is important in multiple linear regression just as it was in simple linear regression. Normality can be understood by examining residuals as well as various diagnostics to examine our data for influential cases. Let us begin by examining the unstandardized residuals for normality. Because the steps for generating normality evidence were presented in previous chapters (see, for example, Chapter 6), they will not be repeated here.

8.5.4.1 Interpreting Normality Evidence

By this point, we are well versed in interpreting quite a range of normality statistics and will do the same for multiple linear regression. The skewness statistic of the residuals is −.336 and kurtosis is .484—both being within the range of what would be considered normal (approximately an absolute value), suggesting some evidence of normality.

Descriptives

			Statistic	Std. Error
Unstandardized Residual	Mean		.0000000	.03039717
	95% Confidence Interval for Mean	Lower Bound	-.0677291	
		Upper Bound	.0677291	
	5% Trimmed Mean		.0015202	
	Median		.0281190	
	Variance		.010	
	Std. Deviation		.10081601	
	Minimum		-.19943	
	Maximum		.17207	
	Range		.37150	
	Interquartile Range		.14051	
	Skewness		-.336	.661
	Kurtosis		.484	1.279

FIGURE 8.20
Normality evidence.

Given the very small sample size, the histogram reflects as normal a distribution as might be expected.

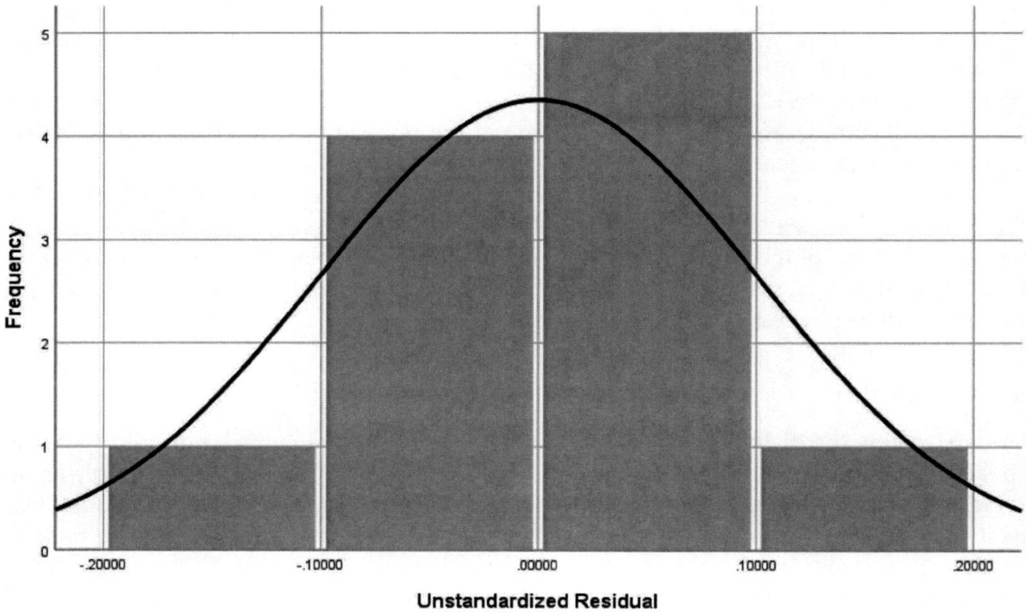

Working in **R**, we can generate a histogram using the *ggplot2* package.

```
install.packages("ggplot2")
```

This command will install the *ggplot2* package which we can use to create various graphs and plots.

```
library(ggplot2)
```

This command will load the *ggplot2* package.

```
qplot(Ch8_GGPA$unstandardizedResiduals,
     geom="histogram",
     main = "Histogram of Unstandardized Residuals",
     xlab = "Unstandardized Residual", ylab = "Count",
     fill=I("gray"),
     col=I("white"))
```

Using the *gplot* command, we create a histogram (i.e., geom = "histogram") from our dataframe (i.e., Ch8_GGPA) using the variable "unstandardizedResiduals." We can add a few commands to change the color of the bars (i.e., *fill=I("gray")*), and outline of the bars (i.e., *col=I("white")*). We can also add a title (i.e., *main = "Histogram of Unstandardized Residuals"*) and change the *X* and *Y* axes (*xlab = "Unstandardized Residual", ylab = "Count"*).

FIGURE 8.21
Histogram.

There are a few other statistics that can be used to gauge normality. The results for the formal test of normality, the Shapiro-Wilk test (*SW*) (Shapiro & Wilk, 1965), is presented below and suggests that our sample distribution for the residual is *not* statistically significantly different than what would be expected from a normal distribution as the *p* value is greater than α (*p* = .918).

Tests of Normality

	Kolmogorov-Smirnov[a]			Shapiro-Wilk		
	Statistic	df	Sig.	Statistic	df	Sig.
Unstandardized Residual	.155	11	.200*	.973	11	.918

*. This is a lower bound of the true significance.

a. Lilliefors Significance Correction

Working in **R**, we can generate various normality statistics as well.

```
install.packages("pastecs")
```

This command will install the *pastecs* package which we will use to generate various forms of normality evidence.

```
library(pastecs)
```

This command will load the *pastecs* package.

```
stat.desc(Ch8_GGPA$unstandardizedResiduals,
          norm = TRUE)
```

This command will generate normality indices on the variable "unstandardizedResiduals" in the dataframe Ch8_GGPA as follows. The *norm=TRUE* command will produce Shapiro-Wilk results (*SW*). We see skew (−.250) and kurtosis (−.694) along with *SW* = .973, *p* = .918 for the "unstandardized residual" variable. All indicate the assumption of normality has been met. As we know, we can divide the skew and kurtosis values by their standard errors to get a standardized value that can be used to determine if the skew and/or kurtosis is statistically different from zero. Since this output provides "2SE," we would simply divide this value by 2 to arrive at the standard error.

Note: You may have noticed that the skewness and kurtosis value that we've just generated differs from what we found in SPSS, which was skew = −.336 and kurtosis = .484. This is because there are different ways to calculate skewness and kurtosis. Let's use another package in **R** to calculate these statistics with different algorithms.

```
      nbr.val       nbr.null         nbr.na            min
 1.100000e+01   0.000000e+00   0.000000e+00  -1.994318e-01
          max          range            sum         median
 1.720684e-01   3.715003e-01  -1.387779e-17   2.811903e-02
         mean        SE.mean   CI.mean.0.95            var
-1.261617e-18   3.039717e-02   6.772912e-02   1.016387e-02
      std.dev       coef.var       skewness      skew.2SE
 1.008160e-01  -7.991015e+16  -2.502561e-01  -1.893907e-01
     kurtosis       kurt.2SE     normtest.W     normtest.p
-6.940916e-01  -2.712533e-01   9.733156e-01   9.178945e-01
```

```
install.packages("e1071")
```

This command will install the e1071 package which we will use to generate skewness and kurtosis.

FIGURE 8.22
Normality evidence: Shapiro-Wilk test.

```
library(e1071)
```

This command will load the e1071 package.

```
skewness(Ch8_GGPA$unstandardizedResiduals, type=3)
skewness(Ch8_GGPA$unstandardizedResiduals, type=2)
skewness(Ch8_GGPA$unstandardizedResiduals, type=1)
```

This command will generate skewness statistics on the variable(s) we specify. The "type=" script defines how skewness is calculated. Specifying "type=2" will use the algorithm that is used by SPSS. Readers interested in learning more, including the algorithms for each of the three methods, are encouraged to review Joanes and Gill (1998). We see that using "type=2," our skew is −.336, the same value as generated using SPSS.

```
# skewness(Ch8_GGPA$unstandardizedResiduals, type=3)
[1] -0.2502561

# skewness(Ch8_GGPA$unstandardizedResiduals, type=2)
[1] -0.3364554

# skewness(Ch8_GGPA$unstandardizedResiduals, type=1)
[1] -0.2887179
```

```
kurtosis(Ch8_GGPA$unstandardizedResiduals, type=3)
kurtosis(Ch8_GGPA$unstandardizedResiduals, type=2)
kurtosis(Ch8_GGPA$unstandardizedResiduals, type=1)
```

This command will generate kurtosis statistics on the variable(s) we specify. The "type=" script defines how kurtosis is calculated. Specifying "type=2" will use the algorithm that is used by SPSS. Readers interested in learning more, including the algorithms for each of the three methods, are encouraged to review Joanes and Gill (1998). We see that using "type=2," our kurtosis is .484, the same value as generated using SPSS.

```
# kurtosis(Ch8_GGPA$unstandardizedResiduals, type=3)
[1] -0.6940916

# kurtosis(Ch8_GGPA$unstandardizedResiduals, type=2)
[1] 0.483582

# kurtosis(Ch8_GGPA$unstandardizedResiduals, type=1)
[1] -0.2098508
```

FIGURE 8.22 (continued)
Normality evidence: Shapiro-Wilk test.

Quantile-quantile (Q-Q) plots are also often examined to determine evidence of normality. The Q-Q plot of residuals suggests relative normality with points that fall on or close to the diagonal line suggesting evidence of normality.

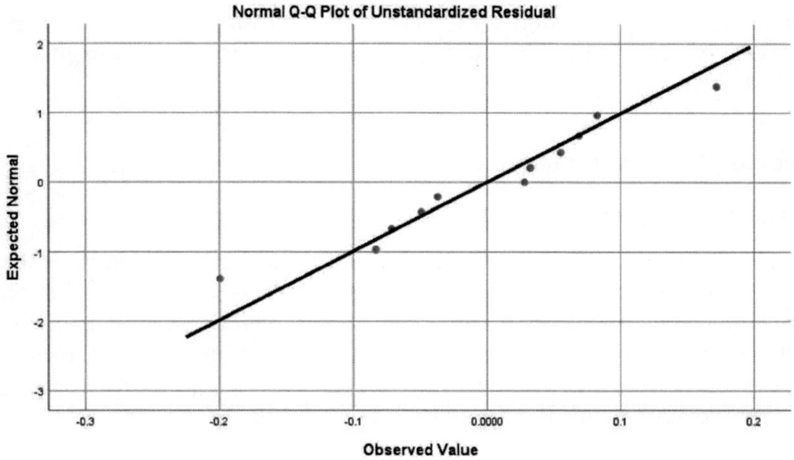

Working in **R**, we can use the *gplot* command to create a Q-Q plot of the variable "unstandardizedResiduals" from the dataframe Ch8_GGPA.

```
qplot(sample=unstandardizedResiduals,
      data = Ch8_GGPA)
```

FIGURE 8.23
Q-Q plot.

The boxplot in Figure 8.24 also suggests a relatively normal distribution of residuals with no outliers.

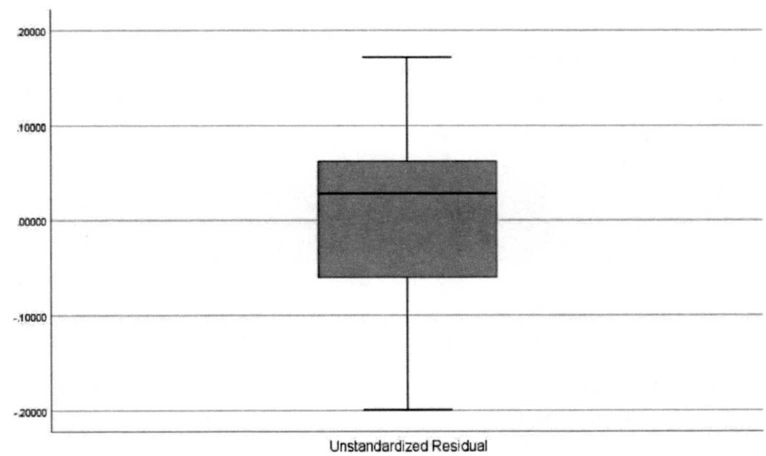

Working in **R**, we can generate a boxplot for unstandardized residuals using the *boxplot* function. To label the *Y* axis, we include the *ylab* command.

```
boxplot(Ch8_GGPA$unstandardizedResiduals,
        ylab="unstandardized residual")
```

FIGURE 8.24
Boxplot.

Considering the forms of evidence we have examined, skewness and kurtosis statistics, the Shapiro-Wilk test, histogram, the Q-Q plot, and the boxplot, all suggest normality is a reasonable assumption.

8.5.5 Screening Data for Influential Points

8.5.5.1 Casewise Diagnostics

Recall that we requested a number of statistics to help in diagnostics. One that we requested was for "Casewise diagnostics." If we had any cases with large values for the standardized residual (outside three standard deviations), information would have been included in our output to indicate the case number, value of the standardized residual, predicted value, and unstandardized residual. This information can be used to more closely examining case(s) with the extreme values on the standardized residuals.

8.5.5.2 Cook's Distance

Cook's distance provides an overall measure for the influence of individual cases. Values greater than one suggest that the case may be problematic in terms of undue influence on the model. Examining the residual statistics in our output, we see that the maximum value for Cook's distance is .260, well under the point at which we should be concerned.

Residuals Statistics[a]

	Minimum	Maximum	Mean	Std. Deviation	N
Predicted Value	3.0714	3.9448	3.5000	.31597	11
Std. Predicted Value	-1.357	1.408	.000	1.000	11
Standard Error of Predicted Value	.038	.079	.058	.011	11
Adjusted Predicted Value	3.0599	3.9117	3.4954	.30917	11
Residual	-.19943	.17207	.00000	.10082	11
Std. Residual	-1.769	1.527	.000	.894	11
Stud. Residual	-1.881	1.716	.017	1.008	11
Deleted Residual	-.22531	.21754	.00458	.12935	11
Stud. Deleted Residual	-2.355	2.020	.000	1.145	11
Mahal. Distance	.240	4.053	1.818	1.048	11
Cook's Distance	.012	.260	.092	.081	11
Centered Leverage Value	.024	.405	.182	.105	11

a. Dependent Variable: Graduate grade point average

Working in **R**, we can create a new variable in our dataframe (i.e., "Ch8_GGPA$largeCook") that notes cases that have a Cook's distance that is greater than 1 using the following command:

```
Ch8_GGPA$largeCook <- Ch8_GGPA$cook > 1
```

We can then run the *sum* command to find out how many large Cook's values there are.

```
sum(Ch8_GGPA$largeCook)
```

We can write similar commands for the centered leverage values.

FIGURE 8.25
Screening data for influential points.

8.5.5.3 Mahalanobis Distances

Mahalanobis distances are measures of the distance from each case to the mean of the independent variable for the remaining cases. We can use the value of Mahalanobis distance as a test statistic value with the chi-square distribution. With two independent variables and one dependent variable, we have three degrees of freedom. Given an alpha level of .05 (alpha of .001 if you want to be a bit more liberal), the chi-square critical value is 7.82. Thus any Mahalanobis distance greater than 7.82 suggests that case is an outlier. With a maximum of 4.053 (see Figure 8.25), there is no evidence to suggest there are outliers in our data.

8.5.5.4 Centered Leverage Values

Centered leverage values less than .20 suggest there are no problems with cases that are exerting undue influence (see Figure 8.25). Values greater than .5 indicate problems.

8.5.5.5 DfBeta

We also asked to save DfBeta values. These values provide another indication of the influence of cases. DfBeta provides information on the change in the predicted value when the case is deleted from the model. For standardized DfBeta values, values greater than an absolute value of 2.0 should be examined more closely. Looking at the minimum and maximum DfBeta values, there are no cases suggestive of undue influence.

Statistics

		Standardized DFBETA Intercept	Standardized DFBETA GRE_Total	Standardized DFBETA UGPA
N	Valid	11	11	11
	Missing	0	0	0
Minimum		-.51278	-.75577	-.32176
Maximum		.63170	.59269	.55938

Working in **R**, we can request DfBetas from our multiple regression model using the following command, and we will name this object "Ch8_dfbeta":

```
Ch8_dfbeta <- dfbetas(GGPA_MultReg)
```

Next, we want to define the range within which there may be influence. Values outside the range of an absolute value of 2 may be influential points. We define the range of our object (i.e., "Ch8_dfbeta") to be < −2 and > 2. We will create an object from this called "Ch8_dfbetasummary."

```
Ch8_dfbetasummary <- Ch8_dfbeta < -2 | Ch8_dfbeta > 2
```

Now, all we need to do is run the *sum* function to see how many DfBeta values are outside this range, and we see there are none.

```
sum(Ch8_dfbetasummary)
```

[1] 0

FIGURE 8.26
Screening for influential points: DfBeta.

8.5.5.6 Diagnostic Plots

A number of diagnostic plots can be generated from the values we saved. For example, a plot of Cook's distance against centered leverage values provides a way to identify influential cases (i.e., cases with leverage of .50 or above and Cook's distance of 1.0 or greater). Here there are no cases that suggest undue influence.

```
plot(GGPA_MultReg)
```

The *plot* command will graph a plot of residuals to fitted values. Note that you have to hit the return key in the RStudio console to generate the plot.

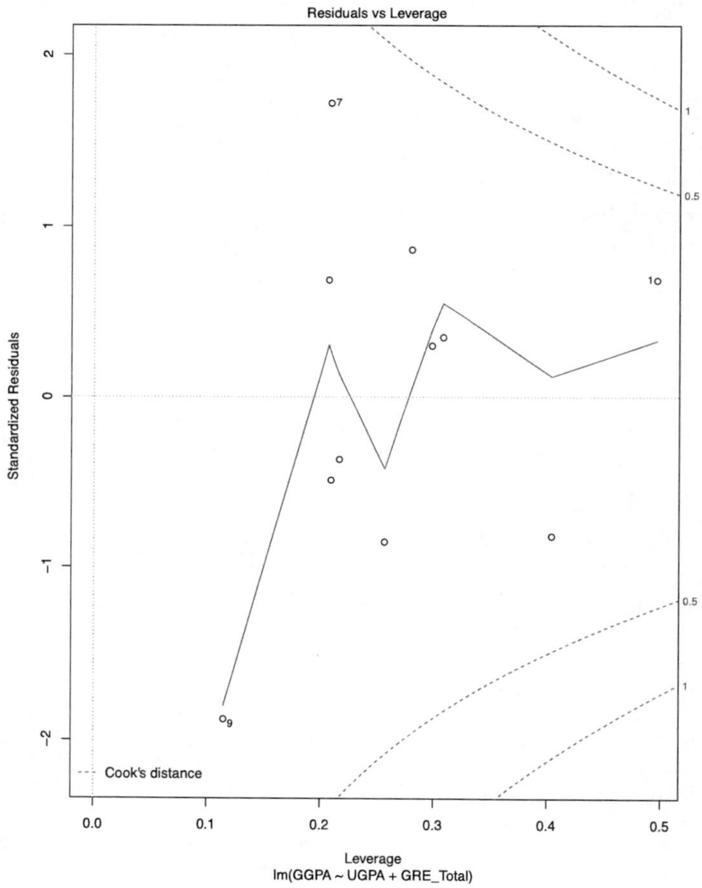


```
layout(matrix(c(1,2,3,4),2,2))
plot(GGPA_MultReg)
```

The *plot* function generates diagnostic plots, and we can use the *layout* command to plot four graphs per page.

FIGURE 8.27
Screening for influential points: diagnostic plots.

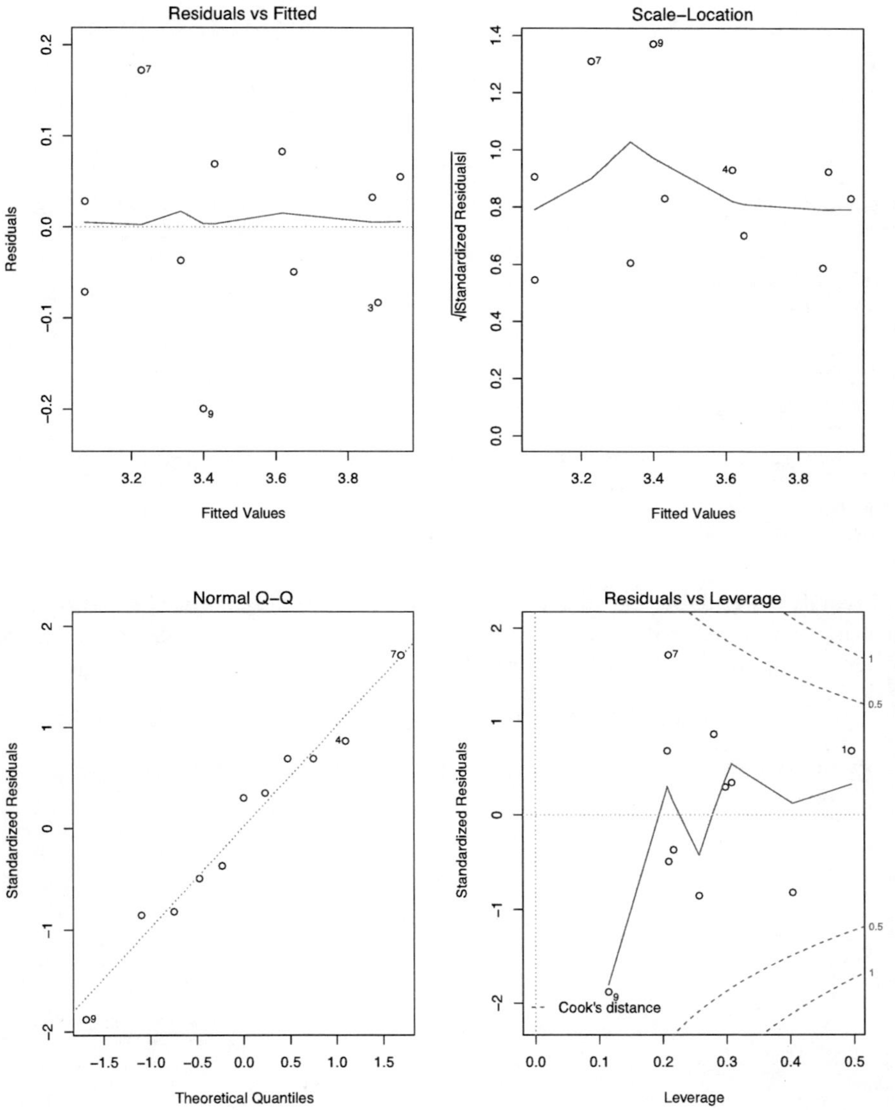

FIGURE 8.27 (continued)
Screening for influential points: diagnostic plots.

8.5.6 Noncollinearity

Detecting multicollinearity can be done by reviewing the **VIF** and **tolerance statistics**. From the table in Figure 8.28, we see tolerance and VIF values. *Tolerance* is calculated as $(1 - R^2)$ and values close to zero (a recommendation is .10 or less) suggest potential multicollinearity problems. Why? A tolerance of .10 suggests that 90% (or more) of the variance in one of the independent variables can be explained by another independent variable. *VIF* is the "variance inflation factor" and is the reciprocal of tolerance where $VIF = \frac{2}{tolerance}$. VIF values greater than 10 (which correspond to a tolerance of .10) suggest potential multicollinearity.

Collinearity Statistics	
Tolerance	VIF
.909	1.100
.909	1.100

Working in **R**, the *car* package can be used to generate VIF statistics. The following command will install *car* and load it into your library. If you've installed *car* previously, you only need to load the package into your library.

```
install.packages(car)
library(car)
```

```
vif(ReadinessLogit)
1/vif(ReadinessLogit)
```

The *vif* and *1/vif* commands will generate the VIF and its reciprocal, which is the tolerance statistic.

FIGURE 8.28
Collinearity statistics.

Collinearity diagnostics can also be reviewed. Multiple *eigenvalues* close to zero indicate independent variables that have strong intercorrelations. The *condition index* is calculated as the square root of the ratio of the largest eigenvalue to each preceding eigenvalue (e.g., $\sqrt{\frac{2.981}{.012}} = 15.76$). A general convention for interpreting condition indices is that values in the range of 10 to 30 should be of concern, greater than 30 indicates trouble, and greater

Collinearity Diagnostics[a]

Model	Dimension	Eigenvalue	Condition Index	Variance Proportions (Constant)	GRE Total Score	Undergraduate grade point average
1	1	2.981	1.000	.00	.00	.00
	2	.012	15.727	.03	.86	.40
	3	.007	20.537	.97	.13	.60

a. Dependent Variable: Graduate grade point average

FIGURE 8.29
Collinearity diagnostics.

than 100 indicates disaster (Belsley, 1991). In this case, both the eigenvalues and condition indices suggest possible problems with multicollinearity.

Noncollinearity can also be examined by computing regression models where each independent variable is considered the outcome and is predicted by the remaining independent variables (the dependent variable is not included in these models). If any of the resultant R_k^2 values are close to one (greater than .9 is a good guideline to follow), then there may be a collinearity problem. For the example data, $R_{12}^2 = .091$, and therefore collinearity is not a concern. Note that in multiple regression situations where there are two independent variables (as in this example with GRE-Total and undergraduate GPA), only one regression needs to be conducted to check for multicollinearity as the results for regressing undergraduate GPA on GRE-Total are the same as regressing GRE-Total on undergraduate GPA.

8.6 Power Using G*Power

A priori and post hoc power can be determined using the specialized software described previously in this text (e.g., G*Power), or you can consult *a priori* power tables (e.g., Cohen, 1988). As an illustration, we use G*Power to first compute the post hoc power of our test. This is followed by an illustration of how to compute *a priori* power.

8.6.1 Post Hoc Power

The first thing that must be done when using G*Power for computing post hoc power is to select the correct test family. In our case, we conducted multiple linear regression. To find regression, we select "Tests" in the top pulldown menu, then "Correlation and regression," and then "Linear multiple regression: Fixed model, R^2 deviation from zero." This will allow us to determine power for the hypothesis that the overall multiple R^2 is equal to zero (i.e., power for the overall regression model). Once that selection is made, the "Test family" automatically changes to "F test."

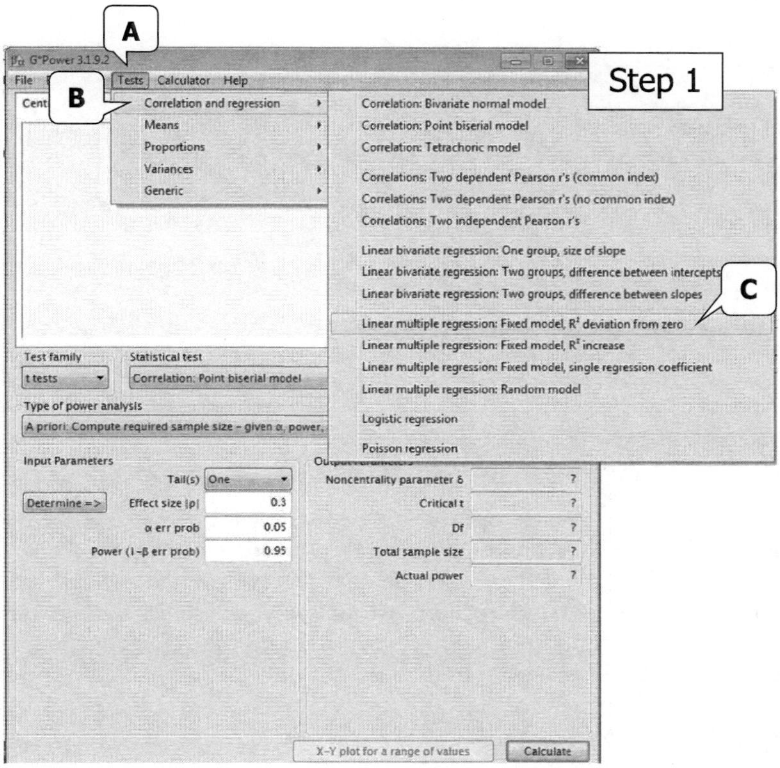

FIGURE 8.30
Post hoc power: Step 1.

The "Type of power analysis" desired needs to be selected. To compute post hoc power, select "Post hoc: Compute achieved power—given α, sample size, and effect size."

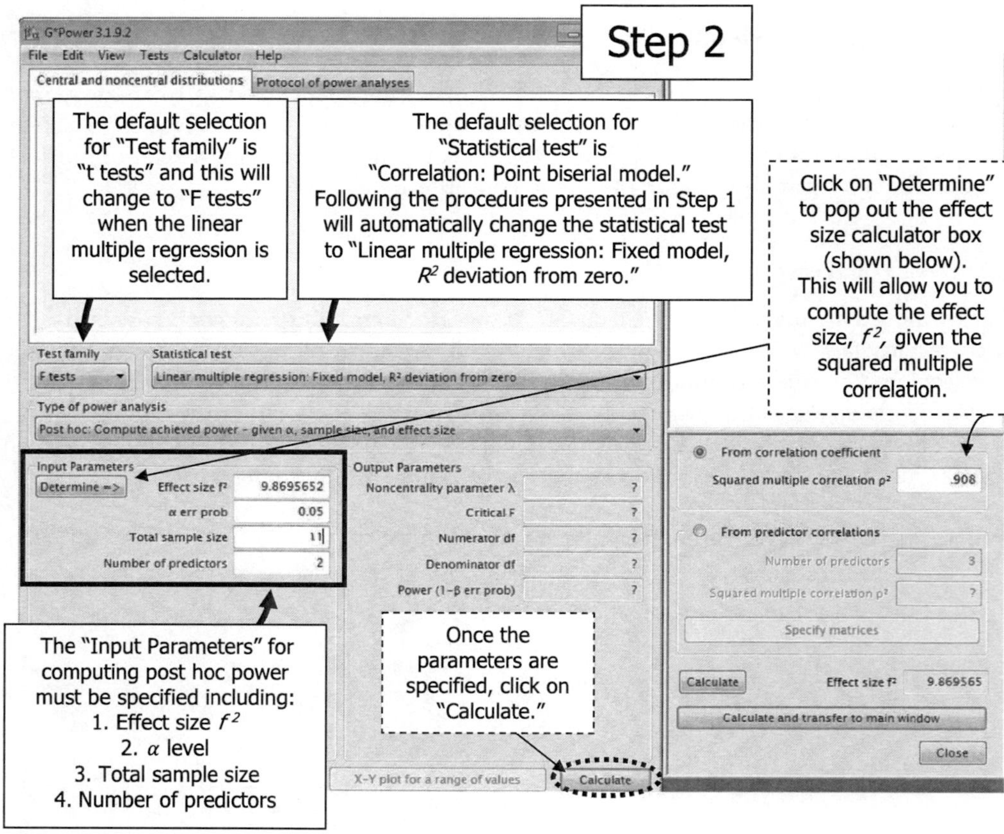

FIGURE 8.31
Post hoc power: Step 2.

The "Input Parameters" must then be specified. We will compute the effect size, f^2, last and so we skip that for the moment. The alpha level which we used was .05, the total sample size was 11 and there were two independent variables. Next, we use the pop out effect size calculator in G*Power to compute the effect size f^2. To do this, click on "Determine" which is displayed under "Input Parameters." In the pop out effect size calculator, input the value for the squared multiple correlation (i.e., the coefficient of determination, R^2). Click on "Calculate" to compute the effect size f^2. Then click on "Calculate and transfer to main window" to transfer the calculated effect size (i.e., 9.8695652) to the "Input Parameters." Once the parameters are specified, click on "Calculate" to find the power statistics.

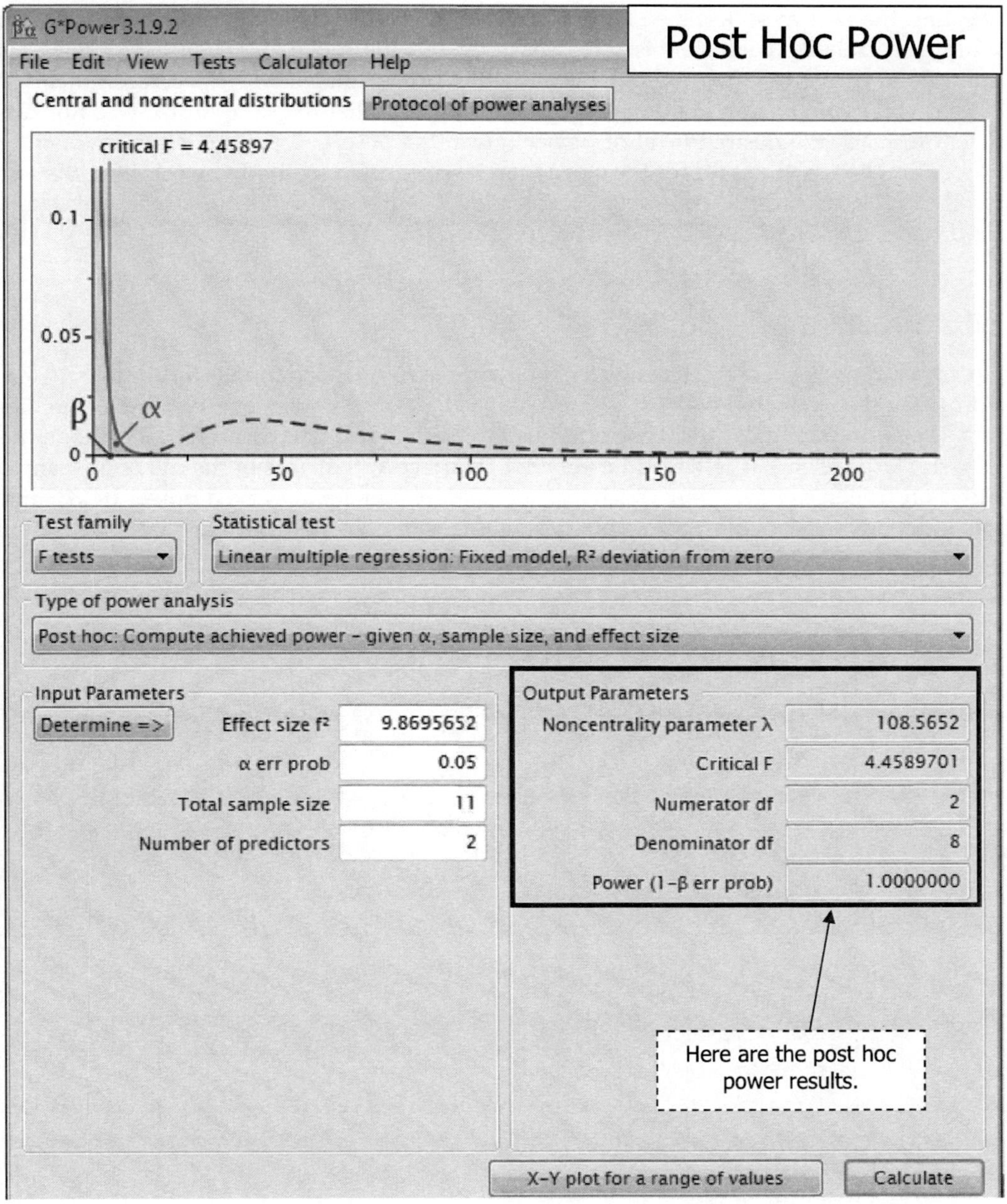

FIGURE 8.32
Post hoc power results.

The "Output Parameters" provide the relevant statistics given the input just specified. Here we were interested in determining post hoc power for a multiple linear regression with a computed effect size f^2 of 9.8695652, an alpha level of .05, total sample size of 11, and two predictors. Based on those criteria, the post hoc power for the overall multiple linear regression model is 1.0000. In other words, given the input parameters, the probability of rejecting the null hypothesis when it is really false (in this case, the probability that the multiple correlation coefficient is zero) was at the maximum (i.e., 1.00) (sufficient power is often .80 or above). Do not forget that conducting power analysis *a priori* is recommended so that you avoid a situation where, post hoc, you find that the sample size was not sufficient to reach the desired level of power (given the observed parameters). Conducting power for change in R^2 and for the slopes can be conducted similarly by selecting the test family of "Linear multiple regression: Fixed model, R^2 increase" or "Linear multiple regression: Fixed model, single regression coefficient," respectively.

8.6.2 *A Priori* Power

For *a priori* power, we can determine the total sample size needed for multiple linear regression given the estimated effect size f^2, alpha level, desired power, and number of predictors. We follow Cohen's (1988) conventions for f^2 effect size (i.e., small R^2 = .02; moderate R^2 = .13; large R^2 = .26). If we had estimated an effect R^2 that was in the moderate range, based on Cohen's conventions, such as f^2 of .15, alpha of .05, observed power of .80, and two independent variables, we would need a total sample size of 68.

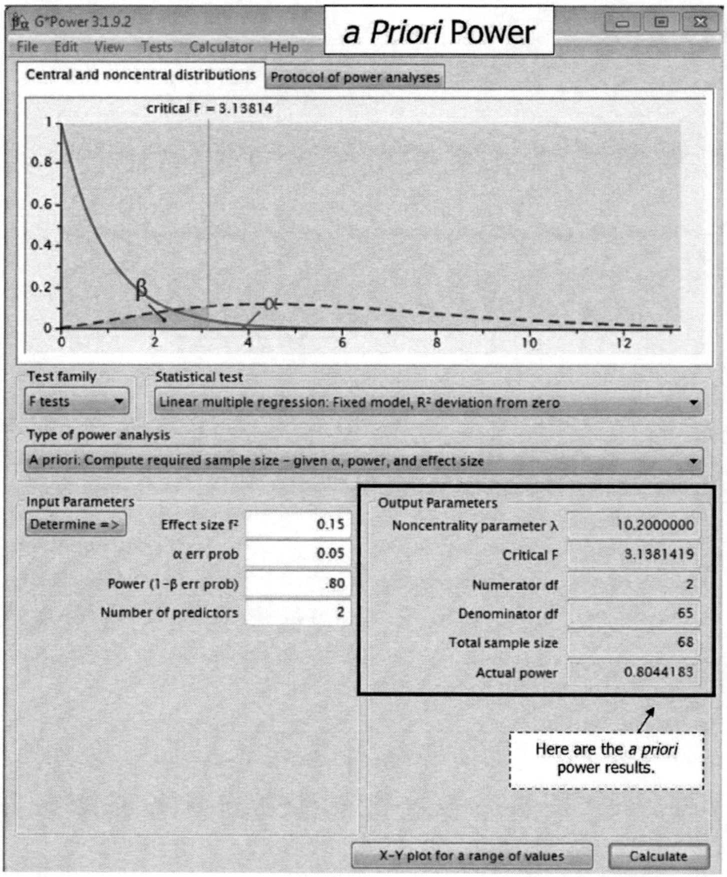

FIGURE 8.33
A priori power results.

8.7 Research Question Template and Example Write-Up

Finally, here is an example write-up for the results of the multiple linear regression analysis. Recall that our graduate research assistant, Addie Venture, was assisting the assistant dean in Graduate Student Services, Dr. Golly. Dr. Golly wanted to know if graduate GPA could be predicted by the total score on the required graduate entrance exam (GRE-Total) and by undergraduate GPA. The research question presented to Dr. Golly by Addie included the following: *Can graduate GPA be predicted from the GRE-Total and undergraduate GPA?*

Addie then assisted Dr. Golly in generating a multiple linear regression as the test of inference, and a template for writing the research question for this design is presented here.

Can [dependent variable] be predicted from [list the set of independent variables]?

It may be helpful to preface the results of the multiple linear regression with information on an examination of the extent to which the assumptions were met. The assumptions include: (a) independence; (b) homoscedasticity; (c) normality; (d) linearity; (e) noncollinearity; and (f) values of X are fixed. Because the last assumption (fixed X) is based on interpretation, it will not be discussed here.

A multiple linear regression model was conducted to determine if graduate GPA (dependent variable) could be predicted from GRE-Total scores and undergraduate GPA (independent variables). The null hypotheses tested were that the multiple R^2 was equal to zero and that the regression coefficients (i.e., the slopes) were equal to zero. The data were screened for missingness and violation of assumptions prior to analysis. There were no missing data.

Linearity. Review of the partial scatterplot of the independent variables (GRE-Total and undergraduate GPA) and the dependent variable (graduate GPA) indicate linearity is a reasonable assumption. Additionally, with a random display of points falling within an absolute value of two, a scatterplot of unstandardized residuals to predicted values provided further evidence of linearity.

Normality. The assumption of normality was tested via examination of the unstandardized residuals. Review of the Shapiro-Wilk test for normality ($SW = .973$, $df = 11$, $p = .918$) and skewness (−.336) and kurtosis (.484) statistics suggested that normality was a reasonable assumption. The boxplot suggested a relatively normal distributional shape (with no outliers) of the residuals. The Q-Q plot and histogram suggested normality was reasonable. Examination of casewise diagnostics, including Mahalanobis distance, Cook's distance, DfBeta values, and centered leverage values, suggested there were no cases exerting undue influence on the model.

Independence. A relatively random display of points in the scatterplots of studentized residuals against values of the independent variables and studentized residuals against predicted values provided evidence of independence. The Durbin-Watson statistic was computed to evaluate independence of errors and was 2.116, which is considered acceptable. This suggests that the assumption of independent errors has been met.

Homoscedasticity. A relatively random display of points, where the spread of residuals appears fairly constant over the range of values of the independent variables (in the scatterplots of studentized residuals against predicted values and studentized residuals against values of the independent variables) provided evidence of homoscedasticity. The nonconstant error test was not statistically significant, $\chi^2 = .539$, $df = 1$, $p = .463$, providing further evidence that there is homogeneity of variance.

Noncollinearity. Tolerance was greater than .10 (.909) and the variance inflation factor was less than 10 (1.100), suggesting that multicollinearity was not an issue. However, the eigenvalues for the predictors were close to zero (.012 and .007) and the respective condition indices were in the range of concern (between 10 and 30, 15.727 and 20.537 respectively). A review of GRE-Total regressed on undergraduate GPA, however, produced R^2 of .091 which suggests noncollinearity. Thus, though there is some isolated cause for concern, the evidence in aggregate suggests that multicollinearity is not an issue.

Here is an example write-up of the results for the multiple linear regression (remember that this will be prefaced by the previous paragraph reporting the extent to which the assumptions of the test were met).

The results of the multiple linear regression suggest that a significant proportion of the total variation in graduate GPA was predicted by GRE-Total and undergraduate GPA, $F(2, 8) = 39.291$, $p < .001$. Additionally, we find:

a. For GRE-Total, the unstandardized partial slope (.012) and standardized partial slope (.614) are statistically significantly different from zero ($t = 5.447$, $df = 8$, $p < .001$). This indicates that with every one-point increase in the GRE-Total score, graduate GPA will increase by approximately 1/100 of one point when controlling for undergraduate GPA.
b. For undergraduate GPA, the unstandardized partial slope (.469) and standardized partial slope (.567) are statistically significantly different from zero ($t = 5.030$, $df = 8$, $p < .001$). This indicates that with every one-point increase in undergraduate GPA, graduate GPA will increase by approximately ½ of one point when controlling for GRE-Total.
c. The confidence intervals around the unstandardized partial slopes do not include zero (GRE-Total, .007, .018; undergraduate GPA, .254, .684) further confirming that these variables are statistically significant predictors of graduate GPA. Thus GRETOT and UGPA were shown to be statistically significant predictors of GGPA, both individually and collectively.
d. The intercept (or average graduate GPA when GRE-Total and undergraduate GPA is zero) was .638, not statistically significantly different from zero ($t = 1.954$, $df = 8$, $p = .087$).
e. R^2 indicates that approximately 91% of the variation in graduate GPA was predicted by the model (i.e., GRE-Total scores and undergraduate GPA). Interpreted according to Cohen (1988), this suggests a large effect. Cohen's f^2, computed as $f^2 = \dfrac{R^2}{(1-R^2)}$, was 9.87, a large effect, and represents the proportion of variation in graduate GPA uniquely explained by the model (i.e., GRE-Total scores and undergraduate GPA) to the proportion of variation in graduate GPA unexplained by the model.
f. Estimated post hoc power to predict multiple R^2 was at the maximum, 1.00.

8.8 Additional Resources

This chapter has provided a preview into conducting multiple linear regression analysis. However, there are a number of areas related to regression and various regression models that space limitations prevent us from delving into. For those of you who are interested in learning more, or if you find yourself in a sticky situation in your analyses, you may wish to look into the following, among many other excellent resources:

- General references related to regression (Olive, 2017; Welc, Esquerdo, & Springer-Link, 2018).
- Bayesian regression (Wang, Faraway, & Yue Ryan, 2018).
- Classification and regression trees (CART), random forests, bagging, boosting, and more (Berk, 2016).
- Nonlinearity in one or more independent variables (Knafl & Ding, 2016).
- Obtaining robust multicollinearity diagnostics when outliers are present (Sinan & Alkan, 2015)
- Quantile regression (Koenker, Chernozhukov, He, & Peng, 2017)
- Regression discontinuity (Hausman & Rapson, 2017; Lee, 2016) and extension of RDD to regression kink design (Card, Lee, Pei, & Weber, 2016)
- Transformations and weighting with heteroscedastic data (Ruppert, 2014)

Problems

Conceptual Problems

1. The correlation of salary and cumulative grade point average controlling for socioeconomic status is an example of which one of the following?

  a. Bivariate correlation

  b. Partial correlation

  c. Regression correlation

  d. Semipartial correlation

2. The most accurate predictions are made when the standard error of estimate equals which one of the following?

  a. $\overline{Y}$

  b. s_Y

  c. 0

  d. 1

3. True or false? The intercept can take on a positive value only.

4. True or false? Adding an additional predictor to a regression equation will necessarily result in an increase in R^2.

5. True or false? The best prediction in multiple regression analysis will result when each predictor has a high correlation with the other predictor variables and a high correlation with the dependent variable.

6. Consider the following two situations:
Situation 1: $r_{Y1} = .6 \quad r_{Y2} = .5 \quad r_{12} = .0$
Situation 2: $r_{Y1} = .6 \quad r_{Y2} = .5 \quad r_{12} = .2$
I assert that the value of R^2 will be greater in Situation 2. Am I correct?

7. Values of variables X_1, X_2, and X_3 are available for a sample of 50 students. The value of $r_{12} = .6$. I assert that if the partial correlation $r_{12.3}$ were calculated, it would be larger than .6. Am I correct?

8. A researcher is building a regression model. There is theory to suggest that science ability can be predicted by literacy skills when controlling for child characteristics (e.g., age and socioeconomic status). Which one of the following variable selection procedures is suggested?
   a. Backward elimination
   b. Forward selection
   c. Hierarchical regression
   d. Stepwise selection

9. I assert that the forward selection, backward elimination, and stepwise regression methods will always arrive at the same final model, given the same dataset and level of significance? Am I correct?

10. I assert the R^2_{adj} will always be larger for the model with the most predictors. Am I correct?

11. True or false? In a two-predictor regression model, if the correlation among the predictors is .95 and VIF is 20, then we should be concerned about collinearity.

12. A researcher is examining how weight is related to age and number of hours exercised per week. The researcher wishes to remove the influence of age from the number of hours exercised per week but not from weight. Which coefficient should the researcher compute?
   a. Bivariate correlation
   b. Partial correlation
   c. Regression correlation
   d. Semipartial correlation

13. Which of the following types of evidence are appropriate for examining the extent to which the assumption of normality has been met?
   a. Maximum value of Cook's distance
   b. Scatterplot of studentized residuals and unstandardized predicted values
   c. Shapiro-Wilk test
   d. Variance inflation factor value

14. A researcher is computing a multiple linear regression model and is interested in including a nominal variable that has four categories. How do you suggest the researcher pursue this?
   a. Create dummy variables and include all four of the dummy variables.
   b. Create dummy variables and include three of the four dummy variables.
   c. Include the nominal variable in the model as is.
   d. Exclude the nominal variable as multiple linear regression cannot deal with variables of this measurement scale.

Answers to Conceptual Problems

1. **b** (partial correlations correlate two variables while holding constant a third)
3. **False** (the intercept can be any value)
5. **False** (best prediction is when there is a high correlation of the predictors with the dependent variable and low correlations among the predictors)
7. **No** (the partial correlation may be larger than, the same as, or smaller than .6)
9. **No** (as discussed, these methods may yield different final models)
11. **True** (that is precisely the situation when we should be very concerned about collinearity)
13. **c** (the Shapiro-Wilk test of normality is one of several types of normality evidence that can be used to examine residuals for meeting the assumption of normality)

Computational Problems

1. You are given the following data, where X_1 (hours of professional development) and X_2 (aptitude test scores) are used to predict Y (annual salary in thousands):

Y	X_1	X_2
40	100	10
50	200	20
50	300	10
70	400	30
65	500	20
65	600	20
80	700	30

Determine the following values: intercept; b_1, b_2, SS_{res}; SS_{reg}; F; s^2_{res}; $s(b_1)$; $s(b_2)$; t_1; t_2.

2. You are given the following data, where X_1 (final percentage in science class) and X_2 (number of absences) are used to predict Y (standardized science test score in third grade):

Y	X_1	X_2
300	65	7
480	98	0
350	70	3
420	80	2
400	82	0
335	70	3
370	75	4
390	80	1
485	99	0
415	95	2
375	88	3

Determine the following values: intercept; b_1, b_2, SS_{res}; SS_{reg}; F; s^2_{res}; $s(b_1)$; $s(b_2)$; t_1; t_2.

3. Complete the missing information for this regression model ($df = 23$).

Y'	=	25.1	+	$1.2\ X_1$	+	$1.0\ X_2$	–	$.50\ X_3$	
		(2.1)		(1.5)		(1.3)		(.06)	standard errors
		(11.9)		( )		( )		( )	*t* ratios
				( )		( )		( )	Significant at .05?

4. Consider a sample of elementary school children. Given that r(strength, weight) = .6, r(strength, age) = .7, and r(weight, age) = .8, what is the first-order partial correlation coefficient between strength and weight holding age constant?
5. For a sample of 100 adults, you are given that $r_{12} = .55$, $r_{13} = .80$, and $r_{23} = .70$. What is the value of $r_{1(2.3)}$?
6. A researcher would like to predict salary from a set of four predictor variables for a sample of 45 subjects. Multiple linear regression analysis was utilized. Complete the following summary table ($\alpha = .05$) for the test of significance of the overall regression model:

Source	*SS*	*df*	*MS*	*F*	Critical Value and Decision
Regression	–	–	20	–	
Residual	400	–	–		
Total	–	–			

7. Calculate the partial correlation $r_{12.3}$ and the part correlation $r_{1(2.3)}$ from the following bivariate correlations: $r_{12} = .5$, $r_{13} = .8$, $r_{23} = .9$.
8. Calculate the partial correlation $r_{13.2}$ and the part correlation $r_{1(3.2)}$ from the following bivariate correlations: $r_{12} = .21$, $r_{13} = .40$, $r_{23} = -.38$.
9. You are given the following data, where X_1 (verbal aptitude) and X_2 (prior reading achievement) are to be used to predict Y (reading achievement):

Y	X_1	X_2
2	2	5
1	2	4
1	1	5
1	1	3
5	3	6
4	4	4
7	5	6
6	5	4
7	7	3
8	6	3
3	4	3
3	3	6
6	6	9
6	6	8

Y	X_1	X_2
10	8	9
9	9	6
6	10	4
6	9	5
9	4	8
10	4	9

Determine the following values: intercept; b_1, b_2, SS_{res}; SS_{reg}; F; s^2_{res}; $s(b_1)$; $s(b_2)$; t_1; t_2

10. You are given the following data, where X_1 (years of teaching experience) and X_2 (salary in thousands) are to be used to predict Y (morale):

Y	X_1	X_2
125	1	24
130	2	30
145	3	32
115	2	28
170	6	40
180	7	38
165	5	48
150	4	42
195	9	56
180	10	52
120	2	33
190	8	50
170	7	49
175	9	53
160	6	49

Determine the following values: intercept; b_1, b_2, SS_{res};SS_{reg}; F; s^2_{res}; $s(b_1)$; $s(b_2)$; t_1; t_2.

11. A researcher has conducted a multiple linear regression. The maximum value for Mahalanobis distance in their model is 8.26. They have tested at alpha of .05 and have three independent variables and one dependent variable. Given this scenario, do the researchers have cause for concern for possible outliers?

Answers to Computational Problems

1. intercept = 28.0952, b_1 = .0381, b_2 = .8333, SS_{res} = 21.4294, SS_{reg} = 1,128.5706, F = 105.3292 (reject at .01), s^2_{res} = 5.3574, $s(b_1)$ = .0058, $s(b_2)$ = .1545, t_1 = 6.5343 (reject at .01), t_2 = 5.3923 (reject at .01).
3. in order, the t values are 0.8 (not significant), 0.77 (not significant), –8.33 (significant).
5. $r_{1(2.3)} = -.0140$
7. $r_{12.3} = -.8412$, $r_{1(2.3)} = -.5047$.
9. intercept = –1.2360, b_1 = .6737, b_2 = .6184, SS_{res} = 58.3275, SS_{reg} = 106.6725, F = 15.5453 (reject at .05), s^2_{res} = 3.4310, $s(b_1)$ = .1611, $s(b_2)$ = .2030, t_1 = 4.1819 (reject at .05), t_2 = 3.0463 (reject at .05).

11. Given alpha of .05, three independent variables, and one dependent variable, there are four degrees of freedom. This results in a chi-square critical value of 9.49. Any Mahalanobis distance value would need to be *greater* than 9.49 to raise concern. Thus, with the maximum value of Mahalanobis distance in their model being 8.26, the researchers do *not* have cause for concern for possible outliers.

Interpretive Problems

1. Using SPSS or **R**, develop a multiple regression model with data supplied for other chapters in this textbook. Write up your results, including interpretation of effect size and testing of assumptions.
2. Use SPSS or **R** to develop a multiple regression model with data available on the textbook's website from the 2017 IPEDS (https://nces.ed.gov/ipeds/). Select one continuous variable as the dependent variable [e.g., 12-month instructional activity credit hours: undergraduates (CDACTUA)] and find at least two strong predictors from among the remaining variables in the dataset. Write up the results in APA style, including testing for the assumptions. Determine and interpret a measure of effect size.
3. Use SPSS or **R** to develop a logistic regression model with data available on the textbook's website from the 2017 NHIS* family file (https://www.cdc.gov/nchs/nhis/). Select one binary variable as the dependent variable [e.g., "# family members receiving Women, Infants, Children (WIC) benefits" (FWICCT)] and find at least two strong predictors from among the remaining variables in the dataset. Write up the results in APA style, including testing for the assumptions. Determine and interpret a measure of effect size.

* It is important to note that we are using only one data file from the NHIS *and* the NHIS is a *complex sample* (i.e., not a simple random sample). Per NHIS (see www.cdc.gov/nchs/nhis/about_nhis.htm#sample_design), "The sampling plan follows a multistage area probability design that permits the representative sampling of households and noninstitutional group quarters (e.g., college dormitories) . . . The current sampling plan was implemented in 2016 . . . [It] is a sample of clusters of addresses that are located in primary sampling units (PSU's). A PSU consists of a county, a small group of contiguous counties, or a metropolitan statistical area." In the NHIS dataset, you will find, for example, a "weight" variable, which is used to adjust for the complex survey design. We won't get into the technical aspects of this, but when the data are analyzed to adjust for the sampling design (including nonsimple random sampling procedure and disproportionate sampling) the end results are then representative of the intended population. The purpose of the text is not to serve as a primer for understanding complex samples, and thus readers interested in learning more about complex survey designs are referred to any number of excellent resources (Hahs-Vaughn, 2005; Hahs-Vaughn, McWayne, Bulotskey-Shearer, Wen, & Faria, 2011a, 2011b; Lee, Forthofer, & Lorimor, 1989; Skinner, Holt, & Smith, 1989). Additionally, so as to not complicate matters any more than necessary, the applications in the textbook do not illustrate how to adjust for the complex sample design. *As such, if you do not adjust for the complex sampling design, the results that you see should not be interpreted to represent any larger population but only that select sample of individuals who actually completed the survey.* I want to stress that the reason why the sampling design has not been illustrated in the textbook applications is because the point of this section of the textbook is to illustrate how

to use statistical software to generate various procedures and how to interpret the output, and not to ensure the results are representative of the intended population. *Please do not let this discount or diminish the need to apply this critical step in your own analyses when using complex survey data as quite a large body of research exists that describes the importance of effectively analyzing complex samples as well as provides evidence of biased results when the complex sample design is not addressed in the analyses* (Hahs-Vaughn, 2005, 2006a, 2006b; Hahs-Vaughn et al., 2011a, 2011b; Kish & Frankel, 1973, 1974; Korn & Graubard, 1995; Lee et al., 1989; Lumley, 2004; Pfeffermann, 1993; Skinner et al., 1989).

9

Logistic Regression

Chapter Outline

Key Concepts

1. Logit
2. Odds
3. Odds ratio

In the previous chapter we examined ordinary least squares (OLS) regression—multiple regression models—that allow us to examine the relationship between one or more predictors when the outcome is continuous. In this chapter, we are introduced to logistic regression, which can also be used when the outcome is categorical and that allows model prediction. Logistic regression and discriminant analysis (which is discussed in an upcoming chapter) share similarities, and there can be confusion on when one is more appropriate than the other. Understanding that you may not be fully familiar with discriminant analysis, we'll offer the condensed version of how the two procedures contrast. The assumptions of multivariate normality and equal variance-covariance matrices, which are required in discriminant analysis, do not hold for logistic regression. Thus, logistic regression is more robust than discriminant analysis when these assumptions are not met. Additionally, logistic regression is oftentimes less interpretatively challenging than discriminant analysis given that it falls within the regression family, more common to many researchers as compared to discriminant analysis.

For the purposes of this chapter, we will concentrate on binary logistic regression which is used when the outcome has only two categories (i.e., dichotomous, binary, or sometimes referred to as a Bernoulli outcome). The logistic regression procedure appropriate for more than two categories is called multinomial (or polytomous) logistic regression. Readers interested in learning more about multinomial logistic regression will be provided some additional references later in this chapter. Also in this chapter, we discuss methods that can be used to enter predictors in logistic regression models. Our objectives are by the end of this chapter you will be able to: (a) understand the concepts underlying logistic regression, (b) determine and interpret the results of logistic regression, (c) understand and evaluate the assumptions of logistic regression, and (d) have a basic understanding of methods of entering the covariates.

9.1 What Logistic Regression Is and How It Works

Oso Wyse, one of the four amazingly talented statistical gurus in the statistics and research lab, has just had a conversation with his faculty advisor. He finds himself embarking on a challenging statistical project.

Oso Wyse finds himself on his final statistical expedition as a graduate research assistant in the stats and research lab. After introduction from his faculty advisor, Oso meets with Dr. Malani, a faculty member in the early childhood department. Dr. Malani has collected data on children who will be entering kindergarten in the fall. Interested in kindergarten readiness issues, Dr. Malani wants to know if scores from a teacher observation scale for social development and family structure (single family versus two-family home) can predict whether children are prepared or unprepared to enter kindergarten. Oso suggests the following research question to Dr. Malani: *Can kindergarten readiness (prepared vs. unprepared) be predicted by social development and family structure (single family vs. two-family home)?* Given that the outcome is dichotomous, Oso determines that binary logistic regression is the appropriate statistical procedure to use to answer Dr. Malani's question. Oso then proceeds with assisting Dr. Malani in analyzing the data.

If the dependent variable is binary (i.e., dichotomous or having only two categories), then ordinary least squares (OLS) regression, described earlier in this text, is inappropriate. Although OLS regression can easily accommodate dichotomous independent variables through dummy coding (i.e., assignment of 1 and 0 to the categories where "1" is traditionally coded as the category of interest, i.e., case outcome; "0" is traditionally coded as the noncase outcome or reference category), it is an entirely different case when the *outcome* is dichotomous. Applying OLS regression to a binary outcome creates problems. For example, a dichotomous outcome violates normality and homogeneity assumptions in OLS regression. In addition, OLS estimates are based on linear relationships between the independent and dependent variables, and forcing a linear relationship (as seen in Figure 9.1) in the case of a binary outcome is erroneous [although we found at least one author (Hellevik, 2009) who argues that OLS regression can be used with dichotomous outcomes]. As seen in this figure, there is obviously not a linear

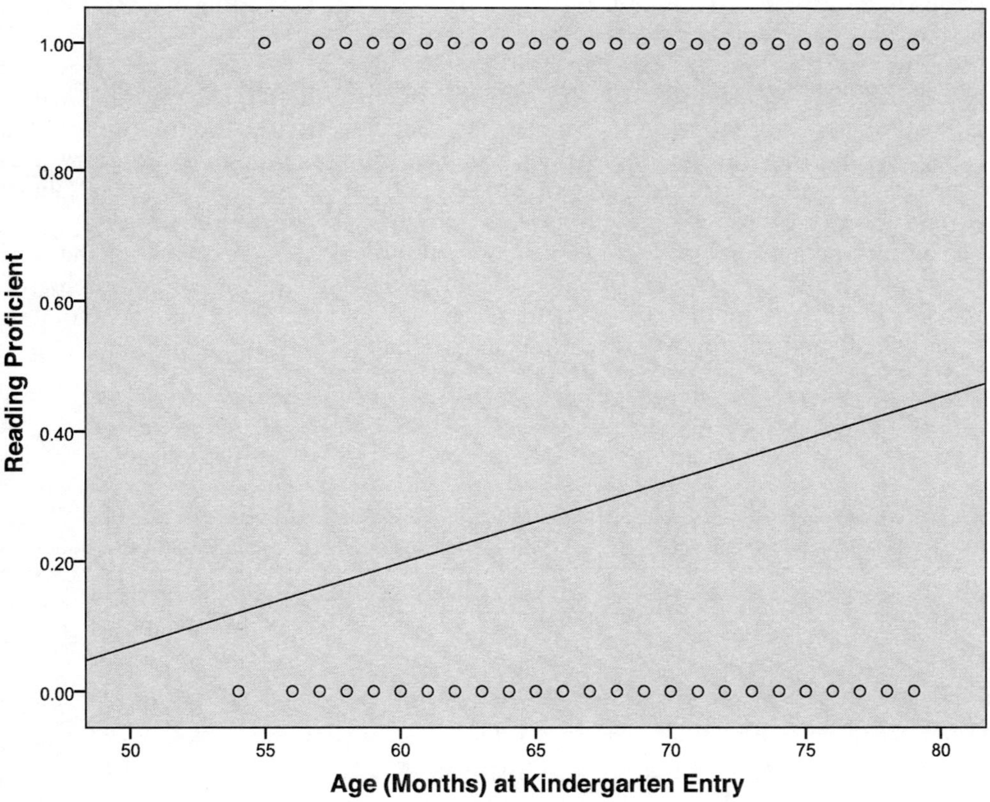

FIGURE 9.1
Nonlinearity of binary outcome.

relationship between age at kindergarten entry and reading proficiency (i.e., proficient or not proficient).

As part of the regression family, logistic regression still allows a prediction to be made; however, now the prediction is whether or not the unit under investigation falls into one of the two categories of the dependent variable. Initially used mostly in the hard sciences, this method has become more broadly popular as there are many situations where researchers want to examine outcomes that are discrete, rather than continuous, in nature. Some examples of dichotomous dependent variables are pass/fail, surviving surgery/not, admit/reject, vote for/against, employ/not, win/lose, or purchase/not. Logistic regression has been applied in a wide variety of situations. As just a few examples, Mehta and colleagues examined public housing and rental assistance and its relationship to asthma (Mehta, Dooley, Kane, Reid, & Shah, 2018). Berg and Brännström (2018) used logistic regression to model the extent to which children who were evicted were then placed in out-of-home care. McGrath and colleagues (McGrath, Hall, Peterson, Kraemer, & Vincent, 2017) used logistic regression to determine whether muscle strength can protect against development of osteoporosis. Cox, Reason, Nix, and Gillman (2016) predicted the likelihood of college graduation based on factors outside of academics using logistic regression.

The idea of using a dichotomous variable was introduced in Chapter 8 on multiple regression as the concept of a *dummy variable*, where the first condition is indicated by a value of 1 (e.g., prepared for kindergarten), whereas a value of 0 indicates the opposite condition (e.g., unprepared for kindergarten). Understanding the coding of 0 and 1 is very important for interpretation purposes. Again, "1" is traditionally the case outcome (with results interpreted in terms of cases) and "0" the non-case or reference category. For the purposes of this text, our discussion will concentrate on dichotomous outcomes where logistic regression is appropriate (i.e., binary logistic regression, referred to throughout this chapter simply as logistic regression). Conditions for which there are more than two possible categories for the dependent variable (e.g., three categories, such as "above satisfactory performance," "satisfactory performance," and "below satisfactory performance"), multinomial logistic regression may be appropriate. An example of the data structure for a logistic regression model with a binary outcome (prepared vs. unprepared for kindergarten), one continuous predictor (social development) and one dichotomous dummy coded predictor (family structure: single parent vs. two-parent home) is presented in Table 9.1.

TABLE 9.1

Kindergarten Readiness Example Data

Child	Social Development (X_1)	Family Structure (X_2)	Kindergarten Readiness (Y)
1	15	Single family (0)	Unprepared (0)
2	12	Single family (0)	Unprepared (0)
3	18	Single family (0)	Prepared (1)
4	20	Single family (0)	Prepared (1)
5	11	Single family (0)	Unprepared (0)
6	17	Single family (0)	Prepared (1)
7	14	Single family (0)	Unprepared (0)
8	18	Single family (0)	Prepared (1)
9	13	Single family (0)	Unprepared (0)

TABLE 9.1 (continued)

Kindergarten Readiness Example Data

Child	Social Development (X_1)	Family Structure (X_2)	Kindergarten Readiness (Y)
10	10	Single family (0)	Unprepared (0)
11	22	Two-parent home (1)	Unprepared (0)
12	25	Two-parent home (1)	Prepared (1)
13	23	Two-parent home (1)	Prepared (1)
14	21	Two-parent home (1)	Prepared (1)
15	30	Two-parent home (1)	Prepared (1)
16	27	Two-parent home (1)	Prepared (1)
17	26	Two-parent home (1)	Prepared (1)
18	28	Two-parent home (1)	Prepared (1)
19	24	Two-parent home (1)	Unprepared (0)
20	30	Two-parent home (1)	Prepared (1)

9.1.1 Characteristics

9.1.1.1 Logistic Regression Equation

As we learned previously with ordinary least squares regression, knowledge of the independent variable(s) provides the information necessary to be able to estimate a precise numerical value of the dependent variable, a predicted value. The following formula recaps the sample multiple regression equation where Y is the predicted outcome for individual i based on: (a) the Y intercept, a, the value of Y when all predictor values are zero; (b) the product of the value of the independent variables, X's, and the regression coefficients, b_k; and (c) the residual, ε_i:

$$Y_i = a + b_1 X_1 + \ldots + b_m X_m + \varepsilon_i$$

As we see, the logistic regression equation is similar in concept to simple and multiple linear regression, but operates much differently. In logistic regression, the binary dependent variable is transformed into a logit variable (which is the natural log of the odds of the dependent variable occurring or not occurring) and the parameters are then estimated using maximum likelihood. The end result is that the odds of an event occurring are estimated through the logistic regression model (whereas OLS estimates a precise numerical value of the dependent variable).

To understand how the logistic regression equation operates, there are three primary computational concepts that must be understood: probability, odds, and the logit. These express the same thing, only in different ways (Menard, 2000). Let us first consider probability.

9.1.1.2 Probability

The overarching difference between OLS regression (i.e., simple and multiple linear regression that we've been learning about) and logistic regression is the measurement scale of the outcome. With OLS regression, our outcome is continuous in scale (i.e., interval or ratio measurement scale). In binary logistic regression, our outcome is dichotomous—one of two categories. Let us use kindergarten readiness ("prepared for kindergarten" coded as "1" vs. "unprepared" coded as "0") as an example of our logistic regression outcome. Therefore, what the regression equation allows us to predict is substantially different for OLS as compared to logistic regression. In comparison to OLS, which allows us to compute a precise numerical value (e.g., a specific predicted score for the dependent variable), the logistic regression equation allows us to compute a *probability*—more specifically, the *probability* that the dependent variable will occur. The logistic regression equation, therefore, generates predicted probabilities that fall between values of 0 and 1. The probability of a case or unit being classified into the lowest numerical category [i.e., $P(Y = 0)$ or, in the case of our example, the probability that a child will be "unprepared" for kindergarten] is equal to 1 minus the probability that it falls within the highest numerical category [i.e., $P(Y = 1)$ or the probability that a child will be "prepared" for kindergarten]. This equates to $P(Y = 0) = 1 - P(Y = 1)$. Applied to our example, the probability that a child will be unprepared for kindergarten is equal to one minus the probability that a child will be prepared for kindergarten. In other words, the knowledge of the probability of one category occurring (e.g., unprepared for kindergarten) allows us to easily determine the probability that the other category will occur (e.g., prepared) as the total probability must equal 1.0. Remember, however, that probabilities have to fall within the range of 0 to 1. As we know, it is not possible to have a negative probability, nor is it possible to have a probability greater than 1 (i.e., greater than 100%). If we try to model the probability as the dependent variable in our OLS equation, it is mathematically possible that the predicted values would be negative or greater than 1—values that are outside the range of what is feasible when considering probabilities. Therefore, this is where our logistic regression equation takes a turn from what we learned with linear regression.

9.1.1.3 Odds and Logit (or Log Odds)

So far, we have talked about the outcome of our logistic regression equation as being a probability, and we also know that predicted probabilities must be between 0 and 1. As we think about how to estimate probabilities, we will see that this takes a few steps to achieve. Rather than the dependent variable being a probability, if it were an *odds value*, then values greater than 1 would be possible and appropriate. **Odds** are simply the ratio of the probability of the dependent variable's two outcomes. The odds that the outcome of a binary variable is 1 (i.e., public school attendance) rather than 0 (or private school attendance), is simply the ratio of the odds that Y equals 1 to the odds that Y does not equal 1. In mathematical terms, this can be written as follows:

$$Odds(Y=1) = \frac{P(Y=1)}{1 - P(Y=1)}$$

As we see in Table 9.2, when the probability that $Y = 1$ (e.g., prepared for kindergarten) equals .50 (column 1 in Table 9.2), then $1 - P(Y = 1)$ (or unprepared for kindergarten) is .50 (column 2) and the odds are equal to 1.00 (column 3). When the probability of $Y = 1$ (e.g., prepared) is very small (say, .100 or less), then the odds for being prepared for kindergarten are also very small and approach zero (i.e., the smaller the probability that a child is prepared for kindergarten). However, as the probability of $Y = 1$ (e.g., being prepared for kindergarten) increases, the odds (column 3) increase tremendously. Thus, the issue that we are faced with when using odds is that while odds can be infinitely large, we are still limited in that the minimum value is zero and we still do not have data that

TABLE 9.2

Illustration of Logged Odds.

$P(Y=1)$	$1-P(Y=1)$	$Odds(Y=1)=\frac{P(Y=1)}{1-P(Y=1)}$	$\ln[Odds(Y=1)]=\ln\left[\frac{P(Y=1)}{1-P(Y=1)}\right]=Logit(Y)$
.001	.999	.001/.999 = .001	*ln*(.001) = −6.908
.100	.900	.100/.900 = .111	*ln*(.111) = −2.198
.300	.700	.300/.700 = .429	*ln*(.429) = −.846
.500	.500	.500/.500 = 1.000	*ln*(1.000) = 0.000
.700	.300	.700/.300 = 2.333	*ln*(2.333) = .847
.900	.100	.900/.100 = 9.000	*ln*(9.000) = 2.197
.999	.001	.999/.001 = 999.000	*ln*(999) = 6.907

can be modeled linearly. When $P(Y = 1) < .5$, the slope below an odds of 1.0 is very steep; yet when $P(Y = 1) > .5$, the slope above odds of 1.0 is much more gradual. It might also be worth noting at this point that the reciprocal odds have the same magnitude of effect but are asymmetrical, and the natural log functions to create a symmetrical outcome variable.

Changing the scale of the odds by taking the natural logarithm of the odds (also called *logit Y* or *log odds*) provides us with a value of the dependent variable that can theoretically range from negative infinity to positive infinity. Thus, taking the log odds of *Y* creates a linear relationship between *X* and the probability of *Y* (Pampel, 2000). The natural log of the odds is calculated as follows, with the residual being the difference between the predicted probability and the actual value of the dependent variable (0 or 1):

$$\ln\frac{P(Y=1)}{1-P(Y=1)}=Logit(Y)$$

In column 4 of Table 9.2, we see what happens when the logit transformation is made. As the odds increase from 1 to positive infinity, the logit (or log odds) of *Y* becomes larger and larger (and remains positive). As the odds decrease from 1 to 0, the logit (or log odds) of *Y* is negative and grows larger and larger (in absolute value).

The logit of *Y* equation is interpreted very similarly to that of OLS. For each one-unit change in the independent variable, the logistic regression coefficients represent the change in the predicted log odds of being in a category. In comparison to OLS regression, the regression coefficients have the exact same interpretation. The difference in interpretation with logistic regression is that the outcome now represents a *log odds* rather than a precise numerical value as we saw with OLS regression. Linking the logit back to probabilities, a one-unit change in the logit equals a bigger change in probabilities that are near the center as compared to the extreme values. This happens because of the linearization once we take the natural log. Taking the natural log stretches the *S*-shaped curve into a linear form; thus, the values at the extreme are stretched less, so to speak, as compared to the values in the middle (Pampel, 2000). By working with log odds, our familiar additive regression equation is applicable:

$$\ln\frac{P(Y=1)}{1-P(Y=1)}=Logit(Y)=\alpha+\beta_1X_1+\beta_2X_2+\ldots+\beta_mX_m$$

It is important to note that although we were accustomed to examining standardized regression coefficients in OLS regression, it is not the norm that standardized coefficients are computed for logistic regression models by statistical software. Standardization is ordinarily accomplished by taking the product of the unstandardized regression coefficient and the ratio of the standard deviation of X to the standard deviation of Y. The interpretation of a standard deviation change in a continuous variable thus makes sense; however, this is not the case for a dichotomous variable, nor is it the case for the log odds (which is the predicted outcome and which does not have a standard deviation).

While interpretation of the logistic equation is relatively straightforward as it holds many similarities to OLS regression, log odds are not a metric that we use often. Therefore, understanding what it means when a predictor, X, has some effect on the log odds, Y, can be difficult. This is where odds come back into the picture.

If we exponentiate the logit (Y) (i.e., the outcome of our logistic regression equation), then it converts back to the odds (see the following equation). Now we can interpret the independent variables as affecting the odds (rather than log odds) of the outcome:

$$Odds(Y=1) = e^{logit(Y)} = e^{ln[Odds(Y=1)]} = e^{\alpha+\beta_1 X_1+\beta_2 X_2+\ldots+\beta_m X_m} = \left(e^{\alpha}\right)\left(e^{\beta_1 X_1}\right)\left(e^{\beta_2 X_2}\right)\ldots\left(e^{\beta_m X_m}\right)$$

As can be seen here, the exponentiation creates an equation that is multiplicative rather than additive, and this then changes the interpretation of the exponentiated coefficients. In previous regression equations we have studied, when the product of the regression coefficient and its predictor is zero, that variable adds nothing to the prediction of the dependent variable. In a multiplicative environment, a value of zero corresponds to a coefficient of 1. In other words, a coefficient of 1 will not change the value of the odds (i.e., the outcome). Coefficients greater than 1 increase the odds, and coefficients less than 1 decrease the odds. In addition, the odds will change more the greater the distance the value is from 1.

Converting the odds back to a probability can be done through the following formula:

$$P(Y=1) = \frac{Odds(Y=1)}{1+Odds(Y=1)} = \frac{e^{\alpha+\beta_1 X_1+\beta_2 X_2+\ldots+\beta_m X_m}}{1+e^{\alpha+\beta_1 X_1+\beta_2 X_2+\ldots+\beta_m X_m}}$$

Probability values close to one indicate increased likelihood of occurrence. In our example, since "1" indicates kindergarten preparedness, a probability close to one would indicate a child was more likely to be prepared for kindergarten. Children with probabilities close to zero suggest a decreased probability of being prepared for kindergarten (and increased probability of not being prepared for kindergarten).

9.1.1.4 Estimation and Model Fit

Now that we understand the logistic regression process and resulting equations a bit better, it is time to turn our attention to how the equation is estimated and how we can determine how well the model fits. We previously learned with multiple regression that the data from the observed values of the independent variables in the sample were used to estimate or predict the values of the dependent variable. In logistic regression, we are also using the knowledge of the values of our predictor(s) to estimate the outcome (i.e., log odds). Now we are using a method called *maximum likelihood estimation* to estimate the values of the parameters (i.e., the logistic coefficients). As we just learned, the dependent variable in a logistic regression model is transformed into a logit value, which is the natural log of the odds of the dependent variable occurring or not occurring. Maximum likelihood estimation is then applied to the model and estimates the odds of occurrence after transformation into the logit. The "likelihood" in maximum likelihood refers to the likelihood of the data occurring given a specific value for population parameters that have been assumed. It is the probability of the data contingent upon a parameter-estimate that is being maximized. Whereas in OLS the sum of squared distance of the observed data to the regression line was minimized, in maximum likelihood the log likelihood is maximized.

The log of the likelihood function (sometimes abbreviated as *LL*) that results from ML estimation then reflects the likelihood of observing the sample statistics given the population parameters. The log likelihood provides an index of how much has not been explained in the model after the parameters have been estimated, and as such, the *LL* can be used as an indicator of model fit. The values of the log likelihood function vary from zero to negative infinity, with values closer to zero suggesting better model fit and larger values (in absolute value terms) indicating poorer fit. The log likelihood value will approach zero the closer the likelihood value is to one. When this happens, this suggests the observed data could be generated from these population parameters. In other words, the smaller the log likelihood, the better the model fit. It follows therefore, that the log likelihood value will grow more negative the closer the likelihood function is to zero. This suggests that the observed data are *less* likely to be generated from these population parameters.

Maximum likelihood estimation performed by statistical software usually begins the estimation process with all regression coefficients equal to the most conservative estimate (i.e., the least squares estimates). Better model fit is accomplished through the use of an algorithm which generates new sets of regression coefficients that produce larger log likelihoods. This is an iterative process that stops when the selection of new parameters creates very little change in the regression coefficients and very small increases in the log likelihood—so small that there is little value in any further estimation.

9.1.1.5 Significance Tests

As with multiple regression, there are two tests of significance in logistic regression. Specifically, these involve testing the significance of the overall logistic regression model and testing the significance of each of the logistic regression coefficients.

9.1.1.5.1 Test of Significance of the Overall Regression Model

The first test is the test of statistical significance to determine overall model fit and provides evidence of the extent to which the predicted values accurately represent the observed values (Xie, Pendergast, & Clarke, 2008). We consider several overall model tests including: (a) change in log likelihood; (b) Hosmer and Lemeshow's goodness of fit test; (c) pseudo-variance explained; and (d) predicted group membership. Additional work (e.g., Xie et al., 2008) has recently been conducted on new methods to assess model fit, but these are not currently available in statistical software, nor easily computed. Also in this section, we briefly address sensitivity, specificity, false positive, false negative, and cross-validation.

9.1.1.5.1.1 Change in Log Likelihood

One way to test overall model fit is the likelihood ratio test. This test is based on the change in the log likelihood function from a smaller model (often the baseline or intercept only model) to a larger model that includes one or more predictors (sometimes referred to as the fitted model). Although we indicate that the smaller model is often the intercept only model, this test can also be used to examine changes in model fit from one fitted model to another fitted model, and we will discuss this in a bit. This likelihood ratio test is similar to the overall *F* test in OLS regression and tests the null hypothesis that all the regression coefficients are equal to zero. Using statistical notation, we can denote the null and alternative hypotheses for the regression coefficients as follows:

$$H_0: \beta_1 = \beta_2 = \ldots = \beta_m = 0$$

$$H_1: Not\,all\,the\;\beta_m = 0$$

For explanation purposes, we assume the smaller model is the baseline or intercept only model. The baseline log likelihood is estimated from a logistic regression model that includes only the constant (i.e., intercept) term. The model log likelihood is estimated from the logistic regression model that includes the constant and the relevant predictor(s). By multiplying the difference in these log likelihood functions by −2, a chi-square test is produced with degrees of freedom equal to the difference in the degrees of freedom of the models ($df = df_{model} - df_{baseline}$) (where "model" refers to the fitted model that includes one or more predictors). In the case of the constant only model, there is only one parameter estimated (i.e., the intercept), so there is only one degree of freedom. In models that include independent variables, the degrees of freedom are equal to the number of independent variables in the model plus one for the constant. The larger the difference between the baseline and model *LL* values, the better the model fit. It is important to note that the log likelihood difference test assumes nested models. In other words, all elements that are included in the baseline or smallest model must also be included in the fitted model. As alluded to previously, the change in log likelihood test can be used for more than just comparing the intercept only model to a fitted model. Researchers often use this test in the model building process to determine if adding predictors (or sets of predictors) aids in model fit by comparing one fitted model to another fitted model. In general, the change in log likelihood is computed as follows:

$$\chi^2 = 2\left(LL_{model} - LL_{baseline}\right)$$

9.1.1.5.1.2 Hosmer-Lemeshow Goodness of Fit Test

The Hosmer-Lemeshow goodness of fit test is another tool that can be used to examine overall model fit. The Hosmer-Lemeshow statistic is computed by dividing cases into deciles (i.e., 10 groups) based on their predicted probabilities. Then a chi-square value is computed based on the observed and expected frequencies. This is a chi-square test for which the researcher does *not* want to find statistical significance. Nonstatistically significant results for the Hosmer-Lemeshow test indicate the model has acceptable fit. In other words, the predicted or estimated model is not statistically significantly different from the observed values. Although the Hosmer-Lemeshow test can easily be requested in SPSS, it has been criticized for being conservative (i.e., lacking sufficient power to detect lack of fit in instances such as nonlinearity of an independent variable), too likely to indicate model fit when five or fewer groups (based on the decile groups created in computing the statistic) are used to calculate the statistic, and offers little diagnostics to assist the researcher when the test indicates poor model fit (Hosmer, Hosmer, LeCessie, & Lemeshow, 1997). Additionally, this test can be overly conservative unless one has very large sample sizes.

9.1.1.5.1.3 Pseudo-Variance Explained

Another overall model fit index for logistic regression is pseudo-variance explained. This index is akin to multiple R^2 (or the coefficient of determination) in OLS regression and can also be considered an effect size measure for the model. The reason these values are considered pseudo-variance explained in logistic regression is that the variance in a dichotomous outcome, as evident in logistic regression, differs as compared to the variance of a continuous outcome, as present in OLS regression.

There are a number of multiple R^2 pseudo-variance explained values that can be computed in logistic regression. Pseudo R^2 measures can be used to interpret one model and as a goodness of fit when comparing multiple models. However, these uses assume there are benchmark values for interpretation and the only influence on the value is the explanatory power provided by the independent variable(s) (Hemmert, Schons, Wieseke, & Schimmelpfennig, 2018). Of these, SPSS automatically computes the Cox and Snell and Nagelkerke indices. There is, however, no consensus on which (if any) of the pseudo-variance explained indices are best and many researchers choose not to report any of them in their published results. In fact, a meta-analysis of studies (1997 to 2011) that had conducted logistic regression and pseudo R^2, Hemmert and colleagues found, among other findings, that the distribution of observations in the outcome and the number of independent variable substantially impact pseudo R^2 (e.g., asymmetrical distributions decreases pseudo R^2, and increasing the number of independent variables increases pseudo R^2) (Hemmert et al., 2018). If you do choose to use and/or report one or more of these values, they should be used only as a guide "without attributing great importance to a precise figure" (Pampel, 2000, p. 50). Additionally, should you use pseudo R^2, review Hemmert et al.'s (2018) study as they identify additional pseudo R^2 values you may wish to consider as well as considerations for interpreting and reporting.

We discuss the following: (a) Cox and Snell (1989); (b) Nagelkerke (1991); (c) Hosmer and Lemeshow (1989); (d) Aldrich and Nelson (1984); (e) Harrell (1986); and (f) traditional R^2.

The Cox and Snell R^2 (1989) is computed as the ratio of the likelihood values raised to the power of $2/n$ (where n is sample size). A problematic is that the computation is such that the theoretical maximum of one cannot be obtained, even when there is perfect prediction:

$$R_{CS}^2 = 1 - \left(\frac{LL_{baseline}}{LL_{model}} \right)^{2/n}$$

Nagelkerke (1991) adjusts the Cox and Snell value so that the maximum value of one can be achieved, and it is computed as follows:

$$R_N^2 = \frac{R_{CS}^2}{1 - \left(LL_{baseline} \right)^{2/n}}$$

Hosmer and Lemeshow's (1989) R^2 is the proportional reduction in the log likelihood (in absolute value terms). Although not provided by SPSS, it can easily be computed by the ratio of the model to baseline $-2LL$. Ranging from zero to one, this value provides an indication of how much the badness of fit of the baseline model is improved by the inclusion of the predictors in the fitted model. Hosmer and Lemeshow's (1989) R^2 is computed as:

$$R_L^2 = \frac{-2LL_{model}}{-2LL_{baseline}}$$

Harrell (1986) proposed that Hosmer and Lemeshow's R^2 be adjusted for the number of parameters (i.e., independent variables) in the model. This adjustment (where m equals the number of independent variables in the model) to the computation makes this R^2 value akin to the adjusted R^2 in OLS regression. It is computed as:

$$R_{LA}^2 = \frac{\left(-2LL_{model}\right) - 2m}{-2LL_{baseline}}$$

Aldrich and Nelson (1984) provided an alternative to the R_L^2 that is equivalent to the squared contingency coefficient. This measure has the same problem as the Cox and Snell R^2; the theoretical maximum of one cannot be obtained even when the independent variable(s) perfectly predict the outcome. It is computed as:

$$pseudo\, R^2 = \frac{-2LL_{model}}{-2LL_{model} + n}$$

The traditional R^2, the coefficient of determination as used in simple and multiple regression, can also be used in logistic regression (only with binary logistic regression, as the mean and variance of a dichotomous variable make sense; however, the mean, for example, in a dummy coded variable situation, is equal to the proportion of cases in the category labeled as 1). R^2 can be computed by correlating the observed values of the binary dependent variable with the predicted values (i.e., predicted probabilities) obtained from the logistic regression model and then squaring the correlated value. Predicted probability values can easily be saved when generating logistic regression models in SPSS.

9.1.1.5.1.4 Predicted Group Membership

Another test of model fit for logistic regression can be accomplished by evaluating predicted to observed group membership. Assuming a cut value of .50, cases with predicted probabilities at .5 or above are predicted as 1 and predicted probabilities below .5 are predicted as 0. A crosstab table of predicted to observed predicted probabilities provides the frequency and percentage of cases correctly classified. Correct classification would be seen in cases that have the same value for both the predicted and observed values. A perfect model produces 100% correctly classified cases. A model that classifies no better than chance would provide 50% correctly classified cases. Press's Q is a chi-square statistic with one degree of freedom that can be used as a formal test of classification accuracy. It is computed as:

$$Q = \frac{\left[N - (nK)^2\right]}{N(K-1)}$$

where N is the total sample size, n represents the number of cases that were correctly classified, and K equals the number of groups. As with other chi-square statistics we have examined, this test is sensitive to sample size. Also, it is important to note that focusing solely on the correct classification overall (as is done with Press's *Q*) may result in overlooking one or more groups that have unacceptable classification. The researcher should evaluate the classification of each group in addition to the overall classification.

Sensitivity is the probability that a case coded as 1 for the dependent variable (aka "positive") is classified correctly. In other words, sensitivity is the percentage of correct predictions of the cases that are coded as 1 for the dependent variable. In the kindergarten readiness example that we will review later, of those 12 children who were prepared for kindergarten (i.e., coded as 1 for the dependent variable), 11 were correctly classified. Thus the sensitivity is 11/12 or about 92%.

Specificity is the probability that a case coded as 0 for the dependent variable (aka "negative") is classified correctly. In other words, specificity is the percentage of correct predictions of the cases that are coded as 0 for the dependent variable. In the kindergarten readiness example that we will review later, of those 8 children who were unprepared for kindergarten (i.e., coded as 0 for the dependent variable), 7 were correctly classified. Thus the specificity is 7/8 or 87.5%.

False positive rate is the probability that a case coded as 0 for the dependent variable (aka "negative") is classified *incorrectly*. In other words, this is the percentage of cases in error where the dependent variable is predicted to be 1 (i.e., prepared), but in fact the observed value is 0 (i.e., unprepared). In the kindergarten readiness example that we will review later, of those 8 children who were unprepared for kindergarten (i.e., coded as 0 for the dependent variable), 1 was incorrectly classified. Thus the false positive rate is 1/8 or 12.5%. The false positive rate is also computed as one minus specificity.

False negative rate is the probability that a case coded as 1 for the dependent variable (aka "positive") is classified *incorrectly*. In other words, this is the percentage of cases in error where the dependent variable is predicted to be 0 (i.e., unprepared), but in fact the observed value is 1 (i.e., prepared). In the kindergarten readiness example that we will review later, of those 12 children who were prepared for kindergarten (i.e., coded as 1 for the dependent variable), 1 was incorrectly classified. Thus the false negative rate is 1/12 or about 8%. The false negative rate is also computed as one minus sensitivity.

9.1.1.5.1.5 Cross-validation

A recommended best practice in logistic regression is to cross-validate the results. If the sample size is sufficient, this can be accomplished by using 75%–80% of the sample to derive the model and then use the remaining cases (the holdout sample) to determine its accuracy. With cross-validation, you are in essence testing the model on two samples—a primary sample (which represents the largest percentage of the sample size) and a holdout sample (that which remains). If classification accuracy of the holdout sample is within 10% of the primary sample, this provides evidence of the utility of the logistic regression model.

9.1.1.6 Test of Significance of the Logistic Regression Coefficients

The second test in logistic regression is the test of the statistical significance of each regression coefficient, b_k. This test allows us to determine if the individual coefficients are statistically significantly different from zero. The null and alternative hypothesis can be illustrated in the same mathematical notation as we used with OLS regression:

$$H_0: \beta_k = 0$$

$$H_1: \beta_k \neq 0$$

Interpreting the test provides evidence of the probability of obtaining the observed sample coefficient by chance if the null hypothesis was true (i.e., if the population regression coefficient value was zero). The Wald statistic, which follows a chi-square distribution, is used as the test statistic for regression coefficients in SPSS. This is calculated by squaring the ratio of the regression coefficient divided by its standard error:

$$W = \frac{\beta_k^2}{SE_{\beta_k^2}}$$

When the logistic regression coefficients are large (in absolute value), rounding error can create imprecision in estimation of the standard errors. This can result in inaccuracies in testing the null hypothesis, and more specifically, increased Type II errors (i.e., failing to reject the null hypothesis when the null hypothesis is false). An alternative to the Wald test, in situations such as this, is the difference in log likelihood test previously described to compare models with and without the variable of interest (Pampel, 2000).

Raftery (1995) proposed a Bayesian information criterion (BIC), computed as the difference between the chi-square value and the natural log of the sample size, that could also be applied to testing logistic regression coefficients:

$$BIC = \chi^2 - \ln n$$

To reject the null hypothesis, the BIC should be positive (i.e., greater than zero). That is, the chi-square value must be greater than the natural log of the sample size. BIC values below zero suggest that the variable contributes little to the model. BIC values between 0 and +2 are considered weak; between 2 and 6, positive; between 6 and 10, strong; and more than 10, very strong.

Beyond determining statistical significance of the individual predictors, you may also want to assess which predictors are adding the most to the model. In OLS regression, we examined the standardized regression coefficients. There are no traditional standardized regression coefficients provided in SPSS for logistic regression, but they are easy to calculate. Simply standardize the predictors before generating the logistic regression model, and then run the model as desired. You can then interpret the logistic regression coefficients as standardized regression coefficients (if necessary, review the multiple regression chapter).

We can also form a confidence interval around the logistic regression coefficient, b_k. The confidence interval formula is the same as in OLS regression: the logistic regression coefficient plus or minus the product of the tabled critical value and the standard error:

$$CI(b_k) = b_k \pm_{(\alpha/2)} t_{(n-m-1)} s_b$$

The null hypothesis that we tested was H_0: $\beta_k = 0$. It follows that if our confidence interval contains zero, then the logistic regression coefficient (b_k) is not statistically significantly different from zero at the specified significance level. We can interpret this to say that β_k will be included in $(1 - \alpha)\%$ of the sample confidence intervals formed from multiple samples.

9.1.1.7 Methods of Predictor Entry

The three categories of model building that will be discussed include: (a) simultaneous logistic regression; (b) stepwise logistic regression; and (c) hierarchical regression.

9.1.1.7.1 Simultaneous Logistic Regression

With simultaneous logistic regression, all the independent variables of interest are included in the model in one set. This method of model building is usually used when the researcher does not hypothesize that some predictors are more important than others. This method of entry allows you to evaluate the contribution of an independent variable over and above that of all other predictors in the model (i.e., each independent variable is evaluated as if it was the last one to enter the equation). One problem that may be encountered with this method of entry is related to strong correlations between the predictor and the outcome. An independent variable that has a strong bivariate correlation with the dependent variable may indicate a weak correlation when entered simultaneously with other predictors. In SPSS, this method of entry is referred to as "Enter."

9.1.1.7.2 Stepwise Logistic Regression

Stepwise logistic regression is a data-driven model building technique where the computer algorithms drive variable entry rather than theory. Issues with this type of technique have previously been outlined in the discussion associated with this method in multiple regression and thus are not rehashed here. If stepwise logistic regression is determined to be the most appropriate strategy to build your model, Hosmer, Lemeshow, and Sturdivant (2000) suggest setting a more liberal criteria for variable inclusion (e.g., α = .15 to .20). They also provide specific recommendations on dealing with interaction terms and scales of variables. Because it is only in unusual instances that this method of model building is appropriate (e.g., exploratory research), additional coverage of the suggestions by Hosmer and Lemeshow is not presented.

SPSS offers forward and backward stepwise methods. For both forward and backward methods, options include conditional, LR, and Wald. The differences between these options are mathematically driven. The LR method of entry uses the $-2LL$ for estimating entry of independent variables. The conditional method also uses the likelihood ratio test, but one that is considered to be computationally quicker. The Wald method applies the Wald test to determining entry of the independent variables. With forward stepwise methods, the model begins with a constant only and, based on some criterion, independent variables are added one at a time until a specified cutoff is achieved (e.g., all independent variables included in the model are statistically significant and any additional variables not included in the model are not statistically significant). Backward stepwise methods work in the reverse fashion where initially all independent variables (and the constant) are included. Independent variables are then removed until only those that are statistically significant remain in the model, and including an omitted independent variable would not improve the model.

9.1.1.7.3 Hierarchical Regression

In hierarchical regression, the researcher specifies *a priori* a sequence for the individual predictor variables (not to be confused with hierarchical linear models, which is a regression approach for analyzing nested data collected at multiple levels, such as child, classroom, and school). The analysis proceeds in a forward selection, backward elimination, or stepwise selection mode according to a researcher specified, theoretically based sequence, rather than an unspecified statistically based sequence. In SPSS, this is conducting by entering predictors in **blocks** and selecting their desired method of entering variables in each block (e.g., simultaneously, forward, backward, stepwise). Because this method was explained in detail in reference to multiple regression and operation of this method of variable selection is the same in logistic regression, additional information will not be presented.

9.1.2 Sample Size

Simulation research suggests that logistic regression is best used with large samples. Samples of size 100 or greater are needed to accurately conduct tests of significance for logistic regression coefficients (Long, 1997). Note that for illustrative purposes, the example in this chapter uses a sample size of 20. We recognize this is insufficient in practice, but have used it for greater ease in presenting the data.

9.1.3 Power

Power in logistic regression can be computed *a priori* (which is ideal) to determine requisite sample size as well as post hoc. It is important to note the relationship between goodness of fit and power in the context of logistic regression. For example, the power of the Hosmer-Lemeshow goodness of fit statistic in detecting ill fit (e.g., nonlinearity in predictor variables) (Xie et al., 2008).

9.1.4 Effect Size

There are a number of effect size indices that may be considered in logistic regression, and a concise summary is presented in Table 9.4. We have already talked about multiple R^2 pseudo-variance explained values which can be used not only to gauge model fit, but also as measures of effect size. Another important statistic in logistic regression is the **odds ratio** (*OR*), also an effect size index that is similar to R^2. The odds ratio is computed by exponentiating the logistic regression coefficient e^{b_k}. Conceptually this is the odds for one category (e.g., prepared for kindergarten) divided by the odds for the other category (e.g., unprepared for kindergarten). The null hypothesis to be tested is that $OR = 1$, which indicates that there is no relationship between a predictor variable and the dependent variable. If an odds ratio of 1 indicates no effect, then an odds ratio *greater than one* indicates higher odds of the outcome occurring. An odds ratio of *less than one* indicates lower odds of the outcome occurring. Thus, assuming we are interesting in finding a relationship between the outcome and some event, we want to find *OR* to be significantly different from 1.

When the independent variable is **continuous**, the odds ratio represents the amount by which the odds change for a one-unit increase in the independent variable. When the odds ratio is greater than one, the independent variable increases the odds of occurrence. When the odds ratio is less than one, the independent variable decreases the odds of occurrence. The odds ratio is provided in SPSS output as "Exp(B)" in the table labeled "Variables in the Equation." In predicting kindergarten readiness, social development is a continuous covariate with a resulting odds ratio of 2.631. We can interpret this odds ratio to be that for every one-unit increase in social development, the odds of being ready for kindergarten (i.e., prepared) increase by 263%, controlling for the other variables in the model.

In the case of **categorical** variables, including dichotomous, multinomial, and ordinal variables, odds ratios are often interpreted in terms of their relative size or the change in odds ratios in comparing models. Consider first the case of a **dichotomous** variable. In the model predicting kindergarten readiness, type of household is one independent variable included in the model where a two-parent home is coded as "1" and a single-parent home as "0." An odds ratio of .002 indicates that the odds of being prepared for kindergarten (compared to unprepared for kindergarten) are decreased by a factor of .002 by being in a single parent home (as opposed to living in a two-family home). We could also state that the odds that a child from a single parent home will be prepared for kindergarten are .998 (i.e., 1 – .002).

In the case of a categorical variable with more than two categories, the odds ratio is interpreted relative to the reference (or left out) category. For example, say we have a predictor in our model that is mother's education level with categories that include: (1) less than high school diploma; (2) high school diploma or GED; and (3) at least some college. Say we set the last category ("at least some college"') as the reference category. An odds ratio of .86 for the category of "high school diploma or GED" for mother's education level suggests that the odds of being prepared for kindergarten (as compared to unprepared) decrease by a factor of .86 when the child's mother has a high school diploma or GED, relative to when the child's mother has at least some college, when the other variables in the model are controlled.

Confidence intervals (CI) can be computed for odds ratios, and these CI reflect precision of the estimated OR. *Larger CI suggest lower precision, and smaller CI reflect higher precision.* An odds ratio with a CI that includes the null value (i.e., $OR = 1$) may provide evidence to suggest a nonstatistically significant relationship between the independent variable and the outcome. There are a number of resources that make for easy computing of effect sizes as well as their confidence intervals. We will illustrate two online calculators for computing odds ratio in the case of two groups (e.g., treatment and control) and their confidence intervals. One is provided by Dr. David B. Wilson and is available through the Campbell Collaboration (see https://campbellcollaboration.org/research-resources/effect-size-calculator.html). Although designed for use when conducting meta-analyses, the online calculator comes in handy whenever an effect size and its CI are desired. Let's take an example using our kindergarten readiness data that will be used later. We see in Table 9.3 a crosstabulation of type of household (which will serve as a proxy for treatment/control) by kindergarten readiness. This is a 2 × 2 frequency table.

Using Campbell's effect size calculator for a 2 × 2 table and treating "two-parent household" as the treatment and "prepared" as the desired (i.e., "yes") outcome, we find an odds ratio of 6, *OR CI* of (.8117, 44.3512). Because the confidence interval contains 1, our null value (i.e., $OR = 1$), this may provide evidence to suggest a nonstatistically significant relationship between the independent variable and the outcome.

We can also use the online effect size calculator by Uanhoro (2017) to compute confidence intervals. This calculator uses the **R** package *epitools* for computing the *OR* and their confidence intervals. Selecting "unconditional maximum likelihood estimation (Wald)" as the method for the *OR* (which uses the normal approximation; see Figure 9.2) will

TABLE 9.3

Type of Household by Kindergarten Readiness Crosstabulation.

		Kindergarten Readiness	
		Unprepared	Prepared
Type of Household	Single-parent household	6	4
	Two-parent household	2	8

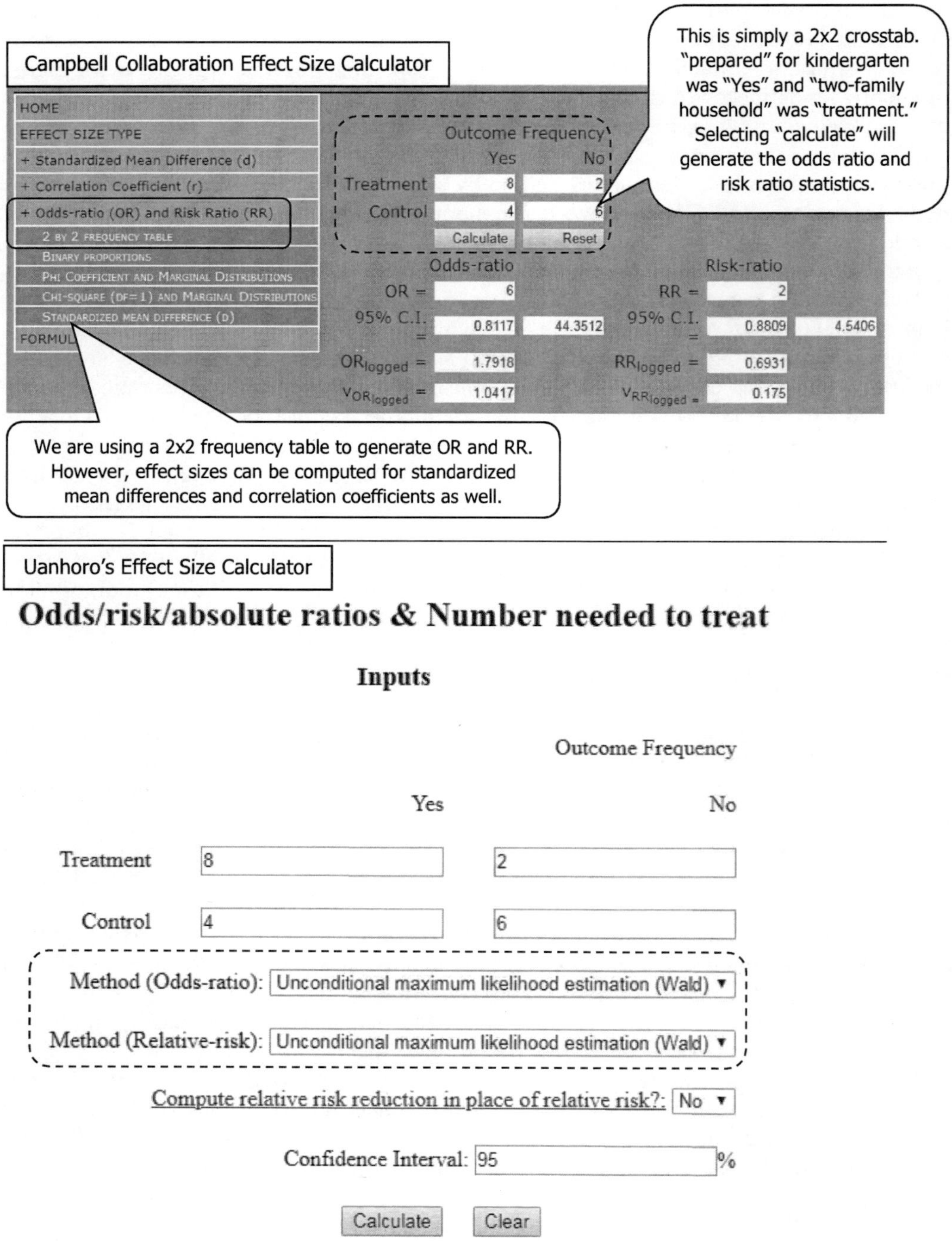

FIGURE 9.2
Computing *OR* CI using the Campbell Collaboration Online Calculator or Uanhoro's Effect Size Calculator.

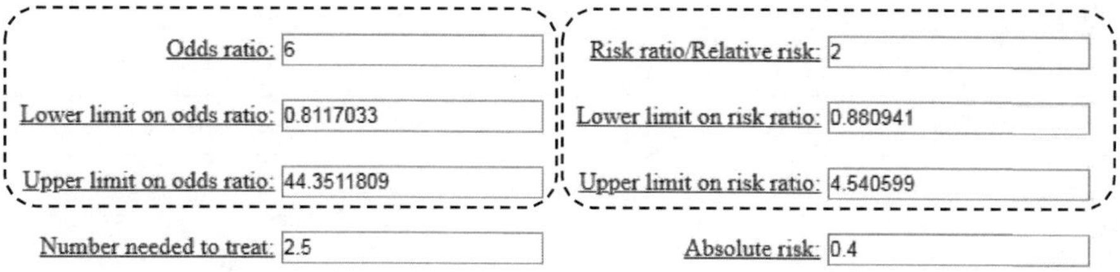

Odds ratio:	6	Risk ratio/Relative risk:	2
Lower limit on odds ratio:	0.8117033	Lower limit on risk ratio:	0.880941
Upper limit on odds ratio:	44.3511809	Upper limit on risk ratio:	4.540599
Number needed to treat:	2.5	Absolute risk:	0.4

Using the small sample adjustment method, the results are:

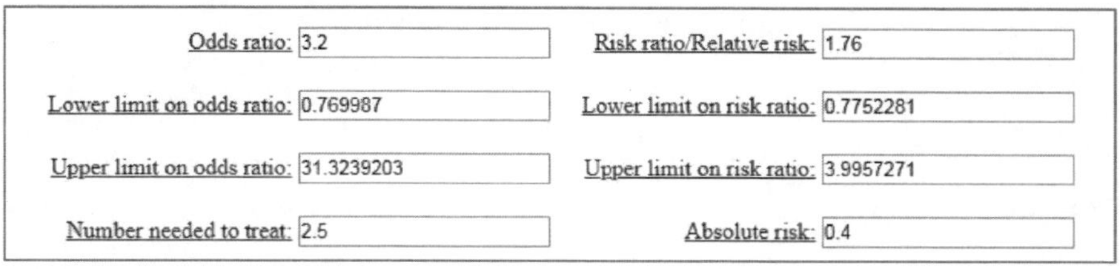

Odds ratio:	3.2	Risk ratio/Relative risk:	1.76
Lower limit on odds ratio:	0.769987	Lower limit on risk ratio:	0.7752281
Upper limit on odds ratio:	31.3239203	Upper limit on risk ratio:	3.9957271
Number needed to treat:	2.5	Absolute risk:	0.4

FIGURE 9.2 (continued)
Computing *OR* CI using the Campbell Collaboration Online Calculator or Uanhoro's Effect Size Calculator.

produce identical results as Campbell's effect size calculator (*OR* = 6; .*CI*, 8117, 44.3512). An added benefit of Uanhoro (2017) is that, in addition to the Wald method for estimation, there are three additional methods provided for calculating the *OR* (https://effect-size-calculator.herokuapp.com/#oddsriskabsolute-ratios-number-needed-to-treat). These include: (1) median-unbiased estimation (mid-*p*); (2) conditional maximum likelihood estimation (Fisher); and (3) small sample adjustment (small). Confidence intervals for mid-*p* and Fisher are computed using exact methods, which is useful in cases where the sample size is small or there is sparse data structure (e.g., rare events) (Hirji, Tsiatis, & Mehta, 1989). Confidence intervals for the "small" method are computed using the normal approximation with small sample adjustment. When using Uanhoro's calculator to calculate relative risk, there are three methods for estimating: (1) unconditional maximum likelihood estimation (Wald); (2) small sample adjustment (small); and (3) bootstrap estimation (boot).

From either of these calculators, we're also provided the **risk ratio**. The risk ratio is computed as the risk of incident in one group divided by the risk of incidence in the other group. In our example of kindergarten preparedness, the "risk" of being "prepared" in the two-parent household was 8/10 or 80%, and the risk of being "prepared" in the single-parent household was 4/10 or 40%. Thus, the risk ratio is simply the incidence of exposure in the "treatment" (i.e., two-parent) group divided by the incidence of exposure in the "control" (i.e., single-parent) group or .80/.40 = 2. A *risk ratio of less than 1* indicates that "exposure" is associated with a reduction in risk; i.e., a decreased risk of the outcome in the exposed group. A *risk ratio greater than 1*, as we see in this example, indicates an increased risk of the outcome in the exposed group. In our illustration, with a *RR* of 2, there is an increased "risk" of being prepared (which is a good thing!) in children from two-parent households. When the *risk ratio is 1, or very near 1*, there is little difference in risk (or incident) between the two groups. The percent *RR* can also be computed by multiplying the *RR* by 100. The interpretation would then be the percent change in the "exposed" group. With a *RR* of 2, the percent relative risk is 200% and indicated that children from two-parent households had a 200% increase in incident (i.e., being prepared) over and above children from single-parent homes.

TABLE 9.4

Effect Sizes and Interpretations.

Effect Size	Interpretation
Multiple R^2 pseudo-variance explained *such as*: • Cox and Snell • Nagelkerke • Hosmer and Lemeshow • Aldrich and Nelson • Harrell • traditional R^2	There is no consensus on which (if any) of the pseudo-variance explained indices are best. Given this, these indices are often not reported in published results. Should you choose to report one or more of these values, they should be used only as a guide "without attributing great importance to a precise figure" (Pampel, 2000, p. 50).
Odds ratio (*OR*)	*OR* is computed by taking the exponent of the logistic regression coefficient, e^{b_k} • $OR = 1$ indicates no relationship between a predictor variable and the dependent variable. • $OR > 1$ = higher odds of the outcome occurring • $OR < 1$ = lower odds of the outcome occurring
Risk ratio (*RR*)	*RR* is computed as the risk of incident in one group divided by the risk of incidence in the other group • $RR = 1$ indicates little or no difference in risk (i.e., incident) between the two groups • $RR > 1$ = increased risk (i.e., incident) of the outcome in the exposed group • $RR < 1$ = decreased risk (i.e., incident) of the outcome in the exposed group
d	*OR* can be converted to Cohen's *d*: $d = \ln OR\left(\frac{\sqrt{3}}{\pi}\right) = \ln OR\left(\frac{\sqrt{3}}{3.1415}\right) = \ln OR(.5513)$

Odds ratio values can also be converted to Cohen's *d* using the following equation:

$$d = \ln OR\left(\frac{\sqrt{3}}{\pi}\right) = \ln OR\left(\frac{\sqrt{3}}{3.1415}\right) = \ln OR(.5513)$$

Guidelines for interpreting Cohen's *d* can be applied. As you may recall, if $d = 1.0$, the sample mean is one standard deviation away from the hypothesized mean. Cohen (1988) has proposed the following subjective standards for the social and behavioral sciences as a convention for interpreting *d*: small effect size, $d = .2$; medium effect size, $d = .5$; large effect size, $d = .8$. Interpretation of effect size can be based on a comparison to similar studies; what is considered a "small" effect using Cohen's rule of thumb may actually be quite large in comparison to other related studies that have been conducted. In lieu of a comparison to other studies, such as in those cases where there are no or minimal related studies, then Cohen's subjective standards may be appropriate.

9.1.5 Assumptions

Compared to OLS regression, the assumptions of logistic regression are somewhat relaxed; however, four primary assumptions must still be considered: (a) noncollinearity;

(b) linearity; (c) independence of errors; and (d) values of X are fixed. In this section, we also discuss conditions that are needed in logistic regression as well as diagnostics that can be performed to more closely examine the data.

9.1.5.1 Noncollinearity

Noncollinearity is applicable to logistic regression models with multiple predictors just as it was in multiple regression (but is not applicable when there is only one predictor in any regression model). This assumption has already been explained in detail in Chapter 8 on multiple regression and thus will not be reiterated other than to explain tools that can be used to detect multicollinearity. Although most standard statistical software does not provide an option to easily generate collinearity statistics in logistic regression, you can generate an OLS regression model (i.e., a traditional multiple linear regression) with the same variables used in the logistic regression model and request collinearity statistics there. Because it is only the collinearity statistics that are of interest, do not be concerned in generating an OLS regression model that violates some of the OLS basic assumptions (e.g., normality). We have previously discussed tolerance and the variance inflation factor as two collinearity diagnostics (where tolerance is computed as $1-R_k^2$ where R_k^2 is the variance in each independent variable, X, explained by the other independent variables and VIF is $\frac{1}{1-R_k^2}$. In reviewing these statistics, tolerance values of less than .20 suggest multicollinearity exists, and values of less than .10 suggest serious multicollinearity. VIF values greater than 10 indicate a violation of noncollinearity.

The effects of a violation of noncollinearity in logistic regression are the same as that in multiple regression. First, it will lead to instability of the regression coefficients across samples, where the estimates will bounce around quite a bit in terms of magnitude, and even occasionally result in changes in sign (perhaps opposite of expectation). This occurs because the standard errors of the regression coefficients become larger, thus making it more difficult to achieve statistical significance. Another result that may occur involves an overall regression that is significant, but none of the individual predictors are significant. Violation will also restrict the utility and generalizability of the estimated regression model.

9.1.5.2 Linearity

In OLS regression, the dependent variable is assumed to have a linear relationship with the continuous independent variable(s), but this does not hold in logistic regression. Because the outcome in logistic regression is a logit, the assumption of linearity in logistic regression refers to linearity between *logit of the dependent variable* and the continuous independent variable(s). Hosmer and Lemeshow (1989) suggest several strategies for detecting nonlinearity, the easiest of which to apply is likely the Box-Tidwell transformation. This strategy is also valuable as it is not overly sensitive to minor violations of linearity. This involves generating a logistic regression model that includes all independent variables of interest along with an interaction term for each—the interaction term being the product of the continuous independent variable and its natural log [i.e., $X * \ln(X)$]. Statistically significant interaction terms suggest nonlinearity. It is important to note that the assumption of linearity is applicable only for continuous predictors. A violation of linearity can result in biased parameters estimates, as well as the expected change in the logit of Y not being constant across the values of X. The Hosmer-Lemeshow test has decreased power in detecting lack of fit in situations where linearity is violated (Xie et al., 2008).

9.1.5.3 Independence of Errors

Independence of errors is applicable to logistic regression models just as it is with OLS regression, and a violation of this assumption can result in underestimated standard errors (and thus overestimated test statistic values and perhaps finding statistical significance more often than is really viable, as well as affecting confidence intervals). This assumption has already been explained in detail during the discussion of multiple regression assumptions and thus additional information will not be provided here.

9.1.5.4 Fixed X

The last assumption is that the values of X_k are **fixed**, where the independent variables X_k are fixed variables rather than random variables. Because this assumption was discussed in detail in relation to multiple regression, we only summarize the main points. When X is fixed, the regression model is valid only for those particular values of X_k that were actually observed and used in the analysis. Thus, the same values of X_k would be used in replications or repeated samples. As discussed in the previous regression chapter (Chapter 8), generally we may not want to make predictions about individuals having combinations of X_k scores outside of the range of values used in developing the prediction model; this is defined as *extrapolating* beyond the sample predictor data. On the other hand, we may not be quite as concerned in making predictions about individuals having combinations of X_k scores within the range of values used in developing the prediction model; this is defined as *interpolating* within the range of the sample predictor data. Table 9.5 summarizes the assumptions of logistic regression and the impact of their violation.

TABLE 9.5

Assumptions and Violation of Assumptions: Logistic Regression Analysis.

Assumption	Effect of Assumption Violation
Noncollinearity of X's	• Regression coefficients can be quite unstable across samples (as standard errors are larger) • Restricted generalizability of the model
Linearity	• Bias in slopes and intercept • Expected change in logit of Y is not a constant and depends on value of X
Independence	• Influences standard errors of the model and thus hypothesis tests and confidence intervals
Values of X's are fixed	• Extrapolating beyond the range of X combinations: prediction errors larger, may also bias slopes and intercept • Interpolating within the range of X combinations: smaller effects than when extrapolating; if other assumptions met, negligible effect

9.1.5.5 Conditions

Although not assumptions, the following conditions should be met with logistic regression: nonzero cell counts; nonseparation of data; lack of influential points; and sufficient sample size.

9.1.5.5.1 Nonzero Cell Counts

The first condition is related to nonzero cell counts in the case of nominal independent variables. A zero cell count occurs when the outcome is constant for one or more categories of a nominal independent variable (e.g., all females pass the course). This results in high standard errors because entire groups of individuals have odds of 0 or 1. Strategies to remove zero cell counts include recoding the categories (e.g., collapsing categories) or adding a constant to each cell of the crosstab table. If the overall model fit is what is of primary interest, then you may choose not to do anything about zero cell counts. The overall relationship between the set of predictors and the dependent variable is not generally impacted by zero cell counts. However, if zero cell counts are retained and the results of the individual predictors are what is of interest, it would be wise to provide a limitation to your results recognizing higher standard errors that are produced due to zero cell counts as well as caution that the values of the individual regression coefficients may be affected. Careful review of the data prior to computing the logistic regression model can help thwart potential problems with zero cell counts.

9.1.5.5.2 Nonseparation of Data

Another condition that should be examined is that of complete or quasi-complete separation. Complete separation arises when the dependent variable is perfectly predicted and results in an inability to estimate the model. Quasi-complete separation occurs when there is less than complete separation and results in extremely large coefficients and standard errors. These conditions may occur when the number of variables equals (or nearly equals) the number of cases in the dataset, such that large coefficients and standard errors result.

9.1.5.5.3 Lack of Influential Points

Outliers and influential cases are problematic in logistic regression analysis just as with OLS regression. Severe outliers can cause the maximum likelihood estimator to reduce to zero (Croux, Flandre, & Haesbroeck, 2002). Residual analysis and other diagnostic tests are equally beneficial for detecting miscoded data and unusual (and potentially influential) cases in logistic regression as it is in OLS regression. SPSS and other statistical software, including **R**, provides the option for saving a number of values including predicted values, residuals, and influence statistics. Both probabilities and group membership predicted values can be saved. Residuals that can be saved include: (a) unstandardized; (b) logit; (c) studentized; (d) standardized; and (e) deviance. The three types of influence values that can be saved include Cook's, leverage values, and DfBetas.

The wide variety of values that can be saved suggests that there are many types of diagnostics that can be performed. Review should be conducted when standardized or studentized residuals are greater than an absolute value of 3.0 and DfBeta values are greater than one. Leverage values greater than $(m + 1)/N$ (where m equals the number of independent variables) indicate an influential case (values closer to 1 suggest problems, while those closer to 0 suggest little influence). If outliers or influential cases are found, it is up to you to decide if removal of the case is warranted. It may be that they, while uncommon, are completely plausible so that they are retained in the model. If they are removed from the model, it is important to report the number of cases that were removed prior to analysis (and evidence to suggest what caused you to remove them). A review of Chapter 8 on multiple regression provides further details on diagnostic analysis of outliers and influential cases.

9.2 Mathematical Introduction Snapshot

To summarize the mathematics that underlie logistic regression, odds are simply the ratio of the probability of the dependent variable's two outcomes and computed as:

$$Odds(Y=1)=\frac{P(Y=1)}{1-P(Y=1)}$$

Changing the scale of the odds by taking the natural logarithm of the odds (aka *logit* Y or *log odds*) provides us with a value of the dependent variable that can theoretically range from negative infinity to positive infinity and thus creates a linear relationship between X and the probability of Y (Pampel, 2000). The natural log of the odds is calculated as follows:

$$\ln\frac{P(Y=1)}{1-P(Y=1)}=Logit(Y)$$

and working with log odds, our familiar additive regression equation is applicable:

$$\ln\frac{P(Y=1)}{1-P(Y=1)}=Logit(Y)=\alpha+\beta_1X_1+\beta_2X_2+\ldots+\beta_mX_m$$

If we exponentiate the logit (Y) (i.e., the outcome of our logistic regression equation), then it converts back to the odds (as noted by the calculation here) which allows us to interpret the independent variables as affecting the odds (rather than log odds) of the outcome:

$$Odds(Y=1)=e^{logit(Y)}=e^{ln[Odds(Y=1)]}=e^{\alpha+\beta_1X_1+\beta_2X_2+\ldots+\beta_mX_m}=(e^{\alpha})(e^{\beta_1X_1})(e^{\beta_2X_2})\ldots(e^{\beta_mX_m})$$

Converting the odds back to a probability can be done through the following formula:

$$P(Y=1)=\frac{Odds(Y=1)}{1+Odds(Y=1)}=\frac{e^{\alpha+\beta_1X_1+\beta_2X_2+\ldots+\beta_mX_m}}{1+e^{\alpha+\beta_1X_1+\beta_2X_2+\ldots+\beta_mX_m}}$$

Probability values close to one indicate increased likelihood of occurrence.

9.3 Computing Logistic Regression Using SPSS

Next we consider SPSS for the logistic regression model. Before we conduct the analysis, let us review the data (Ch9.readiness.sav) (note that we recognize the sample size of 20 does not meet minimum sample size criteria previously specified; however, for illustrative purposes, we felt it important to be able to show the entire dataset, and this would have been more difficult with the recommended sample size for logistic regression). With one dependent variable and two independent variables, the dataset must consist of three variables or columns, one for each independent variable and one for the dependent variable. Each row still represents one individual. As seen in the screenshot in Figure 9.3, the SPSS data are in the form of three columns that represent the two independent variables (a continuous teacher administered social development scale and household—a dichotomous variable, single vs. two-adult household) and one binary dependent variable (kindergarten readiness screening test—prepared vs. not prepared). As our dependent variable is dichotomous, we will conduct binary logistic regression. When the dependent variable consists of more than two categories, multinomial logistic regression is appropriate (although not illustrated here).

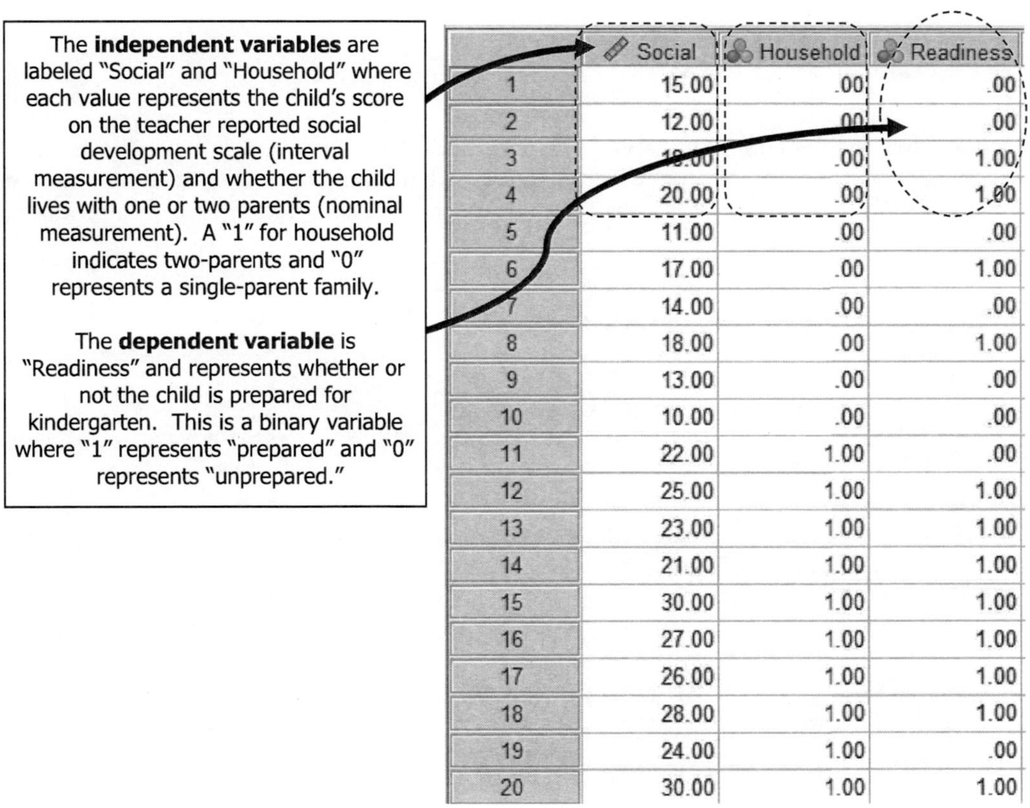

	Social	Household	Readiness
1	15.00	.00	.00
2	12.00	.00	.00
3	[illegible]	.00	1.00
4	20.00	.00	1.00
5	11.00	.00	.00
6	17.00	.00	1.00
7	14.00	.00	.00
8	18.00	.00	1.00
9	13.00	.00	.00
10	10.00	.00	.00
11	22.00	1.00	.00
12	25.00	1.00	1.00
13	23.00	1.00	1.00
14	21.00	1.00	1.00
15	30.00	1.00	1.00
16	27.00	1.00	1.00
17	26.00	1.00	1.00
18	28.00	1.00	1.00
19	24.00	1.00	.00
20	30.00	1.00	1.00

FIGURE 9.3
SPSS data.

Step 1. To conduct a binary logistic regression, go to "Analyze" in the top pulldown menu, then select "Regression," and then select "Binary Logistic." Following the screenshot below (see screenshot for Step 1, Figure 9.4) produces the "Logistic Regression" dialog box.

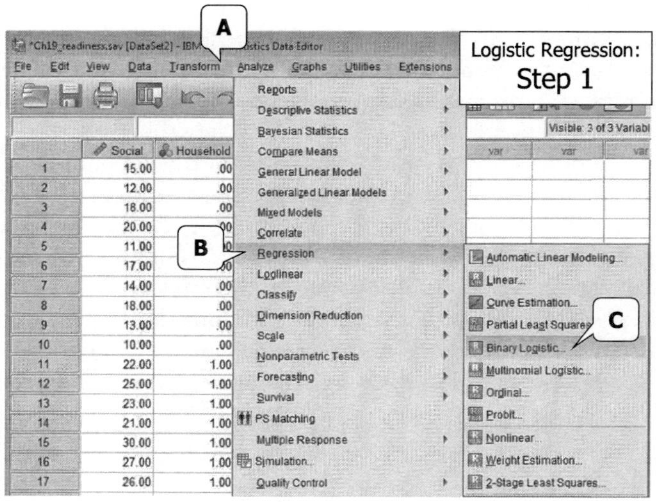

FIGURE 9.4
Step 1.

Step 2. Click the dependent variable (e.g., "Readiness"') and move it into the "Dependent" box by clicking the arrow button. Click the independent variables and move them into the "Covariate(s)" box by clicking the arrow button (see the screenshot for Step 2, Figure 9.5).

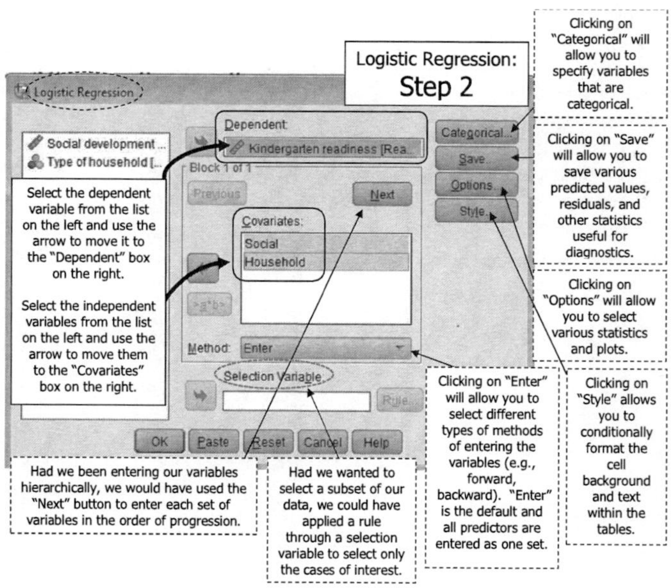

FIGURE 9.5
Step 2.

Step 3. From the Logistics Regression dialog box (see Figure 9.5), clicking on "Categorical" will provide the option to define as categorical those variables that are nominal or ordinal in scale as well as to select which category of the variable is the reference category through the Define Categorical Variables dialog box (see the screenshot for Figure 9.6). From the list of covariates on the left, click the categorical covariate(s) (e.g., "Household") and move it into the "Categorical Covariates" box by clicking the arrow button. By default, "(Indicator)" will appear next to the variable name. Indicator refers to traditional dummy coding and you have the option of selecting which value is the reference category. For binary variables (only two categories), using the "Last" value as the reference category means that the category coded with the largest value will be the category "left out" of the model (or referent), and using the "First" value as the reference category means that the category coded with the smallest value will be the category "left out" of the model. Here two-parent households were coded as 1 and single-parent households as 0. We use single-parent households (coded as 0) as the reference category. Thus we select the radio button for "First" (see Figure 9.6) to define single-parent households as the reference category.

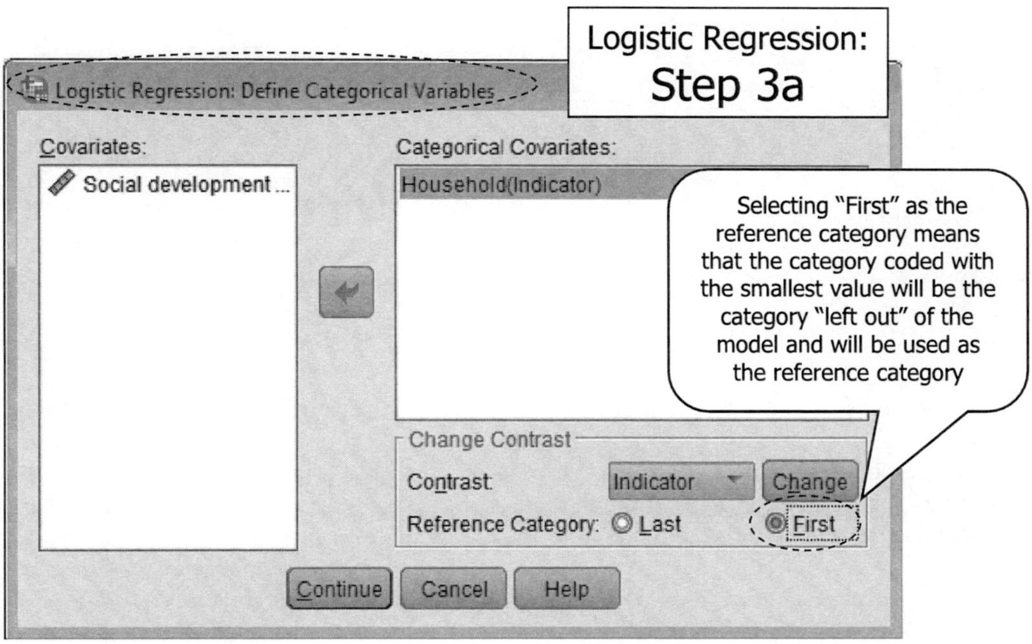

FIGURE 9.6
Step 3a.

Next, we need to click the button labeled "Change" (see the screenshot in Figure 9.7) to define the first value (i.e., zero or single parent household) as the reference (or "left out") category. By doing that, the name of our categorical covariate will now read Household(Indicator(first)). Had we had a categorical variable with more than two categories, we could just define the variable as categorical within logistic regression and select either the first or last value as the reference category. If neither the first or last were what you wanted as the reference category, then some recoding of the data is necessary.

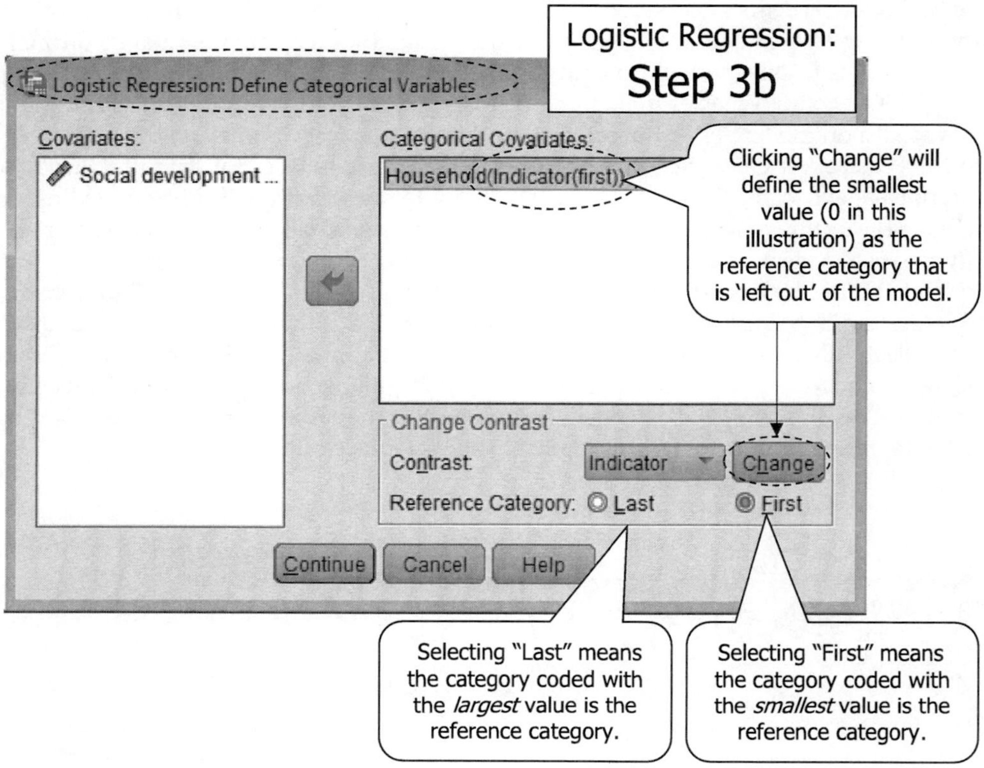

FIGURE 9.7
Step 3b.

Before we move on, notice that the button for "Contrast" is a toggle menu with Indicator as the default option. Selecting the toggle menu allows you to select other types of contrasts often discussed in relation to ANOVA contrasts (e.g., Simple, Difference, Helmert) (see the screenshot for Step 3b contrast shown in Figure 9.8). These will not be reviewed here. Click on "Continue" to return to the Logistic Regression dialog box.

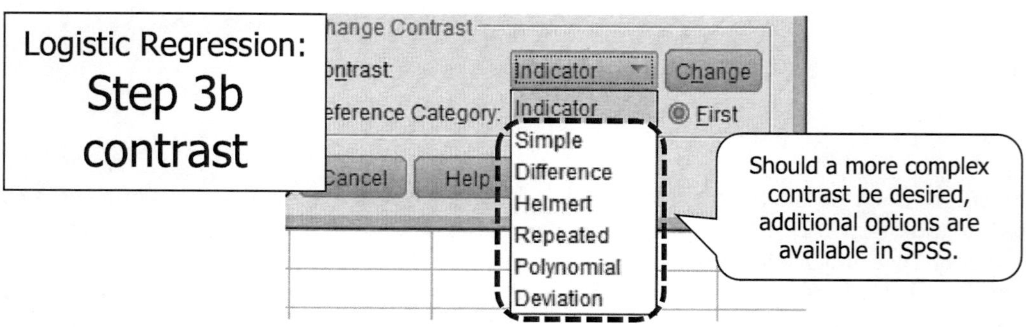

FIGURE 9.8
Step 3b contrast.

Step 4. From the Logistic Regression dialog box (see Figure 9.6), clicking on "Save" will provide the option to save various predicted values, residuals, and statistics that can be used for diagnostic examination (see the screenshot in Figure 9.9). From the Save dialog box under the heading "Predicted Values," place checkmarks in the boxes next to "Probabilities" and "Group membership." Under the heading "Residuals," place a checkmark in the box next to "Standardized." Under the heading "Influences," place checkmarks in the boxes next to "Cook's," "Leverage values," and "DfBeta(s)." Click on "Continue" to return to the original dialog box.

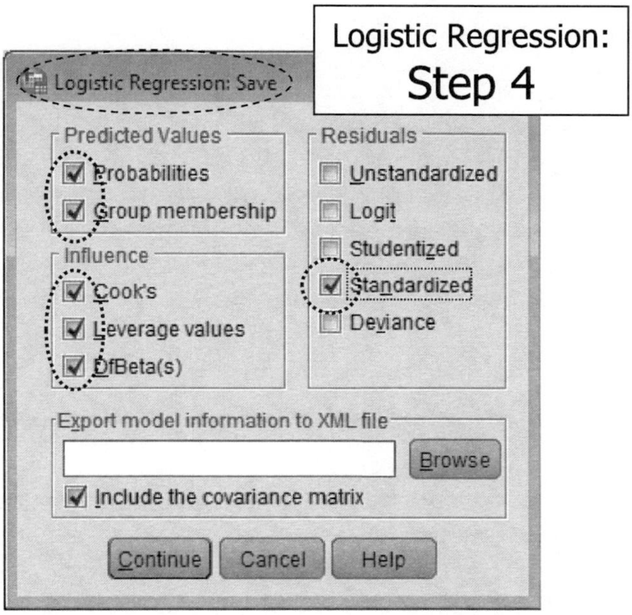

FIGURE 9.9
Step 4.

Step 5. From the Logistic Regression dialog box (see screenshot Step 2 Figure 9.6), clicking on "Options" will allow you to generate various statistics and plots. From the Options dialog box (see the screenshot for Step 5 in Figure 9.10) under the heading "Statistics and Plots," place checkmarks in the boxes next to "Classification plots," "Hosmer-Lemeshow goodness-of-fit," "Casewise listing of residuals," "Outliers outside," and "CI for exp(B)." For Outliers outside, you must specify a numeric value of standard deviations to define what you consider to be an outlier. Common values may be 2 (in a normal distribution, 95% of cases will be within +2 standard deviations), 3 (in a normal distribution, about 99% of cases will be within +3 standard deviations), or 3.29 (in a normal distribution, about 99.9% of cases will be within +3.29 standard deviations). For this illustration, we will use a value of 2. For CI for exp(*B*), you must specify a confidence interval. This should be the complement of the alpha being tested. If you are using an alpha of .05, then the *CI* will be 1 – .05 or 95. All the remaining options in the Options dialog box will be left as the default settings. Click on "Continue" to return to the original dialog box. From the Logistic Regression dialog box, click on "OK" to generate the output.

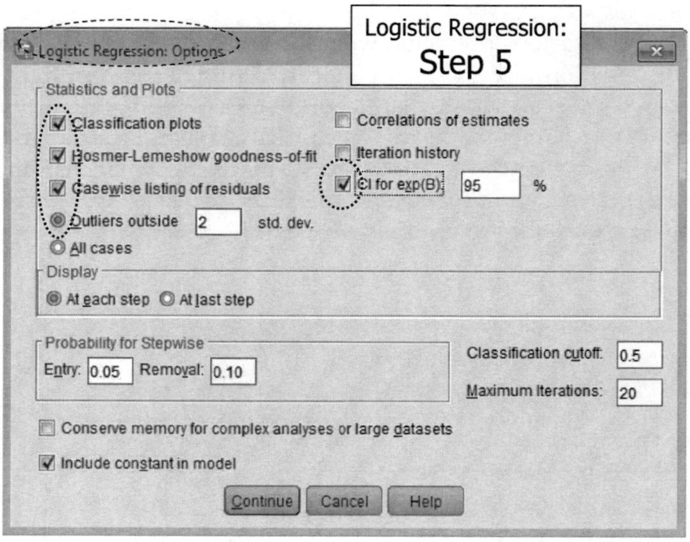

FIGURE 9.10
Step 5.

Interpreting the output. Annotated results are presented in Table 9.6.

TABLE 9.6
SPSS Results for the Binary Logistic Regression Kindergarten Readiness Example

Case Processing Summary

Unweighted Cases[a]		N	Percent
Selected Cases	Included in Analysis	20	100.0
	Missing Cases	0	.0
	Total	20	100.0
Unselected Cases		0	.0
Total		20	100.0

a. If weight is in effect, see classification table for the total number of cases.

This table provides information on sample size and missing data. The sample size is 20 and we have no missing data.

Dependent Variable Encoding

Original Value	Internal Value
Unprepared	0
Prepared	1

Information on how the values of the dependent variable are coded is provided under "internal value." "Unprepared" is coded as 0 and "prepared" is coded as 1.

Categorical Variables Codings

		Frequency	Parameter coding (1)
Type of household	Single parent household	10	.000
	Two-parent household	10	1.000

Information on how the values of the categorical variable(s) are coded is provided as "parameter coding." "Single parent household" is coded as 0 and "two-parent household" is coded as 1. The sample size per group is presented in the "frequency" column.

TABLE 9.6 (continued)

SPSS Results for the Binary Logistic Regression Kindergarten Readiness Example

Block 0 is a summary of the model with the *constant only* (i.e., none of the predictors are included). The **classification table** provides the percentage of cases correctly predicted given the constant only. *Without including covariates,* we can correctly predict children who are prepared for kindergarten 100% of the time but fail to predict any children (0%) who are unprepared. Here all children are predicted to be prepared.

Block 0: Beginning Block

Classification Table[a,b]

Observed			Predicted: Kindergarten readiness: Unprepared	Predicted: Kindergarten readiness: Prepared	Percentage Correct
Step 0	Kindergarten readiness	Unprepared	0	8	.0
		Prepared	0	12	100.0
	Overall Percentage				60.0

a. Constant is included in the model.

b. The cut value is .500

Variables in the Equation

		B	S.E.	Wald	df	Sig.	Exp(B)
Step 0	Constant	.405	.456	.789	1	.374	1.500

Variables Not in the Equation

			Score	df	Sig.
Step 0	Variables	Social development	8.860	1	.003
		Type of household(1)	3.333	1	.068
	Overall Statistics		11.168	2	.004

Variables not in the equation provides an indication of whether each covariate will statistically significantly contribute to predicting the outcome. Only social development (p = .003) is of value in the logistic model. The value of 11.168 for **overall statistics** is a residual chi-square statistic. Since the p value for the residual chi-square statistic indicates statistical significance (p = .004), this indicates that including the two covariates improves the model as compared to the constant only model.

(continued)

TABLE 9.6 (continued)

SPSS Results for the Binary Logistic Regression Kindergarten Readiness Example

Block 1: Method = Enter

Method = Enter indicates that the method of entering the predictors was simultaneous entry (recall this is the default method in SPSS and is called "Enter").

Omnibus Tests of Model Coefficients

		Chi-square	df	Sig.
Step 1	Step	15.793	2	.000
	Block	15.793	2	.000
	Model	15.793	2	.000

The −2LL for the constant only model is computed as the sum of chi-square for the constant only model and −2LL for the full model:

$$\chi^2_{Model} + (-2\text{LL}) = 15.793 + 11.128 = 26.921$$

Model Summary

Step	-2 Log likelihood	Cox & Snell R Square	Nagelkerke R Square
1	11.128[a]	.546	.738

a. Estimation terminated at iteration number 7 because parameter estimates changed by less than .001.

The two R^2 values are pseudo R^2 and are interpreted similarly to multiple R^2. These can be used as effect size indices for logistic regression and Cohen's interpretations for correlation can be used to interpret. Both values indicate a large effect.

Model summary statistics provide overall model fit. For good model fit, the value of −2LL for the full model (11.128) should be less than −2LL for the constant only model (26.921). This is a chi-square value with degrees of freedom equal to the number of parameters in the full model (i.e., two predictors plus one constant) minus the number of parameters in the baseline model (i.e., 1). Thus there are 2 *df*. Using the chi-square table, with an alpha of .05 and 2 *df*, the critical value is 5.99. Since 11.128 is larger than the critical value, we reject the null hypothesis that the best prediction model is the constant only model. *In other words, the full model (with predictors) is better at predicting kindergarten readiness than the constant only model.*

Hosmer and Lemeshow Test

Step	Chi-square	df	Sig.
1	4.691	7	.698

As a measure of classification accuracy, *non-statistical significance (p = .698) indicates good model fit for the Hosmer and Lemeshow test.* This test is affected by small sample size, however; caution should be used when interpreting the results of this test when sample size is less than 50.

Contingency Table for Hosmer and Lemeshow Test

		Kindergarten readiness = Unprepared		Kindergarten readiness = Prepared		
		Observed	Expected	Observed	Expected	Total
Step 1	1	2	1.988	0	.012	2
	2	2	1.922	0	.078	2
	3	1	1.651	1	.349	2
	4	2	1.292	0	.708	2
	5	0	.607	2	1.393	2
	6	1	.404	2	2.596	3
	7	0	.100	2	1.900	2
	8	0	.030	2	1.970	2
	9	0	.005	3	2.995	3

TABLE 9.6 (continued)

SPSS Results for the Binary Logistic Regression Kindergarten Readiness Example

The **classification table** provides information on how well group membership was predicted. Cells on the *diagonal* indicate *correct* classification. For example, children who were prepared for kindergarten were accurately classified 91.7% of the time as compared to unprepared children (87.5%). Overall, 90% of children were correctly classified. This is computed as the number of correctly classified cases divided by total sample size:

$$\frac{7+11}{20} = .90$$

Classification Table[a]

Observed			Predicted: Kindergarten readiness Unprepared	Predicted: Kindergarten readiness Prepared	Percentage Correct
Step 1	Kindergarten readiness	Unprepared	7	1	87.5
		Prepared	1	11	91.7
	Overall Percentage				90.0

a. The cut value is .500

Using Press's *Q* and given the chi-square critical value of 3.841 ($df = 1$), we find:

$$Q = \frac{[N-(nK)]^2}{N(K-1)} = \frac{[20-(18)(2)]^2}{20(2-1)} = 12.8$$

We reject the null hypothesis. There is evidence to suggest that the predictions are statistically significantly better than chance.

NOTE! Interpretations of *B* coefficients are usually done via odds ratios.

The Wald statistic is used to test the statistical significance of each covariate.

The *p* value for "social" ($p = .030$) indicates that the slope is statistically significantly different from zero. This tells us that the independent variable is contributing to predicting kindergarten preparedness. The intercept ($p = .032$) is also statistically significantly different from zero.

Exp(B) values are the odds ratios. The odds ratio of 2.631 for social indicates that the odds for being prepared for kindergarten are over 2-1/2 times greater (or 263%) for every one point increase in social development. The odds for household are nearly zero. This indicates that the odds for being prepared for kindergarten are about the same regardless of the child's household structure (single- versus two-parent home).

Variables in the Equation

		B	S.E.	Wald	df	Sig.	Exp(B)	95% C.I.for EXP(B) Lower	95% C.I.for EXP(B) Upper
Step 1[a]	Social development	.967	.446	4.696	1	.030	2.631	1.097	6.313
	Type of household(1)	-6.216	3.440	3.265	1	.071	.002	.000	1.693
	Constant	-15.404	7.195	4.584	1	.032	.000		

a. Variable(s) entered on step 1: Social development, Type of household.

The *B* coefficient is interpreted as the change in the logit of the dependent variable given a one-unit change in the independent variable. Recall that the logit is the natural log of the dependent variable occurring. With *B* equal to .967, this tells us that a one-unit change in social development will result in nearly a one-unit change in the logit of kindergarten preparedness. The constant is the expected value of the logit of kindergarten readiness for children of single parents (recall this was coded as 0) and when social development is zero.

Since the odds of 1.00 (which indicates similar odds for falling into either category of the outcome) are *not* contained within the interval for social development, this suggests the odds ratio is statistically significantly different from zero. Note that the odds ratio is only computed for the predictors and not for the intercept (i.e., constant).

A negative *B* indicates that an increase in value of that independent variable will result in a *decrease* in the predicted probability of the dependent variable.

A positive *B* indicates that an increase in value of that independent variable will result in an *increase* in the predicted probability of the dependent variable.

(continued)

TABLE 9.6 (continued)

SPSS Results for the Binary Logistic Regression Kindergarten Readiness Example

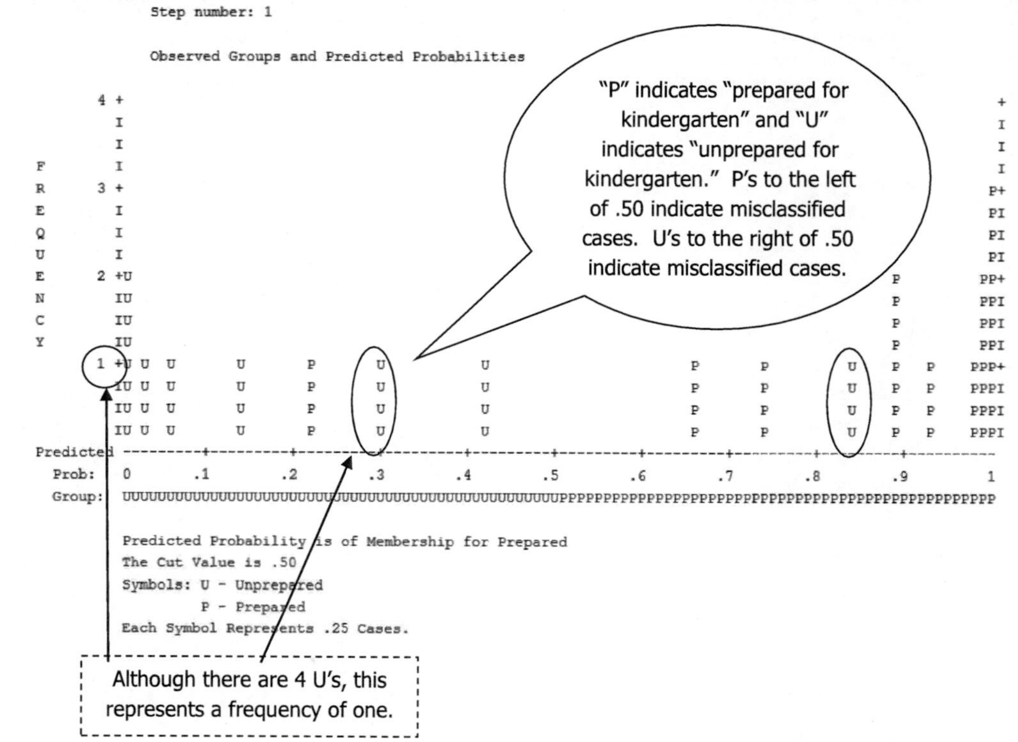

Casewise List[b]

Case	Selected Status[a]	Observed Kindergarten readiness	Predicted	Predicted Group	Temporary Variable Resid	ZResid	SResid
14	S	P**	.214	U	.786	1.918	2.102
19	S	U**	.832	P	-.832	-2.226	-2.106

a. S = Selected, U = Unselected cases, and ** = Misclassified cases.

b. Cases with studentized residuals greater than 2.000 are listed.

Recall we told SPSS to identify residuals that were outside 2 standard deviations. Based on that decision, cases 14 and 19 were identified as potential outliers. We review this output in the discussion on outliers.

9.4 Computing Logistic Regression Using R

Next we consider **R** for the logistic regression model. The commands are provided within the blocks with additional annotation to assist in understanding how the command works. Should you want to write reminder notes and annotation to yourself as you write the commands in **R** (and we highly encourage doing so), remember that any text that follows a hashtag (i.e., #) is annotation only and not part of the **R** code. Thus, you can write annotations directly into **R** with hashtags. We encourage this practice so that when you call up the commands in the future, you'll understand what the various lines of code are doing.

9.4.1 Reading Data Into R

```
getwd()
```

R is always directed to a directory on your computer. To find out which directly it's pointed to, run the *get working directory* command. We will assume that we need to change the working directory, and will use the next line of code to set the working directory to the desired path.

```
setwd("E:/Folder")
```

This command will set your working directory to a specific folder that you name. Change what is in parentheses to your file location. Also, if you are copying the directory name, it will copy in slashes. You will need to change the backslash (i.e., \) to a forward slash (i.e., /) in the **R** command. Also note that you need this in parentheses.

```
Ch9_readiness <- read.csv("Ch9_readiness.csv")
```

This command reads your data into **R**. What's to the left of the "<-" will be what you want to call the dataframe in **R**. In this example, we're calling this **R** dataframe "Ch9_readiness." What's to the right of the "<-" tells **R** to find this particular csv file. In this example, our file is called "Ch9_readiness.csv." Make sure the extension (i.e., .csv) is there. Also note that you need this in quotations.

```
names(Ch9_readiness)
```

This command will produce a list of variable names for the dataframe as follows:

```
[1] "Social" "Household" "Readiness"
```

This is a good check to make sure your data have been read in correctly.

```
View(Ch9_readiness)
```

This command will let you view the dataset in spreadsheet format in RStudio.

```
Ch9_readiness$Household <- factor(Ch9_readiness$Household)
```

This tells **R** to treat the variable "Household" as categorical.

```
Ch9_readiness$Readiness <- factor(Ch9_readiness$Readiness)
```

This tells R to treat the variable "Readiness" as categorical.

FIGURE 9.11
Reading data into **R**.

```
summary(Ch9_readiness)
```

The *summary* command will produce basic descriptive statistics on all the variables in your dataframe. This is a great way to quickly check to see if the data have been read in correctly and get a feel for your data, if you haven't already. The output from the summary statement for this dataframe looks like this. Because we defined Household and Readiness as categorical, we get only a few summary stats.

```
        Social          Household    Readiness
Min.    :10.00          0:10         0: 8
1st Qu. :14.75          1:10         1:12
Median  :20.50
Mean    :20.20
3rd Qu. :25.25
Max.    :30.00
```

FIGURE 9.11 (continued)
Reading data into **R**.

9.4.2 Generating the Logistic Regression Model and Saving Values

With these commands, we will generate the logistic regression model and save variables that can be used for data screening.

```
ReadinessLogit <- glm(formula = Readiness ~ Social + Household,
                      family="binomial",
                      data =Ch9_readiness)
```

The *glm* function will run the logistic regression model. In this example, we're naming our model *ReadinessLogit*. The formula defines our dependent variable as "Readiness," and it is predicted by "Social" and "Household." The command *family="binomial"* tells **R** to compute a logistic regression model using a binomial distribution.

```
summary(ReadinessLogit)
```

The *summary* function will generate the results from the logistic regression model. If you don't run the summary line of code, since we named our model, there won't be any results output!

Deviance Residuals:

```
     Min         1Q     Median        3Q        Max
-1.88892   -0.24308   0.06327   0.41366   1.75662
```

Coefficients:

```
             Estimate  Std. Error  z value  Pr(>|z|)
(Intercept)  -15.4035      7.1941   -2.141    0.0323 *
Social         0.9675      0.4465    2.167    0.0302 *
Household1    -6.2162      3.4402   -1.807    0.0708 .
—
Signif. codes:
0 '***' 0.001 '**' 0.01 '*' 0.05 '.' 0.1 ' ' 1

(Dispersion parameter for binomial family taken to be 1)

    Null deviance: 26.920 on 19 degrees of freedom
Residual deviance: 11.128 on 17 degrees of freedom
AIC: 17.128

Number of Fisher Scoring iterations: 6
```

FIGURE 9.12
Generating the logistic regression model and saving variables.

```
ReadinessLogit2 <- glm(formula = Readiness ~ Social,
                       family="binomial",
                       data =Ch9_readiness)

anova(ReadinessLogit, ReadinessLogit2,
      test = "Chisq") #to compare 2 models
```

There are a number of model fit tests that can be conducted. As an example, if we want to compare one model with fewer predictors (for illustrative purposes, the model has been re-ran as *ReadinessLogit2* with only "Social" as the predictor) to another model, we can do so. The *anova* function with *test = "Chisq"* will generate the likelihood ratio test to compare the two models, *ReadinessLogit* and *ReadinessLogit2*. This test generates the likelihood ratio test to compare the likelihood of the data under the full model (i.e., *ReadinessLogit*) against the likelihood of the data in the reduced model (i.e., *ReadinessLogit2*). A statistically significant likelihood ratio test means we reject the null hypothesis that the reduced model is better than the full model. In other words, a statistically significant likelihood ratio test provides evidence against the reduced model and in favor of the full model. We see $p = .02319$, suggesting the full model, with both predictors, is better model fit than the reduced model with only one predictor.

```
Analysis of Deviance Table

Model 1: Readiness ~ Social + HouseholdF
Model 2: Readiness ~ Social
  Resid. Df Resid. Dev Df Deviance Pr(>Chi)
1        17     11.128
2        18     16.282 -1  -5.1541  0.02319 *
—
Signif. codes: 0 '***' 0.001 '**' 0.01 '*' 0.05 '.' 0.1 ' ' 1
```

```
install.packages("zoo")
library(lmtest)
lrtest(ReadinessLogit, ReadinessLogit2)
```

The likelihood ratio test can also be conducted using the *lrtest* function from the *zoo* package. In parentheses, we input the two models to compare.

```
Likelihood ratio test
Model 1: Readiness ~ Social + HouseholdF
Model 2: Readiness ~ Social
  #Df  LogLik Df   Chisq Pr(>Chisq)
1   3 -5.5638
2   2 -8.1409 -1  5.1541   0.02319 *
—
Signif. codes: 0 '***' 0.001 '**' 0.01 '*' 0.05 '.' 0.1 ' ' 1
```

```
install.packages("survey")
library(survey)
regTermTest(ReadinessLogit, "Social")
regTermTest(ReadinessLogit, "HouseholdF")
```

The Wald test can be generated using the *regTermTest* function from the *survey* package. Within parentheses, we define our logistic regression model (i.e., *ReadinessLogit*) and one of the predictors. Thus, the number of tests generated will equal the number of predictors for which you want to generate the Wald test. The Wald test tests the alternative hypothesis that the coefficient of an independent variable in the model is not equal to zero. Failing to reject the hypothesis provides evidence that removing the variable from the model will not substantially impact the model fit. Not surprising, we see that we could remove the "Household" predictor and our model fit would not be detrimentally impacted.

FIGURE 9.12 (continued)
Generating the logistic regression model and saving variables.

```
# regTermTest(ReadinessLogit, "Social")
Wald test for Social
 in glm(formula = Readiness ~ Social + Household, family = "binomial",
    data = Ch9_readiness)
F =  4.696185 on  1  and   17  df: p= 0.044723
```

```
# regTermTest(ReadinessLogit, "Household")
Wald test for Household
 in glm(formula = Readiness ~ Social + Household, family = "binomial",
    data = Ch9_readiness)
F =  3.264941 on  1  and   17  df: p= 0.088507
```

```
install.packages("caret")
library(caret)
varImp(ReadinessLogit)
```

Using the *varImp* function from the *caret* package our logistic model, *ReadinessLogit*, we can examine variable importance by reviewing the absolute value of the *t* test statistic for each predictor. The measure has a maximum value of 100, with values closer to 100 suggesting greater variable importance.

```
             Overall
Social      2.167068
HouseholdF1 1.806915
```

```
Ch9_readiness$predicted.probabilities <- fitted(ReadinessLogit)
```

The *fitted* function saves the predicted probabilities generated from the "ReadinessLogit" object. Within the parentheses is the name of our logistic regression model (i.e., *ReadinessLogit*). To the left of "<-" is the command that will save the predicted probabilities with the name of *predicted.probabilities* to our dataframe (i.e., *Ch9_readiness*). The remaining variables that we are generating are created and saved to our dataframe similarly.

```
Ch9_readiness$cook <- cooks.distance(ReadinessLogit)
```

The *cooks.distance* function will save Cook's distance, an influence statistic, generated from the "ReadinessLogit" object to our dataframe *Ch9_readiness* and will label the variable "cook."

```
Ch9_readiness$leverage <- hatvalues(ReadinessLogit)
```

The *hatvalues* function saves the leverage values generated from the *ReadinessLogit* object to our dataframe *Ch9_readiness* and will label the variable "leverage."

```
Ch9_readiness$standardized.residuals <- rstandard(ReadinessLogit)
```

The *rstandard* function saves standardized residuals generated from the *ReadinessLogit* object.

```
Ch9_readiness$studentized.residuals <- rstudent(ReadinessLogit)
```

The *rstudent* function saves studentized residuals generated from the *ReadinessLogit* object.

```
Ch9_readiness$dfbeta <- dfbeta(ReadinessLogit)
```

The *dfbeta* function saves DfBeta values generated from the *ReadinessLogit* object.

```
write.csv(Ch9_readiness,"Ch9diag.csv")
```

If you want to save the data that you just created and export to Excel, you can use this command to write a csv file.

FIGURE 9.12 (continued)
Generating the logistic regression model and saving variables.

Comparing our output from **R** to SPSS, we see that, with the exception of small rounding error, the results for the coefficients are the same. There is additional output from **R** that we don't receive from SPSS. For example, the deviance residuals (which are −2*log likelihood) are a model fit measure and can be used to compare the null model (i.e., intercept only model) with the model which includes predictors.

9.4.3 Generating Confidence Intervals of Coefficient Estimates

```
confint(ReadinessLogit)
```

Because we named our model, we can easily request additional stats. With the *confint* command, we can obtain confidence intervals for the coefficient estimates. These CI are based on the profiled log-likelihood function.

```
                    2.5 %      97.5 %
(Intercept)  -35.3699318  -5.1530264
Social         0.3232473   2.1935459
Household1   -15.3437530  -0.7315369
```

```
confint.default(ReadinessLogit)
```

With the *confint.default* statement, we can get *CI* based on just the standard errors.

```
                     2.5 %      97.5 %
(Intercept)  -29.50365894  -1.3034231
Social         0.09246227   1.8425236
Household1   -12.95885210   0.5265212
```

FIGURE 9.13
Generating confidence intervals of coefficient estimates.

9.4.4 Exponentiating Coefficients

```
exp(coef(ReadinessLogit))
```

Use the *exp* command to exponentiate the coefficients and interpret them as odds-ratios. For the intercept and Household variables, we see "−07" and "−03," respectively. This indicates we need to move the decimals that number of places to the left.

```
(Intercept)         Social    Household1
2.043276e-07  2.631339e+00  1.996888e-03
```

FIGURE 9.14
Exponentiating coefficients.

9.4.5 Producing Odds Ratios and Their Confidence Intervals

Earlier, we illustrated two online calculators that can be used for computing *OR* and confidence intervals. However, it's very easy to compute *OR* and confidence intervals using the logistic regression model generated in **R**.

```
exp(cbind(OR=coef(ReadinessLogit),
confint.default(ReadinessLogit)))
```

This statement will produce odds ratios and their confidence intervals based on standard errors. Had we just used the *confint* command, the *CI* produced would be based on the profiled log-likelihood function. Use the *cbind* command to place the coefficients and *CI* in columns.

```
                        OR        2.5 %     97.5 %
(Intercept) 2.043276e-07 1.537176e-13 0.2716005
Social      2.631339e+00 1.096872e+00 6.3124485
Household1  1.996888e-03 2.355277e-06 1.6930324
```

FIGURE 9.15
Producing odds ratios and their confidence intervals.

9.5 Data Screening

Previously we described a number of assumptions used in logistic regression. These included: (a) noncollinearity; (b) linearity between the predictors and logit of the dependent variable; and (c) independence of errors. We also reviewed the data to ensure there are no outliers.

Before we begin to examine assumptions, let us review the values that we requested to be saved to our datafile (see the SPSS dataset screenshot in Figure 9.16, as well as Figure 9.12 for producing the variables earlier in **R**).

1. PRE_1 values are the predicted probabilities.
2. PGR_1 is the predicted group membership (here group membership is either prepared or unprepared for kindergarten).
3. COO_1 values are Cook's influence statistics. As a general guideline, Cook's values greater than one suggest that case is potentially problematic.
4. LEV_1 values are leverage values. As a general guide, leverage values less than .20 suggest there are no problems with cases exerting undue influence. Values greater than .5 indicate problems.
5. ZRE_1 values are standardized residuals computed as the residual divided by an estimate of the standard deviation of the residual. Standardized residuals have a mean of zero and standard deviation of one.
6. DFB0_1, DFB1_1 and DFB2_1 values are DfBeta values and indicate the difference in a beta coefficient if that particular case were excluded from the model.

9.5.1 Noncollinearity

It is not possible to request multicollinearity statistics, such as tolerance and VIF, using logistic regression in SPSS or **R**. We can, however, estimate those values by running the same variables in a multiple regression model and requesting only the collinearity statistics. We are not interested in the parameter estimates of the model—only the collinearity statistics. Tolerance values of less than .10 and VIF values of greater than 10 indicate multicollinearity (Menard, 1995). Because the steps for generating multiple regression were presented previously in the text, we will not reiterate them here. Rather, we will merely present the applicable portion of the output of this model. From the output that follows with a tolerance of .248 and VIF of 4.037, we have evidence that we do not have multicollinearity. In

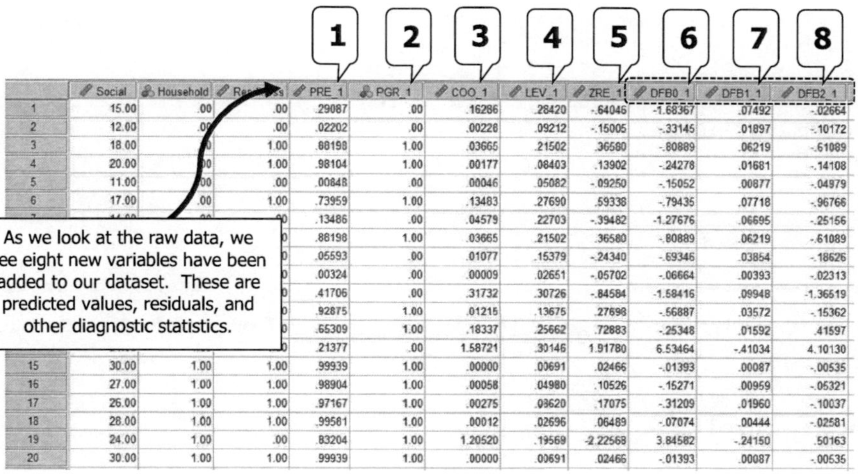

	Social	Household	Re[illegible]ss	PRE_1	PGR_1	COO_1	LEV_1	ZRE_1	DFB0_1	DFB1_1	DFB2_1
1	15.00	.00	.00	.29087	.00	.16286	.28420	-.64046	-1.68367	.07492	-.02664
2	12.00	[illegible]	.00	.02202	.00	.00228	.09212	-.15005	-.33145	.01897	-.10172
3	18.00	[illegible]	1.00	.88198	1.00	.03665	.21502	.36580	-.80889	.06219	-.61089
4	20.00	[illegible]	1.00	.98104	1.00	.00177	.08403	.13902	-.24278	.01681	-.14108
5	11.00	[illegible]	.00	.00848	.00	.00046	.05082	-.09250	-.15052	.00877	-.04979
6	17.00	.00	1.00	.73959	1.00	.13483	.27690	.59338	-.79435	.07718	-.96766
[illegible]	[illegible]	[illegible]	[illegible]	.13486	.00	.04579	.22703	-.39482	-1.27676	.06695	-.25156
[illegible]	[illegible]	[illegible]	[illegible]	.88198	1.00	.03665	.21502	.36580	-.80889	.06219	-.61089
[illegible]	[illegible]	[illegible]	[illegible]	.05593	.00	.01077	.15379	-.24340	-.69346	.03854	-.18626
[illegible]	[illegible]	[illegible]	[illegible]	.00324	.00	.00009	.02651	-.05702	-.06664	.00393	-.02313
[illegible]	[illegible]	[illegible]	[illegible]	.41706	.00	.31732	.30726	-.84584	-1.58416	.09948	-1.36519
[illegible]	[illegible]	[illegible]	[illegible]	.92875	1.00	.01215	.13675	.27698	-.56887	.03572	-.15362
[illegible]	[illegible]	[illegible]	[illegible]	.65309	1.00	.18337	.25662	.72883	-.25348	.01592	.41597
[illegible]	[illegible]	[illegible]	[illegible]	.21377	.00	1.58721	.30146	1.91780	6.53464	-.41034	4.10130
15	30.00	1.00	1.00	.99939	1.00	.00000	.00691	.02466	-.01393	.00087	-.00535
16	27.00	1.00	1.00	.98904	1.00	.00058	.04980	.10526	-.15271	.00959	-.05321
17	26.00	1.00	1.00	.97167	1.00	.00275	.03620	.17075	-.31209	.01960	-.10037
18	28.00	1.00	1.00	.99561	1.00	.00012	.02696	.06489	-.07074	.00444	-.02581
19	24.00	1.00	.00	.83204	1.00	1.20520	.19569	-2.22568	3.84582	-.24150	.50163
20	30.00	1.00	1.00	.99939	1.00	.00000	.00691	.02466	-.01393	.00087	-.00535

FIGURE 9.16
Saved data.

examining collinearity diagnostics, a general guideline for interpreting condition indices is that values in the range of 10 to 30 should be of concern, greater than 30 indicates trouble, and greater than 100 indicates disaster (Belsley, 1991). Here the condition index of dimension three (14.259) is within the range of cause for concern. The last three columns refer to variance proportions. Multiplying these values by 100 provides a percentage of the variance of the regression coefficient that is related to a particular eigenvalue. Multicollinearity is suggested when covariates have high percentages associated with a small eigenvalue (and large condition index). Thus, for purposes of reviewing for multicollinearity, concentrate only on the rows with small eigenvalues. In this example, 100% of the variance of the regression coefficient for social development and 73% for type of household are related to eigenvalue 3 (the dimension with the smallest eigenvalue and largest condition index). This suggests some concern for multicollinearity. In summary, we have met the assumption of noncollinearity with the tolerance and VIF values, but there is some concern for multicollinearity with the condition index and variance proportion values.

Coefficients[a]

		Collinearity Statistics	
Model		Tolerance	VIF
1	Social development	.248	4.037
	Type of household	.248	4.037

a. Dependent Variable: Kindergarten readiness

Collinearity Diagnostics[a]

				Variance Proportions		
Model	Dimension	Eigenvalue	Condition Index	(Constant)	Social development	Type of household
1	1	2.683	1.000	.00	.00	.01
	2	.303	2.974	.05	.00	.25
	3	.013	**14.259**	.95	**1.00**	**.73**

a. Dependent Variable: Kindergarten readiness

FIGURE 9.17
Collinearity output and **R** code.

Working in **R**, the *car* package can be used to generated VIF statistics. The following command will install *car* and load it into your library. If you've installed the package previously, you only need to load the package into your library.

```
install.packages(car)
library(car)
```

```
vif(ReadinessLogit)
1/vif(ReadinessLogit)
```

The *vif* command will generate the VIF and its reciprocal (using the *1/vif* command), which is the tolerance statistic.

FIGURE 9.17 (continued)
Collinearity output and **R** code.

9.5.2 Linearity

Recall that the linearity assumption is applicable only to continuous variables. Thus, we will test this assumption only for social development. The Tidwell transformation test can be used to test that the assumption of linearity has been met. To generate this test, for each *continuous* independent variable we must first create an interaction term that is the product of the independent variable and its natural log (*ln*). Here we have only one continuous independent variable—social development. Thus, only one interaction term will be created.

Step 1. To create an interaction term of our continuous variable and the natural log of this variable, go to "Transform" in the top pulldown menu, then select "Compute Variable." Following the screenshot for Step 1 (Figure 9.18) produces the "Compute Variable" dialog box.

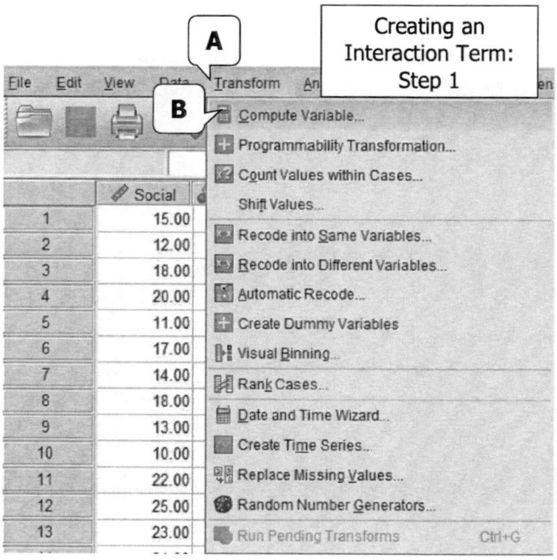

FIGURE 9.18
Creating an interaction term: Step 1.

Step 2. In the Target Variable box in the upper left corner, enter the variable name that you want to appear as the column header (see the screenshot for Step 2, Figure 9.19). Since this is the column header name, this name cannot begin with special characters or numbers and cannot have any spaces. If you wish to define the label for this variable (i.e., what will appear on the output; this *can* include special characters, spaces, and numbers), then click on the "Type & Label" box directly underneath "Target Variable," where additional text to define the name of the variable can be included. Next, click on the continuous covariate (i.e., social development) and move it into the Numeric Expression box by clicking on the arrow in the middle of the screen. Using either the keyboard on screen or your keyboard, click on the asterisks key (i.e.,*). This will be used as the multiplication sign. Next, under Function group, click on "Arithmetic" to display all of the basic mathematical functions. From this alphabetized list click on Ln (natural log). To move this function into the Numeric Expression box, click on the arrow key in the right central part of the dialog box.

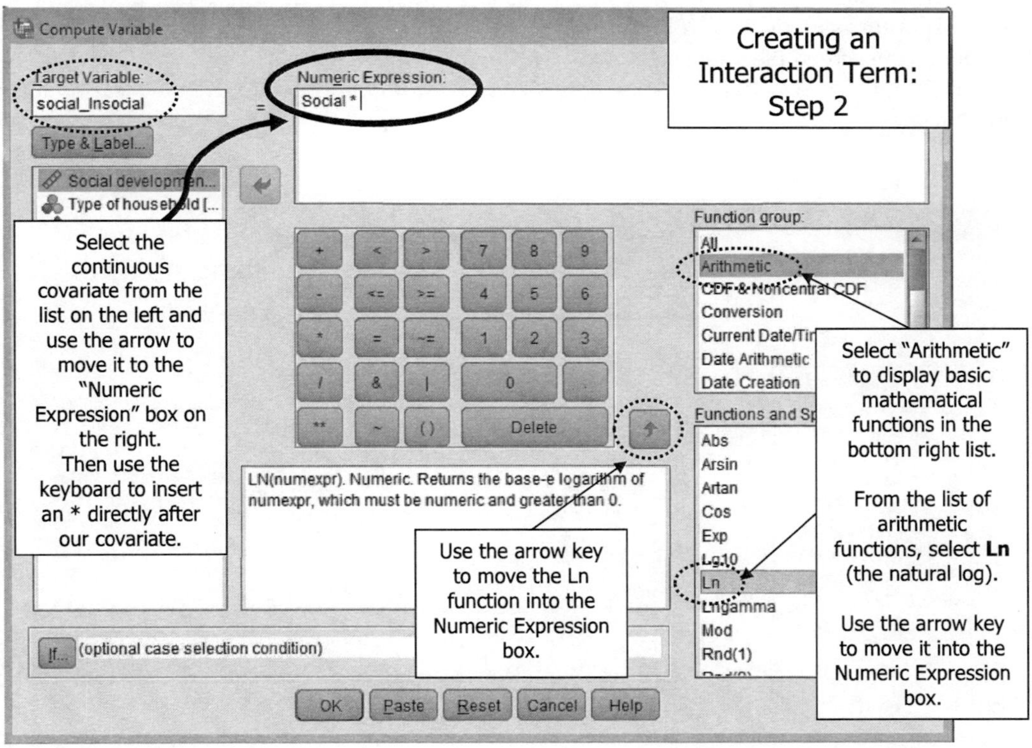

FIGURE 9.19
Creating an interaction term: Step 2.

Step 3. Once the natural log function is displayed in the Numeric Expression box, a question mark enclosed inside parentheses will appear (see screenshot for Step 3a, Figure 9.20). This is SPSS's way of asking for which variable you want the natural log computed. Here it is the continuous covariate, social development.

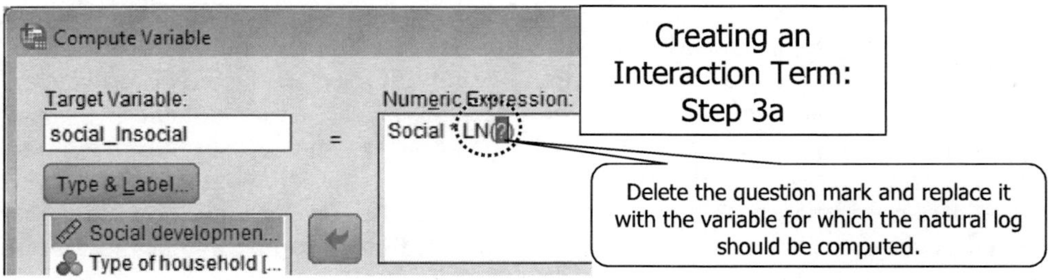

FIGURE 9.20
Creating an interaction term: Step 3a.

Here we want to compute the natural log for the continuous covariate, social development. To move this variable into the parentheses, use the backspace or delete key to remove the question mark. Then, click on the continuous covariate, social development, and move it into the parentheses next to LN in the Numeric Expression box by clicking on the arrow in the middle of the screen (see the screenshot for Step 3b, Figure 9.21). The numeric expression should then read: Social*LN(Social). Click "OK" to compute and create the new variable in the dataset.

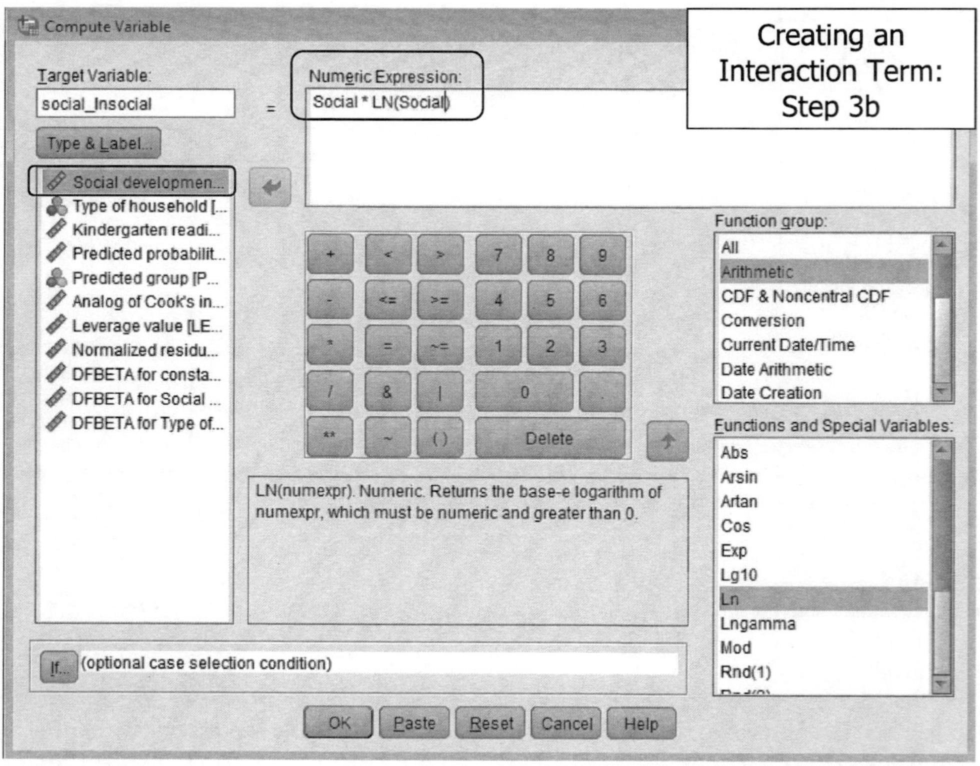

FIGURE 9.21
Creating an interaction term: Step 3b.

Step 4. The next step is to include the newly created variable (i.e., the interaction of the continuous variable with its natural log) into the logistic regression model, along with the other predictors. As those steps have been presented previously, they will not be reiterated here. The output indicates that the interaction term is not statistically significant (p = .300), which suggests we have met the assumption of linearity.

Variables in the Equation

		B	S.E.	Wald	df	Sig.	Exp(B)	95% C.I.for EXP(B)	
								Lower	Upper
Step 1[a]	Social development	12.953	11.897	1.185	1	.276	421981.259	.000	5646804337369759.000
	Type of household(1)	-8.208	5.264	2.432	1	.119	.000	.000	8.236
	social_lnsocial	-2.948	2.845	1.074	1	**.300**	.052	.000	13.845
	Constant	-76.228	64.345	1.403	1	.236	.000		

a. Variable(s) entered on step 1: Social development, Type of household, social_lnsocial.

Working in **R**, we create the natural log of the variable "social development" with the following script and save it to our dataframe, naming the new variable "logsocial."

```
Ch9_readiness$logsocial <-
log(Ch9_readiness$Social)*Ch9_readiness$Social
```

Next, we include the new variable, "logsocial," into the logistic equation with this command. We name the new object "ReadinessLogit2."

```
ReadinessLogit2 <- glm(formula = Readiness ~ Social + Household +logsocial,
                       family="binomial",
                       data =Ch9_readiness)
```

Finally, we review the output of the new model with the *summary* function.

```
summary(ReadinessLogit2)
```

FIGURE 9.22
Interaction output.

9.5.3 Independence

We plot the standardized residuals (which were requested and created through the "Save" option) against the values of X to examine the extent to which independence was met. The general steps for generating a simple scatterplot through "Scatter/dot" have been presented in Chapter 7 in the context of simple regression, and they will not be repeated here. We will create one graph for each independent variable in our model. For the first graph in this example, place the standardized residual (called "normalized residual" in SPSS) on the Y axis and the independent variable (in this case, "social development") on the X axis. For the second graph, repeat these steps, keeping the standardized residual (called "normalized residual") on the Y axis, and move the second independent variable ("household") on the X axis.

Interpreting independence evidence. If the assumption of independence is met, the points should fall randomly within a band of −2.0 to +2.0. Here we have pretty good evidence of independence, especially given the small sample size relative to logistic regression, as all but one point (case 19) is within an absolute value of 2.0.

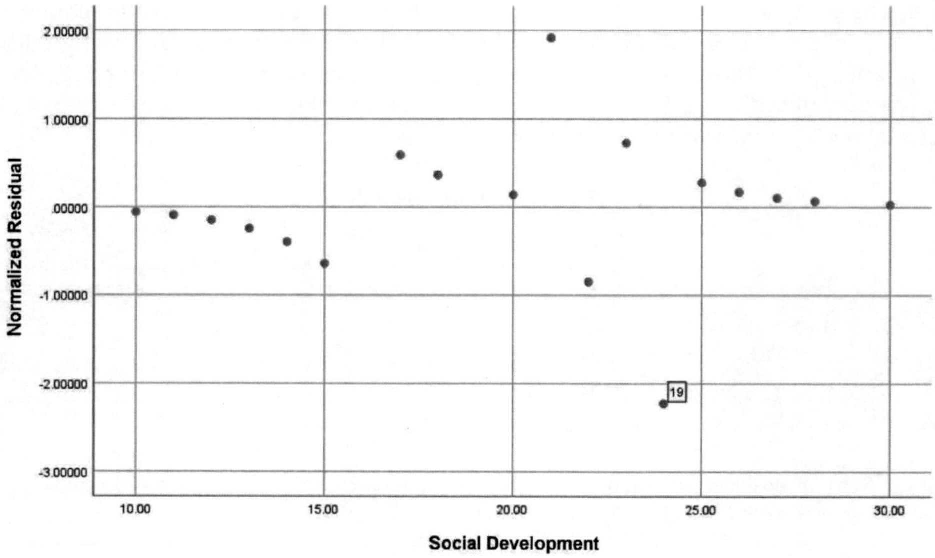

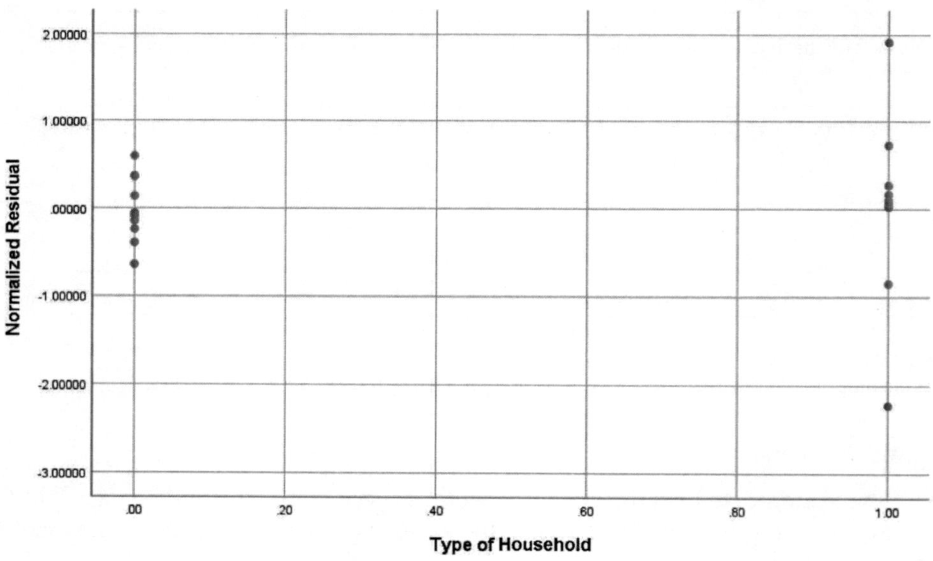

9.5.4 Absence of Outliers

Just as we saw in multiple regression, there are a number of diagnostics that can be used to examine the data for outliers.

9.5.4.1 Cook's Distance

Cook's distance provides an overall measure for the influence of individual cases. Values greater than one suggest that a case may be problematic in terms of undue influence on the model. Examining the residual statistics provided in the binary logistic regression output (see the table in Figure 9.23), we see that the maximum value for Cook's distance is 1.58, which indicates at least one influential point.

9.5.4.2 *Leverage Values*

These values range from 0 to 1, with values close to 1 indicating greater leverage. As a general rule, leverage values greater than $(m + 1)/n$ (where m equals the number of independent variables; here $(2+1)/20 = .15$ indicates an influential case. With a maximum of .307, there is evidence to suggest one or more cases are exerting leverage.

9.5.4.3 *DfBeta*

We saved the DfBeta values as another indication of the influence of a case. The DfBeta provides information on the change in the predicted value when the case is deleted from the model. For logistic regression, the DfBeta values should be smaller than one. Looking at the minimum and maximum DfBeta values for the intercept (labeled "constant") and for household, we have at least one case that is suggestive of undue influence.

Descriptive Statistics

	N	Minimum	Maximum
Analog of Cook's influence statistics	20	.00000	1.58721
Leverage value	20	.00691	.30726
Normalized residual	20	-2.22568	1.91780
DFBETA for constant	20	-1.68367	6.53464
DFBETA for Social development	20	-.41034	.09948
DFBETA for Type of household(1)	20	-1.36519	4.10130
Valid N (listwise)	20		

Working in **R**, we can display the minimum and maximum values (along with other statistics) of all the variables in the dataframe with the *summary* function defined four our dataframe, *Ch9_readiness*. If you have a large dataset and want to review only the variables of interest, they can be listed in parentheses, separated by commas, such as ("Ch9_readiness$cook, Ch9_readiness$leverage")

```
summary(Ch9_readiness)
```

FIGURE 9.23
DfBeta output.

From our logistic regression output, we can review the Casewise List to determine cases with studentized residuals larger than two standard deviations (recall from the Options dialog box that we told SPSS to identify residuals outside two standard deviations). Here there were two cases (cases 14 and 19) that were identified as outliers and the relevant statistics (e.g., observed group, predicted value, predicted group, residual, and standardized residual) are provided. We examine these cases to make sure there was not a data entry error. If the data are correct, then we determine whether to keep or filter out the case(s).

Casewise List[b]

Case	Selected Status[a]	Observed Kindergarten readiness	Predicted	Predicted Group	Temporary Variable Resid	ZResid	SResid
14	S	P**	.214	U	.786	1.918	2.102
19	S	U**	.832	P	-.832	-2.226	-2.106

a. S = Selected, U = Unselected cases, and ** = Misclassified cases.

b. Cases with studentized residuals greater than 2.000 are listed.

FIGURE 9.24
Casewise output.

Since we have a small dataset, we can easily review the values of our diagnostics and see which cases are problematic in terms of exerting undue influence and/or outliers. Those that are circled are values that fall outside of the recommended guidelines and thus are suggestive of outlying or influential cases. Due to the already small sample size, we will not filter out any of these potentially problematic cases. However, in this situation (i.e., with diagnostics that suggest one or more influential cases), you may want to consider filtering out those cases or, at a minimum, reviewing the data to be sure that there was not a data entry error for that case.

PRE_1	PGR_1	COO_1	LEV_1	ZRE_1	DFB0_1	DFB1_1	DFB2_1
.999392	1.00	.000004	.006911	.024661	-.013932	.000875	-.005351
.999392	1.00	.000004	.006911	.024661	-.013932	.000875	-.005351
.995807	1.00	.000117	.026963	.064891	-.070741	.004442	-.025813
.989041	1.00	.000581	.049804	.105262	-.152712	.009590	-.053211
.971670	1.00	.002750	.086198	.170750	-.312087	.019597	-.100367
.928748	1.00	.012153	.136749	.276980	-.568875	.035722	-.153621
.832036	1.00	1.205197	.195686	-2.225679	3.845816	-.241498	.501633
.653086	1.00	.183375	.256624	.728829	-.253484	.015918	.415970
.417058	.00	.317320	.307255	-.845835	-1.584165	.099477	-1.365193
.213769	.00	1.587210	.301455	1.917797	6.534642	-.410342	4.101298
.981041	1.00	.001773	.084029	.139017	-.242779	.016815	-.141077
.881982	1.00	.036653	.215020	.365801	-.808894	.062192	-.610891
.881982	1.00	.036653	.215020	.365801	-.808894	.062192	-.610891
.739590	1.00	.134832	.276901	.593380	-.794347	.077183	-.967663
.290873	.00	.162857	.284197	-.640457	-1.683669	.074919	-.026642
.134862	.00	.045786	.227035	-.394822	-1.276762	.066947	-.251561
.055928	.00	.010766	.153789	-.243396	-.693464	.038535	-.186256
.022018	.00	.002284	.092122	-.150046	-.331452	.018975	-.101723
.008483	.00	.000458	.050824	-.092499	-.150523	.008774	-.049789
.003241	.00	.000089	.026507	-.057023	-.066639	.003932	-.023128

FIGURE 9.25
Reviewing diagnostic values.

9.5.5 Assessing Classification Accuracy

In addition to examining Press's Q for classification accuracy, we can generate a kappa statistic. Kappa is the proportion of agreement above that expected by chance. A kappa statistic of 1.0 indicates perfect agreement whereas a kappa of 0 indicates chance agreement. Negative values can occur and indicate weaker than chance agreement. General rules of interpretation for kappa are: small, < .30; moderate, .30 to .50; and large, > .50.

Step 1. Kappa statistics are generated through the "Crosstab" procedure (go to "Analyze" in the top pulldown menu, then "Descriptive statistics," and then "Crosstabs"). Once the Crosstabs dialog box is open, select the dependent variable from the list on the left and use the arrow key to move it to "Row(s)." Select the predicted group (PGR_1) from the list on the left and use the arrow key to move it to "Column(s)" (see the screenshot for Step 1, Figure 9.26).

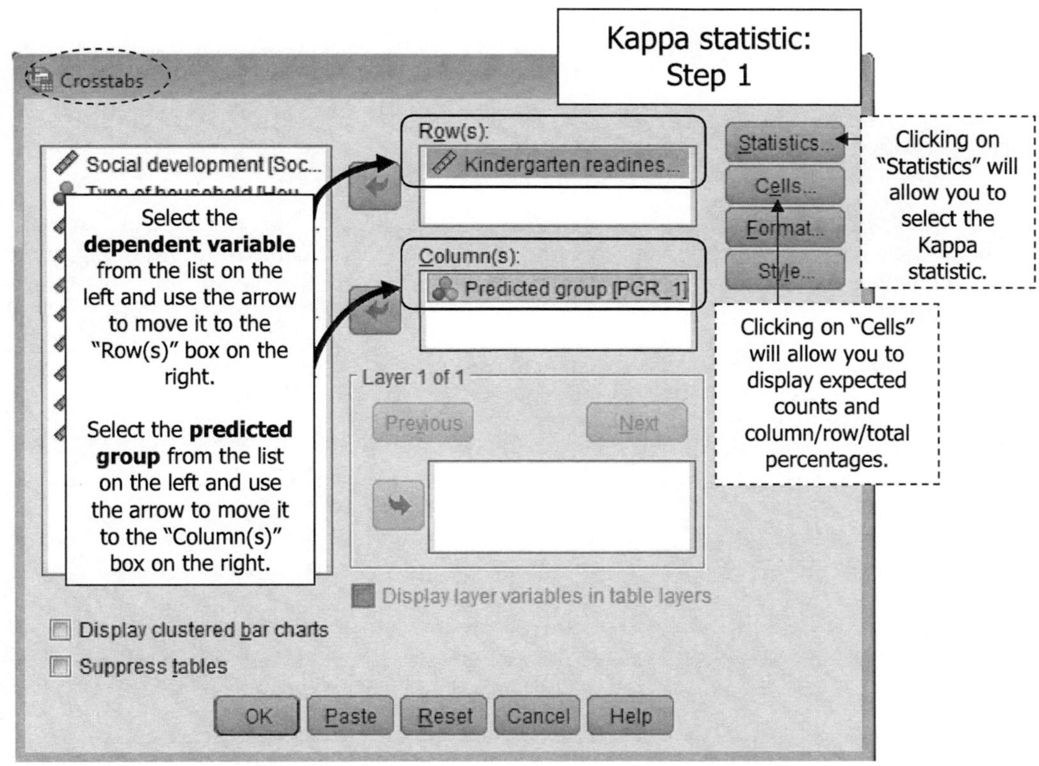

FIGURE 9.26
Kappa statistic: Step 1.

Step 2. Click on the Statistics option button. Place a checkmark in the box next to "Kappa" (see the screenshot for Step 2, Figure 9.27). Then click on "Continue" to return to the main dialog box.

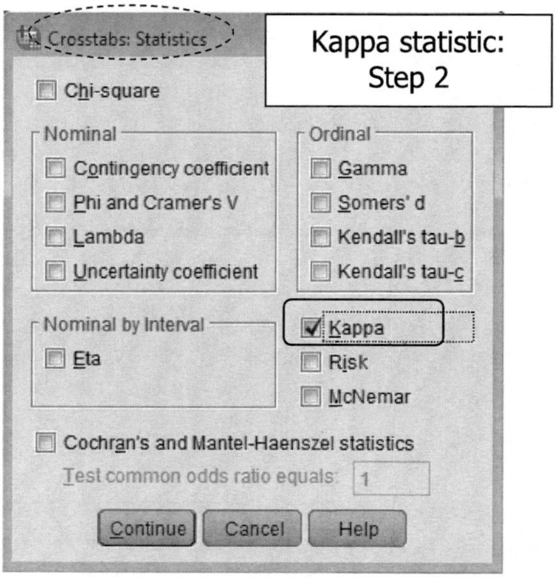

FIGURE 9.27
Kappa statistic: Step 2.

Step 3. Click on the "Cells" option button. In the Cell Display dialog box, place checkmarks in the boxes next to "Observed," "Expected," and "Row" (see the screenshot for Step 3, Figure 9.28). Then click on "Continue" to return to the main dialog box. Then click "OK" to generate the output.

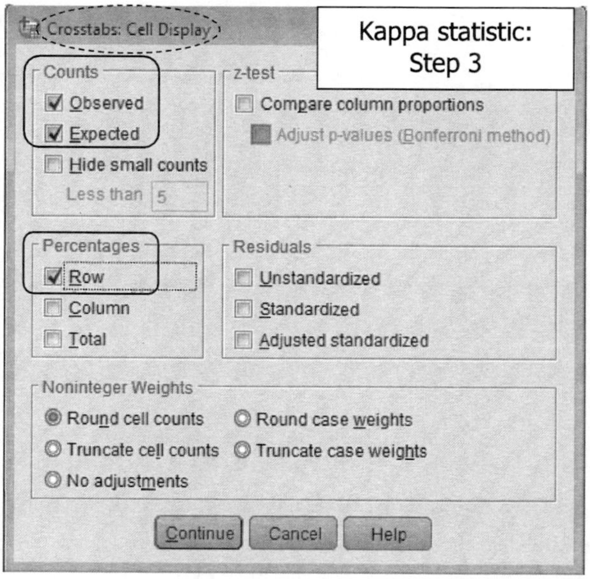

FIGURE 9.28
Kappa statistic: Step 3.

The crosstab table is interpreted as we have seen in the past. The columns represent the predicted group membership and the rows represent the observed group membership. This table should look familiar to the one that was provided to us with the logistic regression results. What is of most interest is the table labeled "Symmetric Measures," as this table contains the Kappa statistic. With a Kappa statistic of .792, and using our conventions for interpretation, this is considered to be a large value, which suggests strong agreement.

Kindergarten readiness * Predicted group Crosstabulation

			Predicted group		
			Unprepared	Prepared	Total
Kindergarten readiness	Unprepared	Count	7	1	8
		Expected Count	3.2	4.8	8.0
		% within Kindergarten readiness	87.5%	12.5%	100.0%
	Prepared	Count	1	11	12
		Expected Count	4.8	7.2	12.0
		% within Kindergarten readiness	8.3%	91.7%	100.0%
Total		Count	8	12	20
		Expected Count	8.0	12.0	20.0
		% within Kindergarten readiness	40.0%	60.0%	100.0%

Symmetric Measures

		Value	Asymptotic Standard Error[a]	Approximate T[b]	Approximate Significance
Measure of Agreement	**Kappa**	.792	.140	3.540	.000
N of Valid Cases		20			

a. Not assuming the null hypothesis.

b. Using the asymptotic standard error assuming the null hypothesis.

Working in **R**, we can use the *caret* package to generate a number of accuracy statistics, including Kappa.

```
install.packages("caret")
library(caret)
```

First, we need to install *caret* and load it into our library.

```
threshold <- 0.5
```

Next, we set our *threshold* level. For this illustration, we will use a threshold of .50.

FIGURE 9.29
Kappa output and ROC curve.

```
confusionMatrix(factor(Ch9_readiness$predicted.probabilities>threshold),
                factor(Ch9_readiness$Readiness==1),
                positive="TRUE")
```

The *confusionMatrix* function will generate the predicted group classification table (called a "confusion matrix") as well as a number of statistics.

```
Confusion Matrix and Statistics

          Reference
Prediction FALSE TRUE
     FALSE     7    1
     TRUE      1   11

               Accuracy : 0.9
                 95% CI : (0.683, 0.9877)
    No Information Rate : 0.6
    P-Value [Acc > NIR] : 0.003611

                  Kappa : 0.7917
 Mcnemar's Test P-Value : 1.000000

            Sensitivity : 0.9167
            Specificity : 0.8750
         Pos Pred Value : 0.9167
         Neg Pred Value : 0.8750
             Prevalence : 0.6000
         Detection Rate : 0.5500
   Detection Prevalence : 0.6000
      Balanced Accuracy : 0.8958

       'Positive' Class : TRUE
```

The model has accuracy of predicting of about 90% ("accuracy: 0.9"). Using the four quadrants of our classification table and labeling the cells as A (upper left), B (upper right), C (bottom left), and D (bottom right) (you may remember this from working with contingency tables in an earlier chapter), specificity and sensitivity can be calculated.

The true negative rate is **specificity** and is calculated as $A/(A+B)$. Denoted in the R output as .8750.

The true positive rate is **sensitivity** and is calculated as $D/(C+D)$. Denoted in the R output as .9167.

```
install.packages("ROCR")
library(ROCR)
```

To generate the ROC curve, we will install and use the *ROCR* package. The *install.packages* and *library* commands will install and call *ROCR* into our library, respectively.

```
predReadiness <- prediction(predict(ReadinessLogit),
                    Ch9_readiness$Readiness)
perfReadiness <- performance(predReadiness, "tpr","fpr")
```

We will create an object called "predReadiness" using our logistic model (i.e., *ReadinessLogit*). The performance measures that we request include the true positive rate (*tpr*) and false positive rate (*fpr*).

```
plot(perfReadiness)
```

Our ROC curve is displayed using the *plot* function.

FIGURE 9.29 (continued)
Kappa output and ROC curve.

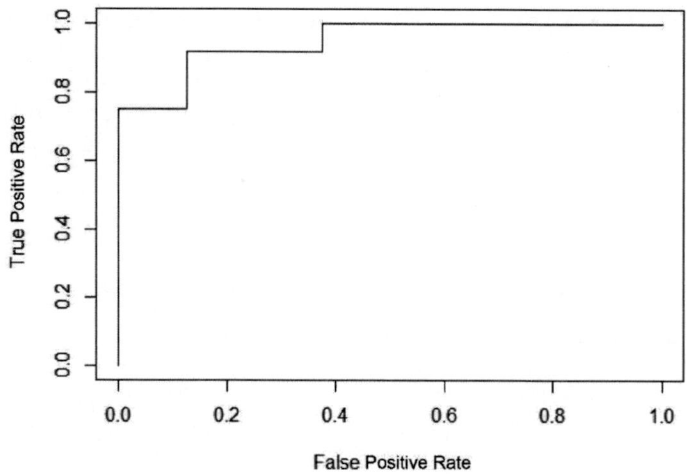


```
performance(predReadiness, 'auc')
```

To find the area under the curve (AUC), we use the *performance* function, inserting our predicted object (predReadiness) and requesting the AUC (*auc*). The output is a scalar, .9479167. AUC ranges from 0 to 1, with 1 indicating 100% specificity and 100% sensitivity. In this example, the AUC is about .95, indicating very good specificity and sensitivity.

```
An object of class "performance"
Slot "x.name":
[1] "None"
```

Slot "y.name":
[1] "Area under the ROC curve"

```
Slot "alpha.name":
[1] "none"

Slot "x.values":
list()
```

Slot "y.values":
[[1]]
[1] 0.9479167

```
Slot "alpha.values":
list()
```

FIGURE 9.29 (continued)
Kappa output and ROC curve.

9.5.5.1 ROC Curves and AUC

Another way to determine classification accuracy is using the **Receiver Operator Characteristic (ROC) curve,** developed during World War II for analyzing radar images and discovered as a useful tool for evaluating medical results in the 1970s. ROC curves plot the true positive rate (sensitivity) to the false positive rate (1-specificity). Each point on the ROC curve is a sensitivity/specificity pair that corresponds to a specific decision threshold. The **area under the curve (AUC)** is a measure of *accuracy*. In other words, how well a parameter can distinguish between the two categories of your outcome. AUC ranges from 0 to 1, with values of .90 to 1.0 indicating excellent accuracy; .80–.89, good; .70–.79, fair; .60–.69, poor; and .59 and less, failing. There are criticisms in using the AUC (Hand, 2009), and thus should you report it, we encourage it to be just one tool to supplement your analysis. **R** commands for generating a ROC curve and the AUC are provided in Figure 9.29.

9.6 Power Using G*Power

A priori and post hoc power can again be determined using the specialized software described previously in this text (e.g., G*Power), or you can consult *a priori* power tables (e.g.,Cohen, 1988). As an illustration, we use G*Power to first compute post hoc power of our example.

9.6.1 Post Hoc Power

The first thing that must be done when using G*Power for computing post hoc power is to select the correct test family. For logistic regression we select "Tests" in the top pulldown menu, then "Correlation and regression," and finally "Logistic regression." Once that selection is made, the "Test family" automatically changes to "z tests."

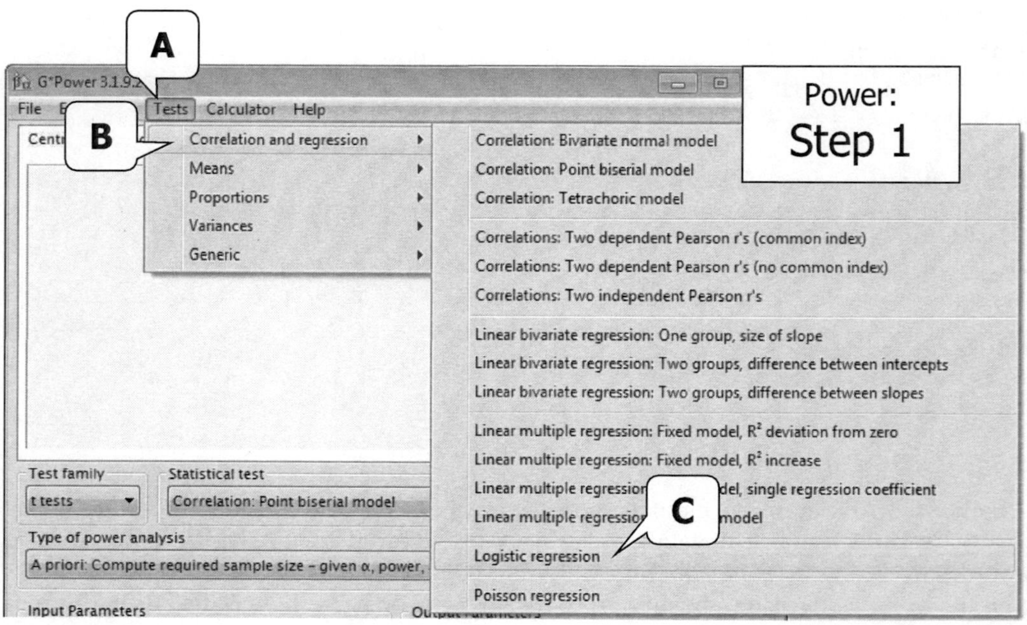

FIGURE 9.30
Post hoc power.

The "Type of power analysis" desired then needs to be selected. To compute post hoc power, select "Post hoc: Compute achieved power—given α, sample size, and effect size." For this illustration, we will compute power for the continuous covariate.

The "Input Parameters" must then be specified. In our example we conducted a two-tailed test. The odds ratio for our continuous variable social development was 2.631. The probability that $Y = 1$ given that $X = 1$ under the null hypothesis is set to .50. The alpha level we used was .05 and the total sample size was 20. "R^2 other X" refers to the squared correlation between social development and our other covariate. In this case, the simple bivariate correlation between these variables is .867 and the squared correlation is .752. Social development is a continuous variable, thus it follows a normal distribution. The last two parameters to be specified are for the mean and standard deviation of our covariate. In this case, the mean of social development was 20.20 and the standard deviation was 6.39. Once the parameters are specified, click on "Calculate" to find the power statistics.

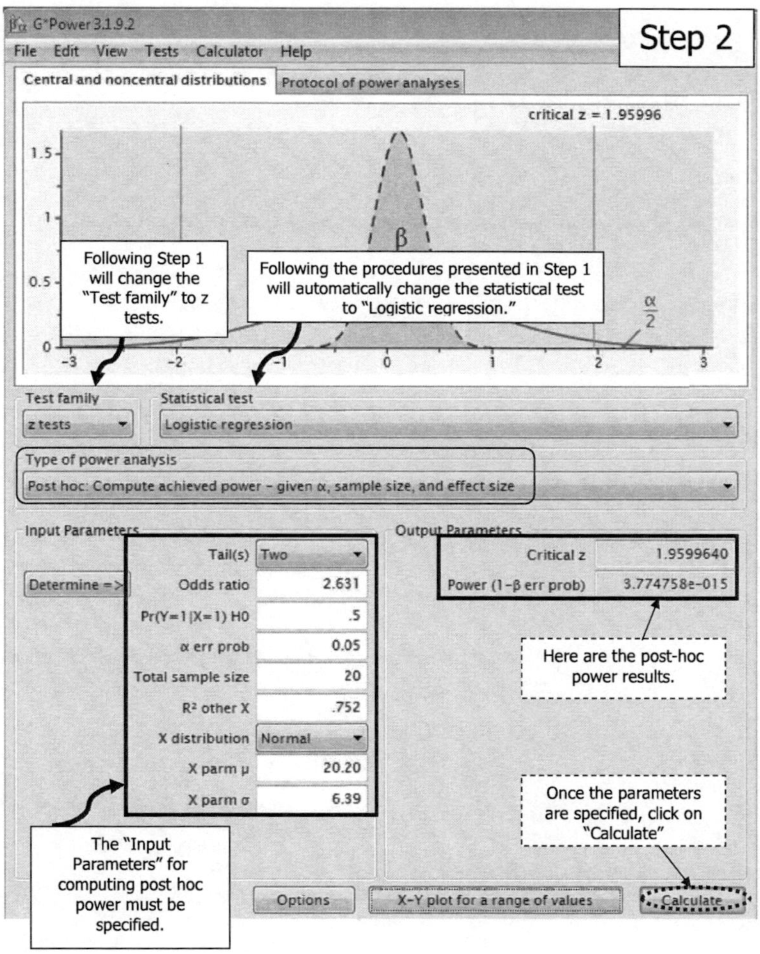

FIGURE 9.31
Post hoc power: Step 2.

The "Output Parameters" provide the relevant statistics for the input just specified. In this example, we were interested in determining post hoc power for a logistic regression model. Based on the criteria specified, the post hoc power was substantially less than 1. In other words, the probability of rejecting the null hypothesis when it is really false was significantly less than 1% (sufficient power is often .80 or above). This finding is not surprising given the very small sample size. Keep in mind that conducting power analysis *a priori* is recommended so that you avoid a situation where, post hoc, you find that the sample size was not sufficient to reach the desired level of power (given the observed parameters).

9.6.2 *A Priori* Power

For *a priori* power, we can determine the total sample size needed for logistic regression given the same parameters just discussed. In this example, had we wanted an *a priori* power of .80 given the same parameters just defined, we would need a total sample size of 7094.

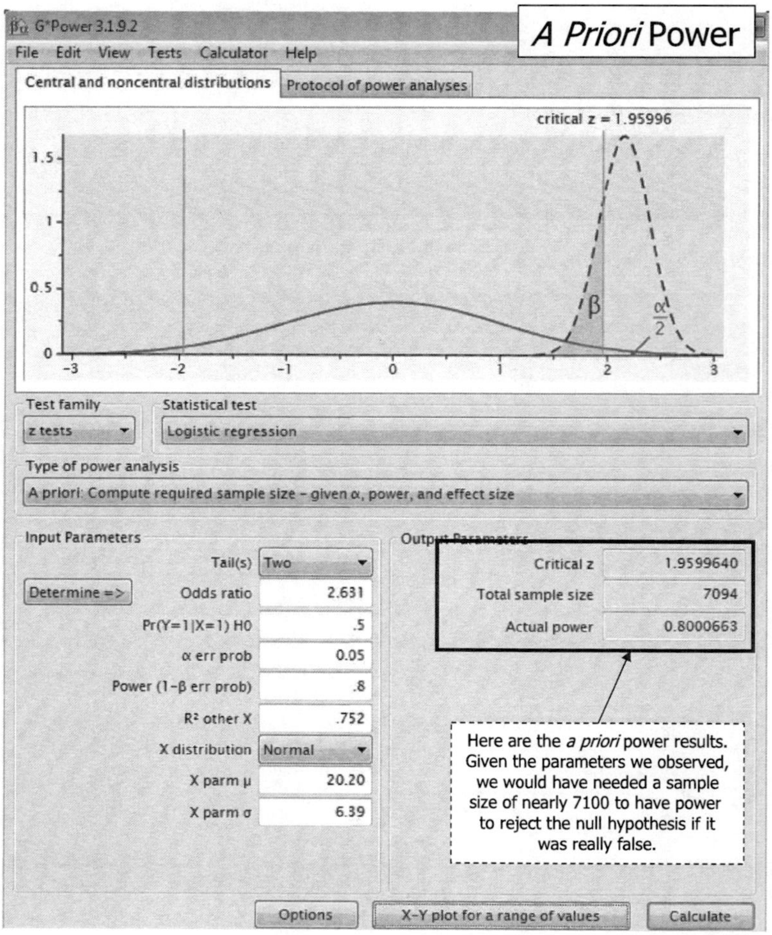

FIGURE 9.32
A priori power.

9.7 Research Question Template and Example Write-Up

Finally, here is an example paragraph for the results of the logistic regression analysis. Recall that our graduate research assistant, Oso Wyse, was assisting Dr. Malani, a faculty member in the early childhood department. Dr. Malani wanted to know if kindergarten readiness (prepared vs. unprepared) could be predicted by social development (a continuous variable) and type of household (single- vs. two-parent home). The research question presented to Dr. Malani from Oso included the following: *Can kindergarten readiness be predicted from social development and type of household?*

Oso then assisted Dr. Malani in generating a logistic regression as the test of inference, and a template for writing the research question for this design is presented as follows:

Can [dependent variable] be predicted from [list independent variables]?

It may be helpful to preface the results of the logistic regression with information on an examination of the extent to which the assumptions were met. The assumptions include: (a) independence; (b) linearity; and (c) noncollinearity. We will also examine the data for outliers and influential points.

Logistic regression was conducted to determine whether social development and type of household (single parent vs. two-parent home) could predict kindergarten readiness.

The assumptions of logistic regression were tested. Specifically, these include: (a) noncollinearity; (b) linearity; and (c) independence of errors.

In terms of **noncollinearity**, a VIF value of 4.037 (below the value of 10.0 which indicates the point of concern) and tolerance of .248 (above the value of .10 which suggests multicollinearity) provided evidence of noncollinearity. However, there was some concern for multicollinearity. In examining the collinearity diagnostics, a condition index value of 14.259 was observed, which falls within the range of concern (specifically 10–30). Review of the variance proportions suggested that 100% of the variance of the regression coefficient for social development and 73% for type of household were related to the smallest eigenvalue. This also suggests concern for multicollinearity. Thu while we met the assumption of noncollinearity with the tolerance and VIF values, but there is some concern for multicollinearity with the condition index and variance proportion values.

Linearity was assessed by re-estimating the model and including, along with the original predictors, an interaction term which was the product of the continuous independent variable (i.e., social development) and its natural logarithm. The interaction term was not statistically significant, thus providing evidence of linearity (social*ln(social), $B = -2.948$, $SE = 2.845$, Wald = 1.074, $df = 1$, $p = .300$).

Independence was assessed by examining a plot of the standardized residuals against values of each independent variable. With the exception of one case which was slightly outside the band, all cases were within an absolute value of 2.0 thus indicating the assumption of independence has been met.

In reviewing for **outliers and influential points**, Cook's distance values were generally within the recommended range of less than 1.0, although the maximum value was 1.587. Leverage values ranged from .007 to .307, well under the recommended .50, suggesting outliers were not problematic. DfBeta values beyond one also suggested cases that may be exerting influence on the model. Based on the evidence reviewed, there are some cases that are suggestive of outlying and influential points. Due to the small sample size however, these cases were retained. Readers are urged to interpret the results with caution given the possible influence of outliers.

Here is an example paragraph of results for the logistic regression (remember that this will be prefaced by the previous paragraph reporting the extent to which the assumptions of the test were met).

Logistic regression analysis was then conducted to determine whether kindergarten readiness (prepared vs. unprepared) could be predicted from social development and type of household (single versus two-parent home). Good model fit was evidenced by nonstatistically significant results on the Hosmer and Lemeshow test, χ^2 (n = 20) = 4.691, df = 7, p =.698, and large effect size indices when interpreted using Cohen (1988) (Cox and Snell R^2 = .546; Nagelkerke R^2 = .738). These results suggest that the predictors, as a set, reliably distinguished between children who are ready for kindergarten (i.e., prepared) versus unprepared. Of the two predictors in the model, only social development was a statistically significant predictor of kindergarten readiness (Wald = 4.696, df = 1, p = .030). The odds ratio for social development suggests that for every one-point increase in social development, the odds are about 2 and 2/3 greater for being prepared for kindergarten as compared to unprepared. Type of household was not statistically significant, which suggests that the odds for being prepared for kindergarten (relative to unprepared) are similar regardless of being raised in a single-parent versus a two-parent household. The table below presents the results for the model including the regression coefficients, Wald statistics, odds ratios, and 95% confidence intervals for the odds ratios. This is followed by a table which presents the group means and standard deviations of each predictor for both children who are prepared and unprepared for kindergarten.

Logistic Regression Results

						95% CI for Exp(*B*)	
	B	***SE***	**Wald**	***p***	**Exp(*B*)**	**Lower**	**Upper**
Intercept (constant)	−15.404	7.195	4.584	.032	NA		
Social development	.967	.446	4.696	.030	2.631	1.097	6.313
Type of household (two-parent home)	−6.216	3.440	3.265	.071	.002	.000	1.693

Group Means (and Standard Deviations) of Predictors

Predictor	Prepared for Kindergarten	Unprepared for Kindergarten
Social development	23.58 (4.74)	15.13 (5.14)
Type of household (two-parent home)	.67 (.49)	.25 (.46)

Overall, the logistic regression model accurately predicted 90% of the children in our sample, with children who are prepared for kindergarten slightly more likely to be classified correctly (91.7% of children prepared for kindergarten and 87.5% of children unprepared correctly classified). To account for chance agreement in classification, the Kappa coefficient was computed and found to be .792, a large value. Additionally, Press's Q was calculated to be 12.8, providing evidence that the predictions based on the logistic regression model are statistically significantly better than chance. The area under the ROC curve was approximately .95, indicating very good specificity and sensitivity. Post hoc power, calculated using G*Power, was less than .01 indicating very weak power.

9.8 Additional Resources

This chapter has provided a preview into conducting logistic regression analysis. However, there are a number of areas that space limitations prevent us from delving into. For those of you who are interested in learning more, or if you find yourself in a sticky situation in your analyses, you may wish to look into the following, among many other excellent resources:

- In-depth coverage of logistic regression (Hilbe, 2016; Osborne, 2015).
- Comprehensive overview of ROC curves, including going beyond the basics with a discussion on Bayesian methods (Krzanowski & Hand, 2009).
- Application of logistic regression with randomized trials and covariate adjustment (Jiang et al., 2017).
- Rare events and imbalanced data (Maalouf, Homouz, & Trafalis, 2018).

Problems

Conceptual Problems

1. Which one of the following represents the primary difference between OLS regression and logistic regression?
   a. Computer processing time to estimate the model
   b. The measurement scales of the independent variables that can be included in the model
   c. The measurement scale of the dependent variable
   d. The statistical software that must be used to estimate the model
2. Which one of the following is NOT an assumption of logistic regression?
   a. Independence
   b. Homogeneity of variance
   c. Linearity
   d. Noncollinearity

3. Which one of the following is NOT an appropriate dependent variable for binary logistic regression?
   a. Bernoulli
   b. Dichotomous
   c. Multinomial
   d. One variable with two categories
4. Which of the following would NOT be appropriate outcomes to examine with binary logistic regression?
   a. Employment status (employed; unemployed not looking for work; unemployed looking for work)
   b. Enlisted member of the military (member vs. non-member)
   c. Marital status (married vs. not married)
   d. Recreational athlete (athlete vs. nonathlete)
5. Which of the following represents what is being predicted in binary logistic regression?
   a. Mean difference between two groups
   b. Odds that the unit of analysis belongs to one of two groups
   c. Precise numerical value
   d. Relationship between one group compared to the other group
6. True or false? While probability, odds, and log odds may be computationally different, they all relay the same basic information.
7. A researcher is studying diet soda drinking habits and has coded "diet soda drinker" as "1" and "non diet soda drinker" as "0." Which of the following is a correct interpretation given a probability value of .52?
   a. The odds of being a diet soda drinker are about equal to those of not being a diet soda drinker.
   b. The odds of being a diet soda drinker are substantially greater than not being a diet soda drinker.
   c. The odds of being a diet soda drinker are substantially less than not being a diet soda drinker.
   d. Cannot be determined from the information provided.
8. A researcher has computed the odds ratio to study the relative odds of participating in family counseling, and has coded "participation" as "1" and "nonparticipation" as "0," based on family stability (a continuous variable). Which of the following is a correct interpretation given an odds ratio of .25?
   a. Families that are more stable participate in family counseling.
   b. The odds of being a stable family are about the same as compared to families that are not stable.
   c. For every one-unit increase in family stability, the odds of participating in family counseling decrease by 75%.
   d. For families that participate in counseling, the odds of family stability are 25% more likely.

9. Which of the following is a correct interpretation of the logit?
   a. The log odds become larger as the odds increase from 1 to 100.
   b. The log odds become smaller as the odds increase from 1 to 100.
   c. The log odds stay relatively stable as the odds decrease from 1 to 0.
   d. The change in log odds becomes larger when the independent variables are categorical rather than continuous.
10. Which of the following correctly contrasts the estimation of OLS regression as compared to logistic regression?
    a. The sum of the squared distance of the observed data to the regression line is minimized in logistic regression. The log likelihood function is maximized in OLS regression.
    b. The sum of the squared distance of the observed data to the regression line is maximized in logistic regression. The log likelihood function is minimized in OLS regression.
    c. The sum of the squared distance of the observed data to the regression line is maximized in OLS regression. The log likelihood function is minimized in logistic regression.
    d. The sum of the squared distance of the observed data to the regression line is minimized in OLS regression. The log likelihood function is maximized in logistic regression.
11. Which of the following is NOT a test that can be used to evaluate overall model fit for logistic regression models?
    a. Change in log likelihood
    b. Hosmer-Lemeshow goodness of fit
    c. Cox and Snell *R* squared
    d. Wald test
12. A researcher is studying diet soda drinking habits and has coded "diet soda drinker" as "1" and "non diet soda drinker" as "0." She has predicted drinking habits based on the individual's weight (measured in pounds). Given this scenario, which of the following is a correct interpretation of an odds ratio of 1.75?
    a. For every one-unit increase in being a diet soda drinker, the odds of putting on an additional pound increase by 75%.
    b. For every one-unit increase in being a diet soda drinker, the odds of putting on an additional pound decrease by 75%.
    c. For every one-pound increase in weight, the odds of attending being a diet soda drinker decrease by 75%.
    d. For every one-pound increase in weight, the odds of attending being a diet soda drinker increase by 75%.
13. A researcher is studying pet ownership and has coded "pet owner" as "1" and "non pet owner" as "0." He has predicted owning a pet based on the individual's household income. Given this scenario, which of the following is a correct interpretation of an odds ratio of 1.90?
    a. For pet owners, the odds of having higher household income increase by 90%.
    b. The odds of being a pet owner, as compared to not being a pet owner, are about the same.

c. For every one-unit increase in household income, the odds of being a pet owner decrease by 90%.

d. For every one-unit increase in household income, the odds of being a pet owner increase by 90%.

Answers to Conceptual Problems

1. **c** (The measurement scale of the dependent variable is the main difference between multiple regression and logistic regression.)
3. **c** (Multinomial)
5. **b** (Odds that the unit of analysis belongs to one of two groups)
7. **a** (The odds of being a diet soda drinker are about equal to those of not being a diet soda drinker, with .50 being exactly equal)
9. **a** (The log odds become larger as the odds increase from 1 to 100)
11. **d** (Wald test (assesses significance of individual predictors))
13. **d** (For every one-unit increase in household income, the odds of pet owner increase by 90%)

Computational Problems

1. You are given the following data, where X_1 (high school cumulative grade point average) and X_2 (participation in school-sponsored athletics; 0 = nonathlete and 1 = athlete; use 0 as the reference category) are used to predict Y (college enrollment immediately after high school, "1," versus delayed college enrollment or no enrollment, "0").

X_1	X_2	Y
4.15	1	1
2.72	0	1
3.16	0	0
3.89	1	1
4.02	1	1
1.89	0	0
2.10	0	1
2.36	1	1
3.55	0	0
1.70	0	0

Determine the following values based on simultaneous entry of independent variables: intercept; $-2LL$; constant; b_1, b_2, $se\ (b_1)$; $se\ (b_2)$; odds ratios; $Wald_1$; $Wald_2$.

2. You are given the following data, where X_1 (participation in high school honors classes; yes = 1, no = 0; use 0 as the reference category) and X_1 (participation in co-op program in college; yes = 1; no = 0; use 0 as the reference category) are used to predict Y (baccalaureate graduation with honors = 1 versus graduation without honors = 0).

X_1	X_2	Y
0	1	1
0	0	1
1	0	0
1	1	1
1	1	1
0	0	0
1	0	1
0	1	1
1	0	0
0	0	0

Determine the following values based on simultaneous entry of independent variables: intercept; $-2LL$; constant; b_1, b_2, $se\ (b_1)$; $se\ (b_2)$; odds ratios; $Wald_1$; $Wald_2$.

3. You are given the following data, where X_1 (high frequency social media user; yes = 1, no = 0; use 0 as the reference category) and X_2 (regularly consume coffee; yes = 1; no = 0; use 0 as the reference category) are used to predict Y (regularly exercise = 1 versus do not regularly exercise = 0).

X_1	X_2	Y
0	1	1
0	1	1
0	1	1
0	0	1
0	0	1
0	1	0
0	1	0
0	1	1
0	1	0
0	0	0
1	0	0
1	0	0
1	1	0
1	1	1
1	1	1
1	1	0
1	1	0
1	1	0

Determine the following values based on simultaneous entry of independent variables: intercept; $-2LL$; constant; b_1, b_2, $se\ (b_1)$; $se\ (b_2)$; odds ratios; $Wald_1$; p.

Answers to Computational Problems

1. $-2LL = 7.558$; $b_{HSGPA} = -.366$; $b_{athlete} = 22.327$; $b_{constant} = .219$; $se(b_{HSGPA}) = 1.309$; $se\ (b_{athlete}) = 20006.861$; odds $\text{ratio}_{HSGPA} = .693$; odds $\text{ratio}_{athlete} < .001$; $Wald_{HSGPA} = .078$; $Wald_{athlete} = .000$
3. $-2LL = 22.342$; $b_{SocMed} = -1.533$; $b_{coffee} = .387$; $b_{constant} = .138$; $se\ b_{SocMed} = 1.050$; $se(b_{coffee}) = 1.145$; odds $\text{ratio}_{SocMed} = .216$; odds $\text{ratio}_{coffee} = 1.472$; $Wald_{SocMed} = 2.132$; $Wald_{coffee} = .114$; $p_{SocMed} = .216$; $p_{coffee} = .736$; $p_{constant} = .893$

Interpretive Problems

1. Use SPSS or **R** to develop a logistic regression model with data available on the website from the Division I-A Football Bowl Subdivision (FBS) obtained from ESPN during January 2016 ($n = 128$; FBS_2015.sav or FBS_2015.csv) (http://espn.go.com/college-football/statistics/team/_/stat/total/sort/totalYards). Utilize "top quartile in overall efficiency" as the dependent (binary) variable to find at least two strong predictors from among the continuous variables in the dataset. Write up the results in APA style, including testing for the assumptions. Determine and interpret a measure of effect size.
2. Use SPSS or **R** to develop a logistic regression model with data available on the textbook's website from the 2017 IPEDS (https://nces.ed.gov/ipeds/). Select one binary variable as the dependent variable [e.g., "institution provides on-campus housing" (ROOM)] and find at least two strong predictors from among the remaining variables in the dataset. Write up the results in APA style, including testing for the assumptions. Determine and interpret a measure of effect size.
3. Use SPSS or **R** to develop a logistic regression model with data available on the textbook's website from the 2017 NHIS* family file (https://www.cdc.gov/nchs/nhis/). Select one binary variable as the dependent variable [e.g., "any family member need help with an activity of daily living (ADL)" (FLAADLYN')] and find at least two strong predictors from among the remaining variables in the dataset. Write up the results in APA style, including testing for the assumptions. Determine and interpret a measure of effect size.

* It is important to note that the NHIS is a *complex sample* (i.e., not a simple random sample). Per NHIS (see www.cdc.gov/nchs/nhis/about_nhis.htm#sample_design), "The sampling plan follows a multistage area probability design that permits the representative sampling of households and noninstitutional group quarters (e.g., college dormitories) . . . The current sampling plan was implemented in 2016 . . . [It] is a sample of clusters of addresses that are located in primary sampling units (PSU's). A PSU consists of a county, a small group of contiguous counties, or a metropolitan statistical area." In the NHIS dataset, you will find, for example, a "weight" variable, which is used to adjust for the complex survey design. We won't get into the technical aspects of this, but when the data are analyzed to adjust for the sampling design (including non-simple random sampling procedure and disproportionate sampling) the end results are then representative of the intended population. The purpose of the text is not to serve as a primer for understanding complex samples, and thus readers interested in learning more about complex survey designs are referred to any number of excellent resources (Hahs-Vaughn, 2005; Hahs-Vaughn, McWayne,

Bulotskey-Shearer, Wen, & Faria, 2011a, 2011b; Lee, Forthofer, & Lorimor, 1989; Skinner, Holt, & Smith, 1989). Additionally, so as to not complicate matters any more than necessary, the applications in the textbook do not illustrate how to adjust for the complex sample design. *As such, if you do not adjust for the complex sampling design, the results that you see should not be interpreted to represent any larger population but only that select sample of individuals who actually completed the survey.* I want to stress that the reason why the sampling design has not been illustrated in the textbook applications is because the point of this section of the textbook is to illustrate how to use statistical software to generate various procedures and how to interpret the output and not to ensure the results are representative of the intended population. *Please do not let this discount or diminish the need to apply this critical step in your own analyses when using complex survey data as quite a large body of research exists that describes the importance of effectively analyzing complex samples as well as provides evidence of biased results when the complex sample design is not addressed in the analyses* (Hahs-Vaughn, 2005, 2006a, 2006b; Hahs-Vaughn et al., 2011a, 2011b; Kish & Frankel, 1973, 1974; Korn & Graubard, 1995; Lee et al., 1989; Lumley, 2004; Pfeffermann, 1993; Skinner et al., 1989).

10

Mediation and Moderation

Chapter Outline

Key Concepts

1. Mediation
2. Moderation
3. Direct effect
4. Indirect effect

In the previous three chapters, we have considered various regression models, specifically looking at using one or more independent variables to predict an outcome. In this chapter we build on our knowledge of regression to examine other ways in which variables can relate in a regression model.

When considering the relationship between two variables (say X and Y), the researcher usually determines some measure of relationship between those variables, such as a correlation coefficient (e.g., r_{XY}, the Pearson product-moment correlation coefficient), as we did in Chapter 10 of the previous volume. Another way of looking at how two variables may be related is through regression analysis, in terms of prediction or explanation. That is, we evaluate the ability of one variable to predict or explain a second variable. With Chapters 8 and 9, we considered the case of multiple predictor variables through multiple linear regression analysis and logistic regression.

In this chapter we consider *differential effects* of predictors. In other words, a predictor may be more or less effective on the outcome in a given situation, where that "given situation" is that the predictor is being mediated or moderated in its relationship to the dependent variable. Our objectives are that by the end of this chapter, you will be able to (a) understand the concepts underlying mediation and moderation, (b) determine and interpret the results of a mediated and moderated model, and (c) understand and evaluate the assumptions of and conditions under which mediation and moderation can be examined.

10.1 What Mediation and Moderation Is and How It Works

Let us consider the basic concepts involved in a simple mediation model. The underlying framework for mediation is to examine the *way* in which the independent variable related to the dependent variable. For example, there may be a direct effect of the independent variable on the dependent variable, but there may also be an indirect effect where the independent variable passes influence through another variable, the mediator, and the mediator then to the dependent variable. We will focus on a simple mediation model, but this can be extended to complex models with multiple mediators.

10.1.1 Characteristics

Before we begin our discussion of mediation, we will have a quick and concise refresher on the simple and multiple regression models. As we learned in simple linear regression, the **population regression model** for the regression of Y, *the criterion*, given X, *the predictor*, often stated as the **regression of Y on X**, although more easily understood as **Y being predicted by X**, is:

$$Y_i = \beta_{YX} X_i + \alpha_{YX} + \varepsilon_i$$

where Y_i is the criterion variable, X_i is the predictor variable, β_{YX} is the population slope for Y_i predicted by X_i, α_{YX} is the population intercept for Y_i predicted by X_i, ε_i are the population residuals or errors of prediction (the part of Y_i not predicted from X_i), and i represents an index for a particular case (an individual or object; in other words, the unit of analysis that has been measured). The index i can take on values from 1 to N, where N is the size of the population, written as $i = 1, \ldots, N$.

The **population prediction model** is

$$Y_i' = \beta_{YX} X_i + \alpha_{YX}$$

where Y_i' is the predicted value of Y for a specific value of X. That is, Y_i is the *actual or observed score* obtained by individual i, while Y_i' is the *predicted score* based on their X score for that same individual (in other words, you are using the value of X to predict what Y will be). Thus, we see that the population prediction error is defined as follows:

$$\varepsilon_i = Y_i - Y_i'$$

There is only one difference between the regression and prediction models. The regression model explicitly includes prediction error as ε_i, whereas the prediction model includes prediction error *implicitly* as part of the predicted score Y_i' (i.e., there is some error in the predicted values).

The **sample multiple linear regression model** for predicting Y_i from m predictors $X_{1,2,\ldots,m}$ is

$$Y_i = b_1 X_{1i} + b_2 X_{2i} + \ldots + b_m X_{mi} + a + e_i$$

where Y_i is the **criterion variable** (also known as the dependent variable); the X_k's are the **predictor (or independent) variables** where $k = 1, \ldots, m$; $\boldsymbol{b_k}$ is the **sample partial slope** of the regression line for Y as predicted by X_k, $\boldsymbol{a}$ is the **sample intercept** of the regression line for Y_i as predicted by the set of X_k's $\boldsymbol{e_i}$ are the **residuals or errors of prediction** (the part of Y_i not predictable from the X_k's); and i represents an index for an individual or object. The index i can take on values from 1 to n where n is the size of the sample (i.e., $i = 1, \ldots, n$). The term **partial slope** is used because it represents the slope of Y for a particular X_k in which we have partialed out the influence of the other X_k's, much as we did with the partial correlation.

The **sample prediction model** is

$$Y_i' = b_1 X_{1i} + b_2 X_{2i} + \ldots + b_m X_{mi} + a$$

Where Y_i' is the predicted value of the outcome for specific values of the X_k's, and the other terms are as before. There is only one difference between the regression and prediction models. The regression model explicitly includes prediction error as e_i whereas the prediction model includes prediction error implicitly as part of the predicted score Y_i' (i.e., there is some error in the predicted values). The goal of the prediction model is to include an independent variable X that minimizes the residual; this means that the independent variable does a nice job of predicting the outcome. We can compute residuals, the e_i, for each of the i individuals or objects by comparing the actual Y values (i.e., Y_i) with the predicted Y values (i.e., Y_i') as

$$e_i = Y_i - Y_i'$$

for all $i = 1, \ldots, n$ individuals or objects in the sample.

Now let's consider the case of multiple predictors, but in the context of **mediation**. Let us first visualize what is happening in mediation. For ease we will drop the subscripts and superscripts with the exception of the direct effect, c'. In Figure 10.1, we have one independent variable, X, one mediator variable, M, and one dependent variable, Y. The arrows show us that the independent variable can be related to the dependent variable by two different paths. One path of influence from X to Y is *direct;* i.e., the arrow which goes directly from X to Y. In this path, X is the antecedent which influences Y, the consequent. The **direct effect** is notated as c' and can be interpreted as follows: two cases that differ by one unit on the independent variable, X, but are equal on M will differ by c' units on the dependent variable, Y. We see this in the following equation:

$$c' = \left[Y' \mid \left(X = x,\ M = m\right)\right] - \left[Y' \mid \left(X = x - 1,\ M = m\right)\right]$$

Where Y' is the estimated outcome which is conditioned on (i.e., $\mid$) the remaining values in parentheses where x is any value of the independent variable, X, and m is any value of the mediator, M. A **positive sign for c'** (i.e., positive direct effect) indicates that the case that is one unit higher on X is estimated to be higher on Y. A **negative sign for c'** (i.e., negative direct effect) indicates that the case that is one unit higher on X is estimated to be lower on Y. Y is the group mean in the case of a binary X, and therefore in situations where the independent variable is dichotomous, c' is estimating the difference between the two group means holding the mediator constant (i.e., the adjusted mean difference in ANCOVA).

The **indirect effect** is the second path of influence from X to Y. It is *indirect* as we see an arrow leads from X to M, then from M to Y. In this indirect path, X is the antecedent which influences M, the consequent, then antecedent M influences Y, the consequent. In other words, the independent variable influences the mediating variable which in turn then influences the dependent variable. In the indirect effect path, path a represents how much two cases which differ by one unit on the independent variable, X, differ on the mediator, M. A positive sign for a indicates that a case higher on the independent variable is higher on the mediator. A negative sign for a indicates that the case higher on the independent variable is lower on the mediator. The b coefficient is interpreted as c' except with the mediator, M, rather than X, as the antecedent. In path b, we find that two cases that differ by one unit on the mediator, M, but are equal on X will differ by b units on the dependent variable, Y. The product, ab, is the *indirect effect* of the independent variable, X, on the dependent variable, Y, through the mediator, M. We can interpret the indirect effect as follows: two cases that differ by one unit on the mediator, M, but are equal on the independent variable,

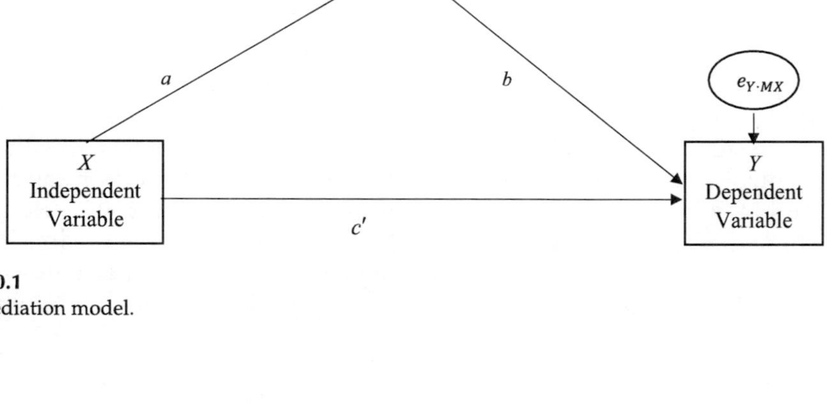

FIGURE 10.1
Simple mediation model.

X, will differ by *b* units on the dependent variable, *Y*. A positive sign for *ab* (i.e., both *a* and *b* are positive *or* both are negative) indicates that the case higher on the independent variable is higher on the dependent variable. A negative sign for *ab* (i.e., either *a* or *b*, but *not both,* are negative) indicates that the case higher on the independent variable is lower on the dependent variable.

The **total effect** of the independent variable on the dependent variable is *c* and indicates how much two cases that differ by one unit on the independent variable will differ on the dependent variable:

$$c = \left[Y' \mid \left(X = x\right)\right] - \left[Y' \mid \left(X = x - 1\right)\right] = c' + ab$$

A summary of the effects is presented in Box 10.1.

BOX 10.1 Summary of Mediating Effects

Path	Effect	Interpretation
a	Indirect effect of the independent variable on the mediator	How much two cases which differ by one unit on the independent variable, *X*, differ on the mediator, *M*
b	Indirect effect of the mediator on the dependent variable	Two cases that differ by one unit on the mediator but are equal on the independent variable will differ by *b* units on the dependent variable
ab	Indirect effect of the independent variable on the dependent variable through the mediator	Two cases that differ by one unit on the mediator but are equal on the independent variable will differ by *b* units on the dependent variable
c′	Direct effect of the independent variable on the dependent variable	Two cases that differ by one unit on the independent variable, *X*, but are equal on *M* will differ by *c′* units on the dependent variable,
c	Total effect of the independent variable on the dependent variable	How much two cases that differ by one unit on the independent variable will differ on the dependent variable

10.1.1.1 Additional Mediation Models

The mediation model in Figure 10.1 is the simplest mediation model that can be conceived. There are many more configurations that may exist, with more *X*'s and more *M*'s as well as mediated moderated models, multilevel mediation, and more. This chapter is designed to provide an overview into mediation and to whet your appetite to learn as you consider more advanced models in your own research.

10.1.2 Sample Size

Estimating sample size for mediation models is more complicated than with multiple linear regression. Although some guidelines for estimating sample size with mediated models have been provided (e.g., Fritz & MacKinnon, 2007), using Monte Carlo simulation to estimate sample size is recommended (Schoemann, Boulton, & Short, 2017). This is detailed further in the discussion of power.

10.1.3 Power

As with sample size, power in mediation models is also more complicated than with multiple linear regression. This is due to the formation of indirect effect as a product of two effects, and as noted by Hayes (2013, p. 141), "with no agreed upon way of quantifying the magnitude of those effects or their product (something you need to do to assess the power to detect an effect of a given size)." The literature on power within mediation is not voluminous; however, there are tables for determining sample sizes needed for detecting indirect effects of a given size (Fritz & MacKinnon, 2007). Additional literature from Zhang (2014) illustrates estimating power in mediation using bootstrap methods through Monte Carlo simulation. Packages in **R**, such as *powerMediation* (Qiu, 2018), can be used to determine power and/or sample size in mediation analysis. *MedPower* is an online tool for computing power and sample size for mediation models (Kenny, n.d.). Schoemann et al. (2017) provide an app that uses Monte Carlo simulation for computing power for indirect effects. Using Schoemann et al. (2017), for example, we see in Figure 10.2, given the correlations and standard deviations from the variables in the model that will be estimated later using SPSS and **R**, along with the default settings for the simulation (i.e., 1000 replications, 20,000 Monte Carlo draws per rep, and 95% confidence level), the power for detecting the indirect effect given a sample size of 44 is *.89*—which is strong power.

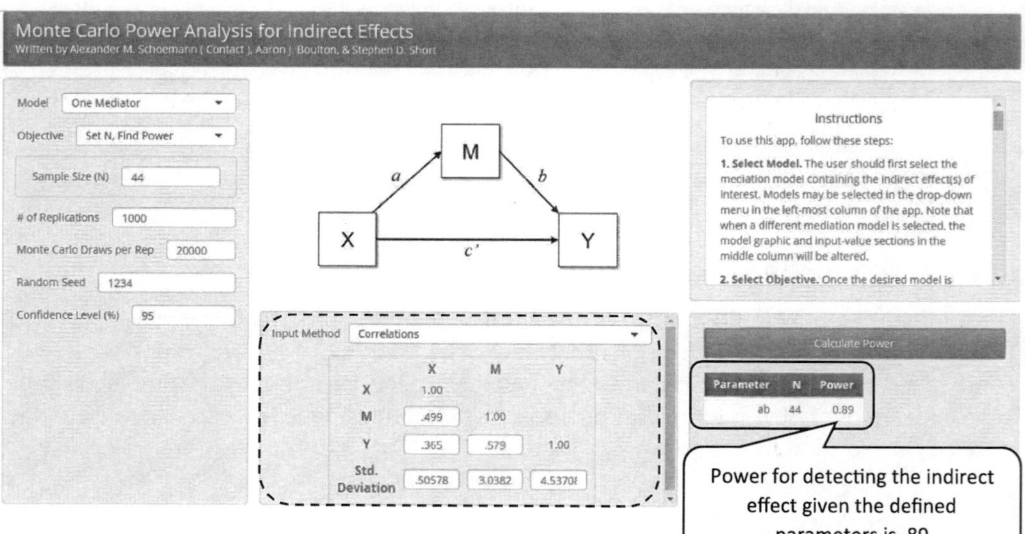

FIGURE 10.2
Power analysis for indirect effects.

10.1.4 Effect Size

As with other elements related to mediation, effect size in mediation analysis is a growing area of research and discussion (e.g., Lachowicz, Preacher, & Kelley, 2018; Preacher & Kelley, 2011) and there are multiple effect size indices that can be considered. We will focus on partially and completely standardized effects using notation from Hayes (2013), summarized in Box 10.2, but will touch on a few other indices that may be encountered in the literature but that are not recommended for use.

BOX 10.2 Effect Sizes in Mediation Models

Effect Size	Formula	Interpretation
Partially standardized direct effect	$c'_{ps} = \dfrac{c'}{SD_Y}$	Independent of the indirect effect (i.e., mediating effect), a unit that is one unit higher on the independent variable will be c'_{ps} standard deviations different on the dependent variable
Partially standardized indirect effect	$ab_{ps} = \dfrac{ab}{SD_Y}$	Two cases that differ by one unit on the independent variable will differ by ab_{ps} standard deviations in the dependent variable as a result of the effect of the independent variable on the mediator
Partially standardized total effect	$c_{ps} = \dfrac{c}{SD_Y} = c'_{ps} + ab_{ps}$	Two cases that differ by one unit on the independent variable will differ by c_{ps} standard deviations on the outcome as a result of the *combined direct and indirect effects* by which the independent variable affects the dependent variable
Completely standardized direct effect	$c'_{ps} = \dfrac{(SD_X)(c')}{SD_Y} = (SD_X)(c'_{ps})$	Independent of the indirect effect (i.e., mediating effect), a unit that is one standard deviation unit higher on the independent variable will be c'_{cs} standard deviations different on the dependent variable
Completely standardized indirect effect (i.e., index of mediation)	$ab_{cs} = \dfrac{(SD_X)(ab)}{SD_Y} = (SD_X)(ab_{ps})$	Two cases that differ by one standard deviation unit on the independent variable will differ by ab_{cs} standard deviations in the dependent variable as a result of the effect of the independent variable on the mediator; in other words, the expected standard deviation change in *the dependent variable* for a one standard deviation increase in the independent variable through the mediating variable
Completely standardized total effect	$c_{cs} = \dfrac{(SD_X)(c)}{SD_Y} = (c'_{cs})(ab_{cs})$	Two cases that differ by one standard deviation unit on the independent variable will differ by c_{cs} standard deviations on the outcome as a result of the *combined direct and indirect effects* by which the independent variable affects the dependent variable

10.1.4.1 Partially Standardized Effect

The **partially standardized effect** is an effect that is relative to the standard deviation (not the original metric) of the outcome, Y. In other words, the independent variable, X, remains in its original metric, but the partially standardized effects are rescaled to the standard deviation of the dependent variable, Y. This means that the size of the partially standardized effect depends on the scale of the independent variable, X.

The **partially standardized direct effect** can be computed as:

$$c'_{ps} = \frac{c'}{SD_Y}$$

The interpretation of the partially standardized *direct effect* is that, independent of the indirect effect (i.e., mediating effect), a unit that is one unit higher on the independent variable will be c'_{ps} standard deviations different on the dependent variable.

The **partially standardized indirect effect** can be computed as:

$$ab_{ps} = \frac{ab}{SD_Y}$$

The interpretation of the partially standardized *indirect effect* is that two cases that differ by one unit on the independent variable will differ by ab_{ps} standard deviations in the dependent variable as a result of the effect of the independent variable on the mediator.

As the total effect of the independent variable, X, is the sum of the direct and indirect effects, the **partially standardized total effect** is the sum of the partially standardized direct and indirect effects, computed as:

$$c_{ps} = \frac{c}{SD_Y} = c'_{ps} + ab_{ps}$$

The interpretation of the partially standardized *total effect* is that two cases that differ by one unit on the independent variable will differ by C_{ps} standard deviations on the outcome as a result of the *combined direct and indirect effects* by which the independent variable affects the dependent variable.

In the case that the independent variable, X, is *dichotomous*, then the partially standardized direct effect and the partially standardized indirect effect are interpreted as the number of standard deviations in the dependent variable, Y, that the groups differ, on average, due to the direct and indirect effects. The direct and indirect effects in the case of binary X sum to the total estimated mean difference in the outcome between the two categories.

10.1.4.2 Completely Standardized Effect

When the scaling of the independent variable, X, is removed from the partially standardized effects, the direct and indirect effects are then expressed in the form of a difference in standard deviations in the dependent variable, Y, between units that differ by one standard deviation on the independent variable, X. The **completely standardized direct effect** then is computed as follows:

$$c'_{cs} = \frac{(SD_X)(c')}{SD_Y} = (SD_X)(c'_{ps})$$

The completely standardized direct effect is interpreted as: independent of the indirect effect (i.e., mediating effect), a unit that is one standard deviation unit higher on the independent variable will be c'_{ps} standard deviations different on the dependent variable.

The **completely standardized indirect effect** is computed as:

$$ab_{cs} = \frac{(SD_X)(ab)}{SD_Y} = (SD_X)(ab_{ps})$$

The completely standardized indirect effect is interpreted as follows: Two cases that differ by one standard deviation unit on the independent variable will differ by ab_{cs} standard deviations in the dependent variable as a result of the effect of the independent variable on the mediator. In other words, the expected standard deviation change in *the dependent variable* for a one standard deviation increase in the independent variable through the mediating variable

Note that when the direct and indirect effects are computed using standardized regression coefficients, *or* when all variables in the model are standardized, they will equate to the completely standardized direct and indirect effects (Preacher & Hayes, 2008).

The **completely standardized total effect** is computed as:

$$c_{cs} = \frac{(SD_X)(c)}{SD_Y} = (c'_{cs})(ab_{cs})$$

The completely standardized total effect is interpreted as follows: Two cases that differ by one standard deviation unit on the independent variable will differ by C_{cs} standard deviations on the outcome as a result of the *combined direct and indirect effects* by which the independent variable affects the dependent variable.

Note that in a simple regression model that estimates the dependent variable from a single independent variable, the completely standardized total effect is equal to the standardized regression coefficient for X. Additionally, when the independent variable is binary, the completely standardized effect is usually not meaningful and is thus not recommended (Hayes, 2013).

10.1.4.3 Other Effect Size Indices for Mediation Models

As is sometimes the case, statistics may be reported even if they are not best practice, and this includes effect sizes in the context of mediation. Thus, we summarize these effects simply because you may find them in the literature; however, we do not encourage their use, as has been recommended by other researchers (e.g., Hayes, 2013).

The **ratio of the indirect effect to total effect** is an effect size for mediation models that is sometimes reported. Problematic with this effect size index is that this proportion may compute to be less than zero (when either but not both *ab* or *c* is less than zero) or greater than one (when *c* is closer in value to zero than *ab*) (Hayes, 2013). Research also suggests that it is unstable from sample to sample (MacKinnon, Warsi, & Dwyer, 1995). Simulation research suggests that a sample of at least 500 is needed for this effect size to produce a trustworthy effect size estimate (MacKinnon et al., 1995).

The **ratio of the indirect effect to the direct effect** is the ratio of the indirect effect, *ab*, to the direct effect, c'. Problematic with this effect size index is that as the direct effect, c', nears zero, the ratio will dramatically increase in size. Thus, from sample to sample, small changes in the indirect effect dramatically alter the value of ratio. Simulation research suggests that a sample of at least 2000 is needed for this effect size index to produce a trustworthy effect size estimate (MacKinnon et al., 1995).

The **proportion of variance** in the dependent variable, *Y*, that is explained by the indirect effect, *ab*, is another effect size index that is sometimes reported. This effect size becomes problematic when the indirect effect, *ab*, exists in the absence of a relationship between the independent variable, *X*, and the dependent variable, *Y*. In other words, the indirect effect, *ab*, is larger, in absolute value terms, than the direct effect, *c*. When this occurs, this proportion of variance effect size can be negative and thus interpretable.

Kappa squared, κ^2, is the ratio of the indirect effect, *ab*, to the maximum possible value of *ab* given the data. While this is a promising effect size index, recent simulation research illustrated that the original derivation of the maximum possible value was computationally in error (Wen & Fan, 2015). Thus software that implemented kappa squared was also in error. Should this be corrected, this effect size may be considered as another appropriate effect size in the future (Hayes, 2013).

10.1.5 Assumptions

By default, moderation and mediation require at least two independent variables, thus the assumptions that must be met are those of multiple regression, including: (a) independence, (b) homoscedasticity, (c) normality, (d) linearity, (e) fixed X, and (f) noncollinearity. Beyond these, there are no further assumptions that must be considered for mediation or moderation. In terms of homoscedasticity, the PROCESS macro that will be illustrated has an option for regression that does not assume homoscedasticity, such as heteroscedasticity-consistent covariance estimators.

10.2 What Moderation Is and How It Works

Moderation is said to occur when the effect of the independent variable, *X*, in terms of size (small or large effect), sign (positive or negative), or strength (weak or strong), on some dependent variable, *Y*, depends on or can be predicted by moderating variable, *W*. In other words, *W* moderates the effect of *X* on *Y*; there is an interaction of *W* and *X* in their influence on *Y*. This is conceptually depicted in Figure 10.3.

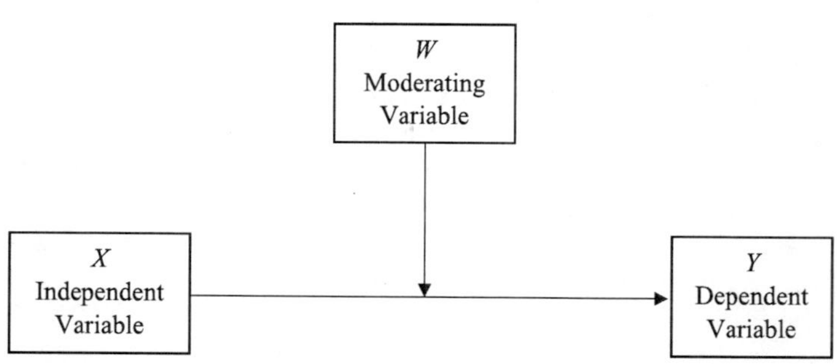

FIGURE 10.3
Conceptual simple moderation model.

While the term *moderation* may be new to you, the concept is likely not as interactions in factorial ANOVA represent moderation. Moderation is simply examining whether the effect of one variable on the dependent variable differs across levels of another variable. While factorial ANOVA assumes categorical variables for both *X* and *W*, we will illustrate moderation via regression, which is not conditioned on the variables being categorical. Additionally, we will work within the framework of ordinary least squares regression using **moderated multiple regression** (MMR). MMR is an inferential approach to examining moderation that consists of comparing two least-squares regression equations (Aiken & West, 1991).

10.2.1 Characteristics

We noted previously the sample multiple linear regression model. Let's consider this model for predicting *Y* from 2 predictors:

$$Y_i = b_1 X + b_2 W + a + e_i$$

where Y_i is the criterion variable (also known as the dependent variable); *X* is one predictor (or independent) variable and *W* is a second antecedent variable; b_k is the sample partial slope of the regression line for Y_i as predicted by *X* or *W*; *a* is the sample intercept of the regression line for Y_i as predicted by the set of predictors; e_i are the residuals or errors of prediction (the part of Y_i not predictable from the predictors); and *i* represents an index for an individual or object. The index *i* can take on values from 1 to *n* where *n* is the size of the sample (i.e., $i = 1, \ldots, n$).

The sample prediction model, therefore, is

$$Y_i' = b_1 X + b_2 W + a$$

Where Y_i' is the predicted value of the dependent variable for specific values of the predictors, and the other terms are as before. We interpret *X*, for example, as a one-unit change in *X* results in a b_1 change in Y_i', and this is *unconditional* on *W*. In other words, the effect of *X* on Y_i' does not depend on the moderating variable, *W*. The value of *X* does not change, i.e., it is *invariant*, across all values of the moderating variable. Similar interpretations can be made for *W* and b_2—the effect of *W* on Y_i' does not depend on the independent variable, *X*; *W* on Y_i' is unconditional on *X*. The value of *W* does not change, i.e., it is *invariant*, across all *X* values.

A **simple linear moderation model** has the following form:

$$Y_i = b_1 X + b_2 W + b_3 XW + a + e_i$$

With the **sample prediction moderation model** being:

$$Y_i' = b_1 X + b_2 W + b_3 XW + a$$

where *XW* is simply the product of *X* and *W* (i.e., there is not mathematical magic needed to construct *XW*; *XW* results from simply multiplying *X* by *W*). Statistically, this model is depicted in Figure 10.4. It is important to note that *X* and *W* should always be included in the model when there is a moderating effect included, even if *X* and/or *W* are not statistically significant (Hayes, 2013).

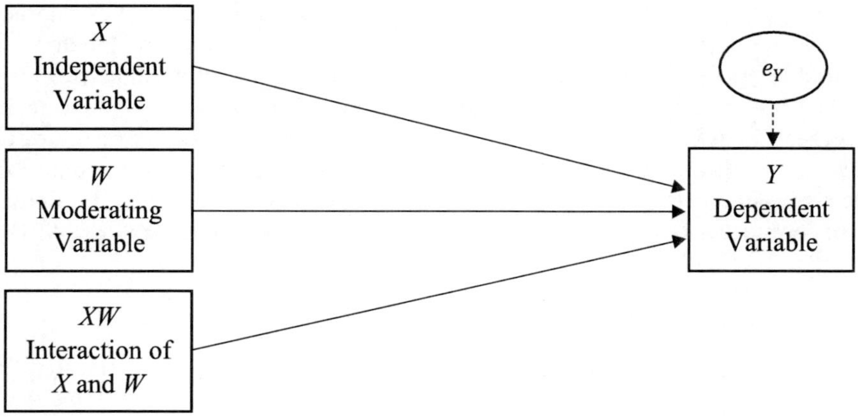

FIGURE 10.4
Simple moderation model.

By including the interaction term, *XW*, we are testing that the effect of *X* on the outcome is **conditional** on *W*. In other words, the effect of *X* on Y_i' is dependent (aka "conditioned") on *W*. The coefficients for *X* and *W*, in a model that includes the moderating term *XW*, are *conditional effects*, with the condition being that the other variable is zero (note that when the moderating term is not included in the model, partial effects—not conditional effects—are estimated).

For two cases that differ on *X* by a unit, the difference in the dependent variable for a one-unit increase (or decrease) in *W* is a change of b_3 units. In other words, as *W* increases by one unit, for two cases that differ on *X* by a unit, the dependent variable will change by b_3 units. These interpretations are predicated on *W* being the moderating variable. If, however, *X* is the moderating variable, then b_3 represents the difference in the dependent variable for a one unit increase (or decrease) in *X* for two cases that differ by a unit on *W*. The degree to which the slopes are not parallel in a moderation model is dependent on b_3. As the absolute value of b_3 increases, the slopes will become increasingly nonparallel.

In the simple moderated model, b_1 is interpreted as the *conditional effect of X* on the dependent variable when the moderating variable is zero. In other words, it is the difference in Y_i' for two cases that differ by *one* on *X* but differ by *zero* on *W*. It is important to note that b_1 is *not* interpreted as a main effect or as the effect of *X* on the dependent variable when controlling for *W*.

A similar interpretation can be made for b_2: It is the *conditional effect of W* on the dependent variable when *X* is zero. In other words, it is the difference in Y_i' for two cases that differ by *one* on *W* but differ by *zero* on *X*. It is important to note that b_2 is *not* interpreted as a main effect or as the effect of *W* on the dependent variable when controlling for *X*.

10.2.1.1 Probing an Interaction

Graphs are an excellent way to visualize an interaction, as we have already seen with factorial ANOVA. However to really understand what's happening when an interaction occurs, the interaction needs to be *probed* more deeply. In other words, does a statistically significant interaction mean that X effects Y for cases that are low on the moderator? High on the moderator? Somewhere in between? The *test of the interaction* (i.e., the inferential test of the coefficient) establishes that an interaction is or is not statistically significant—i.e., whether the relationship between the independent and dependent variables systematically varies as a function of the moderator. If there is a statistically significant interaction, this justifies the next step—*probing*. The *test of the interaction* is *not* the same thing as *probing for an interaction*. Probing for the interaction is necessary to better understand what is happening within the interaction and to understand where the differential variation on the moderator occurs. As researchers, we want to say more than just the effect of X on Y depends on W, and we usually want to say at what point(s) the effect of X on Y depends on W.

A common technique for probing an interaction is the **pick-a-point approach** (Rogosa, 1980). In this approach, the conditional effect of X on Y is computed using one or more values of W, and this is followed by a test of inference. Modern statistical software eliminates the need for computing this by hand, which can be prone to error. If you are so inclined, however, Cohen, Cohen, West, and Aiken (2003) provide an example of hand calculation for the pick-a-point approach. Using the PROCESS macro, as illustrated later, the pick-a-point approach is implemented and output provided automatically. We will illustrating the approach applying the 16th, 50th, and 84th percentiles, as recommended by Hayes (2013) as they correspond to relatively low, moderate, and high values of the moderating variable. In the case that W is *normally distributed*, these values also correspond, respectively, to one standard deviation below the mean (16th percentile), the mean (i.e., 50th percentile in a normal distribution), and one standard deviation above the mean (i.e., 84th percentile) on the moderating variable. And regardless of the distributional shape of W, the 16th, 50th, and 84th percentiles will always be within the range of the observed data.

The challenge with the pick-a-point approach is the selection of often arbitrary values of the moderating variable and the fact that the values selected are sample specific (Hayes, 2013). The **Johnson-Neyman** (JN) technique (Johnson & Neyman, 1936) eliminates these issues. The JN technique was originally proposed as a way to handle violations of homogeneity of regression in ANCOVA mean difference tests for two groups. Bauer and Curran (2005) extended the work to more general regression models. The JN technique can be applied only with continuous moderating variables. The JN technique can be considered as the reverse of the pick-a-point approach in that values of W are derived at the point where the interaction is statistically significant (Hayes, 2013). In other words, the values of W where the conditional effect of X on Y changes from nonstatistically significant to statistically significant.

Hayes (2013, p. 255) refers to the JN technique as identifying "regions of significance" for the effect of X on Y. If the JN technique results in *two solutions*, this suggests that the conditional effect of X on Y is statistically significant in one of the two fashions: $JN_{W1} \leq W \leq JN_{W2}$ *or* $JN_{W1} \geq W$ *and* $W \geq JN_{W2}$. In other words, the region of significance is either contained within two points (i.e., $JN_{W1} \leq W \leq JN_{W2}$) on W or it is outside two points of W (*i.e.*, $JN_{W1} \geq W$ *and* $W \geq JN_{W2}$).

If JN results in only *one solution*, this suggests that the conditional effect of X on Y is statistically significant when $W \geq JN_{W1}$ *or* when $W \leq JN_{W1}$, but not both. In other words, the region of significance is either above (i.e., $W \geq JN_{W1}$) or below (i.e., $\leq JN_{W1}$) some value of W, but not between them, and not in both directions.

It may also be the case that the JN technique results in *no solution*. This can happen when the conditional effect of X on Y is statistically significant across *all* values of W (i.e., the entire range of values of W) *or* when the conditional effect of X on Y is statistically significant across *no* values of W.

10.2.1.2 Centering

Researchers working with moderated models may want to consider centering the X and W variables. This can assist in avoiding multicollinearity (Aiken & West, 1991) and may also increase the interpretability of the regression intercept. As noted by Hayes (2013), however, centering to avoid multicollinearity is largely a myth. In terms of interpretation, on the other hand, centering *is* something that many researchers may want to consider.

When centering is not applied, the intercept is interpreted as the value of the dependent variable when all the predictors are *zero*. If either X or W are not zero, then the intercept has no meaning. In comparison, if the predictors are centered at the average, for example, the intercept becomes the value of the dependent variable when the predictors are at their *average*. In the case of mean centering X and W, b_1 is interpreted as the difference in Y between two cases that differ by one unit on X among cases that are *at the mean* of W. For b_2, we find it is interpreted as the difference in Y between two cases that differ by one unit on W among cases that are *at the mean* of X.

A model estimated without mean centering is mathematically equivalent (e.g., R^2 and $MS_{residual}$) to a model estimated with mean centering. The coefficients and related estimates (t, p, SE) for X and W will differ as they are estimating effects for cases at the average (rather than zero). However, the regression coefficient for the interaction, XW, will be the same regardless of centering. Thus, the test of the moderation will result in the same conclusion regardless of mean centering or not mean centering.

10.2.2 Sample Size

As with multiple regression, there exists conventions for sample size needed for detecting a moderating effect. Stone-Romero and Anderson (1994) found that samples of at least 120 were needed to detect moderate and large moderating effects. Aguinis (2004) recommends a sample size of at least 100 for detecting a moderating effect. Throughout the text, however, we have discouraged the application of conventions for determining sample size, as there are so many factors that need to be considered and applying a one-size-fits-all determination for sample size is thus not best practice. Rather, we suggest estimating sample size with, for example, power software. Shieh (2009), for example, provides SAS IML and **R** code for calculating power and sample size.

10.2.3 Power

Powering a study for a main effect is different from powering a study for an interaction. Aguinis (2004) grouped factors that impact power in moderated multiple regression into five categories: variable distributions; variable operationalization; sample size; predictor variable correlations; and interactive effects of these factors impacting power. We will start our discussion of power within the context of these categories.

The first category relates to the **distribution of the variables**. Aguinis and Stone-Romero (1997) found that power in MMR is dramatically decreased when the variance of the predictor, X, is smaller in the sample than in the population. Range restriction of the independent variable, X, in turn restricts the range of the interaction, XW, and this detrimentally impacts the ability to find a population moderating effect. Another aspect related to variable distribution concerns transformations of outcome variables. Transforming Y, specifically log transformations to correct for nonnormally skewed distributions, has been found to underestimate the moderating effect and decrease power (i.e., which indicates an increased change of a Type II error) (Russell & Dean, 2000).

The second factor impacting power relates to **variable operationalization**, which includes measurement error, operationalization of the dependent variable, and categorizing continuous variables (Aguinis, 2004). The probability of Type II errors increases in the presence of inadequate reliability when testing moderation (Aguinis, 2004). Low reliability of modeled variables is so problematic that measurement error is considered by some researchers to be the most impactful factor on power in MMR (Kromrey & Foster-Johnson, 1999). The measurement scale of the variables included in MMR also impacts power. In particular, the use of Likert items for either or both the independent and/or dependent variable have been shown to decrease power to detect a moderating effect (e.g., Russell & Bobko, 1992). Artificially categorizing (e.g., creating dichotomy or multicategory) a continuous variable has also been found to decrease power in detecting moderating effects.(e.g., Mason & Tu, 1996).

Sample size, both overall and subgroup, can impact power in MMR. Generally in testing hypotheses, regardless of statistical approach, larger sample sizes result in increased power. For MMR, overall sample size is particularly critical. For example, Stone-Romero and Anderson (1994) found that samples of at least 120 were needed to detect moderate and large moderating effects. As noted previously, estimating overall sample size via a power analysis program may mitigate problems with decreased power in MMR. In addition to overall sample size, however, the group sizes within the moderating variable (i.e., subgroups) also impact power. Unequal sizes of the subgroups impact power above and beyond the total sample size (Aguinis, 2004). There is decreased power when one group is substantially smaller than the other group, regardless of the total sample size (Stone-Romero, Alliger, & Aguinis, 1994).

The fourth factor identified by Aguinis (2004) that impacts power in MMR relates to **predictor variable correlations**. Researchers have found that multicollinearity does not detrimentally impact MMR (Cronbach, 1987). However, a weak relationship between the independent and dependent variable (i.e., first-order effect) may limit detection of a moderating effect (Rogers, 2002). In other words, the strength of the relationship between the independent and dependent variable places a cap on the size of the moderating effect.

The last factor relates to **interaction effects between these aspects that impact power**. For example, as noted by Aguinis (2004, p. 78), "the combined effects on power of the simultaneous presence of small total sample size, large measurement error, and unequal sample sizes across the moderator-based subgroups are greater than the sum of the individual effects of these factors." Additionally, the presence of just one factor that detrimentally impacts power can dramatically decrease power even if the other factors are powered sufficiently (Aguinis, 2004).

Researchers interested in assessing power for moderation have a number of resources to consult. For example, Shieh (2009) illustrates power and sample size calculations for detecting moderating effects. Calculating power for moderating effects in cluster randomized designs has also been illustrated (e.g., Dong, Kelcey, & Spybrook, 2018; Dong & Society for Research on Educational, 2014; Spybrook & Kelcey, 2014).

10.2.4 Effect Size

A common effect size for moderated multiple regression is f^2 (Aiken & West, 1991), computed as follows:

$$f^2 = \frac{R_2^2 - R_1^2}{1 - R_2^2}$$

Where R_1^2 is the proportion of variance in the dependent variable that is accounted for by the effects of the independent variable (*X*) and moderating variable (*W*), and R_2^2 is the proportion of variance in the dependent variable that is accounted for by the effects of the independent variable, moderating variable, and interaction term (*XW*). Conventions for interpreting f^2 are offered by Cohen (1988), with small effects of $f^2 = .02$, moderate effects of $f^2 = .15$, and large effects of $f^2 = .35$. Aguinis, Beaty, Boik, and Pierce (2005) proposed a modified f^2 that is appropriate to use when there are categorical moderators when homogeneity of error is violated. An online calculator (see www.hermanaguinis.com/mmr/index.html) is available for computed modified f^2.

When the independent, moderating, and dependent variables have metrics that are interpretable (e.g., number of XYZ, dummy coding), the direction and strength of the conditional effects represent an unstandardized effect size (Bodner, 2017). Standardized regression coefficients can be interpreted as effect size, although this practice is debatable (Smithson & Shou, 2017). With continuous moderators, Bodner (2017) presents an approach for conditional effects expressed in standardized mean differences and semipartial correlations.

10.2.5 Assumptions

The usual assumptions of multiple linear regression are applicable for moderated multiple regression and include: linearity; residuals that are homoscedastic, normally distributed, and independent; and lack of multicollinearity. When there is a categorical moderator, homogeneity of (within-group) error variance assumption—i.e., homoscedasticity—is particularly important to preventing increased probability of Type I and Type II errors.

10.3 Computing Mediation and Moderation Using SPSS

We will first consider SPSS for mediation using the PROCESS macro. This will be followed by illustration for moderation.

10.3.1 Installing the PROCESS Macro

An excellent computational tool for observed variable path analysis-based moderation and mediation is PROCESS (Hayes, n.d.). In addition to estimating coefficients, standard errors (including heteroscedasticity-constant standard errors), and similar calculations, PROCESS provides direct and indirect effects for mediation models (including percent bootstrap and Monte Carlo confidence intervals for indirect effects), conditional effects in moderation models, and conditional indirect effects in conditional process models with a single mediator or with multiple mediators. Additionally, it provides options for probing interactions as well as generates effect size indices for direct, indirect, and total effects. There are a number of templates for estimating models, and models can be custom built as well. These are just a few of the rich tools that PROCESS provides.

PROCESS can be used by writing syntax within SPSS or by installing a custom dialog menu that can be used in the navigational menu within SPSS. We will illustrate using the latter. To install PROCESS as a custom menu tool, visit http://processmacro.org and click "Download" from the top navigational menu. From this page, you will have access to download the latest version of PROCESS (version 3.2.01 at the time of writing) (note, however, that you will need administrator privilege to install). Once installed, PROCESS is provided as an option from the regression menu.

10.3.2 Computing Mediation Analysis Using SPSS

Next we consider SPSS for computing mediation. Before we conduct the analysis, let us review the data. We are using the "Ch10_medmod.sav" data. For this illustration, the dependent variable is "DV," the independent variable is "TRTMT," and the mediating variable is "M1" (see Figure 10.5).

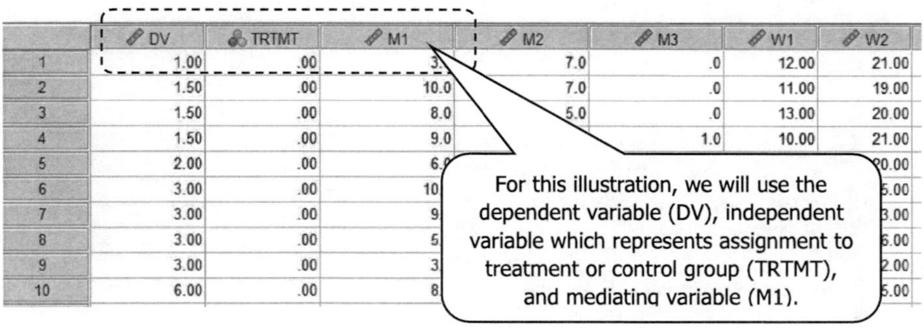

FIGURE 10.5
Mediation data (first 10 cases).

Step 1. To conduct a mediated regression model, go to "**Analyze**" in the top pulldown menu, then select "**Regression**," and then select "**PROCESS**." Following the screenshot for Step 1 (Figure 10.6) produces the "**PROCESS**" dialog box.

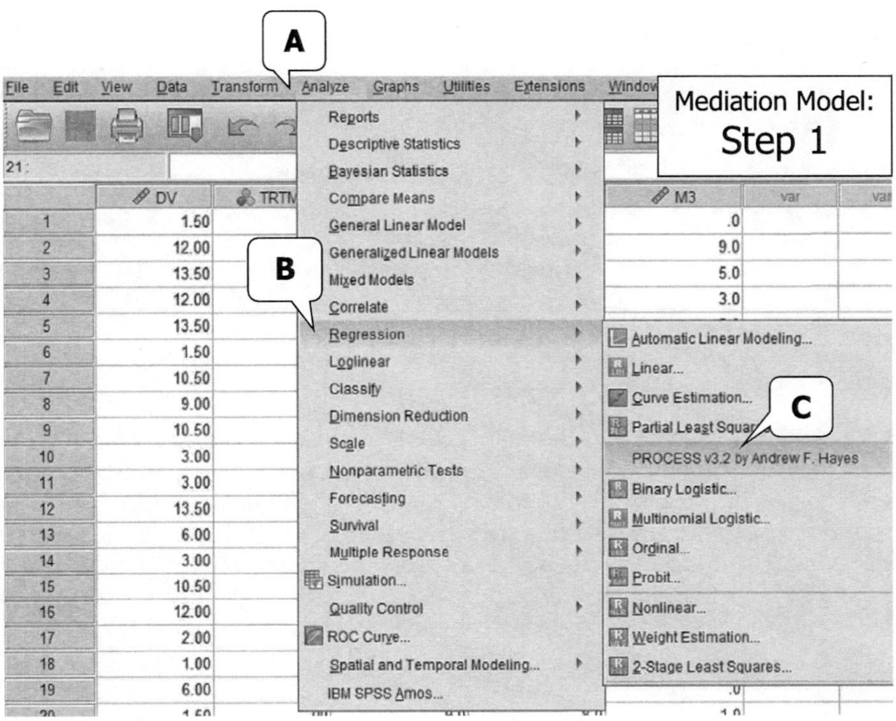

FIGURE 10.6
Mediation model: Step 1.

Step 2. Click the dependent variable (e.g., "DV") and move it into the "Y variable" box by clicking the arrow button. Click the independent variable (e.g., "TRTMT") and move into the "X Variable" box by clicking the arrow button. Click the mediating variable (e.g., "M1") and move to the "Mediator(s) M" box by clicking the arrow button (see the screenshot for Step 2, Figure 10.7). We are using model 4 from the templates, so use the toggle menu to select "4" for "Model number." We will leave the default settings for confidence intervals (i.e., 95) and number of bootstrap samples (e.g., 5000). To obtain the bootstrap inference for model coefficients, place a check in the respective box.

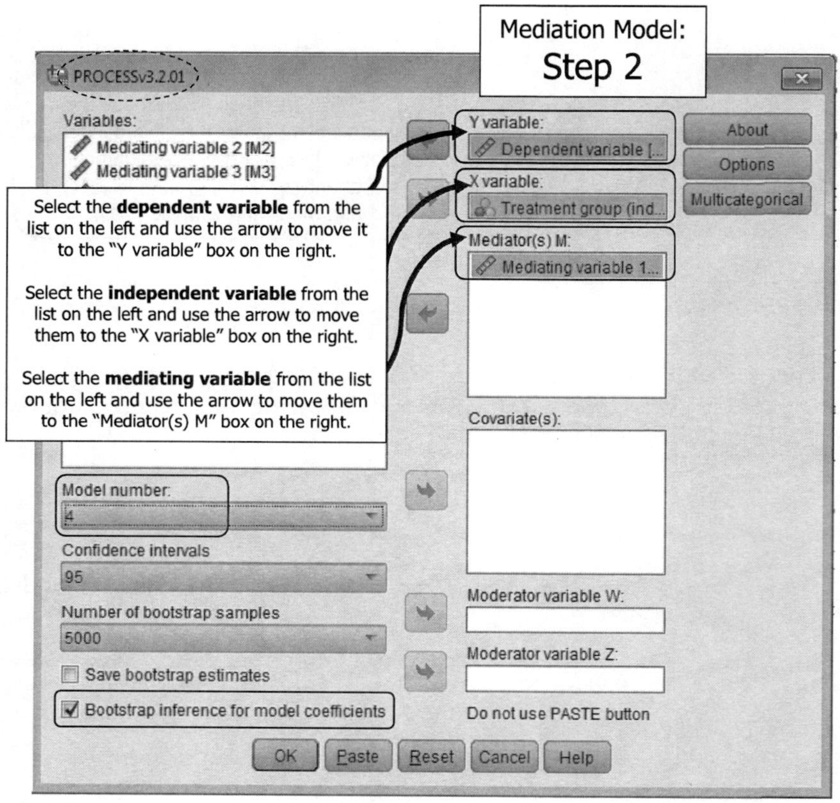

FIGURE 10.7
Mediation Model: Step 2.

Step 3. Clicking on Options from the main dialog box (see Figure 10.7) will produce the dialog box that will allow us to make a number of selections for the output. For this illustration, we place checkmarks for the following: "Show total effect model" (note that we are computing *model 4* from the templates (Hayes, 2013) so this option is appropriate); "Pairwise contrasts of indirect effects"; "Effect size (mediation-only models)"; and "Standardized coefficients (mediation-only models)." Using the toggle menu, we select "HC4" (Cribari-Neto, 2004) for the heteroscedasticity-consistent inference. HC4 takes large leverage values into account when constructing the standard errors, and has been shown to outperform HC3 in the presence of high leverage points and error distributions that are nonnormal (Hayes & Cai, 2007). Using HC3 or HC4 is recommended (Hayes & Cai, 2007). Then click "Continue" to return to the main dialog box.

10.3.2.1 Interpreting Mediation Output

From the output in Table 10.1, we see the mediation analyses is actually a series of models. Among other findings, the mediating variable fully (or completely) mediates the relationship between the independent variable and the outcome.

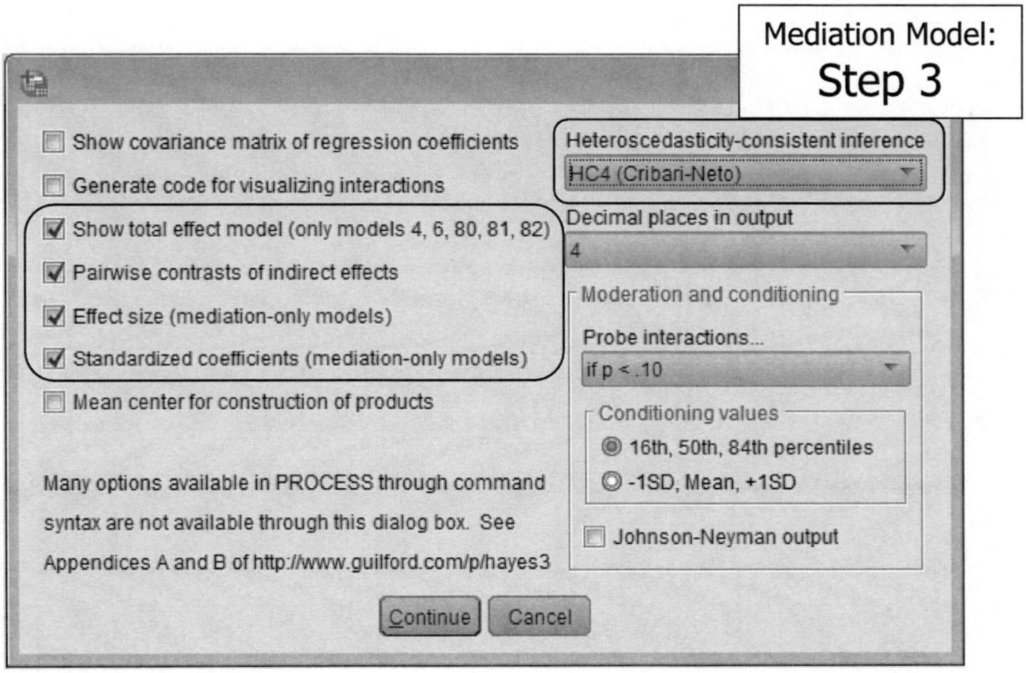

FIGURE 10.8
Mediation model: Step 3.

Note that for the bootstrapped confidence intervals, assumptions about the shape of the sampling distribution are not made. As a result, bootstrapped confidence intervals handle irregularities of the *ab* sampling distribution and yield more accurate inferences than the normal theory approach, which results in increased power (relative to the normal theory approach) (Hayes, 2013). As bootstrapped confidence intervals are derived void of any assumptions of the size of the parameter, it is incorrect to state that intervals that do not include zero are statistically significant. Thus, a bootstrapped confidence interval that is entirely above zero will support a conclusion of a positive indirect effect, but it would technically be incorrect to conclude to reject the null hypothesis that the estimate = 0 with an observed probability of no more than .05 (Hayes, 2013). In practice, however, the interpretation of a bootstrapped confidence interval leads to a similar substantive interpretation—i.e., intervals that do not contain zero provide "evidence that the effect is positive to a 'statistically significant' degree" (Hayes, 2013, p. 101).

TABLE 10.1

Mediation Model SPSS Output

```
Run MATRIX procedure:

************** PROCESS Procedure for SPSS Version 3.2.01 *****************

          Written by Andrew F. Hayes, Ph.D.       www.afhayes.com
    Documentation available in Hayes (2018). www.guilford.com/p/hayes3

**************************************************************************
Model  : 4
    Y  : DV
    X  : TRTMT
    M  : M1

Sample
Size:  44

**************************************************************************
OUTCOME VARIABLE:
 M1

Model Summary
          R       R-sq        MSE     F(HC4)        df1        df2          p
      .4994      .2494     7.0931    13.9573     1.0000    42.0000      .0006

Model
              coeff    se(HC4)          t          p       LLCI       ULCI
constant     7.9545      .5975    13.3123      .0000     6.7487     9.1604
TRTMT        3.0000      .8030     3.7359      .0006     1.3794     4.6206

Standardized coefficients
           coeff
TRTMT      .9874
```

The output provides information on the model template that was used (#4), the variables specified (*Y* is the dependent variable, *X* is the independent variable, and *M* is the mediating variable), and the sample size.

The first set of output relates to the use of the mediating variable (*M*, M1) as the outcome with only the independent variable (*X*, TRTMT) in the model. **This is path *a*.** We want *X* to relate to *M*, as that provides evidence that mediation with *M* is appropriate.

a

This coefficient tells us that two cases that differ by 1 on *X* are estimated to differ by *a* units on *M*. Because *X* in this illustration is dummy coded (where treatment = 1 and control = 0), *a* is the difference between the group means on *M*. The coefficient, therefore, tells us more specifically that two cases that differ by 1 on *X* are estimated to differ, on average, by *a* =3.00 units on *M*. Thus, individuals in the treatment group (*X* = 1) are, on average, 3.00 units higher on the mediating variable than those in the control condition.

The standard errors are constructed using heteroscedasticity consistent methods. For this illustration, we requested HC4. HC4 takes large leverage values into consideration. HC4 has been shown to outperform HC3 in the presence of high leverage points and error distributions that are nonnormal (Cribari-Neto, 2004). Using HC3 or HC4 has been recommended (Hayes & Cai, 2007).

TABLE 10.1 (continued)

Mediation Model SPSS Output

```
************************
OUTCOME VARIABLE:
 DV
```

The second set of output relates to the use of the dependent variable (*Y*, DV) as the outcome with the independent variable (*X*, TRTMT) and mediating variable (*M*, M1) in the model.

```
Model Summary
          R       R-sq        MSE     F(HC4)        df1        df2          p
      .5854      .3427    14.1908    16.1036     2.0000    41.0000      .0000

Model
              coeff    se(HC4)          t          p       LLCI       ULCI
constant      .8819     1.7321      .5091      .6134    -2.6163     4.3800
TRTMT         .9053     1.3870      .6527      .5176    -1.8958     3.7065
M1            .7891      .1928     4.0939      .0002      .3998     1.1784

Standardized coefficients
            coeff
TRTMT       .1995
M1          .5284
```

c′ (points to TRTMT); *b* (points to M1)

Assuming we want to see a mediating effect, we want to see the following: We want to *M* to relate to *Y* but *X* to no longer relate to *Y* (or to relate to a smaller degree). When there is a mediating effect, the relation between *X* and *Y* will decrease or disappear altogether. If the relationship between *X* and *Y* completely disappears, this indicates that there is **full mediation**. In other words, *M* fully mediates the relationship between *X* and *Y*. If some relationship between *X* and *Y* remains after the mediator is included in the model, but that relationship is smaller in magnitude, this indicates that there is **partial mediation**. In other words, *M* partially mediates the relationship between *X* and *Y*. In this illustration, we see that *X* is not statistically significant (p = .5026) but *M* is statistically significant (p = .0001). Thus, there is full mediation as the relationship between the treatment and the outcome has completely disappeared with the inclusion of the mediator. *This suggests that the mediator fully mediates the relationship between the treatment and the outcome.*

The **coefficient for *X*** tells us that two people that differ by one unit on *X* but are equal on *M* are estimated to differ by .9053 units on the outcome. Since *X* is a binary variable (treatment = 1, control = 0), this coefficient suggests that independent of the effect of *M* on *Y*, individuals assigned to the treatment condition are estimated to be nearly 1 point higher (specifically .9053 higher), on average, on the outcome than those assigned to the control condition.

The **coefficient for *M*** tells us that two people who are equal on *X* (i.e., assigned to the same condition) but that differ by one unit on *M* are estimated to differ by .7891 units on the dependent variable. The sign for *b* (i.e., the mediating variable) is positive, which indicates that individuals who are higher on the mediating variable, *M*, are also estimated to be higher on the dependent variable.

```
************************** TOTAL EFFECT MODEL ****************************
OUTCOME VARIABLE:
 DV

Model Summary
          R       R-sq        MSE     F(HC4)        df1        df2          p
      .3648      .1331    18.2700     6.4487     1.0000    42.0000      .0149

Model
              coeff    se(HC4)          t          p       LLCI       ULCI
constant     7.1591     1.0018     7.1460      .0000     5.1373     9.1809
TRTMT        3.2727     1.2888     2.5394      .0149      .6719     5.8736
```

c (points to TRTMT)

(continued)

TABLE 10.1 (continued)

Mediation Model SPSS Output

```
************** TOTAL, DIRECT, AND INDIRECT EFFECTS OF X ON Y **************
```

The total effect of *X* on *Y* is calculated as:

$$c' + ab = .9053 + 2.3674 = 3.2727$$

Which indicates those in the treatment group were, on average, about 3-1/4 units higher on the outcome than those in the control group.

c_ps is the partially standardized direct effect calculated as

$$c_{ps} = \frac{c}{SD_Y} = \frac{3.2727}{4.53708} = .7213$$

This is interpreted as two cases that differ by one unit on *X* will differ by about ¾ of one standard deviation on *Y* as a result of the *combined direct and indirect effects* by which *X* affects *Y*

```
Total effect of X on Y
     Effect     se(HC4)          t          p       LLCI       ULCI       c_ps
     3.2727      1.2888     2.5394      .0149      .6719     5.8736      .7213
```

This provides information on the direct effect of *X* on *Y* (i.e., path c')

```
Direct effect of X on Y
     Effect     se(HC4)          t          p       LLCI       ULCI      c'_ps
      .9053      1.3870      .6527      .5176    -1.8958     3.7065      .1995
```

This provides information on the mediating effect; the *ab* effect—the indirect effect of treatment assignment *(X)* on the dependent variable *(Y)* through the mediator. This is calculated simply as the product of the coefficients (with the difference here due to rounding):

$$ab = (3.00)(.7891) = 2.3673$$

c'_ps is the partially standardized direct effect calculated as

$$c'_{ps} = \frac{c'}{SD_Y} = \frac{.9053}{4.53708} = .1995$$

This is interpreted as, independent of the indirect effect, an individual in the treatment group is estimated to be about 20 standard deviations higher on *Y*

```
Indirect effect(s) of X on Y:
        Effect     BootSE   BootLLCI   BootULCI
M1      2.3674      .7573      .9776     3.9406
```

ab

The indirect effect of *X* on *Y* is the product of the effect of the independent variable, *X*, on the mediating variable, *M*, (i.e., path *a*) and the effect of *M* on the outcome, *Y*, when *X* is held constant (i.e., path *b*). The 95% bootstrap confidence intervals are provided.

The partially standardized indirect effect of *X* on *Y* is a measure of effect size calculated as:

$$ab_{ps} = \frac{ab}{SD_Y} = \frac{2.3674}{4.5371} = .5218$$

```
Partially standardized indirect effect(s) of X on Y:
        Effect     BootSE   BootLLCI   BootULCI
M1       .5218      .1652      .2261      .8722
```

The partially standardized indirect effect rescales *ab* to the stand… the original metric of *X*. Thus, the size of the partially standardized indirect effect depends on the scale of *X*.

The **partially standardized indirect** effect of *X* on *Y* tells us that two individuals who differ by one unit on *X* differ by about # standard deviation units on *Y* as a result of the effect of *M*, which in turn affects *Y*. In this illustration, since *X* is binary, the partially standardized indirect effect of *X* on *Y* tells us that two individuals who differ by one unit on *X* differ by about ½ of one standard deviation unit on *Y*, on average, as a result of the effect of *M*, which in turn affects *Y*.
More specifically, those in the treatment group were, on average, about ½ of one standard deviation higher on the outcome as a result of the indirect effect through the mediating variable, *M*, than those in the control condition.

The bootstrapped confidence interval tells us that the this difference could be as low as about ¼ of one standard deviation and as high as nearly 9/10 of one standard deviation.

TABLE 10.1 (continued)

Mediation Model SPSS Output

```
*********** BOOTSTRAP RESULTS FOR REGRESSION MODEL PARAMETERS ************

OUTCOME VARIABLE:
 M1

              Coeff    BootMean     BootSE    BootLLCI    BootULCI
constant     7.9545      7.9586      .5889      6.8000      9.1111
TRMT         3.0000      2.9919      .7981      1.4060      4.5513

----------

OUTCOME VARIABLE:
 DV

              Coeff    BootMean     BootSE    BootLLCI    BootULCI
constant      .8819       .9200     1.6775     -2.3398      4.2559
TRMT          .9053       .9525     1.3869     -1.7333      3.7262
M1            .7891       .7833      .1888       .3864      1.1410

********************** ANALYSIS NOTES AND ERRORS ************************

Level of confidence for all confidence intervals in output:
  95.0000

Number of bootstrap samples for percentile bootstrap confidence intervals:
  5000

NOTE: Standardized coefficients for dichotomous or multicategorical X are in
      partially standardized form.

NOTE: A heteroscedasticity consistent standard error and covariance matrix
estimator was used.

------ END MATRIX -----
```

Callout labels: *a* (constant/TRMT, outcome M1); *c'* (constant/TRMT, outcome DV); *b* (M1)

Callout: The bootstrapped confidence interval results lend evidence to support the conclusions from the hypothesis tests presented earlier. Among others, that *M* fully mediates the relationship between *X* and *Y* given that 0 is within the interval for *X* and is not within the interval for *M*.

Cribari-Neto, F. (2004). Asymptotic inference under heteroskedasticity of unknown form. *Computational Statistics and Data Analysis, 45*, 215-233. doi:10.1016/S0167-9473(02)00366-3

Hayes, A. F., & Cai, L. (2007). Using heteroskedasticity-consistent standard error estimators in OLS regression: An introduction and software implementation. *Behavior Research Methods, 39*(4), 709-722.

10.3.3 Computing Moderation Analysis Using SPSS

Next we consider SPSS for computing moderation. Before we conduct the analysis, let us review the data. We are using the "Ch10_medmod.sav" data. For this illustration, the dependent variable is "DV," the independent variable is "TRTMT," and the moderating variable is "W1." Note in Figure 10.2 that there is no interaction term, *XW*, that represents the interaction between the moderating variable and independent variable. Using the PROCESS macro, there is no need to compute the interaction.

Step 1. To conduct a simple moderation model, the first step is the same as followed for conducting mediation. Go to "Analyze" in the top pulldown menu, then select "Regression," and then select "PROCESS." Following the screenshot presented earlier in Figure 10.6 (Step 1) produces the "PROCESS" dialog box.

Step 2. Click the dependent variable (e.g., "DV") and move it into the "Y variable" box by clicking the arrow button. Click the independent variable (e.g., "TRTMT") and move into the "X Variable" box by clicking the arrow button. Click the moderating variable (e.g., W) and move to the "Moderator (W)" box by clicking the arrow button (see the screenshot for Step 2, Figure 10.9). We are using model 1 from the templates (Hayes, 2013), so use the toggle menu to select "1" for "Model number." We will leave the default settings for confidence intervals (i.e., 95) and number of bootstrap samples (e.g., 5000). To obtain the bootstrap inference for model coefficients, place a check in the respective box.

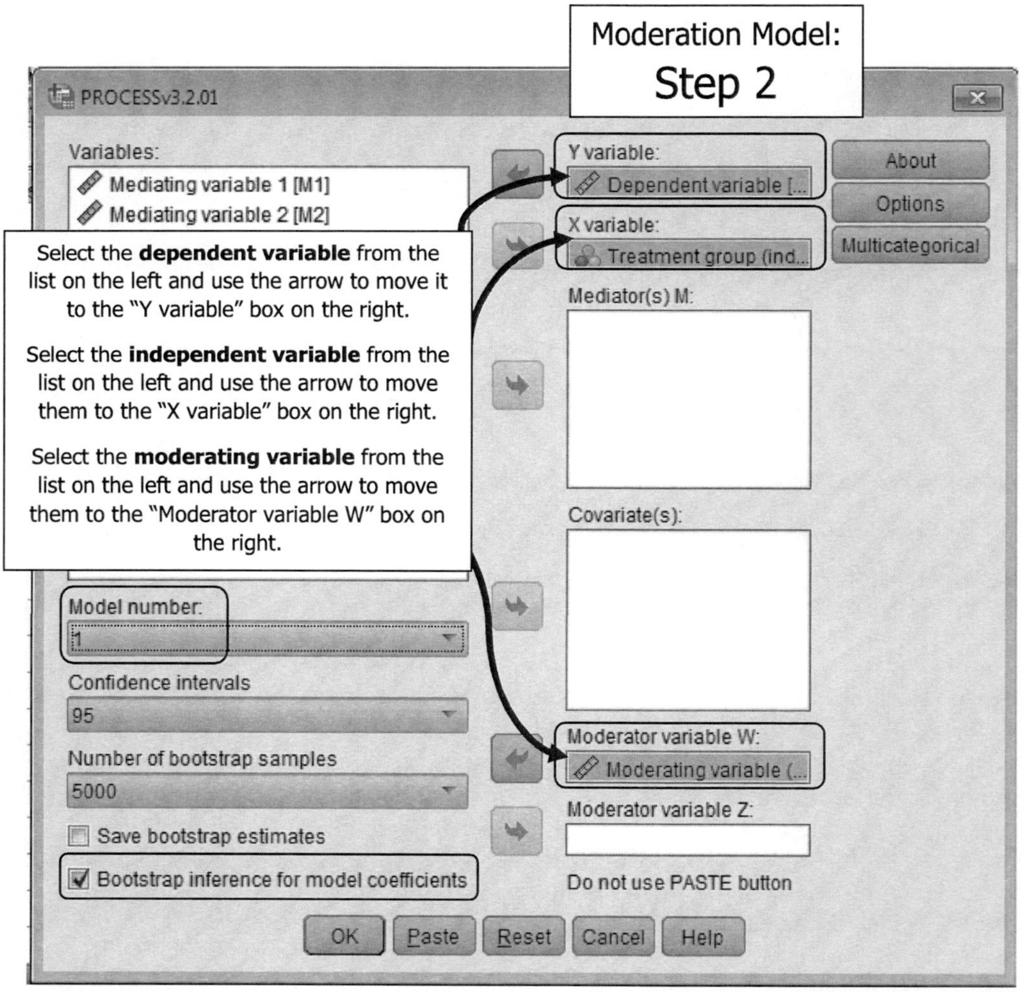

FIGURE 10.9
Simple moderation: Step 2.

Step 3. Clicking on Options from the main dialog box (see Figure 10.9) will produce the dialog box that will allow us to make a number of selections for the output. For this illustration, we place a checkmark in the respective box to generate code for visualizing interactions. Using the toggle menu, we select "HC4" (Cribari-Neto, 2004) for the heteroscedasticity-consistent inference. HC4 takes large leverage values into account when constructing the standard errors, and has been shown to outperform HC3 in the presence of high leverage points and error distributions that are nonnormal (Hayes & Cai, 2007). Using HC3 or HC4 is recommended (Hayes & Cai, 2007). In the "Moderation and conditioning" box, we leave the default settings for "Probe interactions," and place a checkmark for "Johnson-Neyman output." Then click "Continue" to return to the main dialog box. Note that the default conditioning values for probing interactions is the 16th, 50th, and 84th percentiles. When W is normally distributed, this corresponds, respectively, to one standard deviation below the mean, the mean, and one standard deviation above the mean.

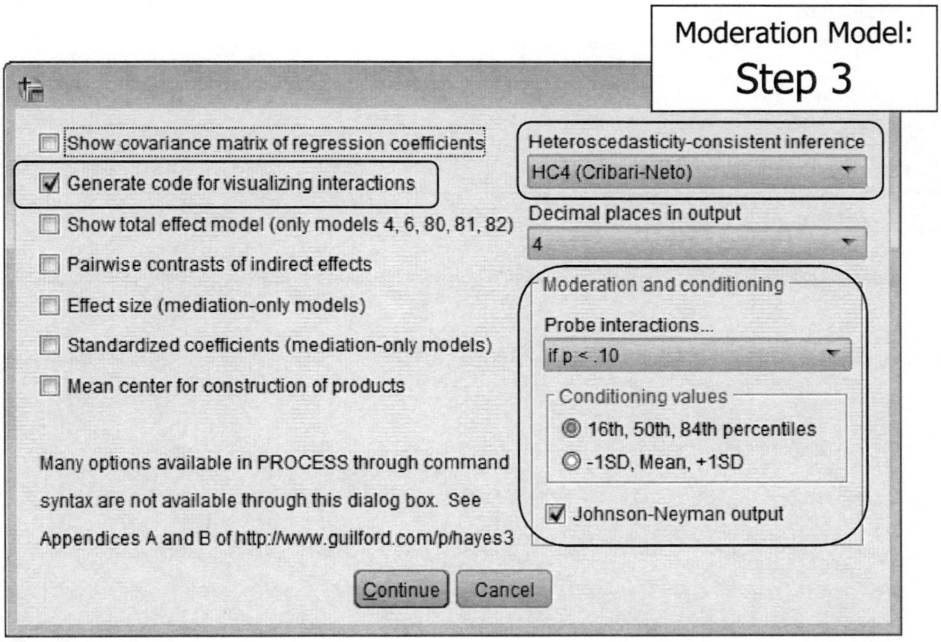

FIGURE 10.10
Moderation model: Step 3.

10.3.3.1 Interpreting Moderation Output

Annotated results are presented in Table 10.2. The OLS regression coefficient for b_1 (i.e., X) is −3.8411. This is interpreted as the difference in the outcome (noted as *DV* on the output) between the treatment and control group among those with a value of 0 on the moderating variable, W (noted as *W1* in the output). Of cases with $W = 0$, those in the treatment group ($X = 1$) had lower values on the dependent variable as noted by the negative sign of the coefficient. Mathematically this is a correct interpretation; however, 0 does not occur in our range of W in this particular example. Thus, in this example, this interpretation interpolates beyond the range of the available data and provides an example where centering makes sense.

The OLS regression coefficient for b_2 (i.e., W) is −1.5811. This coefficient is interpreted as the difference in the dependent variable between two cases that differ by one unit on W

when X is 0. In this illustration, X of zero refers to the control group so this *is* an interpretable coefficient. This is the conditional effect of the moderating variable, W, on the outcome, Y, among those in the control group (i.e., $X = 0$). The sign for the coefficient is negative, which indicates that those higher on W had lower scores on the outcome variable.

The regression coefficient for the interaction, *Int_1*, is .7549 and indicates how the effect of X on the dependent variable changes as W changes by one unit. In this illustration, the interaction is statistically significant, which suggests the effect of X (independent variable) on Y (dependent variable) depends on W (moderating variable). As W increases by one unit, the difference in the dependent variable between those in the treatment and control group increases by .7549 units (i.e., a positive effect, so the effect moves toward larger values on the number as W increases).

As noted previously, we requested probing interactions with the default conditioning values of the 16th, 50th, and 84th percentiles, and when W is normally distributed, this corresponds, respectively, to one standard deviation below the mean, the mean, and one standard deviation above the mean. We see in the graph that the moderating effect of W on X occurs at higher values of the moderator.

TABLE 10.2

Moderation Model SPSS Output

Matrix

```
Run MATRIX procedure:

************** PROCESS Procedure for SPSS Version 3.2.01 *****************

          Written by Andrew F. Hayes, Ph.D.       www.afhayes.com
    Documentation available in Hayes (2018). www.guilford.com/p/hayes3

**************************************************************************
Model  : 1
    Y  : DV
    X  : TRTMT
    W  : W1

Sample
Size:  44

**************************************************************************
OUTCOME VARIABLE:
 DV

Model Summary
          R       R-sq        MSE     F(HC4)        df1        df2          p
      .8320      .6923     6.8099    99.0793     3.0000    40.0000      .0000

Model
              coeff    se(HC4)          t          p       LLCI       ULCI
constant    20.9575      .8565    24.4677      .0000    19.2263    22.6886
TRTMT (b1)  -3.8411     1.4887    -2.5802      .0137    -6.8499     -.8323
W1 (b2)     -1.5811      .1011   -15.6386      .0000    -1.7854    -1.3767
Int_1 (b3)    .7549      .2192     3.4434      .0014      .3118     1.1980

Product terms key:
 Int_1     :        TRTMT       x          W1

Test(s) of highest order unconditional interaction(s):
       R2-chng     F(HC4)        df1        df2          p
X*W      .0477    11.8569     1.0000    40.0000      .0014

----------
    Focal predict: TRTMT      (X)
          Mod var: W1         (W)
```

b_1

b_2

b_3

The output provides information on the model template that was used (#1), the variables specified (*Y* is the dependent variable, *X* is the independent variable, and *M* is the moderating variable), and the sample size.

OLS regression output

The interaction term, *XW*, created by PROCESS is denoted as *Int_1*

Test of the interaction term, *XW*

TABLE 10.2 (continued)

Moderation Model SPSS Output

The **first row** represents the effect of *X* on *Y* conditioned on *W* being low (i.e., 16th percentile or one standard deviation below the mean)—an effect of −.0667. The second row represents the effect of *X* on *Y* conditioned on *W* being moderate (i.e., 50th percentile)—an effect of 2.9528. The third row represents the effect of *X* on *Y* conditioned on *W* being high (i.e., 84th percentile or one standard deviation above the mean)—an effect of 4.4626.

Conditional effects of the focal predictor at values of the moderator(s):

W1	Effect	se(HC4)	t	p	LLCI	ULCI
5.0000	-.0667	.6851	-.0973	.9230	-1.4513	1.3180
9.0000	2.9528	.8956	3.2972	.0021	1.1428	4.7629
11.0000	4.4626	1.2430	3.5902	.0009	1.9504	6.9748

These "conditional effects" values are based on the equation: $\theta_{X \to Y} = b_1 + b_3 W$ and the effect (i.e., regression coefficient) represents the effect of *X* on *Y* among those relatively low (i.e., 16th percentile), moderate (i.e., 50th percentile), and high (i.e., 84th percentile) on *W*, the moderating variable.

Moderator value(s) defining Johnson-Neyman significance region(s):

Value	**% below**	**% above**
6.8479	**27.2727**	**72.7273**

Johnson-Neyman results for probing an interaction. The region of significance for the effect of *X* on *Y*. The Johnson-Neyman technique shows that the relationship between *X* and *Y* is significant when *W* is **greater than 6.8479** but not significant with lower values.

Conditional effect of focal predictor at values of the moderator:

W1	Effect	se(HC4)	t	p	LLCI	ULCI
3.0000	-1.5764	.9379	-1.6807	.1006	-3.4721	.3193
3.5000	-1.1990	.8609	-1.3928	.1714	-2.9389	.5409
4.0000	-.8215	.7915	-1.0379	.3055	-2.4213	.7782
4.5000	-.4441	.7321	-.6067	.5475	-1.9237	1.0354
5.0000	-.0667	.6851	-.0973	.9230	-1.4513	1.3180
5.5000	.3108	.6533	.4757	.6369	-1.0097	1.6312
6.0000	.6882	.6390	1.0770	.2879	-.6033	1.9797
6.5000	1.0657	.6433	1.6565	.1055	-.2346	2.3659
6.8479	1.3283	.6572	2.0211	**.0500**	.0000	2.6565
7.0000	1.4431	.6659	2.1671	.0362	.0972	2.7890
7.5000	1.8205	.7050	2.5823	.0136	.3956	3.2454
8.0000	2.1980	.7581	2.8994	.0060	.6658	3.7301
8.5000	2.5754	.8224	3.1316	.0032	.9133	4.2375
9.0000	2.9528	.8956	3.2972	.0021	1.1428	4.7629
9.5000	3.3303	.9756	3.4137	.0015	1.3586	5.3020
10.0000	3.7077	1.0609	3.4949	.0012	1.5636	5.8519
10.5000	4.0852	1.1503	3.5513	.0010	1.7602	6.4101
11.0000	4.4626	1.2430	3.5902	.0009	1.9504	6.9748
11.5000	4.8400	1.3382	3.6167	.0008	2.1353	7.5448
12.0000	5.2175	1.4355	3.6345	.0008	2.3161	8.1189
12.5000	5.5949	1.5345	3.6461	.0008	2.4936	8.6963
13.0000	5.9724	1.6348	3.6532	.0007	2.6682	9.2765

(continued)

TABLE 10.2 (continued)

Moderation Model SPSS Output

```
Data for visualizing the conditional effect of the focal predictor:
Paste text below into a SPSS syntax window and execute to produce plot.
```

```
DATA LIST FREE/
   TRTMT      W1          DV          .
BEGIN DATA.
      .0000      5.0000     13.0522
     1.0000      5.0000     12.9855
      .0000      9.0000      6.7279
     1.0000      9.0000      9.6807
      .0000     11.0000      3.5658
     1.0000     11.0000      8.0284
END DATA.
GRAPH/SCATTERPLOT=
 W1        WITH      DV        BY        TRTMT      .
```

To visualize the graph, copy and paste into SPSS syntax as noted. The graph, based on this syntax, is pasted here.

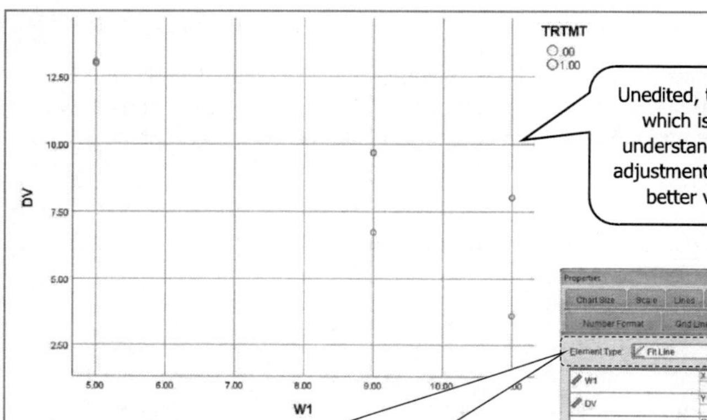

Unedited, the graph appears like this, which isn't completely helpful in understanding the interaction. A few adjustments in chart editor will help to better visualize the interaction.

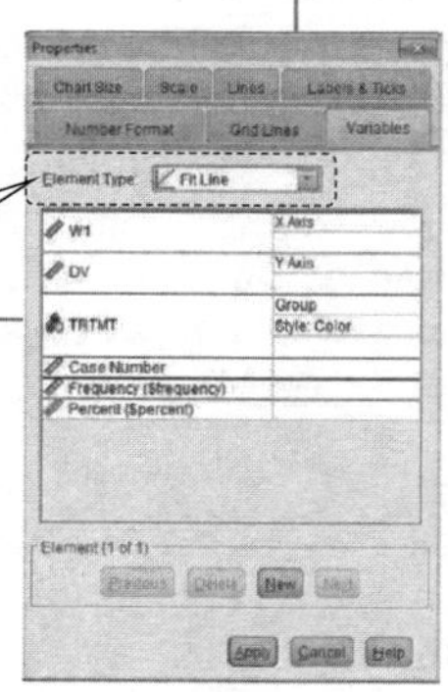

Double click on the graph to enact **chart editor** in SPSS. Go to "Edit" in the top toolbar in chart editor and select "Select *X* axis." This will bring up the **"Properties"** toolbox (shown here). From here, click on "Variables" and change the element type to "Fit line." Click on "Apply" to see lines (rather than dots).

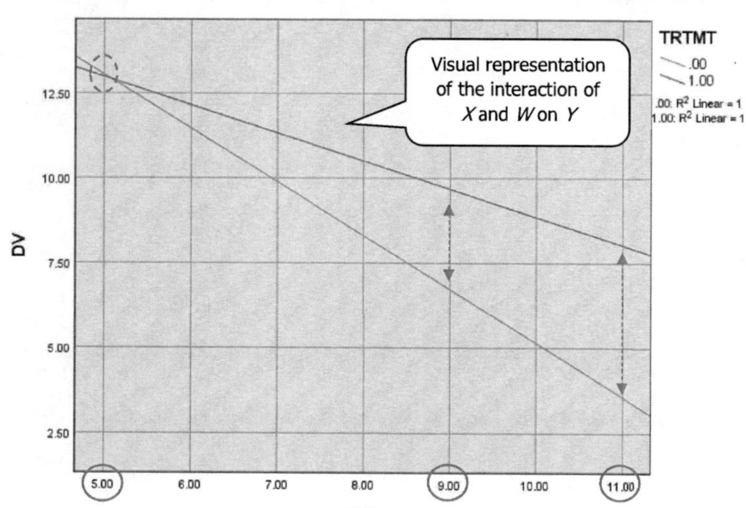

Visual representation of the interaction of *X* and *W* on *Y*

TABLE 10.2 (continued)

Moderation Model SPSS Output

The **first row** represents the effect of *X* on *Y* conditioned on *W* being low (i.e., 16th percentile or one standard deviation below the mean)—an effect of -.0667. The second row represents the effect of *X* on *Y* conditioned on *W* being moderate (i.e., 50th percentile)—2.9528. The third row represents the effect of *X* on *Y* conditioned on *W* being high (i.e., 84th percentile or one standard deviation above the mean)—2.9528.

```
Conditional effects of the focal predictor at values of the moderator(s):

      W1      Effect    se(HC4)        t         p       LLCI       ULCI
  5.0000      -.0667      .6851    -.0973     .9230   -1.4513     1.3180
  9.0000      2.9528      .8956    3.2972     .0021    1.1428     4.7629
 11.0000      4.4626     1.2430    3.5902     .0009    1.9504     6.9748
```

For example, given the conditional effects presented earlier (copied and pasted here in the box above just to make it easier), when ***W* = 5.00**, i.e., *relatively low—one standard deviation below the mean*—**the conditional effect of *X* on *Y* is –.667,** computed as follows. We see at this point, one standard deviation below the mean, there is not a statistically significant conditional effect ($p = .9230$), and that is visualized on the graph by the lines for the two groups *X* in overlaying each other.

$$\theta_{X \to Y} = b_1 + b_3 W = -3.8411 + (.7549)(5.00) = -3.8411 + 3.7745 = -.066$$

```
*********** BOOTSTRAP RESULTS FOR REGRESSION MODEL PARAMETERS ****

OUTCOME VARIABLE:
 DV

              Coeff    BootMean      BootSE    BootLLCI    BootULCI
constant    20.9575     21.0859       .8917     19.7238     23.1469
TRTMT       -3.8411     -3.8229      1.5965     -6.7705      -.3999
W1          -1.5811     -1.5949       .1051     -1.8312     -1.4212
Int_1         .7549       .7500       .2276       .2648      1.1572
```

The bootstrapped confidence interval results lend evidence to support the conclusions from the hypothesis tests presented earlier. Among others, that *W* moderates the relationship between *X* and *Y* given that 0 is not within the interval for *Int_1,* the interaction term.

```
*********************** ANALYSIS NOTES AND ERRORS ************************

Level of confidence for all confidence intervals in output:
  95.0000

Number of bootstrap samples for percentile bootstrap confidence intervals:
  5000

W values in conditional tables are the 16th, 50th, and 84th percentiles.

NOTE: A heteroscedasticity consistent standard error and covariance matrix
estimator was used.
```

```
------ END MATRIX -----
```

10.4 Computing Mediation and Moderation Using R

Next we consider **R** for computing mediation and moderation The commands are provided within the blocks with additional annotation to assist in understanding how the command works. Should you want to write reminder notes and annotation to yourself as you write the commands in **R** (and we highly encourage doing so), remember that any text that follows a hashtag (i.e., #) is annotation only and not part of the **R** code. Thus, you can write annotations directly into **R** with hashtags. We encourage this practice so that when you call up the commands in the future, you'll understand what the various lines of code are doing. You may think you'll remember what you did. However, trust us. There is a good chance that you won't. Thus, consider it best practice when using **R** to annotate heavily!

10.4.1 Reading Data Into R

```
getwd()
```

R is always pointed to a directory on your computer. The *get working directory* function can be used to determine to which directory **R** is pointed. We will assume that we need to change the working directory, and will use the next line of code to set the working directory to the desired path.

```
setwd("E:/FolderName")
```

We use the *setwd* function to establish the working directory. To set the working directory, change what is in quotation marks to your file location. Also, if you are copying the directory name from your properties, you will need to change the backslash (i.e., \) to a forward slash (i.e., /).

```
Ch10_med <- read.csv("Ch10_medmod.csv")
```

The *read.csv* function reads our data into **R**. What's to the left of the "<-" will be what the data will be called in **R**. In this example, we're calling the R dataframe "Ch10_med." What's to the right of the "<-" tells **R** to find this particular csv file. In this example, our file is called "Ch10_medmod.csv." Make sure the extension (i.e., .csv) is included in your script. Also note that the name of your file should be in quotation marks within the parentheses.

```
names(Ch10_med)
```

The *names* function will produce a list of variable names for each dataframe as follows. This is a good check to make sure your data have been read in correctly.

```
[1] "DV"     "TRTMT" "M1"    "M2"    "M3"    "W1"    "W2"
"
```

```
View(Ch10_med)
```

The *View* function will let you view the dataset in spreadsheet format in RStudio.

```
Ch10_med$TRTMTf <- factor(Ch10_med$TRTMT,
labels = c("treatment", "control"))
```

FIGURE 10.11
Reading data into **R**.

This command will create a new variable in our dataframe named "TRTMTf." We use the *factor* function to define the variable TRTMT as nominal with the two groups defined here (i.e., treatment, control). What is to the left of "<-" in the script creates the new TRTMTf variable in our dataframe.

```
summary(Ch10_med)
```

The *summary* function will produce basic descriptive statistics on all the variables in your dataframe. This is a great way to quickly check to see if the data have been read in correctly and to get a feel for your data, if you haven't already. The output from the summary statement for this dataframe looks like this. Because we defined TRTMTf as a factor, we are provided only the frequencies for each category in that variable.

```
       DV              TRTMT          M1               M2
Min.    : 1.000  Min.    :0.0  Min.    : 3.000  Min.    : 4.000
1st Qu.: 5.250  1st Qu.:0.0  1st Qu.: 7.750  1st Qu.: 7.000
Median :10.500  Median :0.5  Median :10.000  Median : 7.000
Mean    : 8.795  Mean    :0.5  Mean    : 9.455  Mean    : 8.136
3rd Qu.:13.500  3rd Qu.:1.0  3rd Qu.:12.000  3rd Qu.: 9.250
Max.    :13.500  Max.    :1.0  Max.    :16.000  Max.    :14.000
       M3               W1               W2            TRTMTf
Min.    : 0.000  Min.    : 3.000  Min.    : 5.00  control   :22
1st Qu.: 0.750  1st Qu.: 6.000  1st Qu.:10.00  treatment :22
Median : 3.000  Median : 9.000  Median :15.00
Mean    : 4.455  Mean    : 8.409  Mean    :14.45
3rd Qu.: 6.250  3rd Qu.:10.000  3rd Qu.:19.00
Max.    :16.000  Max.    :13.000  Max.    :25.00
```

FIGURE 10.11
Reading data into **R**.

10.4.2 Generating a Mediation Model Using R

```
model_a <- lm(M1 ~ TRTMTf,
              Ch10_med)
```

The *lm* function is used to generate a linear regression model with the mediating variable, "M1," as the outcome and the treatment variable, TRTMTf, as the independent variable. The data come from "Ch10_med" and the object is named "model_a." This will produce the coefficient for *a*.

```
summary(model_a)
```

The *summary* function is used to produce the output for model_a.

```
Call:
lm(formula = M1 ~ TRTMTf, data = Ch10_med)

Residuals:
    Min       1Q  Median      3Q     Max
-4.9545  -1.9545  0.0455  2.0455  5.0455

Coefficients:
              Estimate Std.  Error t value Pr(>|t|)
(Intercept)      7.9545  0.5678  14.009  < 2e-16 ***
TRTMTftreatment  3.0000  0.8030  3.736  0.000558 ***
—
Signif. codes: 0 '***' 0.001 '**' 0.01 '*' 0.05 '.' 0.1 ' ' 1
```

FIGURE 10.12
Generating a mediating model in **R**.

```
Residual standard error: 2.663 on 42 degrees of freedom
Multiple R-squared:  0.2494,  Adjusted R-squared:  0.2316
F-statistic: 13.96 on 1 and 42 DF, p-value: 0.0005579
```

```
model_bc <- lm(DV ~ TRTMTf + M1,
              Ch10_med)
```

The *lm* function is used to generate a linear regression model with the dependent variable, DV, as the outcome and the treatment variable, TRTMTf, and mediating variable, M1, as the independent variables. The data come from "Ch10_med" and the object is named "model_bc." This will produce coefficients for c' and b.

```
summary(model_bc)
```

The *summary* function is used to produce the output for model_bc.

```
Call:
lm(formula = DV ~ TRTMTf + M1, data = Ch10_med)

Residuals:
    Min       1Q  Median      3Q     Max
-7.3894  -2.4932  0.5241  3.0323  6.3050

Coefficients:
                Estimate Std.  Error  t value  Pr(>|t|)
(Intercept)       0.8819      1.9129    0.461  0.647223
TRTMTftreatment   0.9053      1.3110    0.691  0.493744
M1                0.7891      0.2183    3.616  0.000812 ***
—
Signif. codes: 0 '***' 0.001 '**' 0.01 '*' 0.05 '.' 0.1 ' ' 1

Residual standard error: 3.767 on 41 degrees of freedom
Multiple R-squared:  0.3427,  Adjusted R-squared:  0.3106
F-statistic: 10.69 on 2 and 41 DF, p-value: 0.0001838
```

```
model_c <- lm(DV ~ TRTMTf,
              Ch10_med)
```

The *lm* function is used to generate a linear regression model with the dependent variable, DV, as the outcome and the treatment variable, TRTMTf, as the independent variable. The data come from Ch10_med and the object is named "model_c." This will produce the coefficient for path *c*.

```
summary(model_c)
```

The *summary* function is used to produce the output for *model_c*.

```
Call:
lm(formula = DV ~ TRTMTf, data = Ch10_med)

Residuals:
    Min       1Q  Median      3Q     Max
-8.9318  -4.1591  0.9545  3.0682  6.3409

Coefficients:
                Estimate  Std. Error  t value  Pr(>|t|)
(Intercept)       7.1591      0.9113    7.856  8.9e-10 ***
TRTMTftreatment   3.2727      1.2888    2.539  0.0149 *
—
Signif. codes: 0 '***' 0.001 '**' 0.01 '*' 0.05 '.' 0.1 ' ' 1

Residual standard error: 4.274 on 42 degrees of freedom
Multiple R-squared: 0.1331, Adjusted R-squared: 0.1125
F-statistic: 6.449 on 1 and 42 DF, p-value: 0.01489
```

FIGURE 10.12 (conitinued)
Generating a mediating model in **R**.

```
install.packages("mediation")
library(mediation)
```

The *install.packages* function is used to install the package, *mediation*. The *library* function is used to load the package into our library.

```
med1 <- mediate(model_a, model_bc,
                treat ='TRTMTf', mediator = 'M1',
                boot = TRUE, sims = 5000)
```

The *mediate* function is used to compute bootstrapped confidence intervals from our models which estimated the coefficients for *a*, *b*, and c', which were model_a and model_bc. The script, "sims = 5000," will generate 5000 bootstrapped samples.

```
summary(med1)
```

The *summary* function is used to produce the output for the bootstrapped results from *med1*. **ACME** is the **average causal mediation effect**, or *ab*, and is 2.367 in this model. The ACME is the mediation effect and is the indirect effect of *X* on *Y* (i.e., the effect of *X* on *Y* through the mediator). An ACME confidence interval that is statistically significant indicates a statistically significant mediating effect. **ADE** is the **average direct effect**, or c', and is .905 in this model. This is the direct effect of *X* on *Y* after taking into account the mediating (or indirect) effect of *M*. From the results, we also see the *total effect* is 3.273. This is the coefficient for *c* and is the total effect of *X* on *Y* without the mediator. It is calculated as the sum of the indirect (i.e., mediation) effect (i.e., 2.367 in this model) and direct effect (i.e., .905 in this model).

```
Causal Mediation Analysis

Nonparametric Bootstrap Confidence Intervals with the Percentile Method
               Estimate 95% CI Lower 95% CI Upper  p-value
ACME              2.367        0.992        3.935     0.00
ADE               0.905       -1.676        3.741     0.49
Total Effect      3.273        0.750        5.654     0.01
Prop. Mediated    0.723        0.244        2.481     0.01

Sample Size Used: 44
```

```
plot(med1)
```

The *plot* function can be used to generate a plot of the confidence intervals. We see that the average direct effect confidence interval crosses zero, indicating that the direct effect of *X* on *Y* after taking the mediator into account is not statistically significant. In comparison, the mediation effect (ACME) and the total effect do not cross zero, indicating those effects are statistically significant.

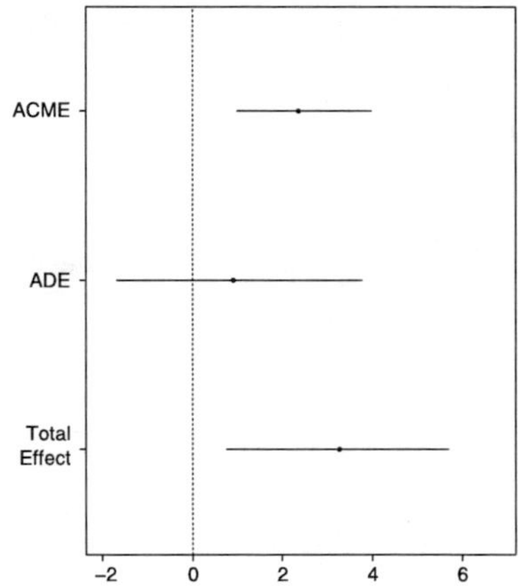

FIGURE 10.12 (conitinued)
Generating a mediating model in **R**.

Mediation models, as with most procedures, can be computed with different packages, and different packages provide different tools and output. Let's look at an example using the *MBESS* package, which provides a number of effect size estimates.

```
install.packages("MBESS")
library(MBESS)
```

The *MBESS* package is first installed, then loaded into our library using the *install.packages* and *library* functions.

```
mediation(x=Ch10_med$TRTMT,
          mediator=Ch10_med$M1,
          dv=Ch10_med$DV)
```

Next, we define our mediation model with *X*, *mediator*, and *dv* from the dataframe Ch10_med.

The estimates for path *c* are provided first.

```
$Y.on.X
$Y.on.X$Regression.Table
                  Estimate Std. Error   t value       p(>|t|)  Low Conf Limit
Intercept.Y_X   7.159091  0.9112933  7.855968  8.900653e-10       5.3200265
c (Regressor)   3.272727  1.2887634  2.539432  1.489384e-02       0.6718975
                Up Conf Limit
Intercept.Y_X        8.998155
c (Regressor)        5.873557

$Y.on.X$Model.Fit
       Residual standard error (RMSE) numerator df denominator df
Values                       4.274345            1             42
       F-Statistic p-value (F)      R^2   Adj R^2 Low Conf Limit
Values    6.448716  0.01489384 0.133104 0.1124636    0.004776991
       Up Conf Limit
Values     0.3505862
```

The estimates for path *a* are provided next.

```
$M.on.X
$M.on.X$Regression.Table
                  Estimate Std. Error    t value       p(>|t|)  Low Conf Limit
Intercept.M_X   7.954545  0.5678137  14.009075  1.986718e-17        6.808651
a (Regressor)   3.000000  0.8030099   3.735944  5.579076e-04        1.379460
                Up Conf Limit
Intercept.M_X         9.10044
a (Regressor)         4.62054

$M.on.X$Model.Fit
       Residual standard error (RMSE) numerator df denominator df
Values                       2.663282            1             42
       F-Statistic  p-value (F)       R^2   Adj R^2 Low Conf Limit
Values    13.95728 0.0005579076 0.2494274 0.2315566     0.05511564
       Up Conf Limit
Values     0.4743653
```

FIGURE 10.12 (conitinued)
Generating a mediating model in **R**.

The estimates for paths c' and b are provided next.

```
$Y.on.X.and.M
$Y.on.X.and.M$Regression.Table

                        Estimate Std. Error   t value       p(>|t|)
Intercept.Y_XM         0.8818695  1.9128789 0.4610169  0.6472228096
c.prime (Regressor)    0.9053181  1.3110235 0.6905430  0.4937437886
b (Mediator)           0.7891364  0.2182535 3.6156865  0.0008122274
                       Low Conf Limit Up Conf Limit
Intercept.Y_XM             -2.9812678      4.745007
c.prime (Regressor)        -1.7423477      3.552984
b (Mediator)                0.3483644      1.229908

$Y.on.X.and.M$Model.Fit
       Residual standard error (RMSE) numerator df denominator df
Values                       3.767066            2             41
       F-Statistic  p-value (F)      R^2   Adj R^2 Low Conf Limit
Values    10.68782 0.0001837648 0.342692 0.3106282      0.1012886
       Up Conf Limit
Values      0.547119
Many effect size values are estimated.
```

Many effect size values are estimated.

```
$Effect.Sizes
                                                             Estimate
Indirect.Effect                                            2.36740922
Indirect.Effect.Partially.Standardized                     0.52179137
Index.of.Mediation                                         0.26391192
R2_4.5                                                     0.12545916
R2_4.6                                                     0.06030367
R2_4.7                                                     0.17597043
Maximum.Possible.Mediation.Effect                          9.35535239
ab.to.Maximum.Possible.Mediation.Effect_kappa.squared      0.25305399
Ratio.of.Indirect.to.Total.Effect                          0.72337504
Ratio.of.Indirect.to.Direct.Effect                         2.61500276
Success.of.Surrogate.Endpoint                              1.09090909
Residual.Based_Gamma                                       0.10865859
Residual.Based.Standardized_gamma                          0.11610848
ES.for.two.groups                                          0.78337195
SOS                                                        0.94256516
```

The *MBESS* package documentation provides a summary of the effect size measures provided in the output (Kelley, 2018, pp. 66–67):

- Indirect.Effect = ab
- Indirect.Effect.Partially.Standardized = $ab_{ps} = \frac{ab}{SD_Y}$ (MacKinnon, 2008)
- Index.of.Mediation = $ab\left(\frac{SD_X}{SD_Y}\right)$ (Preacher & Hayes, 2008)
- R2_4.5 = index of explained variance (MacKinnon, 2008)
- R2_4.6 = index of explained variance (MacKinnon, 2008)
- R2_4.7 = index of explained variance (MacKinnon, 2008)
- Maximum.Possible.Mediation.Effect = "the maximum attainable value of the mediation effect (i.e., the indirect effect), in the direction of the observed indirect effect, that could have been observed, conditional on the sample variances and on the magnitudes of relationships among some of the variables" Kelley, 2018, p. 66

FIGURE 10.12 (conitinued)
Generating a mediating model in **R**.

- ab.to.Maximum.Possible.Mediation.Effect_kappa.squared = the proportion of the maximum possible indirect effect; the indirect effect is the numerator and the maximum possible mediation effect is the denominator (Preacher & Kelley, 2011)
- Ratio.of.Indirect.to.Total.Effect = ratio of the indirect effect to the total effect (Freedman, 2002); this effect size is also referred to as *mediation ratio* (Ditlevsen, Christensen, Lynch, Damsgaard, & Keiding, 2005) and as the *relative indirect effect* (Huang, Sivaganesan, Succop, & Goodman, 2004); "often loosely interpreted as the *relative indirect effect*" Kelley, 2018, p. 66
- Ratio.of.Indirect.to.Direct. Effect = ratio of the indirect effect to the direct effect (Sobel, 1982)
- Success.of.Surrogate.Endpoint = success of a surrogate endpoint (Buyse & Molenberghs, 1998)
- Residual.Based_Gamma = residual based index (Preacher & Kelley, 2011)
- Residual.Based.Standardized_gamma = standardized residual based index, where the scales of M and Y are removed by using standardized values of M and Y (Preacher & Kelley, 2011)
- ES.for.two.groups = Hansen and McNeal (1996) effect size for two groups, applicable when X is binary and coded with values of 0 and 1
- SOS = shared over simple effects (SOS) index; computed as "the ratio of the variance in Y explained by both X and M divided by the variance in Y explained by X" (Lindenberger & Pötter, 1998)

```
upsilon(Ch10_med$TRTMT, Ch10_med$M1, Ch10_med$DV,
        conf.level = 0.95,
        bootstrap = TRUE,
        bootstrap.package = "lavaan",
        bootstrap.type="ordinary", B = 1000,
        boot.data.out=FALSE)
```

To generate the upsilon effect size (Lachowicz et al., 2018), the *upsilon* function is used (note that at the time of writing, this function can be used with simple mediation models only). The first line defines X, M, and Y. Bootstrapped confidence intervals are generate with the "bootstrap = TRUE" script. The default bootstrap package is *lavaan*, and the other option is *boot*. The type of bootstrap confidence interval is *ordinary*, which is the default. When using *lavaan*, the other option is *bollen.stine*. We generate 1000 bootstrap replications with *B = 1000*. Bootstrapped data will be generated only if *boot.data.out = TRUE*. In this case, we have not requested the data by indicating *FALSE*. (Be patient—the bootstrapping may take several minutes to run!)

```
              Estimate 95% ordinary LCL 95% ordinary UCL
Upsilon     0.06964950       0.01251146        0.1860631
Adj Upsilon 0.06026125       0.00619668        0.1714112
```

FIGURE 10.12
Generating a mediating model in **R**.

10.4.3 Generating a Moderation Model Using R

So as not to be confusing with the mediation example, we will read our data in again, but this time call our dataframe a name unique to the moderation illustration, specifically "Ch10_mod."

```
getwd()
setwd("E:\filename")
Ch10_mod <- read.csv("Ch10_medmod.csv")
```

```
install.packages("devtools")
devtools::install_github("markhwhiteii/processr")
library(processr)
```

FIGURE 10.13
Generating a moderating model in **R**.

The *processr* package in **R** allows users to specify models 1, 4, 7, and 14. To access *processr*, we will first install *devtools* (if not already installed), and then run the "install_github" script to download *processr*. The *processr* package runs through *lavaan*, and *lavaan* requires continuous inputs. As such, when using *processr*, any dichotomous variables must be numeric and coded as 0 and 1 (i.e., we will not use the recoded factor variable for the treatment variable).

```
mod1result <- model1(iv="TRTMT", dv="DV",
                    mod="W1", data=Ch10_mod)
mod1result
```

We use the *model1* function to denote Model 1 from Hayes and define the independent variable, "iv," dependent variable, "DV," and moderating variable, "mod." The data come from Ch10_mod. We name the object "mod1result," and use the mod1result script to output our results. The results provide the coefficients along with estimates for three values of the moderating variable allowing us to see the effect of *X* on *Y* conditioned on *W* being low (one standard deviation below the mean)—an effect of 5.756, moderate (the mean)—an effect of 8.409, and high (one standard deviation above the mean)—an effect of 11.062.

```
# A tibble: 7 x 5
  term              estimate std.error statistic  p. value
  <chr>                <dbl>     <dbl>     <dbl>     <dbl>
1 intercept           21.0       1.88      11.1   8.01e-14
2 TRTMT               -3.84      2.66      -1.44  1.57e- 1
3 W1                  -1.58      0.206     -7.67  2.21e- 9
4 interaction          0.755     0.303      2.49  1.70e- 2
5 when W1 = 5.756      0.504     1.12       0.449 6.56e- 1
6 when W1 = 8.409      2.51      0.793      3.16  2.98e- 3
7 when W1 = 11.062     4.51      1.13       3.98   2.86e-4
```

```
mod2result <- model1(iv="W1", dv="DV",
mod="TRTMT", data=Ch10_mod)
mod2result
```

If we swap the roles of the independent and moderating variables, we can see the effect of the moderating variable at levels of the independent variable.

```
# A tibble: 6 x 5
  term              estimate std.error statistic  p. value
  <chr>                <dbl>     <dbl>     <dbl>     <dbl>
1 intercept           21.0       1.88       11.1  8.01e-14
2 W1                  -1.58      0.206      -7.67 2.21e- 9
3 TRTMT               -3.84      2.66       -1.44 1.57e- 1
4 interaction          0.755     0.303       2.49 1.70e- 2
5 when TRTMT = 0      -1.58      0.206      -7.67 2.21e- 9
6 when TRTMT = 1      -0.826     0.222      -3.72 6.17e- 4
```

Now let's look at another way to examine moderation using **R**. We will generate two models: one without the interaction term and one with the term, and will then compare the models.

```
Mod1 <- lm(DV ~ TRTMTf + W1,
           data = Ch10_mod)
```

The *lm* function is used to generate a linear regression model with the outcome, DV, and the treatment variable, TRTMTf, as the independent variable, and the moderating variable, W1. No interaction term is included in this model. The data come from Ch10_mod and the object is named "Mod1."

```
summary(fitMod)
```

FIGURE 10.13 (continued)
Generating a moderating model in **R**.

The *summary* function is used to produce the output.

```
Residuals:
    Min       1Q  Median      3Q     Max
-6.5795  -1.9992  0.4091  2.1770  4.8848

Coefficients:
                 Estimate Std. Error t value Pr(>|t|)
(Intercept)       17.9125     1.5195  11.788 9.49e-15 ***
TRTMTftreatment    2.4886     0.8415   2.958  0.00513 **
W1                -1.2322     0.1604  -7.681 1.83e-09 ***
—
Signif. codes: 0 '***' 0.001 '**' 0.01 '*' 0.05 '.' 0.1 ' ' 1
Residual standard error: 2.77 on 41 degrees of freedom
Multiple R-squared:  0.6445,  Adjusted R-squared:  0.6272
F-statistic: 37.17 on 2 and 41 DF, p-value: 6.181e-10
```

Next, we generate the same model but include the interaction term.

```
Mod2 <- lm(DV ~ TRTMTf + W1 + TRTMTf*W1,
data = Ch10_mod)
```

The *lm* function is used to generate a linear regression model with the moderating variable, *W1*, the outcome, *DV*, and the treatment variable, TRTMTf, as the independent variable, along with an interaction term *XW*, specifically *TRTMTf*W1* in this dataframe. The data come from Ch10_mod and the object is named "Mod2."

```
summary(Mod2)
```

The *summary* function is used to produce the output.

```
Residuals:
    Min       1Q  Median      3Q     Max
-7.3546  -1.1344  0.5217  1.7758  3.8193

Coefficients:
                    Estimate Std. Error t value Pr(>|t|)
(Intercept)          20.9575     1.8825  11.133 8.01e-14 ***
TRTMTftreatment      -3.8411     2.6624  -1.443    0.157
W1                   -1.5811     0.2061  -7.672 2.21e-09 ***
TRTMTftreatment:W1  0.7549     0.3031   2.490 0.017 *
—
Signif. codes: 0 '***' 0.001 '**' 0.01 '*' 0.05 '.' 0.1 ' ' 1
Residual standard error: 2.61 on 40 degrees of freedom
Multiple R-squared: 0.6923, Adjusted R-squared: 0.6692
F-statistic: 29.99 on 3 and 40 DF, p-value: 2.507e-10
```

```
anova(Mod1, Mod2)
```

Then we compare the models using the *ANOVA* function. We see there is a statistically significant difference ($p = .01701$).

```
Analysis of Variance Table

Model 1: DV ~ TRTMTf + W1
Model 2: DV ~ TRTMTf + W1 + TRTMTf * W1
  Res.Df    RSS Df Sum of Sq      F  Pr(>F)
1     41 314.63
2     40 272.40  1    42.236 6.2022 0.01701 *
—
Signif. codes: 0 '***' 0.001 '**' 0.01 '*' 0.05 '.' 0.1 ' ' 1
```

What about centering?

FIGURE 10.13 (continued)
Generating a moderating model in **R**.

```
Xc <- c(scale(Ch10_mod$TRTMT, center=TRUE, scale=FALSE))
```

Should we want to run the model with variables that are centered, this script will create a new variable, *Xc*, that is a *centered predictor*.

```
Wc<- c(scale(Ch10_mod$W1, center=TRUE, scale=FALSE))
```

Should we want to run the model with variables that are centered, this script will create a new variable, "Wc," that is a *centered moderator*.

```
fitMod2 <- lm(DV ~ TRTMTf + Wc + TRTMTf*Wc,
              data = Ch10_mod)
```

Let's first generate a model that centers *W* but not *X*, given that 0 for *X* is an interpretable value (i.e., the control group). The *lm* function is used to generate a linear regression model with the dependent variable, "DV," the centered moderating variable, "Wc," and the uncentered treatment variable, TRTMTf, as the independent variable. The data come from Ch10_mod and the object is named "fitMod2."

```
summary(fitMod2)
```

The *summary* function is used to produce the output. We see that the intercept is now 7.66 and is interpreted as the value of *Y* for those in the control group ($X = 0$) when *W* is at the average.

```
Residuals:
    Min      1Q  Median      3Q     Max
-7.3546 -1.1344  0.5217  1.7758  3.8193

Coefficients:
                    Estimate Std. Error t value Pr(>|t|)
(Intercept)           7.6622     0.5602  13.677  < 2e-16 ***
TRTMTftreatment       2.5068     0.7927   3.162  0.00298 **
Wc                   -1.5811     0.2061  -7.672 2.21e-09 ***
TRTMTftreatment:Wc    0.7549     0.3031   2.490  0.01701 *
---
Signif. codes: 0 '***' 0.001 '**' 0.01 '*' 0.05 '.' 0.1 ' ' 1

Residual standard error: 2.61 on 40 degrees of freedom
Multiple R-squared: 0.6923, Adjusted R-squared: 0.6692
F-statistic: 29.99 on 3 and 40 DF, p-value: 2.507e-10
```

If *both X* and *W* are centered, our script and output appears as such:

```
Xc <- c(scale(Ch10_mod$TRTMT, center=TRUE, scale=FALSE))
Wc<- c(scale(Ch10_mod$W1, center=TRUE, scale=FALSE))
fitMod3 <- lm(DV ~ Xc + Wc + Xc*Wc, data = Ch10_mod)
summary(fitMod3)
```

```
Residuals:
    Min      1Q  Median      3Q     Max
-7.3546 -1.1344  0.5217  1.7758  3.8193

Coefficients:
             Estimate Std. Error t value Pr(>|t|)
(Intercept)    8.9155     0.3964  22.494  < 2e-16 ***
Xc             2.5068     0.7927   3.162  0.00298 **
Wc            -1.2036     0.1516  -7.942 9.48e-10 ***
Xc:Wc          0.7549     0.3031   2.490  0.01701 *
---
Signif. codes: 0 '***' 0.001 '**' 0.01 '*' 0.05 '.' 0.1 ' ' 1
Residual standard error: 2.61 on 40 degrees of freedom
Multiple R-squared: 0.6923, Adjusted R-squared: 0.6692
F-statistic: 29.99 on 3 and 40 DF, p-value: 2.507e-10
```

FIGURE 10.13
Generating a moderating model in **R**.

10.5 Additional Resources

This chapter has provided a preview of conducting moderated and mediated regression. However, once again, space limitations prevent us from delving too deeply into these advanced topics. For those who are interested in a deeper dive, there are quite a number of excellent resources to turn, including the following, among many others:

- A comprehensive overview of moderation and mediation, including details on using the PROCESS macro in SPSS (Hayes, 2013)
- Dr. David A. Kenny's mediation website, http://davidakenny.net/cm/mediate.htm, and moderation website, http://davidakenny.net/cm/moderation.htm
- Dr. Andrew F. Hayes's webpage with links for SPSS, SAS, and Mplus macros and code, among other useful resources, www.afhayes.com/index.html

Problems

Conceptual Problems

1. A researcher is examining team performance and wants to look at the relationship between communication and collaboration. The researcher believes that communication may interact with timing. Which of the following types of models would you recommend the researcher examine?
   a. Mediation
   b. Moderation
   c. Neither
   d. Both
2. A researcher wants to examine the relationship between intelligence in early adulthood and physical performance in late adulthood. However, they believe there may be an indirect effect of intelligence on physical performance through education. Which of the following types of models would you recommend the researcher examine?
   a. Mediation
   b. Moderation
   c. Neither
   d. Both
3. A researcher wants to examine the relationship between stress and high-risk behavior and believes there may be an indirect effect of stress on high-risk behavior through depression. Which of the following types of models would you recommend the researcher examine?
   a. Mediation
   b. Moderation

c. Neither
d. Both

4. A researcher wants to examine the relationship between job demands and health and believes that job demands may interact with cultural values. Which of the following types of models would you recommend the researcher examine?
a. Mediation
b. Moderation
c. Neither
d. Both

5. True or false? Power for moderated multiple regression can be determined in the same way that power for multiple linear regression is determined.

6. A researcher has conducted a moderated multiple regression analysis and finds f^2 of .40. Using Cohen's (1988) conventions, this can be interpreted in which one of the following ways?
a. Small effect
b. Moderate effect
c. Large effect
d. Cannot be determined without additional information

7. A particularly important assumption to consider with moderated multiple regression is which one of the following?
a. Homoscedasticity
b. Lack of multicollinearity
c. Linearity
d. Normality of residuals

8. Which one of the following effect sizes are recommended for mediation analyses?
a. Kappa squared
b. Partially standardized indirect effect
c. Proportion of variance in the independent variable that is explained by the indirect effect
d. Ratio of the indirect to direct effect

9. For which of the following effects is the size of the effect dependent on the scale of the independent variable?
a. Completely standardized indirect effect
b. Completely standardized total effect
c. Partially standardized direct effect
d. None of the above

10. The pick-a-point approach is used for which of the following?
a. To test a direct effect
b. To test an indirect effect
c. To probe an interaction
d. To test an interaction

Answers to Conceptual Problems

1. **b** (an interaction of communication and timing on collaboration suggests a moderating relationship)
3. **a** (an indirect effect of depression suggests that the researcher examine the relationship between stress and high-risk behavior as mediated by depression)
5. **False** (power in multiple moderated regression is more complicated than in multiple linear regression and thus additional factors need to be considered)
7. **a** (homoscedasticity is an especially important assumption in moderated analyses)
9. **c** (in partially standardized effects, the independent variable remains in its original metric; thus the size of the partially standardized effect depends on the scale of *X*)

Computational Problems

1. Using the Ch10_medmod data, conduct a simple mediation model (Figure 10.1) to test the mediating effect of M2 on the relationship between DV and TRTMT. Report the path coefficients and related parameter estimates for a, b, ab, c, and c'. Indicate if there is full, partial, or no mediation.
2. Using the Ch10_medmod data, conduct a simple moderation model (Figure 10.3) to test the moderating effect of W2 on the relationship between DV and TRTMT. Probe interactions using the 16th, 50th, and 84th percentiles and the Johnson-Neyman technique. Report the path coefficients and related parameter estimates for b_1, b_2, and b_3, along with results for probing the interactions.
3. You are given the following pairs of scores on X (number of hours studied) and Y (quiz score).

X	Y
4	5
4	6
3	4
7	8
2	4

   a. Find the linear regression model for predicting Y from X.
   b. Use the prediction model obtained to predict the value of Y for a new person who has a value of 6 for X.

Answers to Computational Problems

1. The path coefficients and related parameter estimates include:
   a. $a = 1.544$, $SE = .6292$, $t = 2.4562$, $p = .0183$
   b. $b = .3338$, $SE = .2481$, $t = 2.7677$, $p = .0084$
   c. $ab = 1.0612$
   d. $c = 3.2727$, $SE = 1.2888$, $t = 2.5394$, $p = .0149$

e. $c' = 2.2116$, $SE = 1.3205$, $t = 1.6748$, $p = .1016$

f. There is full mediation as the effect of TRTMT on the DV is no longer statistically significant when the mediating variable, M1, is included in the model (see paths b and c').

3. a. b (slope) = .8571, a (intercept) = 1.9716; b. Y= (outcome) = 7.1142

Appendix: Tables

TABLE A.1

Standard Unit Normal Distribution.

z	P(z)	z	P(z)	z	P(z)	z	P(z)
0.00	0.5000000	0.32	0.6255158	0.64	0.7389137	0.96	0.8314724
0.01	0.5039894	0.33	0.6293	0.65	0.7421539	0.97	0.8339768
0.02	0.5079783	0.34	0.6330717	0.66	0.7453731	0.98	0.8364569
0.03	0.5119665	0.35	0.6368307	0.67	0.7485711	0.99	0.8389129
0.04	0.5159534	0.36	0.6405764	0.68	0.7517478	1.00	0.8413447
0.05	0.5199388	0.37	0.6443088	0.69	0.7549029	1.01	0.8437524
0.06	0.5239222	0.38	0.6480273	0.70	0.7580363	1.02	0.8461358
0.07	0.5279032	0.39	0.6517317	0.71	0.7611479	1.03	0.848495
0.08	0.5318814	0.40	0.6554217	0.72	0.7642375	1.04	0.85083
0.09	0.5358564	0.41	0.659097	0.73	0.7673049	1.05	0.8531409
0.10	0.5398278	0.42	0.6627573	0.74	0.77035	1.06	0.8554277
0.11	0.5437953	0.43	0.6664022	0.75	0.7733726	1.07	0.8576903
0.12	0.5477584	0.44	0.6700314	0.76	0.7763727	1.08	0.8599289
0.13	0.5517168	0.45	0.6736448	0.77	0.7793501	1.09	0.8621434
0.14	0.55567	0.46	0.6772419	0.78	0.7823046	1.10	0.8643339
0.15	0.5596177	0.47	0.6808225	0.79	0.7852361	1.11	0.8665005
0.16	0.5635595	0.48	0.6843863	0.80	0.7881446	1.12	0.8686431
0.17	0.5674949	0.49	0.6879331	0.81	0.7910299	1.13	0.8707619
0.18	0.5714237	0.50	0.6914625	0.82	0.7938919	1.14	0.8728568
0.19	0.5753454	0.51	0.6949743	0.83	0.7967306	1.15	0.8749281
0.20	0.5792597	0.52	0.6984682	0.84	0.7995458	1.16	0.8769756
0.21	0.5831662	0.53	0.701944	0.85	0.8023375	1.17	0.8789995
0.22	0.5870644	0.54	0.7054015	0.86	0.8051055	1.18	0.8809999
0.23	0.5909541	0.55	0.7088403	0.87	0.8078498	1.19	0.8829768
0.24	0.5948349	0.56	0.7122603	0.88	0.8105703	1.20	0.8849303
0.25	0.5987063	0.57	0.7156612	0.89	0.8132671	1.21	0.8868606
0.26	0.6025681	0.58	0.7190427	0.90	0.8159399	1.22	0.8887676
0.27	0.6064199	0.59	0.7224047	0.91	0.8185887	1.23	0.8906514
0.28	0.6102612	0.60	0.7257469	0.92	0.8212136	1.24	0.8925123
0.29	0.6140919	0.61	0.7290691	0.93	0.8238145	1.25	0.8943502
0.30	0.6179114	0.62	0.7323711	0.94	0.8263912	1.26	0.8961653
0.31	0.6217195	0.63	0.7356527	0.95	0.8289439	1.27	0.8979577

(continued)

TABLE A.1 (continued)

The Standard Unit Normal Distribution

z	*P(z)*	*z*	*P(z)*	*z*	*P(z)*	*z*	*P(z)*
1.28	0.8997274	1.68	0.9535213	2.08	0.9812372	2.48	0.9934309
1.29	0.9014747	1.69	0.954486	2.09	0.9816911	2.49	0.9936128
1.30	0.9031995	1.70	0.9554345	2.10	0.9821356	2.50	0.9937903
1.31	0.9049021	1.71	0.9563671	2.11	0.9825708	2.51	0.9939634
1.32	0.9065825	1.72	0.9572838	2.12	0.982997	2.52	0.9941323
1.33	0.9082409	1.73	0.9581849	2.13	0.9834142	2.53	0.9942969
1.34	0.9098773	1.74	0.9590705	2.14	0.9838226	2.54	0.9944574
1.35	0.911492	1.75	0.9599408	2.15	0.9842224	2.55	0.9946139
1.36	0.913085	1.76	0.9607961	2.16	0.9846137	2.56	0.9947664
1.37	0.9146565	1.77	0.9616364	2.17	0.9849966	2.57	0.9949151
1.38	0.9162067	1.78	0.962462	2.18	0.9853713	2.58	0.99506
1.39	0.9177356	1.79	0.963273	2.19	0.9857379	2.59	0.9952012
1.40	0.9192433	1.80	0.9640697	2.20	0.9860966	2.60	0.9953388
1.41	0.9207302	1.81	0.9648521	2.21	0.9864474	2.61	0.9954729
1.42	0.9221962	1.82	0.9656205	2.22	0.9867906	2.62	0.9956035
1.43	0.9236415	1.83	0.966375	2.23	0.9871263	2.63	0.9957308
1.44	0.9250663	1.84	0.9671159	2.24	0.9874545	2.64	0.9958547
1.45	0.9264707	1.85	0.9678432	2.25	0.9877755	2.65	0.9959754
1.46	0.927855	1.86	0.9685572	2.26	0.9880894	2.66	0.996093
1.47	0.9292191	1.87	0.9692581	2.27	0.9883962	2.67	0.9962074
1.48	0.9305634	1.88	0.969946	2.28	0.9886962	2.68	0.9963189
1.49	0.9318879	1.89	0.970621	2.29	0.9889893	2.69	0.9964274
1.50	0.9331928	1.90	0.9712834	2.30	0.9892759	2.70	0.996533
1.51	0.9344783	1.91	0.9719334	2.31	0.9895559	2.71	0.9966358
1.52	0.9357445	1.92	0.9725711	2.32	0.9898296	2.72	0.9967359
1.53	0.9369916	1.93	0.9731966	2.33	0.9900969	2.73	0.9968333
1.54	0.9382198	1.94	0.9738102	2.34	0.9903581	2.74	0.996928
1.55	0.9394292	1.95	0.9744119	2.35	0.9906133	2.75	0.9970202
1.56	0.9406201	1.96	0.9750021	2.36	0.9908625	2.76	0.9971099
1.57	0.9417924	1.97	0.9755808	2.37	0.991106	2.77	0.9971972
1.58	0.9429466	1.98	0.9761482	2.38	0.9913437	2.78	0.9972821
1.59	0.9440826	1.99	0.9767045	2.39	0.9915758	2.79	0.9973646
1.60	0.9452007	2.00	0.9772499	2.40	0.9918025	2.80	0.9974449
1.61	0.9463011	2.01	0.9777844	2.41	0.9920237	2.81	0.9975229
1.62	0.9473839	2.02	0.9783083	2.42	0.9922397	2.82	0.9975988
1.63	0.9484493	2.03	0.9788217	2.43	0.9924506	2.83	0.9976726
1.64	0.9494974	2.04	0.9793248	2.44	0.9926564	2.84	0.9977443
1.65	0.9505285	2.05	0.9798178	2.45	0.9928572	2.85	0.997814
1.66	0.9515428	2.06	0.9803007	2.46	0.9930531	2.86	0.9978818
1.67	0.9525403	2.07	0.9807738	2.47	0.9932443	2.87	0.9979476

z	$P(z)$	z	$P(z)$	z	$P(z)$	z	$P(z)$
2.88	0.9980116	3.17	0.9992378	3.46	0.9997299	3.75	0.9999116
2.89	0.9980738	3.18	0.9992636	3.47	0.9997398	3.76	0.999915
2.90	0.9981342	3.19	0.9992886	3.48	0.9997493	3.77	0.9999184
2.91	0.9981929	3.20	0.9993129	3.49	0.9997585	3.78	0.9999216
2.92	0.9982498	3.21	0.9993363	3.50	0.9997674	3.79	0.9999247
2.93	0.9983052	3.22	0.999359	3.51	0.9997759	3.80	0.9999277
2.94	0.9983589	3.23	0.999381	3.52	0.9997842	3.81	0.9999305
2.95	0.9984111	3.24	0.9994024	3.53	0.9997922	3.82	0.9999333
2.96	0.9984618	3.25	0.999423	3.54	0.9997999	3.83	0.9999359
2.97	0.998511	3.26	0.9994429	3.55	0.9998074	3.84	0.9999385
2.98	0.9985588	3.27	0.9994623	3.56	0.9998146	3.85	0.9999409
2.99	0.9986051	3.28	0.999481	3.57	0.9998215	3.86	0.9999433
3.00	0.9986501	3.29	0.9994991	3.58	0.9998282	3.87	0.9999456
3.01	0.9986938	3.30	0.9995166	3.59	0.9998347	3.88	0.9999478
3.02	0.9987361	3.31	0.9995335	3.60	0.9998409	3.89	0.9999499
3.03	0.9987772	3.32	0.9995499	3.61	0.9998469	3.90	0.9999519
3.04	0.9988171	3.33	0.9995658	3.62	0.9998527	3.91	0.9999539
3.05	0.9988558	3.34	0.9995811	3.63	0.9998583	3.92	0.9999557
3.06	0.9988933	3.35	0.9995959	3.64	0.9998637	3.93	0.9999575
3.07	0.9989297	3.36	0.9996103	3.65	0.9998689	3.94	0.9999593
3.08	0.998965	3.37	0.9996242	3.66	0.9998739	3.95	0.9999609
3.09	0.9989992	3.38	0.9996376	3.67	0.9998787	3.96	0.9999625
3.10	0.9990324	3.39	0.9996505	3.68	0.9998834	3.97	0.9999641
3.11	0.9990646	3.40	0.9996631	3.69	0.9998879	3,98	0.9999655
3.12	0.9990957	3.41	0.9996752	3.70	0.9998922	3.99	0.999967
3.13	0.999126	3.42	0.9996869	3.71	0.9998964	4.00	0.9999683
3.14	0.9991553	3.43	0.9996982	3.72	0.9999004		
3.15	0.9991836	3.44	0.9997091	3.73	0.9999043		
3.16	0.9992112	3.45	0.9997197	3.74	0.999908		

Values computed by the authors using **R**.

TABLE A.2

Percentage Points of the t Distribution

	$\alpha_1 = .10$	.05	.025	.01	.005	.0025	.001	.0005
v	$\alpha_2 = .20$	.10	.050	.02	.010	.0050	.002	.0010
1	3.077684	6.313752	12.7062	31.82052	63.65674	127.3213	318.3088	636.6192
2	1.885618	2.919986	4.302653	6.964557	9.924843	14.08905	22.32712	31.59905
3	1.637744	2.353363	3.182446	4.540703	5.840909	7.453319	10.21453	12.92398
4	1.533206	2.131847	2.776445	3.746947	4.604095	5.597568	7.173182	8.610302
5	1.475884	2.015048	2.570582	3.36493	4.032143	4.773341	5.89343	6.868827
6	1.439756	1.94318	2.446912	3.142668	3.707428	4.316827	5.207626	5.958816
7	1.414924	1.894579	2.364624	2.997952	3.499483	4.029337	4.78529	5.407883
8	1.396815	1.859548	2.306004	2.896459	3.355387	3.832519	4.500791	5.041305
9	1.383029	1.833113	2.26215	2.821438	3.249836	3.689662	4.296806	4.780913
10	1.372184	1.812461	2.228139	2.763769	3.169273	3.581406	4.1437	4.586894
11	1.36343	1.795885	2.200985	2.718079	3.105807	3.496614	4.024701	4.436979
12	1.356217	1.782288	2.178813	2.680998	3.05454	3.428444	3.929633	4.317791
13	1.350171	1.770933	2.160369	2.650309	3.012276	3.372468	3.851982	4.220832
14	1.34503	1.76131	2.144787	2.624494	2.976843	3.325696	3.78739	4.140454
15	1.340606	1.75305	2.13145	2.60248	2.946713	3.286039	3.732834	4.072765
16	1.336757	1.745884	2.119905	2.583487	2.920782	3.251993	3.686155	4.014996
17	1.333379	1.739607	2.109816	2.566934	2.898231	3.22245	3.645767	3.965126
18	1.330391	1.734064	2.100922	2.55238	2.87844	3.196574	3.610485	3.921646
19	1.327728	1.729133	2.093024	2.539483	2.860935	3.173725	3.5794	3.883406
20	1.325341	1.724718	2.085963	2.527977	2.84534	3.153401	3.551808	3.849516
21	1.323188	1.720743	2.079614	2.517648	2.83136	3.135206	3.527154	3.819277
22	1.321237	1.717144	2.073873	2.508325	2.818756	3.118824	3.504992	3.792131
23	1.31946	1.713872	2.068658	2.499867	2.807336	3.103997	3.484964	3.767627
24	1.317836	1.710882	2.063899	2.492159	2.79694	3.090514	3.466777	3.745399
25	1.316345	1.708141	2.059539	2.485107	2.787436	3.078199	3.450189	3.725144
26	1.314972	1.705618	2.055529	2.47863	2.778715	3.066909	3.434997	3.706612
27	1.313703	1.703288	2.051831	2.47266	2.770683	3.05652	3.421034	3.689592
28	1.312527	1.701131	2.048407	2.46714	2.763262	3.046929	3.408155	3.673906
29	1.311434	1.699127	2.04523	2.462021	2.756386	3.038047	3.39624	3.659405
30	1.310415	1.697261	2.042272	2.457262	2.749996	3.029798	3.385185	3.645959
40	1.303077	1.683851	2.021075	2.423257	2.704459	2.971171	3.306878	3.550966
60	1.295821	1.670649	2.000298	2.390119	2.660283	2.914553	3.231709	3.4602
120	1.288646	1.657651	1.97993	2.357825	2.617421	2.859865	3.159539	3.373454
∞	1.281552	1.644854	1.959964	2.326348	2.575829	2.807034	3.090232	3.290527

Values computed by the authors using **R**.

TABLE A.3

Percentage Points of the χ^2 Distribution

	Alpha							
v	0.990	0.975	0.950	0.900	0.100	0.050	0.025	0.010
1	0.000157088	0.000982069	0.00393214	0.01579077	2.705543	3.841459	5.023886	6.634897
2	0.02010067	0.05063562	0.1025866	0.210721	4.60517	5.991465	7.377759	9.21034
3	0.1148318	0.2157953	0.3518463	0.5843744	6.251389	7.814728	9.348404	11.34487
4	0.2971095	0.4844186	0.710723	1.063623	7.77944	9.487729	11.14329	13.2767
5	0.5542981	0.8312116	1.145476	1.610308	9.236357	11.0705	12.8325	15.08627
6	0.8720903	1.237344	1.635383	2.204131	10.64464	12.59159	14.44938	16.81189
7	1.239042	1.689869	2.16735	2.833107	12.01704	14.06714	16.01276	18.47531
8	1.646497	2.179731	2.732637	3.489539	13.36157	15.50731	17.53455	20.09024
9	2.087901	2.700389	3.325113	4.168159	14.68366	16.91898	19.02277	21.66599
10	2.558212	3.246973	3.940299	4.865182	15.98718	18.30704	20.48318	23.20925
11	3.053484	3.815748	4.574813	5.577785	17.27501	19.67514	21.92005	24.72497
12	3.570569	4.403789	5.226029	6.303796	18.54935	21.02607	23.33666	26.21697
13	4.106915	5.008751	5.891864	7.041505	19.81193	22.36203	24.7356	27.68825
14	4.660425	5.628726	6.570631	7.789534	21.06414	23.68479	26.11895	29.14124
15	5.229349	6.262138	7.260944	8.546756	22.30713	24.99579	27.48839	30.57791
16	5.812212	6.907664	7.961646	9.312236	23.54183	26.29623	28.84535	31.99993
17	6.40776	7.564186	8.67176	10.08519	24.76904	27.58711	30.19101	33.40866
18	7.014911	8.230746	9.390455	10.86494	25.98942	28.8693	31.52638	34.80531
19	7.63273	8.906516	10.11701	11.65091	27.20357	30.14353	32.85233	36.19087
20	8.260398	9.590777	10.85081	12.44261	28.41198	31.41043	34.16961	37.56623
21	8.897198	10.2829	11.59131	13.2396	29.61509	32.67057	35.47888	38.93217
22	9.542492	10.98232	12.33801	14.04149	30.81328	33.92444	36.78071	40.28936
23	10.19572	11.68855	13.09051	14.84796	32.0069	35.17246	38.07563	41.6384
24	10.85636	12.40115	13.84843	15.65868	33.19624	36.41503	39.36408	42.97982
25	11.52398	13.11972	14.61141	16.47341	34.38159	37.65248	40.64647	44.3141
26	12.19815	13.8439	15.37916	17.29188	35.56317	38.88514	41.92317	45.64168
27	12.8785	14.57338	16.1514	18.1139	36.74122	40.11327	43.19451	46.96294
28	13.56471	15.30786	16.92788	18.93924	37.91592	41.33714	44.46079	48.27824
29	14.25645	16.04707	17.70837	19.76774	39.08747	42.55697	45.72229	49.58788
30	14.95346	16.79077	18.49266	20.59923	40.25602	43.77297	46.97924	50.89218
40	22.16426	24.43304	26.5093	29.05052	51.80506	55.75848	59.34171	63.69074
50	29.70668	32.35736	34.76425	37.68865	63.16712	67.50481	71.4202	76.15389
60	37.48485	40.48175	43.18796	46.45889	74.39701	79.08194	83.29767	88.37942
70	45.44172	48.75756	51.73928	55.32894	85.52704	90.53123	95.02318	100.4252
80	53.54008	57.15317	60.39148	64.27784	96.5782	101.8795	106.6286	112.3288
90	61.75408	65.64662	69.12603	73.29109	107.565	113.1453	118.1359	124.1163
100	70.06489	74.22193	77.92947	82.35814	118.498	124.3421	129.5612	135.8067

Values computed by the authors using **R**.

TABLE A.4

Percentage Points of the *F* Distribution

	υ_1																		
υ_2	1	2	3	4	5	6	7	8	9	10	12	15	20	24	30	40	60	120	Infinity
alpha = .10																			
1	39.86346	49.5	53.59324	55.83296	57.24008	58.20442	58.90595	59.43898	59.85759	60.19498	60.70521	61.22034	61.74029	62.00205	62.26497	62.52905	62.79428	63.06064	63.32812
2	8.526316	9	9.16179	9.243416	9.292626	9.32553	9.349081	9.36677	9.380544	9.391573	9.408132	9.424711	9.441309	9.449616	9.457927	9.466244	9.474565	9.482891	9.491222
3	5.538319	5.462383	5.390773	5.342644	5.309157	5.284732	5.266195	5.251671	5.239996	5.230411	5.215618	5.200313	5.184482	5.176365	5.168111	5.159719	5.151187	5.142513	5.133695
4	8.526316	4.324555	4.19086	4.10725	4.050579	4.009749	3.978966	3.95494	3.935671	3.919876	3.895527	3.87036	3.844338	3.830994	3.817422	3.803615	3.789568	3.775275	3.76073
5	4.544771	3.779716	3.619477	3.520196	3.452982	3.404507	3.367899	3.339276	3.316281	3.297402	3.268239	3.238011	3.20665	3.190523	3.174084	3.157324	3.14023	3.122792	3.104996
6	3.77595	3.463304	3.288762	3.180763	3.107512	3.054551	3.014457	2.983036	2.957741	2.936935	2.904721	2.871222	2.83634	2.818345	2.79996	2.781169	2.761952	2.74229	2.722162
7	3.589428	3.257442	3.074072	2.960534	2.883344	2.827392	2.78493	2.75158	2.724678	2.70251	2.668111	2.63223	2.594732	2.575327	2.555457	2.535096	2.514218	2.492792	2.470786
8	3.457919	3.113118	2.923796	2.806426	2.726447	2.668335	2.624135	2.589349	2.561238	2.538037	2.501958	2.464216	2.424637	2.404097	2.383016	2.361362	2.339097	2.316181	2.292566
9	3.360303	3.006452	2.812863	2.69268	2.610613	2.550855	2.505313	2.469406	2.44034	2.416316	2.378885	2.339624	2.298322	2.276827	2.25472	2.231958	2.208493	2.18427	2.159227
10	3.285015	2.924466	2.727673	2.605336	2.521641	2.460582	2.413965	2.37715	2.347306	2.322604	2.284051	2.243515	2.200744	2.178426	2.155426	2.131691	2.107161	2.081765	2.055422
11	3.225202	2.859511	2.660229	2.536188	2.451184	2.389067	2.341566	2.303997	2.273502	2.24823	2.208725	2.167094	2.123046	2.100005	2.076214	2.05161	2.026118	1.999652	1.972109
12	3.176549	2.806796	2.605525	2.480102	2.394022	2.331024	2.28278	2.244575	2.213525	2.187764	2.147437	2.104851	2.059677	2.035993	2.011492	1.986102	1.959732	1.932278	1.903615
13	3.136205	2.763167	2.560273	2.433705	2.346724	2.282979	2.234103	2.19535	2.16382	2.137635	2.096588	2.05316	2.006982	1.982718	1.957575	1.931466	1.904287	1.875915	1.846196
14	3.102213	2.726468	2.522224	2.394692	2.306943	2.242559	2.193134	2.153904	2.121955	2.095396	2.053714	2.009535	1.962453	1.937663	1.911933	1.885163	1.857234	1.828001	1.797283
15	3.073186	2.695173	2.489788	2.361433	2.273022	2.208082	2.158178	2.11853	2.086209	2.059319	2.01707	1.972216	1.924314	1.899044	1.872774	1.845393	1.816764	1.78672	1.755052
16	3.04811	2.668171	2.461811	2.332745	2.243758	2.178329	2.128003	2.087982	2.055331	2.028145	1.985386	1.939921	1.891272	1.865561	1.838792	1.810841	1.781557	1.750747	1.718169
17	3.026232	2.644638	2.437434	2.307747	2.218253	2.152392	2.101689	2.061336	2.028388	2.000936	1.957716	1.911695	1.862361	1.836242	1.80901	1.780528	1.750627	1.71909	1.685641
18	3.006977	2.623947	2.416005	2.285772	2.195827	2.129581	2.078541	2.037889	2.004674	1.97698	1.93334	1.886811	1.836845	1.810348	1.782685	1.753706	1.723222	1.690993	1.656706
19	2.9899	2.605612	2.397022	2.266303	2.175956	2.109364	2.05802	2.017098	1.983639	1.955725	1.911702	1.864705	1.814155	1.787307	1.759241	1.729793	1.698758	1.665869	1.630774
20	2.974653	2.589254	2.380087	2.248934	2.158227	2.091322	2.039703	1.998534	1.964853	1.936738	1.892363	1.844935	1.793843	1.766667	1.738223	1.708334	1.676776	1.643256	1.60738
21	2.960956	2.574569	2.364888	2.233345	2.142311	2.075123	2.023252	1.981858	1.947974	1.919674	1.874975	1.827148	1.775551	1.748068	1.719268	1.688962	1.656907	1.622782	1.586151
22	2.948585	2.561314	2.35117	2.219274	2.127944	2.060497	2.008397	1.966796	1.932725	1.904255	1.859255	1.811057	1.758989	1.731217	1.702083	1.671382	1.638853	1.604147	1.566785
23	2.937356	2.54929	2.338727	2.206512	2.114911	2.047227	1.994915	1.953124	1.91888	1.890252	1.844974	1.796431	1.743921	1.715878	1.686428	1.655352	1.622371	1.587107	1.549035
24	2.927117	2.538332	2.32739	2.194882	2.103033	2.035132	1.982625	1.940658	1.906255	1.87748	1.831942	1.783076	1.730152	1.701854	1.672104	1.640673	1.60726	1.571459	1.532696
25	2.917745	2.528305	2.317017	2.184242	2.092165	2.024062	1.971376	1.929246	1.894693	1.865782	1.820003	1.770834	1.71752	1.688981	1.658947	1.627177	1.59335	1.557031	1.517597
26	2.909132	2.519096	2.307491	2.174469	2.082182	2.013893	1.961039	1.918758	1.884067	1.855028	1.809023	1.759571	1.70589	1.677122	1.646819	1.614725	1.580502	1.543683	1.503595
27	2.901192	2.510609	2.298712	2.165463	2.072981	2.004519	1.95151	1.909087	1.874267	1.845109	1.798891	1.749173	1.695144	1.66616	1.635601	1.603198	1.568595	1.531293	1.490568
28	2.893846	2.502761	2.290595	2.157136	2.064473	1.995851	1.942696	1.900141	1.865199	1.83593	1.789513	1.739543	1.685187	1.655997	1.625193	1.592496	1.557527	1.519759	1.478412
29	2.887033	2.495483	2.283069	2.149415	2.056583	1.987811	1.934521	1.891842	1.856786	1.827412	1.780807	1.7306	1.675932	1.646547	1.615511	1.582531	1.54721	1.50899	1.467036
30	2.880695	2.488716	2.276071	2.142235	2.049246	1.980333	1.926916	1.884121	1.848958	1.819485	1.772704	1.722272	1.667309	1.637737	1.606479	1.573228	1.537569	1.498912	1.456365
40	2.835354	2.440369	2.226092	2.09095	1.99682	1.926879	1.872522	1.828863	1.792902	1.762686	1.714563	1.662411	1.605151	1.574111	1.541076	1.505625	1.467157	1.424757	1.376912
60	2.791068	2.393255	2.177411	2.040986	1.94571	1.87472	1.819393	1.774829	1.73802	1.707009	1.657429	1.603368	1.543486	1.510718	1.475539	1.437342	1.395201	1.347568	1.291464
120	2.747807	2.347338	2.129991	1.992302	1.895875	1.823812	1.767476	1.721959	1.684248	1.652379	1.601204	1.545002	1.482072	1.447226	1.409379	1.367602	1.32034	1.264573	1.192563
Infinity	2.705543	2.302585	2.083796	1.94486	1.847271	1.774107	1.71672	1.670196	1.631517	1.598718	1.545779	1.487142	1.420599	1.383177	1.341867	1.295126	1.23995	1.168605	1.000018

υ_2	υ_1 = 1	2	3	4	5	6	7	8	9	10	12	15	20	24	30	40	60	120	Infinity
alpha = .05																			
1	161.4476	199.5	215.7073	224.5832	230.1619	233.986	236.7684	238.8827	240.5433	241.8817	243.906	245.9499	248.0131	249.0518	250.0951	251.1432	252.1957	253.2529	254.3144
2	18.51282	19	19.16429	19.24679	19.29641	19.32953	19.35322	19.37099	19.38483	19.3959	19.41251	19.42914	19.44577	19.45409	19.46241	19.47074	19.47906	19.48739	19.49573
3	10.12796	9.552094	9.276628	9.117182	9.013455	8.940645	8.886743	8.845238	8.8123	8.785525	8.744641	8.70287	8.66019	8.638501	8.616576	8.594411	8.572004	8.549351	8.52645
4	7.708647	6.944272	6.591382	6.388233	6.256057	6.163132	6.094211	6.041044	5.998779	5.964371	5.911729	5.857805	5.802542	5.774389	5.745877	5.716998	5.687744	5.658105	5.628072
5	6.607891	5.786135	5.409451	5.192168	5.050329	4.950288	4.875872	4.81832	4.772466	4.735063	4.677704	4.618759	4.558131	4.527153	4.495712	4.463793	4.43138	4.398454	4.364997
6	5.987378	5.143253	4.757063	4.533677	4.387374	4.283866	4.206658	4.146804	4.099016	4.059963	3.999935	3.938058	3.874189	3.841457	3.808164	3.774286	3.739797	3.704667	3.668866
7	5.591448	4.737414	4.346831	4.120312	3.971523	3.865969	3.787044	3.725725	3.676675	3.636523	3.574676	3.51074	3.444525	3.410494	3.375808	3.34043	3.304323	3.267445	3.229751
8	5.317655	4.45897	4.066181	3.837853	3.687499	3.58058	3.500464	3.438101	3.38813	3.347163	3.283939	3.218406	3.150324	3.11524	3.079406	3.042778	3.005303	2.966923	3.229751
9	5.117355	4.256495	3.862548	3.633089	3.481659	3.373754	3.292746	3.229583	3.178893	3.13728	3.072947	3.006102	2.936455	2.900474	2.863652	2.825933	2.787249	2.747525	2.927575
10	4.964603	4.102821	3.708265	3.47805	3.325835	3.217175	3.135465	3.071658	3.020383	2.978237	2.912977	2.845017	2.774016	2.737248	2.699551	2.660855	2.621077	2.580122	2.537878
11	4.844336	3.982298	3.587434	3.35669	3.203874	3.094613	3.01233	2.94799	2.896223	2.853625	2.787569	2.71864	2.646445	2.608974	2.570489	2.530905	2.490123	2.448024	2.40447
12	4.747225	3.885294	3.490295	3.259167	3.105875	2.99612	2.913358	2.848565	2.796375	2.753387	2.686637	2.616851	2.543588	2.505482	2.466279	2.42588	2.384166	2.340995	2.296198
13	4.667193	3.805565	3.410534	3.179117	3.025438	2.915269	2.832098	2.766913	2.714356	2.671024	2.603661	2.53311	2.458882	2.420196	2.380334	2.33918	2.296596	2.252414	2.206432
14	4.60011	3.738892	3.343889	3.11225	2.958249	2.847726	2.764199	2.698672	2.645791	2.602155	2.534243	2.463003	2.387896	2.348678	2.308207	2.26635	2.22295	2.177811	2.130693
15	4.543077	3.68232	3.287382	3.055568	2.901295	2.790465	2.706627	2.640797	2.587626	2.543719	2.475313	2.403447	2.327535	2.287826	2.246789	2.204276	2.160105	2.114056	2.065847
16	4.493998	3.633723	3.238872	3.006917	2.852409	2.741311	2.657197	2.591096	2.537667	2.493513	2.42466	2.352223	2.27557	2.235405	2.193841	2.150711	2.105813	2.058895	2.009635
17	4.451322	3.591531	3.196777	2.964708	2.809996	2.69866	2.614299	2.547955	2.494291	2.449916	2.380654	2.307693	2.230354	2.189766	2.147708	2.103998	2.058411	2.010663	1.960386
18	4.413873	3.554557	3.159908	2.927744	2.772853	2.661305	2.576722	2.510158	2.456281	2.411702	2.342067	2.268622	2.190648	2.149665	2.107143	2.062885	2.016643	1.9681	1.91684
19	4.38075	3.521893	3.12735	2.895107	2.740058	2.628318	2.543534	2.47677	2.422699	2.377934	2.307954	2.234063	2.155497	2.114143	2.071186	2.02641	1.979544	1.930237	1.878025
20	4.351244	3.492828	3.098391	2.866081	2.71089	2.598978	2.514011	2.447064	2.392814	2.347878	2.277581	2.203274	2.124155	2.082454	2.039086	1.993819	1.946358	1.896318	1.84318
21	4.324794	3.4668	3.072467	2.8401	2.684781	2.572712	2.487578	2.420462	2.366048	2.320953	2.250362	2.17567	2.096033	2.054004	2.010248	1.964515	1.916486	1.865739	1.811703
22	4.30095	3.443357	3.049125	2.816708	2.661274	2.549061	2.463774	2.396503	2.341937	2.296696	2.225831	2.150778	2.070656	2.028319	1.984195	1.938018	1.889445	1.838018	1.783107
23	4.279344	3.422132	3.027998	2.795539	2.639999	2.527655	2.442226	2.374812	2.320105	2.274728	2.203607	2.128217	2.047638	2.005009	1.960537	1.913938	1.864844	1.81276	1.756997
24	4.259677	3.402826	3.008787	2.776289	2.620654	2.508189	2.422629	2.355081	2.300244	2.254739	2.18338	2.107673	2.026664	1.98376	1.938957	1.891955	1.84236	1.789642	1.733049
25	4.241699	3.38519	2.991241	2.75871	2.602987	2.49041	2.404728	2.337057	2.282097	2.236474	2.164891	2.088887	2.007471	1.964306	1.919188	1.871801	1.821727	1.768395	1.710992
26	4.225201	3.369016	2.975154	2.742594	2.58679	2.474109	2.388314	2.320527	2.265453	2.219718	2.147926	2.071642	1.989842	1.946428	1.90101	1.853255	1.802719	1.748795	1.6906
27	4.210008	3.354131	2.960351	2.727765	2.571886	2.459108	2.373208	2.305313	2.250131	2.204292	2.132303	2.055755	1.97359	1.92994	1.884236	1.836129	1.785149	1.73065	1.671682
28	4.195972	3.340386	2.946685	2.714076	2.558128	2.445259	2.35926	2.291264	2.235982	2.190044	2.117869	2.041071	1.958561	1.914686	1.868709	1.820263	1.768857	1.7138	1.654076
29	4.182964	3.327654	2.93403	2.701399	2.545386	2.432434	2.346342	2.278251	2.222874	2.176844	2.104493	2.027458	1.94462	1.900531	1.854293	1.805523	1.753704	1.698107	1.637644
30	4.170877	3.31583	2.922277	2.689628	2.533555	2.420523	2.334344	2.266163	2.210697	2.16458	2.092063	2.014804	1.931653	1.88736	1.840872	1.79179	1.739574	1.683452	1.622265
40	4.084746	3.231727	2.838745	2.605975	2.449466	2.335852	2.249024	2.18017	2.124029	2.077248	2.003459	1.924463	1.838859	1.792937	1.744432	1.692797	1.637252	1.57661	1.508904
60	4.001191	3.150411	2.758078	2.525215	2.36827	2.254053	2.166541	2.096968	2.040098	1.992592	1.917396	1.836437	1.747984	1.700117	1.649141	1.594273	1.534314	1.467267	1.389276
120	3.920124	3.071779	2.680168	2.447237	2.289851	2.175006	2.08677	2.016426	1.958763	1.910461	1.833695	1.750497	1.65868	1.608437	1.554343	1.495202	1.429013	1.351886	1.253858
Infinity	3.841459	2.995732	2.604909	2.371932	2.2141	2.098598	2.009591	1.938414	1.879886	1.830704	1.752172	1.666386	1.570522	1.517293	1.459099	1.393962	1.318032	1.221395	1.000023

(continued)

υ_2	υ_1 1	2	3	4	5	6	7	8	9	10	12	15	20	24	30	40	60	120	Infinity
alpha = .01																			
1	4052.181	4999.5	5403.352	5624.583	5763.65	5858.986	5928.356	5981.07	6022.473	6055.847	6106.321	6157.285	6208.73	6234.631	6260.649	6286.782	6313.03	6339.391	6365.864
2	98.50251	99	99.1662	99.24937	99.2993	99.33259	99.35637	99.37421	99.38809	99.3992	99.41585	99.43251	99.44917	99.4575	99.46583	99.47416	99.4825	99.49083	99.49916
3	34.11622	30.81652	29.4567	28.7099	28.23708	27.91066	27.6717	27.48918	27.34521	27.22873	27.05182	26.87219	26.68979	26.59752	26.50453	26.41081	26.31635	26.22114	26.12517
4	21.19769	18	16.69437	15.97702	15.52186	15.20686	14.97576	14.79889	14.65913	14.5459	14.37359	14.1982	14.01961	13.92906	13.83766	13.74538	13.6522	13.5581	13.46305
5	16.25818	13.27393	12.05995	11.39193	10.96702	10.67225	10.45551	10.28931	10.15776	10.05102	9.888275	9.722219	9.552646	9.466471	9.379329	9.291189	9.202015	9.111771	9.020417
6	13.74502	10.92477	9.779538	9.148301	8.745895	8.466125	8.259995	8.101651	7.976121	7.874119	7.718333	7.558994	7.395832	7.312721	7.228533	7.143222	7.056737	6.969023	6.880021
7	12.24638	9.546578	8.451285	7.846645	7.460435	7.191405	6.992833	6.840049	6.718752	6.620063	6.469091	6.314331	6.155438	6.074319	5.99201	5.908449	5.823566	5.737286	5.649525
8	11.25862	8.649111	7.590992	7.006077	6.631825	6.370681	6.177624	6.02887	5.910619	5.814294	5.666719	5.515125	5.359095	5.279264	5.19813	5.11561	5.031618	4.946052	4.858799
9	10.56143	8.021517	6.991917	6.422085	6.056941	5.80177	5.612865	5.467123	5.351129	5.256542	5.111431	4.962078	4.807995	4.728998	4.648582	4.566649	4.483087	4.397769	4.31055
10	10.04429	7.559432	6.552313	5.994339	5.636326	5.385811	5.200121	5.056693	4.942421	4.849147	4.70587	4.55814	4.405395	4.326929	4.246933	4.165287	4.081855	3.996481	3.90898
11	9.646034	7.205713	6.21673	5.6683	5.316009	5.06921	4.886072	4.744468	4.63154	4.539282	4.397401	4.250867	4.099046	4.02091	3.941132	3.859573	3.776071	3.690436	3.602442
12	9.330212	6.926608	5.952545	5.411951	5.064343	4.820574	4.639502	4.499365	4.38751	4.296054	4.155258	4.009619	3.858433	3.780485	3.700789	3.619181	3.535473	3.44944	3.360809
13	9.073806	6.700965	5.73938	5.20533	4.861621	4.620363	4.440997	4.302062	4.191078	4.100267	3.960326	3.815365	3.664609	3.586753	3.507042	3.425293	3.341287	3.25476	3.165393
14	8.861593	6.514884	5.563886	5.035378	4.694964	4.45582	4.277882	4.139946	4.02968	3.939396	3.800141	3.655697	3.505222	3.427387	3.347596	3.265641	3.181274	3.094191	3.004018
15	8.683117	6.358873	5.416965	4.89321	4.555614	4.318273	4.141546	4.004453	3.894788	3.80494	3.66624	3.522194	3.371892	3.294029	3.21411	3.131906	3.047135	2.959453	2.868426
16	8.530965	6.226235	5.292214	4.772578	4.43742	4.201634	4.025947	3.889572	3.780415	3.690931	3.552687	3.408947	3.258737	3.180811	3.100733	3.018248	2.933046	2.844737	2.752824
17	8.39974	6.112114	5.185	4.668968	4.335939	4.101505	3.926719	3.790964	3.682242	3.593066	3.455198	3.311694	3.161518	3.083502	3.003241	2.920458	2.834806	2.745852	2.653033
18	8.28542	6.012905	5.09189	4.579036	4.247882	4.014637	3.840639	3.705422	3.597074	3.508162	3.370608	3.227286	3.077097	2.998974	2.918516	2.83542	2.749309	2.659701	2.565963
19	8.184947	5.925879	5.010287	4.500258	4.170767	3.938573	3.765269	3.630525	3.522503	3.433817	3.296527	3.153343	3.003109	2.924866	2.844201	2.760786	2.674211	2.583944	2.48928
20	8.095958	5.848932	4.938193	4.43069	4.102685	3.871427	3.69874	3.564412	3.456676	3.368186	3.23112	3.088041	2.937735	2.859363	2.778485	2.694749	2.607708	2.516783	2.421191
21	8.016597	5.780416	4.874046	4.368815	4.042144	3.811725	3.63959	3.505632	3.398147	3.30983	3.172953	3.029951	2.879556	2.80105	2.719955	2.635896	2.548393	2.456813	2.360294
22	7.945386	5.719022	4.816606	4.313429	3.987963	3.758301	3.58666	3.453034	3.345773	3.257606	3.120891	2.977946	2.827447	2.748802	2.66749	2.583111	2.495149	2.402919	2.305477
23	7.881134	5.663699	4.764877	4.263567	3.939195	3.710218	3.539024	3.405695	3.298634	3.210599	3.074025	2.931118	2.780504	2.70172	2.620191	2.535496	2.447081	2.354209	2.25585
24	7.822871	5.613591	4.718051	4.218445	3.89507	3.666717	3.495928	3.362867	3.255985	3.168069	3.031615	2.888732	2.737997	2.659072	2.577329	2.492321	2.403461	2.309955	2.210685
25	7.769798	5.567997	4.675465	4.17742	3.854957	3.627174	3.456754	3.323937	3.217217	3.129406	2.993056	2.850186	2.699325	2.62026	2.538305	2.45299	2.363691	2.269562	2.16939
26	7.721254	5.526335	4.63657	4.13996	3.818336	3.591075	3.420993	3.288399	3.181824	3.094108	2.957848	2.814982	2.663991	2.584787	2.502624	2.417007	2.327279	2.232536	2.131471
27	7.676684	5.488118	4.600907	4.105622	3.78477	3.557991	3.388219	3.255827	3.149385	3.061754	2.925573	2.782703	2.63158	2.552239	2.469872	2.38396	2.293812	2.198465	2.096517
28	7.635619	5.452937	4.568091	4.074032	3.753895	3.527559	3.358073	3.225868	3.119547	3.031992	2.895881	2.753	2.601744	2.522268	2.439701	2.353501	2.262941	2.167001	2.06418
29	7.597663	5.420445	4.537795	4.044873	3.725399	3.499475	3.330252	3.198219	3.092009	3.004524	2.868472	2.725577	2.574188	2.494579	2.411817	2.325335	2.234372	2.137851	2.034166
30	7.562476	5.390346	4.50974	4.017877	3.699019	3.473477	3.304499	3.172624	3.066516	2.979094	2.843095	2.70018	2.548659	2.468921	2.385967	2.299211	2.207854	2.110762	2.006225
40	7.3141	5.178508	4.312569	3.828294	3.51384	3.291012	3.123757	2.992981	2.88756	2.800545	2.664827	2.521616	2.368876	2.287998	2.203382	2.114232	2.019411	1.917191	1.804707
60	7.077106	4.977432	4.125892	3.649047	3.338884	3.118674	2.953049	2.82328	2.718454	2.631751	2.496116	2.352297	2.197806	2.115364	2.028479	1.936018	1.836259	1.72632	1.600647
120	6.850893	4.78651	3.9491	3.479531	3.173545	2.955854	2.791764	2.662906	2.558574	2.472077	2.3363	2.191504	2.034588	1.950018	1.860005	1.762849	1.655693	1.532992	1.380528
Infinity	6.634897	4.60517	3.781622	3.319176	3.017254	2.801982	2.63933	2.511279	2.407333	2.320925	2.184747	2.038528	1.878312	1.790826	1.696406	1.592268	1.47299	1.324585	1.000033

Values computed by the authors using **R**.

TABLE A.5

Fisher's Z Transformed Values

r	Z	r	Z	r	Z	r	Z
0.00	0.000000	0.25	0.2554128	0.50	0.5493061	0.75	0.9729551
0.01	0.01000033	0.26	0.2661084	0.51	0.5627298	0.76	0.9962151
0.02	0.02000267	0.27	0.2768638	0.52	0.5763398	0.77	1.020328
0.03	0.030009	0.28	0.2876821	0.53	0.5901452	0.78	1.045371
0.04	0.04002135	0.29	0.2985663	0.54	0.6041556	0.79	1.071432
0.05	0.05004173	0.30	0.3095196	0.55	0.6183813	0.80	1.098612
0.06	0.06007216	0.31	0.3205454	0.56	0.6328332	0.81	1.127029
0.07	0.07011467	0.32	0.3316471	0.57	0.6475228	0.82	1.156817
0.08	0.08017133	0.33	0.3428283	0.58	0.6624627	0.83	1.188136
0.09	0.09024419	0.34	0.3540925	0.59	0.6776661	0.84	1.221174
0.10	0.1003353	0.35	0.3654438	0.60	0.6931472	0.85	1.256153
0.11	0.1104469	0.36	0.3768859	0.61	0.7089214	0.86	1.293345
0.12	0.120581	0.37	0.3884231	0.62	0.7250051	0.87	1.33308
0.13	0.1307399	0.38	0.4000597	0.63	0.7414161	0.88	1.375768
0.14	0.1409256	0.39	0.4118	0.64	0.7581737	0.89	1.421926
0.15	0.1511404	0.40	0.4236489	0.65	0.7752987	0.90	1.472219
0.16	0.1613867	0.41	0.4356112	0.66	0.7928136	0.91	1.527524
0.17	0.1716667	0.42	0.447692	0.67	0.8107431	0.92	1.589027
0.18	0.1819827	0.43	0.4598967	0.68	0.829114	0.93	1.65839
0.19	0.1923372	0.44	0.4722308	0.69	0.8479558	0.94	1.738049
0.20	0.2027326	0.45	0.4847003	0.70	0.8673005	0.95	1.831781
0.21	0.2131713	0.46	0.4973113	0.71	0.8871839	0.96	1.94591
0.22	0.2236561	0.47	0.5100703	0.72	0.907645	0.97	2.092296
0.23	0.2341895	0.48	0.5229843	0.73	0.9287274	0.98	2.29756
0.24	0.2447741	0.49	0.5360603	0.74	0.9504794	0.99	2.646652

TABLE A.6

Orthogonal Polynomials

J	Trend	$j=1$	2	3	4	5	6	7	8	9	10	Σc_j^2
$J=3$	Linear	−1	0	1								2
	Quadratic	1	−2	1								6
$J=4$	Linear	−3	−1	1	3							20
	Quadratic	1	−1	−1	1							4
	Cubic	−1	3	−3	1							20
$J=5$	Linear	−2	−1	0	1	2						10
	Quadratic	2	−1	−2	−1	2						14
	Cubic	−1	2	0	−2	1						10
	Quartic	1	−4	6	−4	1						70
$J=6$	Linear	−5	−3	−1	1	3	5					70
	Quadratic	5	−1	−4	−4	−1	5					84
	Cubic	−5	7	4	−4	−7	5					180
	Quartic	1	−3	2	2	−3	1					28
	Quintic	−1	5	−10	10	−5	1					252
$J=7$	Linear	−3	−2	−1	0	1	2	3				28
	Quadratic	5	0	−3	−4	−3	0	5				84
	Cubic	−1	1	1	0	−1	−1	1				6
	Quartic	3	−7	1	6	1	−7	3				154
	Quintic	−1	4	−5	0	5	−4	1				84
$J=8$	Linear	−7	−5	−3	−1	1	3	5	7			168
	Quadratic	7	1	−3	−5	−5	−3	1	7			168
	Cubic	−7	5	7	3	−3	−7	−5	7			264
	Quartic	7	−13	−3	9	9	−3	−13	7			616
	Quintic	−7	23	−17	−15	15	17	−23	7			2184
$J=9$	Linear	−4	−3	−2	−1	0	1	2	3	4		60
	Quadratic	28	7	−8	−17	−20	−17	−8	7	28		2772
	Cubic	−14	7	13	9	0	−9	−13	−7	14		990
	Quartic	14	−21	−11	9	18	9	−11	−21	14		2002
	Quintic	−4	11	−4	−9	0	9	4	−11	4		468
$J=10$	Linear	−9	−7	−5	−3	−1	1	3	5	7	9	330
	Quadratic	6	2	−1	−3	−4	−4	−3	−1	2	6	132
	Cubic	−42	14	35	31	12	−12	−31	−35	−14	42	8580
	Quartic	18	−22	−17	3	18	18	3	−17	−22	18	2860
	Quintic	−6	14	−1	−11	−6	6	11	1	−14	6	780

Source: Reprinted from Pearson, E. S., and Hartley, H. O., *Biometrika Tables for Statisticians*, Cambridge University Press, Cambridge, UK, 1966, Table 47. With permission of Biometrika Trustees.

TABLE A.7

Critical Values for Dunnett's Procedure

df	1	2	3	4	5	6	7	8	9
One tailed, α = .05									
5	2.02	2.44	2.68	2.85	2.98	3.08	3.16	3.24	3.30
6	1.94	2.34	2.56	2.71	2.83	2.92	3.00	3.07	3.12
7	1.89	2.27	2.48	2.62	2.73	2.82	2.89	2.95	3.01
8	1.86	2.22	2.42	2.55	2.66	2.74	2.81	2.87	2.92
9	1.83	2.18	2.37	2.50	2.60	2.68	2.75	2.81	2.86
10	1.81	2.15	2.34	2.47	2.56	2.64	2.70	2.76	2.81
11	1.80	2.13	2.31	2.44	2.53	2.60	2.67	2.72	2.77
12	1.78	2.11	2.29	2.41	2.50	2.58	2.64	2.69	2.74
13	1.77	2.09	2.27	2.39	2.48	2.55	2.61	2.66	2.71
14	1.76	2.08	2.25	2.37	2.46	2.53	2.59	2.64	2.69
15	1.75	2.07	2.24	2.36	2.44	2.51	2.57	2.62	2.67
16	1.75	2.06	2.23	2.34	2.43	2.50	2.56	2.61	2.65
17	1.74	2.05	2.22	2.33	2.42	2.49	2.54	2.59	2.64
18	1.73	2.04	2.21	2.32	2.41	2.48	2.53	2.58	2.62
19	1.73	2.03	2.20	2.31	2.40	2.47	2.52	2.57	2.61
20	1.72	2.03	2.19	2.30	2.39	2.46	2.51	2.56	2.60
24	1.71	2.01	2.17	2.28	2.36	2.43	2.48	2.53	2.57
30	1.70	1.99	2.15	2.25	2.33	2.40	2.45	2.50	2.54
40	1.68	1.97	2.13	2.23	2.31	2.37	2.42	2.47	2.51
60	1.67	1.95	2.10	2.21	2.28	2.35	2.39	2.44	2.48
120	1.66	1.93	2.08	2.18	2.26	2.32	2.37	2.41	2.45
∞	1.64	1.92	2.06	2.16	2.23	2.29	2.34	2.38	2.42
One tailed, α = .01									
5	3.37	3.90	4.21	4.43	4.60	4.73	4.85	4.94	5.03
6	3.14	3.61	3.88	4.07	4.21	4.33	4.43	4.51	4.59
7	3.00	3.42	3.66	3.83	3.96	4.07	4.15	4.23	4.30
8	2.90	3.29	3.51	3.67	3.79	3.88	3.96	4.03	4.09
9	2.82	3.19	3.40	3.55	3.66	3.75	3.82	3.89	3.94
10	2.76	3.11	3.31	3.45	3.56	3.64	3.71	3.78	3.83
11	2.72	3.06	3.25	3.38	3.48	3.56	3.63	3.69	3.74
12	2.68	3.01	3.19	3.32	3.42	3.50	3.56	3.62	3.67
13	2.65	2.97	3.15	3.27	3.37	3.44	3.51	3.56	3.61
14	2.62	2.94	3.11	3.23	3.32	3.40	3.46	3.51	3.56
15	2.60	2.91	3.08	3.20	3.29	3.36	3.42	3.47	3.52
16	2.58	2.88	3.05	3.17	3.26	3.33	3.39	3.44	3.48
17	2.57	2.86	3.03	3.14	3.23	3.30	3.36	3.41	3.45
18	2.55	2.84	3.01	3.12	3.21	3.27	3.33	3.38	3.42
19	2.54	2.83	2.99	3.10	3.18	3.25	3.31	3.36	3.40
20	2.53	2.81	2.97	3.08	3.17	3.23	3.29	3.34	3.38
24	2.49	2.77	2.92	3.03	3.11	3.17	3.22	3.27	3.31

(continued)

TABLE A.7 (continued)

Critical Values for Dunnett's Procedure

df	1	2	3	4	5	6	7	8	9
One tailed, α = .01									
30	2.46	2.72	2.87	2.97	3.05	3.11	3.16	3.21	3.24
40	2.42	2.68	2.82	2.92	2.99	3.05	3.10	3.14	3.18
60	2.39	2.64	2.78	2.87	2.94	3.00	3.04	3.08	3.12
120	2.36	2.60	2.73	2.82	2.89	2.94	2.99	3.03	3.06
∞	2.33	2.56	2.68	2.77	2.84	2.89	2.93	2.97	3.00
Two tailed, α = .05									
5	2.57	3.03	3.29	3.48	3.62	3.73	3.82	3.90	3.97
6	2.45	2.86	3.10	3.26	3.39	3.49	3.57	3.64	3.71
7	2.36	2.75	2.97	3.12	3.24	3.33	3.41	3.47	3.53
8	2.31	2.67	2.88	3.02	3.13	3.22	3.29	3.35	3.41
9	2.26	2.61	2.81	2.95	3.05	3.14	3.20	3.26	3.32
10	2.23	2.57	2.76	2.89	2.99	3.07	3.14	3.19	3.24
11	2.20	2.53	2.72	2.84	2.94	3.02	3.08	3.14	3.19
12	2.18	2.50	2.68	2.81	2.90	2.98	3.04	3.09	3.14
13	2.16	2.48	2.65	2.78	2.87	2.94	3.00	3.06	3.10
14	2.14	2.46	2.63	2.75	2.84	2.91	2.97	3.02	3.07
15	2.13	2.44	2.61	2.73	2.82	2.89	2.95	3.00	3.04
16	2.12	2.42	2.59	2.71	2.80	2.87	2.92	2.97	3.02
17	2.11	2.41	2.58	2.69	2.78	2.85	2.90	2.95	3.00
18	2.10	2.40	2.56	2.68	2.76	2.83	2.89	2.94	2.98
19	2.09	2.39	2.55	2.66	2.75	2.81	2.87	2.92	2.96
20	2.09	2.38	2.54	2.65	2.73	2.80	2.86	2.90	2.95
24	2.06	2.35	2.51	2.61	2.70	2.76	2.81	2.86	2.90
30	2.04	2.32	2.47	2.58	2.66	2.72	2.77	2.82	2.86
40	2.02	2.29	2.44	2.54	2.62	2.68	2.73	2.77	2.81
60	2.00	2.27	2.41	2.51	2.58	2.64	2.69	2.73	2.77
120	1.98	2.24	2.38	2.47	2.55	2.60	2.65	2.69	2.73
∞	1.96	2.21	2.35	2.44	2.51	2.57	2.61	2.65	2.69
Two tailed, α = .01									
5	4.03	4.63	4.98	5.22	5.41	5.56	5.69	5.80	5.89
6	3.71	4.21	4.51	4.71	4.87	5.00	5.10	5.20	5.28
7	3.50	3.95	4.21	4.39	4.53	4.64	4.74	4.82	4.89
8	3.36	3.77	4.00	4.17	4.29	4.40	4.48	4.56	4.62
9	3.25	3.63	3.85	4.01	4.12	4.22	4.30	4.37	4.43
10	3.17	3.53	3.74	3.88	3.99	4.08	4.16	4.22	4.28
11	3.11	3.45	3.65	3.79	3.89	3.98	4.05	4.11	4.16
12	3.05	3.39	3.58	3.71	3.81	3.89	3.96	4.02	4.07
13	3.01	3.33	3.52	3.65	3.74	3.82	3.89	3.94	3.99
14	2.98	3.29	3.47	3.59	3.69	3.76	3.83	3.88	3.93
15	2.95	3.25	3.43	3.55	3.64	3.71	3.78	3.83	3.88
16	2.92	3.22	3.39	3.51	3.60	3.67	3.73	3.78	3.83

df	1	2	3	4	5	6	7	8	9
Two tailed, α = .01									
17	2.90	3.19	3.36	3.47	3.56	3.63	3.69	3.74	3.79
18	2.88	3.17	3.33	3.44	3.53	3.60	3.66	3.71	3.75
19	2.86	3.15	3.31	3.42	3.50	3.57	3.63	3.68	3.72
20	2.85	3.13	3.29	3.40	3.48	3.55	3.60	3.65	3.69
24	2.80	3.07	3.22	3.32	3.40	3.47	3.52	3.57	3.61
30	2.75	3.01	3.15	3.25	3.33	3.39	3.44	3.49	3.52
40	2.70	2.95	3.09	3.19	3.26	3.32	3.37	3.41	3.44
60	2.66	2.90	3.03	3.12	3.19	3.25	3.29	3.33	3.37
120	2.62	2.85	2.97	3.06	3.12	3.18	3.22	3.26	3.29
∞	2.58	2.79	2.92	3.00	3.06	3.11	3.15	3.19	3.22

Source: Reprinted from Dunnett, C.W., *J.Am. Stat. Assoc.*, 50, 1096, 1955 Table 1a and Table 1b With permission of the American Statistical Association; Dunnett, C. W., *Biometrics, 20,* 482, 1964, Table II and Table III. With permission of the Biometric Society.

The columns represent J = number of treatment means (excluding the control).

TABLE A.8

Critical Values for Dunn's (Bonferroni's) Procedure

		Number of Contrasts										
ν	α	2	3	4	5	6	7	8	9	10	15	20
2	0.01	14.071	17.248	19.925	22.282	24.413	26.372	28.196	29.908	31.528	38.620	44.598
	0.05	6.164	7.582	8.774	9.823	10.769	11.639	12.449	13.208	13.927	17.072	19.721
	0.10	4.243	5.243	6.081	6.816	7.480	8.090	8.656	9.188	9.691	11.890	13.741
	0.20	2.828	3.531	4.116	4.628	5.089	5.512	5.904	6.272	6.620	8.138	9.414
3	0.01	7.447	8.565	9.453	10.201	10.853	11.436	11.966	12.453	12.904	14.796	16.300
	0.05	4.156	4.826	5.355	5.799	6.185	6.529	6.842	7.128	7.394	8.505	9.387
	0.10	3.149	3.690	4.115	4.471	4.780	5.055	5.304	5.532	5.744	6.627	7.326
	0.20	2.294	2.734	3.077	3.363	3.610	3.829	4.028	4.209	4.377	5.076	5.626
4	0.01	5.594	6.248	6.751	7.166	7.520	7.832	8.112	8.367	8.600	9.556	10.294
	0.05	3.481	3.941	4.290	4.577	4.822	5.036	5.228	5.402	5.562	6.214	6.714
	0.10	2.751	3.150	3.452	3.699	3.909	4.093	4.257	4.406	4.542	5.097	5.521
	0.20	2.084	2.434	2.697	2.911	3.092	3.250	3.391	3.518	3.635	4.107	4.468
5	0.01	4.771	5.243	5.599	5.888	6.133	6.346	6.535	6.706	6.862	7.491	7.968
	0.05	3.152	3.518	3.791	4.012	4.197	4.358	4.501	4.630	4.747	5.219	5.573
	0.10	2.549	2.882	3.129	3.327	3.493	3.638	3.765	3.880	3.985	4.403	4.718
	0.20	1.973	2.278	2.503	2.683	2.834	2.964	3.079	3.182	3.275	3.649	3.928
6	0.01	4.315	4.695	4.977	5.203	5.394	5.559	5.704	5.835	5.954	6.428	6.782
	0.05	2.959	3.274	3.505	3.690	3.845	3.978	4.095	4.200	4.296	4.675	4.956
	0.10	2.428	2.723	2.939	3.110	3.253	3.376	3.484	3.580	3.668	4.015	4.272
	0.20	1.904	2.184	2.387	2.547	2.681	2.795	2.895	2.985	3.066	3.385	3.620
7	0.01	4.027	4.353	4.591	4.782	4.941	5.078	5.198	5.306	5.404	5.791	6.077
	0.05	2.832	3.115	3.321	3.484	3.620	3.736	3.838	3.929	4.011	4.336	4.574
	0.10	2.347	2.618	2.814	2.969	3.097	3.206	3.302	3.388	3.465	3.768	3.990
	0.20	1.858	2.120	2.309	2.457	2.579	2.684	2.775	2.856	2.929	3.214	3.423
8	0.01	3.831	4.120	4.331	4.498	4.637	4.756	4.860	4.953	5.038	5.370	5.613
	0.05	2.743	3.005	3.193	3.342	3.464	3.589	3.661	3.743	3.816	4.105	4.316
	0.10	2.289	2.544	2.726	2.869	2.967	3.088	3.176	3.254	3.324	3.598	3.798
	0.20	1.824	2.075	2.254	2.393	2.508	2.605	2.690	2.765	2.832	3.095	3.286
9	0.01	3.688	3.952	4.143	4.294	4.419	4.526	4.619	4.703	4.778	5.072	5.287
	0.05	2.677	2.923	3.099	3.237	3.351	3.448	3.532	3.607	3.675	3.938	4.129
	0.10	2.246	2.488	2.661	2.796	2.907	3.001	3.083	3.155	3.221	3.474	3.658
	0.20	1.799	2.041	2.212	2.345	2.454	2.546	2.627	2.696	2.761	3.008	3.185
10	0.01	3.580	3.825	4.002	4.141	4.256	4.354	4.439	4.515	4.584	4.852	5.046
	0.05	2.626	2.860	3.027	3.157	3.264	3.355	3.434	3.505	3.568	3.813	3.989
	0.10	2.213	2.446	2.611	2.739	2.845	2.934	3.012	3.080	3.142	3.380	3.552
	0.20	1.779	2.014	2.180	2.308	2.413	2.501	2.578	2.646	2.706	2.941	3.106
11	0.01	3.495	3.726	3.892	4.022	4.129	4.221	4.300	4.371	4.434	4.682	4.860
	0.05	2.586	2.811	2.970	3.094	3.196	3.283	3.358	3.424	3.484	3.715	3.880
	0.10	2.166	2.412	2.571	2.695	2.796	2.881	2.955	3.021	3.079	3.306	3.468
	0.20	1.763	1.993	2.154	2.279	2.380	2.465	2.539	2.605	2.663	2.888	3.048
12	0.01	3.427	3.647	3.804	3.927	4.029	4.114	4.189	4.256	4.315	4.547	4.714
	0.05	2.553	2.770	2.924	3.044	3.141	3.224	3.296	3.359	3.416	3.636	3.793
	0.10	2.164	2.384	2.539	2.658	2.756	2.838	2.910	2.973	3.029	3.247	3.402
	0.20	1.750	1.975	2.133	2.254	2.353	2.436	2.508	2.571	2.628	2.845	2.999

ν	α	Number of Contrasts 2	3	4	5	6	7	8	9	10	15	20
13	0.01	3.371	3.582	3.733	3.850	3.946	4.028	4.099	4.162	4.218	4.438	4.595
	0.05	2.526	2.737	2.886	3.002	3.096	3.176	3.245	3.306	3.361	3.571	3.722
	0.10	2.146	2.361	2.512	2.628	2.723	2.803	2.872	2.933	2.988	3.198	3.347
	0.20	1.739	1.961	2.116	2.234	2.331	2.412	2.482	2.544	2.599	2.809	2.958
14	0.01	3.324	3.528	3.673	3.785	3.878	3.956	4.024	4.084	4.138	4.347	4.497
	0.05	2.503	2.709	2.854	2.967	3.058	3.135	3.202	3.261	3.314	3.518	3.662
	0.10	2.131	2.342	2.489	2.603	2.696	2.774	2.841	2.900	2.953	3.157	3.301
	0.20	1.730	1.949	2.101	2.217	2.312	2.392	2.460	2.520	2.574	2.779	2.924
15	0.01	3.285	3.482	3.622	3.731	3.820	3.895	3.961	4.019	4.071	4.271	4.414
	0.05	2.483	2.685	2.827	2.937	3.026	3.101	3.166	3.224	3.275	3.472	3.612
	0.10	2.118	2.325	2.470	2.582	2.672	2.748	2.814	2.872	2.924	3.122	3.262
	0.20	1.722	1.938	2.088	2.203	2.296	2.374	2.441	2.500	2.553	2.754	2.896
16	0.01	3.251	3.443	3.579	3.684	3.771	3.844	3.907	3.963	4.013	4.206	4.344
	0.05	2.467	2.665	2.804	2.911	2.998	3.072	3.135	3.191	3.241	3.433	3.569
	0.10	2.106	2.311	2.453	2.563	2.652	2.726	2.791	2.848	2.898	3.092	3.228
	0.20	1.715	1.929	2.077	2.190	2.282	2.359	2.425	2.483	2.535	2.732	2.871
17	0.01	3.221	3.409	3.541	3.644	3.728	3.799	3.860	3.914	3.963	4.150	4.284
	0.05	2.452	2.647	2.783	2.889	2.974	3.046	3.108	3.163	3.212	3.399	3.532
	0.10	2.096	2.296	2.439	2.547	2.634	2.706	2.771	2.826	2.876	3.066	3.199
	0.20	1.709	1.921	2.068	2.179	2.270	2.346	2.411	2.488	2.519	2.713	2.849
18	0.01	3.195	3.379	3.508	3.609	3.691	3.760	3.820	3.872	3.920	4.102	4.231
	0.05	2.439	2.631	2.766	2.869	2.953	3.024	3.085	3.138	3.186	3.370	3.499
	0.10	2.088	2.287	2.426	2.532	2.619	2.691	2.753	2.806	2.857	3.043	3.174
	0.20	1.704	1.914	2.059	2.170	2.259	2.334	2.399	2.455	2.505	2.696	2.830
19	0.01	3.173	3.353	3.479	3.578	3.658	3.725	3.784	3.835	3.881	4.059	4.185
	0.05	2.427	2.617	2.750	2.852	2.934	3.004	3.064	3.116	3.163	3.343	3.470
	0.10	2.080	2.277	2.415	2.520	2.605	2.676	2.738	2.791	2.839	3.023	3.152
	0.20	1.699	1.908	2.052	2.161	2.250	2.324	2.388	2.443	2.493	2.682	2.813
20	0.01	3.152	3.329	3.454	3.550	3.629	3.695	3.752	3.802	3.848	4.021	4.144
	0.05	2.417	2.605	2.736	2.836	2.918	2.986	3.045	3.097	3.143	3.320	3.445
	0.10	2.073	2.269	2.405	2.508	2.593	2.663	2.724	2.777	2.824	3.005	3.132
	0.20	1.695	1.902	2.045	2.154	2.241	2.315	2.378	2.433	2.482	2.668	2.798
21	0.01	3.134	3.308	3.431	3.525	3.602	3.667	3.724	3.773	3.817	3.987	4.108
	0.05	2.408	2.594	2.723	2.822	2.903	2.970	3.028	3.080	3.125	3.300	3.422
	0.10	2.067	2.261	2.396	2.498	2.581	2.651	2.711	2.764	2.810	2.989	3.114
	0.20	1.691	1.897	2.039	2.147	2.234	2.306	2.369	2.424	2.472	2.656	2.785
22	0.01	3.118	3.289	3.410	3.503	3.579	3.643	3.698	3.747	3.790	3.957	4.075
	0.05	2.400	2.584	2.712	2.810	2.889	2.956	3.014	3.064	3.109	3.281	3.402
	0.10	2.061	2.254	2.387	2.489	2.572	2.641	2.700	2.752	2.798	2.974	3.096
	0.20	1.688	1.892	2.033	2.141	2.227	2.299	2.361	2.415	2.463	2.646	2.773

(continued)

TABLE A.8 (continued)

Critical Values for Dunn's (Bonferroni's) Procedure

		Number of Contrasts										
ν	α	2	3	4	5	6	7	8	9	10	15	20
23	0.01	3.103	3.272	3.392	3.483	3.558	3.621	3.675	3.723	3.766	3.930	4.046
	0.05	2.392	2.574	2.701	2.798	2.877	2.943	3.000	3.050	3.094	3.264	3.383
	0.10	2.056	2.247	2.380	2.481	2.563	2.631	2.690	2.741	2.787	2.961	3.083
	0.20	1.685	1.888	2.028	2.135	2.221	2.292	2.354	2.407	2.455	2.636	2.762
24	0.01	3.089	3.257	3.375	3.465	3.539	3.601	3.654	3.702	3.744	3.905	4.019
	0.05	2.385	2.566	2.692	2.788	2.866	2.931	2.988	3.037	3.081	3.249	3.366
	0.10	2.051	2.241	2.373	2.473	2.554	2.622	2.680	2.731	2.777	2.949	3.070
	0.20	1.682	1.884	2.024	2.130	2.215	2.286	2.347	2.400	2.448	2.627	2.752
25	0.01	3.077	3.243	3.359	3.449	3.521	3.583	3.635	3.682	3.723	3.882	3.995
	0.05	2.379	2.558	2.683	2.779	2.856	2.921	2.976	3.025	3.069	3.235	3.351
	0.10	2.047	2.236	2.367	2.466	2.547	2.614	2.672	2.722	2.767	2.938	3.058
	0.20	1.679	1.881	2.020	2.125	2.210	2.280	2.341	2.394	2.441	2.619	2.743
26	0.01	3.066	3.230	3.345	3.433	3.505	3.566	3.618	3.664	3.705	3.862	3.972
	0.05	2.373	2.551	2.675	2.770	2.847	2.911	2.966	3.014	3.058	3.222	3.337
	0.10	2.043	2.231	2.361	2.460	2.540	2.607	2.664	2.714	2.759	2.928	3.047
	0.20	1.677	1.878	2.016	2.121	2.205	2.275	2.335	2.388	2.435	2.612	2.735
27	0.01	3.056	3.218	3.332	3.419	3.491	3.550	3.602	3.647	3.688	3.843	3.952
	0.05	2.368	2.545	2.668	2.762	2.838	2.902	2.956	3.004	3.047	3.210	3.324
	0.10	2.039	2.227	2.356	2.454	2.534	2.600	2.657	2.707	2.751	2.919	3.036
	0.20	1.675	1.875	2.012	2.117	2.201	2.270	2.330	2.383	2.429	2.605	2.727
28	0.01	3.046	3.207	3.320	3.407	3.477	3.536	3.587	3.632	3.672	3.825	3.933
	0.05	2.383	2.539	2.661	2.755	2.830	2.893	2.948	2.995	3.038	3.199	3.312
	0.10	2.036	2.222	2.351	2.449	2.528	2.594	2.650	2.700	2.744	2.911	3.027
	0.20	1.672	1.872	2.009	2.113	2.196	2.266	2.326	2.378	2.424	2.599	2.720
29	0.01	3.037	3.197	3.309	3.395	3.464	3.523	3.574	3.618	3.658	3.809	3.916
	0.05	2.358	2.534	2.655	2.748	2.823	2.886	2.940	2.967	3.029	3.189	3.301
	0.10	2.033	2.218	2.346	2.444	2.522	2.588	2.644	2.693	2.737	2.903	3.018
	0.20	1.671	1.869	2.006	2.110	2.193	2.262	2.321	2.373	2.419	2.593	2.713
30	0.01	3.029	3.188	3.298	3.384	3.453	3.511	3.561	3.605	3.644	3.794	3.900
	0.05	2.354	2.528	2.649	2.742	2.816	2.878	2.932	2.979	3.021	3.180	3.291
	0.10	2.030	2.215	2.342	2.439	2.517	2.582	2.638	2.687	2.731	2.895	3.010
	0.20	1.669	1.867	2.003	2.106	2.189	2.258	2.317	2.369	2.414	2.587	2.707
40	0.01	2.970	3.121	3.225	3.305	3.370	3.425	3.472	3.513	3.549	3.689	3.787
	0.05	2.323	2.492	2.606	2.696	2.768	2.827	2.878	2.923	2.963	3.113	3.218
	0.10	2.009	2.189	2.312	2.406	2.481	2.544	2.597	2.644	2.686	2.843	2.952
	0.20	1.656	1.850	1.983	2.083	2.164	2.231	2.288	2.338	2.382	2.548	2.663
60	0.01	2.914	3.056	3.155	3.230	3.291	3.342	3.386	3.425	3.459	3.589	3.679
	0.05	2.294	2.456	2.568	2.653	2.721	2.777	2.826	2.869	2.906	3.049	3.146
	0.10	1.989	2.163	2.283	2.373	2.446	2.506	2.558	2.603	2.643	2.793	2.897
	0.20	1.643	1.834	1.963	2.061	2.139	2.204	2.259	2.308	2.350	2.511	2.621

		Number of Contrasts										
ν	α	2	3	4	5	6	7	8	9	10	15	20
120	0.01	2.859	2.994	3.067	3.158	3.215	3.263	3.304	3.340	3.372	3.493	3.577
	0.05	2.265	2.422	2.529	2.610	2.675	2.729	2.776	2.816	2.852	2.967	3.081
	0.10	1.968	2.138	2.254	2.342	2.411	2.469	2.519	2.562	2.600	2.744	2.843
	0.20	1.631	1.817	1.944	2.039	2.115	2.178	2.231	2.278	2.319	2.474	2.580
∞	0.01	2.806	2.934	3.022	3.089	3.143	3.188	3.226	3.260	3.289	3.402	3.480
	0.05	2.237	2.388	2.491	2.569	2.631	2.683	2.727	2.766	2.800	2.928	3.016
	0.10	1.949	2.114	2.226	2.311	2.378	2.434	2.482	2.523	2.560	2.697	2.791
	0.20	1.618	1.801	1.925	2.018	2.091	2.152	2.204	2.249	2.289	2.438	2.540

Source: Table 1 reprinted from Games, P. A. (1977), An improved t table for simultaneous control of g contrasts. *Journal of the American Statistical Association, 72*, 531–534. Reprinted with permission of the American Statistical Association, www.amstat.org and by permission of the publisher (Taylor & Francis Ltd., www.tandfonline.com.

TABLE A.9

Critical Values for the Studentized Range Statistic

	J or *r*								
v	2	3	4	5	6	7	8	9	10
$\alpha = .10$									
1	8.929	13.44	16.36	18.49	20.15	21.51	22.64	23.62	24.48
2	4.130	5.733	6.773	7.538	8.139	8.633	9.049	9.409	9.725
3	3.328	4.467	5.199	5.738	6.162	6.511	6.806	7.062	7.287
4	3.015	3.976	4.586	5.035	5.388	5.679	5.926	6.139	6.327
5	2.850	3.717	4.264	4.664	4.979	5.238	5.458	5.648	5.816
6	2.748	3.559	4.065	4.435	4.726	4.966	5.168	5.344	5.499
7	2.680	3.451	3.931	4.280	4.555	4.780	4.972	5.137	5.283
8	2.630	3.374	3.834	4.169	4.431	4.646	4.829	4.987	5.126
9	2.592	3.316	3.761	4.084	4.337	4.545	4.721	4.873	5.007
10	2.563	3.270	3.704	4.018	4.264	4.465	4.636	4.783	4.913
11	2.540	3.234	3.658	3.965	4.205	4.401	4.568	4.711	4.838
12	2.521	3.204	3.621	3.922	4.156	4.349	4.511	4.652	4.776
13	2.505	3.179	3.589	3.885	4.116	4.305	4.464	4.602	4.724
14	2.491	3.158	3.563	3.854	4.081	4.267	4.424	4.560	4.680
15	2.479	3.140	3.540	3.828	4.052	4.235	4.390	4.524	4.641
16	2.469	3.124	3.520	3.804	4.026	4.207	4.360	4.492	4.608
17	2.460	3.110	3.503	3.784	4.004	4.183	4.334	4.464	4.579
18	2.452	3.098	3.488	3.767	3.984	4.161	4.311	4.440	4.554
19	2.445	3.087	3.474	3.751	3.966	4.142	4.290	4.418	4.531
20	2.439	3.078	3.462	3.736	3.950	4.124	4.271	4.398	4.510
24	2.420	3.047	3.423	3.692	3.900	4.070	4.213	4.336	4.445
30	2.400	3.017	3.386	3.648	3.851	4.016	4.155	4.275	4.381
40	2.381	2.988	3.349	3.605	3.803	3.963	4.099	4.215	4.317
60	2.363	2.959	3.312	3.562	3.755	3.911	4.042	4.155	4.254
120	2.344	2.930	3.276	3.520	3.707	3.859	3.987	4.096	4.191
∞	2.326	2.902	3.240	3.478	3.661	3.808	3.931	4.037	4.129

	J or *r*								
v	11	12	13	14	15	16	17	18	19
$\alpha = .10$									
1	25.24	25.92	26.54	27.10	27.62	28.10	28.54	28.96	29.35
2	10.01	10.26	10.49	10.70	10.89	11.07	11.24	11.39	11.54
3	7.487	7.667	7.832	7.982	8.120	8.249	8.368	8.479	8.584
4	6.495	6.645	6.783	6.909	7.025	7.133	7.233	7.327	7.414
5	5.966	6.101	6.223	6.336	6.440	6.536	6.626	6.710	6.789
6	5.637	5.762	5.875	5.979	6.075	6.164	6.247	6.325	6.398
7	5.413	5.530	5.637	5.735	5.826	5.910	5.838	6.061	6.130
8	5.250	5.362	5.464	5.558	5.644	5.724	5.799	5.869	5.935
9	5.127	5.234	5.333	5.423	5.506	5.583	5.655	5.723	5.786
10	5.029	5.134	5.229	5.317	5.397	5.472	5.542	5.607	5.668
11	4.951	5.053	5.146	5.231	5.309	5.382	5.450	5.514	5.573
12	4.886	4.986	5.077	5.160	5.236	5.308	5.374	5.436	5.495
13	4.832	4.930	5.019	5.100	5.176	5.245	5.311	5.372	5.429

v	*J* or *r*								
	11	**12**	**13**	**14**	**15**	**16**	**17**	**18**	**19**
α = .10									
14	4.786	4.882	4.970	5.050	5.124	5.192	5.256	5.316	5.373
15	4.746	4.841	4.927	5.006	5.079	5.147	5.209	5.269	5.324
16	4.712	4.805	4.890	4.968	5.040	5.107	5.169	5.227	5.282
17	4.682	4.774	4.858	4.935	5.005	5.071	5.133	5.190	5.244
18	4.655	4.746	4.829	4.905	4.975	5.040	5.101	5.158	5.211
19	4.631	4.721	4.803	4.879	4.948	5.012	5.073	5.129	5.182
20	4.609	4.699	4.780	4.855	4.924	4.987	5.047	5.103	5.155
24	4.541	4.628	4.708	4.780	4.847	4.909	4.966	5.021	5.071
30	4.474	4.559	4.635	4.706	4.770	4.830	4.886	4.939	4.988
40	4.408	4.490	4.564	4.632	4.695	4.752	4.807	4.857	4.905
60	4.342	4.421	4.493	4.558	4.619	4.675	4.727	4.775	4.821
120	4.276	4.353	4.422	4.485	4.543	4.597	4.647	4.694	4.738
∞	4.211	4.285	4.351	4.412	4.468	4.519	4.568	4.612	4.654

v	*J* or *r*								
	2	**3**	**4**	**5**	**6**	**7**	**8**	**9**	**10**
α = .05									
1	17.97	26.98	32.82	37.08	40.41	43.12	45.40	47.36	49.07
2	6.085	8.331	9.798	10.88	11.74	12.44	13.03	13.54	13.99
3	4.501	5.910	6.825	7.502	8.037	8.478	8.853	9.177	9.462
4	3.927	5.040	5.757	6.287	6.707	7.053	7.347	7.602	7.826
5	3.635	4.602	5.218	5.673	6.033	6.330	6.582	6.802	6.995
6	3.461	4.339	4.896	5.305	5.628	5.895	6.122	6.319	6.493
7	3.344	4.165	4.681	5.060	5.359	5.606	5.815	5.998	6.158
8	3.261	4.041	4.529	4.886	5.167	5.399	5.597	5.767	5.918
9	3.199	3.949	4.415	4.756	5.024	5.244	5.432	5.595	5.739
10	3.151	3.877	4.327	4.654	4.912	5.124	5.305	5.461	5.599
11	3.113	3.820	4.256	4.574	4.823	5.028	5.202	5.353	5.487
12	3.082	3.773	4.199	4.508	4.751	4.950	5.119	5.265	5.395
13	3.055	3.735	4.151	4.453	4.690	4.885	5.049	5.192	5.318
14	3.033	3.702	4.111	4.407	4.639	4.829	4.990	5.131	5.254
15	3.014	3.674	4.076	4.367	4.595	4.782	4.940	5.077	5.198
16	2.998	3.649	4.046	4.333	4.557	4.741	4.897	5.031	5.150
17	2.984	3.628	4.020	4.303	4.524	4.705	4.858	4.991	5.108
18	2.971	3.609	3.997	4.277	4.495	4.673	4.824	4.956	5.071
19	2.960	3.593	3.977	4.253	4.469	4.645	4.794	4.924	5.038
20	2.950	3.578	3.958	4.232	4.445	4.620	4.768	4.896	5.008
24	2.919	3.532	3.901	4.166	4.373	4.541	4.684	4.807	4.915
30	2.888	3.486	3.845	4.102	4.302	4.464	4.602	4.720	4.824
40	2.858	3.442	3.791	4.039	4.232	4.389	4.521	4.635	4.735
60	2.829	3.399	3.737	3.977	4.163	4.314	4.441	4.550	4.646
120	2.800	3.356	3.685	3.917	4.096	4.241	4.363	4.468	4.560
∞	2.772	3.314	3.633	3.858	4.030	4.170	4.286	4.387	4.474

(continued)

TABLE A.9 (continued)

Critical Values for the Studentized Range Statistic

	J or *r*								
v	11	12	13	14	15	16	17	18	19
α = .05									
1	50.59	51.96	53.20	54.33	55.36	56.32	57.22	58.04	58.83
2	14.39	14.75	15.08	15.38	15.65	15.91	16.14	16.37	16.57
3	9.717	9.946	10.15	10.35	10.53	10.69	10.84	10.98	11.11
4	8.027	8.208	8.373	8.525	8.664	8.794	8.914	9.028	9.134
5	7.168	7.324	7.466	7.596	7.717	7.828	7.932	8.030	8.122
6	6.649	6.789	6.917	7.034	7.143	7.244	7.338	7.426	7.508
7	6.302	6.431	6.550	6.658	6.759	6.852	6.939	7.020	7.097
8	6.054	6.175	6.287	6.389	6.483	6.571	6.653	6.729	6.802
9	5.867	5.983	6.089	6.186	6.276	6.359	6.437	6.510	6.579
10	5.722	5.833	5.935	6.028	6.114	6.194	6.269	6.339	6.405
11	5.605	5.713	5.811	5.901	5.984	6.062	6.134	6.202	6.265
12	5.511	5.615	5.710	5.798	5.878	5.953	6.023	6.089	6.151
13	5.431	5.533	5.625	5.711	5.789	5.862	5.931	5.995	6.055
14	5.364	5.463	5.554	5.637	5.714	5.786	5.852	5.915	5.974
15	5.306	5.404	5.493	5.574	5.649	5.720	5.785	5.846	5.904
16	5.256	5.352	5.439	5.520	5.593	5.662	5.720	5.786	5.843
17	5.212	5.307	5.392	5.471	5.544	5.612	5.675	5.734	5.790
18	5.174	5.267	5.352	5.429	5.501	5.568	5.630	5.688	5.743
19	5.140	5.231	5.315	5.391	5.462	5.528	5.589	5.647	5.701
20	5.108	5.199	5.282	5.357	5.427	5.493	5.553	5.610	5.663
24	5.012	5.099	5.179	5.251	5.319	5.381	5.439	5.494	5.545
30	4.917	5.001	5.077	5.147	5.211	5.271	5.327	5.379	5.429
40	4.824	4.904	4.977	5.044	5.106	5.163	5.216	5.266	5.313
60	4.732	4.808	4.878	4.942	5.001	5.056	5.107	5.154	5.199
120	4.641	4.714	4.781	4.842	4.898	4.950	4.998	5.044	5.086
∞	4.552	4.622	4.685	4.743	4.796	4.845	4.891	4.934	4.974

	J or *r*								
v	2	3	4	5	6	7	8	9	10
α = .01									
1	90.03	135.0	164.3	185.6	202.2	215.8	227.2	237.0	245.6
2	14.04	19.02	22.29	24.72	26.63	28.20	29.53	30.68	31.69
3	8.261	10.62	12.17	13.33	14.24	15.00	15.64	16.20	16.69
4	6.512	8.120	9.173	9.958	10.58	11.10	11.55	11.93	12.27
5	5.702	6.976	7.804	8.421	8.913	9.321	9.669	9.972	10.24
6	5.243	6.331	7.033	7.556	7.973	8.318	8.613	8.869	9.097
7	4.949	5.919	6.543	7.005	7.373	7.679	7.939	8.166	8.368
8	4.746	5.635	6.204	6.625	6.960	7.237	7.474	7.681	7.863
9	4.596	5.428	5.957	6.348	6.658	6.915	7.134	7.325	7.495
10	4.482	5.270	5.769	6.136	6.428	6.669	6.875	7.055	7.213
11	4.392	5.146	5.621	5.970	6.247	6.476	6.672	6.842	6.992
12	4.320	5.046	5.502	5.836	6.101	6.321	6.507	6.670	6.814
13	4.260	4.964	5.404	5.727	5.981	6.192	6.372	6.528	6.667

v	J or r 2	3	4	5	6	7	8	9	10
$\alpha = .01$									
14	4.210	4.895	5.322	5.634	5.881	6.085	6.258	6.409	6.543
15	4.168	4.836	5.252	5.556	5.796	5.994	6.162	6.309	6.439
16	4.131	4.786	5.192	5.489	5.722	5.915	6.079	6.222	6.349
17	4.099	4.742	5.140	5.430	5.659	5.847	6.007	6.147	6.270
18	4.071	4.703	5.094	5.379	5.603	5.788	5.944	6.081	6.201
19	4.046	4.670	5.054	5.334	5.554	5.735	5.889	6.022	6.141
20	4.024	4.639	5.018	5.294	5.510	5.688	5.839	5.970	6.087
24	3.956	4.546	4.907	5.168	5.374	5.542	5.685	5.809	5.919
30	3.889	4.455	4.799	5.048	5.242	5.401	5.536	5.653	5.756
40	3.825	4.367	4.696	4.931	5.114	5.265	5.392	5.502	5.599
60	3.762	4.282	4.595	4.818	4.991	5.133	5.253	5.356	5.447
120	3.702	4.200	4.497	4.709	4.872	5.005	5.118	5.214	5.299
∞	3.643	4.120	4.403	4.603	4.757	4.882	4.987	5.078	5.157

v	J or r 11	12	13	14	15	16	17	18	19
$\alpha = .01$									
1	253.2	260.0	266.2	271.8	277.0	281.8	286.3	290.4	294.3
2	32.59	33.40	34.13	34.81	35.43	36.00	36.53	37.03	37.50
3	17.13	17.53	17.89	18.22	18.52	18.81	19.07	19.32	19.55
4	12.57	12.84	13.09	13.32	13.53	13.73	13.91	14.08	14.24
5	10.48	10.70	10.89	11.08	11.24	11.40	11.55	11.68	11.81
6	9.301	9.485	9.653	9.808	9.951	10.08	10.21	10.32	10.43
7	8.548	8.711	8.860	8.997	9.124	9.242	9.353	9.456	9.554
6	8.027	8.176	8.312	8.436	8.552	8.659	8.760	8.854	8.943
9	7.647	7.784	7.910	8.025	8.132	8.232	8.325	3.412	8.495
10	7.356	7.485	7.603	7.712	7.812	7.906	7.993	8.076	8.153
11	7.128	7.250	7.362	7.465	7.560	7.649	7.732	7.809	7.883
12	6.943	7.060	7.167	7.265	7.356	7.441	7.520	7.594	7.665
13	6.791	6.903	7.006	7.101	7.188	7.269	7.345	7.417	7.485
14	6.664	6.772	6.871	6.962	7.047	7.126	7.199	7.268	7.333
15	6.555	6.660	6.757	6.845	6.927	7.003	7.074	7.142	7.204
16	6.462	6.564	6.658	6.744	6.823	6.898	6.967	7.032	7.093
17	6.381	6.480	6.572	6.656	6.734	6.806	6.873	6.937	6.997
18	6.310	6.407	6.497	6.579	6.655	6.725	6.792	6.854	6.912
19	6.247	6.342	6.430	6.510	6.585	6.654	6.719	6.780	6.837
20	6.191	6.285	6.371	6.450	6.523	6.591	6.654	6.714	6.771
24	6.017	6.106	6.186	6.261	6.330	6.394	6.453	6.510	6.563
30	5.849	5.932	6.008	6.078	6.143	6.203	6.259	6.311	6.361
40	5.686	5.764	5.835	5.900	5.961	6.017	6.069	6.119	6.165
60	5.528	5.601	5.667	5.728	5.785	5.837	5.886	5.931	5.974
120	5.375	5.443	5.505	5.562	5.614	5.662	5.708	5.750	5.790
∞	5.227	5.290	5.348	5.400	5.448	5.493	5.535	5.574	5.611

Source: Table 3 from Harter, H. L. (1960). Tables of range and studentized range. *Annals of Mathematical Statistics, 31*, 1122–1147. By permission of the Institute of Mathematical Statistics.

TABLE A.10

Critical Values for the Bryant–Paulson Procedure

$\alpha = .05$

v	$J = 2$	$J = 3$	$J = 4$	$J = 5$	$J = 6$	$J = 7$	$J = 8$	$J = 10$	$J = 12$	$J = 16$	$J = 20$
$X = 1$											
2	7.96	11.00	12.99	14.46	15.61	16.56	17.36	18.65	19.68	21.23	22.40
3	5.42	7.18	8.32	9.17	9.84	10.39	10.86	11.62	12.22	13.14	13.83
4	4.51	5.84	6.69	7.32	7.82	8.23	8.58	9.15	9.61	10.30	10.82
5	4.06	5.17	5.88	6.40	6.82	7.16	7.45	7.93	8.30	8.88	9.32
6	3.79	4.78	5.40	5.86	6.23	6.53	6.78	7.20	7.53	8.04	8.43
7	3.62	4.52	5.09	5.51	5.84	6.11	6.34	6.72	7.03	7.49	7.84
8	3.49	4.34	4.87	5.26	5.57	5.82	6.03	6.39	6.67	7.10	7.43
10	3.32	4.10	4.58	4.93	5.21	5.43	5.63	5.94	6.19	6.58	6.87
12	3.22	3.95	4.40	4.73	4.98	5.19	5.37	5.67	5.90	6.26	6.53
14	3.15	3.85	4.28	4.59	4.83	5.03	5.20	5.48	5.70	6.03	6.29
16	3.10	3.77	4.19	4.49	4.72	4.91	5.07	5.34	5.55	5.87	6.12
18	3.06	3.72	4.12	4.41	4.63	4.82	4.98	5.23	5.44	5.75	5.98
20	3.03	3.67	4.07	4.35	4.57	4.75	4.90	5.15	5.35	5.65	5.88
24	2.98	3.61	3.99	4.26	4.47	4.65	4.79	5.03	5.22	5.51	5.73
30	2.94	3.55	3.91	4.18	4.38	4.54	4.69	4.91	5.09	5.37	5.58
40	2.89	3.49	3.84	4.09	4.29	4.45	4.58	4.80	4.97	5.23	5.43
60	2.85	3.43	3.77	4.01	4.20	4.35	4.48	4.69	4.85	5.10	5.29
120	2.81	3.37	3.70	3.93	4.11	4.26	4.38	4.58	4.73	4.97	5.15
$X = 2$											
2	9.50	13.18	15.59	17.36	18.75	19.89	20.86	22.42	23.66	25.54	26.94
3	6.21	8.27	9.60	10.59	11.37	12.01	12.56	13.44	14.15	15.22	16.02
4	5.04	6.54	7.51	8.23	8.80	9.26	9.66	10.31	10.83	11.61	12.21
5	4.45	5.68	6.48	7.06	7.52	7.90	8.23	8.76	9.18	9.83	10.31
6	4.10	5.18	5.87	6.37	6.77	7.10	7.38	7.84	8.21	8.77	9.20
7	3.87	4.85	5.47	5.92	6.28	6.58	6.83	7.24	7.57	8.08	8.46
8	3.70	4.61	5.19	5.61	5.94	6.21	6.44	6.82	7.12	7.59	7.94
10	3.49	4.31	4.82	5.19	5.49	5.73	5.93	6.27	6.54	6.95	7.26
12	3.35	4.12	4.59	4.93	5.20	5.43	5.62	5.92	6.17	6.55	6.83
14	3.26	3.99	4.44	4.76	5.01	5.22	5.40	5.69	5.92	6.27	6.54
16	3.19	3.90	4.32	4.63	4.88	5.07	5.24	5.52	5.74	6.07	6.33
18	3.14	3.82	4.24	4.54	4.77	4.96	5.13	5.39	5.60	5.92	6.17
20	3.10	3.77	4.17	4.46	4.69	4.88	5.03	5.29	5.49	5.81	6.04
24	3.04	3.69	4.08	4.35	4.57	4.75	4.90	5.14	5.34	5.63	5.86
30	2.99	3.61	3.98	4.25	4.46	4.62	4.77	5.00	5.18	5.46	5.68
40	2.93	3.53	3.89	4.15	4.34	4.50	4.64	4.86	5.04	5.30	5.50
60	2.88	3.46	3.80	4.05	4.24	4.39	4.52	4.73	4.89	5.14	5.33
120	2.82	3.38	3.72	3.95	4.13	4.28	4.40	4.60	4.75	4.99	5.17

α = .05

v	*J* = 2	*J* = 3	*J* = 4	*J* = 5	*J* = 6	*J* = 7	*J* = 8	*J* = 10	*J* = 12	*J* = 16	*J* = 20
X = 3											
2	10.83	15.06	17.82	19.85	21.45	22.76	23.86	25.66	27.08	29.23	30.83
3	6.92	9.23	10.73	11.84	12.72	13.44	14.06	15.05	15.84	17.05	17.95
4	5.51	7.18	8.25	9.05	9.67	10.19	10.63	11.35	11.92	12.79	13.45
5	4.81	6.16	7.02	7.66	8.17	8.58	8.94	9.52	9.98	10.69	11.22
6	4.38	5.55	6.30	6.84	7.28	7.64	7.94	8.44	8.83	9.44	9.90
7	4.11	5.16	5.82	6.31	6.70	7.01	7.29	7.73	8.08	8.63	9.03
8	3.91	4.88	5.49	5.93	6.29	6.58	6.83	7.23	7.55	8.05	8.42
10	3.65	4.51	5.05	5.44	5.75	6.01	6.22	6.58	6.86	7.29	7.62
12	3.48	4.28	4.78	5.14	5.42	5.65	5.85	6.17	6.43	6.82	7.12
14	3.37	4.13	4.59	4.93	5.19	5.41	5.59	5.89	6.13	6.50	6.78
16	3.29	4.01	4.46	4.78	5.03	5.23	5.41	5.69	5.92	6.27	6.53
18	3.23	3.93	4.35	4.66	4.90	5.10	5.27	5.54	5.76	6.09	6.34
20	3.18	3.86	4.28	4.57	4.81	5.00	5.16	5.42	5.63	5.96	6.20
24	3.11	3.76	4.16	4.44	4.67	4.85	5.00	5.25	5.45	5.75	5.98
30	3.04	3.67	4.05	4.32	4.53	4.70	4.85	5.08	5.27	5.56	5.78
40	2.97	3.57	3.94	4.20	4.40	4.56	4.70	4.92	5.10	5.37	5.57
60	2.90	3.49	3.83	4.08	4.27	4.43	4.56	4.77	4.93	5.19	5.38
120	2.84	3.40	3.73	3.97	4.15	4.30	4.42	4.62	4.77	5.01	5.19

α = .01

v	*J* = 2	*J* = 3	*J* = 4	*J* = 5	*J* = 6	*J* = 7	*J* = 8	*J* = 10	*J* = 12	*J* = 16	*J* = 20
X = 1											
2	19.09	26.02	30.57	33.93	36.58	38.76	40.60	43.59	45.95	49.55	52.24
3	10.28	13.32	15.32	16.80	17.98	18.95	19.77	21.12	22.19	23.82	25.05
4	7.68	9.64	10.93	11.89	12.65	13.28	13.82	14.70	15.40	16.48	17.29
5	6.49	7.99	8.97	9.70	10.28	10.76	11.17	11.84	12.38	13.20	13.83
6	5.83	7.08	7.88	8.48	8.96	9.36	9.70	10.25	10.70	11.38	11.90
7	5.41	6.50	7.20	7.72	8.14	8.48	8.77	9.26	9.64	10.24	10.69
8	5.12	6.11	6.74	7.20	7.58	7.88	8.15	8.58	8.92	9.46	9.87
10	4.76	5.61	6.15	6.55	6.86	7.13	7.35	7.72	8.01	8.47	8.82
12	4.54	5.31	5.79	6.15	6.48	6.67	6.87	7.20	7.46	7.87	8.18
14	4.39	5.11	5.56	5.89	6.15	6.36	6.55	6.85	7.09	7.47	7.75
16	4.28	4.96	5.39	5.70	5.95	6.15	6.32	6.60	6.83	7.18	7.45
18	4.20	4.86	5.26	5.56	5.79	5.99	6.15	6.42	6.63	6.96	7.22
20	4.14	4.77	5.17	5.45	5.68	5.86	6.02	6.27	6.48	6.80	7.04
24	4.05	4.65	5.02	5.29	5.50	5.68	5.83	6.07	6.26	6.56	6.78
30	3.96	4.54	4.89	5.14	5.34	5.50	5.64	5.87	6.05	6.32	6.53
40	3.88	4.43	4.76	5.00	5.19	5.34	5.47	5.68	5.85	6.10	6.30
60	3.79	4.32	4.64	4.86	5.04	5.18	5.30	5.50	5.65	5.89	6.07
120	3.72	4.22	4.52	4.73	4.89	5.03	5.14	5.32	5.47	5.69	5.85

(*continued*)

TABLE A.10 (continued)

Critical Values for the Bryant–Paulson Procedure

$\alpha = .01$

v	$J = 2$	$J = 3$	$J = 4$	$J = 5$	$J = 6$	$J = 7$	$J = 8$	$J = 10$	$J = 12$	$J = 16$	$J = 20$
$X = 2$											
2	23.11	31.55	37.09	41.19	44.41	47.06	49.31	52.94	55.82	60.20	63.47
3	11.97	15.56	17.91	19.66	21.05	22.19	23.16	24.75	26.01	27.93	29.38
4	8.69	10.95	12.43	13.54	14.41	15.14	15.76	16.77	17.58	18.81	19.74
5	7.20	8.89	9.99	10.81	11.47	12.01	12.47	13.23	13.84	14.77	15.47
6	6.36	7.75	8.64	9.31	9.85	10.29	10.66	11.28	11.77	12.54	13.11
7	5.84	7.03	7.80	8.37	8.83	9.21	9.53	10.06	10.49	11.14	11.64
8	5.48	6.54	7.23	7.74	8.14	8.48	8.76	9.23	9.61	10.19	10.63
10	5.02	5.93	6.51	6.93	7.27	7.55	7.79	8.19	8.50	8.99	9.36
12	4.74	5.56	6.07	6.45	6.75	7.00	7.21	7.56	7.84	8.27	8.60
14	4.56	5.31	5.78	6.13	6.40	6.63	6.82	7.14	7.40	7.79	8.09
16	4.42	5.14	5.58	5.90	6.16	6.37	6.55	6.85	7.08	7.45	7.73
18	4.32	5.00	5.43	5.73	5.98	6.18	6.35	6.63	6.85	7.19	7.46
20	4.25	4.90	5.31	5.60	5.84	6.03	6.19	6.46	6.67	7.00	7.25
24	4.14	4.76	5.14	5.42	5.63	5.81	5.96	6.21	6.41	6.71	6.95
30	4.03	4.62	4.98	5.24	5.44	5.61	5.75	5.98	6.16	6.44	6.66
40	3.93	4.48	4.82	5.07	5.26	5.41	5.54	5.76	5.93	6.19	6.38
60	3.83	4.36	4.68	4.90	5.08	5.22	5.35	5.54	5.70	5.94	6.12
120	3.73	4.24	4.54	4.75	4.91	5.05	5.16	5.35	5.49	5.71	5.88
$X = 3$											
2	26.54	36.26	42.64	47.36	51.07	54.13	56.71	60.90	64.21	69.25	73.01
3	13.45	17.51	20.17	22.15	23.72	25.01	26.11	27.90	29.32	31.50	33.13
4	9.59	12.11	13.77	15.00	15.98	16.79	17.47	18.60	19.50	20.87	21.91
5	7.83	9.70	10.92	11.82	12.54	13.14	13.65	14.48	15.15	10.17	16.95
6	6.85	8.36	9.34	10.07	10.65	11.13	11.54	12.22	12.75	13.59	14.21
7	6.23	7.52	8.36	8.98	9.47	9.88	10.23	10.80	11.26	11.97	12.51
8	5.81	6.95	7.69	8.23	8.67	9.03	9.33	9.84	10.24	10.87	11.34
10	5.27	6.23	6.84	7.30	7.66	7.96	8.21	8.63	8.96	9.48	9.88
12	4.94	5.80	6.34	6.74	7.05	7.31	7.54	7.90	8.20	8.65	9.00
14	4.72	5.51	6.00	6.36	6.65	6.89	7.09	7.42	7.69	8.10	8.41
16	4.56	5.30	5.76	6.10	6.37	6.59	6.77	7.08	7.33	7.71	8.00
18	4.44	5.15	5.59	5.90	6.16	6.36	6.54	6.83	7.06	7.42	7.69
20	4.35	5.03	5.45	5.75	5.99	6.19	6.36	6.63	6.85	7.19	7.45
24	4.22	4.86	5.25	5.54	5.76	5.94	6.10	6.35	6.55	6.87	7.11
30	4.10	4.70	5.06	5.33	5.54	5.71	5.85	6.08	6.27	6.56	6.78
40	3.98	4.54	4.88	5.13	5.32	5.48	5.61	5.83	6.00	6.27	6.47
60	3.86	4.39	4.72	4.95	5.12	5.27	5.39	5.59	5.75	6.00	6.18
120	3.75	4.25	4.55	4.77	4.94	5.07	5.18	5.37	5.51	5.74	5.90

Source: Tables 1A and 1B from Bryant, J. L. & Paulson, A. S. (1976). An extension of Tukey's method of multiple comparisons to experimental designs with random concomitant variables, *Biometrika, 63*, 631–638. By permission of Oxford University Press.

References

Aberson, C. L. (2010). *Applied power analysis for the behavioral sciences*. New York, NY: Routledge.

Agresti, A. (2018). *Statistical methods for the social sciences* (5th ed.). Upper Saddle River, NJ: Pearson.

Agresti, A., & Pendergast, J. (1986). Comparing mean ranks for repeated measures data. *Communications in Statistics: Theory and Method, 15*, 1417–1433.

Aguinis, H. (2004). *Regression analysis for categorical moderators*. New York, NY: Guilford.

Aguinis, H., Beaty, J. C., Boik, R. J., & Pierce, C. A. (2005). Effect size and power in assessing moderating effects of categorical variables using multiple regression: A 30-year review. *Journal of Applied Psychology, 90*(1), 94–107.

Aguinis, H., & Stone-Romero, E. F. (1997). Methodological artifacts in moderated multiple regression and their effects on statistical power. *Journal of Applied Psychology, 82*(1), 192–206. doi:10.1037/0021-9010.82.1.192

Aiken, L. S., & West, S. G. (1991). *Multiple regression: Testing and interpreting interactions*. Newbury Park, CA: Sage Publications.

Aldrich, J. H., & Nelson, F. D. (1984). *Linear probability, logit, and probit models*. Beverly Hills, CA: Sage.

Algina, J., & Olejnik, S. (2000). Determining sample size for accurate estimation of the squared multiple correlation coefficient. *Multivariate Behavioral Research, 35*, 119–136.

Andrews, D. F. (1971). Significance tests based on residuals. *Biometrika, 58*, 139–148.

Applebaum, M. I., & Cramer, E. M. (1974). Some problems in the nonorthogonal analysis of variance. *Psychological Bulletin, 81*, 335–343.

Atiqullah, M. (1964). The robustness of the covariance analysis of a one-way classification. *Biometrika, 51*(3–4), 365–373.

Atkinson, A. C. (1987). *Plots, transformations, and regression*. Oxford, UK: Oxford University Press.

Barnett, V., & Lewis, T. (1994). *Outliers in statistical data* (3rd ed.). Chichester, UK: Wiley.

Bates, D. M., & Watts, D. G. (1988). *Nonlinear regression analysis and its applications*. New York, NY: Wiley.

Bauer, D. J., & Curran, P. J. (2005). Probing interactions in fixed and multilevel regression: Inferential and graphical techniques. *Multivariate Behavioral Research, 40*(3), 373–400.

Beckman, R. J., & Cook, R. D. (1983). Outliers [in statistical data]. *Technometrics, 25*, 119–149.

Belsley, D. A. (1991). *Conditioning diagnostics: Collinearity and weak data in regression*. New York, NY: Wiley.

Belsley, D. A., Kuh, E., & Welsch, R. E. (1980). *Regression diagnostics*. New York, NY: Wiley.

Benjamini, Y., & Hochberg, Y. (1995). Controlling the false discovery rate: A practical and powerful approach to multiple testing. *Journal of the Royal Statistical Society. Series B, 57*(1), 289–300.

Berg, L., & Brännström, L. (2018). Evicted children and subsequent placement in out-of-home care: A cohort study. *PLoS ONE, 13*(4), 1–13. doi:10.1371/journal.pone.0195295

Berk, R. A. (2016). *Statistical learning from a regression perspective* (2nd ed.). Cham, Switzerland: Springer.

Berry, W. D., & Feldman, S. (1985). *Multiple regression in practice*. Beverly Hills, CA: Sage.

Bodner, T. E. (2017). Standardized effect sizes for moderated conditional fixed effects with continuous moderator variables. *Frontiers in Psychology, 8*. doi:10.3389/fpsyg.2017.00562

Boik, R. J. (1979). Interactions, partial interactions, and interaction contrasts in the analysis of variance. *Psychological Bulletin, 86*(5), 1084–1089. doi:10.1037/0033–2909.86.5.1084

Boik, R. J. (1981). A priori tests in repeated measures designs: Effects of nonsphericity. *Psychometrika, 46*(3), 241–255.

Bonett, D. G., & Seier, E. (2002). A test of normality with high uniform power. *Computational Statistics and Data Analysis, 40*, 435–445. doi:10.1016/S0167–9473(02)00074–9

Borm, G. F., Fransen, J., & Lemmens, W. A. J. G. (2007). A simple sample size formula for analysis of covariance in randomized clinical trials. *Journal of Clinical Epidemiology, 60*, 1234–1238. doi:10.1016/j.jclinepi.2007.02.006

Box, G. E. P. (1954). Some theorems on quadratic forms applied in the study of analysis of variance problems, II: Effects of Inequality of variance and of correlation between errors in the two-way classification. *The Annals of Mathematical Statistics, 25*(3), 484–498.

Box, G. E. P., & Anderson, S. L. (1962). *Robust tests for variances and effect of non-normality and variance heterogeneity on standard tests* (Technical Report Number 7, Ordinance Project Number TB 2–0001 (832)).

Box, G. E. P., & Cox, D. R. (1964). An analysis of transformations. *Journal of the Royal Statistical Society, 26*(Series B), 211–243.

Bradley, J. V. (1978). Robustness? *The British Journal of Mathematical and Statistical Psychology, 31*(2), 144–152.

Bradley, J. V. (1982). The insidious L-shaped distribution. *Bulletin of the Psychonomic Society, 20*(2), 85–88.

Brown, M. B., & Forsythe, A. B. (1974). The ANOVA and multiple comparisons for data with heterogeneous variances. *Biometrics,* (4), 719–724. doi:10.2307/2529238

Brunner, E., Dette, H., & Munk, A. (1997). Box-type approximations in nonparametric factorial designs. *Journal of the American Statistical Association, 92*(440), 1494–1502. doi:10.2307/2965420

Bryant, J. L., & Paulson, A. S. (1976). An extension of Tukey's method of multiple comparisons to experimental designs with random concomitant variables. *Biometrika, 63*(3), 631–638.

Buyse, M., & Molenberghs, G. (1998). Criteria for the validation of surrogate endpoints in randomized experiments. *Biometrics, 54*(3), 1014–1029.

Campbell, D. T., & Stanley, J. C. (1966). *Experimental and non-experimental designs.* Chicago, IL: Rand McNally.

Card, D., Lee, D. S., Pei, Z., & Weber, A. (2016). *Regression kink design: Theory and practice.* Retrieved from Cambridge, MA: https://www.nber.org/papers/w22781.pdf

Carlson, J. E., & Timm, N. H. (1974). Analysis of nonorthogonal fixed-effects designs. *Psychological Bulletin, 81*(9), 563–570. doi:10.1037/h0036936

Carroll, R. J., & Nordholm, L. A. (1975). Sampling characteristics of Kelley's epsilon2 and Hays' omega2. *Educational & Psychological Measurement, 35,* 541–554.

Carroll, R. J., & Ruppert, D. (1982). Robust estimation in heteroscedastic linear models. *Annals of Statistics, 10,* 429–441.

Celik, N., & Senoglu, B. (2018). Robust estimation and testing in one-way ANOVA for type II censored samples: Skew normal error terms. *Journal of Statistical Computation and Simulation, 88*(7), 1382–1393. doi:10.1080/00949655.2018.1433670

Chakravart, I. M., Laha, R. G., & Roy, J. (1967). *Handbook of methods of applied statistics* (Vol. 1). New York, NY: Wiley.

Chandrasekhar, C. K., Bagyalakshmi, H., Srinivasan, M. R., & Gallo, M. (2016). Partial ridge regression under multicollinearity. *Journal of Applied Statistics, 43*(13), 2462–2473. doi:10.1080/02664763.2016.1181726

Chatterjee, S., & Price, B. (1977). *Regression analysis by example.* New York, NY: Wiley.

Clinch, J. J., & Keselman, H. J. (1982). Parametric alternatives to the analysis of variance. *Journal of Educational Statistics, 7.*

Cohen, J. (1968). Multiple regression as a general data-analytic system. *Psychological Bulletin, 70*(6), 426–443. doi:10.1037/h0026714

Cohen, J. (1988). *Statistical power analysis for the behavioral sciences* (2nd ed.). Hillsdale, NJ: Lawrence Erlbaum.

Cohen, J., & Cohen, P. (1983). *Applied multiple regression/correlation analysis for the behavioral sciences* (2nd ed.). Hillsdale, NJ: Erlbaum.

Cohen, J., Cohen, P., West, S. G., & Aiken, L. S. (2003). *Applied multiple regression/correlation analysis for the behavioral sciences* (3rd ed.). Mahwah, NJ: Lawrence Erlbaum Associates.

Conover, W. J., & Iman, R. L. (1982). Analysis of covariance using the rank transformation. *Biometrics,* (3), 715–724. doi:10.2307/2530051

Cook, R. D. (1977). Detection of influential observation in linear regression. *Technometrics, 19*(1), 15.

Cook, R. D. (2000). Detection of influential observation in linear regression. *Technometrics, 42*(1), 65–68. doi:10.2307/1271434

Cook, R. D., & Campbell, D. T. (1979). *Quasi-experimentation: Design and analysis issues for field settings*. Chicago, IL: Rand McNally.

Cook, R. D., & Weisberg, S. (1982). *Residuals and influence in regression*. London, England: Chapman & Hall.

Coombs, W. T., Algina, J., & Oltman, D. O. (1996). Univariate and multivariate omnibus hypothesis tests selected to control type I error rates when population variances are not necessarily equal. *Review of Educational Research, 66*(2), 137–179.

Cotton, J. W. (1998). *Analyzing within-subjects experiments*. Mahwah, NJ: Lawrence Erlbaum.

Cox, B., Reason, R., Nix, S., & Gillman, M. (2016). Life happens (outside of college): Non-college life-events and students' likelihood of graduation. *Research in Higher Education, 57*(7), 823–844. doi:10.1007/s11162-016-9409-z

Cox, D. R., & Snell, E. J. (1989). *Analysis of binary data* (2nd ed.). London: Chapman & Hall.

Cramer, E. M., & Applebaum, M. I. (1908). Nonorthogonal analysis of variance--Once again. *Psychological Bulletin, 87*(51–57).

Cribari-Neto, F. (2004). Asymptotic inference under heteroskedasticity of unknown form. *Computational Statistics and Data Analysis, 45*, 215–233. doi:10.1016/S0167-9473(02)00366-3

Cronbach, L. J. (1987). Statistical tests for moderator variables: Flaws in analyses recently proposed. *Psychological Bulletin, 102*(3), 414–417.

Croux, C., Flandre, C., & Haesbroeck, G. (2002). The breakdown behavior of the maximum likelihood estimator in the logistic regression model. *Statistics and Probability Letters, 60*, 377–386.

D'Agostino, R. B. (1970). Transformation to normality of the null distribution of g1. *Biometrika, 57*(3), 679–681.

Darlington, R. B., & Hayes, A. F. (2017). *Regression analysis and linear models*. New York, NY: Guilford.

David, F. A., & Daryl, P. (1978). Finding the outliers that matter. *Journal of the Royal Statistical Society: Series B (Methodological), 40*(1), 85.

Derksen, S., & Keselman, H. J. (1992). Backward, forward and stepwise automated subset selection algorithms: Frequency of obtaining authentic and noise variables. *British Journal of Mathematical and Statistical Psychology, 45*, 265–282.

Ditlevsen, S., Christensen, U., Lynch, J., Damsgaard, M. T., & Keiding, N. (2005). The mediation proportion: A structural equation approach for estimating the proportion of exposure effect on outcome explained by an intermediate variable. *Epidemiology, 15*(1), 114–120. doi:10.1097/01.ede.0000147107.76079.07

Dong, N., Kelcey, B., & Spybrook, J. (2018). Power analyses for moderator effects in three-level cluster randomized trials. *Journal of Experimental Education, 86*(3), 489–514.

Dong, N., & Society for Research on Educational, E. (2014). *Power analysis to detect the effects of a continuous moderator in 2-level simple cluster random assignment experiments*. Retrieved from Evanston, IL: https://login.ezproxy.net.ucf.edu/login?auth=shibb&url=https://search.ebscohost.com/login.aspx?direct=true&db=eric&AN=ED562787&site=eds-live&scope=site

Dunn, O. J. (1961). Multiple comparisons among means. *Journal of the American Statistical Association*, (293), 52–64. doi:10.2307/2282330

Dunn, O. J. (1974). On multiple tests and confidence intervals. *Communications in Statistics, 3*(1), 101–103.

Dunn, O. J., & Clark, V. A. (1987). *Applied statistics: Analysis of variance and regression* (2nd ed.). New York, NY: Wiley.

Dunnett, C. W. (1955). A multiple comparison procedure for comparing several treatments with a control. *Journal of the American Statistical Association*, (272), 1096–1121. doi:10.2307/2281208

Dunnett, C. W. (1964). New tables for multiple comparisons with a control. *Biometrics*, (3), 482–491. doi:10.2307/2528490

Dunnett, C. W. (1980). Pairwise multiple comparisons in the unequal variance case. *Journal of the American Statistical Association, 75*(372), 796–800. doi:10.2307/2287161

Durbin, J., & Watson, G. S. (1950). Testing for serial correlation in least squares regression: I. *Biometrika, 37*(3–4), 409–428.

Durbin, J., & Watson, G. S. (1951). Testing for serial correlation in least squares regression: II. *Biometrika, 38*(1–2), 159–178.

Durbin, J., & Watson, G. S. (1971). Testing for serial correlation in least squares regression: III. *Biometrika, 58*(1), 1–19.

Egbewale, B. E., Lewis, M., & Sim, J. (2014). Bias, precision and statistical power of analysis of covariance in the analysis of randomized trials with baseline imbalance: A simulation study. *BMC Medical Research Methodology,* 14–49.

Elashoff, J. D. (1969). Analysis of covariance: A delicate instrument. *American Educational Research Journal, 6*(3), 383–401.

Erdfelder, E., Faul, F., & Buchner, A. (1996). GPOWER: A general power analysis program. *Behavior Research Methods, Instruments & Computers, 28*(1), 1–11.

Fahrmeir, L., Kneib, T., Lang, S., & Marx, B. (2013). *Regression: Models, methods, and applications*. Berlin and Heidelberg: Springer.

Faul, F., Erdfelder, E., Buchner, A., & Lang, A-G. (2009). Statistical power analysis using G*Power 3.1: Tests for correlation and regression analyses. *Behavior Research Methods, 41*(4), 1149–1160.

Faul, F., Erdfelder, E., Lang, A-G., & Buchner, A. (2007). G*Power 3: A flexible statistical power analysis program for the social, behavioral, and biomedical sciences. *Behavior Research Methods, 39*(2), 175–191.

Feldt, L. (1958). A comparison of the precision of three experimental designs employing a concomitant variable. *Psychometrika, 23*(4), 335–354.

Ferguson, G. A., & Takane, Y. (1989). *Statistical analysis in psychology and education* (6th ed.). New York, NY: McGraw Hill.

Fern, E. F., & Monroe, K. B. (1996). Effect-size estimates: Issues and problems in interpretation. *Journal of Consumer Research, 23*(2), 89–105.

Festing, M. F. W. (2014). Randomized block experimental designs can increase the power and reproducibility of laboratory animal experiments. *ILAR Journal, 55*(3), 472–476.

Fidler, F., & Thompson, B. (2001). Computing correct confidence intervals for ANOVA fixed-and random-effects effect sizes. *Educational and Psychological Measurement, 61*(4), 575-604.

Fisher, R. A. (1942). *The design of experiments* (3rd ed.). Edinburgh, UK: Oliver and Boyd.

Freedman, L. S. (2002). Confidence intervals and statistical power of the "validation" ratio for surrogate or intermediate endpoints. *Journal of Statistical Planning and Inference, 96*, 143–153.

Friedman, M. (1937). The use of ranks to avoid the assumption of normality implicit in the analysis of variance. *Journal of the American Statistical Association, 32*(200), 675–701. doi:10.2307/2279372

Fritz, M. S., & MacKinnon, D. P. (2007). Required sample size to detect the mediated effect. *Psychological Science, 18*(3), 233–239. doi:10.1111/j.1467-9280.2007.01882.x

Games, P., A., & Howell, J. F. (1976). Pairwise multiple comparison procedures with unequal N's and/or variances: A Monte Carlo study. *Journal of Educational Statistics, 1*(2), 113–125. doi:10.2307/1164979

Geisser, S., & Greenhouse, S. W. (1958). An extension of Box's results on the use of the F distribution in multivariate analysis. *The Annals of Mathematical Statistics, 29*(3), 885.

Glass, G. V., & Hopkins, K. D. (1996). *Statistical methods in education and psychology* (3rd ed.). Boston, MA: Allyn & Bacon.

Glass, G. V., Peckham, P. D., & Sanders, J. R. (1972). Consequences of failure to meet assumptions underlying the fixed effects analyses of variance and covariance. *Review of Educational Research,* (3), 237–288.

Hahs-Vaughn, D. L. (2005). A primer for using and understanding weights with national datasets. *Journal of Experimental Education, 73*(3), 221–248.

Hahs-Vaughn, D. L. (2006a). Analysis of data from complex samples. *International Journal of Research & Method in Education, 29*(2), 163–181.

Hahs-Vaughn, D. L. (2006b). Weighting omissions and best practices when using large-scale data in educational research. *Association for Institutional Research Professional File* (101), 1–9.

Hahs-Vaughn, D. L. (2016). *Applied multivariate statistical concepts*. New York, NY: Routledge and Taylor & Francis.

Hahs-Vaughn, D. L., McWayne, C. M., Bulotskey-Shearer, R. J., Wen, X., & Faria, A. (2011a). Complex sample data recommendations and troubleshooting. *Evaluation Review, 35*(3), 304–313. doi:10.1177/0193841X11412070

Hahs-Vaughn, D. L., McWayne, C. M., Bulotskey-Shearer, R. J., Wen, X., & Faria, A. (2011b). Methodological considerations in using complex survey data: An applied example with the head start family and child experiences survey. *Evaluation Review, 35*(3), 269–303.

Hair, J. F., Black, W. C., Babin, B. J., Anderson, R. E., & Tatham, R. L. (2006). *Multivariate data analysis* (6th ed.). Upper Saddle River, NJ: Pearson Prentice Hall.

Hand, D. J. (2009). Measuring classifier performance: A coherent alternative to the area under the ROC curve. *Machine Learning, 77*(1), 103.

Hansen, W. B., & McNeal, R. B., Jr. (1996). The law of maximum expected potential effect: Constraints placed on program effectiveness by mediator relationships. *Health Education Research, 11*(4), 501–507. doi:10.1093/her/11.4.501

Harrell, F. E. J. (1986). The LOGIST procedure. In I. SAS Institute (Ed.), *SUGI supplemental library user's guide* (5 ed., pp. 269–293). Cary, NC: SAS Institute, Inc.

Harwell, M. R. (1992). Summarizing Monte Carlo results in methodological research. *Journal of Educational Statistics,* (4), 297–313. doi:10.2307/1165126

Harwell, M. R. (2003). Summarizing Monte Carlo results in methodological research: The single-factor, fixed-effects ANCOVA case. *Journal of Educational and Behavioral Statistics, 28*, 45–70.

Hausman, C., & Rapson, D. S. (2017). *Regression discontinuity in time: Considerations for empirical applications*. Cambridge, MA: National Bureau of Economic Research.

Hawkins, D. M. (1980). *Identification of outliers*. London and New York, NY: Chapman & Hall.

Hayes, A. F. (n.d.). The PROCESS macro for SPSS and SAS (Version 3.2). Retrieved from http://processmacro.org/index.html

Hayes, A. F. (2013). *Introduction to mediation, moderation, and conditional process analysis: A regression-based approach*. New York, NY: Guilford.

Hayes, A. F., & Cai, L. (2007). Using heteroskedasticity-consistent standard error estimators in OLS regression: An introduction and software implementation. *Behavior Research Methods, 39*(4), 709–722.

Hays, W. L. (1988). *Statistics* (4th ed.). New York, NY: Holt, Rinehart and Winston.

Hayter, A., J. (1986). The maximum familywise error rate of Fisher's least significant difference test. *Journal of the American Statistical Association, 81*(396), 1000–1004. doi:10.2307/2289074

Heck, R. H., Tabata, L. N., & Thomas, S. L. (2014). *Multilevel and longitudinal modeling with IBM SPSS* (2nd ed.). New York, NY: Routledge.

Hellevik, O. (2009). Linear versus logistic regression when the dependent variable is a dichotomy. *Quality and Quantity, 43*(1), 59–74.

Hemmert, G. A. J., Schons, L. M., Wieseke, J., & Schimmelpfennig, H. (2018). Log-likelihood-based pseudo-R2 in logistic regression. *Sociological Methods & Research, 47*(3), 507–531. doi:10.1177/0049124116638107

Hilbe, J. M. (2016). *Practical guide to logistic regression*. Boca Raton, FL: CRC Press and Taylor & Francis.

Hirji, K. F., Tsiatis, A., A., & Mehta, C. R. (1989). Median unbiased estimation for binary data. *The American Statistician, 43*(1), 7. doi:10.2307/2685158

Hochberg, Y. (1988). A sharper Bonferroni procedure for multiple tests of significance. *Biometrika, 75*(4), 800–802.

Hochberg, Y., & Tamhane, A. C. (1987). *Multiple comparison procedures*. New York, NY: Wiley.

Hochberg, Y., & Varon-Salomon, Y. (1984). On simultaneous pairwise comparisons in analysis of covariance. *Journal of the American Statistical Association,* (388), 863–866. doi:10.2307/2288716

Hocking, R. R. (1976). The analysis and selection of variables in linear regression. *Biometrics, 32*(1), 1–49.

Hoerl, A. E., & Kennard, R. W. (1970a). Ridge regression: Application to non-orthogonal models. *Technometrics, 12*, 591–612.

Hoerl, A. E., & Kennard, R. W. (1970b). Ridge regression: Biased estimation for non-orthogonal models. *Technometrics, 12*, 55–67.

Hosmer, D. W., Hosmer, T., LeCessie, S., & Lemeshow, S. (1997). A comparison of goodness-of-fit tests for the logistic regression model. *Statistics in Medicine, 16*, 965–980.

Hosmer, D. W., & Lemeshow, S. (1989). *Applied logistic regression*. New York, NY: Wiley.

Hosmer, D. W., Lemeshow, S., & Sturdivant, R. X. (2000). *Applied logistic regression* (2nd ed.). Hoboken, NJ: John Wiley & Sons, Inc.

Hox, J. J., Moerbeek, M., & van de Schoot, R. (2017). *Multilevel analysis: Techniques and applications* (3rd ed.). New York, NY: Routledge.

Huang, B., Sivaganesan, S., Succop, P., & Goodman, E. (2004). Statistical assessment of mediational effects for logistic mediational models. *Statistics in Medicine, 23*(17), 2713–2728.

Huberty, C. J. (1989). Problems with stepwise methods—Better alternatives. In B. Thompson (Ed.), *Advances in social science methodology* (Vol. 1, pp. 43–70). Greenwich, CT: JAI Press.

Huck, S. W., & McLean, R. A. (1975). Using a repeated measures ANOVA to analyze the data from a pretest-posttest design: A potentially confusing task. *Psychological Bulletin, 82*(4), 511–518. doi:10.1037/h0076767

Huitema, B. E. (2011). *Analysis of covariance and alternatives statistical methods for experiments, quasi-experiments, and single-case studies* (2nd ed.). Hoboken, NJ: Wiley.

Huynh, H., & Feldt, L. S. (1970). Conditions under which mean square ratios in repeated measurements designs have exact F-distributions. *Journal of the American Statistical Association, 65*(332), 1582. doi:10.2307/2284340

James, G. S. (1951). The comparison of several groups of observations when the ratios of the population variances are unknown. *Biometrika, 38*(3–4), 324–329.

Jennings, E. (1988). Models for pretest-posttest data: Repeated measures ANOVA revisited. *Journal of Educational Statistics, 13*(3), 273. doi:10.2307/1164655

Jiang, H., Kulkarni, P. M., Mallinckrodt, C. H., Shurzinske, L., Molenberghs, G., & Lipkovich, I. (2017). Covariate adjustment for logistic regression analysis of binary clinical trial data. *Statistics in Biopharmaceutical Research, 9*(1), 126–134.

Joanes, D. N., & Gill, C. A. (1998). Comparing measures of sample skewness and kurtosis. *Journal of the Royal Statistical Society. Series D (The Statistician), 47*(1), 183–189.

Johansen, S. (1980). The Welch-James approximation to the distribution of the residual sum of squares in a weighted linear regression. *Biometrika, 67*(1), 85–93.

Johnson, P. O., & Neyman, J. (1936). Tests of certain linear hypotheses and their application to some educational problems. *Statistical Research Memoirs, 1*, 57–93.

Kaiser, L. D., & Bowden, D. C. (1983). Simultaneous confidence intervals for all linear contrasts of means with heterogenous variances. *Communications in Statistics: Theory and Methods, 12*(1), 73–88.

Kelley, K. (2018). Package MBESS (Version 4.4.3). Retrieved from https://cran.r-project.org/web/packages/MBESS/MBESS.pdf

Kenny, D. A. (n.d.). *Power and N computations for mediation*. Retrieved from https://davidakenny.shinyapps.io/MedPower/

Keppel, G. (1982). *Design and analysis: A researcher's handbook* (2nd ed.). Englewood Cliffs, NJ: Prentice-Hall.

Keppel, G. (1991). *Design and analysis: A researcher's handbook* (3rd ed.). Englewood Cliffs, NJ: Prentice-Hall, Inc.

Keppel, G., & Wickens, T. D. (2004). *Design and analysis: A researcher's handbook* (4th ed.). Upper Saddle River, NJ: Pearson Prentice Hall.

Kirk, R. E. (2013). *Experimental design: Procedures for the behavioral sciences* (4th ed.). Thousand Oaks, CA: Sage Publications, Inc.

Kirk, R. E. (2014). *Experimental design: Procedures for the behavioral sciences*. Thousand Oaks, CA: Sage.

Kisbu-Sakarya, Y., MacKinnon, D. P., & Aiken, L. S. (2013). A Monte Carlo comparison study of the power of the analysis of covariance, simple difference, and residual change scores in testing two-wave data. *Educational & Psychological Measurement, 73*(1), 47–62. doi:10.1177/0013164412450574

Kish, L., & Frankel, M. R. (1973, October 17). *Inference from complex samples*. Paper presented at the Annual Meeting of the Royal Statistical Society. https://deepblue.lib.umich.edu/bitstream/handle/2027.42/146969/rssb00981.pdf?sequence=1&isAllowed=y

Kish, L., & Frankel, M. R. (1974). Inference from complex samples. *Journal of the Royal Statistical Society, Series B, 36*, 1–37.

Kleinbaum, D. G., Kupper, L. L., Muller, K. E., & Nizam, A. (1998). *Applied regression analysis and other multivariable models* (3rd ed.). Pacific Grove, CA: Duxbury.

Knafl, G. J., & Ding, K. (2016). *Adaptive regression for modeling nonlinear relationships*. Cham, Switzerland: Springer.

Knofszynski, G. T. (2008). Sample sizes when using multiple linear regression for prediction. *Educational and Psychological Measurement, 68*(3), 431–442.

Koenker, R., Chernozhukov, V., He, X., & Peng, L. (2017). *Handbook of quantile regression* (1st ed.). Boca Raton, FL: CRC Press.

Korn, E. L., & Graubard, B. I. (1995). Examples of differing weighted and unweighted estimates from a sample survey. *American Statistician, 49*, 291–305.

Kramer, C. Y. (1957). Extension of multiple range tests to group correlated adjusted means. *Biometrics*, (1), 13–18. doi:10.2307/3001898

Kreft, I., & de Leeuw, J. (1998). *Introducing multilevel modeling*. Thousand Oaks, CA: Sage.

Kromrey, J. D., & Foster-Johnson, L. (1999). Statistically differentiating between interaction and nonlinearity in multiple regression analysis: A Monte Carlo investigation of a recommended strategy. *Educational & Psychological Measurement, 59*(3), 392–413. doi:10.1177/00131649921969947

Kruskal, W. H., & Wallis, W. A. (1952). Use of ranks in one-criterion variance analysis. *Journal of the American Statistical Association, 47*(260), 583–621. doi:10.2307/2280779

Kruskal, W. H., & Wallis, W. A. (1953). Errata: Use of ranks in one-criterion variance analysis. *Journal of the American Statistical Association, 48*(264), 907–911. doi:10.2307/2281082

Krzanowski, W. J., & Hand, D. J. (2009). *ROC curves for continuous data*. Boca Raton, FL: CRC Press.

Kutner, M., Nachtscheim, C., & Neter, J. (2005). *Applied linear statistical models* (5th ed.). New York, NY: McGraw Hill.

Lachowicz, M. J., Preacher, K. J., & Kelley, K. (2018). A novel measure of effect size for mediation analysis. *Psychological Methods, 23*(2), 244–261. doi:10.1037/met0000165

Larsen, W. A., & McCleary, S. J. (1972). The use of partial residual plots in regression analaysis. *Technometrics, 14*, 781–790.

Lee, E. S., Forthofer, R. N., & Lorimor, R. J. (1989). *Analyzing complex survey data*. Newbury Park, CA: Sage.

Lee, M-J. (2016). *Matching, regression discontinuity, difference in differences, and beyond*. New York, NY: Oxford University Press.

Levin, J. R., Serlin, R. C., & Seaman, M. A. (1994). A controlled, powerful multiple-comparison strategy for several situations. *Psychological Bulletin, 115*(1), 153–159.

Li, J., & Lomax, R. G. (2011). Analysis of variance: What is your statistical software actually doing? *Journal of Experimental Education, 79*, 279–294.

Lilliefors, H. (1967). On the Kolmogorov-Smirnov test for normality with mean and variance unknown. *Journal of the American Statistical Association, 62*, 399–402.

Lindenberger, U., & Pötter, U. (1998). The complex nature of unique and shared effects in hierarchical linear regression: Implications for developmental psychology. *Psychological Methods, 3*(2), 218–230. doi:10.1037/1082–989X.3.2.218

Liu, X. S. (2014). *Statistical power analysis for the social and behavioral sciences: Basic and advanced techniques*. New York, NY: Routledge.

Lomax, R. G., & Surman, S. H. (2007). Factorial ANOVA in SPSS: Fixed-, random-, and mixed-effects models. In S. S. Sawilowsky (Ed.), *Real data analysis*. Greenwich, CT: Information Age.

Long, J. S. (1997). *Regression models for categorical and limited dependent variables*. Thousand Oaks, CA: Sage.

Lord, F. M. (1960). Large-sample covariance analysis when the control variable is fallible. *Journal of the American Statistical Association*, (290), 307–321. doi:10.2307/2281743

Lord, F. M. (1967). A paradox in the interpretation of group comparisons. *Psychological Bulletin, 68*(5), 304–305. doi:10.1037/h0025105

Lord, F. M. (1969). Statistical adjustments when comparing preexisting groups. *Psychological Bulletin, 72*(5), 336–337. doi:10.1037/h0028108

Lumley, T. (2004). Analysis of complex survey samples. *Journal of Statistical Software, 9*(8), 1–19.

Maalouf, M., Homouz, D., & Trafalis, T. B. (2018). Logistic regression in large rare events and imbalanced data: A performance comparison of prior correction and weighting methods. *Computational Intelligence, 34*(1), 161–174. doi:10.1111/coin.12123

Maas, C. J. M., & Hox, J. J. (2005). Sufficient sample size for multilevel modeling. *Methodology: European Journal of Research Methods for the Behavioral and Social Sciences, 1*(3), 86–92.

MacKinnon, D. P. (2008). *Introduction to statistical mediation analysis*. New York, NY: Lawrence Erlbaum Associates.

MacKinnon, D. P., Warsi, G., & Dwyer, J. H. (1995). A simulation study of mediated effect measures. *Multivariate Behavioral Research, 30*(1), 41–62. doi:10.1207/s15327906mbr3001_3

Mansfield, E. R., & Conerly, M. D. (1987). Diagnostic value of residual and partial residual plots. *The American Statistician, 41*, 107–116.

Mansouri, H., & Zhang, F. (2018). Simultaneous rank tests in analysis of covariance based on pairwise ranking. Retrieved from https://arxiv.org/pdf/1802.03884.pdf

Marascuilo, L. A., & Levin, J. R. (1970). Appropriate post hoc comparisons for interaction and nested hypotheses in analysis of variance designs: The elimination of Type IV errors. *American Educational Research Journal, 7*(3), 397–421.

Marascuilo, L. A., & Levin, J. R. (1976). The simultaneous investigation of interaction and nested hypotheses in two-factor analysis of variance designs. *American Educational Research Journal, 13*(1), 61–65.

Marascuilo, L. A., & McSweeney, M. (1977). *Nonparametric and distribution-free methods for the social sciences*. Monterey, CA: Brooks/Cole.

Marascuilo, L. A., & Serlin, R. C. (1988). *Statistical methods for the social and behavioral sciences*. New York, NY: Freeman.

Marquardt, D. W., & Snee, R. D. (1975). Ridge regression in practice. *The American Statistician, 29*, 3–19.

Mason, C. A., & Tu, S. (1996). Assessing moderator variables: Two computer simulation studies. *Educational & Psychological Measurement, 56*(1), 45–62. doi:10.1177/0013164496056001003

Maxwell, S. E. (1980). Pairwise multiple comparisons in repeated measures designs. *Journal of Educational & Behavioral Statistics, 5*(3), 269.

Maxwell, S. E. (2000). Sample size and multiple regression analysis. *Psychological Methods, 5*, 434–458.

Maxwell, S. E., Arvey, R. D., & Camp, C. J. (1981). Measures of strength of association: A comparative examination. *Journal of Applied Psychology, 66*(5), 525–534.

Maxwell, S. E., Delaney, H. D., & Dill, C. A. (1984). Another look at ANCOVA versus blocking. *Psychological Bulletin, 95*, 136–147. doi:10.1037/0033–2909.95.1.136

Maxwell, S. E., Delaney, H. D., & Kelley, K. (2018). *Designing experiments and analyzing data*. New York, NY: Routledge.

McCulloch, C. E. (2005). Repeated measures ANOVA, RIP? *CHANCE, 19*, 29–33.

McGrath, R. P., Hall, O. T., Peterson, M. D., Kraemer, W. J., & Vincent, B. M. (2017). Muscle strength is protective against osteoporosis in an ethnically diverse sample of adults *Journal of Strength & Conditioning Research, 31*(9), 2586–2589.

Mehta, A. J., Dooley, D. P., Kane, J., Reid, M., & Shah, S. N. (2018). Subsidized housing and adult asthma in Boston, 2010–2015. *American Journal of Public Health, 108*(8), 1059–1065. doi:10.2105/AJPH.2018.304468

Menard, S. (1995). *Applied logistic regression analysis*. Thousand Oaks, CA: Sage.

Menard, S. (2000). *Applied logistic regression analysis* (2nd ed.). Thousand Oaks, CA: Sage.

Mendoza, J. L., & Stafford, K. L. (2001). Confidence intervals, power calculation, and sample size estimation for the squared multiple correlation coefficient under the fixed and random regression models: A computer program and useful standard tables. *Educational and Psychological Measurement, 61*, 650–667.

Meuleman, B., Loosveldt, G., & Emonds, V. (2013). Regression analysis: Assumptions and diagnostics. In H. Best & C. Wolf (Eds.), *Handbook of regression analysis and causal inference* (pp. 83–110). London, England: Sage.

Meyers, L. S., Gamst, G., & Guarino, A. J. (2006). *Applied multivariate research: Design and interpretation*. Thousand Oaks, CA: Sage.

Mickey, R. M., Dunn, O. J., & Clark, V. A. (2004). *Applied statistics: Analysis of variance and regression* (3rd ed.). Hoboken, NJ: Wiley.
Miller, A. J. (1984). Selection of subsets of regression variables (with discussion). *Journal of the Royal Statistical Society, A*(147), 389–425.
Miller, A. J. (1990). *Subset selection in regression*. New York, NY: Chapman & Hall.
Miller, R. G. (1997). *Beyond ANOVA: Basics of applied statistics*. Boca Raton, FL: CRC Press.
Morgan, G. A., Leech, N. L., Gloeckner, G. W., & Barrett, K. C. (2012). *IBM SPSS for introductory statistics: Use and interpretation* (5th ed.). Boca Raton, FL: CRC Press/Taylor & Francis Group.
Mosteller, F., & Tukey, J. W. (1977). *Data analysis and regression*. Reading, MA: Addison-Wesley.
Murphy, K. R., Myors, B., & Wolach, A. (2009). *Statistical power analysis: A simple and general model for traditional and modern hypothesis tests* (3rd ed.). New York, NY: Routledge and Taylor & Francis Group.
Murphy, K. R., Myors, B., & Wolach, A. (2014). *Statistical power analysis: A simple and general model for traditional and modern hypothesis tests* (4th ed.). New York, New York: Routledge/Taylor & Francis Group.
Myers, J. L., Lorch, R. F., & Well, A. (2010). *Research design and statistical analysis* (3rd ed.). New York, NY: Routledge.
Myers, J. L., & Well, A. D. (1995). *Research design and statistical analysis*. Mahwah, NJ: Lawrence Erlbaum Associates.
Myers, R. H. (1979). *Fundamentals of experimental design* (4th ed.). Boston, MA: Allyn & Bacon.
Myers, R. H. (1986). *Classical and modern regression with applications*. Boston, MA: Duxbury.
Myers, R. H. (1990). *Classical and modern regression with applications* (2nd ed.). Boston, MA: Duxbury.
Nagelkerke, N. J. D. (1991). A note on a general dievision of the coefficient of determination. *Biometrika, 78*, 691–692.
O'Grady, K. E. (1982). Measures of explained variance: Cautions and limitations. *Psychological Bulletin, 92*(3), 766–777. doi:10.1037/0033–2909.92.3.766
Olejnik, S., & Algina, J. (2000). Measures of effect size for comparative studies: Applications, interpretations, and limitations. *Contemporary Educational Psychology, 25*(3), 241–286.
Olejnik, S., & Algina, J. (2003). Generalized eta and omega squared statistics: Measures of effect size for some common research designs. *Psychological Methods, 8*(4), 434–447.
Olive, D. J. (2017). *Linear regression*. Cham, Switzerland: Springer International Publishing.
Osborne, J. W. (2015). *Best practices in logistic regression*. Los Angeles, CA: Sage.
Overall, J. E., Lee, D. M., & Hornick, C. W. (1981). Comparison of two strategies for analysis of variance in nonorthogonal designs. *Psychological Bulletin, 90*, 367-375. doi:10.1037/0033-2909.90.2.367
Overall, J. E., & Spiegel, D. K. (1969). Concerning least squares analysis of experimental data. *Psychological Bulletin, 72*, 311–322.
Page, M. C., Braver, S. L., & MacKinnon, D. P. (2003). *Levine's guide to SPSS for analysis of variance* (2nd ed.). Mahwah, NJ: Lawrence Erlbaum Associates Publishers.
Pampel, F. C. (2000). *Logistic regression: A primer*. Thousand Oaks, CA: Sage.
Pavur, R. (1988). Type I error rates for multiple comparison procedures with dependent data. *The American Statistician*, (3), 171–173. doi:10.2307/2684994
Peckham, P. D. (1968). *An investigation of the effects of non-homogeneity of regression slopes upon the F-test of analysis of covariance*. Unpublished doctoral dissertation. University of Colorado, Boulder, CO.
Pedhazur, E. J. (1997). *Multiple regression in behavioral research* (3rd ed.). Fort Worth, TX: Harcourt Brace.
Pfeffermann, D. (1993). The role of sampling weights when modeling survey data. *International Statistical Review, 61*(2), 317–337.
Pingel, L. A. (1969). *A comparison of the effects of two methods of block formation on design precision*. Paper presented at the American Educational Research Association, Los Angeles, CA.
Porter, A. C. (1967). *The effects of using fallible variables in the analysis of covariance* (Unpublished doctoral dissertation). University of Wisconsin. Madison, WI.
Porter, A. C., & Raudenbush, S. W. (1987). Analysis of covariance: Its model and use in psychological research. *Journal of Counseling Psychology, 34*, 383–392. doi:10.1037/0022-0167.34.4.383

Preacher, K. J., & Hayes, A. F. (2008). Asymptotic and resampling strategies for assessing and comparing indirect effects in multiple mediator models. *Behavior Research Methods, 40*(3), 879–891.

Preacher, K. J., & Hayes, A. F. (2008). Contemporary approaches to assessing mediation in communication research. In A. F. Hayes, M. D. Slater, & L. B. Snyder (Eds.), *The Sage sourcebook of advanced data analysis methods for communication research.* (pp. 13–54). Thousand Oaks, CA: Sage Publications, Inc.

Preacher, K. J., & Kelley, K. (2011). Effect size measures for mediation models: Quantitative strategies for communicating indirect effects. *Psychological Methods, 16*(2), 93–115.

Puri, M. L., & Sen, P. K. (1969). Analysis of covariance based on general rank scores. *The Annals of Mathematical Statistics,* (2), 610–618.

Qiu, W. (2018). Power mediation: Power/sample size calculation for mediation analysis (Version R package version 3.1.0). Retrieved from https://cran.r-project.org/web/packages/powerMediation/powerMediation.pdf

Quade, D. (1967). Rank analysis of covariance. *Journal of the American Statistical Association,* (320), 1187–1200. doi:10.2307/2283769

Raftery, A. E. (1995). Bayesian model selection in social research. *Sociological Methodology, 25,* 111–163.

Raudenbush, S. W., & Bryk, A. S. (2002). *Hierarchical linear models: Applications and data analysis methods* (2nd ed.). Thousand Oaks, CA: Sage.

Reichardt, C. S. (1979). The statistical analysis of data from nonequivalent control group designs. In T. D. Cook & D. T. Campbell (Eds.), *Quasi-experimentation: Design and analysis issues for field settings*. Chicago, IL: Rand McNally.

Rogers, W. M. (2002). Theoretical and mathematical constraints of interactive regression models. *Organizational Research Methods, 5*(3), 212–230. doi:10.1177/10928102005003002

Rogosa, D. R. (1980). Comparing non-parallel regression lines. *Psychological Bulletin, 88,* 307–321.

Rosenthal, R. (1994). Parametric measures of effect size. In H. Cooper & L. V. Hedges (Eds.), *The handbook of research synthesis* (pp. 231–244). New York, NY: Russell Sage Foundation.

Rosenthal, R., & Rosnow, R. L. (1985). *Contrast analysis: Focused comparisons in the analysis of variance.* Cambridge and New York, NY: Cambridge University Press.

Rosenthal, R., & Rubin, D. B. (1979). A note on percent variance explained as a measure of the importance of effects. *Journal of Applied Social Psychology, 9*(5), 395–396. doi:10.1111/j.1559-1816.1979.tb02713.x

Rosnow, R. L., & Rosenthal, R. (1988). Focused tests of significance and effect size estimation in counseling psychology. *Journal of Counseling Psychology, 35*(2), 203–208.

Rousseeuw, P. J., & Leroy, A. M. (1987). *Robust regression and outlier detection*. New York, NY: Wiley.

Ruppert, D. (2014). Transformation and weighting. In M. Davidian, X. Lin, J. S. Morris, & L. A. Stefanski (Eds.), *The work of Raymond J. Carroll: The impact and influence of a statistician* (pp. 155–161). Cham, Switzerland: Springer.

Ruppert, D., & Carroll, R. J. (1980). Trimmed least squares estimation in the linear model. *Journal of the American Statistical Association, 75,* 828–838.

Russell, C. J., & Bobko, P. (1992). Moderated regression analysis and Likert scales: Too coarse for comfort. *Journal of Applied Psychology, 77*(3), 336–342.

Russell, C. J., & Dean, M. A. (2000). To log or not to log: Bootstrap as an alternative to the parametric estimation of moderation effects in the presence of skewed dependent variables. *Organizational Research Methods, 3*(2), 166–185. doi:10.1177/109442810032002

Rutherford, A. (1992). Alternatives to traditional analysis of covariance. *British Journal of Mathematical and Statistical Psychology, 45*(2), 197–223. doi:10.1111/j.2044–8317.1992.tb00988.x

Rutherford, A. (2011). *ANOVA and ANCOVA: a GLM approach* (2nd ed.). Hoboken, NJ: Wiley.

Scariano, S. M., & Davenport, J. M. (1987). The effects of violation of independence assumptions in the one-way ANOVA. *The American Statistician, 41*(2), 123–129.

Scheffé, H. (1953). A Method for judging all contrasts in the analysis of variance. *Biometrika, 40*(1–2), 87–104.

Schneider, B. A., Avivi-Reich, M., & Mozuraitis, M. (2015). A cautionary note on the use of the analysis

of covariance (ANCOVA) in classification designs with and without within-subject factors. *Frontiers in Psychology, 6*, 1–12. doi:10.3389/fpsyg.2015.00474/full10.3389/fpsyg.2015.00474
Schoemann, A. M., Boulton, A. J., & Short, S. D. (2017). Determining power and sample size for simple and complex mediation models. *Social Psychological and Personality Science, 8*(4), 379–386. doi:10.1177/1948550617715068
Seber, G. A. F., & Wild, C. J. (1989). *Nonlinear regression*. New York, NY: Wiley.
Sechrest, L., & Yeaton, W. H. (1982). Magnitudes of experimental effects in social science research. *Evaluation Review, 6*(5), 579–600.
Shadish, W. R., Cook, T. D., & Campbell, D. T. (2002). *Experimental and quasi-experimental designs for generalized causal inference*. Boston, MA: Houston Mifflin.
Shan, G., & Ma, C. (2014). A comment on sample size calculation for analysis of covariance in parallel arm studies. *Journal of Biometrics and Biostatistics, 5*(1). doi:10.4172/2155–6180.1000184
Shapiro, S. S., & Wilk, M. B. (1965). An analysis of variance test for normality (complete samples). *Biometrika, 52*(3–4), 591–611.
Shavelson, R. J. (1996). *Statistical reasoning for the behavioral sciences* (3rd ed.). Boston, MA: Allyn & Bacon.
Shieh, G. (2009). Detecting interaction effects in moderated multiple regression with continuous variables power and sample size considerations. *Organizational Research Methods, 12*(3), 510–528. doi:10.1177/1094428108320370
Shieh, G. (2017). Power and sample size calculations for contrast analysis in ANCOVA. *Multivariate Behavioral Research, 52*(1), 1–11. doi:10.1080/00273171.2016.1219841
Sidak, Z. (1967). Rectangular confidence regions for the means of multivariate normal distributions. *Journal of the American Statistical Association, 62*(318), 626–633. doi:10.2307/2283989
Sinan, A., & Alkan, B. B. (2015). A useful approach to identify the multicollinearity in the presence of outliers. *Journal of Applied Statistics, 42*(5), 986—993.
Singh, R. (2010). A survey of ridge regression for Improvement over ordinary least squares. *IUP Journal of Computational Mathematics, 3*(4), 54–74.
Skinner, C. J., Holt, D., & Smith, T. M. F. (Eds.). (1989). *Analysis of complex samples*. New York, NY: Wiley.
Smithson, M. (2001). Correct confidence intervals for various regression effect sizes and parameters: The importance of noncentral distributions in computing intervals. *Educational and Psychological Measurement, 61*(4), 605–632.
Smithson, M. (2003). Noncentral confidence intervals for standardized effect sizes. *Confidence intervals* (pp. 33–41). Thousand Oaks, CA: Sage.
Smithson, M., & Shou, Y. (2017). Moderator effects differ on alternative effect-size measures. *Behavior Research Methods, 49*(2), 747–757. doi:10.3758/s13428-016-0735-z
Snijders, T. A. B., & Bosker, R. J. (1999). *Multilevel analysis: An introduction to basic and advanced multilevel modeling*. Thousand Oaks, CA: Sage.
Snijders, T. A. B., & Bosker, R. J. (2012). *Multilevel analysis: An introduction to basic and advanced multilevel modeling* (2nd ed.). Thousand Oaks, CA: Sage.
Sobel, M. E. (1982). Asymptotic confidence intervals for indirect effects in structural equation models. In S. Leinhardt (Ed.), *Sociological methodology* (pp. 290–312). Washington, DC: American Sociological Association.
Spybrook, J., & Kelcey, B. (2014). *Power calculations for binary moderator in cluster randomized trials*. Retrieved from https://login.ezproxy.net.ucf.edu/login?auth=shibb&url=https://search.ebscohost.com/login.aspx?direct=true&db=eric&AN=ED562789&site=eds-live&scope=site
Spybrook, J., Raudenbush, S., Liu, X., Congdon, R., & Martinez, A. (2006). Optimal design (Version 1.76): University of Michigan. Retrieved from http://sitemaker.umich.edu/group-based/optimal_design_software
Steiger, J. H., & Fouladi, R. T. (1992). R2: A computer program in interval estimation power calculation, and hypothesis testing for the squared multiple correlation. *Behavior Research Methods, Instruments & Computers, 4*, 581–582.
Stevens, J. P. (1984). Outliers and influential data points in regression analysis. *Psychological Bulletin, 95*(2), 334–344.

Stevens, J. P. (2009). *Applied multivariate statistics for the social sciences* (5th ed.). New York, NY: Psychology Press.

Stone-Romero, E. F., & Anderson, L. E. (1994). Relative power of moderated multiple regression and the comparison of subgroup correlation coefficients for detecting moderating effects. *Journal of Applied Psychology, 79*(3), 354–359.

Stone-Romero, E. F., Alliger, G. M., & Aguinis, H. (1994). Type II error problems in the use of moderated multiple regression for the detection of moderating effects of dichotomous variables. *Journal of Management, 20*(1), 167–178. doi:10.1177/014920639402000109

Tabachnick, B. G., & Fidell, L. S. (2013). *Using multivariate statistics* (6th ed.). Boston, MA: Pearson.

Tabachnick, B. G., & Fidell, L. S. (2019). *Using multivariate statistics* (7th ed.). Boston, MA: Pearson.

Tabatabai, M. A., & Tan, W. Y. (1985). Some comparative studies on testing parallelism of several straight lines under heteroscedastic variances. *Communications in Statistics: Simulation and Computation, 14*(4), 837–844.

Thompson, B. (2016). The case for using the general linear model as a unifying conceptual framework for teaching statistics and psychometric theory. *Journal of Methods and Measurement in the Social Sciences, 6*(2), 30–41. doi:10.2458/azu_jmmss.v6i2.18801

Thompson, M. L. (1978). Selection of variables in multiple regression. Part I: A review and evaluation. Part II: Chosen procedures, computations and examples. *International Statistical Review, 46*, 1–19, 129–146.

Timm, N. H. (2002). *Applied multivariate analysis*. New York, NY: Springer.

Timm, N. H., & Carlson, J. E. (1975). Analysis of variance through full rank models. *Multivariate Behavioral Research Monographs, 75*(1), 120–143.

Tomarken, A., & Serlin, R. C. (1986). Comparison of ANOVA alternatives under variance heterogeneity and specific noncentrality structures. *Psychological Bulletin, 99*, 90–99.

Tukey, J. W. (1949). One degree of freedom for non-additivity. *Biometrics, 5*(3), 232. doi:10.2307/3001938

Tukey, J. W. (1953). *The problem of multiple comparisons*. Princeton, NJ: Princeton University Press.

Uanhoro, J. O. (2017). *Effect size calculators*. Retrieved from https://effect-size-calculator.herokuapp.com/

Van Breukelen, G. J. P. (2006). ANCOVA versus change from baseline had more power in randomized studies and more bias in nonrandomized studies. *Journal of Clinical Epidemiology, 59*, 920–925. doi:10.1016/j.jclinepi.2006.02.007

Vatcheva, K. P., Lee, M., McCormick, J. B., & Rahbar, M. H. (2016). Multicollinearity in regression analyses conducted in epidemiologic studies. *Epidemiology, 6*(2).

Vaughan, G. M., & Corballis, M. C. (1969). Beyond tests of significance: Estimating strength of effects in selected ANOVA designs. *Psychological Bulletin, 72*(3), 204–213. doi:10.1037/h0027878

Wampold, B. E., & Serlin, R. C. (2000). The consequence of ignoring a nested factor on measures of effect size in analysis of variance. *Psychological Methods, 5*(4), 425–433.

Wang, X., Faraway, J. J., & Yue Ryan, Y. (2018). *Bayesian regression modeling with INLA* (1st ed.). Boca Raton, FL: CRC Press.

Wang, Y., Rodríguez de Gil, P., Chen, Y-H., Kromrey, J. D., Kim, E. S., Pham, T., . . . Romano, J. L. (2017). Comparing the performance of approaches for testing the homogeneity of variance assumption in one-factor ANOVA models. *Educational and Psychological Measurement, 77*(2), 305–329. doi:10.1177/0013164416645162

Weisberg, S. (1985). *Applied linear regression* (2nd ed.). New York, NY: Wiley.

Weisberg, S. (2014). *Applied linear regression* (4th ed.). Hoboken, NJ: Wiley.

Weinberg, S. L., & Abramowitz, S. K. (2002). *Data analysis for the behavioral sciences using SPSS*. Cambridge, UK: Cambridge University Press.

Welc, J., Esquerdo, P. J. R., & SpringerLink. (2018). *Applied regression analysis for business: Tools, traps and applications*. Cham, Switzerland: Springer International Publishing.

Welch, B. L. (1951). On the comparison of several mean values: An alternative approach. *Biometrika, 38*(3–4), 330–336.

Wen, Z., & Fan, X. (2015). Monotonicity of effect sizes: Questioning kappa-squared as mediation effect size measure. *Psychological Methods, 20*(2), 193–203. doi:10.1037/met0000029

Wetherill, G. B. (1986). *Regression analysis with applications*. London, England: Chapman & Hall.

Wilcox, R. R. (1986). Controlling power in a heteroscedastic ANOVA procedure. *British Journal of Mathematical and Statistical Psychology, 39*(1), 65–68. doi:10.1111/j.2044-8317.1986.tb00845.x

Wilcox, R. R. (1987). *New statistical procedures for the social sciences: Modern solutions to basic problems*. Hillsdale, NJ: Lawrence Erlbaum Associates.

Wilcox, R. R. (1988). A new alternative to the ANOVA F and new results on James's second-order method. *British Journal of Mathematical and Statistical Psychology, 41*(1), 109–117. doi:10.1111/j.2044-8317.1988.tb00890.x

Wilcox, R. R. (1989). Adjusting for unequal variances when comparing means in one-way and two-way fixed effects ANOVA models. *Journal of Educational Statistics, 14*(3), 269–278. doi:10.2307/1165019

Wilcox, R. R. (1995). *Statistics for the social sciences*. San Diego, CA: Academic Press.

Wilcox, R. R. (1996). *Statistics for the social sciences*. San Diego, CA: Academic.

Wilcox, R. R. (2003). *Applying comtemporary statistical procedures*. San Diego, CA: Academic.

Wilcox, R. R. (2017). *Introduction to robust estimation and hypothesis testing* (4th ed.). Burlington, MA: Elsevier.

Wonnacott, T. H., & Wonnacott, R. J. (1981). *Regression: A second course in statistics*. New York, NY: Wiley.

Wu, L. L. (1985). Robust M-estimation of location and regression. In N. B. Tuma (Ed.), *Sociological methodology* (pp. 316–388). San Francisco, CA: Jossey-Bass.

Wu, X. W., & Lai, D. (2015). Comparison of statistical methods for pretest—posttest designs in terms of type I error probability and statistical power. *Communications in Statistics: Simulation & Computation, 44*(2), 284–294. doi:10.1080/03610918.2013.7752954

Xie, X-J., Pendergast, J., & Clarke, W. (2008). Increasing the power: A practical approach to goodness-of-fit test for logistic regression models with continuous predictors. *Computational Statistics and Data Analysis, 52*, 2703–2713. doi:10.1016/j.csda.2007.09.027

Zhang, Z. (2014). Monte Carlo based statistical power analysis for mediation models: Methods and software. *Behavior Research Methods, 46*(4), 1184–1198.

Zimmerman, D. W. (1997). A note on interpretation of the paired-samples t test. *Journal of Educational and Behavioral Statistics, 22*(3), 349–360. doi:10.2307/1165289

Name Index

Subject Index

Note: Please note page number in *italics* indicate figures and page number in **bold** indicate tables.